Electromagnetic Fields and Waves

Second Edition

Electromagnetic Fields and Waves

Second Edition

Editor in Chief

Jiao Qixiang

Editors

Li Shufang **Li Li**
Zhang Yang'an **Gao Zehua**
Wang Yafeng **Zhang Xin**

Science Press Alpha Science International Ltd.

Electromagnetic Fields and Waves
Second Edition
452 pgs. | 254 figs. | 12 tbls.

Editor in Chief

Jiao Qixiang

Co-Published by:

Science Press
16 Donghuangchenggen North Street
Beijing 100717, China

and

Alpha Science International Ltd.
7200 The Quorum, Oxford Business Park North
Garsington Road, Oxford OX4 2JZ, U.K.

www.alphasci.com

ISBN 978-1-84265-560-3 (Alpha Science)

Printed in India

Preface to the Second Edition

Compared with the first edition of this book, there is no big change in the text structure of the second edition. It maintains the characteristics of the original book, but partially adds and deletes or modifies the content and examples.

In order to broaden the mind, the second edition still tries to keep the feature of "using diverse analysis methods" to solve some important problems, and give a brief summary before each analysis method. This may provide more options with regard to different teaching hours and reserve some room of thinking for the readers who have the interest and potential.

The modifications in the second edition are as follows:

(1) Text content addition and deletion. For example, a new section "the concept of symmetric dipoles and antenna array" has been added into chapter; individual parts like "low loss transmission line" have been deleted in chapter 9; some content like "the Kirchhoff voltage laws" has been modified; the section "the polarization of the electromagnetic wave" has been reused.

(2) Examples and exercises addition and deletion. Some individual examples which are too complex or too simple have been deleted. Meanwhile, some examples which are very conceptual or have good engineering value have been added into the text. Furthermore, about 30 basic concept problems have been added, and some problems have been deleted or adapted; the order of some problems have been rotated, and problem solving "tips" have been added into some of the problems.

Thanks to Ms. Sun Yongmei, Mr. Qiao Yaojun, Mr. Wu Jian and Mr. Zhang Min who gave valuable suggestions on the modification of this book. These four are all young and middle-aged professors or associate professors; thanks to the precious opinions given by the young teacher, Dr. Shen Yuanmao, and especially thanks to the doctoral supervisor Ms. Gu Wanyi and Professor Wang Wenbo who have been giving strong support and backing to the education of the electromagnetic field and the development of the teaching material for many years. Thanks to Dr. Siamak Sorooshyari and Prof. Xue Quan who have made plenty of efforts to touch up the expression.

We hereby express our heart-felt gratitude to the valuable suggestions given by the readers and other universities of the same profession, and hope for more attention and support from them.

The telecommunications engineering specialty of Beijing University of Posts and Telecommunications is the characteristic specialty development point (TS2055) of the first batch which was authorized by the Ministry of Education of the People's Republic of China. The compiling of this book was subsidized by this development point. The aim of compiling this book is to create an elaborative teaching material in telecommunications engineering around the construction of this development point.

Author

Preface for the First Edition

The theory of Electromagnetic fields and electromagnetic waves is an important basis for specialization subjects. By using strong theories, rigorous logic and conceptual abstraction, it profoundly reveals the basic laws of electromagnetic phenomena. This course will not only deepen our understanding of the electromagnetic laws, but also train the right way of thinking and improve the ability to analyze issues.

As far as problem analysis ability is concerned, we emphasize the principle of diversity of analysis. Scientific analysis is an important part of engineering course in university. Based on this principle, an issue analysis can involve a brief physical concept analysis, a more rigorous mathematics analysis, or a combined analysis from different points of view and in different concepts. Thus, for one thing, students will be provided with further exploration of space, for another, teachers will be provided with more choices in analysis methods and lecture contents, for courses in different professions and different class hours.

In terms of the knowledge structure, the theory of electromagnetic field should be a necessary and indispensable part for college students majoring in electronics and communication engineering. The continuous development of science and technology will further demonstrate its importance.

As is known to everyone, the theory of electromagnetic fields and electromagnetic waves is a basis for specialization subjects, such as microwave technology, antennas, wave propagation, and optical fiber transmission. In fact, its application is far beyond this range.

If you know about high-frequency circuits, you'll know that the leads of resistance, inductance, capacitance should be as short as possible. Because these leads will introduce distributed inductance, capacitance, resistance, which will change the element parameters and do harm to the circuit indicators. we have to use the theory of electromagnetic to deal with the problem of "distributed parameter".

In integrated circuits, particularly in high-speed integrated circuits, it is full of electromagnetic field, for example, the coupling between circuits, the coupling between the earth, the loss of material and dispersion, as well as the reflection caused by connectors, cables, turning, through holes and other discontinuities, they will change the circuit parameters, do harm to the integrity of signal. To do analysis of these issues, theory of electromagnetic field is essential.

Electromagnetic field and wave, as the intersection of disciplines, derived many disciplines. Communications discipline contains mobile communications, satellite communications, optical fiber communication, etc. In addition to communications, radar, radio and television, they are all subject to information carried by electromagnetic wave.

As far as development of students' innovative ability is concerned, as a qualified student of electronics, communications and information engineering, one must take efforts to learn the theory of electromagnetic field to deal with the most basic electromagnetic problem. And because the course is interdisciplinary, it helps students to learn inter-disciplines and improve create capacity.

The content of this book refers to the teaching program made by electromagnetic field theory's Steering Group which belongs to Ministry of Education of the People's Republic of China. And we also consult the excellent teaching books at home and abroad and the teaching experiences of the authors. In this book, there are also many examples which are expected to be helpful to improve the readers' abilities in persuing problems.

The book is made up of 11 chapters. The 1st, 9th, 10th and 11en were compiled in turn by Zehua Gao, Yangan Zhang, Shufang Li and Li Li. Yafeng Wang and Xin Zhang compiled the exercises and answers. These six authors above are all young and middle-aged professors or associate professors who have got the doctor's degree. Qixiang Jiao compiled the chapters from 2 to 8. And he is also the editor in chief of this book and planed the draft as a whole. Professor Wanyi Gu who is also a tutor of a Ph.D. student check and approve this book and thanks a lot to her. We are grateful to Huazhi Wang, Maolin Zhang who gave us many advices about the architecture and content of this book. We also want to show our appreciations to the doctors, masters and undergraduates who have given strong support in clearing up the draft, drawing and so on. Thanks to the leaders of Telecommunications Engineering College, wireless communication center and optical communication center of BUPT who have given us so much help and support.

Thanks to editors Min Kuang and Jiang Yu of Science Press. They worked hard for the book's quality assurance.

The book < Electromagnetic Fields and Waves Problem's fine solutions> which is a support of this book has been published by Science Press at the same time. The PPT has been finished and teachers may get it from the press for free. The readers may get the web resources relating to this book at http://jpkc.bupt.edu.cn:4213/dcc/index.htm. If readers have any suggestion about the content of this book, they may get touch with the author.(wangyf@bupt.edu.cn, lili@bupt.edu.cn, zhang@bupt.edu.cn).

Author

Main Character, Parameters and The Expressions of Gradient, Divergence, Rotation

Main Symbols

Symbols	Name	Unit representation
$\boldsymbol{E}$	electric-field intensity	V/m(volt per meter)
$\boldsymbol{H}$	magnetic field intensity	A/m(Ampere per meter)
$\boldsymbol{D}$	electric displacement (Electric flux density)	C/m^2(Coulomb per square meter)
$\boldsymbol{B}$	magnetic induction (magnetic flux density)	T(Tesla)
φ	electric potential	V(Volt)
Ψ_e	electric flux	C(Coulomb)
Φ	magnetic flux	Wb(Weber)
$\boldsymbol{A}$	magnetic vector potential	Wb/m(Weber per meter)
ρ_l	linear charge density	C/m(Coulomb per meter)
ρ_s	surface charge density	C/m^2(Coulomb per square meter)
ρ	volume charge density	C/m^3(Coulomb per cubic meter)
n	refractive index	
R	reflection coefficient	
T	transmission coefficient, reflection coefficient	
C_0	capacitance per unit length	F/m(Farad per meter)
L_0	inductance per unit length	H/m(Henry per meter)
$\boldsymbol{F}$	force	N(Newton)
$\boldsymbol{T}$	moment	N·m (Newton·meter)
w_e	energy density of electric field	J/m^3(Joule per cubic meter)
w_m	energy density of magnetic field	J/m^3(Joule per cubic meter)
$\boldsymbol{S}$	power density(Poynting vector)	W/m^2(Watt per square meter)
$\boldsymbol{J}_s$	surface current density	A/m(Ampere per meter)
$\boldsymbol{J}$	current density	A/m^2(Ampere per square meter)
γ	propagation constant	1/m(1 per meter)
α	attenuation constant	Np/m, dB/m(Napier per meter, decibel/meter)
β	phase-shift constant	rad/m(radian per meter)
k	wave number, TEM phase-shift constant	rad/m(radian per meter)
η	wave impedance of TEM wave	Ω(Ohm)
η_0	wave impedance of TEM wave in vacuum	Ω(Ohm)
$Z_{W(TE)}$	wave impedance of TE wave	Ω(Ohm)
$Z_{W(TM)}$	wave impedance of TM wave	Ω(Ohm)
Z_c	characteristic impedance	Ω(Ohm)
Z_s	surface impedance	Ω(Ohm)
R_s	surface resistance	Ω(Ohm)
X_s	surface reactance	Ω(Ohm)
R_r	radiation resistance	Ω(Ohm)
λ	wavelength	m(meter)
λ_0	wavelength in vacuum	m(meter)
λ_g	waveguide wavelength	m(meter)
λ_c	cut-off wavelength	m(meter)
ε^e	complex dielectric constant	
$\boldsymbol{P}_e$	electric dipole moment	C·m (Coulomb ·meter)
$\boldsymbol{m}$	magnetic dipole moment	$A\cdot m^2$ (Ampere · square meter)

Common Parameters

$c \approx 3\times10^{8}$m/s	velocity of light(in vacuum)
$\varepsilon_0 = \frac{1}{36\pi}\times10^{-9}$F/m	dielectric constant(in vacuum)
$\mu_0 = 4\pi\times10^{-7}$H/m	permeability (in vacuum)
$\sigma_{\mathrm{Ag}} = 6.17\times10^{7}$S/m	Electroconductibility(silver)
$\sigma_{\mathrm{Cu}}=5.80\times10^{7}$S/m	Electroconductibility(copper)
$\sigma_{\mathrm{Au}}=4.10\times10^{7}$S/m	Electroconductibility(gold)
$\sigma_{\mathrm{Pb}}=3.54\times10^{7}$S/m	Electroconductibility(aluminum)
$\sigma_{\mathrm{Cu*Zn}}=1.57\times10^{7}$S/m	Electroconductibility(brass)
$\sigma_{\mathrm{Fe}} = 1.00\times10^{7}$S/m	Electroconductibility(Ferrum)
$e = -1.602\times10^{-19}$C	quantity of electron charge
$m_{\mathrm{e}} = 9.107\times10^{-31}$kg	rest mass of electron
$R_{\mathrm{e}} = 2.81\times10^{-15}$m	Radius of electron
$m_{\mathrm{p}} = 1.673\times10^{-27}$kg	Rest mass of proton
6.6237×10^{-34}J·s	Planck constant
1.38×10^{-23}J/K	Boltzmann constant

The Expressions of Gradient, Divergence, Rotation and Laplace's Equation in Three Kinds Usual Coordinates

Rectangular coordinate system(x, y, z)

$$\nabla\varphi = \boldsymbol{e}_x\frac{\partial\varphi}{\partial x} + \boldsymbol{e}_y\frac{\partial\varphi}{\partial y} + \boldsymbol{e}_z\frac{\partial\varphi}{\partial z}$$

$$\nabla\cdot\boldsymbol{a} = \frac{\partial a_x}{\partial x} + \frac{\partial a_y}{\partial y} + \frac{\partial a_z}{\partial z}$$

$$\nabla\times\boldsymbol{a} = \boldsymbol{e}_x\left(\frac{\partial a_z}{\partial y} - \frac{\partial a_y}{\partial z}\right) + \boldsymbol{e}_y\left(\frac{\partial a_x}{\partial z} - \frac{\partial a_z}{\partial x}\right) + \boldsymbol{e}_z\left(\frac{\partial a_y}{\partial x} - \frac{\partial a_x}{\partial y}\right)$$

$$\nabla^2\varphi = \Delta\varphi = \frac{\partial^2\varphi}{\partial x^2} + \frac{\partial^2\varphi}{\partial y^2} + \frac{\partial^2\varphi}{\partial z^2}$$

Cylindrical-coordinate system (r, ϕ, z)

$$\nabla\varphi = \boldsymbol{e}_r\frac{\partial\varphi}{\partial r} + \frac{\boldsymbol{e}_\phi}{r}\frac{\partial\varphi}{\partial\phi} + \boldsymbol{e}_z\frac{\partial\varphi}{\partial z}$$

$$\nabla\cdot\boldsymbol{a} = \frac{1}{r}\frac{\partial}{\partial r}(ra_r) + \frac{1}{r}\frac{\partial a_\phi}{\partial\phi} + \frac{\partial a_z}{\partial z}$$

$$\nabla\times\boldsymbol{a} = \boldsymbol{e}_r\left(\frac{1}{r}\frac{\partial a_z}{\partial\phi} - \frac{\partial a_\phi}{\partial z}\right) + \boldsymbol{e}_\phi\left(\frac{\partial a_r}{\partial z} - \frac{\partial a_z}{\partial r}\right) + \boldsymbol{e}_z\left(\frac{1}{r}\frac{\partial}{\partial r}(ra_\phi) - \frac{1}{r}\frac{\partial a_r}{\partial\phi}\right)$$

$$\nabla^2\varphi = \Delta\varphi = \frac{1}{r}\frac{\partial}{\partial r}\left(r\frac{\partial\varphi}{\partial r}\right) + \frac{1}{r^2}\frac{\partial^2\varphi}{\partial\phi^2} + \frac{\partial^2\varphi}{\partial z^2}$$

Spherical coordinate system $(\boldsymbol{r}, \boldsymbol{\theta}, \boldsymbol{\phi})$

$$\nabla\varphi = \boldsymbol{e}_r \frac{\partial\varphi}{\partial r} + \frac{\boldsymbol{e}_\theta}{r}\frac{\partial\varphi}{\partial\theta} + \frac{\boldsymbol{e}_\phi}{r\sin\theta}\frac{\partial\varphi}{\partial\phi}$$

$$\nabla\cdot\boldsymbol{a} = \frac{1}{r^2}\frac{\partial}{\partial r}(r^2 a_r) + \frac{1}{r\sin\theta}\frac{\partial}{\partial\theta}(\sin\theta a_\theta) + \frac{1}{r\sin\theta}\frac{\partial a_\phi}{\partial\phi}$$

$$\nabla\times\boldsymbol{a} = \boldsymbol{e}_r\left(\frac{1}{r\sin\theta}\frac{\partial}{\partial\theta}(\sin\theta a_\phi) - \frac{1}{r\sin\theta}\frac{\partial a_\theta}{\partial\phi}\right) + \boldsymbol{e}_\theta\left(\frac{1}{r\sin\theta}\frac{\partial a_r}{\partial\phi} - \frac{1}{r}\frac{\partial}{\partial r}(r a_\phi)\right) + \boldsymbol{e}_\phi\left(\frac{1}{r}\frac{\partial}{\partial r}(r a_\theta) - \frac{1}{r}\frac{\partial a_r}{\partial\theta}\right)$$

$$\nabla^2\varphi = \Delta\varphi = \frac{1}{r^2}\frac{\partial}{\partial r}\left(r^2\frac{\partial\varphi}{\partial r}\right) + \frac{1}{r^2\sin\theta}\frac{\partial}{\partial\theta}\left(\sin\theta\frac{\partial\varphi}{\partial\theta}\right) + \frac{1}{r^2\sin^2\theta}\frac{\partial^2\varphi}{\partial\phi^2}$$

Vector Identities

$$\boldsymbol{a}\cdot(\boldsymbol{b}\times\boldsymbol{c}) = \boldsymbol{b}\cdot(\boldsymbol{c}\times\boldsymbol{a}) = \boldsymbol{c}\cdot(\boldsymbol{a}\times\boldsymbol{b})$$

$$\boldsymbol{a}\times(\boldsymbol{b}\times\boldsymbol{c}) = (\boldsymbol{a}\cdot\boldsymbol{c})\boldsymbol{b} - (\boldsymbol{a}\cdot\boldsymbol{b})\boldsymbol{c}$$

$$\nabla\cdot\nabla = \nabla^2$$

$$\nabla(\varphi_1\varphi_2) = \varphi_1\nabla\varphi_2 + \varphi_2\nabla\varphi_1$$

$$\nabla\cdot(\varphi\boldsymbol{a}) = \boldsymbol{a}\cdot\nabla\varphi + \varphi\nabla\cdot\boldsymbol{a}$$

$$\nabla\times(\varphi\boldsymbol{a}) = \nabla\varphi\times\boldsymbol{a} + \varphi\nabla\times\boldsymbol{a}$$

$$\nabla(\boldsymbol{a}\cdot\boldsymbol{b}) = (\boldsymbol{a}\cdot\nabla)\boldsymbol{b} + (\boldsymbol{b}\cdot\nabla)\boldsymbol{a} + \boldsymbol{a}\times(\nabla\times\boldsymbol{b}) + \boldsymbol{b}\times(\nabla\times\boldsymbol{a})$$

$$\nabla\cdot(\boldsymbol{a}\times\boldsymbol{b}) = \boldsymbol{b}\cdot\nabla\times\boldsymbol{a} - \boldsymbol{a}\cdot\nabla\times\boldsymbol{b}$$

$$\nabla\times(\boldsymbol{a}\times\boldsymbol{b}) = \boldsymbol{a}\nabla\cdot\boldsymbol{b} - \boldsymbol{b}\nabla\cdot\boldsymbol{a} + (\boldsymbol{b}\cdot\nabla)\boldsymbol{a} - (\boldsymbol{a}\cdot\nabla)\boldsymbol{b}$$

$$\nabla\cdot(\nabla\varphi) = (\nabla\cdot\nabla)\varphi = \nabla^2\varphi = \Delta\varphi$$

$$\nabla\times(\nabla\varphi) = 0$$

$$\nabla\cdot(\nabla\times\boldsymbol{a}) = 0$$

$$\nabla\times(\nabla\times\boldsymbol{a}) = \nabla(\nabla\cdot\boldsymbol{a}) - \nabla^2\boldsymbol{a}$$

$$\oint_S \boldsymbol{a}\cdot\mathrm{d}\boldsymbol{S} = \int_V \nabla\cdot\boldsymbol{a}\mathrm{d}V$$

$$\oint_C \boldsymbol{a}\cdot\mathrm{d}\boldsymbol{l} = \int_S \nabla\times\boldsymbol{a}\cdot\mathrm{d}\boldsymbol{S}$$

$$\oint_S (\boldsymbol{n}\times\boldsymbol{a})\mathrm{d}S = \int_V \nabla\times\boldsymbol{a}\cdot\mathrm{d}V$$

$$\oint_S \varphi\boldsymbol{n}\mathrm{d}S = \int_V \nabla\varphi\mathrm{d}V$$

$$\oint_C \varphi\mathrm{d}\boldsymbol{l} = \int_S \boldsymbol{n}\times\nabla\varphi\mathrm{d}S$$

Contents

Chapter 1

Vector Analysis

The theory of vector analysis for electromagnetic fields and electromagnetic waves is introduced in Chapter 1. The concepts of scalar, vector, scalar field, vector field are also included. We introduce the concepts of the vector operation, the flux of vector, the divergence of vector, Gauss's law, the circumfluence of vector, the curl of vector, the Stokes's law, the gradient of scalar, and the Helmholtz's law.

1.1 Scalar and Vector Fields

1.1.1 Scalar

A scalar is a physical quantity that can be expresed by magnitude and sign. The scalar is a point in the space, and can be called an absolute scalar if it is independent of coordinate system. For example real numbers such as mass, length, area, time,temperature,voltage, electric charge, current and energy are all scalars.

1.1.2 Vector

A vector is a physical quantity that can be specified by direction as well as magnitude. A vector may be denoted by $\boldsymbol{a}$ or by line with direction $\overrightarrow{OA}$. It's magnitude can be denoted with a which is called the magnitude of vector $\boldsymbol{a}$ and marked with:

$$|\boldsymbol{a}| = a$$

Vector $\boldsymbol{a}$ can be denoted by a line with direction in three dimensional space shown in Figure 1.1. The length of the line represent the magnitude of vector $\boldsymbol{a}$. The direction of the line represent the direction of vector $\boldsymbol{a}$.

In the three-dimensional Cartesian case of Figure 1.1 that the magnitude of vector $\boldsymbol{a}$ is,

$$a = |\boldsymbol{a}| = \sqrt{a_x^2 + a_y^2 + a_z^2}$$

The cross angle α, β, γ between vector $\boldsymbol{a}$ and the positve direction of x axis, y axis, z axis correspondingly are called the direction angle of vector $\boldsymbol{a}$. The cosine of orientation angle $\cos\alpha$, $\cos\beta$, $\cos\gamma$ is called the direction cosine of vector $\boldsymbol{a}$. It can be seen from the above figure that,

$$a_x = |\boldsymbol{a}|\cos\alpha, \quad a_y = |\boldsymbol{a}|\cos\beta, \quad a_z = |\boldsymbol{a}|\cos\gamma \tag{1.1}$$

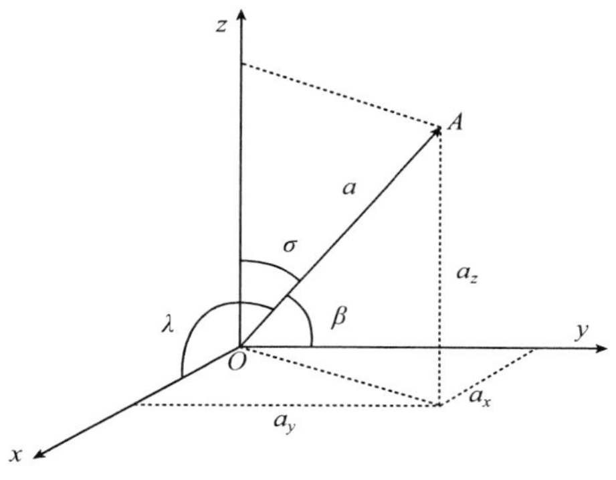

Figure 1.1 Vector $\boldsymbol{a}$ in Cartesian Coordinates

So the orientation cosine satisfy

$$\cos^2\alpha + \cos^2\beta + \cos^2\gamma = 1 \tag{1.2}$$

The vector with a magnitude of 1 is called unit vector and it can be denoted by $\boldsymbol{e}$. The unit vector that has the same direction as vector $\boldsymbol{a}$ with magnitude$a \neq 0$ can be denoted,

$$\boldsymbol{e} = \boldsymbol{a}/a \tag{1.3}$$

The unit vector with the same direction of the positve direction of x axis, y axis, z axis in Cartesian Coordinates is called a base unit vector. The base unit vector can be expressed by $\boldsymbol{e}_x$, $\boldsymbol{e}_y$, $\boldsymbol{e}_z$. The three-dimensional vector $\boldsymbol{a}$ can be decomposed according to the base unit vectors,

$$\boldsymbol{a} = a_x\boldsymbol{e}_x + a_y\boldsymbol{e}_y + a_z\boldsymbol{e}_z$$

When the tail and the head of a vector is superimposed, the vector can be called zero vector. The length of zero vector is zero and the direction of the zero vector is arbitrary. If the magnitude of two vectors are equal and the direction of the two vectors are opposite, the two vector can be called reverse vector. The reverse vector of vector $\boldsymbol{a}$ is denoted $-\boldsymbol{a}$.

If two vectors satisfy: they lien on the same line or they are parallel and their direction is the same; their magnitude is equal, then the two vector is equal vector.

1.1.3 Scalar Field

If the location of scalar φ is a function of time, then the scalar can be described by the function $\varphi(x, y, z, t)$, where x, y, z determine the scalar's position, and t denotes the time. If the range of scalar function is an infinte set, then the set will represent the field of the scalar. As an example, the distribution of space temperature is a temperature field $T(x, y, z, t)$, and the distribution of electric potential is electric $\varphi(x, y, x, t)$.

If the scalar is independent of time, $\varphi(x, y, x, t)$ describes a static field; If the scalar has a relationship with time, $\varphi(x, y, x, t)$ describes a dynamic field.

1.1.4 Vector Field

If a vector $\boldsymbol{F}$ is a function of space, position, and time, it can be expressed by $\boldsymbol{F}(x, y, z, t)$, where x, y, z is space position and t is time. If the range of a vector function is an infinite set, this set represent the field of vector. For instance, the distribution of electric field intensity is given by an electric field vector $\boldsymbol{E}(x, y, z, t)$.

In three-dimensional space, a vector can be represented by three weights in a three-dimensional coordinate system. Without dimension, the three weight would correspond to three scalar values. Thus, one vector can be represented by three scalars. A vector can be written as follows:

$$\boldsymbol{F}(x, y, z) = F_x(x, y, z)\boldsymbol{e}_x + F_y(x, y, z)\boldsymbol{e}_y + F_z(x, y, z)\boldsymbol{e}_z$$

where $F_x(x, y, z), F_y(x, y, z)$ and $F_z(x, y, z)$ are three scalar values.

1.2 Operation of Vector

1.2.1 Vector Addition

In physics, two forces can be decomposed into a superposition of forces via a parallelogram law. Thus, we can define vector addition as follows.

Suppose there are two vectors $\boldsymbol{a}$ and $\boldsymbol{b}$(such as in Figure 1.2), their start point is O, take vector $\boldsymbol{a}$ and $\boldsymbol{b}$ as the side of parallelogram, then vector $\boldsymbol{c}$ corresponding with the diagonal is the resultant vector $\boldsymbol{a}$ and $\boldsymbol{b}$, denote as $\boldsymbol{c}=\boldsymbol{a}+\boldsymbol{b}$. This law is the plus of vector's parallelogram law.

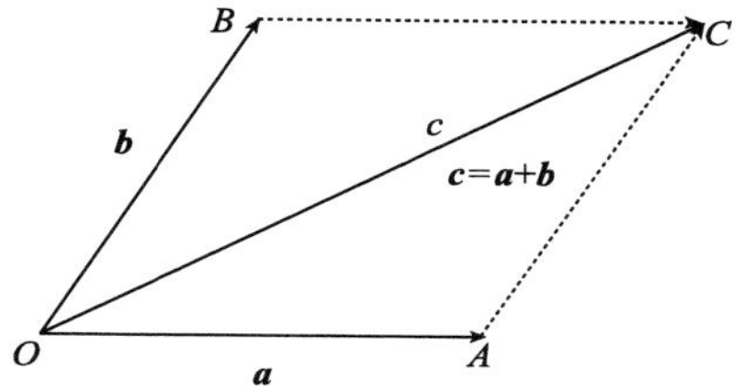

Figure 1.2 Parallelogram law of vector's plus

The plus of vector also can be denoted by basic vector.

If

$$\boldsymbol{a} = a_x \boldsymbol{e}_x + a_y \boldsymbol{e}_y + a_z \boldsymbol{e}_z$$

and

$$\boldsymbol{b} = b_x \boldsymbol{e}_x + b_y \boldsymbol{e}_y + b_z \boldsymbol{e}_z$$

then

$$\boldsymbol{a} + \boldsymbol{b} = (a_x + b_x)\boldsymbol{e}_x + (a_y + b_y)\boldsymbol{e}_y + (a_z + b_z)\boldsymbol{e}_z$$

From the above we can see that the coordinate components of the vector plus are the plus of the corresponding coordinate components of the two vectors

The subtraction of vector is the inverse operation of vector plus. Suppose we have $\boldsymbol{a}+\boldsymbol{b}=\boldsymbol{c}$, then

$$\boldsymbol{a} = \boldsymbol{c} - \boldsymbol{b} = \boldsymbol{c} + (-\boldsymbol{b})$$

Vector addition follows commutativity and associativity:

$$\boldsymbol{a} + \boldsymbol{b} = \boldsymbol{b} + \boldsymbol{a} \tag{1.4}$$

$$\boldsymbol{a}+(\boldsymbol{b}+\boldsymbol{c})=(\boldsymbol{a}+\boldsymbol{b})+\boldsymbol{c} \tag{1.5}$$

1.2.2 Scalar Vector Multiplication

Multiplication between real number λ and vector $\boldsymbol{a}$ defined as multiply of vector between real number λ and vector $\boldsymbol{a}$, the result must be a vector, can be written $\lambda\boldsymbol{a}$, and its magnitude is:

$$|\lambda\boldsymbol{a}| = |\lambda||\boldsymbol{a}| \tag{1.6}$$

The vector's direction is determined as follows: if $\lambda > 0$, direction of $\lambda\boldsymbol{a}$ is the same of $\boldsymbol{a}$; if $\lambda < 0$, $\lambda\boldsymbol{a}$ is right-about with $\boldsymbol{a}$; if $\lambda = 0$, $\lambda\boldsymbol{a}= 0$.

when $\lambda = -1$, $\lambda\boldsymbol{a} = -\boldsymbol{a}$ is defined as reverse vector of vector $\boldsymbol{a}$.

The multiplication of vector $\boldsymbol{a}$ by a scalar λ can be expressed as:

$$\lambda\boldsymbol{a} = \lambda a_x \boldsymbol{e}_x + \lambda a_y \boldsymbol{e}_y + \lambda a_z \boldsymbol{e}_z$$

Supposed $\boldsymbol{a}$ and $\boldsymbol{b}$ are vectors, λ_1 and λ_2 are scalar, then we have

$$(\lambda_1 + \lambda_2)\boldsymbol{a} = \lambda_1\boldsymbol{a} + \lambda_2\boldsymbol{a}$$

$$\lambda_1(\boldsymbol{a} + \boldsymbol{b}) = \lambda_1\boldsymbol{a} + \lambda_1\boldsymbol{b}$$

1.2.3 Scalar Product of Vector (number product, inner product)

In physics, objects have a diaplacement $\boldsymbol{l}$ under force $\boldsymbol{F}$, if angle between $\boldsymbol{F}$ and $\boldsymbol{l}$ is θ, then the force $\boldsymbol{F}$'s power on objects is:

$$W = |\boldsymbol{F}||\boldsymbol{l}| \cos\theta \tag{1.7}$$

According to the above calculation, we can define one kind of product of vector: suppose we have two vectors $\boldsymbol{a}$ and $\boldsymbol{b}$, θ is the angle between them, the scalar product of vectors $\boldsymbol{a}$ and $\boldsymbol{b}$ is denoted by $\boldsymbol{a} \cdot \boldsymbol{b}$. We equate this to the product of magnitudes and the cosine of the angle between the two vectors via

$$\boldsymbol{a} \cdot \boldsymbol{b} = |\boldsymbol{a}||\boldsymbol{b}| \cos\theta \tag{1.8}$$

The scalar product of a vector is also called number product, dot product or inner product.

According to this define, W, power of $\boldsymbol{F}$, can be denoted as:

$$W = \boldsymbol{F} \cdot \boldsymbol{l}$$

The scalar product of vector $\boldsymbol{a}$ and $\boldsymbol{b}$, $\boldsymbol{a} \cdot \boldsymbol{b}$ can be denoted by basic unit vector

$$\boldsymbol{a} \cdot \boldsymbol{b} = a_x b_x + a_y b_y + a_z b_z \tag{1.9}$$

According to last expression, the scalar of vector equal to the sum of product of corresponding weight of two vector.

According to $\boldsymbol{a} \cdot \boldsymbol{b} = |\boldsymbol{a}||\boldsymbol{b}| \cos\theta$, we have

$$\cos\theta = \frac{\boldsymbol{a} \cdot \boldsymbol{b}}{|\boldsymbol{a}|\,|\boldsymbol{b}|}$$

Thinking about expression $\boldsymbol{a} \cdot \boldsymbol{b} = a_x b_x + a_y b_y + a_z b_z$, then

$$\cos\theta = \frac{a_x b_x + a_y b_y + a_z b_z}{\sqrt{a_x^2 + a_y^2 + a_z^2}\sqrt{b_x^2 + b_y^2 + b_z^2}}$$

A Scalar Product will satisfy the following properties:

$$\left.\begin{array}{l} \boldsymbol{a} \cdot \boldsymbol{b} = \boldsymbol{b} \cdot \boldsymbol{a} \\ \boldsymbol{a} \cdot (\boldsymbol{b} + \boldsymbol{c}) = \boldsymbol{a} \cdot \boldsymbol{b} + \boldsymbol{a} \cdot \boldsymbol{c} \\ \lambda_1(\boldsymbol{a} \cdot \boldsymbol{b}) = (\lambda_1 \boldsymbol{a}) \cdot \boldsymbol{b} = \boldsymbol{a} \cdot (\lambda_1 \boldsymbol{b}) = (\boldsymbol{a} \cdot \boldsymbol{b})\lambda_1 \end{array}\right\} \tag{1.10}$$

1.2.4 Vector Product of vector(cross product, external product)

In the study of turning,moment is a physical quantity that is often used.

Supposed force $\boldsymbol{F}$ is working at a point on object A, a moment is produced to a fulcrum with a value that is equal to the product of quantity of the force and distance between O and line of action. This is illustrated in Figure 1.3. Suppose the angle between $\boldsymbol{F}$ and $\overrightarrow{OA}$ is θ, so the moment is:

$$\boldsymbol{T} = \overrightarrow{OA} \times \boldsymbol{F} \tag{1.11}$$

The moment is a vector, and its direction is perpendicular to the plane which is specifyed by $\boldsymbol{F}$ and $\overrightarrow{OA}$. And $\overrightarrow{OA}$, $\boldsymbol{F}$ and $\boldsymbol{T}$ meet right-hand system, means if the fingers of right hand are curled from $\overrightarrow{OA}$'s direction toward the control and $\boldsymbol{F}$, the thumb points in the direction of $\boldsymbol{T}$.

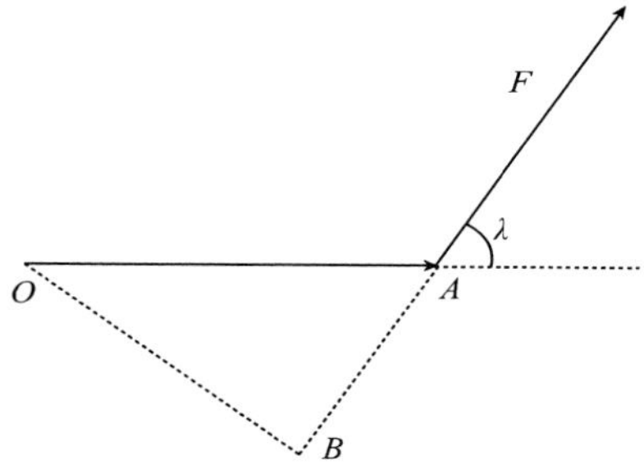

Figure 1.3 Force and moment

According to the above operation, we define another kind of vector operation: suppose there are two vector $\boldsymbol{a}$ and $\boldsymbol{b}$, their included angle is θ, and vector $\boldsymbol{c}$ is vector product of vector $\boldsymbol{a}$ and $\boldsymbol{b}$, denoted as $\boldsymbol{a} \times \boldsymbol{b}$, its value equal to the product of two vector's magnitude and direction sine, namely

$$|\boldsymbol{c}| = |\boldsymbol{a} \times \boldsymbol{b}| = |\boldsymbol{a}||\boldsymbol{b}| \sin\theta$$

Vector $\boldsymbol{c}$ is perpendicular to vector $\boldsymbol{a}$ and $\boldsymbol{b}$, and vector $\boldsymbol{a}$,$\boldsymbol{b}$ and $\boldsymbol{c}$ form the right-handed system.

Vector product ia also named cross product or external product.

According to the above definition, the moment $\boldsymbol{T}$ made by force $\boldsymbol{F}$ also can be denoted as

$$\boldsymbol{T} = \overrightarrow{OA} \times \boldsymbol{F}$$

From the definition of vector product, we can see

$$\left.\begin{array}{l}\boldsymbol{a} \times \boldsymbol{a} = 0 \\ \boldsymbol{a} \times \boldsymbol{b} = -\boldsymbol{b} \times \boldsymbol{a} \\ (\boldsymbol{a}+\boldsymbol{b}) \times \boldsymbol{c} = \boldsymbol{a} \times \boldsymbol{c} + \boldsymbol{b} \times \boldsymbol{a} \\ \lambda_1(\boldsymbol{a} \times \boldsymbol{b}) = (\lambda_1 \boldsymbol{a}) \times \boldsymbol{b} = \boldsymbol{a} \times (\lambda_1 \boldsymbol{b}) = (\boldsymbol{a} \times \boldsymbol{b})\lambda_1\end{array}\right\} \tag{1.12}$$

We infer the following about a vector product expression with basic unit:

According to the vector computation rules,we get

$$\boldsymbol{a} \times \boldsymbol{b} = (a_x \boldsymbol{e}_x + a_y \boldsymbol{e}_y + a_z \boldsymbol{e}_z) \times (b_x \boldsymbol{e}_x + b_y \boldsymbol{e}_y + b_z \boldsymbol{e}_z)$$

Expand the expression, we get

$$\boldsymbol{a} \times \boldsymbol{b} = (a_y b_z - a_z b_y)\boldsymbol{e}_x + (a_z b_x - a_x b_z)\boldsymbol{e}_y + (a_x b_y - a_y b_x)\boldsymbol{e}_z \tag{1.13}$$

The cross product can be expressed in terms of a determinant as follows

$$\boldsymbol{a} \times \boldsymbol{b} = \begin{vmatrix} a_y & a_z \\ b_y & b_z \end{vmatrix} \boldsymbol{e}_x + \begin{vmatrix} a_z & a_x \\ b_z & b_x \end{vmatrix} \boldsymbol{e}_y + \begin{vmatrix} a_x & a_y \\ b_x & b_y \end{vmatrix} \boldsymbol{e}_x$$

or

$$\boldsymbol{a} \times \boldsymbol{b} = \begin{vmatrix} \boldsymbol{e}_x & \boldsymbol{e}_y & \boldsymbol{e}_z \\ a_x & a_y & a_z \\ b_x & b_y & b_z \end{vmatrix} \tag{1.14}$$

Example 1.1 The vectors $\boldsymbol{a}=2\boldsymbol{e}_x - 6\boldsymbol{e}_y - 3\boldsymbol{e}_z$ and $\boldsymbol{b}=4\boldsymbol{e}_x+3\boldsymbol{e}_y-\boldsymbol{e}_z$ specify a plane. What is the unit normal vector?

Method1 basic normal vector is perpendicular to this plane, so this vector must be perpendicular to vectors $\boldsymbol{a}$ and $\boldsymbol{b}$. Suppose the unit vector $\boldsymbol{c}=c_x\boldsymbol{e}_x + c_y\boldsymbol{e}_y + c_z\boldsymbol{e}_z$, so

$$\boldsymbol{c} \cdot \boldsymbol{a} = 2c_x - 6c_y - 3c_z = 0$$

$$\boldsymbol{c} \cdot \boldsymbol{b} = 4c_x + 3c_y - c_z = 0$$

And the same, thinking about $\boldsymbol{c}=c_x\boldsymbol{e}_x + c_y\boldsymbol{e}_y + c_z\boldsymbol{e}_z$ is an unit vector, so $c_x^2 + c_y^2 + c_z^2 = 1$. Finally, the solution of this unit vector is

$$\boldsymbol{c} = \pm\left(\frac{3}{7}\boldsymbol{e}_x - \frac{2}{7}\boldsymbol{e}_y + \frac{6}{7}\boldsymbol{e}_z\right)$$

Method2 because $\boldsymbol{a}\times\boldsymbol{b}$ is perpendicular to the plane specifid by $\boldsymbol{a}$ and $\boldsymbol{b}$, and that

$$\boldsymbol{a} \times \boldsymbol{b} = \begin{vmatrix} \boldsymbol{e}_x & \boldsymbol{e}_y & \boldsymbol{e}_z \\ 2 & -6 & -3 \\ 4 & 3 & -1 \end{vmatrix} = 15\boldsymbol{e}_x - 10\boldsymbol{e}_y + 30\boldsymbol{e}_z$$

Because unit vector parallel $\boldsymbol{a}\times\boldsymbol{b}$, this unit vector is

$$\boldsymbol{c} = \frac{\boldsymbol{a} \times \boldsymbol{b}}{|\boldsymbol{a} \times \boldsymbol{b}|} = -\frac{3}{7}\boldsymbol{e}_x + \frac{2}{7}\boldsymbol{e}_y - \frac{6}{7}\boldsymbol{e}_z$$

The reserve direction unit vector is

$$c = -\frac{3}{7}\boldsymbol{e}_x + \frac{2}{7}\boldsymbol{e}_y - \frac{6}{7}\boldsymbol{e}_z$$

According to $|\boldsymbol{a}\times\boldsymbol{b}| = |\boldsymbol{a}||\boldsymbol{b}|\ \sin\theta$, there are

$$\sin\theta = \frac{|\boldsymbol{a}\times\boldsymbol{b}|}{|\boldsymbol{a}|\,|\boldsymbol{b}|}$$

Thinking about expression $\boldsymbol{a}\times\boldsymbol{b}=(a_y b_z - a_z b_y)\boldsymbol{e}_x+(a_z b_x - a_x b_z)\boldsymbol{e}_y+(a_x b_y - a_y b_x)\boldsymbol{e}_z$, so

$$\sin\theta = \frac{\sqrt{(a_y b_z - a_z b_y)^2 + (a_z b_x - a_x b_z)^2 + (a_x b_y - a_y b_x)^2}}{\sqrt{a_x^2 + a_y^2 + a_z^2}\sqrt{b_x^2 + b_y^2 + b_z^2}}$$

1.2.5 Mixed-Product of Vector

Suppose there are three vectors $\boldsymbol{a}=a_x\boldsymbol{e}_x + a_y\boldsymbol{e}_y + a_z\boldsymbol{e}_z$, $\boldsymbol{b}=b_x\boldsymbol{e}_x + b_y\boldsymbol{e}_y + b_z\boldsymbol{e}_z$ and $\boldsymbol{c}=c_x\boldsymbol{e}_x + c_y\boldsymbol{e}_y + c_z\boldsymbol{e}_z$, we first consider $\boldsymbol{a}\times\boldsymbol{b}$, then take the scalar product of $\boldsymbol{a}\times\boldsymbol{b}$ and $\boldsymbol{c}$, to mean $(\boldsymbol{a}\times\boldsymbol{b})\cdot\boldsymbol{c}$. This is a mixed-product of a vector.

According to the definition of $(\boldsymbol{a}\times\boldsymbol{b})$, we have

$$\boldsymbol{a}\times\boldsymbol{b} = \begin{vmatrix} a_y & a_z \\ b_y & b_z \end{vmatrix}\boldsymbol{e}_x + \begin{vmatrix} a_z & a_x \\ b_z & b_x \end{vmatrix}\boldsymbol{e}_y + \begin{vmatrix} a_x & a_y \\ b_x & b_y \end{vmatrix}\boldsymbol{e}_z$$

so

$$(\boldsymbol{a}\times\boldsymbol{b})\cdot c = \begin{vmatrix} a_y & a_z \\ b_y & b_z \end{vmatrix}c_x + \begin{vmatrix} a_z & a_x \\ b_z & b_x \end{vmatrix}c_y + \begin{vmatrix} a_x & a_y \\ b_x & b_y \end{vmatrix}c_z \tag{1.15}$$

Written as a third order determinant

$$(\boldsymbol{a}\times\boldsymbol{b})\cdot\boldsymbol{c} = \begin{vmatrix} a_x & a_y & a_z \\ b_x & b_y & b_z \\ c_x & c_y & c_z \end{vmatrix}$$

According to the nature of determinant, two straight swap does not change the value and sign of the determinant. So we can easily get

$$(\boldsymbol{a}\times\boldsymbol{b})\cdot\boldsymbol{c} = (\boldsymbol{b}\times\boldsymbol{c})\cdot\boldsymbol{a} = (\boldsymbol{c}\times\boldsymbol{a})\cdot\boldsymbol{b} \tag{1.16}$$

Example 1.2 Prove the following identity

$$(\boldsymbol{a}\times\boldsymbol{b})\times\boldsymbol{c} = (\boldsymbol{a}\cdot\boldsymbol{c})\boldsymbol{b} - (\boldsymbol{b}\cdot\boldsymbol{c})\boldsymbol{a}$$

Prove Suppose $\boldsymbol{a}=a_x\boldsymbol{e}_x+a_y\boldsymbol{e}_y+a_z\boldsymbol{e}_z$, $\boldsymbol{b}=b_x\boldsymbol{e}_x+b_y\boldsymbol{e}_y+b_z\boldsymbol{e}_z$ and $\boldsymbol{c}=c_x\boldsymbol{e}_x+c_y\boldsymbol{e}_y+c_z\boldsymbol{e}_z$, then

$$\boldsymbol{a}\times\boldsymbol{b} = \begin{vmatrix} a_y & a_z \\ b_y & b_z \end{vmatrix}\boldsymbol{e}_x - \begin{vmatrix} a_x & a_z \\ b_x & b_z \end{vmatrix}\boldsymbol{e}_y + \begin{vmatrix} a_x & a_y \\ b_x & b_y \end{vmatrix}\boldsymbol{e}_z$$

Supposed $(\boldsymbol{a}\times\boldsymbol{b})\times\boldsymbol{c}=x\boldsymbol{e}_x + x\boldsymbol{e}_y + x\boldsymbol{e}_z$, then

$$\begin{aligned} x &= -\begin{vmatrix} a_x & a_z \\ b_x & b_z \end{vmatrix}c_z - \begin{vmatrix} a_x & a_y \\ b_x & b_y \end{vmatrix}c_y \\ &= -a_x b_z c_z + a_z b_x c_z - a_x b_y c_y + a_y b_x c_y \\ &= a_x b_x c_x + a_y b_x c_y + a_z b_x c_z - a_x b_x c_x - a_x b_y c_y - a_x b_z c_z \\ &= (a_x c_x + a_y c_y + a_z c_z)b_x - (b_x c_x + b_y c_y + b_z c_z)a_x \\ &= (\boldsymbol{a}\cdot\boldsymbol{c})b_x - (\boldsymbol{b}\cdot\boldsymbol{c})a_x \end{aligned}$$

furthermore

$$y = (\boldsymbol{a}\cdot\boldsymbol{c})b_y - (\boldsymbol{b}\cdot\boldsymbol{c})a_y$$

$$z = (\boldsymbol{a}\cdot\boldsymbol{c})b_z - (\boldsymbol{b}\cdot\boldsymbol{c})a_z$$

thus we have

$$(\boldsymbol{a} \times \boldsymbol{b}) \times \boldsymbol{c} = (\boldsymbol{a} \cdot \boldsymbol{c})\boldsymbol{b} - (\boldsymbol{b} \cdot \boldsymbol{c})\boldsymbol{a}$$

1.3 Flux and Divergence of Vector

1.3.1 Flux

A space surface contains unilateral and bilateral surfaces. The one we often meet is the bilateral .According to the specific problem statement, one side of the bilateral surface must be selected. Once we have selected the side, the bilateral surface is called orientable surface. The direction of normal vector of any point on this orientable surface point to the side we select. One of two methods are used to select a side: one corresponds to the surface S which is an open surface and it is enclosed by a closed curve C, as show in Figure1.4. After selecting the direction of closed curve C, we can specify the direction of an area dSin this curve by the right-handed rule, namely if four fingers take along with the direction of closed curve C, then the direction the thumb will point in a direction perpendicular to this area, and provide the direction $\boldsymbol{n}$ of the area dS. Another methodology corrponds to a closed-surface, generally we adopt outside normal vector as direction of this closed-surface. d$\boldsymbol{S}$=$\boldsymbol{n}$dS is called as vector area unit.

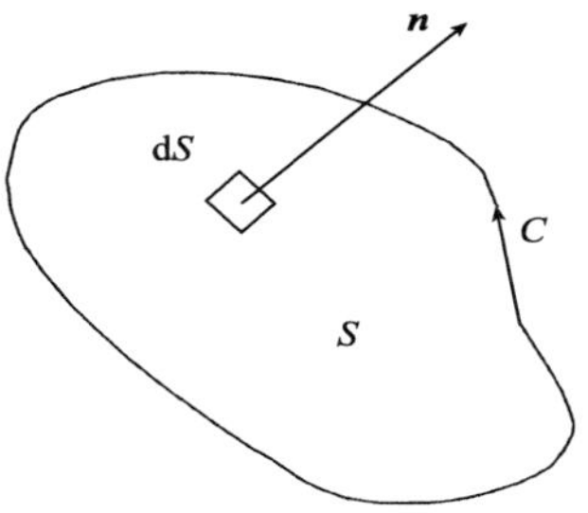

Figure 1.4 The direction of opened-surface S and little area

We consider a vector $\boldsymbol{a}= a_x\boldsymbol{e}_x + a_y\boldsymbol{e}_y + a_z\boldsymbol{e}_z$ going through vector area unit d$\boldsymbol{S}$=$\boldsymbol{n}$dS, because dS is very small, the vector on it can be seen steadily. The scalar product between vector $\boldsymbol{a}$ and vector area unit d$\boldsymbol{S}$ is defined as:

$$\boldsymbol{a} \cdot \mathrm{d}\boldsymbol{S} = \boldsymbol{a} \cdot \boldsymbol{n}\mathrm{d}S = a\mathrm{d}S\cos\theta$$

as flux made by vector $\boldsymbol{a}$ going through vector area unit d$\boldsymbol{S}$, where θ is angle between $\boldsymbol{a}$ and $\boldsymbol{n}$.

When vector $\boldsymbol{a}$ goes through a surface S, the flux will be given by

$$\int_S \boldsymbol{a} \cdot \mathrm{d}\boldsymbol{S} = \int_S \boldsymbol{a} \cdot \boldsymbol{n}\mathrm{d}\boldsymbol{S} = \int_S a\mathrm{d}S\cos\theta$$

In this section, we give attention to a situation where the surface is closed-surface, it has a special meaning. The flux made by vector $\boldsymbol{a}$ go through closed-surface can be denoted as

$$\oint_S \boldsymbol{a} \cdot \mathrm{d}\boldsymbol{S} = \oint_S \boldsymbol{a} \cdot \boldsymbol{n}\mathrm{d}\boldsymbol{S} = \oint_S a\mathrm{d}S\cos\theta \tag{1.17}$$

Explain this equation by fluid-field: if flux go through closed-surface is greater than zero, namely $\oint_S \boldsymbol{a} \cdot \mathrm{d}\boldsymbol{S} > 0$, it means there are net flow come from inside of closed-surface; if flux travels through a closed-surface less than zero, namely $\oint_S \boldsymbol{a} \cdot \mathrm{d}\boldsymbol{S} < 0$, it means there are net flow come from the outside of volume surrounded by the closed-surface into the inside, at this time, there are fluid "cave"(also be called as negative source), which can absorb flow, in the inner of closed-surface. In the first situation, there may have negative source, but the surface integral of negative source is less than positive; in the second, there may have positive source, but the integral of positive source is less than negative. If the flux go through closed-surface S is 0, namely $\oint_S \boldsymbol{a} \cdot \mathrm{d}\boldsymbol{S} = 0$, this means that the flow out from the closed-surface is the same as the flow into the closed-surface, namely the integral of positive source equal to negative, or there are't any positive and negative source.

1.3.2 Divergence

Flux describes a large area integral, the value of integral can't reveal the real value of everywhere. To investigate the vector field $\boldsymbol{a}$, we let a closed-surface shrink by taking the limit

$$\lim_{\Delta V \to 0} \frac{\oint_S \boldsymbol{a} \cdot \mathrm{d}\boldsymbol{S}}{\Delta V}$$

This limit is the real value of this point, called as the intensity of this source, and denoted by $\mathrm{div}\boldsymbol{a}$. Thus, the definition of intensity can be expressed by

$$\mathrm{div}\boldsymbol{a} = \lim_{\Delta V \to 0} \frac{\oint_S \boldsymbol{a} \cdot \mathrm{d}\boldsymbol{S}}{\Delta V} \tag{1.18}$$

Indicating that the divergence coming from a unit volume on this point. We note the following from the last expression: the divergence of vector field is scalar. If $\mathrm{div}\boldsymbol{a} > 0$, we can state that there is a source on this point; if $\mathrm{div}\boldsymbol{a} < 0$, there is cave; if $\mathrm{div}\boldsymbol{a} = 0$, we can say this point is no-divergent, if $\mathrm{div}\boldsymbol{a} = 0$ is true for everywhere, then the field is no-divergent field.

According to the definition of divergence, $\mathrm{div}\boldsymbol{a}$ states that the value on a point, is irrelative with shape of volume element ΔV, when we take a limit to zero, all the range is near zero, when infering the expression of divergence, we adopt Cartesian Coordinates and take parallelepiped as volume element. In Cartesian Coordinates, we consider (x,y,z) as one point on a parallelepiped, as shown in Figure 1.5. respectively, we can calculate the flux going through the six surfaces before departing through the back sides as

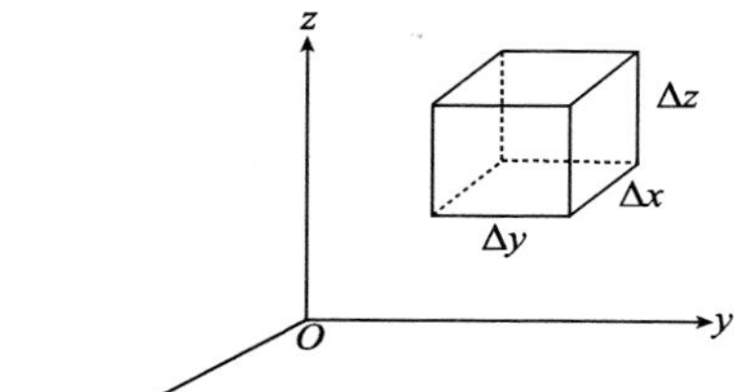

Figure 1.5 Calculate flux by parallelepiped

$$\left(a_x + \frac{\partial a_x}{\partial x}\Delta x\right)\Delta y \Delta z - a_x \Delta y \Delta z = \frac{\partial a_x}{\partial x}\Delta x \Delta y \Delta z$$

The flux going through right and left sides is

$$-a_y \Delta x \Delta z + \left(a_y + \frac{\partial a_y}{\partial y}\Delta y\right)\Delta x \Delta z = \frac{\partial a_y}{\partial y}\Delta x \Delta y \Delta z$$

The flux going through upward and down sides is

$$\left(a_z + \frac{\partial a_z}{\partial z}\Delta z\right)\Delta x \Delta y - a_z \Delta x \Delta y = \frac{\partial a_z}{\partial z}\Delta x \Delta y \Delta z$$

From the definition of divergence, we have

$$\mathrm{div}\boldsymbol{a} = \lim_{\Delta V \to 0} \frac{\oint_S \boldsymbol{a} \cdot \mathrm{d}\boldsymbol{S}}{\Delta V} = \lim_{\Delta V \to 0} \frac{\left(\dfrac{\partial a_x}{\partial x} + \dfrac{\partial a_y}{\partial y} + \dfrac{\partial a_z}{\partial z}\right)\Delta x \Delta y \Delta z}{\Delta x \Delta y \Delta z} = \frac{\partial a_x}{\partial x} + \frac{\partial a_y}{\partial y} + \frac{\partial a_z}{\partial z}$$

The Hamiltonian operator (also referred to as the vector operator), written as "∇", is a differential operator, and should be treated as a vector. Under Cartesian Coordinate system, the Hamiltonian can be denoted as

$$\nabla = \boldsymbol{e}_x\frac{\partial}{\partial x} + \boldsymbol{e}_y\frac{\partial}{\partial y} + \boldsymbol{e}_z\frac{\partial}{\partial z} \tag{1.19}$$

By use of the Hamiltonian operator, the divergence can be written as

$$\text{div}\boldsymbol{a} = \frac{\partial a_x}{\partial x} + \frac{\partial a_y}{\partial y} + \frac{\partial a_z}{\partial z} = \nabla \cdot \boldsymbol{a} \tag{1.20}$$

The divergence of a vector field that is the sum of two vector will equal the sum of each vector field's divergence. More specifically, we have the property

$$\nabla \cdot (\boldsymbol{a} + \boldsymbol{b}) = \nabla \cdot \boldsymbol{a} + \nabla \cdot \boldsymbol{b} \tag{1.21}$$

1.4 Gauss's Theorem

Gauss's Theorem makes use of the divergence property discussed above. The flux generated by vector $\boldsymbol{a}$ through a closed-surface is equal to the integral of vector $\boldsymbol{a}$'s divergence in the volume V surrounded by closed-surface S, namely

$$\oint_S \boldsymbol{a} \cdot \mathrm{d}\boldsymbol{S} = \int_V \nabla \cdot \boldsymbol{a}\mathrm{d}\boldsymbol{V} \tag{1.22}$$

where vector $\boldsymbol{a}$ is supposed to be successive in area of S, and to have successive 1st-order partial derivative. Gauss's theorem is frequently referred to as the divergence throrem.

Proof of Gauss's Theorem: Divide the volume V surrounded by closed-surface S into the volume element: $\mathrm{d}V_1$, $\mathrm{d}V_2, \cdots$, first we calculate the flux made by vector $\boldsymbol{a}$ going through each volume element's closed-surface. According to definition, there is

$$\oint_{S_1} \boldsymbol{a} \cdot \mathrm{d}\boldsymbol{S} = \nabla \cdot \boldsymbol{a}\mathrm{d}\boldsymbol{V}_1, \quad \oint_{S_2} \boldsymbol{a} \cdot \mathrm{d}\boldsymbol{S} = \nabla \cdot \boldsymbol{a}\mathrm{d}\boldsymbol{V}_2, \cdots$$

Since the two near volume element have a common surface, sum of flux made by vector $\boldsymbol{a}$ going through this common surface is zero. Except volume element near closed-surface S, other volume elements are all surrounded by common surface with other volume elements, so the sum of flux going through these inner volume element is zero. About volume element near surface S, there are only flux going through surface S and can't be cancelled, sum of these flux are equal to flux made by vector $\boldsymbol{a}$ going through surface S. That is

$$\oint_{S_1} \boldsymbol{a} \cdot \mathrm{d}\boldsymbol{S} + \oint_{S_2} \boldsymbol{a} \cdot \mathrm{d}\boldsymbol{S} + \cdots = \oint_S \boldsymbol{a} \cdot \mathrm{d}\boldsymbol{S}$$

so

$$\oint_S \boldsymbol{a} \cdot \mathrm{d}\boldsymbol{S} = \nabla \cdot \boldsymbol{a}\mathrm{d}\boldsymbol{V}_1 + \nabla \cdot \boldsymbol{a}\mathrm{d}\boldsymbol{V}_2 + \cdots = \int_V \nabla \cdot \boldsymbol{a}\mathrm{d}\boldsymbol{V}$$

Q.E.D.

Example 1.3 Calculate the following quantities:

(1) the flux of vector $\boldsymbol{a}=x^2\boldsymbol{e}_x+(xy)^2\boldsymbol{e}_y+24x^2y^2z^3\boldsymbol{e}_z$;

(2) the integral of $\nabla\cdot\boldsymbol{a}$ to a unit volume with center on the origin of coordinate system;

(3) the integral of $\boldsymbol{a}$ to surface of this volume.

Answer (1) the flux of vector $\boldsymbol{a}=x^2\boldsymbol{e}_x+(xy)^2\boldsymbol{e}_y+24x^2y^2z^3\boldsymbol{e}_z$ is

$$\nabla \cdot \boldsymbol{a} = \frac{\partial a_x}{\partial x} + \frac{\partial a_y}{\partial y} + \frac{\partial a_z}{\partial z} = 2x + 2x^2y + 72x^2y^2z^2$$

(2) the integral of $\nabla\cdot\boldsymbol{a}$ to a unit volume with center on origin of coordinates is

$$\int_V \nabla\cdot\boldsymbol{a}\mathrm{d}V = \int_{-\frac{1}{2}}^{\frac{1}{2}}\int_{-\frac{1}{2}}^{\frac{1}{2}}\int_{-\frac{1}{2}}^{\frac{1}{2}}(2x+2x^2y+72x^2y^2z^2)\mathrm{d}x\mathrm{d}y\mathrm{d}z$$

$$=\int_{-\frac{1}{2}}^{\frac{1}{2}}\int_{-\frac{1}{2}}^{\frac{1}{2}}(2x+2x^2y+6x^2y^2)\mathrm{d}x\mathrm{d}y=\int_{-\frac{1}{2}}^{\frac{1}{2}}\left(2x+\frac{1}{2}x^2\right)\mathrm{d}x=\frac{1}{24}$$

(3) the integral of $\boldsymbol{a}$ to surface of this volume is

$$\oint_S \boldsymbol{a}\cdot\mathrm{d}\boldsymbol{S}$$

$$=\int_{-\frac{1}{2}}^{\frac{1}{2}}\int_{-\frac{1}{2}}^{\frac{1}{2}} a_z\Big|_{z=\frac{1}{2}}\mathrm{d}x\mathrm{d}y-\int_{-\frac{1}{2}}^{\frac{1}{2}}\int_{-\frac{1}{2}}^{\frac{1}{2}} a_z\Big|_{z=-\frac{1}{2}}\mathrm{d}x\mathrm{d}y$$

$$+\int_{-\frac{1}{2}}^{\frac{1}{2}}\int_{-\frac{1}{2}}^{\frac{1}{2}} a_x\Big|_{x=\frac{1}{2}}\mathrm{d}y\mathrm{d}z-\int_{-\frac{1}{2}}^{\frac{1}{2}}\int_{-\frac{1}{2}}^{\frac{1}{2}} a_x\Big|_{x=-\frac{1}{2}}\mathrm{d}y\mathrm{d}z$$

$$+\int_{-\frac{1}{2}}^{\frac{1}{2}}\int_{-\frac{1}{2}}^{\frac{1}{2}} a_y\Big|_{y=\frac{1}{2}}\mathrm{d}z\mathrm{d}x-\int_{-\frac{1}{2}}^{\frac{1}{2}}\int_{-\frac{1}{2}}^{\frac{1}{2}} a_y\Big|_{y=-\frac{1}{2}}\mathrm{d}z\mathrm{d}x$$

$$=\int_{-\frac{1}{2}}^{\frac{1}{2}}\int_{-\frac{1}{2}}^{\frac{1}{2}}(24x^2y^2z^3)\Big|_{z=\frac{1}{2}}\mathrm{d}x\mathrm{d}y-\int_{-\frac{1}{2}}^{\frac{1}{2}}\int_{-\frac{1}{2}}^{\frac{1}{2}}(24x^2y^2z^3)\Big|_{z=-\frac{1}{2}}\mathrm{d}x\mathrm{d}y$$

$$+\int_{-\frac{1}{2}}^{\frac{1}{2}}\int_{-\frac{1}{2}}^{\frac{1}{2}} x^2\Big|_{x=\frac{1}{2}}\mathrm{d}y\mathrm{d}z-\int_{-\frac{1}{2}}^{\frac{1}{2}}\int_{-\frac{1}{2}}^{\frac{1}{2}} x^2\Big|_{x=-\frac{1}{2}}\mathrm{d}y\mathrm{d}z$$

$$+\int_{-\frac{1}{2}}^{\frac{1}{2}}\int_{-\frac{1}{2}}^{\frac{1}{2}}(xy)^2\Big|_{y=\frac{1}{2}}\mathrm{d}z\mathrm{d}x-\int_{-\frac{1}{2}}^{\frac{1}{2}}\int_{-\frac{1}{2}}^{\frac{1}{2}}(xy)^2\Big|_{y=-\frac{1}{2}}\mathrm{d}z\mathrm{d}x$$

$$=\left(\frac{1}{48}+\frac{1}{48}\right)+0+0=\frac{1}{24}$$

We can see from expression $\oint_S \boldsymbol{a}\cdot\mathrm{d}\boldsymbol{S}=\int_V\nabla\cdot\boldsymbol{a}\mathrm{d}V$, this example has proveed the Divergence theorem.

1.5 Vector Circulation and Rotation

We consider the integral of vector $\boldsymbol{a}= a_x\boldsymbol{e}_x+a_y\boldsymbol{e}_y+a_z\boldsymbol{e}_z$ on a curve C. The direction of curve C is the same as the line element on the path of integral. The vector of line element is denoted by $\mathrm{d}\boldsymbol{l}$, and the line integral of vector $\boldsymbol{a}$ on curve C can be written as

$$\int_C \boldsymbol{a}\cdot\mathrm{d}\boldsymbol{l}$$

In field theory (specifically, in the electromagnetic field theory) an important situation is the curve C is closed curve. In such a case the integral of vector $\boldsymbol{a}$ on closed curve Ccan be denoted as

$$\oint_C \boldsymbol{a}\cdot\mathrm{d}\boldsymbol{l} \tag{1.23}$$

The above scenario is referred to as a loop integral (or circulation).

Circulation is a macroscopic value, and its integral value can't reflect the circulation state at each spot. In order to research vector $\boldsymbol{a}$'s circulation state on everywhere in space, we shrink the closed curve, make the area element ΔS_n surrounded by this curve approach zero, and take limit

$$\lim_{\Delta S_n \to 0} \frac{\oint_C \boldsymbol{a} \cdot \mathrm{d}\boldsymbol{l}}{\Delta S_n} \tag{1.24}$$

The final expression above can reflect the circulation state at each spot in terms of area density of circulation, the direction will satisfy the right-handed rule, and is named weight of vector $\boldsymbol{a}$'s rotation in $\boldsymbol{n}$'s direction, denoted $\mathrm{rot}_n \boldsymbol{a}$ or $\mathrm{curl}_n \boldsymbol{a}$. Thus, we have

$$\mathrm{rot}_n \boldsymbol{a} = \lim_{\Delta S_n \to 0} \frac{\oint_C a \cdot \mathrm{d}l}{\Delta S_n} \tag{1.25}$$

Rotation is max circulation in unit area at a point.

It should be noted that if $\mathrm{rot}\boldsymbol{a}=0$ is always true, then the vector field is a non-rotational field.

In the following, we analyse a rotations expression in the Cartesian Coordinate system.

According to Definition(1.25), $\mathrm{rot}\boldsymbol{a}$ is the value of one point, independent of the shape of area element ΔS. When we take a limit to zero, all length of the area approaches zero, where area element is rectangular surface element, as shown in Figure1.6, adopt an area element parallel to axis x and y, its area is ΔS_z. At point $A(x, y, z)$, vector $\boldsymbol{a}=a_x \boldsymbol{e}_x + a_y \boldsymbol{e}_y + a_z \boldsymbol{e}_z$, the integrals of vector $\boldsymbol{a}$ on loop l_1, l_2, l_3, l_4 are

$$\begin{aligned}\oint_C \boldsymbol{a} \cdot \mathrm{d}\boldsymbol{l} &= a_x \Delta x + \left(a_y + \frac{\partial a_y}{\partial x}\Delta x\right)\Delta y - \left(a_x + \frac{\partial a_x}{\partial y}\Delta y\right)\Delta x - a_y \Delta y \\ &= \left(\frac{\partial a_y}{\partial x} - \frac{\partial a_x}{\partial y}\right)\Delta x \Delta y\end{aligned}$$

so

$$\lim_{\Delta S_z \to 0} \frac{\oint_C \boldsymbol{a} \cdot \mathrm{d}\boldsymbol{l}}{\Delta S_z} = \frac{\partial a_y}{\partial x} - \frac{\partial a_x}{\partial y} = \mathrm{rot}_z \boldsymbol{a} \tag{1.26}$$

This limit is given by the rotation $\mathrm{rot}\boldsymbol{a}$'s on ΔS_z $\mathrm{rot}_z \boldsymbol{a}$, namely the weight of $\mathrm{rot}\boldsymbol{a}$ on axis z.

Following the same analysis, we adopt an area element ΔS_x, ΔS_y, respectively parallels to axis x and y, then the weight of $\mathrm{rot}\boldsymbol{a}$ on axis x is

$$\lim_{\Delta S_x \to 0} \frac{\oint_C \boldsymbol{a} \cdot \mathrm{d}\boldsymbol{l}}{\Delta S_x} = \frac{\partial a_z}{\partial y} - \frac{\partial a_y}{\partial z} = \mathrm{rot}_x \boldsymbol{a} \tag{1.27}$$

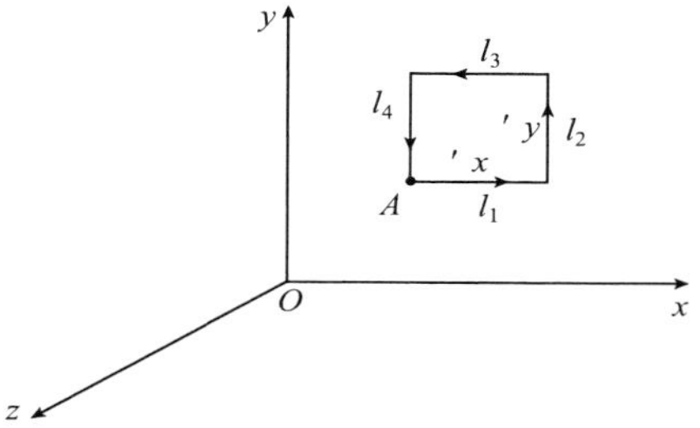

Figure 1.6 Area element ΔS_z used in calculating $\mathrm{rot}_z \boldsymbol{a}$

The weight of $\mathrm{rot}\boldsymbol{a}$ on axis y is given by

$$\lim_{\Delta S_y \to 0} \frac{\oint_C \boldsymbol{a} \cdot \mathrm{d}\boldsymbol{l}}{\Delta S_y} = \frac{\partial a_x}{\partial z} - \frac{\partial a_z}{\partial x} = \mathrm{rot}_y \boldsymbol{a} \tag{1.28}$$

The following operation

$$\begin{aligned}\mathrm{rot}\boldsymbol{a} &= \boldsymbol{e}_x \mathrm{rot}_x \boldsymbol{a} + \boldsymbol{e}_y \mathrm{rot}_y \boldsymbol{a} + \boldsymbol{e}_z \mathrm{rot}_z \boldsymbol{a} \\ &= \left(\frac{\partial a_z}{\partial y} - \frac{\partial a_y}{\partial z}\right)\boldsymbol{e}_x + \left(\frac{\partial a_x}{\partial z} - \frac{\partial a_z}{\partial x}\right)\boldsymbol{e}_y + \left(\frac{\partial a_y}{\partial x} - \frac{\partial a_x}{\partial y}\right)\boldsymbol{e}_z\end{aligned} \tag{1.29}$$

Indicates that the rotation of $\boldsymbol{a}$ equals to the sum of circulation's weight: $\mathrm{rot}_x\boldsymbol{a}$, $\mathrm{rot}_y\boldsymbol{a}$ and $\mathrm{rot}_z\boldsymbol{a}$.

Use of the Hamiltonian operator allows the rotation to be written as

$$\mathrm{rot}\boldsymbol{a} = \nabla \times \boldsymbol{a} = \left(\boldsymbol{e}_x\frac{\partial}{\partial x} + \boldsymbol{e}_y\frac{\partial}{\partial y} + \boldsymbol{e}_z\frac{\partial}{\partial z}\right) \times (a_x\boldsymbol{e}_x + a_y\boldsymbol{e}_y + a_z\boldsymbol{e}_z) \tag{1.30}$$

By using the determinant operator, the rotation can be written as

$$\nabla \times \boldsymbol{a} = \begin{vmatrix} \boldsymbol{e}_x & \boldsymbol{e}_y & \boldsymbol{e}_z \\ \dfrac{\partial}{\partial x} & \dfrac{\partial}{\partial y} & \dfrac{\partial}{\partial z} \\ a_x & a_y & a_z \end{vmatrix} \tag{1.31}$$

Rotation of two vector's sum equals to the sum of each vector's rotation, means

$$\nabla \times (\boldsymbol{a} + \boldsymbol{b}) = \nabla \times \boldsymbol{a} + \nabla \times \boldsymbol{b}$$

A useful rotation quality states that a rotation's divergence is always zero.
Prove

$$\begin{aligned}
\nabla \cdot (\nabla \times \boldsymbol{a}) &= \left(\boldsymbol{e}_x\frac{\partial}{\partial x} + \boldsymbol{e}_y\frac{\partial}{\partial y} + \boldsymbol{e}_z\frac{\partial}{\partial z}\right) \cdot \left[\left(\boldsymbol{e}_x\frac{\partial}{\partial x} + \boldsymbol{e}_y\frac{\partial}{\partial y} + \boldsymbol{e}_z\frac{\partial}{\partial z}\right)\right. \\
&\left.\times (a_x\boldsymbol{e}_x + a_y\boldsymbol{e}_y + a_z\boldsymbol{e}_z)\right] = \left(\boldsymbol{e}_x\frac{\partial}{\partial x} + \boldsymbol{e}_y\frac{\partial}{\partial y} + \boldsymbol{e}_z\frac{\partial}{\partial z}\right) \\
&\left[\left(\frac{\partial a_z}{\partial y} - \frac{\partial a_y}{\partial z}\right)\boldsymbol{e}_x + \left(\frac{\partial a_x}{\partial z} - \frac{\partial a_z}{\partial x}\right)\boldsymbol{e}_y + \left(\frac{\partial a_y}{\partial x} - \frac{\partial a_x}{\partial y}\right)\boldsymbol{e}_z\right] \\
&= \frac{\partial}{\partial x}\left(\frac{\partial a_z}{\partial y} - \frac{\partial a_y}{\partial z}\right) + \frac{\partial}{\partial y}\left(\frac{\partial a_x}{\partial z} - \frac{\partial a_z}{\partial x}\right) + \frac{\partial}{\partial z}\left(\frac{\partial a_y}{\partial x} - \frac{\partial a_x}{\partial y}\right) = 0
\end{aligned}$$

The above can be written as

$$\nabla \cdot (\nabla \times \boldsymbol{a}) = 0 \tag{1.32}$$

The fact that a rotation's divergence is always zero is a very important property. Alternatively, if the flux of vector filed $\boldsymbol{b}$ is zero at anywhere, we can see this vector field as another vector $\boldsymbol{a}$'s rotation, namely if $\nabla\cdot\boldsymbol{b}=0$, then $\boldsymbol{b}=\nabla\times\boldsymbol{a}$.

1.6 Stockes' Theorem

The loop integral of vector $\boldsymbol{a}$ on a closed-curve is equal to the flux of vector $\boldsymbol{a}$'s rotation through any surface S with a border that is a closed-curve C, namely

$$\oint_C a \cdot \mathrm{d}l = \int_S \nabla \times \boldsymbol{a} \cdot \mathrm{d}\boldsymbol{S} \tag{1.33}$$

The above is a mathematical expression referred to as Stockes' Theorem.

Proof Surface S is shown in Figure 1.7, divide this surface to many little area element, make them minimizing to zero and denoted by $\mathrm{d}S_k$, C_k is a closed curve surrounding this area element. According to the definition of rotation, the loop integral of vector $\boldsymbol{a}$ on closed curve C_k is

$$\oint_{C_k} \boldsymbol{a} \cdot \mathrm{d}\boldsymbol{l} = \nabla \times \boldsymbol{a} \cdot \mathrm{d}\boldsymbol{S}_k$$

Plus these circulation on all little surface, then the integral on common curve cancel each other(as shown in figure, the integral on common curve is reverse), so there is only partial integral on non-common curve be reversed. As a result, the sum of circulation on each small closed curve equals the circulation on closed curve C, namely

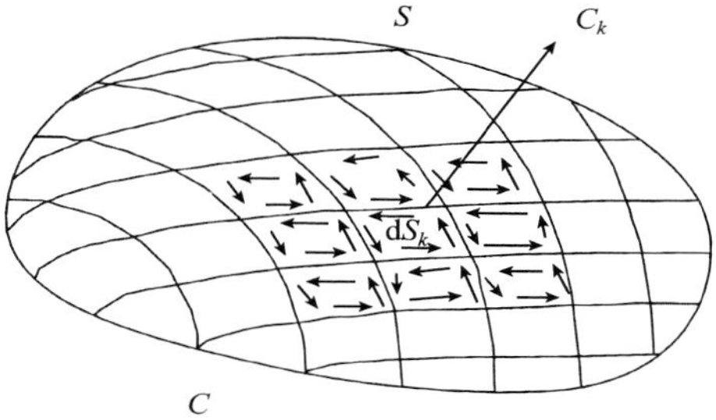

Figure 1.7 Divide surface S into little area element

$$\oint_{C_1} \boldsymbol{a}\cdot\mathrm{d}\boldsymbol{l}+\oint_{C_2} \boldsymbol{a}\cdot\mathrm{d}\boldsymbol{l}+\cdots+\oint_{C_k} \boldsymbol{a}\cdot\mathrm{d}\boldsymbol{l}+\cdots=\oint_{C} \boldsymbol{a}\cdot\mathrm{d}\boldsymbol{l}$$

Because of $\oint_{C_k} \boldsymbol{a}\cdot\mathrm{d}\boldsymbol{l} = \nabla\times\boldsymbol{a}\cdot\mathrm{d}\boldsymbol{S}_k$, so

$$\oint_{C_1} \boldsymbol{a}\cdot\mathrm{d}\boldsymbol{l}+\oint_{C_2} \boldsymbol{a}\cdot\mathrm{d}\boldsymbol{l}+\cdots+\oint_{C_k} \boldsymbol{a}\cdot\mathrm{d}\boldsymbol{l}+\cdots$$
$$=\nabla\times\boldsymbol{a}\cdot\mathrm{d}\boldsymbol{S}_1+\nabla\times\boldsymbol{a}\cdot\mathrm{d}\boldsymbol{S}_2+\cdots+\nabla\times\boldsymbol{a}\cdot\mathrm{d}\boldsymbol{S}_k+\cdots$$

Due to the area element $\mathrm{d}S_k$ approaching to zero, we have

$$\nabla\times\boldsymbol{a}\cdot\mathrm{d}\boldsymbol{S}_1+\nabla\times\boldsymbol{a}\cdot\mathrm{d}\boldsymbol{S}_2+\cdots+\nabla\times\boldsymbol{a}\cdot\mathrm{d}\boldsymbol{S}_k+\cdots=\int_S \nabla\times\boldsymbol{a}\cdot\mathrm{d}\boldsymbol{S}$$

which leads to Stockes' Theorem

$$\oint_C \boldsymbol{a}\cdot\mathrm{d}l=\int_S \nabla\times a\cdot\mathrm{d}\boldsymbol{S}$$

Example 1.4 Given the vector field $\boldsymbol{a}=z\boldsymbol{e}_x+x\boldsymbol{e}_y+y\boldsymbol{e}_z$, and take hemispherical surface $S(x^2+y^2+z^2=1, z\geqslant 0)$, certify Stockes Theorem.

Answer On the $x-y$ plane the closed curve is $x^2+y^2=1(z=0)$, $\mathrm{d}\boldsymbol{l}=\mathrm{d}x\boldsymbol{e}_x+\mathrm{d}y\boldsymbol{e}_y$, there is

$$\oint_C \boldsymbol{a}\cdot\mathrm{d}\boldsymbol{l}=\oint_C (z\mathrm{d}x+x\mathrm{d}y)=\oint_C x\mathrm{d}y=\int_{-1}^{1}\sqrt{1-y^2}\mathrm{d}y+\int_{1}^{-1}(-\sqrt{1-y^2})\mathrm{d}y=\pi$$

and

$$\nabla\times\boldsymbol{a}=\left(\frac{\partial a_z}{\partial y}-\frac{\partial a_y}{\partial z}\right)\boldsymbol{e}_x+\left(\frac{\partial a_x}{\partial z}-\frac{\partial a_z}{\partial x}\right)\boldsymbol{e}_y+\left(\frac{\partial a_y}{\partial x}-\frac{\partial a_x}{\partial y}\right)\boldsymbol{e}_z$$
$$=\boldsymbol{e}_x+\boldsymbol{e}_y+\boldsymbol{e}_z$$

so

$$\int_S \nabla\times\boldsymbol{a}\cdot\mathrm{d}\boldsymbol{S}=\int_S(\boldsymbol{e}_x+\boldsymbol{e}_y+\boldsymbol{e}_z)\cdot\boldsymbol{e}_r r^2\sin\theta\mathrm{d}\theta\mathrm{d}\phi$$
$$=\int_S(\boldsymbol{e}_x\cdot\boldsymbol{e}_r+\boldsymbol{e}_y\cdot\boldsymbol{e}_r+\boldsymbol{e}_z\cdot\boldsymbol{e}_r)\sin\theta\mathrm{d}\theta\mathrm{d}\phi$$
$$=\int_0^{2\pi}\int_0^{\frac{\pi}{2}}(\sin\theta\cos\phi+\sin\theta\sin\phi+\cos\theta)\sin\theta\mathrm{d}\theta\mathrm{d}\phi$$
$$=\int_0^{2\pi}\cos\phi\mathrm{d}\phi\int_0^{\frac{\pi}{2}}\sin^2\theta\mathrm{d}\theta+\int_0^{2\pi}\sin\phi\mathrm{d}\phi\int_0^{\frac{\pi}{2}}\sin^2\theta\mathrm{d}\theta+\int_0^{2\pi}\mathrm{d}\phi\int_0^{\frac{\pi}{2}}\sin\theta\cos\theta\mathrm{d}\theta=\pi$$

as a result

$$\oint_C \boldsymbol{a}\cdot\mathrm{d}l=\int_S \nabla\times\boldsymbol{a}\cdot\mathrm{d}\boldsymbol{S}$$

holds with Stockes theorem being verified.

1.7 Gradient of a Scalar Field

In this part, the concept of gradient is introduced, in order to learn more about the change of scalar φ in scalar field, suppose there is a vector $\boldsymbol{b}$ in a scalar field, which meets the two requirements: its direction is equal to that at where the change in the function $\varphi(x, y, z)$ is maximal; and its modulus is the value of this maximal change. Then we call vector $\boldsymbol{b}$ is the gradient of function $\varphi(x, y, z)$ at point A, denoted as grad φ, namely

$$\text{grad}\varphi = \boldsymbol{b} = \frac{\partial \varphi}{\partial x}\boldsymbol{e}_x + \frac{\partial \varphi}{\partial y}\boldsymbol{e}_y + \frac{\partial \varphi}{\partial z}\boldsymbol{e}_z$$

By using of Hamiltonian operateor, gradient can be written in

$$\text{grad}\varphi = \frac{\partial \varphi}{\partial x}\boldsymbol{e}_x + \frac{\partial \varphi}{\partial y}\boldsymbol{e}_y + \frac{\partial \varphi}{\partial z}\boldsymbol{e}_z = \nabla\varphi \tag{1.34}$$

In scalar field, the gradient of function $\varphi(x, y, z)$ on point is a vector, and its direction is that at where fuction $\varphi(x, y, z)$'s change is maximal at A, its modularity is the value of this maximal change, namely function φ's maimum change rate.

The Gradient of the sum of a vector is equal to the sum of each vector's gradient, namely

$$\nabla(\varphi_1 + \varphi_2) = \nabla\varphi_1 + \nabla\varphi_2$$

There exists a useful gradient's quality: the rotation of gradient is always zero, namely

$$\text{rot}(\text{grad}\varphi) = \nabla \times (\nabla\varphi) = 0 \tag{1.35}$$

Prove

$$\begin{aligned}
&\nabla \times (\nabla\varphi)\\
&= \left(\boldsymbol{e}_x\frac{\partial}{\partial x} + \boldsymbol{e}_y\frac{\partial}{\partial y} + \boldsymbol{e}_z\frac{\partial}{\partial z}\right) \times \left(\frac{\partial \varphi}{\partial x}\boldsymbol{e}_x + \frac{\partial \varphi}{\partial y}\boldsymbol{e}_y + \frac{\partial \varphi}{\partial z}\boldsymbol{e}_z\right)\\
&= \left(\frac{\partial}{\partial y}\frac{\partial \varphi}{\partial z} - \frac{\partial}{\partial z}\frac{\partial \varphi}{\partial y}\right)\boldsymbol{e}_x + \left(\frac{\partial}{\partial z}\frac{\partial \varphi}{\partial x} - \frac{\partial}{\partial x}\frac{\partial \varphi}{\partial z}\right)\boldsymbol{e}_y + \left(\frac{\partial}{\partial x}\frac{\partial \varphi}{\partial y} - \frac{\partial}{\partial y}\frac{\partial \varphi}{\partial x}\right)\boldsymbol{e}_z = 0
\end{aligned}$$

The above expressing can be written as

$$\nabla \times (\nabla\varphi) = 0$$

A very important quality is that for a scalar, the rotation of gradient is always zero. Alternatively, if the rotation of vector $\boldsymbol{b}$ is always zero, namely vector field $\boldsymbol{b}$ is potential field, we can see vector $\boldsymbol{b}$ as the gradient of scalar field φ; namely if $\nabla \times \boldsymbol{b} = 0$, then we can suppose $\boldsymbol{b} = \nabla\varphi$.

Example 1.5 Prove

$$\nabla^2\left(\frac{1}{r}\right) = \nabla \cdot \nabla\left(\frac{1}{r}\right) = 0$$

Prove

$$\nabla^2\left(\frac{1}{r}\right) = \left(\frac{\partial^2}{\partial x^2} + \frac{\partial^2}{\partial y^2} + \frac{\partial^2}{\partial z^2}\right)\left(\frac{1}{\sqrt{x^2 + y^2 + z^2}}\right)$$

$$\frac{\partial}{\partial x}\left(\frac{1}{\sqrt{x^2 + y^2 + z^2}}\right) = \frac{\partial}{\partial x}(x^2 + y^2 + z^2)^{-\frac{1}{2}} = -x(x^2 + y^2 + z^2)^{-\frac{3}{2}}$$

$$\frac{\partial^2}{\partial x^2}\left(\frac{1}{\sqrt{x^2 + y^2 + z^2}}\right) = \frac{\partial}{\partial x}\left[-x(x^2 + y^2 + z^2)^{-\frac{3}{2}}\right]$$

$$= 3x^2(x^2 + y^2 + z^2)^{-\frac{5}{2}} - (x^2 + y^2 + z^2)^{-\frac{3}{2}} = \frac{2x^2 - y^2 - z^2}{(x^2 + y^2 + z^2)^{\frac{5}{2}}}$$

Similarly, we have

$$\frac{\partial^2}{\partial y^2}\left(\frac{1}{\sqrt{x^2+y^2+z^2}}\right)=\frac{2y^2-z^2-x^2}{(x^2+y^2+z^2)^{\frac{5}{2}}}$$
$$\frac{\partial^2}{\partial z^2}\left(\frac{1}{\sqrt{x^2+y^2+z^2}}\right)=\frac{2z^2-x^2-y^2}{(x^2+y^2+z^2)^{\frac{5}{2}}}$$

so

$$\nabla^2\left(\frac{1}{r}\right)=\left(\frac{\partial^2}{\partial x^2}+\frac{\partial^2}{\partial y^2}+\frac{\partial^2}{\partial z^2}\right)\left(\frac{1}{\sqrt{x^2+y^2+z^2}}\right)$$
$$=\frac{2x^2-y^2-z^2}{(x^2+y^2+z^2)^{\frac{5}{2}}}+\frac{2y^2-y^2-z^2}{(x^2+y^2+z^2)^{\frac{5}{2}}}+\frac{2z^2-y^2-z^2}{(x^2+y^2+z^2)^{\frac{5}{2}}}=0$$

The Laplace equation has been verified, $\varphi=\dfrac{1}{r}$ is the solution to this equation, with ∇^2 being the Laplace operator, which can also be denoted Δ. In Cartesian Coordinates, it can be definition as

$$\nabla^2=\Delta=\nabla\cdot\nabla=\frac{\partial^2}{\partial x^2}+\frac{\partial^2}{\partial y^2}+\frac{\partial^2}{\partial z^2} \tag{1.36}$$

1.8 The Helmholtz Theorem

We have described basic definitions of electric fields. We can describe a vector field with its divergence and rotation, scalar with gradient.

For a scalar field φ, the gradient is a vector field $\nabla\varphi=\boldsymbol{b}$. If vector $\boldsymbol{b}$ is known, one must calculate the scalar field φ, since $\nabla\times(\nabla\varphi)=\nabla\times\boldsymbol{b}=0$, $\boldsymbol{b}$ is a non-rotation vector field, which is known as conservatism vector field. And at this time, scalar field φ is named position field or potential field. A gravity field $\boldsymbol{G}$ is a vector with conservatism, corresponding scalar field is potential field U, of object, there has $\nabla U=\boldsymbol{G}$. the power made gravity on closed path is zero, namely $\oint_C G\cdot \mathrm{d}\boldsymbol{l}=\oint_C \nabla U\cdot \mathrm{d}\boldsymbol{l}=0$, as shown in Figure 1.8, loop integral also can written in

$$\oint_C G\cdot \mathrm{d}\boldsymbol{l}=\oint_C \nabla U\cdot \mathrm{d}\boldsymbol{l}=\int_{C_1}\nabla U\cdot \mathrm{d}\boldsymbol{l}+\int_{C_2}\nabla U\cdot \mathrm{d}\boldsymbol{l}=0$$

Where the integral path C_1 is from point A to point B and goes through point C_1. The integral path C_2 is from point B to point A, and goes through point C_1. Last expression also can be written in

$$\int_{C_1}\nabla U\cdot \mathrm{d}\boldsymbol{l}=-\int_{C_2}\nabla U\cdot \mathrm{d}\boldsymbol{l}=\int_{-C_2}\nabla U\cdot \mathrm{d}\boldsymbol{l}$$

where the integral path is from point A to point B and goes through $-C_2$. We can see from this derivation that the value of integral is dependent on the start point A and the end point B, and independent of the integral path.

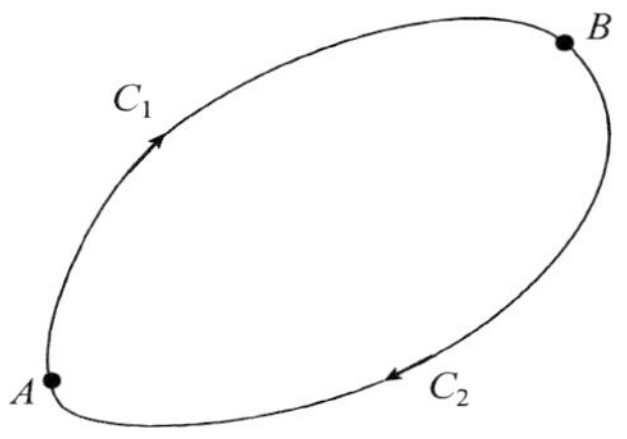

Figure 1.8 Integral of gravity G along closed path

$$\int_A^B \nabla U\cdot \mathrm{d}\boldsymbol{l}=\int_A^B \frac{\mathrm{d}U}{\mathrm{d}l}\mathrm{d}\boldsymbol{l}=U(B)-U(A)$$

Suppose point A is a fixed point and point B is a point in the field; the above expression can be written as

$$U(x,y,z)=\int_A^B \nabla U\cdot \mathrm{d}\boldsymbol{l}+U(A)$$

because point A is a fixed point, then $U(A)$ is a constant, which can be denoted by C, there is

$$U(x,y,z)=\int_A^B \nabla U\cdot \mathrm{d}\boldsymbol{l}+C$$

Suppose point A is the reference point, constant C is decided by the selection of reference point.

For a non-rotation vector field, its divergence can't be zero at anywhere, otherwise this field is without rotation and divergence, then this field don't exist; For a non-divergence vector field, its rotation can't be zero at anywhere, otherwise this field does not exist either. A general vector field $\boldsymbol{a}$ may have divergence and rotation at the same time. Such a vector field can be denoted by the sum of non-rotation (when there is only divergence) weight and non-divergence (when there is only rotation) field, denoted by the expression

$$\boldsymbol{a}(x,y,z)=\boldsymbol{a}_l(x,y,z)+\boldsymbol{a}_s(x,y,z) \tag{1.37}$$

where $\boldsymbol{a}_l(x,y,z)$ is the non-rotation weight, and its divergence is not zero everywhere. Suppose that the divergence is $\rho(x,y,z)$, $\boldsymbol{a}_s(x,y,z)$ is the non-divergence weight, and the vector's rotation isn't zero everywhere. Also, suppose that the rotation is $\boldsymbol{J}\ (x,y,z)$, so there is

$$\left.\begin{array}{l}\nabla\cdot\boldsymbol{a}=\nabla\cdot(\boldsymbol{a}_l+\boldsymbol{a}_s)=\nabla\cdot\boldsymbol{a}_l=\rho\\ \nabla\times\boldsymbol{a}=\nabla\times(\boldsymbol{a}_l+\boldsymbol{a}_s)=\nabla\times\boldsymbol{a}_s=\boldsymbol{J}\end{array}\right\} \tag{1.38}$$

We can infer from the above expression that the divergence and rotation describe the source in vector field respectively. Thinking about the divergence and rotation contain derivatives with respect to space, so choosing divergence and rotation must be in successive field, if the field isn't successive, then we can't analyse with divergence and rotation at the border of this field. So when the distribution of field and border condition of vector's divergence $\rho(x,y,z)$ and rotation $\boldsymbol{J}(x,y,z)$ have been confirmed, the vector field also has been confirmed, this rule is Helmholtz Theorem.

Knowledge of the Helmholtz Theorem, can also extend to infinite space. If vector approach zero at infinity, then divergence and rotation of vector are certainty, namely the property of vector approaching zero at infinity can be seen as a border condition.

We can infer from such discussion that firstly, knowledge pertaining to vector fields is necessary to research a vector's divergence and rotation to obtain its general solution, or by the way of researching integral, namely flux on closed surface and loop flow on closed curve., Then take border condition in hand at the same time, we can get a fixed solution of vector field.

Exercises

1.1 Prove that $(\boldsymbol{a}\times\boldsymbol{b})\times(\boldsymbol{c}\times\boldsymbol{d})=\boldsymbol{b}[\boldsymbol{a}\cdot(\boldsymbol{c}\times\boldsymbol{d})]-\boldsymbol{a}[\boldsymbol{b}\cdot(\boldsymbol{c}\times\boldsymbol{d})]=\boldsymbol{c}[\boldsymbol{a}\cdot(\boldsymbol{b}\times\boldsymbol{d})]-\boldsymbol{d}[\boldsymbol{a}\cdot(\boldsymbol{b}\times\boldsymbol{c})]$.

1.2 Supposed $\boldsymbol{a}=xz^3\boldsymbol{e}_x-2x^2yz\boldsymbol{e}_y+2yz^4\boldsymbol{e}_z$, calculate rotation on point $M(1,-1,-1)$.

1.3 Supposed $\varphi(x,y,z)=3x^2y-y^3z^2$,calculate $\nabla\varphi$ on point $M(1,-2,1)$.

1.4 calculate $\nabla\left(\dfrac{1}{r}\right)$.

1.5 There are three vector $\boldsymbol{A}$,$\boldsymbol{B}$ and $\boldsymbol{C}$:

$$\begin{aligned}\boldsymbol{A}&=\boldsymbol{e}_x+\boldsymbol{e}_y2-\boldsymbol{e}_z3\\ \boldsymbol{B}&=-\boldsymbol{e}_y4+\boldsymbol{e}_z\\ \boldsymbol{C}&=\boldsymbol{e}_x5-\boldsymbol{e}_y2\end{aligned}$$

calculate: (1) $\boldsymbol{e}_A$; (2) $\boldsymbol{A}\cdot\boldsymbol{B}$; (3) $\boldsymbol{A}\times\boldsymbol{C}$.

1.6 Suppose $|\boldsymbol{R}| = \left[(x-x')^2+(y-y')^2+(z-z')^2\right]^{\frac{1}{2}}$ is the distance between source point $\boldsymbol{r}'$ and field point $\boldsymbol{r}$, $\boldsymbol{R}$'s direction is from source to field point. By using of Cartesian-coordinates, try to prove

$$\nabla^2\left(\frac{1}{\boldsymbol{R}}\right) = -4\pi\delta(r-r')$$

1.7 Calculate the line integral of vector $\boldsymbol{A} = \boldsymbol{e}_x x^2 + \boldsymbol{e}_y xy^2$ on circuit $x^2+y^2+a^2$. Then calculate the integral of $\nabla\times\boldsymbol{A}$ on this circuit, and verify Stockes' Theorem.

1.8 In sphere coordinates, prove $\boldsymbol{A} = \frac{1}{r^3}\boldsymbol{r}$ is potential field, and calculate its potential function v.

1.9 There are vectors $\boldsymbol{A}$ and $\boldsymbol{B}$, they satisfy

$$\boldsymbol{A} = \boldsymbol{e}_r z^2\sin\phi + \boldsymbol{e}_\phi z^2\cos\phi + \boldsymbol{e}_z 2rz\sin\phi$$
$$\boldsymbol{B} = \boldsymbol{e}_x(3y^2-2x) + \boldsymbol{e}_y x^2 + \boldsymbol{e}_z 2z$$

(1) which vector can be denoted as gradient of scalar function? And which vector can be denoted as rotation of vector function?

(2) calculate the distribution of vector's source.

1.10 In cylindrical space $r=5, z=0, z=4$, verity divergence theorem for $\boldsymbol{A} = \boldsymbol{e}_r r^2 + \boldsymbol{e}_z 2z$.

1.11 In Cartesian-coordinates, verify the following identity:

$$\nabla\times(f\boldsymbol{G}) = f\nabla\times\boldsymbol{G} + \nabla f\times\boldsymbol{G}$$

1.12 Prove the following: If $\boldsymbol{A}\cdot\boldsymbol{B}=\boldsymbol{A}\cdot\boldsymbol{C}$ and $\boldsymbol{A}\times\boldsymbol{B}=\boldsymbol{A}\times\boldsymbol{C}$, then $\boldsymbol{B}=\boldsymbol{C}$.

1.13 If there are scalar product and vector product between a known vector and unknown vector, then we can certify the unknown vector. Suppose $\boldsymbol{A}$ is a known vector, $P=\boldsymbol{A}\cdot\boldsymbol{X}$ and $\boldsymbol{P}=\boldsymbol{A}\times\boldsymbol{X}$, P and $\boldsymbol{P}$ are known vector, try to calculate $\boldsymbol{X}$.

1.14 Calculate the integral of vector $\boldsymbol{r}$ with respect to a sphere surface which center on origin point and radius is a, and calculate integral of $\nabla\cdot\boldsymbol{r}$ with respect to sphere volume.

1.15 Calculate the loop line integral of vector $\boldsymbol{A} = \boldsymbol{e}_x x + \boldsymbol{e}_y x^2 + \boldsymbol{e}_z y^2 z$ with respect to a square whose side is 2 and it has two side coincide with axis x and y. Then calculate surface integral of $\nabla\times\boldsymbol{A}$ with respect to this square, and verify Stockes Theorem.

1.16 Suppose $r = \sqrt{x^2+y^2+z^2}$ is the magnitude of vector $M(x,y,z)$; prove $\nabla r = \frac{\boldsymbol{r}}{r} = \boldsymbol{r}_0$.

1.17 By using of Cartesian-coordinates, prove: $\nabla\cdot(f\boldsymbol{A}) = f\nabla\cdot\boldsymbol{A} + \boldsymbol{A}\cdot\nabla f$.

1.18 Prove that the vectors $\boldsymbol{A} = \boldsymbol{e}_x 3 + \boldsymbol{e}_y\frac{1}{3} - \boldsymbol{e}_z 2$ and $\boldsymbol{B} = \boldsymbol{e}_x 4 + \boldsymbol{e}_y 6 + \boldsymbol{e}_z 5$ are perpendicular to one another.

1.19 Prove that the vectors $\boldsymbol{A} = \boldsymbol{e}_x 4 + \boldsymbol{e}_y 10 + \boldsymbol{e}_z 5$ and $\boldsymbol{B} = \boldsymbol{e}_x 8 + \boldsymbol{e}_y 20 + \boldsymbol{e}_z 10$ are parallel to one another.

1.20 Consider the vectors $\boldsymbol{A} = \boldsymbol{e}_x 2 + \boldsymbol{e}_y 3 + \boldsymbol{e}_z 4$ and $\boldsymbol{B} = -\boldsymbol{e}_x 6 - \boldsymbol{e}_y 4 + \boldsymbol{e}_z$. Calculate the weight of $\boldsymbol{A}\times\boldsymbol{B}$ with respect to $\boldsymbol{C} = \boldsymbol{e}_x - \boldsymbol{e}_y + \boldsymbol{e}_z$.

1.21 Calculate the gradient of vector $\boldsymbol{A} = \boldsymbol{e}_x x^3 + \boldsymbol{e}_y(xy)^5 + \boldsymbol{e}_z x^2y^3z^5$.

1.22 In a unit cube which center is origin, verify divergence with respect to vector $\boldsymbol{A} = \boldsymbol{e}_x x^2 + \boldsymbol{e}_y(xy)^2 + \boldsymbol{e}_z 24x^2y^2z^3$.

1.23 Calculate directional derivative of function $u = 3x^2+z^2-2yz+2xz$ at direction of vector $\boldsymbol{A} = \boldsymbol{e}_x yz + \boldsymbol{e}_y xz + \boldsymbol{e}_z xy$ on point $M(1,2,3)$.

1.24 Given function $\boldsymbol{E} = \boldsymbol{e}_x y + \boldsymbol{e}_y x$, calculate line integral $\int E\cdot \mathbf{d}\boldsymbol{l}$ from point $P_1(2,1,-1)$ to point $P_2(8,2,-1)$, and the integral path is:

(1) parabola $x = 2y^2$;

(2) straight line between these two point.

and $\boldsymbol{E}$ is a conservative field?

Chapter 2

Electrostatic Fields

An electrostatic field is an electric field that is static to the observer and is generated by a charge which does not vary with time. An electrostatic field is vector field. Helmholtz's theorem states that to uniquely determined a vector, we must know its rotation, divergence and boundary condition. So, in this chapter, we will set out from the electrostatic's basic experiment law (coulomb's law), and then derive the fundamental equation for the electrostatic field's differential form and integral form. They describe the electrostatic field's irrotationality, divergencevortex sources and divergence-producing types of sources of one point in the space. The integral form is used for the study of the field in large-scale, and so is the relationship with the sources. The fundamental equation of the electrostatic field's integral form can also be used for analyzing, the electrostatic field's boundary condition. In the current chapter, we will also discuss potential gradiant, electric dipole, energy of electric field, electrostatic force and system of conductors' self-capacitance and mutual capacitance.

2.1 Electrostatic Field's Divergence Equation and Rotation Equation

We shall introduce coulomb's law and then derive the electrostatic field's irrotationality and the divergence fundamental equation. That is also the fundamental equation of electrostatic field. There are two forms that must be considered: a differential form and an integral form. The differential equations consist of the divergence equation and the rotation equation.

2.1.1 Coulomb's Law and Electric Field Intensity $\boldsymbol{E}$

1. Coulomb's Law

The law is encompasses the rules summed up by the experiments of forces between two point charges in the vacuum. Namely, we have that

$$\boldsymbol{F}_{10} = \frac{q_1 q_0}{4\pi\varepsilon_0 R^2}\boldsymbol{e}_R = \frac{q_1 q_0}{4\pi\varepsilon_0 R^3} \tag{2.1}$$

in the vacuum, between stationary point charge q_1, q_0's force is $\boldsymbol{F}_{10}$, the size is proportional to the product of q_1 and q_0, and is inversely proportional to the square of the distance R. The direction of the force is the same of their connecting line. Where,

$$\boldsymbol{R} = \boldsymbol{e}_R R = \boldsymbol{r} - \boldsymbol{r}\prime$$

2. Electric Field Intensity $\boldsymbol{E}$

Unit test charge was affected by an electric force, and this force is known as the electric field intensity $\boldsymbol{E}$. It is

$$\boldsymbol{E} = \lim_{q_0 \to 0} \frac{\boldsymbol{F}}{q_0} \tag{2.2}$$

The test charge $q_0 \rightarrow 0$ is to prevent the affection to the original field distribution because of the introduction of the test charge q_0.

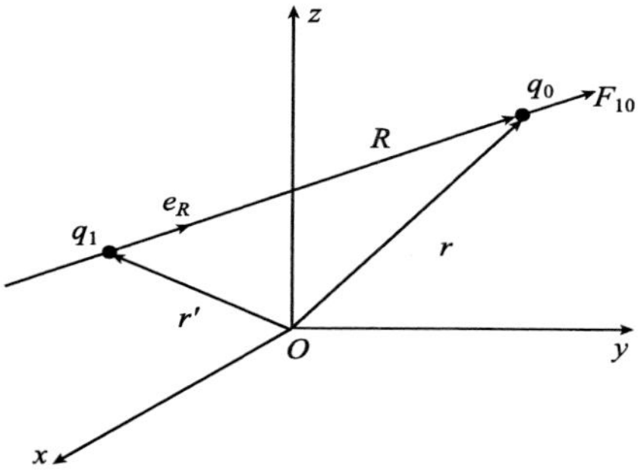

Figure 2.1 Forces between two point charges

Customarily, the charge of field source's location is called the "source point" and the observation point is known as the "field point". The field point's location is expressed by the coordinates x, y, z and its position vector by r. Source point coordinates x', y', z' are used with position vector given by r'. If q_1 in Figure 2.1 (indicated below by q) is the field source charge, and q_0 is the field point, then the electric field intensity $\boldsymbol{E}$ of field point is given by

$$\boldsymbol{E} = \lim_{q_0 \to 0} \frac{\boldsymbol{F}}{q_0} = \frac{q}{4\pi\varepsilon_0 R^3}\boldsymbol{R} = \frac{q}{4\pi\varepsilon_0 R^2}\boldsymbol{e}_R \tag{2.3}$$

where we have

$$R = |\boldsymbol{r} - \boldsymbol{r}'| = \left[(x - x')^2 + (y - y')^2 + (z - z')^2\right]^{1/2}$$

3. Superposition of the Electric Field Vector

Electric field generated by point charges is proportional to the power of the charges and the total electric field generated by a number of points can take advantage of the superposition principle. Specifically, we seek each point charge's electric field and then sum the vectors.

$$\boldsymbol{E} = \sum_{i=1}^{N} = \frac{q_i}{4\pi\varepsilon_0 R_i^2}\boldsymbol{e}_{R_i} \tag{2.4}$$

R_i is the distance between q_i to the field point, and $\boldsymbol{e}_{R_i}$ is the unit vector on the R_i.

For an electric field that is generated by continuous distribution charge, we can make use of integral summation method and integrate the regional meta-charge electric field generated by the charge in the distribution of regional to get the total electric field. Figure 2.2 shows the distribution of body-charge with the electric field generated by the volume element given by

$$\mathrm{d}\boldsymbol{E} = \frac{\rho \mathrm{d}V'}{4\pi\varepsilon_0 R^2}\boldsymbol{e}_R \tag{2.5}$$

To obtain the total electric field generated by a charged body, we may get the (2.5)'s volume integral of charge distribution in the whole region. It is

$$\boldsymbol{E} = \int_V \frac{\rho \mathrm{d}V'}{4\pi\varepsilon_0 R^2}\boldsymbol{e}_R \tag{2.6}$$

If the charged region has a surface charge distribution, then the total electric field is computed by integrating the ρ_s's surface area. It is

$$\boldsymbol{E} = \int_{S'} \frac{\boldsymbol{e}_R \rho_s \mathrm{d}S'}{4\pi\varepsilon_0 R^2} \tag{2.7}$$

Similarly, as shown in Figure 2.3 for the line charge distribution, the total electric field is the line integral of the distributed charge ρ_l.

$$\boldsymbol{E} = \int_{l'} \frac{\boldsymbol{e}_R \rho_l \mathrm{d}l'}{4\pi\varepsilon_0 R^2} \tag{2.8}$$

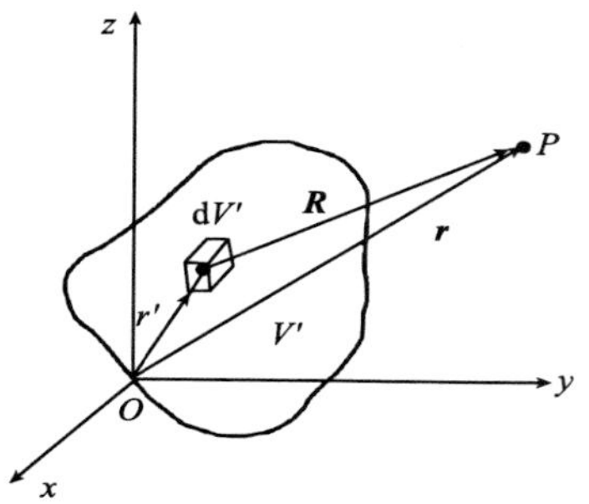

Figure 2.2 The electric field distribution of body-charge

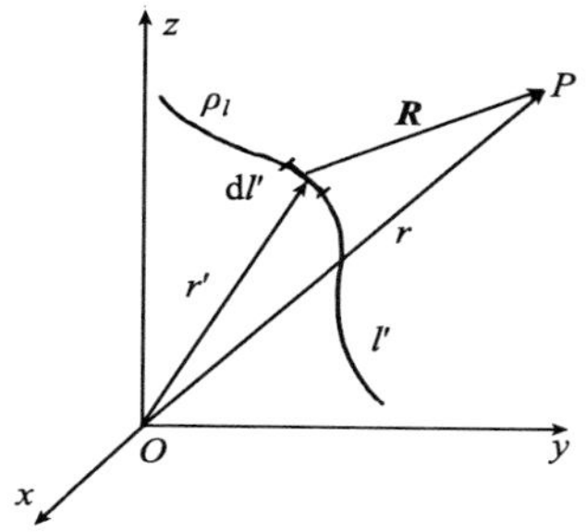

Figure 2.3 The electric field of line charge distribution

Example 2.1 An infinitely long straight line with uniformly charged in free-space, and its line charge density is ρ_l. Please seek the electric field intensity $\boldsymbol{E}$ of a point out of the line.

Answer choose cylindrical coordinates, according to the structure of the symmetry axis, the electric field generated by the infinitely long straight line with uniformly charged in free-space must be radial, and its electric field amplitude on the cylindrical surface which is on a axis of the electrical line are equal. Therefore, Gauss's law can be used. Construct a unit length of cylindrical surface with radius r, such as shown in Figure of Example 2.1.

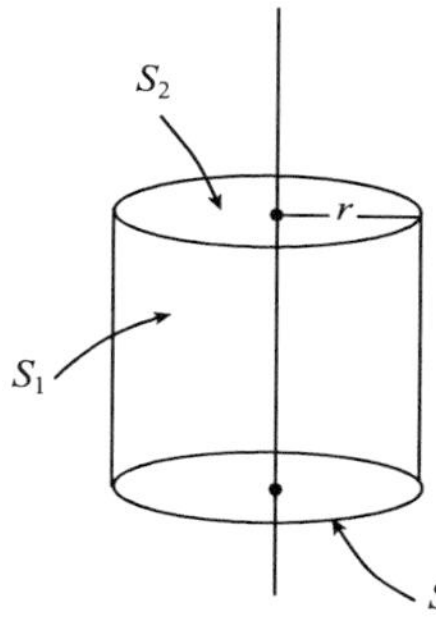

Figure of Example 2.1

We know from the Gauss's law that

$$\oint_S \boldsymbol{E}\cdot \mathrm{d}\boldsymbol{S} = \int_{S_1} \boldsymbol{E}\cdot \mathrm{d}\boldsymbol{S} + \int_{S_2} \boldsymbol{E}\cdot \mathrm{d}\boldsymbol{S} + \int_{S_3} \boldsymbol{E}\cdot \mathrm{d}\boldsymbol{S} = q/\varepsilon_0$$

The direction of top surface S_2 and bottom surface S_3's normal is vertical with the electric field's, then the integral of S_2, S_3 above is zero, while the side of S_1 surface normal direction is parallel to the electric field.

$$\int_{S_1} \boldsymbol{E}\cdot \mathrm{d}\boldsymbol{S} = \int_{S_1} E_r \mathrm{d}S = 2\pi r E_r$$

So the total electric quantity q contained in the sealing face is ρ_i and thus

$$E = E_r = \frac{\rho l}{2\pi\varepsilon_0 r} \quad \mathrm{V/m}$$

Example 2.2 A $2l$ long straight line with uniformly charged in free-space has a total charge of q, Please seek the electric field intensity $\boldsymbol{E}$ of a point out of the line.

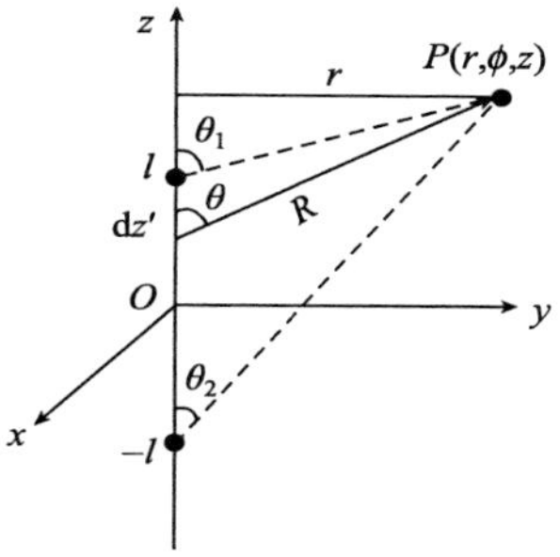

Figure of Example 2.2

Answer we choose a cylindrical coordinate system, whose z axis coincides with the straight line and let the coordinate origin be located in the midpoint of line segment. It is shown in Figure of Example 2.2. According to the axial symmetry of charge distribution, we may know that the electric field intensity is coordinate-free with the sites ϕ and there is noe_ϕ, but its size and direction are concerned with the z coordinates. In fact, there is not a surface that is perpendicular with the electric field's direction and of equal magnitude everywhere, and therefore it should not be sensible to apply Gauss's law. We may make use of the relationship of field and the Source (2.8) to seek the electric field.

It is known that charge q is uniformly distributed in a straight line segment, so the line charge density of the straight line segment is $\rho_l = q/2l$, and electric field in point P generated by the line charge $\rho_l \mathrm{d}z'$ is:

$$\mathrm{d}\boldsymbol{E} = \frac{\rho_l \mathrm{d}z'}{4\pi\varepsilon_0 R^2}\boldsymbol{e}_R$$

Suppose that the angle between Path-based vector R and the z-axis is θ, then

$$\mathrm{d}\boldsymbol{E} = \frac{\rho_l \mathrm{d}z'}{4\pi\varepsilon_0 R^2}(\boldsymbol{e}_r \sin\theta + \boldsymbol{e}_z \cos\theta) = \boldsymbol{e}_r \mathrm{d}E_r + \boldsymbol{e}_z \mathrm{d}E_z$$

The electric field intensity of points P that is generated by the entire segment is

$$E_r = \int_{-l}^{l} \frac{\rho_l \sin\theta \mathrm{d}z'}{4\pi\varepsilon_0 R^2}$$

$$E_z = \int_{-l}^{l} \frac{\rho_l \cos\theta \mathrm{d}z'}{4\pi\varepsilon_0 R^2}$$

According to the Figure of Example 2.2 we can see that the distance R, the angle θ and the coordinates r's relationship are as follows

$$R = \frac{r}{\sin\theta}, \quad z' = z - R\cos\theta = z - r\tan\theta, \quad \mathrm{d}z' = \frac{r\mathrm{d}\theta}{\sin^2\theta}$$

Combining the above relationships into the integral relation equation, we obtain: θ_1 and θ_2 are included angle between arrays of upper and lower ends to the scene point and the z axis. This result can be discussed by two points:

(1) If the length of segment $2l$ is much larger than the distance r between the field points to the z axis, then the electric field near the middle of the line segment should be nearly the same as Example 1.1. Because if the length $2l$ approaches infinity, that is $l \to \infty$, then we may know that $\theta_1 = \pi$, $\theta_2 = 0$ from Figure of Example 2.2. Substituting this relationship into the E_r, E_z, we may get that:

$$E_r = \frac{\rho_l}{2\pi\varepsilon_0 r}, \quad E_z = 0$$

It's the same as Example 2.1.

(2) If the distance R is much larger than the length of the straight-line $2l$, then the solution of the electric field at a distant point is given by

$$\begin{aligned} E_r &= \frac{\rho_l}{4\pi\varepsilon_0 r}(\cos\theta_2 - \cos\theta_1) \\ &= \frac{\rho_l}{4\pi\varepsilon_0 r}\left(\frac{l+z}{\sqrt{(l+z)^2 + r^2}} - \frac{z-l}{\sqrt{(l-z)^2 + r^2}}\right) \\ E_z &= \frac{\rho_l}{4\pi\varepsilon_0 r}(\sin\theta_1 - \sin\theta_2) \\ &= \frac{\rho_l}{4\pi\varepsilon_0 r}\left(\frac{r}{\sqrt{(l-z)^2 + r^2}} - \frac{r}{\sqrt{(l+z)^2 + r^2}}\right) \end{aligned}$$

Use $\dfrac{l}{\sqrt{z^2 + r^2}} \ll 1$, and binomial expansion theorem. Then take an entry similar to the terms with power 1, at last we get:

$$\boldsymbol{E} = E_r \boldsymbol{e}_r + E_z \boldsymbol{e}_z = \frac{q}{4\pi\varepsilon_0 R^3}(r\boldsymbol{e}_r + z\boldsymbol{e}_z) = \frac{q\boldsymbol{R}}{4\pi\varepsilon_0 R^3} = \frac{q\boldsymbol{e}_r}{4\pi\varepsilon_0 R^2}$$

It is the same as point charge.

It can be seen that, despite the fact that there is no real point charge, because even the smallest charge's capacity is not zero, and there is no infinitely long straight line, but these assumptions in ideal case can be used in the practical application to make the problems simple.

2.1.2 Electrostatic Field's Divergence Equation

In this section, we first analyze the force of the unit charge that is given by Coulomb's law, $\boldsymbol{E}=\boldsymbol{e}_r q/4\pi\varepsilon_0 r^2$, and then seek the differential and integral forms of Gauss's law.

$$\oint_S \boldsymbol{D}\cdot \mathrm{d}\boldsymbol{S} = q$$
$$\nabla\cdot\boldsymbol{D}=\rho$$

The above expressions describe divergence of an electrostatic field. That is, the electric flux through the closed surface equals to the quantity of the charges surrounded by the surface.

1. Gauss's Law in Integral Form

Method 1 set a sphere with a point charge q as the center, and seek the electric flux ψ_e through the spherical surface, then derive Gauss's law.

The electric flux $\mathrm{d}\psi_e$ through the spherical surface $\mathrm{d}\boldsymbol{S}$ is

$$\mathrm{d}\psi_e = \boldsymbol{D}\cdot\mathrm{d}\boldsymbol{S} = D\mathrm{d}S\cdot\cos 0^\circ = D\mathrm{d}S$$

Because the ball has symmetric properties of center to the charge q, any point on the ball surface has the same size of electric flux density D, as shown in Figure 2.4. Therefore, the electric flux ψ_e through the entire sphere can be directly given as:

$$\psi_e = \oint_S \mathrm{d}\psi_e = \oint_S \boldsymbol{D}\cdot\mathrm{d}\boldsymbol{S} = \oint_S D\mathrm{d}S = D\oint_S \mathrm{d}S$$
$$= \frac{q}{4\pi r^2}\cdot 4\pi r^2 = q$$

That is

$$\oint_S \boldsymbol{D}\cdot\mathrm{d}\boldsymbol{S} = q \tag{2.9}$$

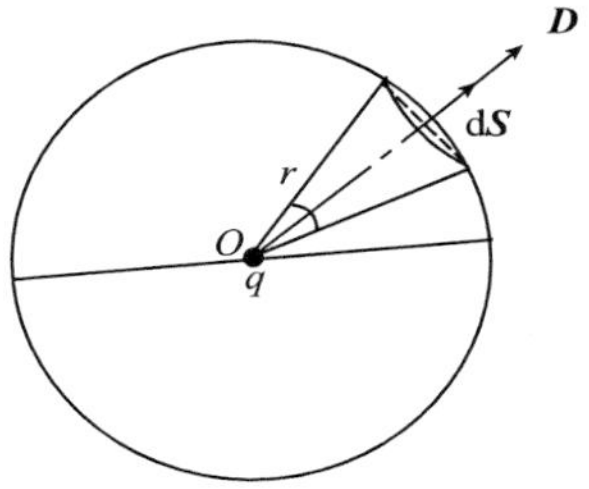

Figure 2.4 The electric flux

That is Gauss's law in integral form.

Method 2 Using the solid angle Ω, seeking electric flux ψ_e, and then derive the Gauss's law. There is a closed surface S with arbitrary shape and a point charge q on point O in the surface, as shown in Figure 2.5. The electric flux d ψ_e through any area element $\mathrm{d}S$ is

$$\mathrm{d}\psi_e = \boldsymbol{D}\cdot\mathrm{d}\boldsymbol{S} = D\mathrm{d}S\cdot\cos\theta = D\mathrm{d}S'$$

with $\mathrm{d}S'(=\mathrm{d}S\cdot\cos\theta)$ as the unit area corresponding to the solid angle $\mathrm{d}\Omega$ on the surface of a ball with O-point as the center r as the radius . The relationship between them is

$$\mathrm{d}\Omega = \frac{\mathrm{d}\boldsymbol{S}\cdot\boldsymbol{e}_r}{r^2} = \frac{\mathrm{d}S'}{r^2}$$

then the electric flux through the surface $\mathrm{d}S$ (that is, through $\mathrm{d}S'$ surface)

$$\mathrm{d}\psi_e = D\mathrm{d}S' = Dr^2\mathrm{d}\Omega = \frac{q}{4\pi}\mathrm{d}\Omega$$

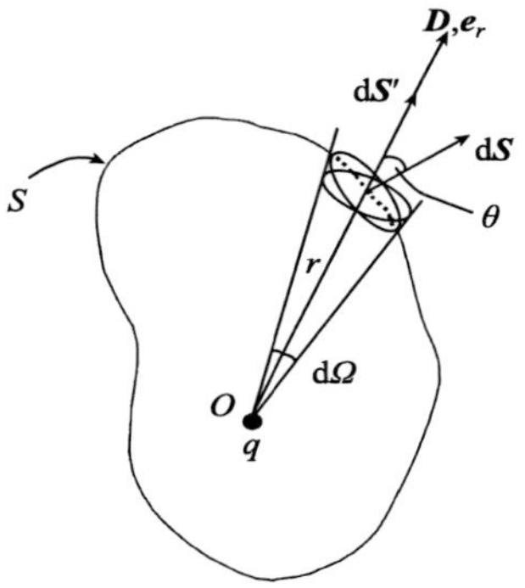

Figure 2.5 The electric flux through any area element

The total electric flux ψ_e through the closed surface S of arbitrary shape is

$$\psi_e = \oint_S \boldsymbol{D} \cdot \mathrm{d}\boldsymbol{S} = \oint_S \mathrm{d}\psi_e = \frac{q}{4\pi} \oint_S \mathrm{d}\Omega$$

A O point's solid angle Ω in a arbitrary closed surface S is the same as a ball's surface to its center O's solid angle Ω. And the ball set O-point as the center and r as the radius. The whole face of the ball to its center's solid angle Ω is

$$\Omega = \oint_S \mathrm{d}\Omega = \oint_S \frac{\mathrm{d}S}{r^2} = \frac{1}{r^2} \cdot 4\pi r^2 = 4\pi \quad \text{(sphere angle)}$$

So the solid angle that an arbitrary closed surface $\boldsymbol{S}$ makes to the point O is 4π. Therefore, the total electric flux ψ_e through an arbitrary closed surface is

$$\psi_e = \oint_S \boldsymbol{D} \cdot \mathrm{d}\boldsymbol{S} = \frac{q}{4\pi} \oint_S \mathrm{d}\Omega = q$$

The more general form

$$\oint_S \boldsymbol{D} \cdot \mathrm{d}\boldsymbol{S} = \sum q$$

which is referred to as the integral form of Gauss's law. It shows that the electric flux through any closed surface is equal to the quantity of the total charge in the closed surface.

We investigate the charge given by Gauss's law if q is not within the closed surface $\boldsymbol{S}$ (as what shows in the figure).

Since the electric flux through $\mathrm{d}\boldsymbol{S}$ is given by

$$\mathrm{d}\psi_e = \boldsymbol{D} \cdot \mathrm{d}\boldsymbol{S} = D\mathrm{d}S' = Dr^2\mathrm{d}\Omega = \frac{q}{4\pi}\mathrm{d}\Omega$$

As shown in Figure 2.6, $\boldsymbol{D}\cdot\mathrm{d}\boldsymbol{S}_1$ is positive and $\boldsymbol{D}\cdot\mathrm{d}\boldsymbol{S}_2$ is negative. The solid angle $\mathrm{d}\Omega$ of $\mathrm{d}\boldsymbol{S}_1$, $\mathrm{d}\boldsymbol{S}_2$ to the O-point will have the same size, but the electric flux through $\mathrm{d}\boldsymbol{S}_1$, $\mathrm{d}\boldsymbol{S}_2$ is equivalent with opposite and sign; thus the net flux is zero. The conclusion is that the electric flux that through a closed surface without any charge is zero. That is

$$\oint_S \boldsymbol{D} \cdot \mathrm{d}\boldsymbol{S} = 0$$

Gauss's Law shows that, the electric flux through an arbitrary closed surface is equal to the quantity of the free charge in the closed surfaces. It shows that electrostatic field has a flux source. Using Gauss's law, one can easily calculate the electrostatic field problems with a plane symmetry, axis symmetry, hot core symmetry. Once using the symmetry of the above-mentioned characteristics in the closed surface (Gauss surface), one may get two points as follows. First, one may suppose that portions of the closed surface's flux are zero. Second, some parts of the electric field E will be the same. Thus, it will not be difficult to derive the electric field intensity $\boldsymbol{E}$.

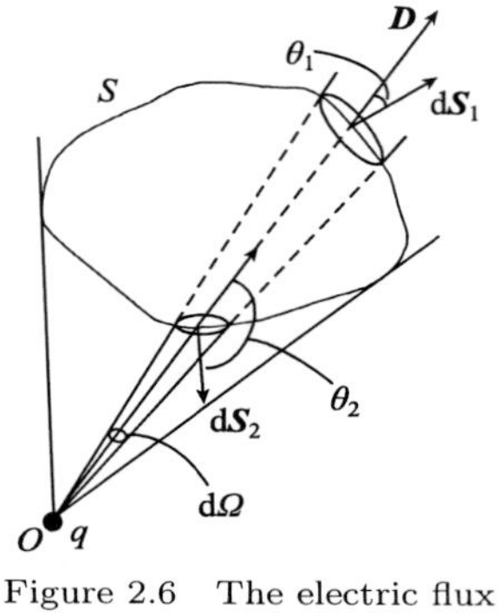

Figure 2.6 The electric flux with charge out of the close surface

2. Differential form of Gauss's Law, Namely, $\boldsymbol{D}$'s Divergence Equation

$$\Delta \cdot \boldsymbol{D} = \rho$$

Method 1 Using the divergence theorem and the integral form of Gauss's law of the electric field, we can derive the divergence equation of $\boldsymbol{D}$.

The mathematical expressions of divergence theorem is

$$\int_V \nabla \cdot \boldsymbol{D} \mathrm{d}V = \oint_S \boldsymbol{D} \cdot \mathrm{d}\boldsymbol{S} \tag{2.10}$$

when the distribution of electric charge is volume charge density ρ, the integral form of Gauss's law is

$$\oint_S \boldsymbol{D} \cdot \mathrm{d}\boldsymbol{S} = \sum q = \int_V \rho \mathrm{d}V$$

substituting into Equation (2.10) gives

$$\int_V \nabla \cdot \boldsymbol{D} \mathrm{d}V = \int_V \rho \mathrm{d}V \tag{2.11}$$

the volume V is random. Therefore, from the above analysis we obtain

$$\Delta \cdot \boldsymbol{D} = \rho \tag{2.12}$$

This is the differential form of Gauss's law. The divergence of $\boldsymbol{D}$ is equal to the point charge's density. This shows that the electrostatic field is divergent and its divergence source is electric field.

Method 2 seek the divergence of electric field intensity $\boldsymbol{E}$ and derive the related nature with δ function, then we get $\Delta \cdot \boldsymbol{D} = \rho$.

Suppose that there is volume charge density $\rho(r')$ located in the volume V', as shown in Figure 2.7. Thus any point P's electric field intensity $\boldsymbol{E}$ in space is given by

$$\boldsymbol{E} = \frac{1}{4\pi\varepsilon_0} \int_{V'} \frac{\rho \boldsymbol{R}}{R^3} \mathrm{d}V' \tag{2.13}$$

Here the vector $\boldsymbol{R}$ is

$$\begin{aligned} \boldsymbol{R} &= \boldsymbol{r} - \boldsymbol{r}' \\ &= \boldsymbol{e}_x(x - x') + \boldsymbol{e}_y(y - y') + \boldsymbol{e}_z(z - z') \end{aligned}$$

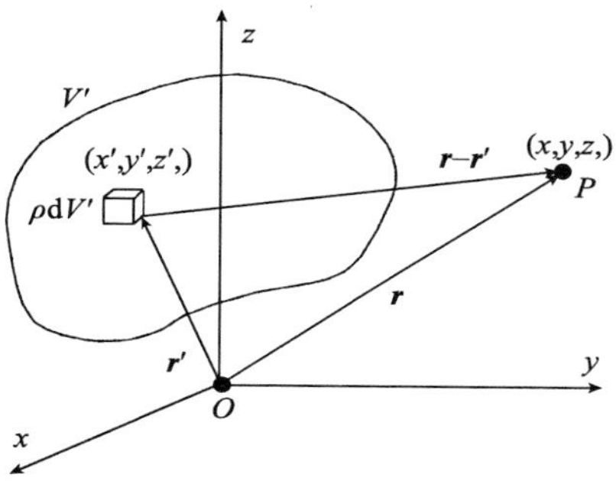

Figure 2.7 The electric field of volume charge distribute

The scalar R is

$$\begin{aligned} R &= |\boldsymbol{r} - \boldsymbol{r}'| \\ &= \left[(x - x')^2 + (y - y')^2 + (z - z')^2\right]^{\frac{1}{2}} \end{aligned}$$

The unit vector $\boldsymbol{e}_R$ is

$$\boldsymbol{e}_R = \boldsymbol{R}/R$$

we now set the field point coordinates x, y, z as variables, and take the divergence of Equation (2.13), to obtain

$$\nabla \cdot \boldsymbol{E} = \frac{1}{4\pi\varepsilon_0} \int_{V'} \left(\nabla \cdot \frac{\boldsymbol{R}}{R^3}\right) \rho \mathrm{d}V' \tag{2.14}$$

Since taking divergence ($\nabla \cdot$) is on the field point (x, y, z), while volume integral is on the origin (x', y', z'), so we may take the divergence (i.e. $\nabla \cdot$) inside the integral.

Next we discuss the characteristics of $\nabla \cdot \dfrac{\boldsymbol{R}}{R^3}$ in the Equation (2.14), by using vector identities

$$\nabla \cdot \left(\frac{\boldsymbol{A}}{f}\right) = \nabla\left(\frac{1}{f}\right) \cdot \boldsymbol{A} + \left(\frac{1}{f}\right)\nabla \cdot \boldsymbol{A}$$

after operation we can get

$$\nabla \cdot \frac{\boldsymbol{R}}{R^3} = \nabla\left(\frac{1}{R^3}\right) \cdot \boldsymbol{R} + \frac{1}{R^3}\nabla \cdot \boldsymbol{R} = -\left(\frac{3}{R^3} - \frac{3}{R^3}\right) \tag{2.15}$$

in this formula, gradient and divergence operators is carried out to the field point in the rectangular coordinate system (x, y, z).

An analysis of R in Equation (2.15) can derive two situations:

(1) the field point P is outside the source area, namely, $R \neq 0$. Then vector r (vector of the scene coordinate and origin point) and r' (vector of coordinate origin to the source point) do not overlap, so $R = |\boldsymbol{r} - \boldsymbol{r}'| \neq 0$. then the Equation (2.15) is zero, $\nabla \cdot \dfrac{\boldsymbol{R}}{R^3} = 0$, Thus Equation (2.14) is zero, that is,

$$\nabla \cdot \boldsymbol{E} = 0$$

(2) Any point in the area between the field point P and point source. We note that $\boldsymbol{r}$ and $\boldsymbol{r}'$ will tend to overlap leading to $R \approx 0$, and eventually leading Equation (2.15) to become a indeterminate without a definite solution, that is,

$$\nabla \cdot \frac{\boldsymbol{R}}{R^3} \to -\infty + \infty$$

Equation (2.14) is also not fixed solution. This issue is addressed by using two properties of the δ function (2.10: δ function and its related properties)

$$\nabla^2\left(\frac{1}{R}\right) = -4\pi\delta \tag{2.16}$$

$$\int_{V'} f(r)\delta(r - r')\mathrm{d}V' = f(r') \tag{2.17}$$

as well as the relationship

$$\nabla\left(\frac{1}{R}\right) = -\frac{\boldsymbol{e}_R}{R^2} = -\frac{\boldsymbol{R}}{R^3}$$

then Equation (2.14) can be expressed as:

$$\begin{aligned}
\nabla \cdot \boldsymbol{E}(r) &= \frac{1}{4\pi\varepsilon_0}\int_{V'}\left(\nabla \cdot \frac{\boldsymbol{R}}{R^3}\right)\rho \mathrm{d}V' \\
&= \frac{-1}{4\pi\varepsilon_0}\int_{V'}\left(\nabla \cdot \nabla\frac{1}{R}\right)\rho \mathrm{d}V' = \frac{-1}{4\pi\varepsilon_0}\int_{V'}\nabla^2\left(\frac{1}{R}\right)\rho \mathrm{d}V' \\
&= \frac{1}{4\pi\varepsilon_0}\int_{V'}\rho 4\pi\delta \mathrm{d}V' = \frac{1}{\varepsilon}\int_{V'}\rho(r')\delta(r - r')\mathrm{d}V' = \frac{\rho(r)}{\varepsilon_0}
\end{aligned}$$

That is

$$\nabla \cdot \boldsymbol{D} = \rho \tag{2.18}$$

the evolution of the Equations (2.12), (2.18) shows that divergence of the static field is the point charge density divided by ε_0. This shows that an electrostatic field has a divergence field and a divergent source. Besides, the source of this divergence is the charge density at that point.

Example 2.3 It is known that in free-space the spherical coordinate distribution of an electric field is given as

$$\begin{aligned} \boldsymbol{E} &= E_0(r/a)^2\boldsymbol{e}_r, \quad 0 < r < a \\ \boldsymbol{E} &= E_0(a/r)^2\boldsymbol{e}_r, \quad r > a \end{aligned}$$

please seek any point in space's charge density distribution.

Analysis We use the relation of electric field and charge source, we can obtain the charge density distribution from the given distribution of electric field. According the questions $E_\theta = E_\varphi = 0$, only E_r component exists, thus

$$\nabla \cdot \boldsymbol{E} = \frac{1}{r^2}\frac{\partial}{\partial r}(r^2 E_r) = \begin{cases} \dfrac{4rE_0}{a^2}, & 0 < r < a \\ 0, & r > a \end{cases}$$

so

$$\begin{aligned} \rho &= \varepsilon_0 \nabla \cdot \boldsymbol{E} = \frac{4\varepsilon_0 E_0 r}{a^2}, \quad 0 < r < a \\ \rho &= 0, \quad r > a \end{aligned}$$

2.1.3 Electrostatic Field's Curl Equations in Electrostatic Field

In this section, we will export that the loop flow along any close way in the electrostatic field is zero from the electric field intensity $\boldsymbol{E} = \boldsymbol{e}_r q/4\pi\varepsilon_0 r^2$ in Coulomb's law, and arrive at the conclusion that the electrostatic field is a non-rotational and conservative field. The power of the electric field force or the force outside is independent of the size and path, but dependent upon the location of the starting and ending points. Electrostatic field integral in the loop flow shows is that

$$\oint_l \boldsymbol{E} \cdot \mathrm{d}\boldsymbol{l} = 0$$

it shows that large-scale electric field loop flow is zero. Its differential form is

$$\nabla \times \boldsymbol{E} = 0$$

This shows that any point's curl of electric field in space is zero.

1. Electrostatic Field Loop Flow's Integral Form

At the electric field of the point charge q, we seek the closed line integral of the electric field $\boldsymbol{E}$ from a point along the curve through the b, c and back to a point. First we get the curve integral from a point to b point. We can learn that from Figure 2.8

$$\int_l \boldsymbol{E}\cdot \mathrm{d}\boldsymbol{l} = \int_l E\mathrm{d}l \cdot \cos\theta = \int_{r_a}^{r_b} E\mathrm{d}r = \frac{q}{4\pi\varepsilon_0}\int_{r_a}^{r_b}\frac{\mathrm{d}r}{r^2} = \frac{q}{4\pi\varepsilon_0}\left(\frac{1}{r_a} - \frac{1}{r_b}\right) \tag{2.19}$$

when the path is start from a point and through b, c, then back to a point, that closed the path is clearly

$$\oint_l \boldsymbol{E} \cdot \mathrm{d}\boldsymbol{l} = 0 \tag{2.20}$$

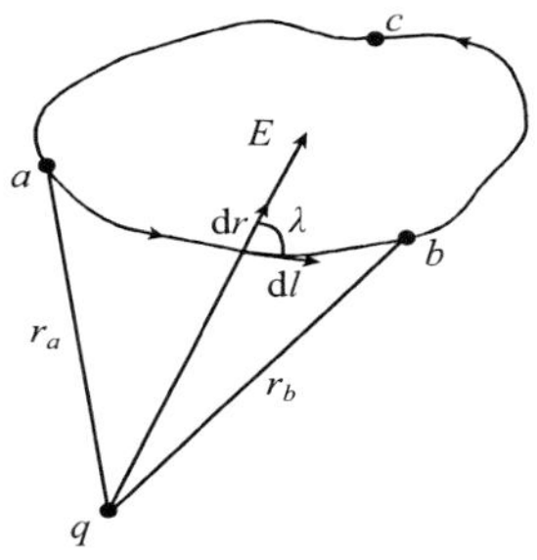

Figure 2.8 Curve integral of electric field

2. Electrostatic Field Ring Flow's Differential Form (electrostatic field's curl equation)

Method 1 Using Stokes theorem and the fact that the electrostatic field's loop flow is zero, we derive the electrostatic field curl equation. Recall that Stokes theorem is given by

$$\oint_l \boldsymbol{E} \cdot \mathrm{d}l = \int_S \nabla \times \boldsymbol{E} \cdot \mathrm{d}\boldsymbol{S}$$

Since in the electrostatic field $\oint_l \boldsymbol{E} \cdot \mathrm{d}\boldsymbol{l} = 0$, we have

$$\oint_S \nabla \times \boldsymbol{E} \cdot \mathrm{d}\boldsymbol{S} = 0$$

as closed lines l and its surrounding area S are arbitrary, so the electrostatic field rotation is zero

$$\nabla \times \boldsymbol{E} = 0 \tag{2.21}$$

Method 2 We can directly obtain the curl of electric field intensity $\boldsymbol{E}$ of the electrostatic field, and then take $\nabla\times\boldsymbol{E}=0$. The electric field intensity $\boldsymbol{E}$ of point charge q is

$$\boldsymbol{E} = e_r \frac{q}{4\pi\varepsilon_0 r^2} = \boldsymbol{e}_r E_r$$

in the electric field spherical coordinates (the coordinate origin's and the point charge q's positions are overlap) there is only one component $\boldsymbol{E}_r$. The $\boldsymbol{E}_r$ component is only a function of r and unrelated to θ, φ. Taking curl of the electric field $\boldsymbol{E}$, we obtain

$$\nabla \times \boldsymbol{E} = \boldsymbol{e}_\theta \frac{1}{r\sin\theta} \cdot \frac{\partial E_r}{\partial \phi} - \boldsymbol{e}_\phi \frac{1}{r}\frac{\partial E_r}{\partial \theta} = 0$$

That is

$$\nabla \times \boldsymbol{E} = 0 \tag{2.22}$$

So far, we have the electrostatic field's integral form and differential form of expression to describe divergent characteristics:

$$\oint_S \boldsymbol{D} \cdot \mathrm{d}\boldsymbol{S} = q$$
$$\nabla \cdot \boldsymbol{D} = \rho$$

We also obtain an expression for the electrostatic field characteristics of non-integral spin form and differential form as follows

$$\oint_l \boldsymbol{E} \cdot \mathrm{d}\boldsymbol{l} = 0$$
$$\nabla \times \boldsymbol{E} = 0$$

These equations that describe the basic characteristics of electrostatic field are called the basic equations of electrostatic field theory. As a vector field, electrostatic field's basic differential equations contain the divergence and curl equations. Based on Helmholtz theorem, in the unbounded space, the only condition to make sure a vector is knowing its rotation and divergence. These mathematical expressions are complete since the basic equations have the electrostatic field's rotation and divergence. From a physics point of view, the electrostatic field's rotation and divergence, respectively, describe its curl source and source of divergence, and thus the description of the source is complete.

2.2 Electric Potential and Electric Potential Gradient

2.2.1 Electric Potential

As shown in the electric field of the charge q, the power that the electric field force used in moving the unit positive charge from a point to the b point, as shown in Figure 2.8, is that

$$\int_l \boldsymbol{E} \cdot \mathrm{d}\boldsymbol{l} = \int_l E\mathrm{d}l \cos\theta = \int_{r_a}^{r_b} E\mathrm{d}r = \frac{q}{4\pi\varepsilon_0} \int_{r_a}^{r_b} \frac{\mathrm{d}r}{r^2} = \frac{q}{4\pi\varepsilon_0}\left(\frac{1}{r_a} - \frac{1}{r_b}\right) \tag{2.23}$$

we discover the following: the power of electric field force is only related with the start and end position, and unrelated with the passed path. Therefore, the electrostatic field's integral along any closed loop will be zero. This shows that the electrostatic field is the same as gravitational field. Both fields belong to a conservative field, or potential field. Supposed a point in Electrostatic field's potential, its meaning in the physics is the following: the power of electric field force to move unit positive charge from a point to the reference point (usually chosen at the zero potential) is the unit positive charge's potential energy. That point's potential is φ.

If we choose the infinity is zero potential reference point, then in the infinite space, in homogeneous medium conditions, the point which is r_a far way from the source charge q has potential φ_a

$$\varphi_a = \int_{r_a}^{\infty} \boldsymbol{E} \cdot \mathrm{d}\boldsymbol{r} = \frac{q}{4\pi\varepsilon_0} \int_{r_a}^{\infty} \frac{\boldsymbol{e}_r \cdot \boldsymbol{e}_r \mathrm{d}r}{r^2} = \frac{q}{4\pi\varepsilon_0 r_a} \tag{2.24}$$

Obviously, the difference of a's and b's potential is φ_{ab}. And the power of electric field force to move unit charge from a to b is

$$\varphi_{ab} = \varphi_a - \varphi_b = \int_{r_a}^{r_b} E\mathrm{d}r = \frac{q}{4\pi\varepsilon_0}\left(\frac{1}{r_a} - \frac{1}{r_b}\right) \tag{2.25}$$

The potential and charge q have a linear relationship and will satisfy the superposition principle. The total potential made by a number of point charges in space is equal to the sum of the potential from each point charge individually.

$$\varphi = \sum_{i=1}^{N} \frac{q_i}{4\pi\varepsilon_0 R_i} \tag{2.26}$$

The potential made by the continuous distribute charges in a particular area can be obtained by integrating the entire electric potential of the charges in this area.
volume charge distribution

$$\varphi = \int_{V'} \frac{\rho \mathrm{d}V'}{4\pi\varepsilon_0 R} \tag{2.27}$$

surface charge distribution

$$\varphi = \int_{S'} \frac{\rho_s \mathrm{d}S'}{4\pi\varepsilon_0 R} \tag{2.28}$$

line charge distribution

$$\varphi = \int_{L'} \frac{\rho_l \mathrm{d}l'}{4\pi\varepsilon_0 R} \tag{2.29}$$

2.2.2 Electric Potential Gradient

The potential φ is a scalar potential gradient, if you can find the relationship between potential and electric field strength $\boldsymbol{E}$, then you could seek electric field (vector) through potential (scalar), and this will be convenient to the calculation of electric field. Next we will find the relationship between the electric field and the electric potential.

Method 1 we may directly use the non-rotating nature of electric field and vector identity: $\nabla\times\nabla\varphi=0$, to find the relationship between the electric field and the electric potential.

Since

$$\begin{cases}\nabla\times\nabla\varphi=0\\ \nabla\times\boldsymbol{E}=0\end{cases}$$

we have

$$\boldsymbol{E}=-\nabla\varphi$$

From the above analysis we obtain

$$\nabla\times\boldsymbol{E}=-\nabla\times\nabla\varphi=0$$

the above analysis shows that if assume $\boldsymbol{E}=-\nabla\varphi$, $\boldsymbol{E}$ still has the free rotation nature. That asserts the assumption that the electric field strength is equal to the negative potential gradient.

$$\boldsymbol{E}=-\nabla\varphi \tag{2.30}$$

Method 2 Express the potential incremental $\mathrm{d}\varphi$ along the length of the element $\mathrm{d}\boldsymbol{l}$ respectively as the electric field force doing work and in the mathematics by total differential (Figure 2.9), then we can prove that $\boldsymbol{E}=-\nabla\varphi$.

First is the physical concept of the potential increment

$$\mathrm{d}\varphi=-\boldsymbol{E}\cdot\mathrm{d}\boldsymbol{l} \tag{2.31}$$

$\boldsymbol{E}\cdot\mathrm{d}\boldsymbol{l}$ denotes the power of electric field force along $\mathrm{d}\boldsymbol{l}$, where the negative sign indicates the electric field strength along the $\mathrm{d}\boldsymbol{l}$'s work is reactive, the potential is reduced. that is also to say the potential increment is negative.

Then, assuming that $\mathrm{d}\varphi$ is the total differential of a potential φ, the total differential is the sum of three partial differentials. That is

$$\mathrm{d}\varphi=\frac{\partial\varphi}{\partial x}\mathrm{d}x+\frac{\partial\varphi}{\partial y}\mathrm{d}y+\frac{\partial\varphi}{\partial z}\mathrm{d}z \tag{2.32}$$

then put $-\boldsymbol{E}\cdot\mathrm{d}\boldsymbol{l}$ into the cartesian coordinates

$$\begin{aligned}-\boldsymbol{E}\cdot\mathrm{d}l&=-(\boldsymbol{e}_xE_x+\boldsymbol{e}_yE_y+\boldsymbol{e}_zE_z)\cdot(\boldsymbol{e}_x\mathrm{d}x+\boldsymbol{e}_y\mathrm{d}y+\boldsymbol{e}_z\mathrm{d}z)\\&=-(E_x\mathrm{d}x+E_y\mathrm{d}y+E_z\mathrm{d}z)\\ \mathrm{d}\varphi&=-\boldsymbol{E}\cdot\mathrm{d}\boldsymbol{l}=-(E_x\mathrm{d}x+E_y\mathrm{d}y+E_z\mathrm{d}z)\end{aligned} \tag{2.33}$$

Comparing Equations (2.32), (2.33), we observe that

$$\begin{aligned}E_x&=-\frac{\partial\varphi}{\partial x},\quad E_y=-\frac{\partial\varphi}{\partial y},\quad E_z=-\frac{\partial\varphi}{\partial z}\\ \boldsymbol{E}&=\boldsymbol{e}_xE_x+\boldsymbol{e}_yE_y+\boldsymbol{e}_zE_z\\&=-\left(\boldsymbol{e}_x\frac{\partial\varphi}{\partial x}+\boldsymbol{e}_y\frac{\partial\varphi}{\partial y}+\boldsymbol{e}_z\frac{\partial\varphi}{\partial z}\right)\\&=-\left(\boldsymbol{e}_x\frac{\partial}{\partial x}+\boldsymbol{e}_y\frac{\partial}{\partial y}+\boldsymbol{e}_z\frac{\partial}{\partial z}\right)\varphi\\&=-\nabla\varphi\end{aligned} \tag{2.34}$$

Equation (2.30) or (2.34) shows the relationship between the electric field and potential. It shows that the size of the electric field intensity $\boldsymbol{E}$ is equal to the maximum potential change rate at the point of the space (that is, the point's potential gradient), the direction of the electric field E is to the direction of lower potential (expressed as the negative sign).

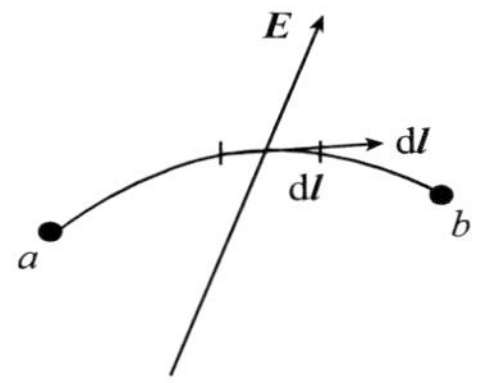

Figure 2.9 The potential incremental along d$\boldsymbol{l}$

Example 2.4 The Figure of Example 2.4 shows a parallel-plate capacitor, the bipolar plate distance d, the potential distribution between the plate is $\varphi = \dfrac{U_0}{d}x$. Find the electric field strength of capacitor.

Solution $\varphi(x) = \dfrac{U_0}{d}x \ (0 \leqslant x \leqslant d)$ is known.

From Equation (2.34) we observe that by seeking potential gradient we can get $\boldsymbol{E} = -\nabla\varphi = -\boldsymbol{e}_x\dfrac{U_0}{d}$. The electric field direction points to the direction where the electric potential decline.

Example 2.5 Let's have a look at the Figure of Example 2.5. Assume a radius of a and a uniformly charged disc with a surface charge density of ρ_s. Please seek any point's electric field intensity of the disc's outside axis.

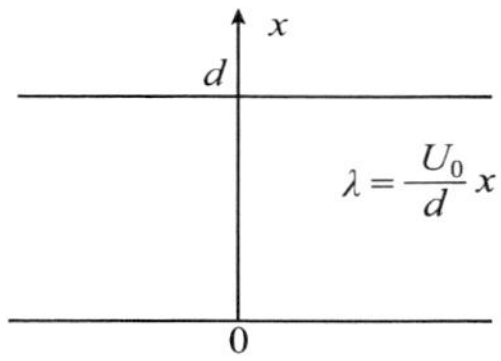

Figure of Example 2.4

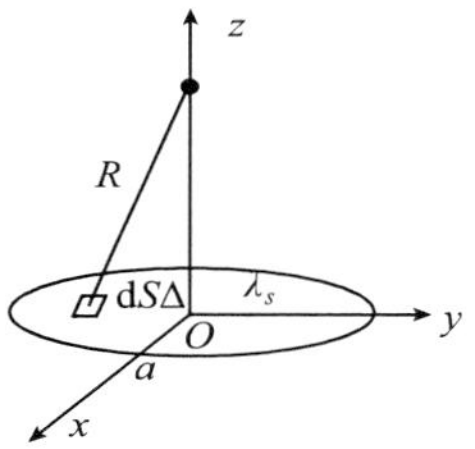

Figure of Example 2.5

Solution Gauss's law is rather inconvenient to use in this case. We use Equation (2.28), first calculate the potential of any point on the axis, and then get the electric field from the electric field gradient.

Set infinity as the zero point of the electric potential reference point, the field point (0, 0, z)'s potential is:

$$\varphi = \int_{s'} \frac{\rho_s \mathrm{d}S'}{4\pi\varepsilon_0 R}$$

The distribution area of charge is expressed by (r', ϕ', z'), the area we seek is expressed by (r, ϕ, z). The integral is in the charge distribution of the active region.

$$\varphi = \frac{\rho_s}{4\pi\varepsilon_0}\int_0^a\int_0^{2\pi}\frac{r'\mathrm{d}r'\mathrm{d}\phi}{\sqrt{z^2+r'^2}} = \frac{\rho_s}{2\varepsilon_0}\left(\sqrt{z^2+a^2}-|z|\right)$$

so

$$\boldsymbol{E} = -\nabla\varphi = \begin{cases} \boldsymbol{e}_z\dfrac{\rho_s s}{2\varepsilon_0}\left[1-\dfrac{z}{\sqrt{z^2+a^2}}\right], & z>0 \\ -\boldsymbol{e}_z\dfrac{\rho_s}{2\varepsilon_0}\left[1+\dfrac{z}{\sqrt{z^2+a^2}}\right], & z<0 \end{cases}$$

The direction of the electric field on z-axis should only have $\boldsymbol{e}_z$components by the symmetry of charge distribution on the disc.

2.3 Laplace's equation and Poisson's equation

In a static field, it is much more convenient to find the electric potential (scalar) than to find the electric field (vector). Once we have the potential function, the electric field will be straight forward to obtain. In order to calculate the electric potential we need to establish the differential equation of the electric potential on the basis of the differential expressions of the two basic properties of the electrostatic field (the properties of rotation and divergence).

First, the electrostatic field is a nonzero divergence field, that is

$$\nabla \cdot \boldsymbol{E} = \rho/\varepsilon \quad \text{(when the medium is uniform)} \tag{2.35}$$

The electrostatic field is also a potential field and an irrotational field, so that

$$\boldsymbol{E} = -\nabla\varphi \quad (\text{because} \nabla \times \nabla\varphi \equiv 0) \tag{2.36}$$

By inserting Formula (2.36) into Formula (2.35), we obtain

$$\nabla^2\varphi = -\rho/\varepsilon \tag{2.37}$$

The above constitutes Poisson's equation of the potential function φ.

In the region with no charge, i.e. $\rho = 0$, the Formula (2.37) changes into

$$\nabla^2\varphi = 0 \tag{2.38}$$

This is the Laplace's equation of the potential function φ.

In Cartesian coordinates, the Laplace operator ∇^2 is given by

$$\nabla^2 = \nabla \cdot \nabla = \frac{\partial^2}{\partial x^2} + \frac{\partial^2}{\partial y^2} + \frac{\partial^2}{\partial z^2}$$

When electric potential φ is known, we can find the volume density distribution of the charge ρ by using the Poisson's equation. When the charges follow the volume distribution, the expression of electric potential is given by

$$\varphi = \int_{V'} \frac{\rho \mathrm{d}V'}{4\pi\varepsilon_0 R} \tag{2.39}$$

This formula and the Poisson's equation are both used for describing the relation between the electric potential and the volume density of charges, and they mutually represent each other from two aspects in the forms of integral and differential. Therefore, this integral expression of electric potential is actually the solution of the differential equation (the Poisson's equation) on the condition of infinite space.

From the derivation process above we can clearly see that Laplace's equation and Poisson's equation are derived from the two basic properties of the electrostatic field: divergence ($\nabla\cdot\boldsymbol{D}= \rho$), irrotationality ($\nabla\times\boldsymbol{E}= 0$, i.e. $\boldsymbol{E}= -\nabla\varphi$). Laplace's equation and Poisson's equation are the summarization of the two basic properties of the electrostatic field. These constitute the basic equations of a static field.

Example 2.6 In a given region of space, the distribution the electric potential is $\varphi = ax^2\sin(2y)\text{ch}(3z)$. Calculate the volume charge distribution and the electric field strength $\boldsymbol{E}$, and check the correctness of ρ and $\boldsymbol{E}$.

(1) Calculate the charge volume density ρ.

By inserting the known electric potential distribution $\varphi = ax^2\sin(2y)\text{ch}(3z)$ into the Poisson's equation, we get

$$\nabla^2\varphi = (2 + 5x^2)a\sin(2y)\text{ch}(3z) = -\rho/\varepsilon_0$$
$$\rho = -\varepsilon_0(5x^2 + 2)a\sin(2y)\text{ch}(3z)$$

(2) Calculate the electric field strength $\boldsymbol{E}$.

To calculate $\boldsymbol{E}$ when φ is known, use of the formula $\boldsymbol{E}= -\nabla\varphi$ will correspond to the most sensible methodology. From

$$\begin{aligned}\boldsymbol{E} &= -\nabla\varphi = -\left[\boldsymbol{e}_x\frac{\partial\varphi}{\partial x} + \boldsymbol{e}_y\frac{\partial\varphi}{\partial y} + \boldsymbol{e}_z\frac{\partial\varphi}{\partial z}\right] \\ &= \boldsymbol{e}_x E_x + \boldsymbol{e}_y E_y + \boldsymbol{e}_z E_z \\ E_x &= -\frac{\partial\varphi}{\partial x} = -2ax\cdot\sin(2y)\cdot\mathrm{ch}(3z) \\ E_y &= -\frac{\partial\varphi}{\partial y} = -2ax^2\cdot\cos(2y)\cdot\mathrm{ch}(3z) \\ E_z &= -\frac{\partial\varphi}{\partial z} = -3ax^2\cdot\sin(2y)\cdot\mathrm{ch}(3z)\end{aligned}$$

(3) Check the correctness of ρ and $\boldsymbol{E}$.

By inserting the charge volume density ρ and the electric field strength $\boldsymbol{E}$ into the formula $\nabla\cdot\boldsymbol{E} = \rho/\varepsilon_0$, we shall check the accuracy of the result.

From

$$\nabla\cdot\boldsymbol{E} = \frac{\partial E_x}{\partial x} + \frac{\partial E_y}{\partial y} + \frac{\partial E_z}{\partial z}$$

By inserting the resulting E_x, E_y, E_z into the formula above, we obtain

$$\begin{aligned}\nabla\cdot\boldsymbol{E} &= [-2a\sin(2y)\mathrm{ch}(3z)] + \left[4ax^2\sin(2y)\mathrm{ch}(3z)\right] \\ &\quad + \left[-9ax^2\sin(2y)\mathrm{ch}(3z)\right] \\ &= -\left(5x^2+2\right)a\sin(2y)\mathrm{ch}(3z)\end{aligned}$$

Due to the relation

$$\nabla\cdot\boldsymbol{E} = \rho/\varepsilon_0$$

we have

$$\rho = -\varepsilon_0\left[5x^2+2\right]a\sin(2y)\cdot\mathrm{ch}(3z)$$

It can be seen that the results ρ and $\boldsymbol{E}$ derived from φ are correct.

2.4 Electric Dipole

An electricaly charged system, which consists of a pair of point electric charges of equal magnitude but opposite sign, separated by small distance, is called electric dipole, as shown in Figure 2.10. It is the far field of the electric dipole that is to be discussed, namely the field at $r \gg l$. This is also very useful to the analysis of the electric field in the medium. When discussing the radiation of electromagnetic wave later, we may encounter the problems of the electric dipole again.

The easiest way to calculate the electric field of electric dipole consists of finding the scalar electric potential φ first, and then taking the gradient. The calculation process uses the spherical coordinates.

The electric potential produced by a pair of point electric charges of equal magnitude but opposite sign at a certain point in the space is given by

$$\varphi = \frac{q}{4\pi\varepsilon_0}\left(\frac{1}{r_+} - \frac{1}{r_-}\right) = \frac{q(r_- - r_+)}{4\pi\varepsilon_0 r_+\cdot r_-} \quad (2.40)$$

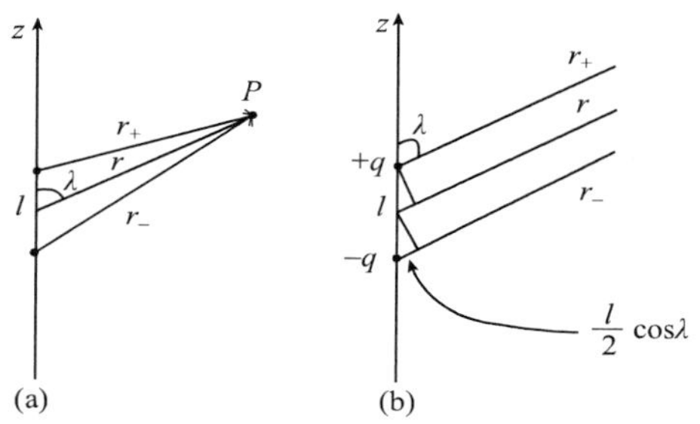

Figure 2.10

(a) Several geometric dimensions of the electric dipoles of which the P point is in the near region;

(b) When P point is in the far region ($r \gg l$), these three lines, r_+, r_-, r, are almost parallel to each other.

2.4.1 Find the Expression of the Electric Potential of the Electric Dipole in the Spherical Coordinate System

Method 1 On the condition of $r \gg l$, by reference to Figure 2.10(b), we can directly write out the r_+, r_-, and thus find φ.

When $r \gg l$ (far point), these three lines, r_+, r_-, r, are almost parallel to each other, so

$$\left.\begin{aligned} r_+ &\approx r - \frac{l}{2}\cos\theta \\ r_- &\approx r + \frac{l}{2}\cos\theta \end{aligned}\right\} \tag{2.41}$$

$$\left.\begin{aligned} r_- - r_+ &\approx l\cos\theta \\ r_+ \cdot r_- &\approx r^2 \end{aligned}\right\} \tag{2.42}$$

By inserting Formula (2.42) into Formula (2.40), we get the expression for the electric potential of the electric dipole

$$\varphi = \frac{q(r_- - r_+)}{4\pi\varepsilon_0 r_+ r_-} = \frac{ql\cos\theta}{4\pi\varepsilon_0 r^2} = \frac{P_e\cos\theta}{4\pi\varepsilon_0 r^2} \tag{2.43}$$

where $\boldsymbol{P}_e$ is the electric dipole moment. That is

$$\boldsymbol{P}_e = ql(\mathrm{C}\cdot\mathrm{m}) \tag{2.44}$$

$\boldsymbol{P}_e$ is equal to the product of the quantity of the electric charges and the distance between the charges. And it points in the direction which is from the negative charge to the positive charge. Therefore, the electric potential of electric dipole can also be written in the following form

$$\varphi = \frac{P_e\cos\theta}{4\pi\varepsilon_0 r^2} = \frac{\boldsymbol{P}_e\cdot\boldsymbol{e}_r}{4\pi\varepsilon_0 r^2} \tag{2.45}$$

Method 2 By reference to Figure 2.10(a), by using the law of cosines and the power series expression of binomial, find the r_+, r_-, and thus find the electric potential φ.

Based on the law of cosines, from Figure 2.10(a), we know that

$$\left.\begin{aligned} r_+ &= \sqrt{r^2 + \left(\frac{l}{2}\right)^2 - rl\cos\theta} \\ r_- &= \sqrt{r^2 + \left(\frac{l}{2}\right)^2 + rl\cos\theta} \end{aligned}\right\} \tag{2.46}$$

Writing r_+, r_-, in the form of binomial $\left[(r^{2-} + rl\cos\theta) + \left(\frac{l}{2}\right)^2\right]^{\frac{1}{2}}$, and expanding it in the form of power series. Then, by using the condition $r \gg l$, and omitting the higher-order items, we get the exactly same expression of r_+, r_- with Formula (2.41), and thus obtain the expression of the electric potential φ. The calculative process is omitted.

2.4.2 Find the Expression of the Electric Field of the Electric Dipole

By calculating the gradient of Formula (2.43) in spherical coordinates, we find the electric field of electric dipole

$$\boldsymbol{E} = -\nabla\varphi = \boldsymbol{e}_r \frac{P_\mathrm{e}\cos\theta}{2\pi\varepsilon_0 r^3} + \boldsymbol{e}_\theta \frac{P_\mathrm{e}\sin\theta}{4\pi\varepsilon_0 r^3} \tag{2.47}$$

i.e.

$$E_r = \frac{P_\mathrm{e}\cos\theta}{2\pi\varepsilon_0 r^3}, \quad E_\theta = \frac{P_\mathrm{e}\sin\theta}{4\pi\varepsilon_0 r^3}$$

From Formula (2.47), can we see two features of the electric field of the electric dipole:

(1) The electric field strength descends at the speed of r^{-3}. Compared with the electric field produced by single point charge whose strength descends at the speed of r^{-2}, the electric field strength of electric dipole descends faster. This is because $\pm q$ are so close to each other that the electric fields respectively produced by them cancel each other to some extent.

(2) The electric field only has the r and θ direction components, and is independent of the coordinate ϕ, and the distribution of it is axially symmetric.

2.4.3 The Equipotential Lines and Electric Force Lines of the Electric Dipole

Curves with constant space potentials are called the equipotential surfaces. The equipotential surface equation of the electric dipole can be obtained by fixating φ in Formula (2.43) as a constant. Since q and l are both constant, the equipotential surface equation is

$$r = \left(\frac{P_\mathrm{e}}{4\pi\varepsilon_0\varphi}\right)^{1/2} \sqrt{\cos\theta} = C_V\sqrt{\cos\theta} \tag{2.48}$$

Corresponding to the different electric potentials φ, C_V can have different values. Hereby, we can draw a family of r-θ curves, as shown by solid lines in Figure 2.11.

In order to describe the distribution characteristic of the electric field strength visually, a family of curves in the space, where the electric field exits, are used for representing the electric field. The tangential direction of each point on the curves is in the direction of the electric field of the point, i.e. the $\mathrm{d}\boldsymbol{l}$ and $\boldsymbol{E}$ have the same direction. The density of the curves is in direct proportion to the strength of electric field. Such curves are referred to the lines of electric force. The vector equation of the lines of the electric force is given by $\boldsymbol{E}\times\mathrm{d}\boldsymbol{l}=0$. That the product of these two vectors equals zero means these two vectors are parallel.

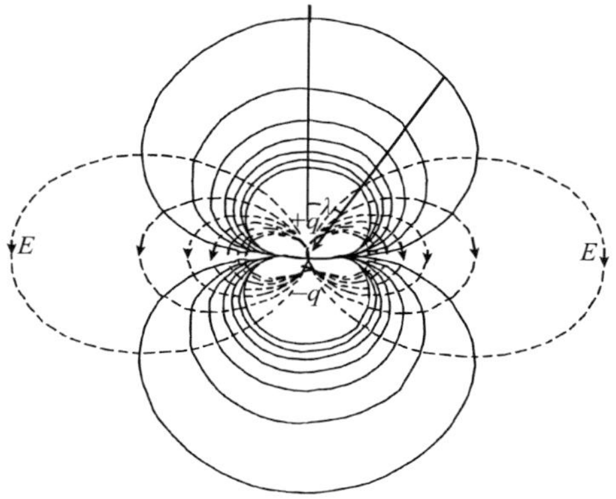

Figure 2.11 The equipotential lines (solid lines) and electric force lines (dotted lines) of the electric dipole

The electric field $\boldsymbol{E}$ only has two components which are E_r and E_θ, and all the lines of electric force are distributed on the surface which is composed by $\boldsymbol{e}_r$ and $\boldsymbol{e}_\theta$, i.e. the lines of electric force are on the meridian plane. And the $\boldsymbol{E}$ is not related to the coordinate ϕ, so the distribution of the lines of electric focused on any meridian plane is the same as the others.

From

$$\boldsymbol{E} \times \mathrm{d}\boldsymbol{l} = (\boldsymbol{e}_r E_r + \boldsymbol{e}_\theta E_\theta) \times (\boldsymbol{e}_r \mathrm{d}r + \boldsymbol{e}_\theta r \mathrm{d}\theta) = 0$$

we get

$$\frac{\mathrm{d}r}{r} = \frac{E_r}{E_\theta}\mathrm{d}\theta$$

By inserting the expressions of E_r and E_θ of the electric field of electric dipole into the formula above, we obtain

$$\frac{\mathrm{d}r}{r} = \frac{2\cos\theta}{\sin\theta}\mathrm{d}\theta$$

after calculating the integral on each side of the equation above, we get

$$r = c' \sin^2\theta \tag{2.49}$$

This is the equation of the lines of the electric force of the electric dipole.

In the Formula (2.49) the c' is a constant. By selecting different values of the c', we obtain different lines of electric force. The dotted lines in Figure 2.11 represent the lines of electric force. The lines of electric force are perpendicular to the equipotential lines.

2.5 Conductors in the Electrostatic Field

In the preceding sections we have analyzed the electrostatic field in free space. We need to discuss the electrostatic field in different substances, thereby obtaining the general laws of the electrostatic field in general substances.

A conductor is a substance which contains large amounts of movable electric charges. If the electric field exists in a conductor, the electric charges inside will positively be acted upon by the electric field force. This is contradictory to the precondition that in the electrostatic field the distribution of electric charges keeps static. Therefore, on the condition of electrostatic field, all the electric charges in a conductor have been under a steady state of electrostatic balance, and the electric field at every point in a conductor is zero. By applying the Gauss law in the conductor, the electric field flux through any closed surface in the conductor is zero, which means that there is no electric charge in the conductor. All of the electric charge of the conductor can only be distributed on the surface. Meanwhile, a conductor in electrostatic field is an equipotential body, and the surface of the conductor is an equipotential surface. Since the equipotential surface is perpendicular to the lines of electric force, we can conclude that the electric field is positively perpendicular to the surface of the conductor.

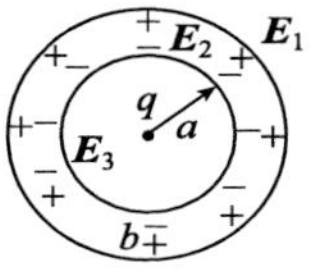

Figure of Example 2.7

Example 2.7 A positive point charge q is at the center of a conductor spherical shell whose inner radius is a, and outer radius is b, as shown in Figure of Example 2.7. Calculate the electric field strength and the electric potential of all the points in the space.

Solution On the basis of the structure of the system in the example and the features of the distribution of electric charges, we can judge that the distribution of electric field is spherically symmetric. The direction of the electric field is the direction of the radius. The spherical shell divides the space into three regions ($r > b$, $b > r > a$, $r < a$). Next we will calculate the electric field and electric potential in these three regions respectively.

(1) $r > b$ Since the distribution of the electric field is spherically symmetric, we can use the Gauss law to find the electric field E_1. Consider an integral surface with the radius of $r > b$, The center of it is the center of the spherical shell. As the conductor spherical shell itself is not charged, so

$$E_1 = \frac{q}{4\pi\varepsilon_0 r^2}, \qquad r > b$$

$$\varphi = \int_r^{\infty} E_1 \mathrm{d}r = \frac{q}{4\pi\varepsilon_0 r}, \qquad r > b$$

(2) $b > r > a$ The electric field in the conductor is zero

$$E_2 = 0$$

$$\varphi = \int_r^{\infty} E \mathrm{d}r = \int_r^{b} E_2 \mathrm{d}r + \int_b^{\infty} E_1 \mathrm{d}r = \frac{q}{4\pi\varepsilon_0 b}, \qquad a < r < b$$

Consider an integral surface inside the conductor with the radius of $r(a < r < b)$. The electric field in the conductor is zero, and the electric field flux through this closed surface is zero, which means there is no electric charges surrounded by this closed surface. Therefore, there are negative charges $-q$ distributed on the inner surface of the conductor that have the equal quantity but the opposite sign to those at the centre of the spherical shell. And since the conductor itself is not charged, there must be corresponding charges $+q$distributed on the outer surface of the conductor spherical shell. As shown in Figure of Example 2.7.

(3) $r < a$ Likewise, by using the Gauss's Law in this region, we can obtain

$$E_3 = \frac{q}{4\pi\varepsilon_0 r^2}, \qquad r < a$$

$$\varphi = \int_r^{\infty} E \mathrm{d}r = \int_r^{a} E_3 \mathrm{d}r + \int_a^{b} E_2 \mathrm{d}r + \int_b^{\infty} E_1 \mathrm{d}r$$

$$= \frac{q}{4\pi\varepsilon_0}\left(\frac{1}{r} - \frac{1}{a} + \frac{1}{b}\right), \qquad r < a$$

From this example it can be seen that when a conductor is put into the electric field, distributed electric charges will be induced on the surface of the conductor. The distribution of these induced charges is related to the shape of the surface of the conductor and the outer electric field. The electric field produced by these distributed electric charges is called the secondary electric field. The total electric field in the conductor is the result of the joint actions of the outer electric field and the secondary electric field produced by distributed electric charges. In steady state, two kinds of fields always have the same amplitude but the opposite direction and cancel out each other. This ensures that the total electric field in the conductor stays zero.

2.6 Dielectrics in the Electrostatic Field

2.6.1 Three Types of Dielectrics Polarization

There are large amounts of free electric charges in the conductor, but there is no free electric charge in the perfect dielectrics. When the dielectrics are put into the electric field, the electric field in the dielectrics will also change. That is because the electric field has polarized the dielectrics, where the bound charges are formed (Figure 2.12). The bound charges can also produce the electric field, and consequently change the total electric field in the dielectrics.

The phenomenon of dielectrics polarization is related to the atoms and molecules that constitute the matter. Three causes of the phenomenon can be concluded as follows: Under the influence of the outer electric field, the electron cloud which revolves around the atomic nucleus shifts a little relative to the nucleus, which consequently makes the electrically neutral atom become a very small electric dipole. This phenomenon is called electronic

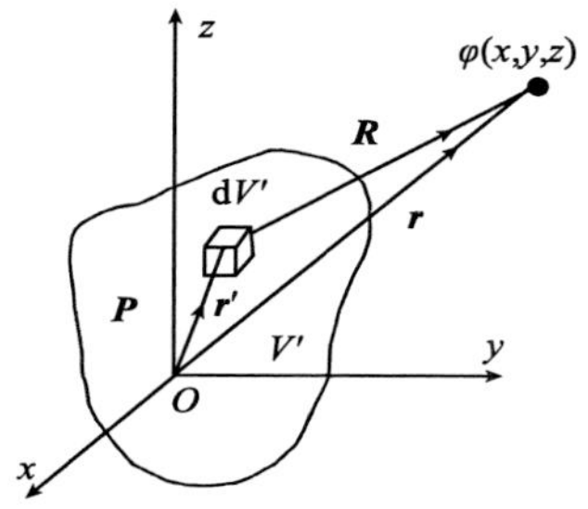

Figure 2.12 Electric potential procerant by polarized dielectric

polarization; Under the influence of the electric field, the positive and negative ions which constitute the molecule shifts a little. And consequently, the molecule becomes a very small electric dipole. This phenomenon is called ionic polarization. There are polar molecules in the dielectrics, whose center of the positive charges and center of the negative charges do not coincide, which means the polar molecule is still a electric dipole, even if the outer electric field is zero. But under the condition of no outer electric field, the orientation of the polar molecules is disordered, and the total resultant electric moment of it is zero. When the outer electric field exists, under the influence of the outer electric field, the rotation direction of the electric moment of the polar molecules is the same with the direction of the outer electric field. Consequently, the resultant electric moment forms. This phenomenon is called orientation polarization. Any one of the polarization phenomena above will produce many electric dipoles inside the dielectrics whose orientations are approximately the same. These electric dipoles produce the secondary electric field. In the isotropic dielectrics, the direction of the secondary electric field is always opposite to that of the outer electric field, and the result of dielectrics polarization is always to make the total electric field in the dielectrics smaller.

In order to analyze the macroeffect of dielectrics polarization (a macroscopic physical quantity) the intensity of polarization $\boldsymbol{P}$, is introduced. It is defined as: the vector sum of the electric moments of electric dipoles per unit volume, namely

$$\boldsymbol{P} = \lim_{\Delta V \to 0} \frac{\sum_{i=1}^{N} \boldsymbol{P}_{e_i}}{\Delta V} \quad (\mathrm{C/m})^2 \tag{2.50}$$

where N is the number of the electric moments in the volume of ΔV, $\boldsymbol{P}_{e_i}$ is the electric moment of the ith electric dipole in ΔV. From this formula it can be seen that the intensity of polarization $\boldsymbol{P}$ is the volume density of electric dipole moments. The units of the intensity of polarization $\boldsymbol{P}$ is: Coulomb/ square meter ($\mathrm{C/m^2}$).

2.6.2 The Relationship between Intensity of Polarization $\boldsymbol{P}$ and The Volume Density of the Bound Charges ρ_P, ρ_{PS}

There are secondary electric fields in the polarized dielectric. These electric fields are produced by the electric dipole moments induced in the electric fields. Considering that electric charges are the ultimate source that produces the electrostatic field, these electric dipole moments in the dielectrics correspond to the equivalent charge distribution. In order to be distinguished from the free electric charges in the dielectrics, we call these electric charges in the dielectrics the bound charges. Next we will seek the relation between the intensity of polarization $\boldsymbol{P}$ and the volume density of the bound charges ρ_p by calculating and comparing.

As shown in Figure 2.13, the electric dipole moment $\mathrm{d}\boldsymbol{P}_\mathrm{e}$ within the volume element $\mathrm{d}V'$ should equal the product of the intensity of polarization $\boldsymbol{P}$ and the volume element $\mathrm{d}V'$, i.e. $\mathrm{d}\boldsymbol{P}_\mathrm{e}=\boldsymbol{P}\mathrm{d}V'$. The electric potential $\mathrm{d}\varphi$ produced by this electric dipole moment, according to the formula of the electric potential of electric dipole Formula (2.45), we get

$$\mathrm{d}\varphi = \frac{\boldsymbol{P} \cdot \boldsymbol{e}_R \mathrm{d}V'}{4\pi\varepsilon_0 R^2}$$

where R is the distance between the point of source $\mathrm{d}V'$ and the point of field. The electric potential produced by the whole polarized dielectric can be obtained by calculating the integral of the V':

$$\varphi = \int_{V'} \frac{\boldsymbol{P} \cdot \boldsymbol{e}_R \mathrm{d}V'}{4\pi\varepsilon_0 R^2} \tag{2.51}$$

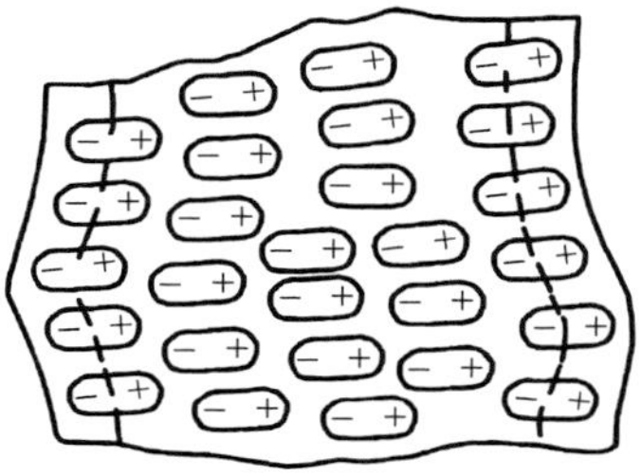

Figure 2.13 The bound charges in dielectrics

Due to the relation

$$\nabla'\left(\frac{1}{R}\right) = \nabla'\left(\frac{1}{|\boldsymbol{r} - \boldsymbol{r}'|}\right) = \frac{\boldsymbol{r} - \boldsymbol{r}'}{|\boldsymbol{r} - \boldsymbol{r}'|^3} = \frac{1}{R^2}\boldsymbol{e}_R \tag{2.52}$$

Equation (2.51) can be expressed as

$$\varphi = \frac{1}{4\pi\varepsilon_0} \int_{V'} \boldsymbol{P} \cdot \nabla'\left(\frac{1}{R}\right) \mathrm{d}V'$$

through the use of the vector identity

$$\nabla \cdot (f\boldsymbol{A}) = f\nabla \cdot \boldsymbol{A} + \boldsymbol{A} \cdot \nabla f$$

The integral above is divided into two terms

$$\varphi = \frac{1}{4\pi\varepsilon_0} \int_{V'} \nabla' \cdot \left(\frac{\boldsymbol{P}}{R}\right) \mathrm{d}V' - \frac{1}{4\pi\varepsilon_0} \int_{V'} \frac{\nabla' \cdot \boldsymbol{P}}{R} \mathrm{d}V'$$

After applying the divergence theorem to the first term, the formula above changes to

$$\varphi = \frac{1}{4\pi\varepsilon_0} \oint_{S'} \frac{\boldsymbol{P} \cdot \boldsymbol{e}_n \mathrm{d}S'}{R} - \frac{1}{4\pi\varepsilon_0} \int_{V'} \frac{\nabla' \cdot \boldsymbol{P}}{R} \mathrm{d}V' \tag{2.53}$$

where $\boldsymbol{e}_n$ is the unit vector in the normal direction outside the closed surface.

By comparing the formula above with the formula of the electric potential produced by the surface charges Formula (2.28) and the formula of the electric potential produced by the volume charges Formula (2.27), it can be seen that the $\boldsymbol{P} \cdot \boldsymbol{e}_n$ and $\nabla \cdot \boldsymbol{P}$ in the formula above correspond to ρ_s and ρ. So the relations between the intensity of polarization $\boldsymbol{P}$ and the volume density, surface density of the equivalent bound charges in the dielectrics are

$$\rho_p = -\nabla \cdot \boldsymbol{P} \tag{2.54}$$

$$\rho_{ps} = \boldsymbol{P} \cdot \boldsymbol{e}_n \tag{2.55}$$

From the formula above, it can be seen that when dielectrics exist in the electric field, bound charges will be produced inside and on the edge of the dielectrics, and these bound charges also produce electric field as a secondary source. Therefore, the calculation of the total electric field in the dielectrics is different from that of the electric field in a vacuum. The basic equation of the divergence of the electric field in the dielectric should be amended as follows

$$\nabla \cdot \boldsymbol{E} = \frac{\rho + \rho_p}{\varepsilon_0} \tag{2.56}$$

This formula shows the relation between the field and the source in the dielectrics.

When the volume density of free electric charges and bound charges are known, we can find the divergence of the electric field in the dielectrics. However, it is still not convenient to calculate the electric field in the dielectrics through this formula. This is because the volume density of bound charges needs to be known. In order to make the actual analysis and calculation simpler and more convenient, new quantities need to be introduced as substitutes for the influence of the ρ_p. Insert $\nabla \cdot \boldsymbol{P} = -\rho_p$ into Formula (2.56). We obtain

$$\nabla \cdot (\boldsymbol{E}\varepsilon_0 + \boldsymbol{P}) = \rho \tag{2.57}$$

It can be seen that the divergence of the resultant vector $\boldsymbol{E}\varepsilon_0+\boldsymbol{P}$ is only related to the volume density of free electric charges, and has nothing to do with the volume density of bound charges. We call this resultant vector the electric displacement vector, or electric flux density. The vector is represented by the symbol $\boldsymbol{D}$. The unit of $\boldsymbol{D}$ is Coulomb/ square meter (C/m^2).

$$\boldsymbol{D} = \boldsymbol{E}\varepsilon_0 + \boldsymbol{P} \tag{2.58}$$

$$\nabla \cdot \boldsymbol{D} = \rho \tag{2.59}$$

By calculating the volume integral of both sides of Formula (2.59), and using the divergence theorem, we can obtain the Gauss law in the dielectrics

$$\oint_S \boldsymbol{D} \cdot \mathrm{d}\boldsymbol{S} = q \tag{2.60}$$

The above formula shows that the flux of the electric displacement vector through any closed surface in the dielectrics is equal to the total amount of free charge surrounded by this closed surface, and independent of the bound charges in this closed surface. The result of the experiment shows that in terms of the isotropic linear dielectrics, the intensity of polarization $\boldsymbol{P}$ is proportional to the strength of the electric field in the dielectrics

$$\boldsymbol{P} = \varepsilon_0 \chi_e E \tag{2.61}$$

where χ_e is a dimensionless constant termed the dielectric susceptibility. By inserting Formula (2.61) into Formula (2.58), we get

$$\boldsymbol{D} = \varepsilon_0(1+\chi_e)\boldsymbol{E} = \varepsilon_r\varepsilon_0\boldsymbol{E} = \varepsilon\boldsymbol{E} \tag{2.62}$$

where

$$\varepsilon_r = 1 + \chi_e = \frac{\varepsilon}{\varepsilon_0} \tag{2.63}$$

In the above equation ε is the dielectric constant of the dielectrics (permittivity). ε_r is a dimensionless constant, and is called the relative dielectric constant (or relative permittivity). In all the isotropic dielectrics the directions of $\boldsymbol{P}$ and $\boldsymbol{E}$ are always the same. χ_e is a nonnegative constant. The relative dielectric constant of any dielectrics is greater than or equal to 1. In general dielectrics, the relation among $\boldsymbol{E}$, $\boldsymbol{P}$, $\boldsymbol{D}$ is linearity, and if the proportional constant χ_e or ε is independent of space coordinates, then the dielectric is uniform. In some dielectrics, the relation among $\boldsymbol{E}$, $\boldsymbol{P}$, $\boldsymbol{D}$ is not directly proportional. This kind of dielectrics is called the nonlinear dielectrics. There are some other dielectrics, in which the relation among $\boldsymbol{E}$, $\boldsymbol{P}$, $\boldsymbol{D}$ is related to the direction of the impressed electric field $\boldsymbol{E}$, and the χ_e and ε here are constantThis type of dielectric is referred to as an anisotropic dielectric.

After the electric displacement vector $\boldsymbol{D}$ is introduced, it is very convenient to find the distribution law of the bound charges in the polarized dielectrics. From Formula (2.61), (2.62), we obtain

$$\begin{aligned}\nabla \cdot \boldsymbol{P} = \nabla \cdot (\varepsilon_0 \chi_e \boldsymbol{E}) &= \nabla \cdot \left(\frac{\varepsilon_r - 1}{\varepsilon_r}\boldsymbol{D}\right) \\ &= \frac{\varepsilon_r - 1}{\varepsilon_r}\nabla \cdot \boldsymbol{D} + \boldsymbol{D} \cdot \nabla\left(\frac{\varepsilon_r - 1}{\varepsilon_r}\right)\end{aligned}$$

By inserting $\rho_p = -\nabla \cdot \boldsymbol{P}$ and $\nabla \cdot \boldsymbol{D} = \rho$ into the formula above, we obtain

$$\rho_p = -\nabla \cdot \boldsymbol{P} = -\left(\frac{\varepsilon_r - 1}{\varepsilon_r}\right)\rho - \boldsymbol{D} \cdot \nabla\left(\frac{\varepsilon_r - 1}{\varepsilon_r}\right) \tag{2.64}$$

It can be seen that in uniform dielectrics with no free electric charges, the density of bound charges is zero. However, in the nonuniform dielectrics or the dielectrics with free electric charges, the density of bound charges is not zero.

We now analyze the surface density of bound charges ρ_{ps}.

$$\boldsymbol{D} = \varepsilon_0\varepsilon_r\boldsymbol{E} = \varepsilon_0\boldsymbol{E} + \boldsymbol{P}$$

$$\boldsymbol{P} = \varepsilon_0\boldsymbol{E}(\varepsilon_r - 1) = \varepsilon_0(\boldsymbol{D}/\varepsilon_0\varepsilon_r)\cdot(\varepsilon_r - 1) = \left(\frac{\varepsilon_r - 1}{\varepsilon_r}\right)\boldsymbol{D}$$

From Formula (2.55), we derive

$$\rho_{ps} = \boldsymbol{P}\cdot\boldsymbol{e}_n = \left(\frac{\varepsilon_r - 1}{\varepsilon_r}\right)\boldsymbol{D}\cdot\boldsymbol{e}_n = \varepsilon_0(\varepsilon_r - 1)\boldsymbol{E}\cdot\boldsymbol{e}_n \tag{2.65}$$

which indicates that only in the region where the electric displacement vector is zero or parallel to the surface of the dielectrics, the surface density of bound charges on the surface of the dielectrics is equal to zero.

Under the condition of a general electric field strength, the intensity of polarization in the linear dielectrics $\boldsymbol{P}$ is proportional to the strength of the electric field inside the dielectrics $\boldsymbol{E}$. The small displacement between the positive and negative charges in molecules and atoms inside the dielectrics increases with the electric field in the dielectrics. However, if the electric field in the dielectrics is so large that the bound charges go beyond the control of the molecules and become free electric charges, then the insulating dielectrics may change into conductors. This phenomenon is called the dielectric breakdown, and the maximum field strength that the dielectric material can bear before breakdown is called the breakdown strength of the dielectric.

Example 2.8 A dielectric sphere whose radius is a and dielectric constant is ε, is filled with electric charges of which the volume density is ρ_0. Calculate the following:

(1) The $\boldsymbol{E}$, $\boldsymbol{P}$ in and out of the dielectric sphere;

(2) The volume density of the bound charges in the dielectric sphere, and the density of the bound charges on the surface of the dielectric sphere.

Solution (1) Due to the symmetric property of the dielectric sphere and the distribution of the electric charges, the $\boldsymbol{E}$, $\boldsymbol{P}$, $\boldsymbol{D}$ in and out of the sphere are all in the radial direction of the sphere and respectively have the same amplitude on the circles of equal radius which have the same center with the dielectric sphere. Through the use of Gauss's law in dielectrics, we have

$$\oint_S \boldsymbol{D}\cdot \mathrm{d}\boldsymbol{S} = q$$

and

$$\boldsymbol{D} = \varepsilon\boldsymbol{E} = \varepsilon_0\boldsymbol{E} + \boldsymbol{P}$$

from which we get

when $r < a$,

$$\boldsymbol{E} = \frac{r\rho_0}{3\varepsilon}\boldsymbol{e}_r$$

$$\boldsymbol{P} = \varepsilon_0(\varepsilon_r - 1)\boldsymbol{E} = \frac{(\varepsilon - \varepsilon_0)r\rho_0}{3\varepsilon}\boldsymbol{e}_r$$

when $r > a$,

$$\boldsymbol{E} = \frac{a^3\rho_0}{3\varepsilon_0 r^2}\boldsymbol{e}_r$$

$$\boldsymbol{P} = 0,$$

(2) The ρ_p in the dielectric sphere ($r < a$)

$$\rho_p = -\nabla \cdot \boldsymbol{P} = -\frac{1}{r^2}\frac{\partial}{\partial r}(r^2 P_r) = \frac{(\varepsilon_0 - \varepsilon)\rho_0}{\varepsilon}$$

On the surface of the dielectric sphere ρ_{ps}

$$\rho_{ps} = \boldsymbol{P} \cdot \boldsymbol{e}_r \,|_{r=a} = \frac{(\varepsilon_0 - \varepsilon)a\rho_0}{3\varepsilon}$$

2.7 The Boundary Conditions of the Electrostatic Field

The boundary in this context indicates the interface between different dielectrics in the electrostatic field. Different dielectrics have different polarizations in the electric field. Thereby, when the electric field passes through the interface between different dielectrics, the strength and direction of it will change under ordinary circumstances. The boundary conditions of the electrostatic field are used for discussing the laws which the electric field $\boldsymbol{E}$ and the electric displacement vector $\boldsymbol{D}$ follow at the interface. These boundary conditions will play an important role when in solving the field problems later in this text.

Since the boundary conditions are used for studying the laws which the field follows at the interface between the dielectrics, we need to analyze them by using the basic equations which describe the divergence and irrotationality of the electrostatic field. However, at the interface between the dielectrics, the strength and direction of the field will change abruptly. Thus, it is necessary to use the basic equations in the integral (and not differential) form to find the boundary conditions.

The strength and direction of the electric field in the dielectrics are random to a great extent. But in terms of the interface, all the fields can be expressed as (resolved into) the tangential field and the normal field, as shown in Figure 2.14. Subsequently, by using the basic equations in integral form, respectively find the laws which the tangential and normal fields follow at the interface between the dielectrics and the boundary conditions.

Resolve the $\boldsymbol{D}$ and $\boldsymbol{E}$ into two components, one of which is perpendicular to the interface and the other is parallel to the interface

$$\boldsymbol{D} = D_n\boldsymbol{e}_n + D_t\boldsymbol{e}_t$$
$$\boldsymbol{E} = E_n\boldsymbol{e}_n + E_t\boldsymbol{e}_t$$

where $\boldsymbol{e}_n$ is the normal unit vector out of the surface of the dielectrics, and $\boldsymbol{e}_t$ is the unit vector tangential to the interface.

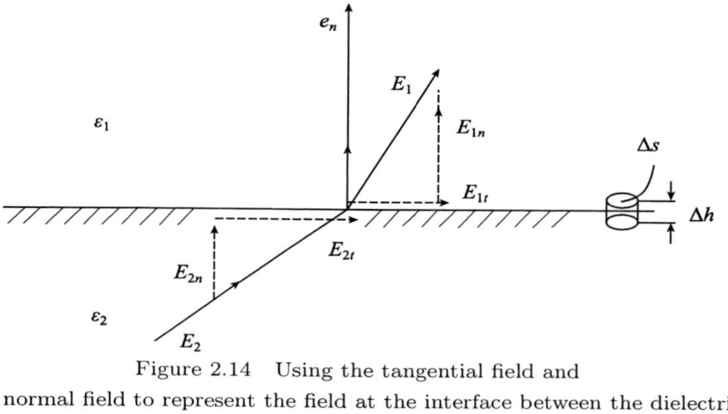

Figure 2.14 Using the tangential field and the normal field to represent the field at the interface between the dielectrics

2.7.1 The Boundary Conditions of the Normal Field

To study the normal boundary, the Gauss's law that describes the flux properties needs to be used. Accordingly, we must draw-upon to make a Gaussian surface. A closed cylindrical surface, as shown in Figure 2.15, of which the top and bottom surfaces are both ΔS and parallel to the interface between the dielectrics. ΔS must be small enough so the field on it can be considered as a uniform field. The height of the cylinder is Δh. Let $\Delta h \to 0$, to insure that the fields calculated from it are on both sides of the dielectrics, but infinitely close to each other. This is the manner by which the changing laws of the field at the boundary be reflected. As $\Delta h \to 0$, the cylindrical surface makes no contribution to the electric flux. Namely, the top and bottom surfaces of a closed cylindrical surface which are small enough to make a contribution to the electric flux. Thus, by using the Gauss law, we have

$$\begin{aligned}\oint_S \boldsymbol{D} \cdot \mathrm{d}\boldsymbol{S} &= \boldsymbol{D}_1 \cdot \Delta \boldsymbol{S}_1 + \boldsymbol{D}_2 \cdot \Delta \boldsymbol{S}_2 \\ &= D_{1n}\Delta S - D_{2n}\Delta S \\ &= \rho_s \Delta S \\ D_{1n} - D_{2n} &= \rho_s \end{aligned} \tag{2.66}$$

The above formula shows that when there is the surface density of free charges ρ_s distributed at the interface between the dielectrics, the normal component of the electric displacement vector is discontinuous from one side of the interface to another with the difference between them being equal to ρ_s. The boundary conditions under two different circumstances below are especially important.

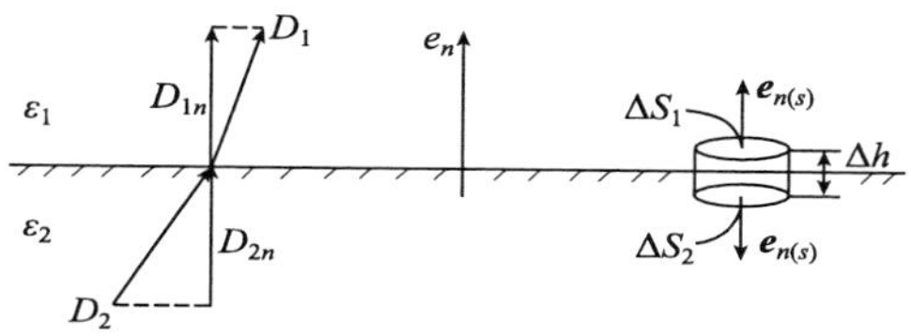

Figure 2.15 The boundary conditions of the normal electric displacement $\boldsymbol{D}$

1. The Boundary Conditions of the Normal Electric Displacement D_n at the Interface Between the Dielectrics

When there is no surface density of free charges at the interface between two dielectrics, i.e. $\rho_s = 0$, we get

$$D_{1n} = D_{2n} \tag{2.67}$$

Namely, the normal electric displacement D_n on the two sides of the dielectrics is continuous.

2. The Boundary Conditions of the Normal Electric Displacement D_n of Conductors

If the dielectric 2 is a conductor, as the electric displacement vector $\boldsymbol{D}$ in the conductor is zero, the normal component of the $\boldsymbol{D}$ at any point on the surface of the conductor is equal to the surface density of the free charges at this point, i.e. $D_{1n} = \rho_s$. This is usually written as

$$D_n = \rho_s \tag{2.68}$$

and indicates that if there is ρ_s on the surface of the conductor, it will produce the $\boldsymbol{D}$ vector which is perpendicular to the surface of the conductor. Similarly, consider there is an outer electric field in the dielectric 1. No matter which direction it is along, when it arrives at the surface of the conductor, its direction has to be perpendicular to the surface of the conductor, and its strength satisfies $D_n = \rho_s$.

2.7.2 The Boundary Conditions of the Tangential Field

When studying the tangential boundary, the conservative property of the electrostatic field needs to be used, i.e. $\oint_l \boldsymbol{E} \cdot \mathrm{d}\boldsymbol{l} = 0$. So we need to make a rectangular closed loop, as shown in Figure 2.16. The length of the upper and lower sides of it is Δl, and they are parallel to the interface between the dielectrics. Δl must be small enough so that the field on it can be considered as a uniform field; The heights of the left and right sides of the closed loop is Δh. We let $\Delta h \to 0$, to ensure that the fields calculated from it are respectively on both sides of the interface between the dielectrics, but infinitely close to the interface. Only in this way, can the changing laws of the field on the boundary be reflected correctly. As the $\Delta h \to 0$, the left and right sides of the rectangular loop make no contribution to the circulation of the electric field, namely the closed line integral of the electric field, but only the upper and lower sides whose Δl is small enough can make contribution to it. Thus we have

$$\begin{aligned}\oint_l \boldsymbol{E} \cdot \mathrm{d}\boldsymbol{l} &= \boldsymbol{E}_1 \cdot \Delta \boldsymbol{l}_1 + \boldsymbol{E}_2 \cdot \Delta \boldsymbol{l}_2 \\ &= E_{1t}\Delta l - E_{2t}\Delta l \\ &= 0\end{aligned}$$

$$E_{1t} = E_{2t} \tag{2.69}$$

This shows that, at the interface between the dielectrics, the tangential electric field E_t is continuous. The boundary conditions under two different circumstances below are especially important.

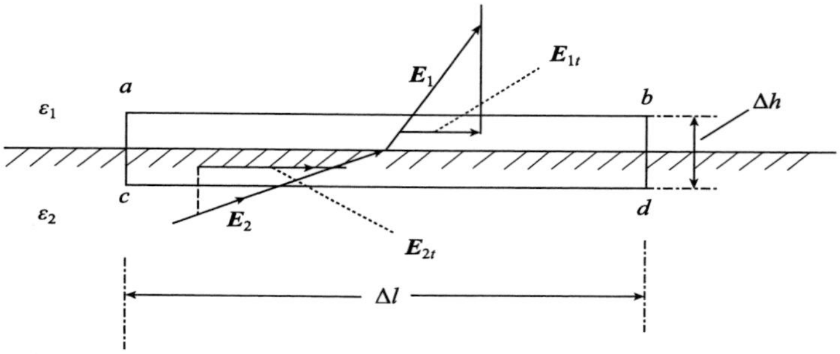

Figure 2.16 The boundary conditions of the tangential electric field E_t

1. The Boundary Conditions of the Tangential Electric Field E_t at the Interface between the Dielectrics

In terms of the interface between two kinds of dielectrics, it satisfies

$$E_{1t} = E_{2t} \tag{2.70}$$

2. The Boundary Conditions of the Tangential Electric Field on the Surface of the Conductor E_{1t}

If the dielectric 1 is a conductor and the electrostatic field in the conductor is zero, then the $E_{2t} = 0$, so the $E_{1t} = 0$. It is usually written as

$$E_t = 0 \tag{2.71}$$

It shows that the tangential electric field on the surface of the conductor is zero, and namely the surface of the conductor is an equipotential surface.

2.7.3 The Continuity of the Electric Potential at the Interface between the Dielectrics

At the interface between the dielectrics, the sudden change of the dielectric parameters will certainly cause a sudden change in the field. But the sudden change of the field quantity won't result in the sudden change of the electric potential φ at the interface between the dielectrics (on the boundary). As shown in Figure 2.17, take two symmetric points a, f on both sides of the interface between the dielectrics along the normal line, but the two points are infinitely close to each other. The distance Δh is approaching zero, i.e. $\overline{af} = \Delta h \rightarrow 0$. Let the electric potential at point a be φ_a, the electric potential at point f be φ_f, and the electric potential difference $\Delta\varphi$ between a and f be

$$\Delta\varphi = \int_a^f \boldsymbol{E} \cdot \mathrm{d}\boldsymbol{l}$$

The line integral path of the E·dl in the formula above can be $a \rightarrow b \rightarrow c \rightarrow d \rightarrow e \rightarrow f$, where the $\overline{bc}$, $\overline{de}$ are parallel to the interface between the dielectrics, $\overline{ab}$, $\overline{cd}$, $\overline{ef}$ are perpendicular to the interface between the dielectrics. As the $E_{1t} = E_{2t}$, the sum of $\boldsymbol{E}\cdot\mathrm{d}\boldsymbol{l}$ along the segment $\overline{bc}$ and $\overline{de}$ is zero. As the $D_{1n} = D_{2n}$, and it is of infinite value, the $\Delta h \rightarrow 0$, and $\overline{ab} + \overline{cd} + \overline{ef} = \Delta h$, so the sum of the $\boldsymbol{E}\cdot\mathrm{d}\boldsymbol{l}$ along the three segments $\overline{ab}$, $\overline{cd}$, $\overline{ef}$ is zero. Thus we have

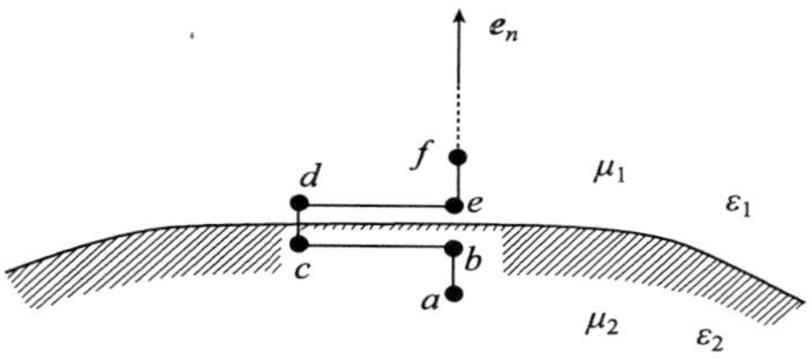

Figure 2.17 The continuity of electric potential φ

$$\varphi_a = \varphi_f \tag{2.72}$$

indicating that the electric potential φ at the interface between the dielectrics is continuous.

Example 2.9 As shown in Figure of Example 2.9, there is an infinitely large uniform dielectric board with the thickness of d in the uniform electric field $\boldsymbol{E}_0$ in the free space, where the relative dielectric constant is ε_r. The angle between the direction of the normal line of the dielectric board and the direction of the outer electric field is θ_0. If the angle between the direction of the electric field and the direction of the normal line of the board in the dielectric board θ_1 is equal to 45°, try to find the angle θ_0 and the surface density of the bound charges on the two sides of the dielectric board.

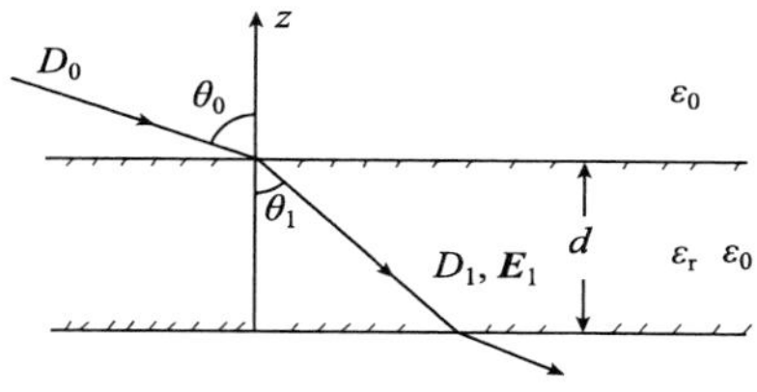

Figure of Example 2.9

Solution If there are no free charges on the surface of the dielectric board, then the boundary conditions are: the tangential electric field is continuous

$$E_0 \sin\theta_0 = E_1 \sin\theta_1$$

As the ρ_s on the surface is equal to zero, then the normal electric displacement vector is continuous

$$D_0 \cos\theta_0 = D_1 \cos\theta_1$$

By calculating the quotient of the two formula above, we obtain

$$\varepsilon_1 \tan\theta_0 = \varepsilon_0 \tan\theta_1 = \varepsilon_0$$

The derivation of the solution to the above equation leads to

$$\theta_0 = \arctan\left[\left(\frac{\varepsilon_0}{\varepsilon_1}\right)\tan\theta_1\right] = \arctan\left(\frac{1}{\varepsilon_\mathrm{r}}\right)$$

From Formula (2.65), we know that

$$\begin{aligned}
\rho_\mathrm{ps} &= \boldsymbol{P}_1 \cdot \boldsymbol{e}_n|_\mathrm{boundary} = \varepsilon_0(\varepsilon_\mathrm{r} - 1)\boldsymbol{E}_1 \cdot \boldsymbol{e}_n|_\mathrm{boundary} \\
\rho_\mathrm{ps\text{-}up} &= \varepsilon_0(\varepsilon_\mathrm{r} - 1)\boldsymbol{E}_1 \cdot \boldsymbol{e}_z = -\varepsilon_0(\varepsilon_\mathrm{r} - 1)\boldsymbol{E}_1 \cos\theta_1 \\
&= -\varepsilon_0(\varepsilon_\mathrm{r} - 1)\boldsymbol{E}_0 \sin\theta_0 \cot\theta_1 = -\varepsilon_0(\varepsilon_\mathrm{r} - 1)\boldsymbol{E}_0 \sin\theta_0 \\
\rho_\mathrm{ps\text{-}down} &= \varepsilon_0(\varepsilon_\mathrm{r} - 1)\boldsymbol{E}_1 \cdot (-\boldsymbol{e}_z) = \varepsilon_0(\varepsilon_\mathrm{r} - 1)\boldsymbol{E}_1 \cos\theta_1 \\
&= \varepsilon_0(\varepsilon_\mathrm{r} - 1)\boldsymbol{E}_0 \sin\theta_0
\end{aligned}$$

2.8 Capacitance of Conductor System

We consider two conductors which carry charges $+q$ and $-q$. The ratio of charge q and voltage U between them is called the capacitance of the conductor system. The capacitance of conductor system relates to its own shape and size and the permittivity of the surrounding medium, and has no relationship with the charge q which the conductor carries. The system whose own positive and negative electric charges are equal and all electricity lines start from the internal positive charges and end at the internal negative charges is called electrostatic stand-alone systems.

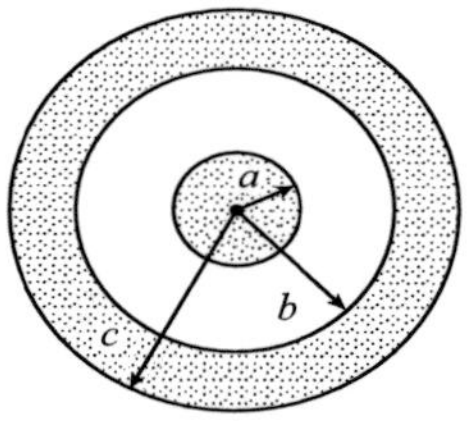

Figure 2.18 Concentric metal sphere and spherical shell

Taking concentric metal spherical shell for example, we will analyse the concepts and calculation methods of potential factors P_{ij}. Capacitance coefficient β_{ij} and partial capacity C_{ij}. Shown in Figure 2.18, the radius of the inner conductor sphere is a, the radii of inside and outside of outer conductor spherical shell respectively are b and c. The permittivity between the conductor sphere and the conductor spherical shell is ε_0, so is the permittivity outside the spherical conductor. Seeking the capacitance of the conductor system.

If inner conductor sphere carries charge q, outer conductor spherical shell carries charge $-q$, then the electric field around the space is

$$\begin{aligned}
E &= \frac{q}{4\pi\varepsilon_0 r^2}, \quad && a < r < b \\
E &= 0, && r > b
\end{aligned}$$

The voltage between the conductor sphere and conductor spherical shell is given by the potential difference

$$U = \int_a^b E\mathrm{d}r = \frac{q}{4\pi\varepsilon_0}\left(\frac{1}{a} - \frac{1}{b}\right)$$

The capacitance between two conductor spherical shells is given by

$$C = \frac{q}{U} = \frac{4\pi\varepsilon_0}{\dfrac{1}{a} - \dfrac{1}{b}}$$

As the ball and the spherical shell carry charges of $+q$ and $-q$, respectively, the electric field entirely distributes between the ball and the shell. When $r > b$, the electric field is zero everywhere, which is equal to that spherical shell the grounded. And there is no spherical shell-to-ground capacitance.

The more general situation involves the conductor ball carrying charge q_1, the conductor spherical shell carries charge q_2, and the electric field throughout the system being given by

$$\begin{aligned} E &= \frac{q_1 \boldsymbol{e}_r}{4\pi\varepsilon_0 r^2}, \qquad && a < r < b \\ E &= 0, && b < r < c \\ E &= \frac{(q_1 + q_2)\boldsymbol{e}_r}{4\pi\varepsilon_0 r^2}, && r > c \end{aligned}$$

The potential on the conductor ball will be given by

$$\varphi_1 = \int_a^\infty E\mathrm{d}r = \int_a^b \frac{q_1 \mathrm{d}r}{4\pi\varepsilon_0 r^2} + \int_c^\infty \frac{(q_1 + q_2)\mathrm{d}r}{4\pi\varepsilon_0 r^2} \tag{2.73}$$

while the potential on the conductor spherical shell is given by

$$\varphi_2 = \frac{q_1 + q_2}{4\pi\varepsilon_0 c} \tag{2.74}$$

The potential difference between the conductor sphere and spherical shell is

$$\varphi_1 - \varphi_2 = \left(\frac{1}{a} - \frac{1}{b}\right)\frac{q_1}{4\pi\varepsilon_0}$$

This indicates that under normal circumstances, potential difference exists between conductor spherical shell and conductor ball, the conductor ball and infinity, and the conductor spherical shell and the infinity. So there is capacitance between conductor spherical shell and conductor ball, the conductor ball and infinity, the conductor spherical shell and the infinity.

In order to analyze and describe such a system with has multiple capacitors, we need to express φ_1, φ_2 with the charge q and the potential coefficient P as

$$\begin{aligned} \varphi_1 &= P_{11}q_1 + P_{12}q_2 \\ \varphi_2 &= P_{21}q_1 + P_{22}q_2 \end{aligned} \tag{2.75}$$

P_{ij} is called potential modulus (V/C) and only relates to the conductor system's shape and size and the permittivity of the surrounding medium.

Comparing Equations (2.73), (2.74) and (2.75), we get

$$\begin{aligned} P_{11} &= \frac{\left(\frac{1}{a} - \frac{1}{b} + \frac{1}{c}\right)}{4\pi\varepsilon_0}, \qquad && P_{12} = \frac{1}{4\pi\varepsilon_0 c} \\ P_{21} &= \frac{1}{4\pi\varepsilon_0 c}, && P_{22} = \frac{1}{4\pi\varepsilon_0 c} \end{aligned} \tag{2.76}$$

and note that

$$P_{12} = P_{21}$$

We can also express q in Equation (2.75) with the capacitance coefficient β and the potential φ

$$\begin{aligned} q_1 &= \beta_{11}\varphi_1 + \beta_{12}\varphi_2 \\ q_2 &= \beta_{21}\varphi_1 + \beta_{22}\varphi_2 \end{aligned} \tag{2.77}$$

the quantities β_{11}, β_{22} in the above equation are the capacitance coefficients, and β_{12}, β_{21} are the inductances. From linear algebra, we know that the coefficient matrix P is β's inverse matrix. In Equations (2.75), (2.77), it is

$$[\beta] = [P]^{-1} \quad \text{or} \quad [P] = [\beta]^{-1}$$

Through the matrix inversion calculation we can get that

$$\begin{aligned}\beta_{11} &= \frac{P_{22}}{\Delta}, \quad \beta_{12} = \frac{-P_{21}}{\Delta}\\ \beta_{21} &= \frac{-P_{12}}{\Delta}, \quad \beta_{22} = \frac{P_{11}}{\Delta}\end{aligned} \tag{2.78}$$

$\Delta = P_{11}P_{22} - P_{12}^2$ in the equation.

Inserting each P value in Equation (2.76) yeilds

$$\begin{aligned}\beta_{11} &= \frac{4\pi\varepsilon_0 ab}{b-a}, \qquad \beta_{12} = \frac{-4\pi\varepsilon_0 ab}{b-a}\\ \beta_{21} &= \frac{-4\pi\varepsilon_0 ab}{b-a}, \quad \beta_{22} = 4\pi\varepsilon_0\left(c + \frac{ab}{b-a}\right)\end{aligned} \tag{2.79}$$

In order to express the capacitance relationship between the each conductor in system, we must use the potential of the conductors and the potential difference between conductors to indicate the charge q. Subsequently, Equation (2.77) can be rewritten into the following form

$$\left.\begin{aligned}q_1 &= (\beta_{11} + \beta_{12})\varphi_1 - \beta_{12}(\varphi_1 - \varphi_2)\\ q_2 &= -\beta_{21}(\varphi_2 - \varphi_1) + (\beta_{21} + \beta_{22})\varphi_2\end{aligned}\right\} \tag{2.80}$$

or as

$$\left.\begin{aligned}q_1 &= C_{11}\varphi_1 + C_{12}(\varphi_1 - \varphi_2)\\ q_2 &= C_{21}(\varphi_2 - \varphi_1)\varphi_1 + C_{22}\varphi_2\end{aligned}\right\} \tag{2.81}$$

The coefficient C_{ij} is called partial capacity, where C_{11}, C_{22} is its own partial capacity, C_{12}, C_{21} is each other partial capacity. Comparing Equation (2.81) and equation (2.80), we can get

$$\left.\begin{aligned}C_{11} &= \beta_{11} + \beta_{12}, \quad C_{12} = -\beta_{12}\\ C_{21} &= -\beta_{21}, \qquad C_{22} = \beta_{21} + \beta_{22}\end{aligned}\right\} \tag{2.82}$$

Inserting each of the β values in the equation above yields

$$\left.\begin{aligned}C_{11} &= 0, \qquad C_{12} = \frac{4\pi\varepsilon_0 ab}{b-a}\\ C_{21} &= \frac{4\pi\varepsilon_0 ab}{b-a}, \quad C_{22} = 4\pi\varepsilon_0 c\end{aligned}\right\} \tag{2.83}$$

As shown above, C_{12}, C_{21} is the mutual capacitance between the ball and the spherical shell,C_{22} is the self-capacitance between the conductor spherical shell and the ground.

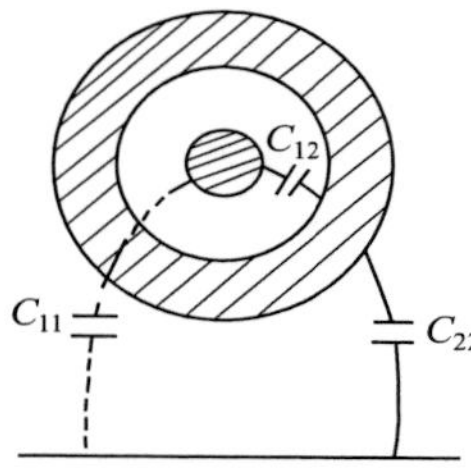

Figure 2.19 The partcapacitance between concentric metal sphere

It should be noted that when the self-capacitance C_{11} of the conductor ball is zero, it does not mean that the capacitor between the conductor ball and the ground is zero. As shown in Figure 2.19, in the case of the self-capacitance C_{11} of the conductor ball being zero, the capacitance between the conductor ball and the ground is equal toC_{12} in series with C_{22}.

The reason why $C_{11} = 0$ is that C_{11} is the self-capacitance of the metal ball in the conductor system, it is the metal ball-to-ground capacitance in the conductor system. As the metal ball is enclosed by concentric metal spherical shell, the electric field which can connect the metal ball with ground is zero. This means that this part of the stored energy is zero causing the capacitor C_{11} to be zero. One should not confuse C_{11} with the ground capacitance of the isolated metal ball.

Example 2.10 there are two metal balls whose radius are respectively a_1 and a_2, the distance between two centers of the balls is d, and $d \gg a$. This scenario is illustrated in Figure 2.10. Obtain the mutual capacitance of the conductor system C_{12}, C_{21} and the self-capacitance C_{11}, C_{22}.

Solution Let the two metal balls respectively carry charge q_1 and q_2, potentials are respectively φ_1, φ_1, as $d \gg a$, it may be approximately considered that charges uniformly distribute on the surface of the metal ball. The potential of each ball is the sum of the potential generated by q_1 and q_2. Suppose that the potential at infinity is zero. Then there we have

$$\begin{cases} \varphi_1 = \dfrac{1}{4\pi\varepsilon_0 a_1} q_1 + \dfrac{1}{4\pi\varepsilon_0 d} q_2 \\ \varphi_2 = \dfrac{1}{4\pi\varepsilon_0 d} q_1 + \dfrac{1}{4\pi\varepsilon_0 a_2} q_2 \end{cases}$$

It is known that

$$\begin{cases} \varphi_1 = P_{11} q_1 + P_{12} q_2 \\ \varphi_2 = P_{21} q_1 + P_{22} q_2 \end{cases}$$

$$\begin{cases} P_{11} = \dfrac{1}{4\pi\varepsilon_0 a_1}, \quad P_{12} = P_{21} = \dfrac{1}{4\pi\varepsilon_0 d} \\ P_{22} = \dfrac{1}{4\pi\varepsilon_0 a_2} \end{cases}$$

Through Equation (2.78) we get a relationship between the capacitance coefficient β_{ij} and the potential modulus P_{ij}, inserting P_{11}, P_{22}, P_{12} and P_{21} in Equation (2.78) we can get each capacitance coefficient β_{ij}

$$\left.\begin{aligned} \beta_{11} &= \frac{P_{22}}{P_{11}P_{22} - P_{12}^2} = \frac{4\pi\varepsilon_0 d^2 a_1}{d^2 - a_1 a_2} \\ \beta_{12} &= \beta_{21} \frac{-P_{12}}{P_{11}P_{22} - P_{12}^2} = -\frac{4\pi\varepsilon_0 a_1 a_2 d}{d^2 - a_1 a_2} \\ \beta_{22} &= \frac{P_{11}}{P_{11}P_{22} - P_{12}^2} = d\frac{4\pi\varepsilon_0 d^2 a_2}{d^2 - a_1 a_2} \end{aligned}\right\}$$

Through Equation (2.80) we obtain

$$\begin{cases} q_1 = \beta_{11}\varphi_1 + \beta_{12}\varphi_2 = (\beta_{11} + \beta_{12})\varphi_1 - \beta_{12}(\varphi_1 - \varphi_2) \\ q_2 = \beta_{21}\varphi_1 + \beta_{22}\varphi_2 = -\beta_{21}(\varphi_2 - \varphi_2) + (\beta_{21} + \beta_{22})\varphi_2 \end{cases}$$

and via Equation (2.81) we attain

$$\begin{cases} q_1 = C_{11}\varphi_1 + C_{12}(\varphi_1 - \varphi_2) \\ q_2 = C_{21}(\varphi_2 - \varphi_1) + C_{22}\varphi_2 \end{cases}$$

We can get each C by comparing the two expressions for q_1 and q_2

$$\begin{aligned} C_{11} &= \beta_{11} + \beta_{12} = \frac{4\pi\varepsilon_0 a_1 d^2}{d^2 - a_1 a_2} - \frac{4\pi\varepsilon_0 a_1 a_2 d}{d^2 - a_1 a_2} = \frac{4\pi\varepsilon_0 a_1 (d^2 - a_2 d)}{d^2 - a_1 a_2} \\ C_{22} &= \beta_{22} + \beta_{21} = \frac{4\pi\varepsilon_0 a_2 (d^2 - a_1 d)}{d^2 - a_1 a_2} \\ C_{12} &= C_{21} = -\beta_{12} = \frac{4\pi\varepsilon_0 a_1 a_2 d}{d^2 - a_1 a_2} \end{aligned}$$

It can also be seen from the expressions of the partial capacity that when the two conductor balls are very far from each other, that is, $d \gg a_1, a_2$, the two balls become isolated conductor balls. We know from the above analysis that

$$\begin{aligned} C_{11} &\approx 4\pi\varepsilon_0 a_1 \\ C_{22} &\approx 4\pi\varepsilon_0 a_2 \\ C_{12} &= C_{21} \approx 0 \end{aligned}$$

The expressions for C_{11} and C_{22} at this point equal the capacitance expressions when C_{11} and C_{22} are isolated balls. And the mutual capacitance becomes zero.

We have analyzed the system formed by two conductors and infinity (or earth) in the above, the same method can apply to the electrostatic system formed by N conductors and infinity, if the potential and the charge of the i-conductor are φ_i and q_i. In such a case, each conductor in the system can be expressed as

$$\begin{aligned} \varphi_1 &= P_{11}q_1 + P_{12}q_2 + \cdots + P_{1n}q_n \\ \varphi_2 &= P_{21}q_1 + P_{22}q_2 + \cdots + P_{2n}q_n \\ &\vdots \\ \varphi_n &= P_{n1}q_1 + P_{n2}q_2 + \cdots + P_{nn}q_n \end{aligned} \tag{2.84}$$

P_{ij} is the potential modulus. We can express q with β_{ij} and φ in Equation (2.84) as

$$\begin{aligned} q_1 &= \beta_{11}\varphi_1 + \beta_{12}\varphi_2 + \cdots + \beta_{1n}\varphi_n \\ q_2 &= \beta_{21}\varphi_1 + \beta_{22}\varphi_2 + \cdots + \beta_{2n}\varphi_n \\ &\vdots \\ q_n &= \beta_{n1}\varphi_1 + \beta_{n2}\varphi_2 + \cdots + \beta_{nn}\varphi_n \end{aligned} \tag{2.85}$$

where β_{ii} is the capacitance coefficient, and β_{ij} for $i \neq j$ are the inductances. From linear algebra we know that the coefficient matrix P and the matrix β are each other's inverses. Specifically, in Equations (2.84), (2.85)we have

$$[\beta] = [P]^{-1} \tag{2.86}$$

Converting the potential in Equation (2.85) of each conductor to the potential difference between the different bodies, then we attain

$$\begin{aligned} q_1 &= C_{11}\varphi_1 + C_{12}(\varphi_1 - \varphi_2) + \cdots + C_{1n}(\varphi_1 - \varphi_n) \\ q_2 &= C_{21}(\varphi_2 - \varphi_1) + C_{22}\varphi_2 + \cdots + C_{2n}(\varphi_2 - \varphi_n) \\ &\vdots \\ q_n &= C_{n1}(\varphi_n - \varphi_1) + C_{n2}(\varphi_n - \varphi_2) + \cdots + C_{nn}\varphi_n \end{aligned} \tag{2.87}$$

In the equation

$$C_{ii} = \sum_{j=1}^{N} \beta_{ij} \tag{2.88}$$

C_{ii} is called own partial capacity.

$$C_{ij} = -\beta_{ij}, \quad i \neq j \tag{2.89}$$

C_{ij} is called each other partial capacity

Figure 2.20 shows the equivalent diagram of the partial capacity of multi-conductor system. When there are more than two conductors in the real problems, calculating the capacitance between two conductors needs consideration of the effects of other conductors. A two-wire transmission line which is close to the ground, is shown in Figure 2.21. We must use the calculation method of partial capacity to calculate the capacitance between the two lines. This will be equivalent to the capacitance C_{11} in series with C_{22} and subsequently in parallel with C_{12}. However, this situation is different since the coaxial line is near the ground; the outer conductor of the coaxial line has the shielding effect, so the capacitance of the coaxial is free from the influence of the outside conductor and ground.

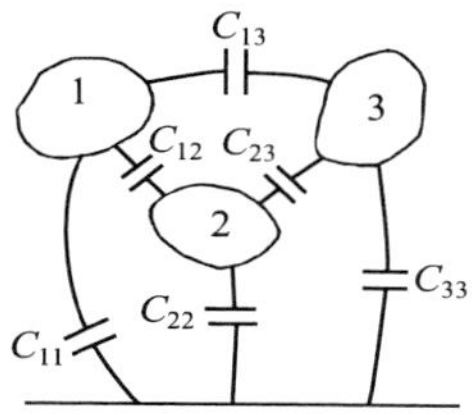

Figure 2.20 Partial capacity of multi-conductor system

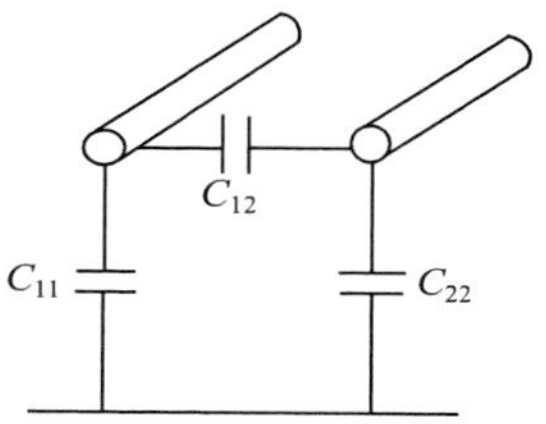

Figure 2.21 Partial capacity of the two-wire transmission line which is close to the ground

2.9 Energy of Electrostatic Field and Electrostatic Force

An electrified body in the electrostatic field will be affected by the electric field force. If there exists no other external forces to form a balance, an electric field force will move the electrified body and do work. This shows that the energy will be stored in the electrostatic field in such a scenario. Energy accumulates in the process of building the electric field. In order to move a charge from infinity to a static electric field, on one hand an external force must do work against the electric force, on the other hand the new charge introduces the electric field changes. Thus, the work which external force makes will store in the electrostatic field.

2.9.1 Energy of Electrostatic Field

In an electric field which is generated by q_1, moving another charge q_2 from infinity to a place which is a distance R_{12} away from the point charge q_1, the external force against the electric field force will make

$$W_2 = q_2\varphi_2 = \frac{q_1 q_2}{4\pi\varepsilon_0 R_{12}} \tag{2.90}$$

where φ_2 stands for the potential charge q_1 generates on place q_2.

In an electric field which is generated by q_2, moving another charge q_1 from infinity to a place which is R_{21} away from the point charge q_2, the external force against the electric field force will make

$$W_1 = q_1\varphi_1 = \frac{q_1 q_2}{4\pi\varepsilon_0 R_{21}} \tag{2.91}$$

φ_1 stands for the potential charge q_2 generates on place q_1.In linear medium, the potential is unrelated to the establishment methods and processes of the charge, so the results of the above two ways are exactly the same. The electric field energy is stored in the conjunct system constituted by q_1,q_2, according to Equations (2.90), (2.91) the electric field energy can be expressed as

$$W = \frac{1}{2}W_1 + \frac{1}{2}W_2 = \frac{1}{2}(q_1\varphi_1 + q_2\varphi_2) \tag{2.92}$$

If moving a point charge q_3 from infinity to a location which is R_{13} away from the point charge q_1, and a distanceR_{23} away from q_2 in this system, this will make

$$W_3 = q_3\varphi_3 = q_3\left(\frac{q_1}{4\pi\varepsilon_0 R_{13}} + \frac{q_2}{4\pi\varepsilon_0 R_{23}}\right) \tag{2.93}$$

φ_3 stands for the potential q_1 and q_2 generates on place q_3. Then the energy of the system will equal

$$W = \frac{q_1 q_2}{4\pi\varepsilon_0 R_{21}} + \frac{q_1 q_3}{4\pi\varepsilon_0 R_{13}} + \frac{q_2 q_3}{4\pi\varepsilon_0 R_{23}} \tag{2.94}$$

Along the lines of Equation (2.92), we can express Equation (2.94) as

$$\begin{aligned} W =& \frac{1}{2}\left[q_1\left(\frac{q_2}{4\pi\varepsilon_0 R_{12}} + \frac{q_3}{4\pi\varepsilon_0 R_{13}}\right) + q_2\left(\frac{q_1}{4\pi\varepsilon_0 R_{12}} + \frac{q_3}{4\pi\varepsilon_0 R_{23}}\right)\right. \\ &\left. + q_3\left(\frac{q_1}{4\pi\varepsilon_0 R_{13}} + \frac{q_2}{4\pi\varepsilon_0 R_{23}}\right)\right] \\ =& \frac{1}{2}(q_1\varphi_1 + q_2\varphi_2 + q_3\varphi_3) \end{aligned} \tag{2.95}$$

The potential φ_1 at location q_1 is generated by point charges q_2 and q_3. This is the same as φ_2, φ_3.

The above discussion can be extended to a more general situation. In the system generated by N point charges, the electric field energy can be expressed as

$$W_e = \frac{1}{2}\sum_{i=1}^{N} q_i\varphi_i \tag{2.96}$$

where φ_i stands for the potential generated by all other point charges (except q_i) on place q_i,

$$\varphi_i = \sum_{\substack{j=1 \\ j\neq i}}^{N} \frac{q_i}{4\pi\varepsilon_0 R_{ij}} \tag{2.97}$$

In the above system for energy expression, we have not considered q_i's own potential impact on the places of each point charge q_i. This is because here we have only discussed the work the external force do in moving N-point to a common electric field. The work does not include the energy which accumulates in each point charge's own process of formation .Or we can say Equation (2.96) only gives interaction energy in the point charge system, but the equation does not include the point charge's own inherent energy.

2.9.2 Total Energy of Charged System

First, calculate the energy of an isolated charged body. If a conductor carry charge q, its potential is φ. From the analysis above, it can be seen that an isolated conductor has certain potential. The potential of the conductor is proportional to the conductor charge. If at some point as a conductor with a charge αq, then the corresponding potential of conductor must be $\alpha\varphi$ where α stands for the coefficient from the $0 \to 1$. At this point, moving the small charge $dq\prime = q d\alpha$ from infinity to the conductor, the external forces are doing work

$$dW_e = \alpha\varphi dq' = \alpha\varphi q d\alpha$$

In the process of an isolated conductor's charge changing from 0 to q, it will accumulate a total energy of

$$W_e = \int_0^1 \alpha\varphi q d\alpha = \frac{1}{2}q\varphi \tag{2.98}$$

The potential φ is generated by conductor's own charges. The energy which the equation expresses is the isolated conductor's own energy.

For a system comprised of N conductors with charges $q_1, q_2, \cdots, q_n$, and potentials φ_1, $\varphi_2, \cdots, \varphi_n$; the total system energy can be derived in the same manner as above

$$W_e = \frac{1}{2}\sum_{i=1}^{N} q_i\varphi_i \tag{2.99}$$

Equation (2.99) and Equation (2.96) are identical in form, however, the φ_i is generated by all charges (including q_i) in the system. Therefore, the equation stands for the total electric field energy, which involves not only each conductor's own inherent energy, but also the interaction energy between each conductor.

When the charge under consideration is a continuous body-charge, a surface charge or a line charge, they can be divided into the charge distribution with a charge element $\rho \mathrm{d}V'$, $\rho_s \mathrm{d}s'$, $\rho_l \mathrm{d}l'$. According to the integral in Equation (2.99), the energy of distributed charge system will be

$$W_e = \int_{V'} \frac{\varphi\rho}{2} \mathrm{d}V' \tag{2.100}$$

$$W_e = \int_{S'} \frac{\varphi\rho_s}{2} \mathrm{d}S' \tag{2.101}$$

$$W_e = \int_{l'} \frac{\varphi\rho_l}{2} \mathrm{d}l' \tag{2.102}$$

The integration area in the equations above is the space occupied by the distributed charge.

2.9.3 Volume Energy Density of Electric Field

In order to describe the relationship between the energy of an electrostatic field and an electric field strength $\boldsymbol{E}$, we need to study the electric field energy in unit volume in space. That is, the relationship between the electric field energy density w_e and the electric field strength $\boldsymbol{E}$ $\left(w_e = \frac{1}{2}\varepsilon E^2\right)$. There are several different methods that can be followed.

Method 1 Using the charging energy DC power supplies for the capacitor, based on the conservation of energy principle, we can find the electric field energy density is $\frac{1}{2}\varepsilon E^2$.

Plate capacitor C, plate area is S, the distance between plates is d, if S is very large, and d is small, we may neglect the edge effect of electric field and assume that E is evenly distributed in the capacitor.

The space between plates is filled with a homogeneous medium having a dielectric constant ε. The capacitor voltage u is changing, so is the current i. Throughout the charging process, the charging energy DC power supplies for the capacitor in time $\mathrm{d}t$ is $ui\mathrm{d}t$. When the charging has completed, the total energy w_e the DC power supplied is given by

$$\begin{aligned} W_e &= \int_0^t ui\mathrm{d}t = \int_0^t \left(\frac{q}{c}\right)\left(\frac{\mathrm{d}q}{\mathrm{d}t}\right)\mathrm{d}t = \int_0^Q \frac{q}{c}\mathrm{d}q \\ &= \frac{1}{2}\cdot\frac{Q^2}{C} = \frac{1}{2}\cdot\frac{(CU)^2}{C} = \frac{1}{2}CU^2 \\ &= \frac{1}{2}\left(\frac{\varepsilon S}{d}\right)(Ed)^2 = \frac{1}{2}\varepsilon E^2\cdot(Sd) \end{aligned} \tag{2.103}$$

This is the total energy stored in the capacitor, and is essentially equal to $\frac{1}{2}\varepsilon E^2$ by volume (Sd). It should be obvious that $\frac{1}{2}\varepsilon E^2$ is the unit volume electric energy, that is, the electric field energy density. Furthermore, we have two methods to express the energy W_e stored in capacitor (Field and Road—distribution and concentration parameter), more specifically

$$W_e = \frac{1}{2}\varepsilon E^2(S\cdot d) = \frac{1}{2}CU^2 \tag{2.104}$$

Method 2 Use the electric field total energy Equation (2.100) (which we have already derived) of charged system whose volume is V', $W_e = \frac{1}{2}\int_{V'} \varphi\rho \mathrm{d}V'$. Then use the divergence theorem and the vector identities. Doing so allows us to arrive at the electric field energy density as $\frac{1}{2}\varepsilon E^2$.

Volume V' has charge distribution within the body, $\nabla \cdot \boldsymbol{D} = \rho$, and there is no charge out of V', $\nabla \cdot \boldsymbol{D} = 0$. But there is electric field and electric field energy in the internal and external of V' (total volume V).

First replace the volume integral of W_e on V' with an integral on the total volume V. Since $\rho = 0$ outside V', the two integrals will be equal leading to

$$W_e = \frac{1}{2}\int_{V'} \varphi\rho \mathrm{d}V' = \frac{1}{2}\int_{V} \varphi\rho \mathrm{d}V \tag{2.105}$$

Substitute $\nabla \cdot \boldsymbol{D} = \rho$, $\boldsymbol{E} = -\nabla\varphi$, the divergence theorem and the vector identities

$$\nabla \cdot \varphi\boldsymbol{D} = \varphi\nabla \cdot \boldsymbol{D} + \boldsymbol{D} \cdot \nabla\varphi$$

in Equation (2.105)

$$\begin{aligned} W_e &= \frac{1}{2}\int_{V'} \varphi\rho \mathrm{d}V' = \frac{1}{2}\int_{V'} \varphi\nabla \cdot \boldsymbol{D}\mathrm{d}V' \\ &= \frac{1}{2}\int_{V'} [\nabla \cdot \varphi\boldsymbol{D} - \boldsymbol{D} \cdot \nabla\varphi]\,\mathrm{d}V' \\ &= \frac{1}{2}\oint_{S'} \varphi\boldsymbol{D} \cdot \mathrm{d}\boldsymbol{S}' + \frac{1}{2}\int_{V'} \boldsymbol{D} \cdot \boldsymbol{E}\mathrm{d}V' \end{aligned} \tag{2.106}$$

The equation shows the following: the total electric power W_e is the sum of the internal and external electric energy of V', and the second part of the equation: $\int_{V'} \frac{1}{2}\varepsilon E^2 \mathrm{d}V'$, is the internal electric energy of V'. Obviously, $\frac{1}{2}\varepsilon E^2$is the electric energy in unit volume, that is, the electric field energy density.

In order to better understand this equation, replace the area S', the volume V' in the Equation (2.106) with total area S, total volume V according to Equation (2.105), so as to obtain

$$\begin{aligned} W_e &= \frac{1}{2}\oint_{S} \varphi\boldsymbol{D} \cdot \mathrm{d}\boldsymbol{S} + \frac{1}{2}\int_{V} \boldsymbol{D} \cdot \boldsymbol{E}\mathrm{d}V \\ &= \frac{1}{2}\int_{V} \boldsymbol{D} \cdot \boldsymbol{E}\mathrm{d}V = \int_{V} \frac{1}{2}\varepsilon E^2 \mathrm{d}V \end{aligned} \tag{2.107}$$

In the derivation of Equation (2.107), taking $\varphi \propto \frac{1}{r}$, $D \propto \frac{1}{r^2}$, $S \propto r^2$ into account. Obviously, when the area S is very large, $\oint_S \varphi\boldsymbol{D} \cdot \mathrm{d}\boldsymbol{S} = 0$, therefore, the electric field energy W_e in the entire volume V is given by

$$W_e = \int_{V} \frac{1}{2}\varepsilon E^2 \mathrm{d}V$$

The electric field energy density will be

$$w_e = \frac{1}{2}\varepsilon E^2 \tag{2.108}$$

Example 2.11 There is a metal ball, its radius is a, its electrical charge is q, seeking the total electric field energy W_e of this isolated charged metal ball.

Method 1 Using the electric field energy density $\frac{1}{2}\varepsilon E^2$, derive the total energy W_e.

Electric field energy distributes in the entire space outside the metal balls. We can derive W_e by integral of $\frac{1}{2}\varepsilon E^2$ on this space:

$$\begin{aligned} W_e &= \int_V \frac{1}{2}\varepsilon_0 E^2 \mathrm{d}V = \int_a^\infty \frac{1}{2}\varepsilon_0 E^2 \cdot 4\pi r^2 \cdot \mathrm{d}r \\ &= \int_a^\infty \frac{1}{2}\varepsilon_0 \left(\frac{q}{4\pi\varepsilon_0 r^2}\right)^2 \cdot 4\pi r^2 \mathrm{d}r = \frac{q^2}{8\pi\varepsilon_0}\int_a^\infty \frac{\mathrm{d}r}{r^2} \\ &= \frac{q^2}{8\pi\varepsilon_0 a} \end{aligned} \tag{a}$$

Method 2 Using $W_e = \frac{1}{2}\sum q_i\varphi_i$.

The above question pertains to the isolated metal ball, thus q_i is the charge q of metal ball and φ_i is the potential φ of the metal ball.

$$W_e = \frac{1}{2}q\varphi = \frac{1}{2}q\int_a^\infty E_r \mathrm{d}r = \frac{q^2}{2\times 4\pi\varepsilon_0}\int_a^\infty \frac{\mathrm{d}r}{r^2} = \frac{q^2}{8\pi\varepsilon_0 a}$$

Example 2.12 The metal ball and the concentric metal spherical shell form an electrostatic system. The first part of the system is given by a metal ball with radius a and charge q. The second part of the system is given by a metal spherical shell without charge. An Example of this is shown in Figure of Example 2.12. Seeking the total electric energy W_e of the system.

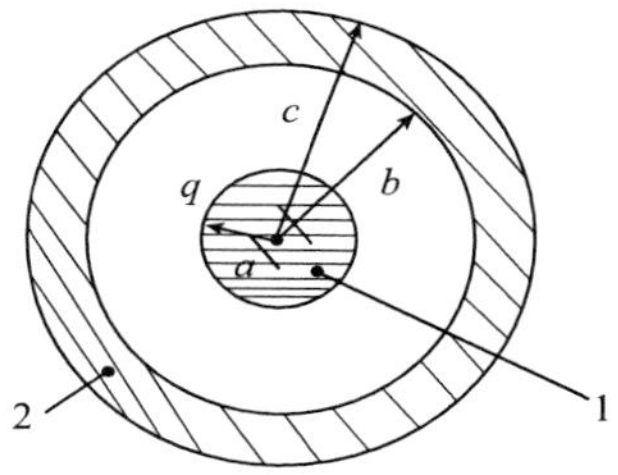

Figure of Example 2.12

Method 1 We can obtain W_e via the volume integral of $\frac{1}{2}\varepsilon E^2$.

The metal ball has charge q and the metal spherical shell has no charge. The electric field distributes in all of the space outside the metal ball (excluding spherical shell), then the total electric energy will be given by

$$\begin{aligned} W_e = \int_V \frac{1}{2}\varepsilon E^2 \mathrm{d}V &= \int_a^b \left(\frac{1}{2}\varepsilon_0 E^2\right) 4\pi r^2 \cdot \mathrm{d}r \\ &+ \int_c^\infty \left(\frac{1}{2}\varepsilon_0 E^2\right) 4\pi r^2 \cdot \mathrm{d}r \end{aligned}$$

Here $E_r = \frac{q^2}{4\pi\varepsilon_0 r^2}$. Subsequently, we insert E_r in the equation above and compute the integral to arrive at

$$W = \frac{q^2}{8\pi\varepsilon_0}\left(\frac{1}{a} - \frac{1}{b} + \frac{1}{c}\right) \tag{a}$$

Method 2 Using $W = \frac{1}{2}\sum_{i=1}^{2} q_i\varphi_i$ (see Equation (2.99)) get the total electric field energy W_e of system.

The total electric field energy of the system includes the conductor's own inherent energy and the interaction energy between conductors. Therefore, the φ_i should be the sum of potential that is generated by all of the charges (including the self-charge q_i).

$$W_{\mathrm{e}} = \frac{1}{2}q_1\varphi_1 + \frac{1}{2}q_2\varphi_2$$

Since $q_1 = q$ (the charge of metal ball), $q_2 = 0$ (metal spherical shell with no charge)

$$W_{\mathrm{e}} = \frac{1}{2}q_1\varphi_1 = \frac{1}{2}q\left(\int_a^b E_r \mathrm{d}r + \int_c^\infty E_r \mathrm{d}r\right)$$

Here $E_r = \dfrac{q^2}{4\pi\varepsilon_0 r^2}$ and insertion of E_r in the above equation and calculation of the integral leads to

$$W_{\mathrm{e}} = \frac{q^2}{8\pi\varepsilon_0}\left(\frac{1}{a} - \frac{1}{b} + \frac{1}{c}\right) \tag{b}$$

Method 3 By the expression of system energy

$$W_{\mathrm{e}} = \frac{1}{2}\int_{V'} \boldsymbol{D}\cdot\boldsymbol{E}\mathrm{d}V' + \frac{1}{2}\oint_{S'}\varphi\boldsymbol{D}\cdot\mathrm{d}\boldsymbol{S}' \tag{2.106}$$

to get the total electric energy W_{e}.

The equation is divided into two parts:

The first is the total energy within the spherical shell whose $r = c$

$$\frac{1}{2}\int_{V'} \boldsymbol{D}\cdot\boldsymbol{E}\mathrm{d}V' = \frac{1}{2}\int_a^b \varepsilon_0 E^2(4\pi r^2)\mathrm{d}r \tag{c}$$

The second is the total energy outside the spherical shell whose $r = c$

$$\frac{1}{2}\oint_{S'}\varphi\boldsymbol{D}\cdot\mathrm{d}\boldsymbol{S}' = \frac{1}{2}\oint_{S'}\varphi D\mathrm{d}S' \tag{d}$$

Calculating Formula (c) we obtain

$$\frac{1}{2}\int_a^b \varepsilon_0\left(\frac{q}{4\pi\varepsilon_0 r^2}\right)^2\cdot 4\pi r^2\mathrm{d}r = \frac{q^2}{8\pi\varepsilon_0}\int_a^b\frac{\mathrm{d}r}{r^2} = \frac{q^2}{8\pi\varepsilon_0}\left(\frac{1}{a} - \frac{1}{b}\right)$$

Calculating Formula (d) we obtain

$$\frac{1}{2}\oint_{S'}\varphi D\mathrm{d}S' = \frac{1}{2}\varphi D\oint_{S'}\mathrm{d}S' = \frac{1}{2}\left(\frac{q}{4\pi\varepsilon_0 c}\right)\cdot\left(\frac{q}{4\pi c^2}\right)\cdot 4\pi c^2 = \frac{q^2}{8\pi\varepsilon_0 c}$$

The sum of the two equations above is the total electric energy of system

$$W_{\mathrm{e}} = \frac{q^2}{8\pi\varepsilon_0}\left(\frac{1}{a} - \frac{1}{b} + \frac{1}{c}\right) \tag{e}$$

From Formula (a), (b), and (e) we observe that when the metal spherical shell is particularly thin, that is $b \approx c$, then

$$W_{\mathrm{e}} = \frac{q^2}{8\pi\varepsilon_0 a}$$

This is consistent with Equation (a) in Example 2.12.

Example 2.13 There is a point charge q and a medium ($\varepsilon = \varepsilon_{\mathrm{r}}\varepsilon_0$) spherical shell (inner and outer radiuses are a and b) in infinite space. Move this spherical shell, from the infinity to the position where the center of the spherical shell and the point charge q coincides, seeking the work done by external force.

Solution The work done by external force will change the energy of system. Namely, if the external forces do positive work, then the system energy will increase, and vice-versa.

No matter the dielectric spherical shell is at infinity, or moving it to the point charge, system's electric field strength $\boldsymbol{E}$ only changes in the dielectric layer of spherical shell (from E_0 to E_ε). The rest of the electric field doesn't change. This part of the changed electrical field has a corresponding change of energy given by:

$$\begin{aligned}\int_V \left(\frac{1}{2}\varepsilon_0 E_0^2 - \frac{1}{2}\varepsilon E_\varepsilon^2\right) \mathrm{d}V &= \int_a^b \left[\frac{1}{2}\varepsilon_0 \left(\frac{q}{4\pi\varepsilon_0 r^2}\right)^2 - \frac{1}{2}\varepsilon\left(\frac{q}{4\pi\varepsilon r^2}\right)^2\right] 4\pi r^2 \mathrm{d}r \\ &= \frac{q^2}{8\pi}\cdot\frac{(b-a)}{ab}\cdot\frac{(\varepsilon-\varepsilon_0)}{\varepsilon_0\varepsilon} \\ &= \frac{q^2(b-a)(\varepsilon_\mathrm{r}-1)}{8\pi ab\varepsilon_\mathrm{r}\varepsilon_0}\end{aligned}$$

It can be seen that moving the medium spherical shell closer causes the system energy to become smaller and the external forces to do negative work. That is, the electric field does positive work.

2.9.4 Electrostatic Force

A charged body in an electrostatic field will be given an electrostatic force. In a multi-conductor charged system, the force a conductor carries is the resultant force generated by the electric field resulting from the charges on the other conductors. For the interactive force between the point charges, Coulomb's Law provides a quantitative expression. For the electric force a continuous distribution charged body received in the electric field, although we can use Coulomb's law in principle and calculate the integral of the forces of each elementary charge, this method needs to determine the charge distribution on a charged body. It is very difficult to do such thing in many cases. So we usually use another relatively simple approach, using the principle of work-energy transformation, to calculate the electric force through electric potential gradient.

Method 1 Virtual displacement method.

Based on conservation of energy principle, the work the electric field forces done in the system are equal to the decrease of total energy in the system. Consider the plate capacitor shown in Figure 2.22. Two plates have respectively $+q$ and $-q$, plate space is x, the positive and negative charges on these two plates attract each other with force. As the charges are distributed on the conductor plate, the force is passed to the conductor plate, assuming that the conductor plate moves the distance $\mathrm{d}x$ under the electric field force $\boldsymbol{F}$, the work the electric force do is $\boldsymbol{F}\cdot\mathrm{d}\boldsymbol{x}$. If the amount of charges which capacitor carries are constant, capacitor is a stand-alone system which has no energy exchange with the outside world.

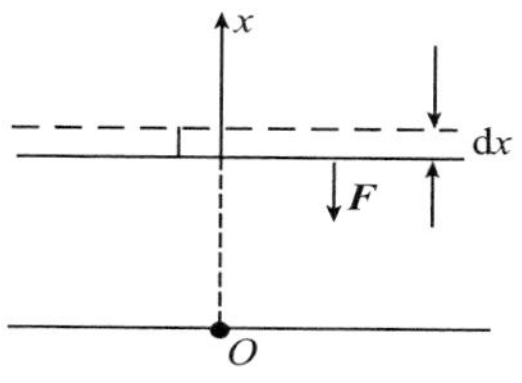

Figure 2.22 Virtual displacement method for plate capacitor

According to the principle of conservation of energy, the work the electric field forces do in the system is equal to the decrease of total energy of the system. That is

$$\boldsymbol{F}\cdot\mathrm{d}\boldsymbol{x} = -\mathrm{d}W_\mathrm{e}$$

When the force $\boldsymbol{F}$ is consistent with the direction of displacement $\mathrm{d}x$, then we have

$$\boldsymbol{F} = -\boldsymbol{e}_x \mathrm{d}W_\mathrm{e}/\mathrm{d}x|_{q=\mathrm{const.}} \tag{2.109}$$

The electric field energy in plate capacitor is known as

$$W_e = \frac{1}{2}CU^2 = \frac{1}{2}\frac{q^2}{C}$$

The expression of the plate capacitor's capacitance, excluding edge effects, is

$$C = \varepsilon S/x$$

Inserting the above result into $W_e = \frac{1}{2}\frac{q^2}{C}$, we obtain the electric field energy W_e as

$$W_e = \frac{q^2 x}{2\varepsilon S}$$

Similarly, inserting the above Equation into Equation (2.109), we can derive $\boldsymbol{F}$ as

$$\boldsymbol{F} = -\boldsymbol{e}_x \frac{\mathrm{d}}{\mathrm{d}x}\left(\frac{q^2 x}{2\varepsilon S}\right)\bigg|_{q=\text{const.}} = -\boldsymbol{e}_x \frac{q^2}{2\varepsilon S} \tag{2.110}$$

The negative sign describes the fact that the direction of the real force is to reduce the distance between plates. If the amount of charge is expressed as

$$q = \rho_s S = \varepsilon E S$$

then

$$F = -\frac{1}{2}\varepsilon E^2 S \tag{2.111}$$

The electric field force which the conductor plate of capacitor is tensile force, and the force which acts on a per unit area of the conductor surface is given by

$$f = -\frac{1}{2}\varepsilon E^2 \tag{2.112}$$

Hence, we obtain the electric force that the plate conductor receives using the principle of work-energy transformation. The method to seek the electric field force is based on the law of conservation of energy (and assume the conductor board displacement), and is known as virtual displacement method.

Method 2 Describe the electric field force with the gradient of the electric potential energy.

In the electrostatic field: $\boldsymbol{E} = -\nabla\varphi$, either side of the equals sign is multiplied by the charge q (constant) and we shall arrive at

$$q\boldsymbol{E} = \boldsymbol{F} = -\nabla(q\varphi) = -\nabla W_e \tag{2.113}$$

then we have

$$\boldsymbol{F} = -\nabla W_e|_{q=\text{const.}} \tag{2.114}$$

Electrostatic force $\boldsymbol{F}$ is equal to the negative gradient of the electrical potential energy W_e (the maximum reduce rate of potential). Taking the plate capacitor in "Method 1" as an example, the total potential energy of plate capacitor, the total electric field energy W_e is given by

$$W_e = \frac{1}{2}\varepsilon E^2(Sx) \tag{2.115}$$

In the equation S is the area of the plate, x is the distance between the plates. Inserting Equation (2.115) into Equation (2.114) allows us to obtain the electrostatic force $\boldsymbol{F}$ as

$$\boldsymbol{F} = -\nabla W_e|_{q=\text{const.}} = -\boldsymbol{e}_x\frac{1}{2}\varepsilon E^2 S \tag{2.116}$$

The force which the capacitor plate receives is

$$F = \frac{1}{2}\varepsilon E^2 S \tag{2.117}$$

The force which per unit area of the capacitor plate receives is

$$f = \frac{1}{2}\varepsilon E^2 \tag{2.118}$$

2.10 δ Function and Its Related Properties

2.10.1 δ Function

The distribution character of charge can be expressed by electric charge volume density $\rho(r)$ which is typically a continuous function of space coordinates. However, the point charge situation often happens in electromagnetic field problems. If the point charge q is also considered as the distributed charge, there will be such a special nature (case): its charge density $\rho\ (r)$ is infinite in the place where point charge is located. However, we will get a finite value q when calculating the volume integral of the charge density of a point charge -a special distributed charge. In order to describe the special nature of the point charge, we need to introduce a δ function.

δ function is a function which is used to describe the charge distribution of a unit point charge. Regarding the unit point charge as the distributed charge, the charge density $\rho(r)$ is zero outside of the tiny ball (the ball includes a unit point charge), and is not zero inside of the tiny ball. The volume density $\rho(r)$ increases when reducing the size of the tiny ball. And when the radius of the tiny ball tends to be zero the electric charge volume density $\rho(r)$ becomes infinite. However, the total charge will still remain unchanged since it is a unit (point) charge.

The δ function has important applications in electromagnetic field theory, transport theory, and communication theory.

2.10.2 The Definition of δ Function

For the unit point charge, its electric charge volume density distribution can be expressed as

$$\rho(\boldsymbol{r}) = \delta(\boldsymbol{r}-\boldsymbol{r}') = \begin{cases} 0, & \boldsymbol{r} \neq \boldsymbol{r}' \\ \infty, & \boldsymbol{r} = \boldsymbol{r}' \end{cases} \tag{2.119}$$

$$\int_V \rho(\boldsymbol{r})\mathrm{d}V = \int_V \delta(\boldsymbol{r}-\boldsymbol{r}')\mathrm{d}V = \begin{cases} 0, & \boldsymbol{r} \neq \boldsymbol{r}' \\ 1, & \boldsymbol{r} = \boldsymbol{r}' \end{cases} \tag{2.120}$$

The position vector $\boldsymbol{r}'$ in the equation is the vector from the origin of coordinates to the source point (where the unit point charge locates), and the position vector $\boldsymbol{r}$ is the vector from the origin of coordinates to the field point. In the Equation (2.120), $\boldsymbol{r}$=$\boldsymbol{r}'$ indicates that if the range of the volume integral contains the location of the unit point charge, then the volume integral of the δ function is equal to 1.

Similarly, for the point charge whose quantity of electric charge is q, the density function will be given by

$$\rho(r) = q\delta(\boldsymbol{r}-\boldsymbol{r}') = \begin{cases} 0, & \boldsymbol{r} \neq \boldsymbol{r}' \\ \infty, & \boldsymbol{r} = \boldsymbol{r}' \end{cases} \tag{2.121}$$

$$\int_V \rho(r)\mathrm{d}V \int_V q\delta(\boldsymbol{r}-\boldsymbol{r}')\mathrm{d}V = \begin{cases} 0, & \boldsymbol{r} \neq \boldsymbol{r}' \\ q, & \boldsymbol{r} = \boldsymbol{r}' \end{cases} \tag{2.122}$$

δ function is also called impulse function. δ function can be described by the impulse that is infinite at altitude, zero at which and 1 at area. Then

$$\delta(t) = \begin{cases} 0, & t \neq 0 \\ \infty, & t = 0 \end{cases}$$

$$\int_{-\infty}^{\infty} \delta(t)\mathrm{d}t = 1 \text{ or } \int \delta(t)\mathrm{d}t = \begin{cases} 0, & t \neq 0 \\ 1, & t = 0 \end{cases}$$

2.10.3 The Related Properties of δ Function

1. The Screening (Sampling) of δ Function

If $f(r)$ is a continuous function, then we have

$$\int_V f(r)\delta(\boldsymbol{r}-\boldsymbol{r}')\mathrm{d}V = f(\boldsymbol{r}') \tag{2.123}$$

Using the $\delta(\boldsymbol{r}-\boldsymbol{r}')$ function, we can screen the $f(r\prime)$ at $\boldsymbol{r}=\boldsymbol{r}'$.

2. $\nabla^2\left(\frac{1}{r}\right)$ Can be Described by the δ Function, Namely, $\nabla^2\left(\frac{1}{r}\right) = -4\pi\delta(x,y,z)$

If the source point (unit point charge) coincides with the origin of coordinates, then

$$r^2 = x^2 + y^2 + z^2$$

The potential φ is inversely proportional to r, that is, $\varphi \propto \frac{1}{r}$. It can be proved that

$$\nabla^2\left(\frac{1}{r}\right) = \begin{cases} 0, & r \neq 0, \text{outside the source} \\ \infty, & r = r' = 0, \text{at the unit point charge} \end{cases} \tag{2.124}$$

Then this feature of the function $\nabla^2\left(\frac{1}{r}\right)$ can be described by the δ function, it is

$$\nabla^2\left(\frac{1}{r}\right) = -4\pi\delta(x,y,z) \tag{2.125}$$

It is straight-forward to prove that this expression is correct. we can prove this by calculating the volume integral in Cartesian coordinates in the range which contains the source (unit point charge).

Calculating the volume integral of the right side of the equals sign in Equation (2.125), using the screening of δ function we obtain

$$\int_V -4\pi\delta(x,y,z)\mathrm{d}V = -4\pi \tag{2.126}$$

Calculate the volume integral of the left side of the equals sign in Equation (2.125), and apply the Gauss divergence theorem. The quantities $\boldsymbol{e}_r$, $\mathrm{d}\boldsymbol{S}$, $\mathrm{d}\boldsymbol{S}'$, θ and the solid angle $\mathrm{d}\Omega$ which appear in the derivation have been marked on Figure 2.5.

$$\begin{aligned}\int_V \nabla^2\left(\frac{1}{r}\right)\mathrm{d}V &= \int_V \nabla\cdot\nabla\left(\frac{1}{r}\right)\mathrm{d}V \\ &= \oint_S \nabla\left(\frac{1}{r}\right)\cdot \mathrm{d}\boldsymbol{S} = -\oint_S \frac{\boldsymbol{e}_r}{r^2}\cdot\mathrm{d}\boldsymbol{S} \\ &= -\oint_S \frac{\mathrm{d}S\cos\theta}{r^2} = -\oint_S \frac{\mathrm{d}S'}{r^2} \\ &= -\oint_S \mathrm{d}\Omega = -4\pi \end{aligned} \tag{2.127}$$

Equations (2.126), (2.127) shows that Equation (2.125) is correct.

If the unit point source charge is not at the origin of coordinates, then the distance R from source point(x', y', z') to the field points(x, y, z) is given by

$$R^2 = (x-x')^2 + (y-y')^2 + (z-z')^2$$

It can also be proven that $\nabla^2\left(\frac{1}{R}\right)$ can be described by the δ function, and indicates that the following result is correct

$$\nabla^2\left(\frac{1}{R}\right) = -4\pi\delta(x, y, z) \tag{2.128}$$

Exercises

2.1 The electric field strength $E = e_x(yz-2x) + e_y xz + e_z xy$, find: (1) Can the electric field be the solution of electrostatic field? (2) If it is an electrostatic field, find the potential corresponding to the electric field strength.

2.2 There is a ring in the $x - y$ plane, it has radius a and its center is at the origin. There is a line charge whose density is $\rho_l(\phi)$. find each $\boldsymbol{E}$ on the axis in following cases

(1) $\rho_l(\phi) = \rho_l$;

(2) $\rho_l(\phi) = \rho_l \cos\phi$;

(3) $\rho_l(\phi) = \rho_l \sin\phi$.

2.3 Find the $\boldsymbol{E}$ on the axis of uniformly charged disc. Disk radius is a, surface charge density is ρ_s.

2.4 The radii of the inside and outside of the spherical conductor capacitor are a and b. If we keep the potential difference U between the inside and outside of the conductor unchanged, prove that the electric field amplitude on the surface of the inside ball reach a minimum value in the condition that the two radii of the inside and outside of the conductor satisfy $a = b/2$. Find the minimum electric field intensity $\boldsymbol{E}_{\min}$.

2.5 A very long semi-cylinder with a radius of a, and surface charge density ρ_s is uniformly distributed on the surface of the cylindrical (Figure of Exercise 2.5). Find the field strength on the cylinder axis.

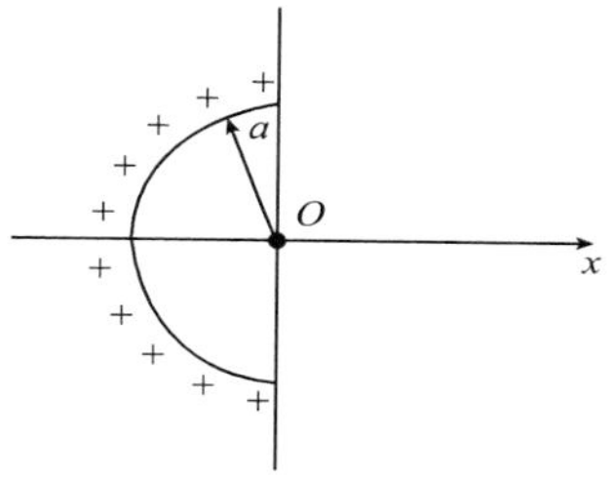

Figure of Exercise 2.5

2.6 A cylinder whose radius is a, and its length is infinity, and surface charge density ρ_s is uniformly distributed on the surface of the cylindrical. Find the potential and the electric field inside and outside the cylinder.

2.7 There is a sphere whose electric charge volume density is ρ(radius is a), the permittivity both inside and outside the sphere are ε_0. Find: (1) the $\boldsymbol{D}$ and $\boldsymbol{E}$ inside and outside the sphere; (2) the potential φ inside and outside the sphere; (3) electrostatic field energy.

2.8 There are concentric conductor spherical shells whose inner and outer radii are respectively a and b, the voltage between the two spherical shells is U. Find this electric field strength between the two spherical shells.

2.9 There is a spherical capacitor whose inner and outer radii are respectively a and b, the space between the two balls is filled with homogeneous medium whose permittivity is ε, the inner ball carries a charge q. Find:

(1) The electric field distribution in the capacitor;

(2) The potential difference between inner and outer conductor;

(3) The capacitance C.

2.10 There is a coaxial line, the radius of inner conductor is a, the outer is b, its thickness is negligible. The space between the inner and outer conductors is filled with homogeneous medium whose permittivity is ε. Find the capacitance of unit length.

2.11 There is an electric field inside and outside of a spherical region whose radius is a

$$\boldsymbol{E} = \begin{cases} \boldsymbol{e}_r A\left(\dfrac{r}{3\varepsilon_0} - \dfrac{r}{3a^2\varepsilon_0}\right), & r < a \\ \boldsymbol{e}_r \dfrac{Ba^2}{\varepsilon_0 r^2}, & r > a \end{cases}$$

Find the charge distribution which generates the electric field.

2.12 A conductor ball whose radius is a carries a charge q, the center of the ball locates at the boundary surface of two kinds of media (Figure of Exercise 2.12). Find:

(1) Electric field distribution;

(2) The electrostatic charge distribution on the spherical surface;

(3) Electrostatic energy of the whole system.

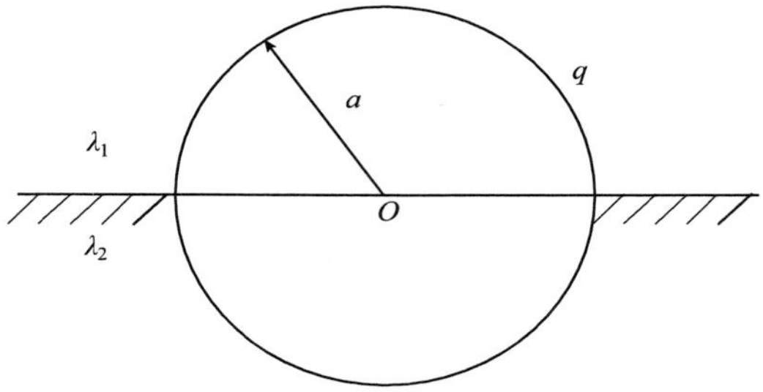

Figure of Exercise 2.12

2.13 There is a coaxial cable, the radius of inner conductor is a, the outer is b, the cable is filled with homogeneous medium whose breakdown strength is constant. Assume that b is constant. Find the value of a where the cable conductors suffers the maximum voltage.

2.14 A concentric sphere capacitor is formed by a conductor ball whose radius is a and a conductor concentric spherical shell, the inner radius of the shell is b, the space between the ball and half of the shell (separated along the radial) is filled with uniform medium whose permittivity is ε_1, The other half is filled with uniform medium whose permittivity is ε_2 (Figure of Exercise 2.14). Find the capacitance of the spherical capacitor.

2.15 As shown in Figure of Exercise 2.15, there are two uniformly charged straight lines, their lengths are $l$and they are parallel to each other, these two lines carry respectively charge $\pm q$, they are separated by distance l, find the electric field at the center of the charged system.

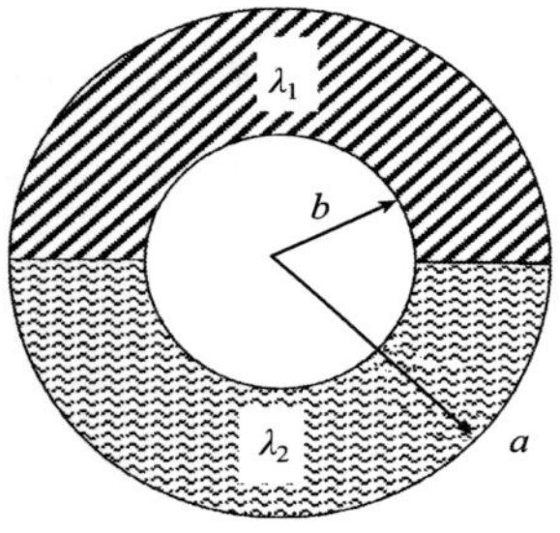

Figure of Exercise 2.14

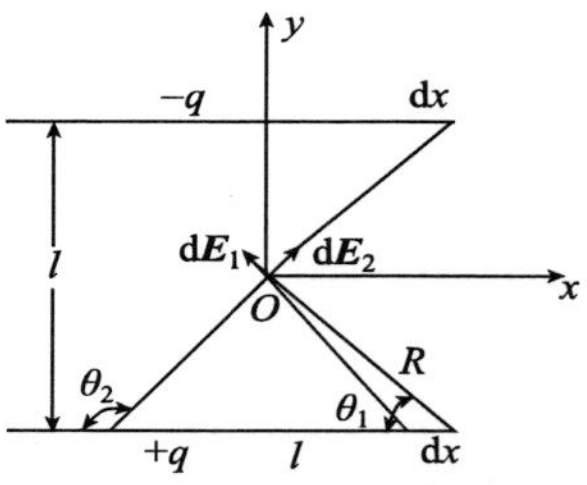

Figure of Exercise 2.15

2.16 There is a uniformly charged plane, it is infinitely big and very thin, and its surface charge density is ρ_s. Proof: when z-axis is perpendicular to the plane, half of the electric field strength $\boldsymbol{E}$ at $z = z_0$ is generated by the charges which are distributed in the circle whose radius is $\sqrt{3}z_0$ on the plane.

2.17 As shown in Figure of Exercise 2.17, there is a sphere with the radius a in a vacuum. The poles of the sphere are set to point charges $+q$ and $-q$, try to calculate the flux of electric displacement vector on the equatorial plane of the sphere.

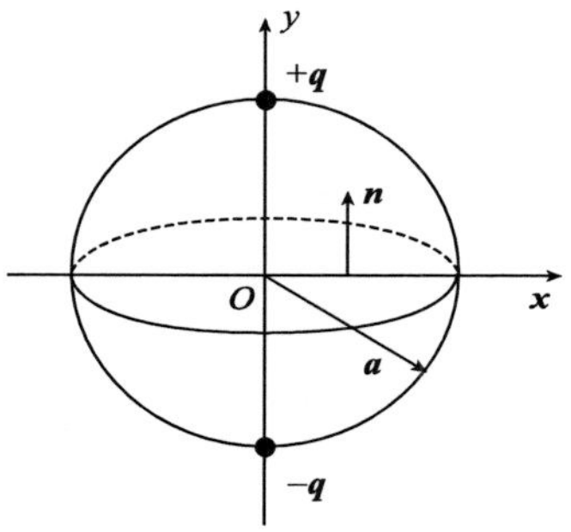

Figure of Exercise 2.17

2.18 The ball with the radius a is filled with the charge whose density is $\rho(r)$, the electric field is known as

$$D_r = \begin{cases} \varepsilon_0 \left(r^3 + Ar^2\right), & r \leqslant a \\ \varepsilon_0 \left(a^5 + Aa^4\right)/r^2, & r \geqslant a \end{cases}$$

Find charge density $\rho(r)$ and the charge surface density ρ_s at $r = a$.

2.19 There is a thin conductor spherical shell with the radius a and an inner surface coated with a thin insulating film. Inside of the ball is filled with a total charge Q, the spherical shell is filled with another charge Q. The internal electric field $\boldsymbol{E}=\boldsymbol{e}_r(r/a)^4$, assume that the medium of the inside of the ball is a vacuum. Find: (1) charge distribution within the ball; (2) the charge distribution on outer surface on the spherical shell; (3) potential of the spherical shell; (4) potential of the center of the ball.

2.20 There are two infinitely long coaxial cylinders, their radii are respectively a and $r = b$(with $b > a$). These surface charge densities are respectively ρ_{s1} and ρ_{s2}. Find: (1) electric field strength $\boldsymbol{E}$; (2) if $\boldsymbol{E} = 0$ at $r > b$, what relationship should the ρ_{s1} and ρ_{s2} have?

2.21 There are concentric conductor spheres whose radii are respectively a and b. The charges are evenly distributed on their surfaces. These surface charge densities are respectively ρ_{s1} and ρ_{s2}. Find: (1) electric field strength $\boldsymbol{E}$ in space; (2) the voltage $\boldsymbol{U}$ between the two spheres.

2.22 There is the line charge with length l and uniform charge density ρ_l. Find: (1) the potential φ on the diagonal plane of the line charge; (2) calculate the electric field strength $\boldsymbol{E}$ on the diagonal plane by Coulomb's law, and check it with $-\nabla\varphi$.

2.23 There is a coaxial capacitor, the radius of inner conductor is a, the outer is c, the part of $a < r < b$(with $b < c$) is filled with dielectric medium whose dielectric constant is ε. Find the capacitance of unit length.

2.24 Uniformly charged disc with radius a and surface charge density ρ_s. Find the potential of any point P at the edge of the disc.

2.25 A point charge q is placed at the axis of a thin grounded ring whose radius is a, the distance between this charge and the center of the ring is b, the inductive charge on the ring is $-q$, if the dielectric of the surrounding is air. Determine the capacitance of this ring.

2.26 There is a coaxial, its radius of inner conductor is a, the outer is b, applied voltage is U_0, the middle is filled with dielectric whose $\varepsilon_{\mathrm{r}} = \dfrac{r}{a}$. Find:

(1) $\boldsymbol{E}$, $\boldsymbol{D}$, $\boldsymbol{P}$ and φ in the medium;
(2) The polarization charge volume density and polarization charge surface density;
(3) The capacitance of unit length.

2.27 The polarization vector in the dielectric cube whose center is located at the

$$\boldsymbol{P} = P_0(\boldsymbol{e}_x x + \boldsymbol{e}_y y + \boldsymbol{e}_z z)$$

(1) Calculate the bound charge volume density and surface density;
(2) Prove that the total bound charge is zero.

Chapter 3

Constant Magnetic Field

A constant current generates a constant magnetic field. A constant magnetic field is a vector field, thus the analysis of a constant magnetic field can make use of basic experiment theorems such as the Ampere Force law and the Biot-Savart law. Subsequently, the differential and integral forms of basic equations of constant magnetic field can be derived. This will be similar to the analysis of static electric field. The differential form is the curl equation and divergence equation of constant magnetic field. This reveals that the constant magnetic field is solenoidal field (divergence is zero at any point), which means that constant magnetic field has no divergence-producing types of sources but vortex sources. The integral form helps to describe the attributes of a magnetic field in large-scale and the relationship between fields and sources, and its basic equation is utilized to analyze the boundary conditions of constant magnetic field.

The notion of a magnetic vector potential, magnetic dipole, and the energy and force of a magnetic field are also discussed. Lastly, the concepts of the calculation of self-inductance and co-inductance are presented.

3.1 The Curl Equation and Divergence Equation of Constant Magnetic Field

In this section, the Biot-Savart Law will be introduced and used to derive basic equations for describing the attributes of constant current magnetic field. The basic equations include differential and integral forms. The differential forms are the divergence equation and curl equation respectively, which give the attributes of constant magnetic field at a certain point as well as the relationship between fields and sources.

3.1.1 The Ampere (Force) Law and the Biot-Savart Law

1. The Ampere Force Law

The Ampere Force law describes the forces between two constant current circuits. As illustrated in Figure 3.1, experiments conducted by Ampere helped to prove the Ampere Force law. A circuit $l_1$has a current of I_1in vacuum, and it has a force over circuit $l_2$which have a current of I_2. The Ampere Force law can be described as follows

$$\boldsymbol{F}_{12} = \frac{\mu_0}{4\pi} \oint_{l_2} \oint_{l_1} \frac{\boldsymbol{I}_2 \mathrm{d}\boldsymbol{l}_2 \times (\boldsymbol{I}_1 \mathrm{d}\boldsymbol{l}_1 \times \boldsymbol{e}_R)}{\boldsymbol{R}^2} \tag{3.1}$$

which is the Ampere (Force) Law. In the equation

$$\boldsymbol{R} = \boldsymbol{e}_R R = \boldsymbol{r} - \boldsymbol{r}' = \boldsymbol{e}_R \left| \boldsymbol{r} - \boldsymbol{r}' \right| \tag{3.2}$$

Transform $\boldsymbol{F}_{12}$ as below

$$\boldsymbol{F}_{12} = \oint_{l_2} \boldsymbol{I}_2 \mathrm{d}\boldsymbol{l}_2 \times \left(\frac{\mu_0}{4\pi} \oint_{l_1} \frac{\boldsymbol{I}_1 \mathrm{d}\boldsymbol{l}_1 \times \boldsymbol{e}_R}{R^2} \right) = \oint_{l_2} \boldsymbol{I}_2 \mathrm{d}\boldsymbol{l}_2 \times \boldsymbol{B} = \oint_{l_2} \mathrm{d}\boldsymbol{F}_{12} \tag{3.3}$$

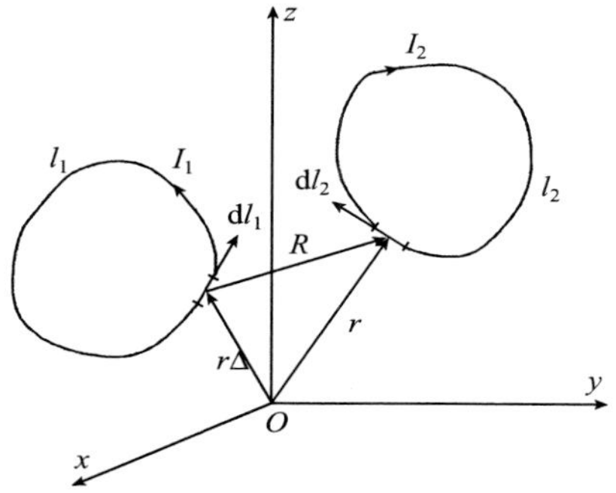

Figure 3.1 Ampere force between two current-carrying circuits

In the equation

$$dF = Idl \times B \tag{3.4}$$

the Ampere force is a force that is imposed on a moving charge by a magneticfield. Thus dF can be given as below

$$dF = Idl \times B = \left(\frac{dq}{dt}\right)(vdt) \times B = dqv \times B \tag{3.5}$$

This above equation states that a moving charge dq before with speed v has been imposed a force dF over in magnetic field B. The force is called the Lorentz force.

If both magnetic field B and electric field E exist, the moving charge q before with a speed of v has been imposed the force F given by

$$F = qE + qv \times B \tag{3.6}$$

2. The Biot-Savart Law

B denotes the magnetic flux intensity at dl_2in circuit $l_2$generated by the current I_1in circuit l_1. A general formula is given below

$$B = \frac{\mu_0}{4\pi}\oint_{l'}\frac{Idl' \times e_R}{R^2} \tag{3.7a}$$

The integration is conducted over coordinates in the source area where current exists.

In the case that current is distributed over a surface S' (J_S denotes the surface current density) or a volume V' (J denotes the volume current density), B can be given as

$$B = \frac{\mu_0}{4\pi}\int_{S'}\frac{J_S \times e_R}{R_2}dS' \tag{3.7b}$$

$$B = \frac{\mu_0}{4\pi}\int_{V'}\frac{J_S \times e_R}{R_2}dV' \tag{3.7c}$$

Comparison of the above result with Formulas (3.7a), it is obvious that for the line current distribution, the magnetic flux intensity dB generated by the current element Idl' in the circuit is

$$dB = \frac{\mu_0}{4\pi}\frac{Idl' \times (r - r')}{|r - r'|^3} = \frac{\mu_0}{4\pi}\frac{Idl' \times e_R}{R^2}. \tag{3.8}$$

It is refered to as the Biot-Savart Law. It reveals the relationship between current element Idl' and the magnetic flux intensity dB generated by the current element. This is illustrated in Figure 3.2.

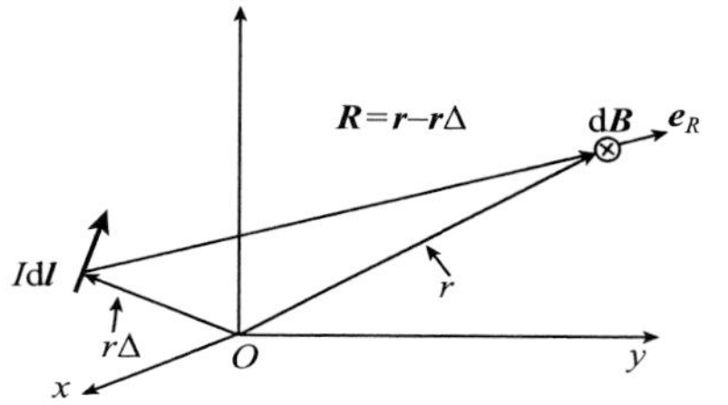

Figure 3.2 Relationship between Idl and its generated dB

Example 3.1 Given an infinite straight conductor with current I, calculate the magnetic field intensity $\boldsymbol{H}$ at point P which is a vertical distance r away to the conductor.

Solution As shown in the Figure of Example 3.1, the magnetic field intensity can be calculated according to the Biot-Savart Law by integrating over the current element $I\mathrm{d}\boldsymbol{l}'$.

According to the Biot-Savart law of Equation (3.8) we have

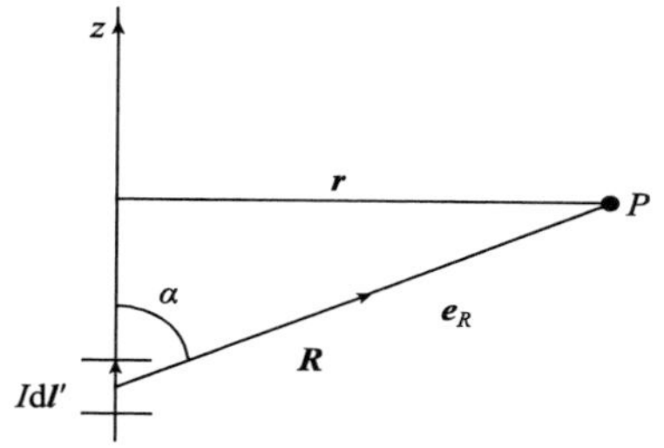

Figure of Example 3.1

$$\mathrm{d}\boldsymbol{B} = \frac{\mu_0}{4\pi}\frac{I\mathrm{d}\boldsymbol{l}' \times \boldsymbol{e}_R}{R^2}$$

and

$$\boldsymbol{H} = \int \mathrm{d}\boldsymbol{H} = \int_{-\infty}^{\infty} \frac{I\mathrm{d}\boldsymbol{l}' \times \boldsymbol{e}_R}{4\pi R^2} = \boldsymbol{e}_\phi \int_{-\infty}^{\infty} \frac{I\sin\alpha}{4\pi R^2}\mathrm{d}z$$

$$= \boldsymbol{e}_\phi \int_{-\infty}^{\infty} \frac{I(r/\sqrt{r^2+z^2})}{4\pi(r^2+z^2)}\mathrm{d}z = \boldsymbol{e}_\phi \frac{I}{2\pi r}$$

$$H_\phi = \frac{I}{2\pi r}$$

H_ϕ is axially symmetric, and is not related with coordinates ϕ. It's only decided by r.

Example 3.2 As shown in the Figure of Example 3.2, an x-axis parralleled metal bar with length of d is moving along the conductive orbit (neglect electrical resistance) at a speed of $\boldsymbol{v}$ (neglect friction) in a uniform constant magnetic field $\boldsymbol{e}_y B$. A resistor of R is connected at the other side of the conductive orbit. Calculate the external mechanical power P_m required for the metal bar (neglect electrical resistance) to move at the speed of $\boldsymbol{v}$.

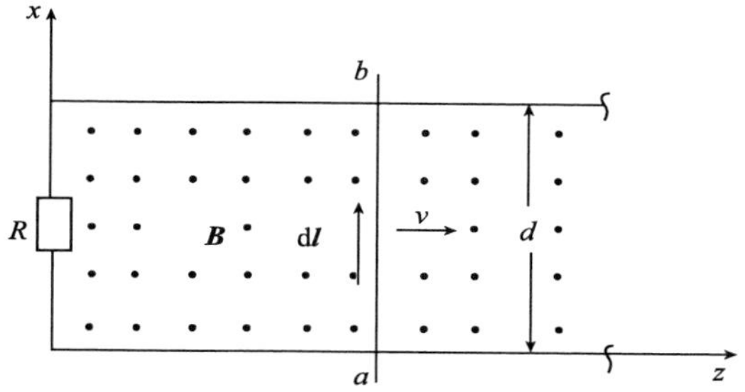

Figure of Example 3.2

Solution The metal bar moves at the speed of v and cuts the magnetic line of force, and thus generates electromotive force over the bar. Upon connection with the resistor R, there is a flow of current in the resistor R and the metal bar. It is known that a conductor with current in the magnetic field will be imposed over a magnetic force $\boldsymbol{F}$. An external force $\boldsymbol{F}'$ is required to overcome the magnetic force imposed over the current-carrying conductor if the metal bar is to be kept moving at a speed of $\boldsymbol{v}$. The mechanical power equals $\boldsymbol{F}' \times \boldsymbol{v}$.

From the above analysis, the key for solving such a problem is to calculate the magnetic force $\boldsymbol{F}$ imposed over the current-carrying metal bar in magnetic field $\boldsymbol{B}$. It is known that

$$\boldsymbol{F} = \int_d I\mathrm{d}\boldsymbol{l} \times \boldsymbol{B} \tag{a}$$

We must determine the magnitude and direction of the current inside the integration symbol. The open circuit voltage between a, b on the metal bar is

$$U_{ab} = \int_a^b \boldsymbol{E}\cdot \mathrm{d}\boldsymbol{l} = \int_a^b (\boldsymbol{v}\times\boldsymbol{B})\cdot \mathrm{d}\boldsymbol{l} = \int_a^b (\boldsymbol{e}_z\times\boldsymbol{e}_y)vB\cdot\boldsymbol{e}_x\mathrm{d}x = -vBd \tag{b}$$

The direction of the current can be determined directly through the magnetic force imposed over the charge

$$\boldsymbol{v}\times\boldsymbol{B} = \boldsymbol{e}_z\times\boldsymbol{e}_y vB = -\boldsymbol{e}_x vB$$

That is, current flux along the x direction, which is opposite to the assumed direction of $I\mathrm{d}\boldsymbol{l}$ in the figure.

The value of the current I along $-x$ direction in the metal bar is given by

$$I = \frac{U_{ab}}{R} = \frac{vBd}{R} \tag{c}$$

Substituting Formula (a) in the equation above yields

$$\boldsymbol{F} = (-\boldsymbol{e}_x\times\boldsymbol{e}_y)IBd = -\boldsymbol{e}_z\frac{vB^2d^2}{R} \tag{d}$$

Thus, the required external mechanical power for the moving metal bar at a speed of $\boldsymbol{v}$ on the conductive orbit is given by

$$\boldsymbol{F}'\cdot\boldsymbol{v} = -\boldsymbol{F}\cdot\boldsymbol{v} = \frac{v^2B^2d^2}{R} \tag{e}$$

The negative sign in front of $\boldsymbol{F}\cdot\boldsymbol{v}$ in Equation (e) denotes that the external force $\boldsymbol{F}'$ is in the opposite direction to the magnetic force $\boldsymbol{F}$ imposed over the metal bar.

The friction of moving on the conductive orbit is neglected. According to the conservation of energy, external mechanical power is converted into thermal loss power. The heat is emitted from the resistor R and can be given as

$$\frac{v^2B^2d^2}{R} = \left(\frac{vBd}{R}\right)^2 R = I^2R$$

3.1.2 Divergence Equation of Constant Magnetic Field

In this subsection we use the Biot-Savart Law to derive the magnetic continuity equation of a constant magnetic field. The equation includes a differential and integral form.

1. Differential Form of Magnetic Continuity Equation of Constant Magnetic Field, Which is $\nabla\cdot\boldsymbol{B}=0$

Method 1 Calculate the divergence of the magnetic induction $\boldsymbol{B}$ of the Biot-Savart law and integrate the vector identity to derive $\nabla\cdot\boldsymbol{B}=0$.

Directly calculate the divergence of $\boldsymbol{B}$ in the Biot-Savart Law as given in Equation (3.7c) and thus we have

$$\nabla\cdot\boldsymbol{B} = \frac{\mu_0}{4\pi}\nabla\cdot\int_{V'}\frac{\boldsymbol{J}\times\boldsymbol{e}_R}{R^2}\mathrm{d}V' = \frac{\mu_0}{4\pi}\int_{V'}\nabla\cdot\left[\nabla\frac{1}{R}\times\boldsymbol{J}\right]\mathrm{d}V' \tag{3.9}$$

When deriving the equation above, the divergence operator ($\nabla\cdot$) can be moved into the sign of integral. This is because the calculation of divergence is imposed on the coordinate of field point. The second reason is that the volume integral is imposed on the coordinate of source point. At the same time, the equation below is used

$$\frac{\boldsymbol{e}_R}{R^2} = -\nabla\left(\frac{1}{R}\right) \tag{3.10}$$

Then we use the vector identity

$$\nabla\cdot(\boldsymbol{A}\times\boldsymbol{F}) = \boldsymbol{F}\cdot\nabla\times\boldsymbol{A} - \boldsymbol{A}\cdot\nabla\times\boldsymbol{F}$$

The variable inside the sign of integral in Equation (3.9) can be transformed as

$$\nabla\cdot\left(\nabla\frac{1}{R}\times\boldsymbol{J}\right) = \boldsymbol{J}\cdot\nabla\times\nabla\frac{1}{R} - \nabla\frac{1}{R}\cdot\nabla\times\boldsymbol{J} \tag{3.11}$$

In the equation above, $\nabla\times\nabla\left(\frac{1}{R}\right)\equiv 0$ is always correct, and the calculation of curl of the field point coordinate imposed over J equals 0. That is $\nabla\times\boldsymbol{J}=0$, because the $\boldsymbol{J}$ in source point is not a function of field point coordinate variable. Thus, Equation (3.11) equals 0, that is

$$\nabla\cdot\left(\nabla\frac{1}{R}\times\boldsymbol{J}\right) = 0 \tag{3.12}$$

The integration in Equation (3.9) equals 0, and as a result, Equation (3.9) is transformed as

$$\nabla\cdot\boldsymbol{B} = 0 \tag{3.13}$$

The equation states that a *constant magnetic field is a non-divergence field. The divergence equals 0 at any point, which means that the constant magnetic field is field with no source of divergence.* This is totally different from the divergence field ($\nabla\cdot\boldsymbol{D}=\rho$) of static electric field.

Method 2 Using the vector identity, $\boldsymbol{B}$ from the Biot-Savart Law can be transformed into a curl of a vector. Calculate the divergence of $\boldsymbol{B}$. On the basis that the divergence of a variable always equals 0 if the variable is a curl of a vector, $\nabla\cdot\boldsymbol{B}=0$ can be derived.

In circuit l', the current is I, and the generated $\boldsymbol{B}$ at the field point is

$$\begin{aligned}\boldsymbol{B} &= \frac{\mu_0 I}{4\pi}\oint_{l'}\frac{\mathrm{d}\boldsymbol{l}'\times\boldsymbol{e}_R}{R^2} = \frac{\mu_0 I}{4\pi}\oint_{l'}\mathrm{d}\boldsymbol{l}'\times(-)\nabla\frac{1}{R}\\ &= \frac{\mu_0 I}{4\pi}\oint_{l'}\nabla\left(\frac{1}{R}\right)\times\mathrm{d}\boldsymbol{l}'\end{aligned} \tag{3.14a}$$

Equation (3.10) is utilized in the derivation.

Use of the vector identity

$$\nabla\times f\boldsymbol{A} = \nabla f\times\boldsymbol{A} + f\nabla\times\boldsymbol{A}$$

The variable inside the sign of integral in equation (3.14a) can be transformed as

$$\nabla\left(\frac{1}{R}\right)\times\mathrm{d}\boldsymbol{l}' = \nabla\times\frac{\mathrm{d}\boldsymbol{l}'}{R} - \frac{1}{R}\nabla\times\mathrm{d}\boldsymbol{l}' = \nabla\times\frac{\mathrm{d}\boldsymbol{l}'}{R} \tag{3.14b}$$

In the equation, $\nabla\times\mathrm{d}\boldsymbol{l}'=0$ because the calculation of curl is imposed on the coordinate variable of field point and $\mathrm{d}\boldsymbol{l}'$ is only a coordinate function of the source point. Transformed

Equation (3.14a) and according to (3.14b), we have

$$\begin{aligned} \boldsymbol{B} =& \frac{\mu_0 I}{4\pi} \oint_{l'} \nabla \left(\frac{1}{R}\right) \times \mathrm{d}\boldsymbol{l}' \\ =& \frac{\mu_0 I}{4\pi} \oint_{l'} \nabla \times \frac{\mathrm{d}\boldsymbol{l}'}{R} = \nabla \times \left(\frac{\mu_0 I}{4\pi} \oint_{l'} \frac{\mathrm{d}\boldsymbol{l}'}{R}\right) \end{aligned} \tag{3.15}$$

Because the calculation of the curl ($\nabla\times$) is imposed over the field point and the integration ($\mathrm{d}\boldsymbol{l}'$) is imposed over the source point, the divergence operation can be moved out of the integral.

Through the vector identity

$$\nabla \cdot \nabla \times \boldsymbol{A} = 0$$

We calculate the divergence of Equation (3.15)

$$\begin{aligned} \nabla \cdot \boldsymbol{B} =& \nabla \cdot \nabla \times \left(\frac{\eta_0 I}{4\pi} \oint_{l'} \frac{\mathrm{d}\boldsymbol{l}'}{R}\right) = 0 \\ \nabla \cdot \boldsymbol{B} =& 0 \end{aligned} \tag{3.16}$$

The above equation reveals that the divergence of constant magnetic flux density $\boldsymbol{B}$ is equal at any point. This means that a constant magnetic field is without divergence sources. This is in contrast with the divergence field ($\nabla \cdot \boldsymbol{D} = \rho$) of static electric field.

2. The Integral Form of the Continuity of Magnetic Flux

Analysis Method Through the use of the divergence theorem and the differential form of continuity of magnetic flux $\nabla \cdot \boldsymbol{B} = 0$, equation $\oint_S \boldsymbol{B} \cdot \mathrm{d}\boldsymbol{S} = 0$ can be derived.

The expression of the divergence theorem is

$$\oint_S \boldsymbol{B} \cdot \mathrm{d}\boldsymbol{S} = \int_V \nabla \cdot \boldsymbol{B} \mathrm{d}V \tag{3.17}$$

From the equation $\nabla \cdot \boldsymbol{B} = 0$ the volume of integration of the equation is 0, and we have

$$\oint_S \boldsymbol{B} \cdot \mathrm{d}\boldsymbol{S} = 0 \tag{3.18}$$

The above equation reveals that the total value of magnetic flux through any closed plane equals 0. This is in contrast from the case in the static electric field where the total value of electric flux through any close plane equals the free charge q inside the close plane. The electric flux line in static electric field has starting point (starts from positive charge) and termination point (ends in negative charge). It is significantly different from the magnetic line of force which has neither starting point nor termination point.

Equations (3.18) and (3.13) represent the continuity of magnetic flux in integral and differential form. This is the basic equation that expresses the solenoidal field of a magnetic field.

3.1.3 The Curl Equation of Constant Current Magnetic Field

We use the magnetic field intensity $\boldsymbol{H}$ generated from the Biot-Savart law, and calculate the closed line integral of $\boldsymbol{H}$. The Ampere Circuital Theorem can be gained to calculate the curl of constant magnetic field.

1. The Integral Form of Constant Current Magnetic Field, the Ampere Circuital Theorem

Use the equation $\mathrm{d}\boldsymbol{B} = \dfrac{\mu_0 I \mathrm{d}\boldsymbol{l}' \times \boldsymbol{e}_R}{4\pi R^2}$ generated from Biot-Savart law, the magnetic field intensity of a long straight conductor (the current is I) in vacuum is $H = \dfrac{I}{2\pi r}$. Impose close line integration over $\boldsymbol{H}$, we can have the Ampere Circuital Theorem.

From the Figure 3.3,

$$\begin{aligned}\oint_l \boldsymbol{H} \cdot \mathrm{d}\boldsymbol{l} &= \oint_l H \mathrm{d}l \cos\theta = \int_0^{2\pi} H r \mathrm{d}\phi \\ &= \int_0^{2\pi} \frac{I}{2\pi r} r \mathrm{d}\phi = I\end{aligned}$$

then

$$\oint_l \boldsymbol{H} \cdot \mathrm{d}\boldsymbol{l} = I \tag{3.19}$$

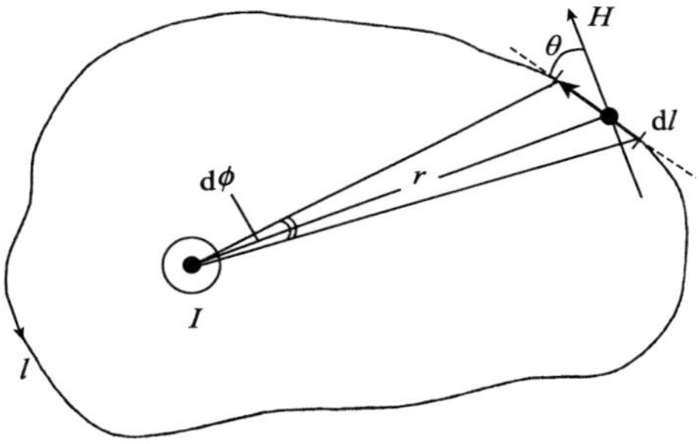

Figure 3.3 Line integration of H with any close path

The equation is the integral form of the Ampere Circuital Theorem and states that the line integral of magnetic field intensity $\boldsymbol{H}$ along a closed circuit in vacuum equals the value of electric current through the plane determined by the circuit. If the electric current through the plane equals 0, then we shall have

$$\oint_l \boldsymbol{H} \cdot \mathrm{d}\boldsymbol{l} = 0 \tag{3.20}$$

2. The Differential Form of Circulation Flow of Constant Current Magnetic Field, Which is the Curl Equation of Magnetic Field

Method 1 Calculate the curl value of magnetic field intensity $\boldsymbol{H}$ inside and outside of a long straight round conductor of radius a with current I. This is equivalent to calculating the curl value of H where $r < a$ and $r > a$, to help find out the physical meaning of $\nabla \times \boldsymbol{H}$.

Select cylindrical coordinates since the conductor is round. Let the axis of the round conductor coincidents with the z-axis of the cylindrical coordinates, so that the $\boldsymbol{H}$ inside and outside the conductor respectively equals

$$\begin{aligned}\boldsymbol{H} &= \boldsymbol{e}_\phi H_\phi = \boldsymbol{e}_\phi \frac{I'}{2\pi r} = \boldsymbol{e}_\phi \frac{(1/\pi a^2)\pi r^2}{2\pi r} \\ &= \boldsymbol{e}_\phi \frac{J\pi r^2}{2\pi r} = \boldsymbol{e}_\phi \frac{Jr}{2}, \quad r \leqslant a\end{aligned} \tag{3.21}$$

$$\boldsymbol{H} = \boldsymbol{e}_\phi H_\phi = \boldsymbol{e}_\phi \frac{I}{2\pi r}, \qquad r \geqslant a \tag{3.22}$$

In Equation (3.21) the current density is given by

$$\boldsymbol{J} = \boldsymbol{e}_z J = \boldsymbol{e}_z \left(\frac{I}{\pi a^2}\right) \tag{3.23}$$

(1) Calculate $\nabla \times \boldsymbol{H}$ where $r < a$ (inside the conductor)
From Equation (3.21), we have

$$\begin{aligned}\nabla \times \boldsymbol{H} &= \boldsymbol{e}_z \frac{1}{r}\frac{\partial}{\partial r}(rH_\phi) = \boldsymbol{e}_z \frac{1}{r}\frac{\partial}{\partial r}\left(\frac{Jr^2}{2}\right)\\ &= \boldsymbol{e}_z J = \boldsymbol{J}, \qquad r < a\end{aligned} \tag{3.24}$$

(2) Calculate $\nabla \times \boldsymbol{H}$ where $r > a$ (outside the conductor)
From Equation (3.22), we have

$$\nabla \times \boldsymbol{H} = \boldsymbol{e}_z \frac{1}{r}\frac{\partial}{\partial r}(rH_\phi) = \boldsymbol{e}_z \frac{1}{r}\frac{\partial}{\partial r}\left(\frac{I}{2\pi}\right) = 0, \qquad r > a \tag{3.25}$$

Which indicates that *the curl of magnetic field equals the current density* $\boldsymbol{J}$.

$$\nabla \times \boldsymbol{H} = \boldsymbol{J} \tag{3.26}$$

If there is no current

$$\nabla \times \boldsymbol{H} = 0 \tag{3.27}$$

Method 2 Through the use of Stokes formula and the integral form of the Ampere Circuital Theorem, the curl expression of magnetic field intensity $\boldsymbol{H}$ can be generated.
Stokes formula is given by

$$\oint_l \boldsymbol{H} \cdot \mathrm{d}\boldsymbol{l} = \int_S \nabla \times \boldsymbol{H} \cdot \mathrm{d}\boldsymbol{S}$$

And the integral form of the Ampere Circuital Theorem is given by

$$\oint_l \boldsymbol{H} \cdot \mathrm{d}\boldsymbol{l} = I = \int_S \boldsymbol{J} \cdot \mathrm{d}\boldsymbol{S} \tag{3.28}$$

Comparing the two formula, we obtain

$$\int_S \nabla \times \boldsymbol{H} \cdot \mathrm{d}\boldsymbol{S} = \int_S \boldsymbol{J} \cdot \mathrm{d}\boldsymbol{S} \tag{3.29}$$

Due to the arbitrary nature of the area S in the formula, we have the curl expression of $\boldsymbol{H}$ as below

$$\nabla \times \boldsymbol{H} = \boldsymbol{J} \tag{3.30}$$

The equation $\nabla \times \boldsymbol{H} = \boldsymbol{J}$ represents that constant magnetic field is a swirl vector field. The source of the curl is the current density J at the point, which is completely different from the case in static electric field where the curl equals 0 meaning that the static electric field is a non-swirl field (with no curl source).

The differential form (3.26) and integral form (3.19) of the Ampere Circuital Theorem *are the other basic equation of constant current magnetic field, which represent the magnetic field is a swirl field.*

3.2 Magnetic Vector Potential $\boldsymbol{A}$ and Scalar Magnetic Potential φ_m

3.2.1 Magnetic Vector Potential $\boldsymbol{A}$

The introduction of the magnetic vector potential $\boldsymbol{A}$ helps to conveniently analyze upcoming problems. Whether introducing magnetic vector potential or scalar magnetic potential, the foundation is based upon Maxwell's equations. As we know, a constant current magnetic field is a non-divergent field

$$\nabla \cdot \boldsymbol{B} = 0$$

and we have the vector identity

$$\nabla \cdot \nabla \times \boldsymbol{A} = 0$$

Comparing the above two equations, we obtain

$$\boldsymbol{B} = \nabla \times \boldsymbol{A}$$

The equation describes the relationship between a vector magnetic potential $\boldsymbol{A}$ and a magnetic induction $\boldsymbol{B}$. If $\boldsymbol{A}$ is known, $\boldsymbol{B}$ can be generated by calculating the curl of $\boldsymbol{A}$. Now we find how to calculate the vector magnetic potential $\boldsymbol{A}$.

Method 1 Transform the expression of $\mathrm{d}\boldsymbol{B}$ from the Biot-Savart Law into the form of calculating the curl of a vector. The vector is $\mathrm{d}\boldsymbol{A}$, and expression of $\mathrm{d}\boldsymbol{B}$ is transformed into $\mathrm{d}\boldsymbol{B} = \nabla\times\mathrm{d}\boldsymbol{A}$ and $\boldsymbol{A}$ can be calculated.

The $\mathrm{d}\boldsymbol{B}$ given in the Biot-Savart Law is

$$\mathrm{d}\boldsymbol{B} = \frac{\mu_0}{4\pi}\frac{\boldsymbol{J} \times \boldsymbol{e}_R}{R^2}\mathrm{d}V' \tag{3.31}$$

with $\boldsymbol{J}\mathrm{d}V'$ denoting the volume current element. After imposing a mathematical process over $\dfrac{\boldsymbol{J} \times \boldsymbol{e}_R}{R^2}$ in the equation and we have

$$\boldsymbol{J} \times \left(\frac{\boldsymbol{e}_R}{R^2}\right) = \nabla \times \frac{\boldsymbol{J}}{R} \tag{3.32}$$

Substitude Equation (3.32) into Equation (3.31)

$$\mathrm{d}\boldsymbol{B} = \nabla \times \left(\frac{\mu_0}{4\pi}\frac{\boldsymbol{J}}{R}\mathrm{d}V'\right) = \nabla \times \mathrm{d}\boldsymbol{A} \tag{3.33}$$

$$\mathrm{d}\boldsymbol{A} = \frac{\mu_0 \boldsymbol{J}}{4\pi R}\mathrm{d}V' \tag{3.34}$$

Calculate the integral of the equation above to obtain

$$\boldsymbol{A} = \frac{\mu_0}{4\pi}\int_V^{\prime} \frac{\boldsymbol{J}}{R}\mathrm{d}V' \tag{3.35}$$

Method 2 First, establish the differential equation of $\boldsymbol{A}$: $\nabla^2 \boldsymbol{A} = -\mu_0 \boldsymbol{J}$, and calculate $\boldsymbol{A}$. Due to

$$\nabla \times \boldsymbol{B} = \nabla \times \nabla \times \boldsymbol{A} = \mu_0 \boldsymbol{J} \tag{3.36}$$

and

$$\nabla \times \nabla \times \boldsymbol{A} = \nabla(\nabla \cdot \boldsymbol{A}) - \nabla^2 \boldsymbol{A}$$

use the Coulomb standard [see Equation (6.92)]:

$$\nabla \cdot \boldsymbol{A} = 0 \tag{3.37}$$

we have

$$\nabla^2 \boldsymbol{A} = -\mu_0 \boldsymbol{J} \tag{3.38}$$

This is a vector Poisson equation. The x component (scalar) is $\nabla^2 A_x = -\mu_0 J_x$, which is similar to the Poisson's equation [Equation (2.37)] in static electric field. Hence, the solution form should be the same as Equation (2.39), and the solution of A_x is

$$A_x = \frac{\mu_0}{4\pi} \int_{V'} \frac{J_x}{R} \mathrm{d}V' \tag{3.39}$$

Thus the vector solution of Equation (3.37) is given by

$$\boldsymbol{A} = \frac{\mu_0}{4\pi} \int_{V'} \frac{\boldsymbol{J}}{R} \mathrm{d}V' \tag{3.40}$$

In case of surface current distribution and line current distribution, change the volume integration in Equation (3.39) into surface integration and line integration respectively:

$$\boldsymbol{A} = \frac{\mu_0}{4\pi} \int_{S'} \frac{\boldsymbol{J}_S}{R} \mathrm{d}S' \tag{3.41}$$

$$\boldsymbol{A} = \frac{\mu_0}{4\pi} \int_{l'} \frac{I}{R} \mathrm{d}\boldsymbol{l}' \tag{3.42}$$

The expression of magnetic vector potential $\mathrm{d}\boldsymbol{A}$ generated by line current element $I\mathrm{d}\boldsymbol{l}$ is

$$\mathrm{d}\boldsymbol{A} = \frac{\mu_0 I \mathrm{d}\boldsymbol{l}}{4\pi R} \tag{3.43}$$

Obviously, magnetic vector potential $\mathrm{d}\boldsymbol{A}$ has the same direction with current source $I\mathrm{d}\boldsymbol{l}$. There exists a simple linear relation between vector $\mathrm{d}\boldsymbol{A}$ and $I\mathrm{d}\boldsymbol{l}$.

3.2.2 Scalar Magnetic Potential φ_m

1. The Introduction of Scalar Magnetic Potential φ_m

At the point where there is no current, we have $\nabla \times \boldsymbol{H} = 0$. Use of the vector identity $\nabla \times \nabla\varphi_m = 0$ allows us to arrive at

$$\boldsymbol{H} = -\nabla\varphi_m \tag{3.44}$$

2. Scalar Magnetic Potential φ_m *Satisfies Laplace Equation*

From the equation

$$\nabla \cdot \boldsymbol{B} = \mu_0 \nabla \cdot \boldsymbol{H} = -\mu_0 \nabla \cdot \nabla\varphi_m = -\mu_0 \nabla^2 \varphi_m = 0$$

we have

$$\nabla^2 \varphi_m = 0 \tag{3.45}$$

Example 3.3 There is a straight conductor which has length l and current I. Calculate the magnetic vector potential $\boldsymbol{A}$ and magnetic induction $\boldsymbol{B}$.

Solution Choose cylindrical coordinates. Make the conductor as z-axis and the middle of the conductor situated at the origin of the coordinate system. It can be seen from Figure of Example 3.3 that the distance between position $\mathrm{d}z'$ on the conductor and the field point P is given by

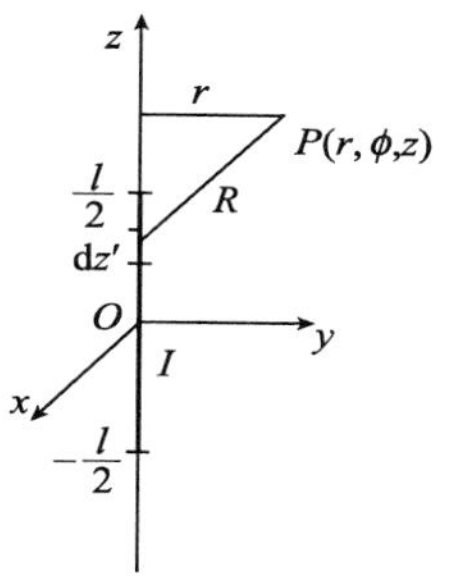

Figure of Example 3.3

$$R = \sqrt{(z-z')^2 + r^2}$$

$$A_z = \int_{-\frac{l}{2}}^{\frac{l}{2}} \frac{\mu_0 I \mathrm{d}z'}{4\pi\sqrt{(z-z')^2 + r^2}}$$

$$= \frac{\mu_0 I}{4\pi} \ln \left[\frac{\sqrt{(l/2-z)^2 + r^2} + (l/2-z)}{\sqrt{(l/2+z)^2 + r^2} - (l/2+z)} \right]$$

Calculate the curl of $\boldsymbol{A}$ to get the magnetic induction $\boldsymbol{B}$

$$\boldsymbol{B} = \nabla \times \boldsymbol{A} = -\boldsymbol{e}_\phi \frac{\partial A_z}{\partial r}$$

$$= \frac{\mu_0 I}{4\pi} \left[\frac{l/2 - z}{\sqrt{(l/2-z)^2 + r^2}} + \frac{l/2+z}{\sqrt{(l/2+z)^2 + r^2}} \right] \boldsymbol{e}_\phi$$

If the length l is much larger than the distance $(r^2+z^2)^{\frac{1}{2}}$ between the field point P and the origin coordinate, which means $l \gg (r^2+z^2)^{\frac{1}{2}}$, then

$$A_z \approx \frac{\mu_0 I}{4\pi} \ln \left(\frac{\sqrt{(l/2)^2 + r^2} + l/2}{\sqrt{(l/2)^2 + r^2} - l/2} \right)$$

$$\approx \frac{\mu_0 I}{4\pi} \ln \left(\frac{l}{r} \right)^2 = \frac{\mu_0 I}{4\pi} \ln \left(\frac{l}{r} \right) \tag{a}$$

Obviously, when $l \to \infty$, A_z approach infinity, which is similar to the case of calculating the electric potential generated by long straight conductor in static electric field. So we choose the *zero vector potential* $\boldsymbol{A}$ *at the reference point of* r_0. The result can be transformed as

$$A_z = \frac{\mu_0 I}{2\pi} \ln \left(\frac{r_0}{r} \right) \tag{b}$$

The change of reference point will not influence the expression of $\boldsymbol{B}$, because the change of reference point only adds a constant value of which the derivative (calculate the curl) equals zero. On the condition that $l \gg (r^2+z^2)^{\frac{1}{2}}$, $\boldsymbol{B}$ can be approximately expressed as

$$\boldsymbol{B} = \nabla \times \boldsymbol{A} = -\boldsymbol{e}_\phi \frac{\partial A_z}{\partial_r} = \boldsymbol{e}_\phi \frac{\mu_0 I}{2\pi r}$$

As expected, the above result is the same as the result calculated through the Ampere Circuital Theorem.

Example 3.4 Calculate the magnetic vector potential $\boldsymbol{A}$ generated by two parallel transmission lines at any point P in space.

Solution $r$denotes the vertical distance between the conductor and a point randomly chosen. r^+ and r^- respectively denote the vertical distance between the point and the conductor of which current direction is along z-axis. In Equation (b) of Example 3.3 the

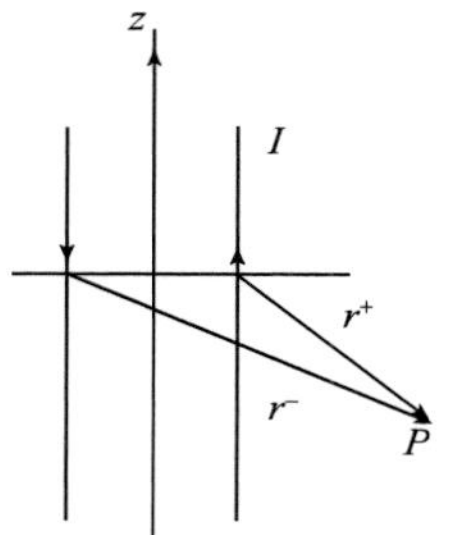

Figure of Example 3.4

magnetic vector potential $\boldsymbol{A}$ generated by long straight single conductor at any point (with vertical distance r) is given by

$$\boldsymbol{A} = \frac{\mu_0 I}{2\pi} \ln\left(\frac{r_0}{r}\right) \boldsymbol{e}_z$$

According to the above equation and Figure of Example 3.4, $\boldsymbol{A}$ generated by two parallel transmission line is

$$\begin{aligned}\boldsymbol{A} =& \boldsymbol{A}^+ + \boldsymbol{A}^- = \frac{\mu_0 I}{2\pi}\left[\ln\left(\frac{r_0}{r^+}\right) - \ln\left(\frac{r_0}{r^-}\right)\right]\boldsymbol{e}_z \\ =& \boldsymbol{e}_z \frac{\mu_0 I}{2\pi} \ln\left(\frac{r^-}{r^+}\right)\end{aligned}$$

In the above equation, the actual direction of $\boldsymbol{A}$ is determined by the value of (r^-/r^+). When $r^- > r^+$, $\boldsymbol{A}$ will be along the positive z-axis. Otherwise, $\boldsymbol{A}$ will be along the negative z-axis.

3.3 Magnetic Dipole

A small current-carrying circuit is called magnetic dipole. This concept is very important in the research of magnetic fields and radiation.

Calculate the magnetic vector potential $\boldsymbol{A}$ and the magnetic induction $\boldsymbol{B}$ generated by a magnetic dipole for which the radius is a and the current is I at any point. As shown in Figure 3.4, we shall select a spherical coordinates system. We select the circuit coincident with the plane constructed by Ox axis and Oy axis and the center of circle as the origin of the coordinates. According to the symmetric structure of the circuit, magnetic vector potential $\boldsymbol{A}$ generated by current on the circuit is of axis symmetry at any point of the plane. This indicates that the magnetic vector potential is independent of the ϕ coordinates. Hence, there is generality in the calculation of magnetic vector potential at point $P(r,\ \theta,\ 0)$. Choose a current element $I\mathrm{d}\boldsymbol{l}_1$, $I\mathrm{d}\boldsymbol{l}_2$ which are symmetric with the x axis. Since the direction of $\boldsymbol{A}$ and the direction of the current are the same, the two $x$components are of the same value and contrary direction generated thus canceling each other. Furthermore, the two $y$components are added up, at point P(on plane constructed by Ox axis and Oz axis), which means there is only A_y component at point P. Due to the use of the spherical coordinates, there is only an A_ϕ component present. Thus, the $A_\phi$generated by current element $I\mathrm{d}\boldsymbol{l}$ is

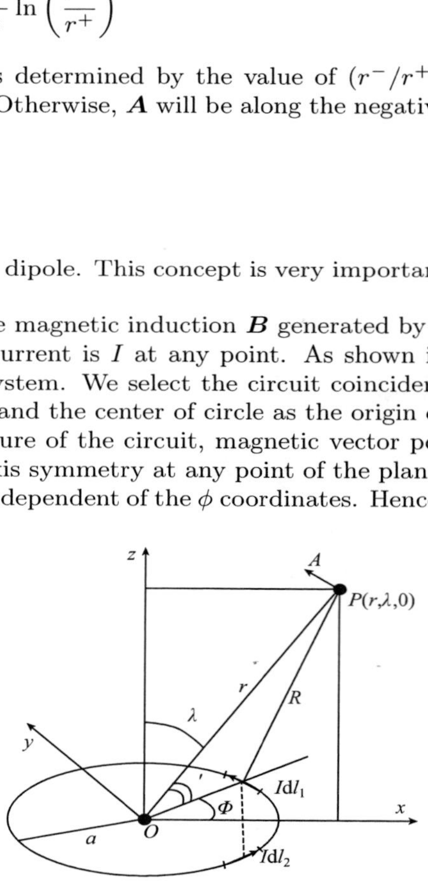

Figure 3.4 Calculate the magnetic vector potential A and the magnetic field intensity H of magnetic dipole

$$\mathrm{d}A_\phi = \mathrm{d}A\cos\phi = \frac{\mu_0 I(a\mathrm{d}\phi)}{4\pi R}\cos\phi \tag{3.46}$$

The total existing $\boldsymbol{A}$ generated by circuital current at any point is

$$\boldsymbol{A} = \boldsymbol{e}_\phi A_\phi = \boldsymbol{e}_\phi \int \mathrm{d}A_\phi = \boldsymbol{e}_\phi \int_0^{2\pi} \frac{\mu_0 I a\mathrm{d}\phi}{4\pi R}\cos\phi \tag{3.47}$$

The problem involves the expression of Rvia spherical coordinates. We shall refer to Figure 3.4 and the cosine theorem

$$R^2 = r^2 + a^2 - 2ra\cos\alpha \tag{3.48}$$

The $\cos\alpha$ in the above equation must be expressed through spherical coordinate variable. This can be solved by introducing in the two form of expression of $\boldsymbol{r}\cdot\boldsymbol{a}$. Refer to Figure 3.4, and

$$\boldsymbol{r}\cdot\boldsymbol{a} = ra\cos$$

$$\begin{aligned}\boldsymbol{r}\cdot\boldsymbol{a} =&(\boldsymbol{e}_x x + \boldsymbol{e}_z z)\cdot(\boldsymbol{e}_x a\cos\phi + \boldsymbol{e}_y a\sin\phi)\\ =&xa\cos\phi\end{aligned}$$

Compare the above two equation of $\boldsymbol{r}\cdot\boldsymbol{a}$ and

$$\cos\alpha = \frac{x}{r}\cos\phi = \sin\theta\cos\phi$$

Substituting the above equation into Equation (3.48) gives

$$R = (r^2 + a^2 - 2ra\sin\theta\cos\phi)^{\frac{1}{2}} \tag{3.49}$$

When $r \gg a$, we shall use the binomial theorem and ignore high-order terms to obtain

$$\begin{aligned}\frac{1}{R} =&\frac{1}{r}\left(1 + \frac{a^2}{r^2} - z\frac{a}{r}\sin\theta\cos\phi\right)^{-\frac{1}{2}}\\ \approx&\frac{1}{r}\left(1 + \frac{a}{r}\sin\theta\cos\phi\right)\end{aligned}$$

Substitute the equation into Equation (3.47), we have

$$\begin{aligned}A_\phi =&\frac{\mu_0 I a}{4\pi r}\times 2\int_0^\pi\left(1 + \frac{a}{r}\sin\theta\cos\phi\right)\cos\phi \mathrm{d}\phi\\ =&\frac{\mu_0 I a^2}{2\pi r^2}\cdot\sin\theta\int_0^\pi\cos^2\phi \mathrm{d}\phi\\ =&\frac{\mu_0 m}{4\pi r^2}\sin\theta\end{aligned} \tag{3.50}$$

m is referred to as magnetic dipole moment. Express it as vector, the direction is vertical to the plane constructed by the circuit and is right-handed spiral with current.

$$m = IS = I(\pi a^2) \tag{3.51}$$

Magnetic vector potential $\boldsymbol{A}$ of magnetic dipole can be expressed as

$$\boldsymbol{A} = \boldsymbol{e}_\phi\frac{\mu_0 m}{4\pi r^2}\sin\theta = \frac{\mu_0\boldsymbol{m}\times\boldsymbol{e}_r}{4\pi r^2} \tag{3.52}$$

With magnetic vector potential $\boldsymbol{A}$, magnetic induction $\boldsymbol{B}$ in spherical coordinate can be expressed as

$$\boldsymbol{B} = \nabla\times\boldsymbol{A} = \frac{\mu_0 m}{4\pi r^3}(\boldsymbol{e}_r 2\cos\theta + \boldsymbol{e}_\theta 2\sin\theta)$$

The above analysis indicates that

$$\boldsymbol{B} = \boldsymbol{e}_r\frac{\mu_0 m}{2\pi r^3}\cos\theta + \boldsymbol{e}_\theta\frac{\mu_0 m}{4\pi r^3}\sin\theta, \quad r \gg a \tag{3.53}$$

The electric field expression of electric dipole in static electric field is given by

$$\boldsymbol{E} = \boldsymbol{e}_r\frac{P}{2\pi\varepsilon_0 r^3}\cos\theta + \boldsymbol{e}_\theta\frac{P}{4\pi\varepsilon_0 r^3}\sin\theta, \quad r \gg l \tag{3.54}$$

Obviously, the magnetic induction expression $\boldsymbol{B}$ in Equation (3.53) of magnetic dipole is due to the electric field intensity $\boldsymbol{E}$ in Equation (3.54).

3.4 Medium in Constant Magnetic Field

The discussion above pertains to magnetic field in free space. In the case that the medium is in a magnetic field, the medium will be magnetized and will generate appending magnetic field. This will change the origin of the magnetic field.

3.4.1 Medium Magnetized Intensity $\boldsymbol{P}_{\rm m}$ and the Relationship Between $\boldsymbol{P}_{\rm m}$ and Magnetized Current Density $\boldsymbol{J}_{\rm m}$, $\boldsymbol{J}_{\rm ms}$

In order to describe the influence of a magnetized medium imposed on the magnetic induction $\boldsymbol{B}$, we introduce the magnetized intensity vector $\boldsymbol{P}_{\rm m}$. This quantity is defined as the vector sum of magnetic dipole moment $\boldsymbol{m}_i$ in unit volume. Specifically, we have

$$\boldsymbol{P}_{\rm m} = \lim_{\Delta V \to 0} \frac{\sum\limits_{i=1}^{N} \boldsymbol{m}_i}{\Delta V} \tag{3.55}$$

where N represents the number of magnetic dipole in ΔV volume, $\boldsymbol{m}_i$ represents the the magnetic moment of the ith magnetic dipole and $\boldsymbol{P}_{\rm m}$ represents the volume density of magnetic dipole. A magnetic dipole moment generates a magnetic field as a second source. It adds the total magnetic field in the medium with the external magnetic field. Whether external magnetic field or second magnetic field generated by magnetized medium, the source is current. So there is relationship between magnetic dipole and equivalent macro current generated by molecular and atom current. According to Equation (3.50), magnetic vector potential $\mathrm{d}\boldsymbol{A}$ generated by magnetic dipole moment $\boldsymbol{P}_{\rm m}\mathrm{d}V'$ in micro volume element $\mathrm{d}V'$ is expressed as

$$\mathrm{d}\boldsymbol{A} = \frac{\mu_0 \boldsymbol{P}_{\rm m} \times \boldsymbol{e}_R \mathrm{d}V'}{4\pi R^2} \tag{3.56}$$

The magnetic vector potential generated by the whole medium can be calculated through integration of the equation above over the volume V' of the medium

$$\boldsymbol{A} = \int_{V'} \frac{\mu_0 \boldsymbol{P}_{\rm m} \times \boldsymbol{e}_R \mathrm{d}V'}{4\pi R^2} \tag{3.57}$$

Substitution of Equation (2.52) into the above expression yields

$$\boldsymbol{A} = \frac{\mu_0}{4\pi} \int_{V'} \boldsymbol{P}_{\rm m} \times \nabla' \left(\frac{1}{r}\right) \mathrm{d}V' \tag{3.58}$$

Using the vector identity: $\nabla \times (f\boldsymbol{A}) = f\nabla \times \boldsymbol{A} + \nabla f \times \boldsymbol{A}$, and dividing Equation (3.58) into two parts yields

$$\boldsymbol{A} = \frac{\mu_0}{4\pi} \left(\int_{V'} \frac{\nabla' \times \boldsymbol{P}_{\rm m}}{R} \mathrm{d}V' - \int_{V'} \nabla' \times \left(\frac{\boldsymbol{P}_{\rm m}}{R}\right) \mathrm{d}V' \right) \tag{3.59}$$

As for the second part, use divergence theorem: $\int_{V'} \nabla \times \boldsymbol{P}_{\rm m} \mathrm{d}V' = -\oint_{S'} \boldsymbol{P}_{\rm m} \times \mathrm{d}\boldsymbol{S}'$, to convert the volume integral into a surface integral

$$\boldsymbol{A} = \frac{\mu_0}{4\pi} \left(\int_{V'} \frac{\nabla' \times \boldsymbol{P}_{\rm m}}{R} \mathrm{d}V' + \frac{\mu_0}{4\pi} \oint_{S'} \frac{\boldsymbol{P}_{\rm m} \times \mathrm{d}\boldsymbol{S}'}{R} \right) \tag{3.60}$$

Comparing this equation with former magnetic vector potential (3.40), (3.41) generated by volume current density and surface current density, equivalent relations can be derived.

$$\boldsymbol{J}_{\mathrm{m}} = \nabla \times \boldsymbol{P}_{\mathrm{m}} \tag{3.61}$$

$$\boldsymbol{J}_{\mathrm{ms}} = \boldsymbol{P}_{\mathrm{m}} \times \boldsymbol{e}_n \tag{3.62}$$

Changing the curl equation ($\nabla' \times$) into ($\nabla \times$) in Equation (3.61), this is similar to the process to Equation (2.54).

$\boldsymbol{J}_{\mathrm{m}}$ is called magnetized volume current density, and $\boldsymbol{J}_{\mathrm{ms}}$ is called magnetized surface current density. The quantities reflect the effective macro current density of a molecule current and atom current. It can be seen that the total magnetic induction $\boldsymbol{B}$ inside the external magnetic field is the sum of the conduction current and magnetized current. Thus, the equation to describe the relationship between the curl of the magnetic field and the source of magnetic field is updated as

$$\nabla \times \boldsymbol{B} = \mu_0(\boldsymbol{J} + \boldsymbol{J}_{\mathrm{m}}) \tag{3.63}$$

The magnetized medium is shown in Figure 3.5, where the direction of magnetic moment is the same to that of magnetized intensity $\boldsymbol{P}_{\mathrm{m}}$. It can be seen that the direction of magnetized surface current generated by molecular current on the surface of medium is in agreement with Equation (3.62). Inside the medium, if the magnetization is uniform and the molecule current cancel each other, then the effective magnetized current density will equal zero. According to Equation (3.61), if $\boldsymbol{P}_{\mathrm{m}}$ is constant to the change of the coordinates and the spatial derivatives equals zero, a same conclusion will be derived. If the medium magnetization is not uniform, the molecule currents will not cancel each other and there will be a macro volume magnetized current. The value and direction of the macro volume magnetized current will be decided according to Equation (3.61).

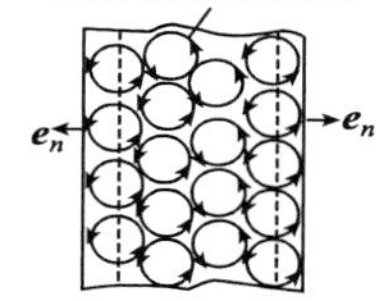

Figure 3.5 Restraint current in magnetized medium

3.4.2 Magnetic Field Intensity H in Medium and the Principle H and B Follows in Medium

Substitute Equation (3.61) into Equation (3.63), then

$$\nabla \times (\boldsymbol{B}/\mu_0 - \boldsymbol{P}_{\mathrm{m}}) = \boldsymbol{J} \tag{3.64}$$

The curl of the total vector $\boldsymbol{B}/\mu_0 - \boldsymbol{P}_{\mathrm{m}}$ is only related to the current density, and is independent of the magnetized current density. Use this total vector to analyse magnetic field in medium. We note that, only the free current density needs to be considered which makes the problem easier. Specifically, we define a new vector $\boldsymbol{H}$ to express magnetic field intensity.

$$\boldsymbol{H} = \boldsymbol{B}/\mu_0 - \boldsymbol{P}_{\mathrm{m}} \tag{3.65}$$

Substituting Equation (3.65) into Equation (3.64), we obtain

$$\nabla \times \boldsymbol{H} = \boldsymbol{J} \tag{3.66}$$

The magnetized intensity $\boldsymbol{P}_{\mathrm{m}}$ is proportional to magnetic field intensity $\boldsymbol{H}$ in a linear medium.

$$\boldsymbol{P}_{\mathrm{m}} = \chi_{\mathrm{m}}\boldsymbol{H} \tag{3.67}$$

In the above equation, χ_m is called magnetic permeability and is a constant. According to theoretical analysis and experiments, all of the magnetic mediums are linear. An exception is the ferromagnetic medium for which χ_m is a complicated function of $\boldsymbol{H}$ while $\boldsymbol{P}_m$ and $\boldsymbol{B}$ are both not linear to $\boldsymbol{H}$.

Substitute Equation (3.67) into Equation (3.65) to obtain

$$\boldsymbol{B} = \mu_0(1 + \chi_m)\boldsymbol{H} = \mu_0\mu_r\boldsymbol{H} = \mu\boldsymbol{H} \tag{3.68}$$

In the equation

$$\mu_r = 1 + \chi_m \tag{3.69}$$

$$\mu = \mu_r\mu_0 \tag{3.70}$$

μ_r is called relative permeability and is a dimensionless constant.

Imposing a surface integral over Equation (3.66) and use of Stokes law to evaluate the left side of the equation, and the integral result of rightside equals current. From our discussion we note that the Ampere Circuital law in medium can be derived as

$$\oint_l \boldsymbol{H} \cdot \mathrm{d}\boldsymbol{l} = I \tag{3.71}$$

$$\nabla \times \boldsymbol{H} = \boldsymbol{J} \tag{3.72}$$

Since no single magnetic charge is discovered, the magnetic continuity equation is still correct. More precisely, we have

$$\begin{aligned} &\oint_S \boldsymbol{B} \cdot \mathrm{d}\boldsymbol{S} = 0 \\ &\nabla \cdot \boldsymbol{B} = 0 \end{aligned} \tag{3.73}$$

Example 3.5 Consider an immeasurable hollow cylinder conductor with Magnetic permeability$\mu$with inner and outer diameter of a and b, respectively. The axis directional uniform current density is $\boldsymbol{J}$. Calculate the magnetic field intensity $\boldsymbol{H}$ and the magnetized current density $\boldsymbol{J}_m$.

Solution According to the Ampere Circuital Law in a medium

$$\oint_l \boldsymbol{H} \cdot \mathrm{d}\boldsymbol{l} = \int_S \boldsymbol{J} \cdot \mathrm{d}\boldsymbol{S}$$

$$\boldsymbol{H} = 0, \quad 0 < r < a$$

then

$$2\pi r\boldsymbol{H} = \int_S \boldsymbol{J} \cdot \mathrm{d}\boldsymbol{S} = J(\pi r^2 - \pi a^2)$$

and hence

$$\boldsymbol{H} = \frac{J(r^2 - a^2)}{2r}\boldsymbol{e}_\phi, \quad a < r < b \tag{a}$$

$$\boldsymbol{H} = \frac{J(b^2 - a^2)}{2r}\boldsymbol{e}_\phi, \quad r > b \tag{b}$$

Calculate the magnetized current density

$$\boldsymbol{J}_m = \nabla \times \boldsymbol{P}_m = \nabla \times (\mu_r - 1)\boldsymbol{H} = (\mu_r - 1) \cdot \nabla \times \boldsymbol{H}$$

$$\begin{cases} \boldsymbol{J}_m = (\mu_r - 1)\boldsymbol{J}\boldsymbol{e}_z, & a < r < b \\ \boldsymbol{J}_m = 0, & 0 < r < a \text{ or } r > b \end{cases}$$

The surface magnetized current density at the border is

$$\boldsymbol{J}_{\rm ms} = \boldsymbol{P}_{\rm m} \times \boldsymbol{e}_n \,|_{\rm border} = (\mu_{\rm r} - 1)\boldsymbol{H} \times \boldsymbol{e}_n|_{\rm border}$$

Then at the point where $r = a^+$

$$\boldsymbol{J}_{\rm ms} = (\mu_{\rm r} - 1)H\boldsymbol{e}_\phi \times (-\boldsymbol{e}_r)\,|_{r=a^+} = (\mu_{\rm r} - 1)H\boldsymbol{e}_z|_{r=a^+} = 0 \tag{c}$$

At the point where $r = b^-$

$$\boldsymbol{J}_{\rm ms} = (\mu_{\rm r} - 1)H\boldsymbol{e}_\phi \times \boldsymbol{e}_r\,|_{r=b^-} = -(\mu_{\rm r} - 1)\,\frac{J(b^2 - a^2)}{2r}\boldsymbol{e}_z \tag{d}$$

3.5 Boundary Condition of Constant Magnetic Field

The boundary condition of a constant magnetic field discuss the change rule of magnetic induction $\boldsymbol{B}$ and magnetic field intensity $\boldsymbol{H}$ at the two side of the boundary. Similar to our previous analysis, we shall use the integral form of basic equation of constant magnetic field.

3.5.1 The Boundary Condition of Normal Field $\boldsymbol{B}_n$

The research of boundary conditions of normal fields requires the use of the integral form of magnetic flux continuity

$$\oint_S \boldsymbol{B} \cdot {\rm d}\boldsymbol{S} = 0 \tag{3.74}$$

We consider a closed cylinder surface randomly at the boundary of two kinds of medium. The height is Δh. The top surface and bottom surface ΔS are both parallel to the boundary surface of the mediums, as displayed in Figure 3.6. Assume that $\Delta h \to 0$, and ΔS is rather small (*the magnetic flux density* $\boldsymbol{B}$ *is uniform, and of cause it will be rigorous if* $\Delta S \to 0$ *when understanding the boundary condition*). The close surface is constructed by the top and bottom surface as well as the side surface of a cylinder, of which the area approaches zero. According to Equation (3.74),

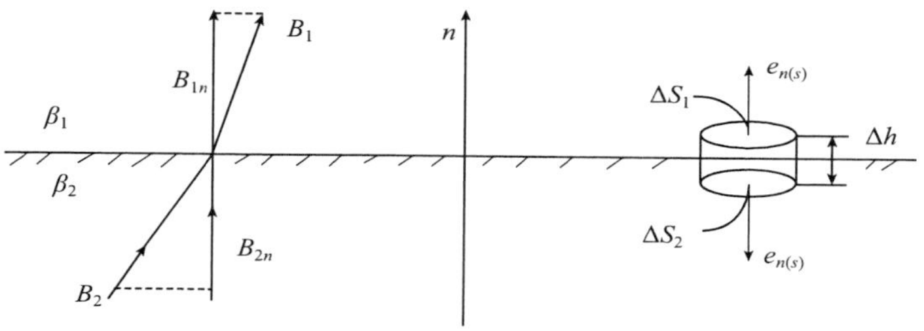

Figure 3.6 Boundary condition of normal field B_n

$B_{1n}\Delta S = B_{2n}\Delta S$, which represents the normal continuity of magnetic induction $\boldsymbol{B}$ at the boundary, that is

$$B_{1n} = B_{2n} \tag{3.75}$$

3.5.2 The Boundary Condition of Tangential Field H_t

The research of boundary conditions of tangential fields requires the use of the integral form of magnetic circumfluence component. This is the Ampere circuital law given by

$$\oint_l \boldsymbol{H} \cdot \mathrm{d}\boldsymbol{l} = I \tag{3.76}$$

As shown in Figure 3.7, we consider a rectangular closed circuit along the boundary surface of the medium with length Δl. This will be parallel to the boundary surface. The height is Δh, which approaches zero. And Δl is rather small (*the magnetic intensity* $\boldsymbol{H}$ *is uniform, and of cause it will be rigorous if* $\Delta l \to 0$ *when understanding the boundary condition*). We impose Equation (3.76) to this circuit, to obtain

$$\oint_l \boldsymbol{H} \cdot \mathrm{d}\boldsymbol{l} = H_{1t}\Delta l + 0 - H_{2t}\Delta l + 0 = J_S \Delta l \tag{3.77}$$

$$H_{1t} - H_{2t} = J_{\mathrm{s}}$$

J_{s} in the equation is the free surface current density at the boundary surface of mediums. This means that the tangential component of magnetic intensity is not continuous while there is free surface current at the boundary surface. If the surface current density is equal to zero at the boundary surface, the tangential component of magnetic intensity will be continuous. This will provide the following relation

$$H_{1t} = H_{2t} \tag{3.78}$$

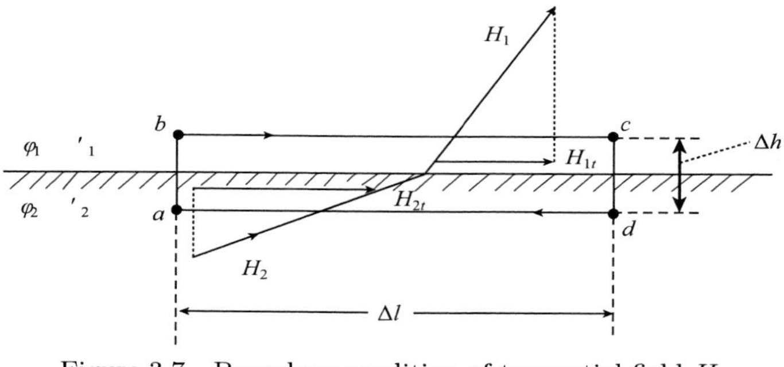

Figure 3.7 Boundary condition of tangential field H_t

3.5.3 Boundary Condition of Magnetic Vector Potential A and Scalar Vector Potential φ_{m}

1. The Boundary Condition of Magnetic Vector Potential $\boldsymbol{A}$

Due to $\nabla \times \boldsymbol{A} = \boldsymbol{B}$,

$$\int_S \nabla \times \boldsymbol{A} \cdot \mathrm{d}\boldsymbol{S} = \oint_l \boldsymbol{A} \cdot \mathrm{d}\boldsymbol{l} = \int_S \boldsymbol{B} \cdot \mathrm{d}\boldsymbol{S} = \varPhi \tag{3.79}$$

Stokes formula is used when deducing the equation above, and $\varPhi$ is the magnetic flux through the surface S. Consider the rectangular loop made for working out the tangential field boundary conditions. Since its width Δl is so small and $\Delta h \to 0$, the area of the rectangular $(\Delta l \cdot \Delta h) \approx 0$. Therefore the passed magnetic flux Φ is zero. It is known from Equation (3.79) that

$$\oint_l \boldsymbol{A} \cdot \mathrm{d}\boldsymbol{l} = 0 \tag{3.80}$$

Since the closed-loop height $\Delta h \to 0$, and $\boldsymbol{A}$ has a limited value, we derive $\boldsymbol{A}$'s tangential component as being continuous. That is

$$A_{1t} = A_{2t} \tag{3.81}$$

and$\nabla \cdot \boldsymbol{A} = 0$, thus

$$\int_V \nabla \cdot \boldsymbol{A} \mathrm{d}V = \oint_S \boldsymbol{A} \cdot \mathrm{d}\boldsymbol{S} = 0$$

The above equality has used the divergence theorem. Since we have $\oint_S \boldsymbol{A} \cdot \mathrm{d}\boldsymbol{S} = 0$, and considering the discussion of the boundary conditions addressed above and the height of the cylindrical surface$\Delta h \to 0$, it can be derived that the normal component of $\boldsymbol{A}$ is continuous, that is

$$\boldsymbol{A}_{1n} = \boldsymbol{A}_{2n} \tag{3.82}$$

Due to the fact that $A_{1t} = A_{2t}, A_{1n} = A_{2n}$, in the medium interface, the magnetic vector potential $\boldsymbol{A}$ is continuous, i.e.

$$A_1 = A_2$$

2. The Boundary Condition of Scalar Magnetic Potential φ_{m}

$$\left.\begin{aligned} &\varphi_{\mathrm{m1}} = \varphi_{\mathrm{m2}} \\ &\mu_1 \frac{\partial \varphi_{\mathrm{m1}}}{\partial n} = \mu_2 \frac{\partial \varphi_{\mathrm{m2}}}{\partial n} \quad (\text{viz } B_{1n} = B_{2n}) \end{aligned}\right\} \tag{3.83}$$

3.5.4 Vector Expression of Constant Magnetic Field's Boundary Condition

If we use $\boldsymbol{n}$ to express the unit vector in the normal direction of the media interface, then the scalar and vector representation of the boundary conditions correspondent to each other are given by

$$\left\{\begin{aligned} &H_{1t} - H_{2t} = J_{\mathrm{s}} \\ &\boldsymbol{n} \times (\boldsymbol{H}_1 - \boldsymbol{H}_2) = \boldsymbol{J}_{\mathrm{s}} \end{aligned}\right. \qquad \left\{\begin{aligned} &H_{1t} = H_{2t} \\ &\boldsymbol{n} \times \boldsymbol{H}_1 = \boldsymbol{n} \times \boldsymbol{H}_2 \end{aligned}\right. \tag{3.84}$$

$$\left\{\begin{aligned} &B_{1n} = B_{2n} \\ &\boldsymbol{n} \cdot \boldsymbol{B}_1 = \boldsymbol{n} \cdot \boldsymbol{B}_2 \end{aligned}\right. \tag{3.85}$$

3.6 Self Inductance and Mutual Inductance

Alternating currents of the conductor loop will generate an alternating magnetic field. The alternating magnetic field will generate an induced electromotive force in its own loop which belongs to a self-induced phenomenon. We shall denote the self inductance as L. When there are two or more conductor circuit loops, for any one of them, it will not only generate the inducing electromotive force within itself but also yield the inducing electromotive force in other loops, which belongs to mutual-induced phenomenon.

3.6.1 External Inductance $\boldsymbol{L}$

The external inductance is defined as below

$$L = \frac{\psi}{I}$$

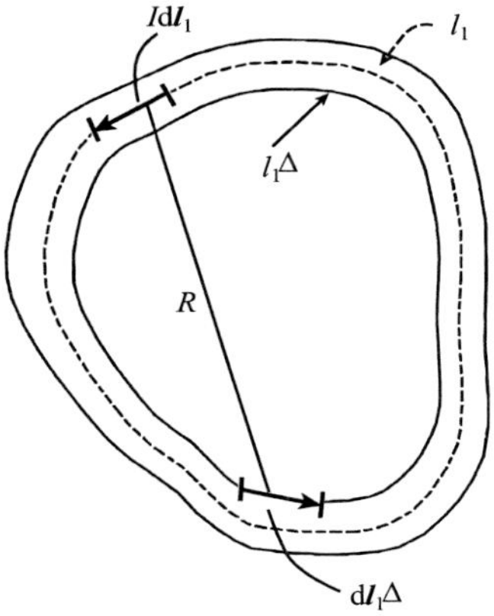

Figure 3.8 Seek the external inductance L

ψ is the outer magnetic chain of the current loop as shown in Figure 3.8, that is, the magnetic chain of the surrounded area within the inner circuit (along the closed-loop l_1') of the wire loop. Because the circuit is only a circle, the magnetic flux is equal to the flux

$$\psi = \Phi = \oint_{l_1'} \boldsymbol{A}_1 \cdot \mathrm{d}\boldsymbol{l}_1' \tag{3.86}$$

If the wire loop size is much larger than the wire radius, we can consider that loop current as being focused at the center line of wire as shown in Figure 3.8. In Equation (3.86) the magnetic flux Φ and the current I are cross-linking. $\boldsymbol{A}_1$ is the magnetic vector potential generated by the current of l_1 on the loop of l_1'. Therefore, $\boldsymbol{A}_1$ is given by

$$\boldsymbol{A}_1 = \oint_{l_1} \mathrm{d}\boldsymbol{A}_1 = \oint_{l_1} \frac{\mu_0 I \mathrm{d}\boldsymbol{l}_1}{4\pi R} \tag{3.87}$$

The magnetic flux can be derived when we substitute the above equation into Equation (3.86)

$$\Phi = \frac{\mu_0 I}{4\pi} \oint_{l_1'} \oint_{l_1} \frac{\mathrm{d}\boldsymbol{l}_1 \cdot \mathrm{d}\boldsymbol{l}_1'}{R} \tag{3.88}$$

Subsequently, the outer inductance of the wire loop is given by

$$L = \Phi / I = \frac{\mu_0}{4\pi} \oint_{l_1'} \oint_{l_1} \frac{\mathrm{d}\boldsymbol{l}_1 \cdot \mathrm{d}\boldsymbol{l}_1'}{R} \tag{3.89}$$

When the turn of the wire loop is N and all of the turns are close to each other, the magnetic flux of each turn can be considered as being the same. The magnetic flux of each turn will be given by

$$\Phi' = N\Phi = \frac{\mu_0 I N}{4\pi} \oint_{l_1'} \oint_{l_1} \frac{\mathrm{d}\boldsymbol{l}_1 \cdot \mathrm{d}\boldsymbol{l}_1'}{R} \tag{3.90}$$

The flux ψ is

$$\psi = N\Phi' = \frac{\mu_0 I N^2}{4\pi} \oint_{l_1'} \oint_{l_1} \frac{\mathrm{d}\boldsymbol{l}_1 \cdot \mathrm{d}\boldsymbol{l}_1'}{R} \tag{3.91}$$

Then

$$L = \psi / I = \frac{\mu_0 N^2}{4\pi} \oint_{l_1'} \oint_{l_1} \frac{\mathrm{d}\boldsymbol{l}_1 \cdot \mathrm{d}\boldsymbol{l}_1'}{R} \tag{3.92}$$

It can be seen that the self-inductance L is proportional to the square of turn N.

3.6.2 Inner Inductance L_i

We consider a long straight wire of radius as an example, and seek to derive the inner inductance L_i per unit length. We assume that the current in the wire is evenly distributed. Since we have the magnetic field within the wire, there exist inner magnetic flux and inner magnetic chain, and thereby the inner inductor. We now use several different methods to obtain the inner inductance L_i per unit length of wire.

Method 1 The *magnetic field energy* per unit length of the circle wire is expressed in the parameters of "field" and "loop" as

$$W_{\mathrm{m}} = \frac{1}{2}L_i I^2 = \int_V \frac{1}{2}\mu_0 H^2 \mathrm{d}V \tag{3.93}$$

We shall use the above equation to find the inner inductance L_i per unit length.

Assume that the radius of circle-based wire is a, the magnetic field strength on the radius r of the wire is expressed as

$$H_\phi = \frac{I'}{2\pi r} = \frac{(I/\pi a^2)\pi r^2}{2\pi r} = \frac{Ir}{2\pi a^2} \tag{3.94}$$

The total magnetic energy per unit length of wire will be given by

$$\begin{aligned} W_{\mathrm{m}} &= \int_V \frac{1}{2}\mu H^2 \mathrm{d}V = \int_0^a \frac{1}{2}\mu_0 \left(\frac{Ir}{2\pi a^2}\right)^2 (2\pi r \mathrm{d}r \times l) \\ &= \frac{\mu_0}{8\pi} \times \frac{1}{2}I^2 = \frac{1}{2}L_i I^2 \end{aligned} \tag{3.95}$$

Subsequently, we obtain the inner inductance per unit length of wire as

$$L_i = \frac{\mu_0}{8\pi} \quad (\mathrm{H\ /\ m}) \tag{3.96}$$

Method 2 We can firstly find the magnetic flux density to magnetic flux and magnetic chains using traditional methods, and then work out the inner inductance L_i per unit length of wire.

Assume current is uniformly distributed in Figure 3.9. Through the use of the Ampere circuit law we can find that in the magnetic flux passing through the area $\mathrm{d}S = \mathrm{d}r \times 1$ (i.e. the area unit formed by unit length and the width of $\mathrm{d}r$) at the r department, is

$$\mathrm{d}\Phi = \boldsymbol{B} \cdot \mathrm{d}\boldsymbol{S} = B\mathrm{d}S = \frac{\mu_0 I r}{2\pi a^2}\mathrm{d}r \tag{3.97}$$

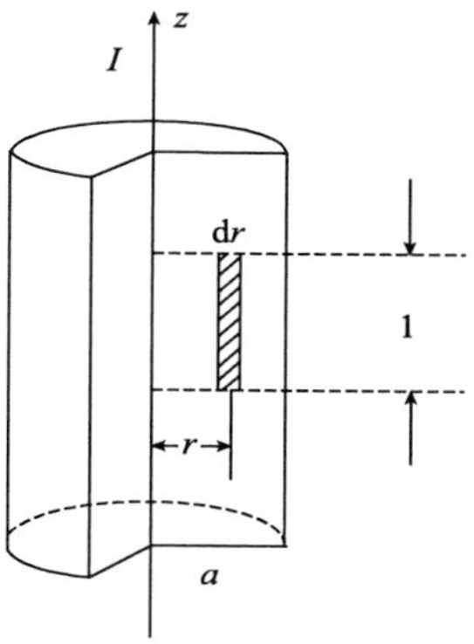

Figure 3.9 Seek the inner inductance of the wire L_i

The above portion of the magnetic flux does not intersect with all currents, but only intersects with the current that is within the radius r. That is, the N turns of this part of the magnetic flux $\mathrm{d}\Phi$ is less than 1, N is equal to the ratio of the current within the radius r and the total current, i.e.

$$N = r^2/a^2 \tag{3.98}$$

Therefore, the magnetic chain through the area $\mathrm{d}r \times 1$ is

$$\mathrm{d}\psi = N\mathrm{d}\Phi = \frac{\mu_0 I r^3 \mathrm{d}r}{2\pi a^4} \tag{3.99}$$

$$\psi = \int d\psi = \int_0^a \frac{\mu_0 I r^3 dr}{2\pi a^4} = \frac{\mu_0 I}{8\pi}$$

and we have

$$L_i = \psi/I = \frac{\mu_0}{8\pi} \tag{3.100}$$

It can be seen that the inner inductance per unit length of wire does not have an impact on the wire radius.

Example 3.6 2 long straight parallel conductors with a radius of a and axis spacing $D(D \gg a)$ are shown in Figure of Example 3.6. The permeability between the wires and the surrounding space is μ_0. Derive the self-inductance per unit length between the parallel pairs of wires.

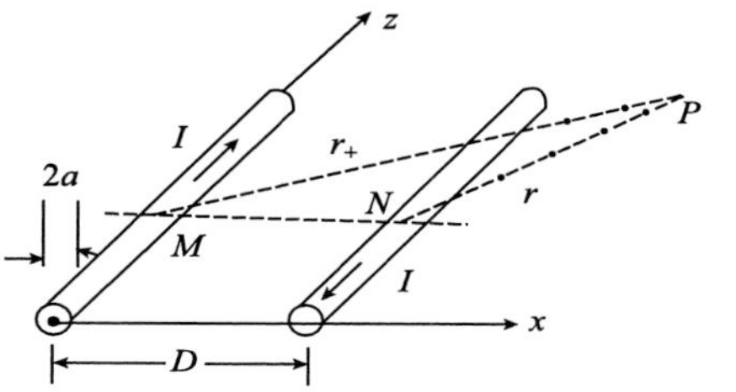

Figure of Example 3.6

Solution The self-inductance per unit length includes the inner and outer inductance. The inner wire inductance per unit length is a known constant $\mu_0/8\pi$. Since there are parallel *pairs of wires* with inner inductance $L_i = 2 \times \frac{\mu_0}{8\pi} = \frac{\mu_0}{4\pi}$, we can directly utilize this result. The remaining problem is to strive for its external inductance.

The current directions of the two lines are shown in the figure above. When $D \gg a$, we can approximate that the current focuses on the geometric axis of the wire. Seeking external inductor needs to find out the magnetic flux Φ_0, this is solved via two different methods.

Method 1 Use the Ampere circuit law to seek the external magnetic flux Φ_0, and thus find the external inductance per unit length L_0.

Based on the Ampere circuit law, the magnetic flux density in the axis connecting the two wires is

$$B = \frac{\mu_0 I}{2\pi}\left(\frac{1}{x} + \frac{1}{D-x}\right)$$

The magnetic flux between the two wires is given by

$$\Phi_0 = \int_a^{D-a} B dx = \frac{\mu_0 I}{\pi} \ln\left(\frac{D-a}{a}\right)$$

and the external inductance per unit length is

$$L_0 = \Phi_0/I = \frac{\mu_0}{\pi} \ln\left(\frac{D-a}{a}\right) \approx \frac{\mu_0}{\pi} \ln\left(\frac{D}{a}\right)$$

Then the total inductance per unit length of the two wires will be

$$L \approx 2 \times \frac{\mu_0}{8\pi} + \frac{\mu_0}{\pi} \ln\left(\frac{D}{a}\right) = \frac{\mu_0}{4\pi} + \frac{\mu_0}{\pi} \ln\left(\frac{D}{a}\right)$$

Method 2 Use magnetic vector potential $\boldsymbol{A}$ to find the outer magnetic flux Φ_0, and thus find the external inductance per unit length L_0.

We know the expression of magnetic vector potential $\boldsymbol{A}$ of a parallel pair of wires (see Example 3.4) as

$$\boldsymbol{A} = \boldsymbol{e}_z \frac{\mu_0 I}{2\pi} \ln\left(\frac{r-}{r+}\right)$$

where r^-, r^+ are the vertical distance for any space point to reach the wire axis (where the current are in negative direction of z and positive direction of z respectively). The external magnetic flux per unit length of the pair of the wires Φ_0 will be

$$\Phi_0 = \oint_l \boldsymbol{A} \cdot \mathrm{d}\boldsymbol{l} = A_M - A_N = \frac{\mu_0 I}{2\pi}\left[\ln\left(\frac{D-a}{a}\right) - \ln\left(\frac{a}{D-a}\right)\right]$$
$$= \frac{\mu_0 I}{\pi}\ln\left(\frac{D-a}{a}\right) \approx \frac{\mu_0 I}{\pi}\ln\left(\frac{D}{a}\right)$$

and the outer inductance per unit length will be given by

$$L_0 \approx \frac{\Phi_0}{I} = \frac{\mu_0}{\pi}\ln\left(\frac{D}{a}\right)$$

Finally, the total inductance per unit length of the pair of wires will be

$$L \approx 2 \times \frac{\mu_0}{8\pi} + \frac{\mu_0}{\pi}\ln\left(\frac{D}{a}\right) = \frac{\mu_0}{4\pi} + \frac{\mu_0}{\pi}\ln\left(\frac{D}{a}\right)$$

3.6.3 Mutual Inductance M

Please find the mutual inductance M between the pair of wire loops l_1 and l_2 drawn in Figure 3.10.

Suppose the current in wire l_1 is I_1 and the magnetic field generated in current I_1 will generate the mutual magnetic flux in loop l_2. That is

$$\Phi_{12} = \oint_{l_2} \boldsymbol{A}_1 \cdot \mathrm{d}\boldsymbol{l}_2 \tag{3.101}$$

and

$$\boldsymbol{A}_1 = \int \mathrm{d}\boldsymbol{A}_1 = \oint_{l_1} \frac{\mu_0 I_1 \mathrm{d}\boldsymbol{l}_1}{4\pi R} \tag{3.102}$$

$$\Phi_{12} = \oint_{l_2}\oint_{l_1} \frac{\mu_0 I_1 \mathrm{d}\boldsymbol{l}_1 \cdot \mathrm{d}\boldsymbol{l}_2}{4\pi R} \tag{3.103}$$

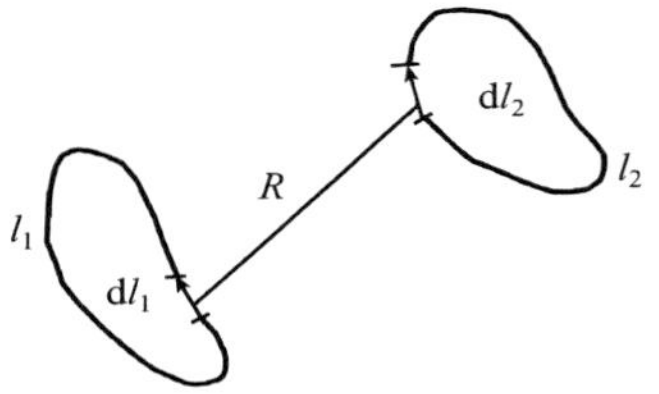

Figure 3.10 The mutual inductance between two wires

Then the mutual inductance M_{12} is

$$M_{12} = \frac{\Phi_{12}}{I_1} = \frac{\mu_0}{4\pi}\oint_{l_2}\oint_{l_1} \frac{\mathrm{d}\boldsymbol{l}_1 \cdot \mathrm{d}\boldsymbol{l}_2}{R} \tag{3.104}$$

Similarly, the magnetic field generated by the current I_2 will generate a mutual inductance M_{21} in loop l_1, that is

$$M_{21} = \frac{\Phi_{21}}{I_2} = \frac{\mu_0}{4\pi}\oint_{l_2}\oint_{l_1} \frac{\mathrm{d}\boldsymbol{l}_2 \cdot \mathrm{d}\boldsymbol{l}_1}{R} \tag{3.105}$$

Since exchange of integration order of $\mathrm{d}l_1$ and $\mathrm{d}l_2$ does not affect integration results, we have

$$M = M_{12} = M_{21} \tag{3.106}$$

Suppose loop l_1 has N_1 turns and loop l_2 has N_2 turns then the mutual inductance between loop l_1 and loop l_2 is

$$M = \frac{N_1 N_2 \mu_0}{4\pi}\oint_{l_2}\oint_{l_1} \frac{\mathrm{d}\boldsymbol{l}_1 \cdot \mathrm{d}\boldsymbol{l}_2}{R} \tag{3.107}$$

It can be seen that the mutual inductance M and the product of N_1 and N_2 are proportional.

Example 3.7 A long straight wire and an $a \times b$ rectangular coil are in the same plane. The side b of the rectangular coil is parellel to the straight wire. The distance from the side b which is close to the straight wire is d, such as the cases shown in Figure of Example 3.7. Derive the mutual inductance between the two wires.

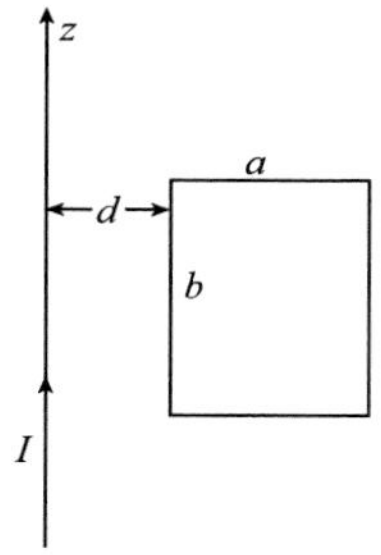

Figure of Example 3.7

Method 1 Find the mutual magnetic flux Φ by using the magnetic flux density B, and then finding the mutual inductance M. Let the current in straight wire be I, the mutual magnetic flux Φ generated in the open-face enclosed by the rectangular coil is

$$\Phi = \int_S \boldsymbol{B} \cdot \mathrm{d}\boldsymbol{S} = \int_d^{d+a} \left(\frac{\mu_0 I}{2\pi r}\right) b\mathrm{d}r = \frac{\mu_0 I}{2\pi} b \ln\left(\frac{d+a}{d}\right)$$

The the mutual inductance is given by

$$M = \frac{\Phi}{I} = \frac{\mu_0}{2\pi} b \ln\left(\frac{d+a}{d}\right)$$

Method 2 Use magnetic vector potential $\boldsymbol{A}$ to find magnetic flux, and then find the mutual inductance M.

The long straight wire with current I will generate a magnetic vector potential in a point in space. The magnetic vector potential $\boldsymbol{A}$ is

$$A_z = \frac{\mu_0 I}{2\pi} \ln(r_0/r)$$

In the above equation r_0 is the reference point for $\boldsymbol{A} = 0$. That is, when $r = r_0$, A_z=0.

The mutual magnetic flux Φ through the open-face enclosed by the rectangular coil is expressed as

$$\Phi = \oint_l \boldsymbol{A} \cdot \mathrm{d}\boldsymbol{l} = b\frac{\mu_0 I}{2\pi} \ln\left(\frac{r_0}{d}\right) - b\frac{\mu_0 I}{2\pi} \ln\left(\frac{r_0}{d+a}\right) = \frac{\mu_0 I}{2\pi} b \ln\left(\frac{d+a}{d}\right)$$

and the mutual inductance is given by

$$M = \Phi/I = \frac{\mu_0}{2\pi} b \ln\left(\frac{d+a}{d}\right)$$

Example 3.8 2 coils whose number of turn are n_1 and n_2 are in the ring around the magnetic core. As shown in Figure of Example 3.8, the core's average radius is r_0, cross-sectional radius is a, permeability is $\mu(\mu_r \mu_0)$. Please find mutual inductance M between the two coils.

Solution Supose $r_0 \gg a$, then let the magnetic line of force be constrained in a magnetic ring and neglect its leakage. First we find the magnetic flux density B_1 generated in the magnetic core by coil 1

$$2\pi r_0 B_1 = \mu \sum I = \mu n_1 I_1$$

$$B_1 = \frac{\mu n_1 I_1}{2\pi r_0}$$

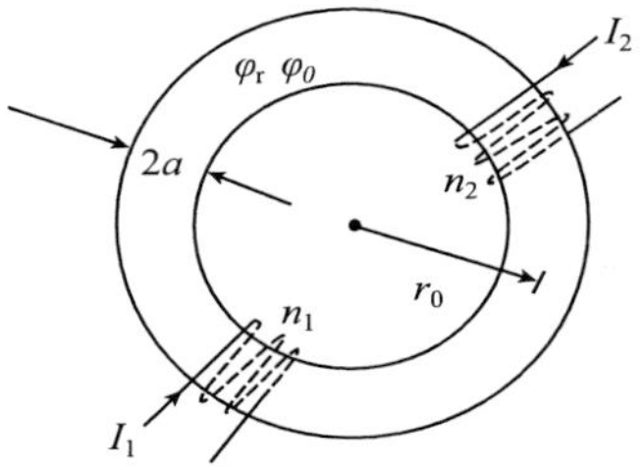

Figure of Example 3.8 The mutual inductance between the two coils of the toroidal core

Next, we shall derive the mutual inductance M_{12}. Since $r_0 >> a$, the mutual magnetic flux linkage is

$$\begin{aligned}\psi_{12} &= N_2\Phi_{12} = n_2\int_S \boldsymbol{B}_1\cdot \mathrm{d}\boldsymbol{S} = n_2B_1\pi a^2\\ &= \frac{\mu n_1n_2I_1\pi a^2}{2\pi r_0} = \frac{\mu n_1n_2I_1a^2}{2r_0}\end{aligned}$$

$$M_{12} = \frac{\psi_{12}}{I_1} = \frac{\mu n_1n_2a^2}{2r_0}$$

$$M = M_{12} = M_{21} = \frac{\mu n_1n_2a^2}{2r_0}, \quad r_0 >> a$$

3.7 Magnetic Energy and Magnetic Force

A magnetic field produces an acting force on moving charges. Therefore, a constant magnetic field also produces an acting force on the current. There are many similarities between a constant magnetic field and an electrostatic field when we consider energy distribution and power issues. It is known that when external factors change the current in the wire, the wire loop will stop the current changes by inducing an electromotive force. Thus, in the process of generating the current, the power supply must provide the energy to overcome the induced electromotive force and work. And the power generated by working will be stored in the magnetic field in the form of magnetic energy.

3.7.1 Magnetic Energy

1. Magnetic Energy of A Closed Loop System

We now analyze the magnetic energy possessed by two current loop systems in free space. Two wire loops l_1 and l_2 are shown in Figure 3.11. Under the action of outside power supply, the current in l_1 and l_2 will be increased from zero to I_1 and I_2, if the wire has zero resistance. The work done by an external power supply should all be converted to magnetic energy.

(1) Loop l_2 is an open circuit with current i_2=0. Under the action of the outer power supply, while the current i_1 in loop l_1 increases by $\mathrm{d}i_1$, the flux linkage ψ_1 that intersects with the loop l_1 will change and thereby generating the electromotive force $\mathcal{E}_1$ to prevent the current from increasing.

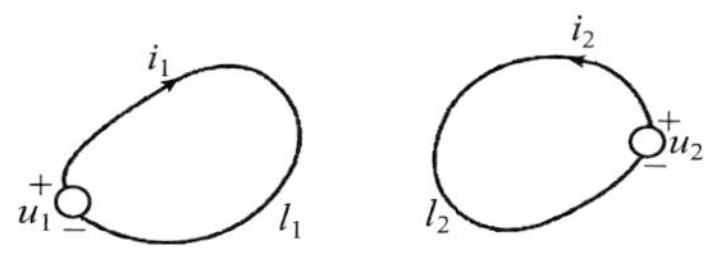

Figure 3.11 2 magnetic coupling coils including voltage source

$$\mathcal{E}_1 = -\frac{\partial\psi_1}{\partial t} \tag{3.108}$$

To maintain the increment of the current i_1, the outer power supply must provide voltage u_1 against the induced electromotive force, which is

$$u_1 = -\mathcal{E}_1 = \frac{\partial\psi_1}{\partial t} \tag{3.109}$$

then, while the loop current is i_1, the work done by the outer power supply in time dt is given by

$$\mathrm{d}W_1 = U_1I_1\mathrm{d}t = I_1\mathrm{d}\psi_1 = L_1I_1\mathrm{d}I_1 \tag{3.110}$$

While I_2 =0, in the process of current i_1 in loop l_1 changing from zero to I_1, the total work done by the outer power supply is

$$W_1 = \int \mathrm{d}W_1 = \int_0^{I_1} L_1 I_1 \mathrm{d}I_1 = \frac{L_1 I_1^2}{2} \tag{3.111}$$

This work (W_1) transforms to the energy of the magnetic field generated by current I_1.

(2) While the current I_1 in loop l_1 is invariant, the work done by outer power supply in the process of the current i_2 in loop l_2 will increase from zero to I_2.

Under the action of the outside power supply, a change in the current i_2 will induce electromotive forces $\mathcal{E}_1 = -\partial\psi_2/\partial t$, and $\mathcal{E}_2 = -\partial\psi_{21}/\partial t$ in loops l_1 and l_2. For example, in order to maintain the increase of current i_2 and maintain the l_1 loop current I_1, there must be outside power supply U_1 and U_2 to offset induced electromotive force. In duration dt, the work done by the two power supply is

$$\mathrm{d}W_2 = u_2 i_2 \mathrm{d}t = i_2 \mathrm{d}\psi_2 = L_2 i_2 \mathrm{d}i_2$$
$$\mathrm{d}W_{21} = I_1 \mathrm{d}\psi_{21} = M_{21} I_1 \mathrm{d}i_2$$

Thus, in the process that I_1 doesn't change and the current of l_2 changes from zeros to I_2, the total work made by the two outer power supplies is

$$W_2 = \int \mathrm{d}W_2 = \int_0^{I_2} L_2 i_2 \mathrm{d}i_2 = \frac{L_2 I_2^2}{2} \tag{3.112}$$

$$W_{21} = \int_0^{I_2} \mathrm{d}W_{21} = \int_0^{I_2} M_{21} I_1 \mathrm{d}i_2 = M_{21} I_1 I_2 \tag{3.113}$$

(3) While the currents in the two loops l_1 and l_2 change from zero to I_1 and I_2, the total magnetic energy owed by this system is

$$\begin{aligned} W_\mathrm{m} =& W_1 + W_2 + W_{21} \\ =& \frac{L_1 I_1^2}{2} + \frac{L_2 I_2^2}{2} + M_{21} I_1 I_2 = \frac{1}{2}\sum_{i=1}^{2}\sum_{j=1}^{2} M_{ij} I_i I_j \end{aligned} \tag{3.114}$$

In the above quantity W_1, W_2 represent the self energy of l_1, l_2, W_{21} represents the interaction energy between loop l_1, l_2. The $M_{ii} = L_i$, Equation (3.114) can be expressed in the alternative manner

$$\begin{aligned} W_\mathrm{m} =& \frac{1}{2}(L_1 I_1 + M I_2) I_1 + \frac{1}{2}(L_2 I_2 + M I_2) I_2 \\ =& \frac{\psi_1 I_1}{2} + \frac{\psi_2 I_2}{2} = \frac{1}{2}\sum_{i=1}^{2} \psi_i I_i \end{aligned} \tag{3.115}$$

In the above equality ψ_i expresses the sum of the self magnetic chain and the mutual magnetic chain of the chain that is intersected with the loop number i, i.e.

$$\psi_1 = L_1 I_1 + M I_2$$
$$\psi_2 = L_2 I_2 + M I_1$$

For a system with N current loops, the total magnetic energy will be given by

$$W_\mathrm{m} = \frac{1}{2}\sum_{i=1}^{N} \psi_i I_i = \frac{1}{2}\sum_{i=1}^{N} L_i I_i^2 + \frac{1}{2}\sum_{i=1}^{N}\sum_{\substack{j=1 \\ i\neq j}}^{N} M_{ij} I_i I_j \tag{3.116}$$

We can transform the Equation (3.114) to

$$
\begin{aligned}
W_\mathrm{m} =& \frac{1}{2}L_1I_1^2 + \frac{1}{2}L_1I_1^2 + M_{21}I_1I_2 \\
=& \frac{1}{2}L_2\left(I_2 + \frac{M_{21}}{L_2}I_1\right)^2 + \frac{1}{2}\left(I_1 - \frac{M_{21}^2}{L_2}\right)I_1^2
\end{aligned}
$$

The first term of this equation will always be positive, and the second term will always satisfy the relation $I_1 - \dfrac{M_{21}^2}{L_2} \geqslant 0$ so as to guarantee that the magnetic energy be positive. Consequently, we obtain the following equation

$$M_{21} \leqslant \sqrt{L_1L_2}$$

The maximum possible mutual inductance is given by $M_{21(\max)} = \sqrt{L_1L_2}$.

The inductive coupling between the two circuits can be expressed by the coupling coefficient k

$$k = \frac{M_{21}}{\sqrt{L_1L_2}} \tag{3.117}$$

with $0 \leqslant k \leqslant 1$.

2. Magnetic Energy Density

Equation (3.115) represents the magnetic energy expressed by the current, self-inductance and mutual inductance. This is the total magnetic energy of the system which is given in terms of the circuit. We shall now find the relationship between the magnetic intensity and magnetic energy density. That is to find that the magnetic energy density equals to $\dfrac{1}{2}\mu H^2$. This can be done by one of two methods.

Method 1 In the analysis of electrostatic field energy, exploiting the method used by a DC power supply for charging a capacitor, we find that the electric energy density is given by $\dfrac{1}{2}\varepsilon E^2$. A dual approach involves having the DC power supply charge the magnetic energy into the solenoid (inductor). This dual approach allows us to arrive at a magnetic energy density of $\dfrac{1}{2}\mu H^2$.

Consider a long straight solenoid with a cross section of S as shown in Figure 3.12. The power supply overcomes the reaction force. In the time dt, the energy provided to the inductance will be $-eidt$. When magnetization is completed (i.e. current is increased from zero to a constant DC value), the total energy provided by the power supply will be given by

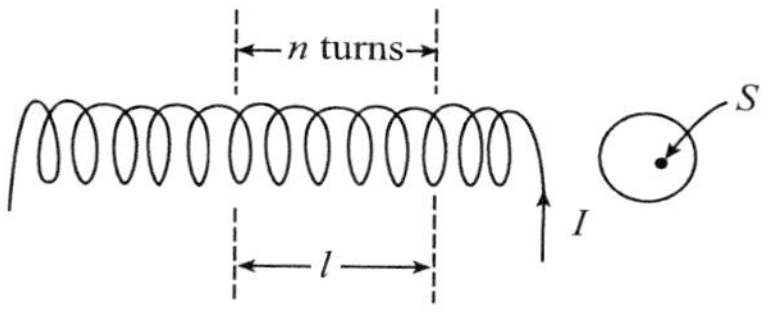

Figure 3.12 The outer power supply provides the magnetic energy for the solenoid

$$W_\mathrm{m} = -\int_0^\infty eidt = \int_0^\infty \frac{\mathrm{d}\psi}{\mathrm{d}t}idt = \int_0^\psi id\psi \tag{3.118}$$

From the perspective of the circuit and electromagnetic fields, we can describe the magnetic energy W_m in the solenoid with length l and n turns.

From the perspective of the circuit we have

$$W_{\mathrm{m}} = \int_0^{\psi} i\mathrm{d}\psi = \int_0^{I} i\mathrm{d}(Li) = \int_0^{I} Li\mathrm{d}i = \frac{1}{2}LI^2 \tag{3.119}$$

From the perspective of the field we have

$$\begin{aligned} W_{\mathrm{m}} &= \int_0^{\psi} i\mathrm{d}\psi = \int_0^{H} i\mathrm{d}(n\mu HS) = \int_0^{H} in\mu S\mathrm{d}H \\ &= \int_0^{H} \left(\frac{Hl}{n}\right) n\mu S\mathrm{d}H = \frac{1}{2}\mu H^2(Sl) \end{aligned} \tag{3.120}$$

Since we are considering a long and straight solenoid, the magnetic field in the solenoid will be evenly distributed. Clearly, the total magnetic field energy should be equal to the magnetic energy density multiplied by the volume of this solenoid Sl. Therefore, $\frac{1}{2}\mu H^2$in the above equation should be the magnetic energy density, that is

$$w_{\mathrm{m}} = \frac{1}{2}\mu H^2 \tag{3.121}$$

The total magnetic energy in the solenoid of length l is given by

$$W_{\mathrm{m}} = \frac{1}{2}LI^2 = \frac{1}{2}\mu H^2(Sl)$$

Method 2 In a system composed of N current loops, we let the total magnetic energy $W_{\mathrm{m}} = \frac{1}{2}\sum_{i=1}^{N} \Psi_i I_i$ be a starting point. With the use of related concepts and vector identities we derive the magnetic energy density as $\frac{1}{2}\mu H^2$.

Inserting $\Psi = \oint \boldsymbol{A}\cdot \mathrm{d}\boldsymbol{l}$ into the expression for the system magnetic energy, we arrive at

$$W_{\mathrm{m}} = \frac{1}{2}\sum_{i=1}^{N} \psi_i I_i = \frac{1}{2}\sum_{i=1}^{N} I_i \oint_{l_i} \boldsymbol{A}\cdot \mathrm{d}\boldsymbol{l}_i \tag{3.122}$$

Suppose system conductor section is $\mathrm{d}S$, then $I_i\mathrm{d}l_i = J\mathrm{d}S\mathrm{d}l_i = J\mathrm{d}V$ change the sum in Equation (3.122) into integral. We now obtain

$$W_{\mathrm{m}} = \frac{1}{2}\int_{V'} \boldsymbol{A}\cdot \boldsymbol{J}\mathrm{d}V' = \frac{1}{2}\int_{V'} \boldsymbol{A}\cdot \nabla\times \boldsymbol{H}\mathrm{d}V' \tag{3.123}$$

The integral volume in this equation can be extended to infinity. When $\boldsymbol{J} = 0, W_{\mathrm{m}}$ does not change since the space integral is zero. Then we use the vector identities

$$\nabla\cdot(\boldsymbol{H}\times\boldsymbol{A}) = \boldsymbol{A}\cdot\nabla\times\boldsymbol{H} - \boldsymbol{H}\cdot\nabla\times\boldsymbol{A}$$

and note that Equation (3.123) will charge to

$$\begin{aligned} W_{\mathrm{m}} =& \frac{1}{2}\int_{V'} \nabla\cdot(\boldsymbol{H}\times\boldsymbol{A})\mathrm{d}V' + \frac{1}{2}\int_{V'} \boldsymbol{H}\cdot\nabla\times\boldsymbol{A}\mathrm{d}V' \\ =& \frac{1}{2}\oint_S (\boldsymbol{H}\times\boldsymbol{A})\cdot \mathrm{d}S + \frac{1}{2}\int_V (\boldsymbol{H}\cdot\boldsymbol{B})\mathrm{d}V \end{aligned} \tag{3.124}$$

The divergence theorem is used in the derivation of the above equation. In the surface S, $\boldsymbol{H}$ decreases at a rate of $1/R^2$. The vector potential $\boldsymbol{A}$ decreases at a rate of $1/R$, and the integral area increases at a rate of R^2. Thus, the first item in Equation (3.124) decreases at a rate of $1/R$, especially when R tends to be infinity the first integral is zero. Therefore the magnetic energy of the entire space is

$$W_{\mathrm{m}} = \frac{1}{2}\int_V \boldsymbol{H}\cdot\boldsymbol{B}\mathrm{d}V \tag{3.125}$$

When there is a magnetic field, there will be magnetic energy. The magnetic energy density is given by

$$w_{\mathrm{m}} = \frac{1}{2}\boldsymbol{H}\cdot\boldsymbol{B} = \frac{1}{2}\mu H^2 \tag{3.126}$$

The magnetic energy density is $\frac{1}{2}\mu H^2$. Recall from chapter 2 that the current energy density will be $\frac{1}{2}\varepsilon E^2$. The magnetic energy density and the current energy density share a dual relationship.

3.7.2 Magnetic Force

Between the current loops l_1 and l_2, there is a magnetic field strength interacting with them. The magnetic force is calculated according to $\boldsymbol{F} = q\boldsymbol{v}\times\boldsymbol{B}$ and the Biot-Savart law. The magnetic force in the current element of length $\mathrm{d}l$ is given by

$$\begin{aligned}\mathrm{d}\boldsymbol{F} &= \mathrm{d}q\boldsymbol{v}\times\boldsymbol{B} = \mathrm{d}q\frac{\mathrm{d}\boldsymbol{l}}{\mathrm{d}t}\times B \\ &= \frac{\mathrm{d}q}{\mathrm{d}t}\mathrm{d}\boldsymbol{l}\times\boldsymbol{B} = I\mathrm{d}\boldsymbol{l}\times\boldsymbol{B}\end{aligned} \tag{3.127}$$

In the magnetic field $\boldsymbol{B}_1$ generated by current I_1, the magnetic force in current element $I_2\mathrm{d}\boldsymbol{l}_2$ is

$$\boldsymbol{F}_{12} = \oint_{l_2} I_2\mathrm{d}\boldsymbol{l}_2\times\boldsymbol{B}_1 \tag{3.128}$$

Substituting Equation (3.7a) from the Biot-Savart law into the above equation provides

$$\boldsymbol{F}_{12} = \oint_{l_2}\oint_{l_1}\frac{\mu_0 I_1 I_2\mathrm{d}\boldsymbol{l}_2\times\mathrm{d}l_1\times\boldsymbol{e}_R}{4\pi R^2} \tag{3.129}$$

The above expression represents the law of Ampere force. We can use this relation to calculate the interaction of the magnetic field between the loop current. However, the calculation is often very difficult and inconvenient to perform. A relatively simple technique is to calculate the magnetic force by means of virtual displacement method.

Consider the current system consisting of N closed loops. *Assuming a loop in the magnetic field strength produce displacement* $\mathrm{d}l$, and the loop flux in the system changes by $\mathrm{d}\psi_i$. Then, the work $\mathrm{d}W$ done by the external power supply is equal to the sum of incremental energy $\mathrm{d}W_{\mathrm{m}}$ and the work done by the magnetic force

$$\mathrm{d}W = \mathrm{d}W_{\mathrm{m}} + F\mathrm{d}l \tag{3.130}$$

The work done by the external power supply can prevent the magnetic flux from changing. This work can be expressed via the following equation

$$\mathrm{d}W = \sum_{i=1}^{N} i_i u_i \mathrm{d}t = \sum_{i=1}^{N} i_i\frac{\mathrm{d}\Psi_i}{\mathrm{d}t}\mathrm{d}t = \sum_{i=1}^{N} I_i\mathrm{d}\Psi_i \tag{3.131}$$

The change of the storage $W_{\rm m}$ of the system is expressed by Equation (3.116)

$$\mathrm{d}W_{\rm m} = \mathrm{d}\left[\sum_{i=1}^{N}\frac{1}{2}i_i\,\Psi_i\right] = \frac{1}{2}\sum_{i=1}^{N}i_i\mathrm{d}\,\Psi_i + \frac{1}{2}\sum_{i=1}^{N}\Psi_i\mathrm{d}i_i \tag{3.132}$$

We shall now discuss two cases:

(1) The system current remains unchanged.

When a system loop displacement occurs, the system electromotive force will change. Then the external power supply should work to offset the power of this electromotive force and maintain each loop current of the system unchanged, that is, $\mathrm{d}(i_i)$=0. From Equation (3.131) and Equation (3.132), we can get that

$$\mathrm{d}W_{\rm m} = \frac{1}{2}\sum_{i=1}^{N}I_i\mathrm{d}\,\Psi_i = \frac{1}{2}\mathrm{d}W \tag{3.133}$$

In the above equation, I_i is the unchanged current. Substituting Equation (3.132) into Equation (3.130) provides $F\mathrm{d}l$=$\mathrm{d}W_{\rm m}$. Thus, we now have

$$F = \mathrm{d}W_{\rm m}/\mathrm{d}l|_{I_i=\rm const}. \tag{3.134}$$

We can obtain from Equation (3.133) that half of the external power supply will provide for system energy storage and the other half for work.

(2) The system magnetic flux maintenance is unchanged.

From the condition of a constant flux the induced electromotive force is zero and the external power supply doesn't work for the system. This leads to $\mathrm{d}W$=0, and the work of the mobile loop power is provided by the magnetic field energy storage. From Equation (3.130), we note that

$$F\mathrm{d}l = -\mathrm{d}W_{\rm m}$$

$$F = -\mathrm{d}W_{\rm m}/\mathrm{d}l|_{\Psi_i=\rm const}. \tag{3.135}$$

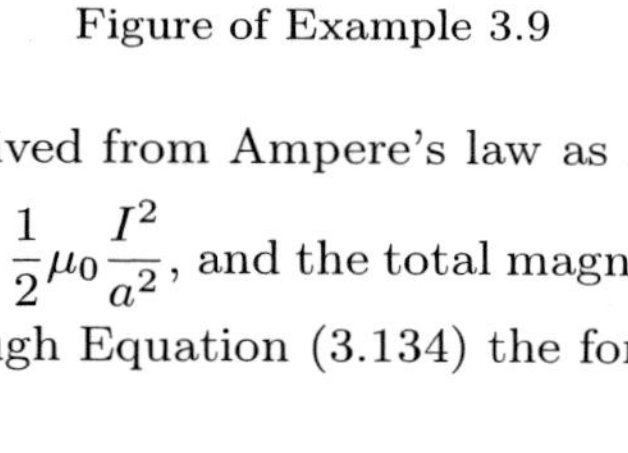

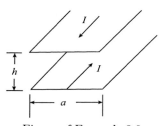

Figure of Example 3.9

Example 3.9 There are parallel strip transmission lines whose width is a and distance h, which have a direction opposite to the current flow I. This is shown in Figure of Example 3.9. If the bandwidth $a \gg h$, ignoring the edge effects, please seek the force per unit length between the lines.

Solution: From the condition $a \gg h$ and neglecting the edge effects, magnetic field between the two strip lines can be seen as uniformly distributed. The magnetic field between the wires is derived from Ampere's law as $H = \frac{I}{a}$. The magnetic energy density between the wires is $w_{\rm m} = \frac{1}{2}\mu_0\frac{I^2}{a^2}$, and the total magnetic energy per unit length is given by $W_{\rm m}$=$w_{\rm m}ha$=$\frac{\mu_0 I^2 h}{2a}$. Through Equation (3.134) the force per unit length between the strip lines will be

$$F = \frac{\mathrm{d}W_{\rm m}}{\mathrm{d}h} = \frac{\mu_0 I^2}{2a}$$

The force between the two boards is repulsive.

Example 3.10 In the z=0 plane, there is a large electrically charged sheet. The sheet's charge surface density is ρ_s=4(c/m^2). The sheet moves at a constant speed v_x, where v_x=2(m/s). In the $z = 0$ plane at various points exists the electric field of $\boldsymbol{E} = 2\boldsymbol{e}_x - 3\boldsymbol{e}_z$(V/m) and magnetic field $\boldsymbol{B} = 3\boldsymbol{e}_y$T. Derive the force $\boldsymbol{F}$ received by the plate area per unit.

Solution The charge surface density ρ_s receives two forces. One is the electric field force $\boldsymbol{F}_\mathrm{e}$, the other is the magnetic force $\boldsymbol{F}_\mathrm{m}$, that is the magnetic field strength received from the moving charge. By virtue of Equation (3.6) we have

$$\boldsymbol{F} = \rho_s\boldsymbol{E} + \rho_s\boldsymbol{v} \times \boldsymbol{B} = \boldsymbol{F}_\mathrm{e} + \boldsymbol{F}_\mathrm{m}$$

We can see that the movement ρ_s receiving the force $\boldsymbol{F}$ that is experienced by a force plate per unit area is given by

$$\boldsymbol{F} = \rho_s\boldsymbol{E} + \rho_s\boldsymbol{v} \times \boldsymbol{B} = 4(2\boldsymbol{e}_x - 3\boldsymbol{e}_z) + 4(2\boldsymbol{e}_x \times 3\boldsymbol{e}_y) = 8\boldsymbol{e}_x + 12\boldsymbol{e}_z$$

Exercises

3.1 Derive the two basic equations of constant magnetic field (respectively describe the constant magnetic field for the vortex field and non-finish at a) in differential form. Derive the expression in rectangular coordinates.

3.2 Consider a wire ring with radius a and current I (Figure of Exercise 3.2). Qualitatively answer following questions about the center of the circle O:

(1) Why the magnetic vector potential at point O is $\boldsymbol{A} = 0$?

(2) Why is the magnetic induction B at point O not equal to zero?

(3) Use $\boldsymbol{B} = \nabla \times \boldsymbol{A}$ to demonstrate that at point O we have $\boldsymbol{A} = 0$, and $\boldsymbol{B} \neq 0$?

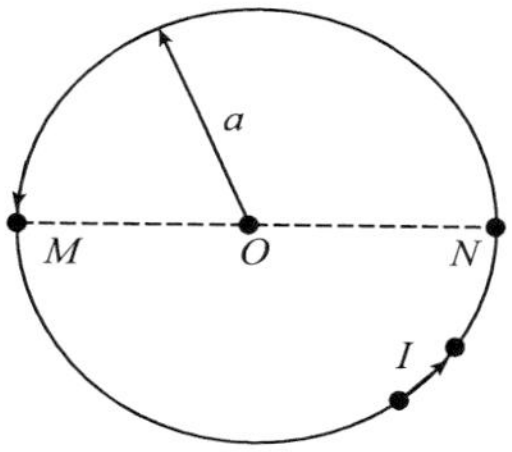

Figure of Exercise 3.2

3.3 In cylindrical coordinates, in the regions $r \leqslant a$, the current density $\boldsymbol{J} = \boldsymbol{e}_z J$. Derive

(1) $\nabla \times \boldsymbol{H}$ in area $r \leqslant a$;

(2) $\nabla \times \boldsymbol{H}$ in area $r \geqslant a$;

(3) From (1), (2), what is the rotation source of the constant magnetic field?

3.4 In cylindrical coordinates, the current density is given by $\boldsymbol{J} = kr^2\boldsymbol{e}_z(r \leqslant a)$.

(1) Seek the magnetic induction $\boldsymbol{B}$;

(2) Demonstrate that $\nabla \times \boldsymbol{B} = \mu_0\boldsymbol{J}$.

3.5 Consider an infinitely long conductor cylinder with radius a, in which current is uniformly distributed along the cross section. The total current is I. Derive the magnetic field strength inside and outside the cylinder $\boldsymbol{H}$.

3.6 The distribution of magnetic induction intensity $\boldsymbol{B}$ is known in cylindrical coordinate system. $\boldsymbol{B} = 0(0 < r < a)$, $\boldsymbol{B} = \dfrac{\mu_0 I}{2\pi r}[(r^2 - a^2)/(b^2 - a^2)]\boldsymbol{e}_\phi(a < r < b)$ and $\boldsymbol{B} = \dfrac{\mu_0 I}{2\pi r}\boldsymbol{e}_\phi(r > b)$. Please derive the current density $\boldsymbol{J}$ in the space.

3.7 As shown in Figure of Exercise 3.7, there is an infinitely long thin strip line with a width of b. A current I flowing in the negative direction of z is uniformly distributed along the strip. Derive $\boldsymbol{B}$ at point P with the belt-line distance of a in the center line.

3.8 Consider a current I is passing through a wire of length L. Please seek the magnetic vector potential $\boldsymbol{A}$ and magnetic induction intensity $\boldsymbol{B}$ in space at any point.

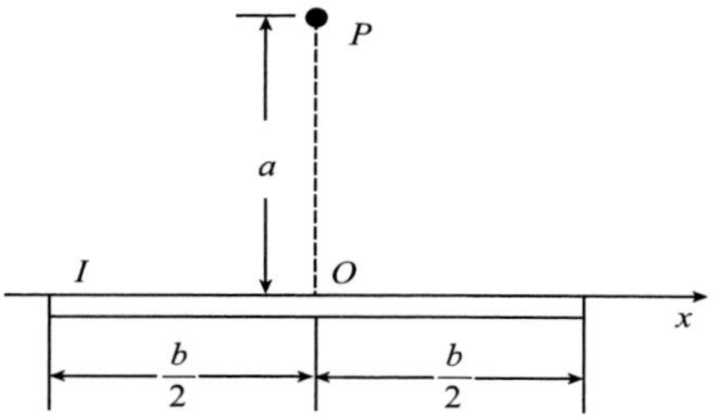

Figure of Exercise 3.7

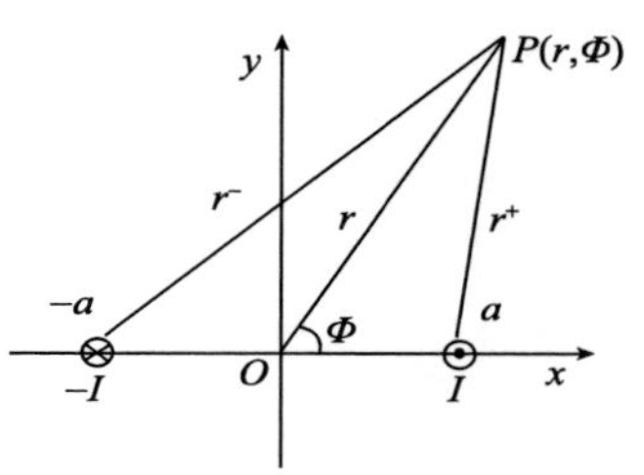

Figure of Exercise 3.9

3.9 Consider a pair of infinitely long parallel transmission lines as shown in Figure of Exercise 3.9. There distances are $2a$ and, the current in the two lines are I (running in the opposite direction for each line) . Derive the magnetic vector potential $\boldsymbol{A}$.

3.10 Consider an infinite long cylindrical conductor with $\boldsymbol{J} = \boldsymbol{e}_z J_0 \left(1 - \dfrac{2r}{a}\right)$. The conductor column axis is the z-axis. Derive $\boldsymbol{B}$.

3.11 Derive $\boldsymbol{B}$ at point P of each current loop in the Figure of Exercise 3.11.

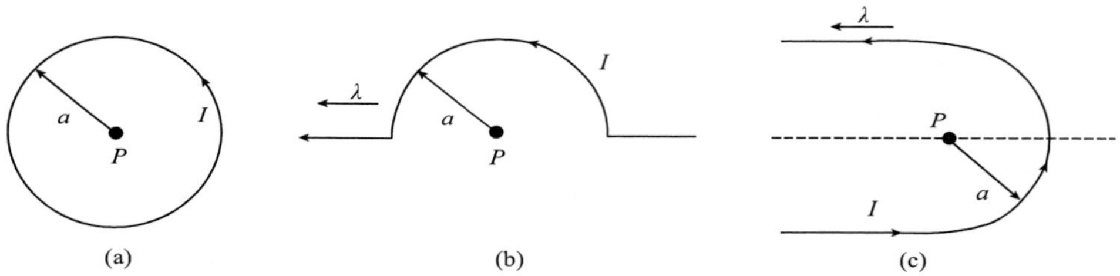

Figure of Exercise 3.11

3.12 Consider a current of z-distribution $J_z = r^2 + 4r (r \leqslant a)$. Use Ampere's law to derive $\boldsymbol{B}$.

3.13 Located in $x < 0$ half-space, the space is full of the homogeneous medium of permeability μ, the space of $x > 0$ is vacuum, there is a line current moving along the z axis, seek the magnetic field strength (Figure of Exercise 3.13).

3.14 There are infinite parallel metal plates of distance d flowing through the same size and opposite current density $\boldsymbol{J}_s$. Derive the magnetic field strength $\boldsymbol{H}$ in the space around (Figure of Exercise 3.14).

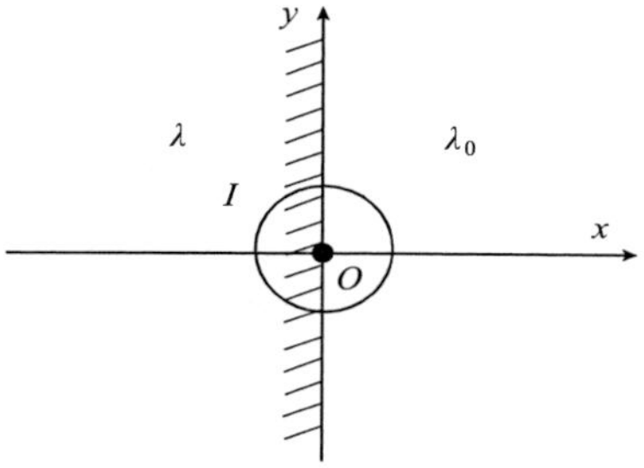

Figure of Exercise 3.13

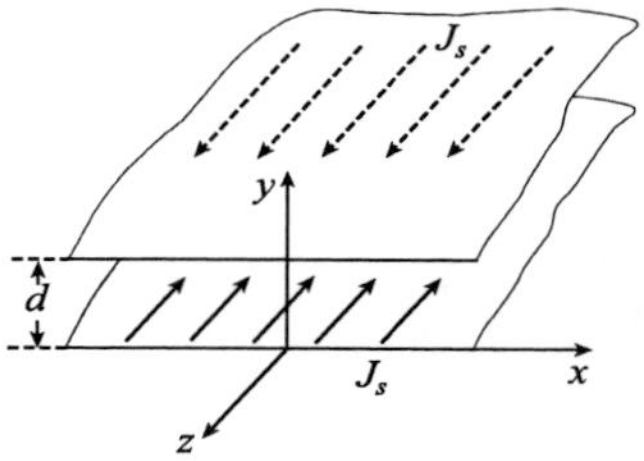

Figure of Exercise 3.14

3.15 Consider a long, hollow, straight copper tube of radius R_0 and thickness d, as shown in Figure of Exercise 3.15. Current I will pass through the copper pipe. Derive $\boldsymbol{B}$ and $\boldsymbol{H}$ in the area $0 \leqslant r \leqslant \infty$. Verify that $\boldsymbol{B}$ and $\boldsymbol{H}$ respectively satisfy the corresponding boundary parts.

3.16 Consider an infinitely long straight current I perpendicular to the interface between two kinds of magnetic media whose magnetic permeability μ_1 and μ_2 as shown in Figure of Exercise 3.16. Derive the magnetic flux density $\boldsymbol{B}_1$ and $\boldsymbol{B}_2$ in these kinds of media.

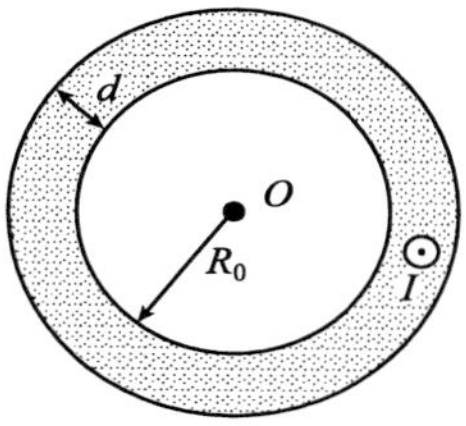

Figure of Exercise 3.15

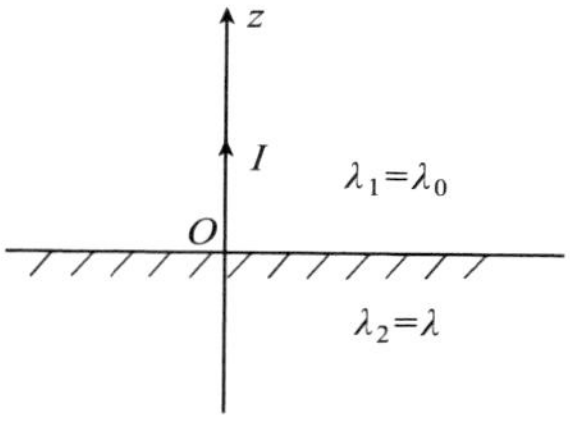

Figure of Exercise 3.16

3.17 Consider that a column, coaxial conductor radius is a, outer conductor radius is c (ignoring its thickness), and a middle with two layers of different magnetic media. This is shown in Figure of Exercise 3.17. The internal and external conductors have the same values, but the opposite direction of current I. Derive $\boldsymbol{H}$, $\boldsymbol{B}$, $\boldsymbol{M}$ for any point in this conductor.

3.18 An infinitely long straight wire is co-planar with a ring of radius a, the distance from the circle center to the wire is $d(d \gg a)$. Derive the mutual inductance M between the ring and the wire (Figure of Exercise 3.18).

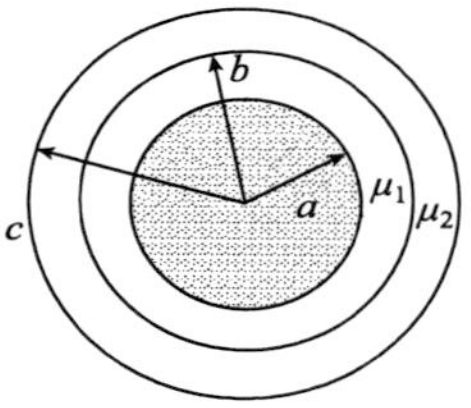

Figure of Exercise 3.17

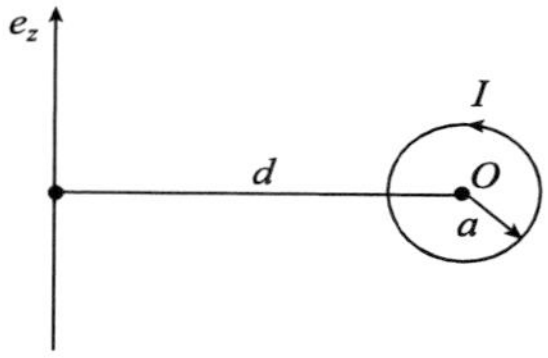

Figure of Exercise 3.18

3.19 Consider a pair of parallel wires of radius a, and an axis spacing $D(D \gg a)$. The permeability between the wire and the surrounding space is μ_0. Derive the self-inductance per unit length between the parallel wires.

3.20 Consider an infinite cylindrical conductor of radius a. Derive the inner inductance per unit length.

3.21 A long straight wire is in the same plane as a rectangular coil whose side lengths are a and b. The wide edge of the coil is parallel to the straight wire as shown in Figure of Exercise 3.21. Derive the interaction between the coil and the wire.

3.22 As shown in Figure of Exercise 3.22, we have a vacuum with a long-line current of I. The wire is co-planar with the loop of equilateral triangle. Derive the magnetic flux in the loop of the triangular.

3.23 Two long rectangular coil are placed on the same plane. The lengths are l_1 and l_2, the widths are w_1 and w_2, and the nearest distance between the two coils is S as shown in Figure of Exercise 3.23. Verify that the mutual inductance of the two coil is $M = \dfrac{\mu_0 l_2}{2\pi} \cdot \ln \dfrac{S + w_2}{S\left(1 + \dfrac{w_2}{S + w_1}\right)}$.
Suppose that $l_1 \gg l_2$, $l_1 \gg S$, and that both of the coils have only one turn. In addition, we neglect the end effect.

3.24 Consider two parallel circular coaxial coils. The distance between them is d, one coil's radius is $a(a \gg d)$, and the other coil's radius is b. Seek the mutual inductance between the two coils.

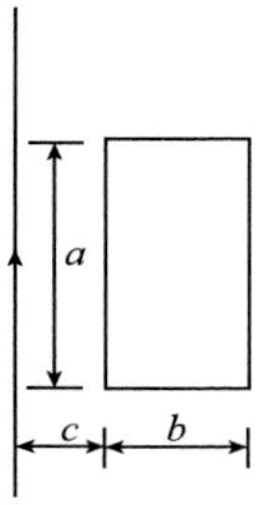

Figure of Exercise 3.21

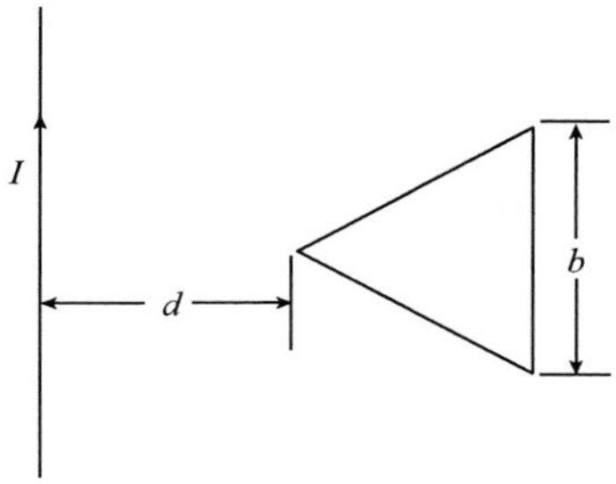

Figure of Exercise 3.22

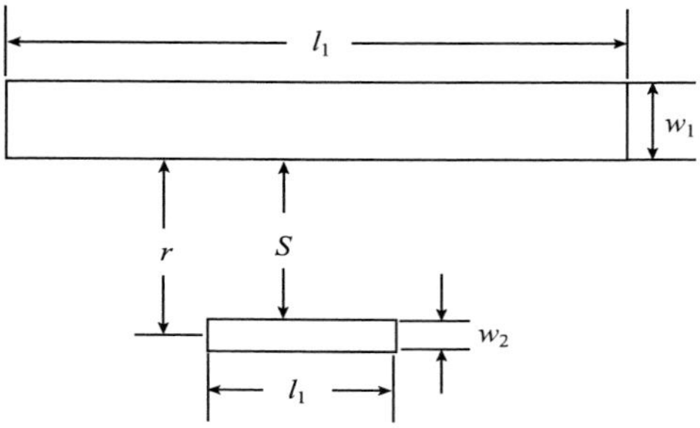

Figure of Exercise 3.23

3.25 Consider an air-insulated coaxial line. The inner conductor radius is a, and the outer conductor radius is b. The current through the coaxial line is I. Suppose that the outer conductor is thin and thus the stored energy can be taken as negligible. Find the magnetic energy stored per unit length of coaxial line. Calculate the inductance per unit length by magnetic energy.

3.26 Consider a long, straight solenoid with a radius of a. Assume the solenoid is coiled with N-turn per unit length. At the solenoid axis, there is a radius of b, small single-turn coil. The angle between the $\boldsymbol{e}_n$ normal of the coil plane and the axis is θ. Find the mutual inductance between the solenoid and the small circle.

Chapter 4

Steady Electric Field

Electrostatic fields will not exist in conductors. Although steady electric fields may exist within a conductor; it must be maintained by an external force. The inside of the conductor is no longer an equipotential volume and the surface is not an equipotential surface under the action of an external power supply. However, in the presence of the constant current, the charge distribution in the medium both in and outside the conductor is unchanged and the steady electric field created by these unchanged distribution of charges has some similarities with the electrostatic field.

In this chapter, we will analyze the two equations $\nabla \cdot \boldsymbol{J} = 0$ and $\nabla \times \boldsymbol{E} = 0$. These equations provide an analogy between the steady electric field and the electrostatic field.

4.1 Current Density

The current I is defined as the charges that pass through any cross-sections in unit time. If the charge density in the conducting medium is ρ and the moving velocity of the charged particle is v, then according to the definition of currents, the current passing through the bin ΔS perpendicular to the moving direction of the particles is given by

$$\Delta I = \Delta q / \Delta t = \rho \boldsymbol{v} \cdot \Delta \boldsymbol{S} = \rho v \Delta S \tag{4.1}$$

*1. Current Density **J***

In the conducting medium, the currents in different positions may have distinct values. In order to describe the inhomogeneity of the distribution of charges, the vector current density $\boldsymbol{J}$ is defined by

$$\boldsymbol{J} = \rho \boldsymbol{v} \quad \mathrm{A} \ / \ \mathrm{m}^2 \tag{4.2}$$

The direction of the above vector is the moving direction of positive charges, and its magnitude is the quantity of charges which perpendicularly pass through a unit area in unit time.

Equation (4.1) can now be written as $\Delta \boldsymbol{I} = \boldsymbol{J} \cdot \Delta \boldsymbol{S}$, and the current passing through any section can be obtained from the surface integral to current density $\boldsymbol{J}$ via

$$I = \int_S \boldsymbol{J} \cdot \mathrm{d}\boldsymbol{S} \tag{4.3}$$

2. Ohm's law

Through experimentation it has been determined that the current density $\boldsymbol{J}$ in a conducting medium with conductivity σ is proportional to the electric field intensity $\boldsymbol{E}$

$$\boldsymbol{J} = \sigma \boldsymbol{E} \tag{4.4}$$

The above equation is called Ohm's law in differential form.

We shall now analyze the relationship between the voltage and current in areas of the conducting medium from Equation (4.4). A uniform conducting medium cylinder with cross sectional area S, length l and conductivity σ will be considered, as shown in Figure 4.1. Assume that the electric field intensity $\boldsymbol{E}$ and the current density $\boldsymbol{J}$ are uniform and $\boldsymbol{J}$ is vertical to the cross section S. The total current passing through the cross section of the conducting medium will be given by

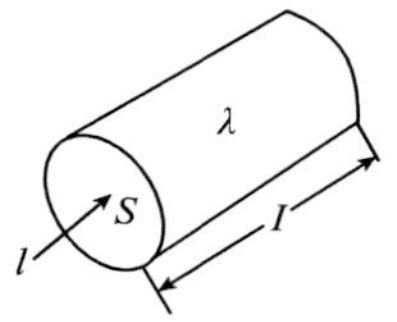

Figure 4.1 The conducting medium which cross sectional area is S and σ distributes uniformly

$$I = \int_S \boldsymbol{J} \cdot \mathrm{d}\boldsymbol{S} = JS \tag{4.5}$$

The voltage drop in length l caused by electric field intensity $\boldsymbol{E}$ is given by

$$U = \int_l \boldsymbol{E} \cdot \mathrm{d}\boldsymbol{l} = El = \frac{SJl}{\sigma S} = \frac{Il}{\sigma S} = IR \tag{4.6}$$

Substituting Equations (4.5) and (4.6) into Equation (4.4), a relationship for the voltage and current in the conducting medium is obtained.

$$U = Il/(\sigma S) = IR \tag{4.7}$$

$$R = l/(\sigma S) \tag{4.8}$$

Equation (4.7) is Ohm's law in linear circuit theory, and Equation (4.4) is its differential form.

Example 4.1 The radii of the inside and outside conductors of a spherical capacitor are a and b respectively. The parameters of the medium between the two conductors are μ_0, ε and σ. Determine the drain conductance of this capacitor.

Solution Suppose that the charge of the inside conductor of the capacitor is q, then the electric field intensity can be calculated from Guass law as

$$E = \frac{q}{4\pi\varepsilon r^2}, \quad a < r < b$$

According to Ohm's law, the current density in the medium is

$$J = \sigma E = \frac{\sigma q}{4\pi\varepsilon r^2}$$

The total leakage current is equal to the integral of the current density along the sphere surface. Namely,

$$I = \oint_S \boldsymbol{J} \cdot \mathrm{d}\boldsymbol{S} = \int_0^{2\pi} \int_0^{\pi} \frac{\sigma q \sin\theta}{4\pi\varepsilon r^2} r^2 \mathrm{d}\theta \mathrm{d}\varphi = \frac{\sigma q}{\varepsilon}$$

The potential difference between the two conductors is given by

$$U = \int_a^b E\mathrm{d}r = \left(\frac{1}{a} - \frac{1}{b}\right) \frac{q}{4\pi\varepsilon}$$

The drain conductance of the capacitor is attained as

$$G = I/U = 4\pi\sigma \Big/ \left(\frac{1}{a} - \frac{1}{b}\right)$$

4.2 Current Continuity Equation

The law of conservation of charge is one of the basic laws in nature. Charge can neither be produced nor be destroyed. The fact that the quantity of charges on an object is zero can only show that the amount of positive charges will be equal to that of the negative charges. The process of charging is actually a process of transportation of the charges. According to the law of conservation of charge, the charge flowing out of a closed surface in unit time is equal to the reduction rate of the total charges in the closed surface . Based on the law of conservation of charge we obtain the current continuity equation

$$\oint_S \boldsymbol{J} \cdot \mathrm{d}\boldsymbol{S} = -\frac{\partial q}{\partial t} = -\frac{\partial}{\partial t}\int_V \rho \mathrm{d}V \tag{4.9}$$

In the above equation, V is the volume surrounded by the closed surface S. On the left side of the equality the surface integral on current density is changed to a volume integral according to the divergence theorem. The differential $\frac{\partial}{\partial t}$ on the right side of the equal sign can be moved inside the integral since the volume V is not the function of time. Thus, we obtain

$$\int_V \nabla \cdot \boldsymbol{J} \mathrm{d}V = -\int_V \frac{\partial \rho}{\partial t}\mathrm{d}V \tag{4.10}$$

The above equation is suitable for any volume V. Hence, the integrands of both sides of the equation are equal to each other, indicating that

$$\nabla \cdot \boldsymbol{J} = -\frac{\partial \rho}{\partial t} \tag{4.11}$$

This equation is the current continuity equation in differential form. Under the condition of a steady electric field, the charge in a closed surface S will not change. Therefore, the following two equations can be obtained from Equations (4.11) and (4.9) respectively

$$\nabla \cdot \boldsymbol{J} = 0 \tag{4.12}$$

$$\oint_S \boldsymbol{J} \cdot \mathrm{d}\boldsymbol{S} = 0 \tag{4.13}$$

Equations (4.12) and (4.13) are the differential form and the integral form of the current continuity equation, respectively. This indicates that the flux of current density $\boldsymbol{J}$ passing through a closed surface is zero under the condition of a steady electric field. Furthermore, this indicates that steady currents will be continuous.

If what the closed surface surrounds is a node in a circuit, we obtain kirchhoff's law which means that the algebraic sum of the currents flowing into a node at any time is zero.

4.3 Steady Electric Fields are Irrotational Fields

4.3.1 Steady Electric Fields are Irrotational Fields

In a conducting medium the steady currents must be maintained by external power supplies. The non-electrostatic force inside the power supplies move positive charges from the negative electrode to the positive electrode. Alternately, they may move negative charges from the negative electrode to the positive electrode so as to form an electric field. The electric field will produce currents in conducting medium. the charges accumulated on the positive and negative electrodes of the power supply are constantly moving to produce the current. Under the stability conditions, the charges accumulated on the positive and negative electrodes of the power supply are not time-varying. So it is the electric field they produce. Under the condition of homeostasis, the electric field produced by constant charges is equal to the electric field produced by static electric charges. Therefore, the steady electric field is a conservative field, which is not dependent on whether the medium is uniform or not. It indicates that

$$\nabla \times \boldsymbol{E} = 0 \tag{4.14}$$

$$\oint_l \boldsymbol{E} \cdot \mathrm{d}\boldsymbol{l} = 0 \tag{4.15}$$

Equations (4.14) and (4.15) illustrate that the steady electric field is an irrotational field, and provide another set of the fundamental equations for the steady electric field.

4.3.2 Steady Electric Field is A Solenoidal Field in Uniform Conducting Mediums

Substituting $\boldsymbol{J} = \boldsymbol{\sigma E}$ into Equation (4.12), we obtain an expression for the electric field divergence in a conducting medium as

$$\nabla \cdot (\sigma \boldsymbol{E}) = \sigma \nabla \cdot \boldsymbol{E} + \boldsymbol{E} \cdot \nabla \sigma = 0$$

$$\nabla \cdot \boldsymbol{E} = -\boldsymbol{E} \cdot \frac{\nabla \sigma}{\sigma} \tag{4.16}$$

In a uniform conducting medium, $\nabla \sigma = 0$, so we obtain

$$\nabla \cdot \boldsymbol{E} = 0 \tag{4.17}$$

The electric field divergence is related with the volume charge density. Equation (4.17) shows that under the condition of steady electric fields, the volume charge density in uniform conducting mediums is zero, and all the charges are only distributed on the surface of the conducting medium.

4.3.3 The Potential of the Steady Electric Fields in A Uniform Medium Satisfies the Laplace Equation

As $\nabla \times \boldsymbol{E} = 0$, so

$$\boldsymbol{E} = -\nabla \varphi$$

And because $\nabla \cdot \boldsymbol{J} = 0$, for uniform medium $\nabla \cdot \boldsymbol{J} = \nabla \cdot \sigma \boldsymbol{E} = \sigma \nabla \cdot \boldsymbol{E} = 0$, so

$$\nabla \cdot \boldsymbol{E} = 0$$

Substituting $\boldsymbol{E} = -\nabla \varphi$ into $\nabla \cdot \boldsymbol{E} = 0$, we obtain

$$\nabla^2 \varphi = 0 \tag{4.18}$$

This indicates that an electric field of the steady currents is a potential field, and its potential φ in a uniform medium satisfies the Laplace's equation.

4.3.4 Kirchhoff's Voltage Law

The relationship between the electromotive force and the internal and external resistors of a power supply will be analyzed. The equivalent electric field $\boldsymbol{E}_1'$ and the steady electric field $\boldsymbol{E}_1$ exist within the power supply. $\boldsymbol{E}_1'$ and $\boldsymbol{E}_1$are produced by non-electrostatic force and charges separately. $\boldsymbol{E}_T = \boldsymbol{E}_1 + \boldsymbol{E}_1'$ is the total field within the power supply; $\boldsymbol{E}_2$ is the steady electric field outside the power supply.

In order to keep the state of the constant currents, the charges transported to the electrodes in unit time by the electrostatic and non-electrostatic force inside the power supply is equal to the current flowing into the conducting medium outside the power supply. As the steady electric field is a conservative field and its closed line integral is zero, kirchhoff's voltage law can be obtained.

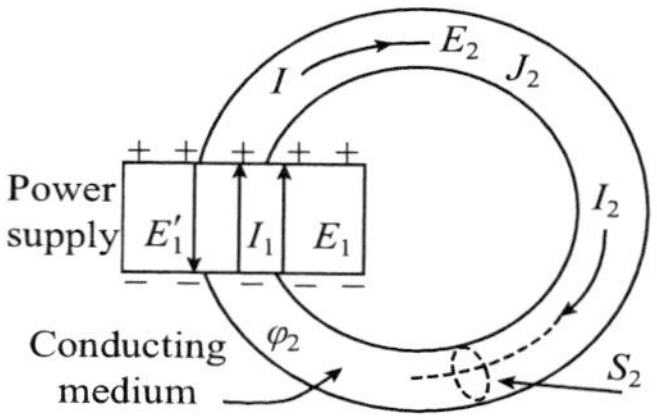

Figure 4.2 Steady electric field including the source

If we use $\boldsymbol{E}_T$, l_1, σ_1, s_1 to represent the parameters in the power supply, and use $\boldsymbol{E}_2$, l_2, σ_2, s_2 to represent the parameters in the medium outside the power supply, then the integral of the closed line of $\boldsymbol{E}$ passing through the inside and outside of the power supply is

$$\oint_l \boldsymbol{E}\cdot \mathrm{d}\boldsymbol{l} = \oint_{l_1} \boldsymbol{E}_T \cdot \mathrm{d}\boldsymbol{l} + \oint_{l_2} \boldsymbol{E}_2 \cdot \mathrm{d}\boldsymbol{l} = \oint_{l_1} \frac{J_1}{\sigma_1}\cdot \mathrm{d}l + \oint_{l_2} \frac{J_2}{\sigma_2}\cdot \mathrm{d}l$$
$$= I\left(\int_{l_1} \frac{\mathrm{d}l}{\sigma_1 s_1} + \int_{l_2} \frac{\mathrm{d}l}{\sigma_2 s_2}\right) = I(R_1 + R_2)$$

where R_1 and R_2 are the internal and external resistances of the power supply. It should be noted that $\boldsymbol{E}_T = \boldsymbol{J}_1/\sigma_1$ is an equivalent electric field expression of the non-electrostatic force $\boldsymbol{F}$ because $\boldsymbol{E}_1'(= \boldsymbol{F}/q)$ inside the power supply .The power supply induced by $\oint_l \boldsymbol{E}\cdot \mathrm{d}\boldsymbol{l}$ can be represented as follows

$$\oint_l \boldsymbol{E}\cdot \mathrm{d}l = \int_{l_1} \boldsymbol{E}_1' \cdot \mathrm{d}l + \left(\int_{l_1} \boldsymbol{E}_1 \cdot \mathrm{d}\boldsymbol{l} + \int_{l_2} \boldsymbol{E}_2 \cdot \mathrm{d}\boldsymbol{l}\right)$$
$$= \int_{l_1} \boldsymbol{E}_1' \cdot \mathrm{d}l = \mathcal{E}$$

The sum of the two integrals in the parentheses above is the closed line integral of the steady electric field. The value of the closed line integral is zero. The quantity $\mathcal{E}$ is the electromotive force of the power supply, and it is supplied by the work of the non-electrostatic force. From the two formulas above we obtain

$$\mathcal{E} = I(R_1 + R_2) \tag{4.19}$$

In the scenario of multiple power supplies and multiple resistances existing in a circuit, we obtain

$$\sum_{i=1}^{N} \mathcal{E}_i = \sum_{j=1}^{M} I_j R_j \tag{4.20}$$

The above formula is referred to as Kichhoff's voltage law and illustrates that the total electromotive force in a circuit is equal to the sum of the individual voltage drops of the resistances in a circuit.

4.4 Loss of Energy in A Conducting Medium

Free electrons in a conducting medium move in a certain direction under the action of the electric field force. The electrons ceaselessly collide with atom crystal lattices in the process of moving, and transfer the kinetic energy to the thermal vibration of atoms. This will cause a loss of energy. In order to maintain the directional migration of free electrons in a conducting medium, the external electric field force must work and provide energy for them. If the external electric field force moves the charge $\mathrm{d}q$ a distance of $\mathrm{d}l$ in a time $\mathrm{d}t$, the work done by the external electric field force is $\mathrm{d}W = \mathrm{d}q\boldsymbol{E}\cdot\mathrm{d}\boldsymbol{l}$. The work supplements the energy loss of free electrons in the time $\mathrm{d}t$. The corresponding power loss of the electric field in the conducting medium is

$$\begin{aligned} P &= \frac{\mathrm{d}W}{\mathrm{d}t} = \frac{\mathrm{d}q\boldsymbol{E}\cdot\mathrm{d}l}{\mathrm{d}t} = I\boldsymbol{E}\cdot\mathrm{d}\boldsymbol{l} \\ &= \boldsymbol{J}\cdot\boldsymbol{E}\mathrm{d}l S = \boldsymbol{J}\cdot\boldsymbol{E}\mathrm{d}V \end{aligned}$$

Thus, the power loss in a unit volume is given by

$$p = \boldsymbol{J}\cdot\boldsymbol{E} \tag{4.21}$$

The power loss in a uniform conducting medium with the cross sectional area $S$and length l is

$$P = pSl = JESl = IU = I^2R = U^2/R \tag{4.22}$$

Where I is the total current through the area S, U is the voltage drop across the length l, and R is the resistance of the conducting medium. The above equation corresponds to the well-known *Joule's law* in the circuit theory. More specifically, Equation (4.21) is known as the differential form of Joule's law.

Example 4.2 The interior and exterior radii of a coaxial line are a and b, respectively. The conductivity of the medium inside the coaxial line is σ. The voltage between the inside and outside conductor is U. Determine the power loss of the unit length coaxial line caused by the medium

Solution Assume that the leakage current of the coaxial line from the inside conductor to the outside conductor is I, thus the current density in the location where the radius is r is given by

$$J = I/(2\pi r), \quad a < r < b$$

$$U = \int_a^b \boldsymbol{E}\cdot\mathrm{d}\boldsymbol{r} = \int_a^b \frac{J\mathrm{d}r}{\sigma} = \frac{I}{2\pi\sigma}\ln\left(\frac{b}{a}\right) = \frac{Jr}{\sigma}\ln\left(\frac{b}{a}\right)$$

Thus, we have

$$J = \frac{U\sigma}{r\ln\left(\frac{b}{a}\right)}$$

The power loss of the unit length coaxial line is

$$P = \int \frac{J^2}{\sigma}\mathrm{d}V = \int_0^{2\pi}\int_a^b \frac{U^2\sigma r\mathrm{d}r\mathrm{d}\varphi}{\left[r\ln\left(\frac{b}{a}\right)\right]^2} = \frac{2\pi U^2\sigma}{\ln\left(\frac{b}{a}\right)}$$

It can be observed from the above equation that the radial leakage resistance of the unit length coaxial line is

$$R = \ln\frac{b}{a}\bigg/ 2\pi\sigma$$

4.5 Boundary Condition of the Steady Electric Field

The magnitude and the direction of the current density vector and the electric field vector will be changed on each side of the interface of two distinct conducting mediums when the steady electric current flows through it. The boundary conditions of the steady electric field are the rule that the current density vector and the steady electric field vector must obey on the interface.

The integral form of the steady electric field is given by

$$\oint_S \boldsymbol{J} \cdot \mathrm{d}\boldsymbol{S} = 0, \quad \oint_l \boldsymbol{E} \cdot \mathrm{d}\boldsymbol{l} = 0$$

The relationship between $\boldsymbol{J}$ and $\boldsymbol{E}$ in a conducting medium is $\boldsymbol{J} = \sigma\boldsymbol{E}$. From the discussion of the boundary conditions of the electrostatic field we know that the tangential component of the steady electric field vector on the interface is continuous according to $\oint_l \boldsymbol{E} \cdot \mathrm{d}\boldsymbol{l} = 0$. From the relation $\oint_S \boldsymbol{J} \cdot \mathrm{d}\boldsymbol{S} = 0$, it is known that the normal component of the current density vector is continuous on the interface. This can be proven via the same method used in Section 2.7. More specifically, the boundary conditions of the steady electric field are

$$J_{1n} = J_{2n} \tag{4.23}$$

$$E_{1t} = E_{2t} \tag{4.24}$$

According to Ohm's law in differential form we obtain

$$\sigma_1 E_{1n} = \sigma_2 E_{2n} \tag{4.25}$$

$$\sigma_2 J_{1t} = \sigma_1 J_{2t} \tag{4.26}$$

The rule of the change of the current density and electric field intensity in the interface can be determined according to the boundary conditions. Assume that the angles between the direction of the current density and the normal direction of the interface of conducting mediums 1 and 2 are given by θ_1 and θ_2, respectively. This scenario is depicted in Figure 4.3.

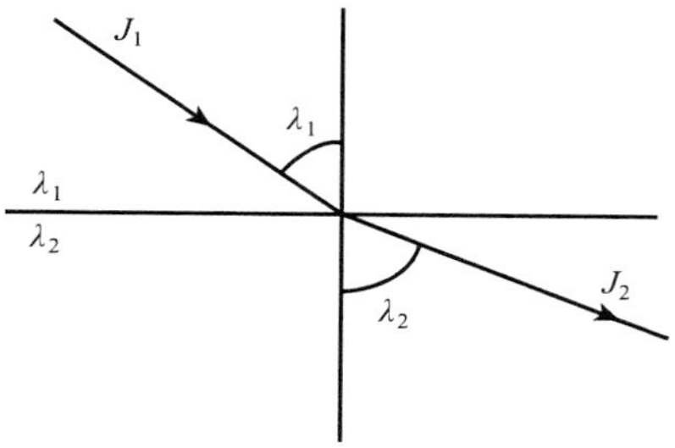

Figure 4.3 Current density vector in the interface

Since the normal component of $\boldsymbol{J}$ is continuous we obtain

$$J_1 \cos\theta_1 = J_2 \cos\theta_2$$

The fact that the tangential component of $\boldsymbol{E}$ is continuous allows us to arrive at

$$\sigma_2 J_1 \sin\theta_1 = \sigma_1 J_2 \sin\theta_2$$

This is followed by the derivation

$$\sigma_2 \tan\theta_1 = \sigma_1 \tan\theta_2 \tag{4.27}$$

4.6 Analogy of the Steady Electric Field and the Electrostatic Field

A comparison of the equations for a steady electric field $\boldsymbol{E}$ and a steady current density $\boldsymbol{J}$ with the equations for electrostatic field $\boldsymbol{E}$ and displacement vector $\boldsymbol{D}$ enables us to derive similarities between the equations and the parameters involved in the equations. The similarities are illustrated in the Table below.

Table 4.1 Comparison of the steady electric field and the electrostatic field

Comparison content	Electrostatic field	Steady electric field	Corresponding parameters
Basic equations in the differential form	$\nabla\times\boldsymbol{E}=0$ $\nabla\cdot\boldsymbol{D}=0$ $\boldsymbol{E}=-\nabla\varphi$	$\nabla\times\boldsymbol{E}=0$ $\nabla\cdot\boldsymbol{J}=0$ $E=-\nabla\varphi$	$E\leftrightarrow E$ $D\leftrightarrow J$
Boundary conditions	$E_{1t}=E_{2t}$ $D_{1n}=D_{2n}$ $\varphi_1=\varphi_2$	$E_{1t}=E_{2t}$ $J_{1n}=J_{2n}$ $\varphi_1=\varphi_2$	
Potential equation	$\nabla^2\varphi=0$	$\nabla^2\varphi=0$	$\varphi\leftrightarrow\varphi$
Basic equation in the integral form and the expression of C and G	$\int_S D\cdot \mathrm{d}\boldsymbol{S}=\int_S \rho s\mathrm{d}\boldsymbol{S}=q$ $C=\frac{q}{U}$	$\int_S J\cdot \mathrm{d}S=I$ $G=\frac{I}{U}$	$q\leftrightarrow I$ $C\leftrightarrow G$
Field and medium	$\boldsymbol{D}=\varepsilon\boldsymbol{E}$	$\boldsymbol{J}=\sigma\boldsymbol{E}$	$\varepsilon\leftrightarrow\sigma$

From Table 4.1 we note that the equations pertaining to an electrostatic field can be altered into the equations for the steady electric field by changing $\boldsymbol{D}$ and ε in the electrostatic field equation to $\boldsymbol{J}$ and σ.

As the potentials of both the steady electric field and the electrostatic field satisfy the Laplace equation, the other field will not need to be solved over again when we know the solution of one field. It can be obtained by replacing the correspondence quantities of the field, which is solved at the same boundary conditions.

The corresponding relationship of fields causes the homologous relationship of parameters such as the correspondence of $\boldsymbol{D}$ and $\boldsymbol{J}$, q and I and the capacitance $C\left(=\frac{q}{U}\right)$ and the conductance $G\left(=\frac{I}{U}\right)$. For the same system we can obtain the conductance G in the steady electric field by replacing ε in the capacitance formula in electrostatic field with σ, and vice versa.

Example 4.3 The radius of the conductor of a two-conductor transmission line is a, the distance between the lines is d, and the conductivity of the medium surrounding the two-conductor transmission line is σ. Determine the drain conductance of the two-conductor transmission line.

Solution As $d >> a$, the per-unit-length capacitance C_0 of the parallel two-conductor transmission line is known as

$$C_0=\frac{\pi\varepsilon}{\ln\left(\frac{d}{a}\right)}\ (\text{F/m})$$

According to the corresponding relationships of $C_0 \leftrightarrow G_0$ and $\varepsilon \leftrightarrow \sigma$ the per-unit-length drain conductance G_0 of the two-conductor transmission line can be obtained by replacing

C_0 and ε in above equation with G_0 and σ, respectively.

$$G_0 = \frac{\pi\sigma}{\ln\left(\frac{d}{a}\right)} \text{ (S/m)}$$

Example 4.4 A conductor sphere with radius a is deeply buried in the soil as a ground installation, and the conductivity of the soil is σ. Determine the ground resistance of the installation.

Solution As shown in Figure of Example 4.4, the conductor is deeply buried, thus the influence of the ground surface to the distribution of the current can be ignored. After flowing into the conductor sphere the current will flow into the soil vertical to the surface of the conductor sphere. This will lead to the soil acting as a resistance. The ground resistance of the metal sphere must be determined when the resistances of the metal lines and the conductor sphere are ignored.

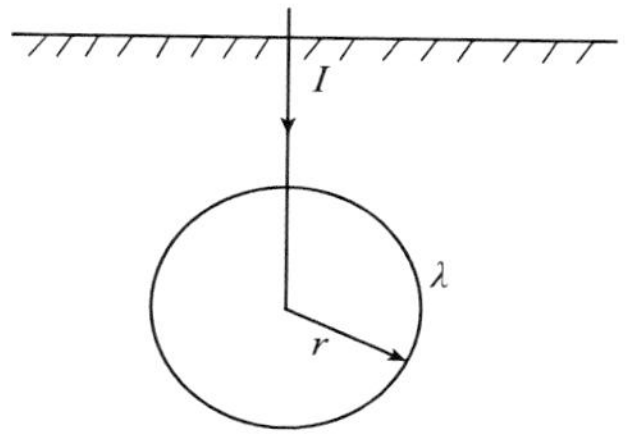

Figure of Example 4.4

Method 1 We determine the current density J_r according to the current I and determine E_r according to $J_r = \sigma E_r$. Subsequently, we obtain the voltage U and the ground resistance R.

Since the influence of the ground surface is ignored, the current flowing out the conductor sphere is in the radial direction and the distribution of the current passing through the sphere surface is uniform in the soil. Hence, the current density J_r is

$$J_r = \frac{I}{4\pi r^2}$$

Where r is the distance from the location of J_r to the center of the conductor sphere. The electric field intensity E_r is given by

$$E_r = \frac{J_r}{\sigma} = \frac{I}{4\pi\sigma r^2}$$

The voltage U of the conductor sphere is

$$U = \int_a^\infty \boldsymbol{E}\cdot \mathrm{d}\boldsymbol{r} = \int_a^\infty E_r \mathrm{d}r = \frac{I}{4\pi\sigma a}$$

The ground resistance R of the conductor sphere is

$$R = \frac{U}{I} = \frac{1}{4\pi\sigma a}$$

Method 2 The electric field intensity E_r is determined according to the charge q. This is followed by a determination of the voltage U and current I. At Last, the ground resistance R can be determined from the aforementioned quantities.

Assume that the total charge q on the sphere surface is uniformly distributed. The influence of the ground surface can be ignored as the metal is deeply buried, then the electric field intensity E_r created by q outside the sphere (in the soil) is given by

$$E_r = \frac{q}{4\pi\varepsilon r^2}$$

The voltage U of the metal sphere is

$$U = \int_a^\infty \boldsymbol{E} \cdot \mathrm{d}\boldsymbol{r} = \int_a^\infty E_r \mathrm{d}r = \frac{q}{4\pi\varepsilon a}$$

The current I flowing into the soil from the metal sphere surface in the radial direction (the direction of e_r) is obtained via

$$\begin{aligned} I &= \oint_S \boldsymbol{J} \cdot \mathrm{d}\boldsymbol{S} = \oint_S J_r \mathrm{d}S = \sigma \oint_S E_r \mathrm{d}s \\ &= \sigma \left(\frac{q}{4\pi\varepsilon r^2} \right) 4\pi r^2 = \frac{\sigma q}{\varepsilon} \end{aligned}$$

The ground resistance R in the soil of the metal sphere is

$$R = \frac{U}{I} = \frac{1}{4\pi\sigma a}$$

Method 3 The resistance R is determined directly according to the definition of resistance

$$\mathrm{d}R = \frac{\mathrm{d}l}{\sigma S} = \frac{\mathrm{d}r}{\sigma 4\pi r^2}$$

Replacing $\mathrm{d}l$, in the above formula with $\mathrm{d}r$ since the current is in the radial direction (i.e. in the direction of r). Then the ground resistance R of the metal sphere in the soil is

$$R = \int \mathrm{d}R = \int_a^\infty \frac{\mathrm{d}r}{\sigma 4\pi r^2} = \frac{1}{4\pi\sigma a}$$

Method 4 The ground resistance R is determined via the analogy between the steady electric field and the electrostatic field.

For a steady electric field, the conductivity of a system in the soil is expressed as

$$G = \frac{I}{U} = \frac{\sigma \oint_S \boldsymbol{E} \cdot \mathrm{d}\boldsymbol{S}}{\int_l \boldsymbol{E} \cdot \mathrm{d}\boldsymbol{l}}$$

Where I is just the current in the soil, the capacitance C of electrostatic is

$$C = \frac{q}{U} = \frac{\varepsilon \oint_S \boldsymbol{E} \cdot \mathrm{d}\boldsymbol{S}}{\int_l \boldsymbol{E} \cdot \mathrm{d}\boldsymbol{l}}$$

For the same system we have

$$G = C \left(\frac{\sigma}{\varepsilon} \right)$$

The capacitance C of an isolated conductor sphere with radius a in an electrostatic field is

$$C = \frac{q}{U} = \frac{q}{\int_a^\infty E_r \mathrm{d}r} = \frac{q}{\int_a^\infty \frac{q}{4\pi\varepsilon r^2} \mathrm{d}r} = 4\pi\varepsilon a$$

Now, the ground resistance R of the conductor sphere in the soil is attained via

$$R = \frac{1}{G} = \frac{1}{C\left(\frac{\sigma}{\varepsilon}\right)} = \frac{1}{4\pi\sigma a}$$

Example 4.5 A half metal sphere with radius a is buried under the ground surface as shown in Figure of Example 4.5. Determine the ground resistance R.

Solution Let's analyze the problem from the concept first. The resistance of the conductor sphere itself can be ignored as it is very small.

The current flowing out the half conductor sphere is in the radial direction (the sphere center is the origin of the spherical coordinates). There are currents in a horizontal direction to the ground surface as shown in Figure of Example 4.5(a). The current distribution is similar to the current distribution of an entire metal sphere in a conducting medium. Thus, the current distribution outside a metal sphere can keep the current on the ground surface in the radial direction (horizontal direction).So the primary boundary conditions are unchanged. As shown in Figure of Example 4.5(b).

Determine the ground resistance R of the half metal sphere buried under the ground surface in Figure of Example 4.5(b).

Method 1 Determine the ground resistance R of the half metal sphere buried under the ground surface using the electrostatic analogy method. (The capacitance C in electrostatic field is corresponding to the conductance G in steady electric field.)

The capacitance of the isolated metal sphere C' in the electrostatic system in Figure of Example 4.5(b) is expressed as

$$C' = \frac{q}{U} = \frac{q}{\displaystyle\int_a^{\infty} E_r \mathrm{d}r} = \frac{q}{\displaystyle\frac{1}{4\pi\varepsilon}\int_a^{\infty}\frac{q}{r^2}\mathrm{d}r} = 4\pi\varepsilon a$$

Replacing ε with the correspondence parameter σ and C' with G' allows the ground conductance G' of the metal sphere to be expressed by

$$G' = 4\pi\sigma a$$

The ground resistance R' of the metal sphere is given by

$$R' = 1/G' = \frac{1}{4\pi\sigma a}$$

The ground resistance R of the half metal sphere buried under the ground surface is

$$R = 2R' = \frac{1}{2\pi\sigma a}$$

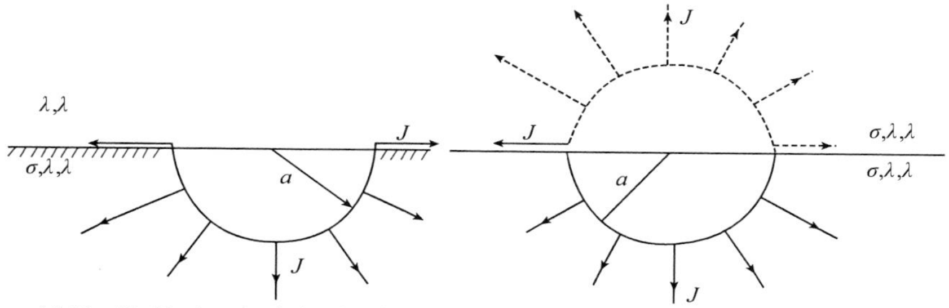

(a) Metal half sphere buried under the ground surface

(b) The mirror image of the metal half sphere

Figure of Example 4.5 Determine the ground resistance of the half metal sphere under the ground surface

Method 2 The ground resistance R of the metal sphere is determined according to $R' = \dfrac{U}{I}$ As shown in Figure of Example 4.5(b), for the entire metal sphere we have

$$U = \int_a^{\infty} E_r \mathrm{d}r = \frac{q}{4\pi\varepsilon}\int_a^{\infty}\frac{\mathrm{d}r}{r^2} = \frac{q}{4\pi\varepsilon a}$$

$$I = \oint_S \boldsymbol{J} \cdot \mathrm{d}\boldsymbol{S} = \oint_S J_r \mathrm{d}s = \int_0^{\pi} \int_0^{2\pi} \sigma E_r \cdot r^2 \sin\theta \mathrm{d}\theta \mathrm{d}\varphi = \frac{\sigma q}{\varepsilon}$$

$$R' = \frac{U}{I} = \frac{q}{4\pi\varepsilon a} \times \frac{\varepsilon}{\sigma q} = \frac{1}{4\pi\sigma a}$$

where I is the current in the soil, and the ground resistance of the half metal sphere buried under the ground surface is given by

$$R = 2R' = \frac{1}{2\pi\sigma a}$$

4.7 Capacitor Considering the Loss of Medium

If the material of the medium filled in the capacitor is lossy, conductivity $\sigma \neq 0$, try to determine the equivalent circuit of the capacitor and the expression of its parameters (R, C).

In order to analyze the equivalent circuit of the capacitor filled with a lossy medium, time varying voltage is applied to the capacitor. The practical conditions of the operating frequency being very low or the distance of the two plates being smaller than a wavelength will cause the magnetic energy to be much smaller than the electric energy. Then the magnetic energy can be ignored, that is the equivalent inductance can be ignored. This equivalent inductance (very small) is series connected with the equivalent resistance R of the lossy medium. The equivalent resistance R should be very big or else the loss is so big that the capacitor may not be used. The series connection of L and R will cause a voltage drop. The equivalent circuit of the lossy capacitor becomes a parallel connection of R and C after ignoring the equivalent inductance. This is depicted in Figure 4.4.

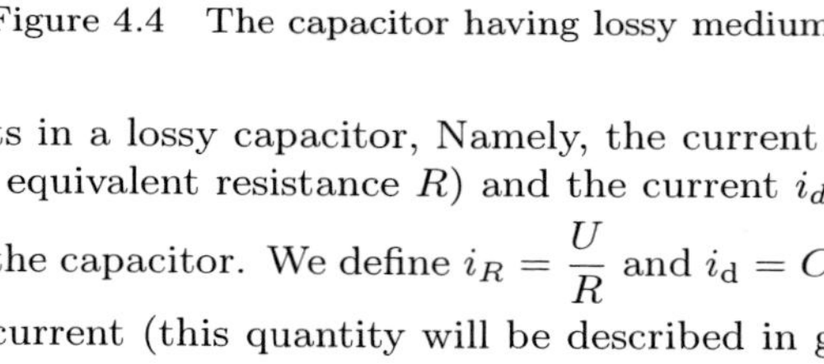

(a) The skeich map of a capacitor (b) Equivalent circuit

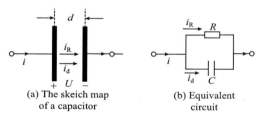

Figure 4.4 The capacitor having lossy medium

There are two currents in a lossy capacitor, Namely, the current i_R which flows through the lossy medium (with equivalent resistance R) and the current i_d which is caused by the time varying voltage in the capacitor. We define $i_R = \dfrac{U}{R}$ and $i_\mathrm{d} = C\dfrac{\mathrm{d}U}{\mathrm{d}t}$ where i_d is referred to as the displacement current (this quantity will be described in greater detail in Section 6.3.3).

In light of the definitions above, the expression for R and C can be expressed as

$$R = \frac{U}{i_R} = \frac{\displaystyle\int_d \boldsymbol{E} \cdot \mathrm{d}\boldsymbol{l}}{\displaystyle\oint_{S+} \sigma \boldsymbol{E} \cdot \mathrm{d}\boldsymbol{S}} \tag{4.28}$$

$$C = \frac{q}{U} = \frac{\oint_{S+} \varepsilon \boldsymbol{E} \cdot \mathrm{d}\boldsymbol{S}}{\int_d \boldsymbol{E} \cdot \mathrm{d}\boldsymbol{l}} \tag{4.29}$$

In the above expressions S_+ is the area of the positive plate of the capacitor, d is the distance of the two plates, σ, ε are the parameters of the medium filled in the capacitor, $\boldsymbol{E}$ is the electric field intensity in the medium filled in the capacitor. Since the medium filled in the capacitor is uniform, and ε and σ can be taken out of the above integrals, we obtain

$$RC = \frac{\oint_{S_+} \varepsilon \boldsymbol{E} \cdot \mathrm{d}\boldsymbol{S}}{\oint_{S_+} \sigma \boldsymbol{E} \cdot \mathrm{d}\boldsymbol{S}} = \frac{\varepsilon}{\sigma} \tag{4.30}$$

Through the relation

$$RC = \frac{\varepsilon}{\sigma} \tag{4.31}$$

According to the formula above, if R is already given, C can be obtained and R can be obtained if C is given.

Example 4.6 Consider a plate capacitor with the plate area $S = (0.01)^2\mathrm{m}^2$ and the distance $d = 10^{-4}\mathrm{m}$ between two plates. The permittivity of the medium between the plates is $\varepsilon = 2.55\varepsilon_0$ and the equivalent resistance of the medium between the plates is measured as $R = 10^4\Omega$. Determine the conductivity σ of the medium.

Solution Different methods can be used to determine the conductivity σ.

Method 1 σ is determined using the equivalent resistance R of the medium.

If the fringing is ignored, the electric field will be uniformly distributed in the capacitor. Then we shall have

$$R = \frac{U}{i_R} = \frac{Ed}{JS} = \frac{d}{\sigma S}$$

with

$$\sigma = \frac{d}{RS} = \frac{10^{-4}}{10^4 \times (0.01)^2} = 10^{-4}(\mathrm{s/m})$$

Method 2 σ is determined using Equation (4.31).

The capacitance of the plate capacitor is determined first.

$$C = \frac{\varepsilon S}{d} = \frac{2.55\varepsilon_0 \times 10^{-4}}{10^{-4}} = 22.58(\mathrm{pF})$$

According to Equation (4.31) we obtain

$$\sigma = \frac{\varepsilon}{RC} = \frac{2.55\varepsilon_0}{10^4 \times 22.58 \times 10^{-12}} = 10^{-4}(\mathrm{s/m})$$

Exercises

4.1 Write the differential form of two basic equations of the steady electric field and the expressions of them in cartesian coordinates.

4.2 Given the integral form of the two basic equations in the steady electric field: $\oint_S J \cdot \mathrm{d}S = 0$, $\oint_l E \cdot \mathrm{d}l = 0$, write J_n and the boundary conditions of E_t determined by the equations.

4.3 The two basic equations for the steady electric field are correct whether the distribution of medium is uniform or not. Please explain why this is so.

4.4 The outer radius of the inside conductor and the inside radius of the outside conductor of a coaxial line are a and b respectively. The medium filled in the coaxial line is imperfect as quantified by a drain conductivity of σ. Determine the per-unit-length drain resistance R of the coaxial line given that the amount of the steady charges of the inside and outside conductors of the unit length are $+\rho_L$ and $-\rho_L$ respectively.

4.5 Given $\boldsymbol{J} = \boldsymbol{e}_x 10y^2z - \boldsymbol{e}_y 2x^2y + \boldsymbol{e}_z 2x^2z(\text{A/m})^2$, determine:

(1) The total current passing through the area $x = 3$, $2 \leqslant y \leqslant 3$, $3.8 \leqslant z \leqslant 5.2$ in the direction of $\boldsymbol{e}_x$;

(2) The current density at the center of the area above;

(3) The mean value of J_x in the area above.

4.6 The current I passing through the conducting thin-line flows downward along the z-axis to a thin conductor layer which is vertical to z-axis. The conductor layer has its center located at $z = 0$. Determine the current density of the thin conductor layer and the current of the 60 degree sector of the thin layer with axis z.

4.7 Consider a steady current flowing through a conducting medium. Given an electric field intensity $\boldsymbol{E}$ in the conducting medium, conductivity $\sigma = \sigma(x, y, z)$, and permittivity $\varepsilon = \varepsilon(x,y,z)$; determine the volume free charge density in the medium.

4.8 A material with conductivity $\sigma_0 + \left(1 + \dfrac{K}{r}\right)$ is filled in the space between two conductor concentric sphere surfaces with radius R_1 and $R_2 (R_1 < R_2)$, respectively. The quantity K is a constant. Determine the resistance of the two perfect conductor sphere surfaces.

4.9 Consider two layers of a medium in a plate capacitor with conductivities σ_1 and σ_2. Given that the thickness of the first layer is d_1, determine the thickness of the second layer d_2 if the power losses are equal in the two layers.

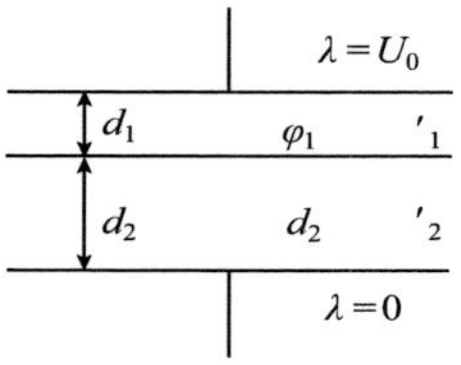

Figure of Exercise 4.10

4.10 As shown in Figure of Exercise 4.10, two pieces of the conducting medium are filled between two plates of the parallel plate capacitor. A voltage of U_0 is applied between the plates. Determine:

(1) $\boldsymbol{E}$, $\boldsymbol{J}$ in the two mediums;

(2) The voltage of each medium piece;

(3) The surface free charge density on the interface of the two mediums and the two plates.

4.11 Consider a conductor plate with thickness d. A sector volume is cut by two circular arcs with radius of r_1 and r_2 and angle α. Given its conductivity σ, determine:

(1) The resistance along the direction of the thickness;

(2) The resistance between the two circular arc areas;

(3) The resistance of the two electrodes in the direction of α.

4.12 Consider a concentric sphere capacitor with the outside radius of the inside sphere equal to a and the inside radius of the outside sphere equal to c. Two layers of medium are filled between the two spheres with interface $r = b$. The permittivity and conductivity of the medium for the inside and outside layer are ε_1, σ_1 and ε_2, σ_2, respectively.

(1) Determine $\boldsymbol{J}$ in the mediums of the two spherical layers and the surface free charge density at $r = a$, b, c, if a voltage of U_0 is applied on the inside and the outside sphere;

(2) Determine the drain resistance of the capacitor.

4.13 Consider a coaxial cable with two medium layers. The interface of the mediums is a coaxial cylinder surface. The radii of the inside conductor, the interface, and the outside conductor are a, b and c, respectively. The permittivity of the two medium layers are ε_1 and ε_2 from inside to outside, and the drain conductivity are given by σ_1 and σ_2. When a voltage V_0 is applied, determine:

(1) The electric field intensity;

(2) The surface free charge density on the interface;

(3) The capacitance and drain conductivity in unit length.

4.14 Consider a volume limited by the sphere surfaces $r = a$, $r = b$ and the conical surfaces $\theta = \theta_0$, $\theta = \pi - \theta_0$ with a medium conductivity of σ. Determine the resistance between the two conical surfaces.

4.15 Two small perfect conductor spheres are located in a limitless uniform medium with conductivity σ. The radii of the spheres are R_1 and R_2, and the distance between the two spheres is d with $d \gg R_1$, $d \gg R_2$. Determine the resistance between the two small spheres.

Chapter 5

Solutions of Electrostatic Field Boundary Value Problem

Boundary value problem consists of a differential equation along with a set of restraints which are referred to as boundary conditions. The boundary conditions correspond to surfaces on which either the surface charge density or the potential is specified. Solution to a boundary value problem is the solution to this differential equation which also satisfies the boundary conditions.

5.1 Electrostatic Field Boundary Value Problems

Generally there are two kinds of electrostatic field problems. One is called a distribution problem, which pertains to electric field intensity and potential distribution directly. For instance, the electrostatic field problems discussed in Chapter 4 can be solved using either Coulomb's law or Gauss's law for the electrostatic field. However, a large number of problems with practical application belong to the second kind of electrostatic field problems, which are called boundary value problems.

Due to materials discontinuities, the equations governing the behavior of the electric fields on either side of an interface are known as the boundary conditions. In this section we consider certain boundary surfaces for which either the surface charge density or the potential is specified. The electric field intensity and potential distribution in materials can be investigated by solving boundary value problems.

There are three kinds of boundary value problems, which are listed follows:

(1) Dirichlet boundary conditions. In this condition, the potential on all boundary of the domains are known. Such problems are referred to as a Dirichlet problem, or a First-type boundary value problem.

(2) Neumann boundary conditions, the orthogonal directional derivative of potential, or called normal derivative, functions on all boundary problems whose domains are known. It is also called the second class boundary value problem. For conductor, the second class boundary problem means that the distribution of surface electric charge density (ρ_s) is known.

(3) Hybrid boundary conditions, which is also be called the third class boundary value problem. Here the potential of some boundary is known, while the normal derivative of potential functions on the rest boundary is known. A normal derivative is a directional derivative taken in the direction normal (that is, orthogonal) to some surface in space.

This is a mixed boundary conditions of the first and second class boundary conditions.

In this chapter, we will discuss how to solve boundary condition problems using Laplace's equation or using the image method for those symmetrical boundaries. Furthermore, conformal mopping method will be introduced to solve those electric field problems with plane-parallel boundaries for two-dimensional field problems. Solution based on numerical finite difference method will be briefly introduced at the end of this chapter

5.2 Uniqueness Theorem

Electrostatic field problems can be solved by a variety of methods. The solution can even be "guessed" based on people's experience. However how to make sure that this obtained solution is correct? The electrostatic field uniqueness theorem states that if one solution satisfies both Laplace's equation (or Poisson's equation) and the given boundary conditions, this solution is unique and accurate.

The term "boundary conditions" mentioned above is generic. To take electrostatic field for an example, boundary conditions generally refer to the following conditions:

(1) Sources on boundary surface such as, electric charge distribution, value of electric charge, electric potential and its distribution.

(2) Boundary conditions, such as, (i)electric field and electric potential on boundary e.g. $E_{1t} = E_{2t}, \varphi_1 = \varphi_2$; (ii) natural boundary e.g. infinitely distant position from source, both electric field intensity and potential would be degraded to zero.

Apagogic method would be used to prove the uniqueness theorem. First, we assume that there are two solutions φ_1 and φ_2 at least which can meet both the Laplace's equation and the given boundary conditions, then we will prove that these two solutions are actually the same one Due to $\varphi_1 = \varphi_2$, solution would be unique.

Green's first identity will be derived below, which will be applied later.

Using the identity

$$\nabla \cdot (\varphi_0 \nabla \varphi_0) = \varphi_0 \nabla^2 \varphi_0 + \nabla \varphi_0 \cdot \nabla \varphi_0$$

Its volume integrals can be simplified using the Divergence theorem, equation can be rewritten as

$$\begin{aligned}\int_V \nabla \cdot (\varphi_\nabla \varphi_0)\mathrm{d}V &= \int_V (\varphi_0 \nabla^2 \varphi_0 + \nabla \varphi_0 \cdot \nabla \varphi_0)\mathrm{d}V \\ &= \oint_S \varphi_0 \nabla \varphi_0 \cdot \mathrm{d}S\end{aligned} \tag{5.1}$$

It is noticed that

$$\nabla \varphi_0 \cdot \mathrm{d}S = \frac{\partial \varphi_0}{\partial n} e_n \cdot e_n \mathrm{d}S = \frac{\partial \varphi_0}{\partial n}\mathrm{d}S \tag{5.2}$$

Then we get

$$\int_V (\varphi_0 \nabla^2 \varphi_0 + |\nabla \varphi_0|^2)\mathrm{d}V = \oint_S \varphi_0 \frac{\partial \varphi_0}{\partial n}\mathrm{d}S \tag{5.3}$$

Expression(5.3)is Green's first identity mentioned above.

Then we will prove the uniqueness theorem.

Method 1 For the first class of boundary value problems, which is the most common boundary problem, electric potential value is given. Green's first identity will be used to prove the uniqueness theorem.

Electric potential on the boundary surface is known, we can use the Apagogic method, which is also called "Reductio ad absurdum" method to prove the uniqueness theorem.

Suppose that there are two solutions φ_1 and φ_2, define

$$\varphi_0 = \varphi_1 - \varphi_2 \tag{5.4}$$

φ_0 stands for the difference of these two solutions. φ_1 and φ_2 satisfy Poisson's equation (or Laplace's equation) we can get

$$\nabla^2 \varphi_0 = \nabla^2 \varphi_1 - \nabla^2 \varphi_2$$

or

$$\nabla^2\varphi_0 = 0 \tag{5.5}$$

Noticing that $\nabla^2\varphi_0 = 0$ and on the boundary surface S we have $\varphi_0 = 0$ (because of same electric potential on the boundary), we can use Green's first identity to obtain the following expression

$$\int_V \nabla\varphi_0 \cdot \nabla\varphi_0 \mathrm{d}V = \int_V |\nabla\varphi_0|^2 \, \mathrm{d}V = 0$$

That is

$$\int_V |\nabla\varphi_0|^2 \, \mathrm{d}V = 0 \tag{5.6}$$

Since $|\nabla\varphi_0|^2 > 0$ using Equation(5.6)we obtain

$$\nabla\varphi_0 = 0 \tag{5.7}$$

which indicates that φ_0 is constant and therefore $\varphi_0 = \varphi_1 - \varphi_2 = \text{const.}$ On the boundary surface $S\varphi_0 = 0$, therefore, $\varphi_1 - \varphi_2 = \text{const.} = 0$ and we have that

$$\varphi_1 = \varphi_2 \tag{5.8}$$

Equation (5.8)states that the solution which satisfies the Laplace's equation and boundary condition is unique.

Method 2 For the mixed boundary value problem, we use Green's first identity to prove the uniqueness theorem.

Given the charge density distribution ρ in the space with volume V, there are n pieces of energized conductors in one limited area. If the electric potential of the surface of conductors $S_1, \cdots, S_k$ are given and the charge density of the rest of the surfaces such as $S_{k+1}, \cdots, S_n$ are given too. This means that the normal derivative of the potential functions are given is the mixed boundary value problem as shown in Figure 5.1.

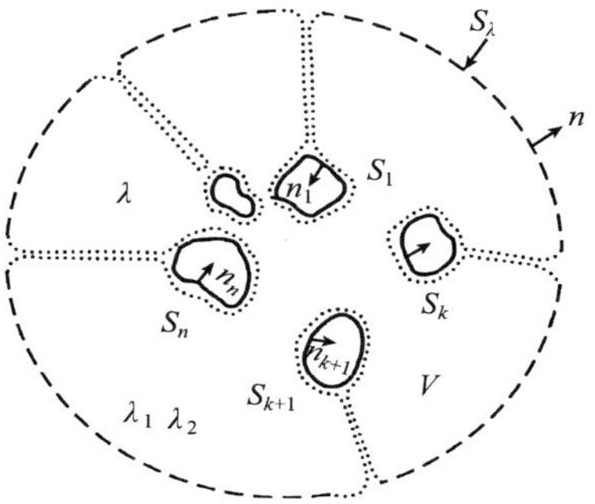

Figure 5.1 Uniqueness theorem under a hybrid boundary condition

We can use reductio ad absurdum method to prove this conclusion. Assuming that there are two electric potentials φ_1 and φ_2 at the same point. Both φ_1 and φ_2 satisfiy the Poisson's equation.

We first write

$$\varphi_0 = \varphi_1 - \varphi_2 \tag{5.9}$$

and note that since

$$\nabla^2\varphi_1 = -\rho/\varepsilon, \quad \nabla^2\varphi_2 = -\rho/\varepsilon$$

$$\nabla^2 \varphi_0 = 0 \tag{5.10}$$

Green's first identity provides that

$$\oint_S \varphi_0 \frac{\partial \varphi_0}{\partial n} \mathrm{d}S = \int_V |\nabla \varphi_0|^2 \, \mathrm{d}V \tag{5.11}$$

The surface integral in the Equation(5.11)covers all of the surfaces surrounding the spatial area. Parameter V is the total space volume extended to the infinity, but except all the conductors. Therefore the surface integral in the Equation(5.11)is composed of three parts. One of the parts is the surface integral of $S_1, \cdots, S_k$, and the second part is the surface integral of $S_{k+1}, \cdots, S_n$. The final part is the surface integral of the infinite surface S_∞which is expressed as

$$\oint_S \varphi_0 \frac{\partial \varphi_0}{\partial n} \mathrm{d}S = \int_{S_1, \cdots, S_k} \frac{\partial \varphi_0}{\partial n} \mathrm{d}S + \int_{S_{k+1}, \cdots, S_n} \varphi_0 \frac{\partial \varphi_0}{\partial n} \mathrm{d}S + \int_{S_\infty} \varphi_0 \frac{\partial \varphi_0}{\partial n} \mathrm{d}S \tag{5.12}$$

We first discuss the surface integral of S_∞ as $r \to \infty$. So long as the surface is large enough, the energized system can be treated as a point charge. Under such a condition we have

$$\varphi_0 \propto \frac{1}{r}, \quad \frac{\partial \varphi_0}{\partial n} \propto \frac{1}{r^2}, \quad \mathrm{d}S \propto r^2$$

Thus as $r \to \infty$ the integral of S_∞ in Equation(5.12)becomes zero. Since the electric potential of the surface $S_1, \cdots, S_k$ are given, φ_1 and φ_2 has no difference on such surface. That is to say $\varphi_0 = \varphi_1 - \varphi_2 = 0$ and the surface integral of these parts is also zero.

The final expression in Equation(5.12)represents the surface integral of $S_{k+1}, \cdots, S_n$. Since the normal derivative of potential functions are given, we note that the surface charge density will be given. Therefore, there will be no difference between the the expressions $\frac{\partial \varphi_1}{\partial n}$ and $\frac{\partial \varphi_2}{\partial n}$, thus

$$\frac{\partial \varphi_0}{\partial n} = \frac{\partial \varphi_1}{\partial n} - \frac{\partial \varphi_2}{\partial n} = 0$$

The surface integral of $S_{k+1}, \cdots, S_n$ is zero, causing Equation(5.12)to equal zero. From Equation(5.11)we have that

$$\int_V |\nabla \varphi_0|^2 \, \mathrm{d}V = 0 \tag{5.13}$$

Since $|\nabla \varphi_0|^2$ is nonnegative we have $\nabla \varphi_0 = 0$ indicating that

$$\varphi_0 = \varphi_1 - \varphi_2 = \text{const} \tag{5.14}$$

Since the electric potential on the surface $S_1, \cdots, S_k$ are given, the constant equals to zero. Therefore, we have

$$\varphi_0 = \varphi_1 - \varphi_2 = \text{const} = 0 \tag{5.15}$$

$$\varphi_1 = \varphi_2 \tag{5.16}$$

5.3 Solving the One-Dimension Field by Integral

For symmetrical problems such as symmetry of flat plane, the axial symmetry of cylindrical surface (body), the hot core symmetry of spherical (body), the selection of an appropriate coordinate system may lead to two-stage three-dimensional partial differential equations being dramatically simplified. Furthermore, in the one-dimensional problem the partial differential equations may simplify to ordinary differential equations. Now we can use integration to solve these equations. The point of this section is to discuss how to find the proper boundary conditions. Then how to use these conditions which we find to solve the constant in equations. And finally solve the equation. We shall now provide several examples.

Example 5.1 The inner radius of a coaxial-line is given by a and the outer radius of is given by b. The coaxial-line is infinite along the z axis. The outer conductor is grounded. The electric potential of the inner conductor is U_0 as shown in Figure of Example 5.1. Find the electric potential and electric field between the coaxial-line.

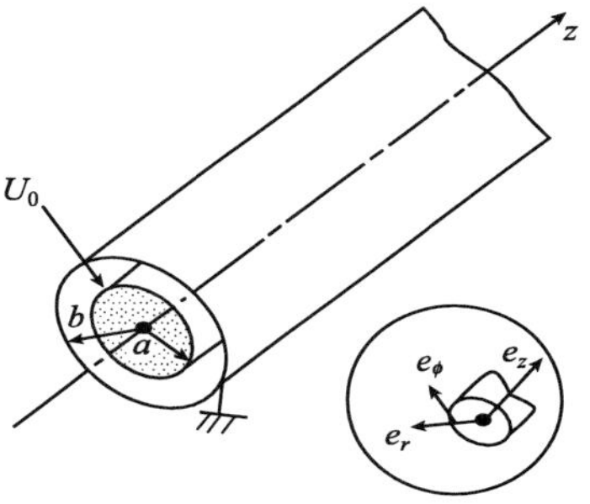

Figure of Example 5.1 Find out the potential field inside the coaxial line

This is the axial symmetry boundary value problem, thus we choose a cylindrical coordinate system. Since the coaxial-line is long enough the electric potential φ is independent of the z coordinate. Furthermore φ is independent of unrelated ϕ coordinate because of the axial symmetry. Therefore φ is related to r then we have

$$\nabla^2\varphi = \frac{1}{r}\frac{\mathrm{d}}{\mathrm{d}r}\left(r\frac{\mathrm{d}\varphi}{\mathrm{d}r}\right) = 0$$

Take integration twice for this equation we have

$$\varphi = C_1 \ln r + C_2 \tag{a}$$

where C_1, C_2 is the undetermined constants.

Using the two given boundary conditions to find C_1, C_2. The two boundary conditions are given by

$$r = a, \quad \varphi = U_0$$

and

$$r = b, \quad \varphi = 0$$

Applying the two boundary conditions in Equation (a) we have

$$C_1 = U_0 / \ln\left(\frac{a}{b}\right)$$

$$C_2 = -U_0 \ln b / \ln\left(\frac{a}{b}\right)$$

Using C_1 and C_2 in Equation (a) we derive the electric potential distribution given by

$$\varphi = \frac{U_0}{\ln \frac{a}{b}} \ln \frac{r}{b}$$

The electric field between the coaxial-line is

$$\boldsymbol{E} = -\nabla\varphi = -\boldsymbol{e}_r \frac{\partial \varphi}{\partial r} = \boldsymbol{e}_r \frac{-U_0}{r \ln \frac{a}{b}}$$

That is

$$E_r = \frac{U_0}{r \ln\left(\frac{b}{a}\right)}$$

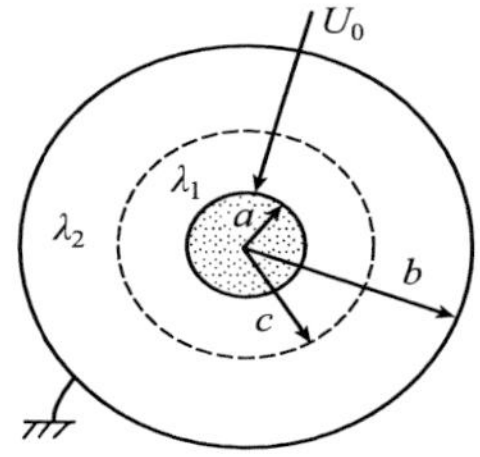

Figure of Example 5.2 A coaxial line that is filled with two kinds of medium

Example 5.2 The inner radius of a coaxial-line is given by a, and the outer radius of is given by b. There are two different media ε_1 and ε_2 between the coaxial-line. The outer conductor is grounded. The electric potential of the inner conductor is U_0 as shown in Figure of Example 5.2. Find the electric potential distribution between the outer and inner conductors.

Since the coaxial-line is infinitely long, the electric potential is unrelated to the z coordinate. The outer and inner conductors and the two media layer are coaxial hence we are faced with an axial symmetry problem. We note that the electric potential φ is unrelated to the ϕ coordinate and φ is related to r. Thus, we have

$$\frac{1}{r}\frac{\mathrm{d}}{\mathrm{d}r}\left(r\frac{\mathrm{d}\varphi}{\mathrm{d}r}\right) = 0 \tag{a}$$

Make integration twice for Equation (a) we have

$$\varphi = C \ln r + C' \tag{b}$$

where C and C' are undetermined constants.

Consider the media ε_1 where $a \leqslant r \leqslant c$, and the media ε_2 where $c \leqslant r \leqslant b$. The electric potential will satisfy the Equation (a) and the solutions of the Laplace equation are given by

$$\varphi_1 = C_1 \ln r + C_2, \quad a \leqslant r \leqslant c \tag{c}$$

$$\varphi_2 = C_3 \ln r + C_4, \quad c \leqslant r \leqslant b \tag{d}$$

There are four undetermined constants in Equation (c) and Equation(d). The constants can be determined via the boundary conditions. There are two obvious boundary conditions.

$$\text{When } r = a, \quad \varphi_1 = U_0 \tag{e}$$

and

$$\text{When } r = b, \quad \varphi_2 = 0 \tag{f}$$

The other two boundary conditions can be obtained on the interface between the two media where $r = c$. One boundary condition is

$$\text{When } r = c, \quad \varphi_1 = \varphi_2 \tag{g}$$

The second boundary condition is when $r = c$, $D_{1n} = D_{2n}$, or $\varepsilon_1 E_{1n} = \varepsilon_2 E_{2n}$or equivalently

$$\text{when } r = c, \quad \varepsilon_1 \frac{\partial \varphi_1}{\partial r} = \varepsilon_2 \frac{\partial \varphi_2}{\partial r} \tag{h}$$

In this problem the normal of interface and the radial cylindrical coordinate have the same direction.

Using the four boundary conditions we have the following relations

(1) via $r = a$, $\varphi_1 = U_0$ and (c),

$$U_0 = C_1 \ln a + C_2 \quad \text{or} C_2 = U_0 - C_1 \ln a \tag{i}$$

(2) via $r = b$, φ_2=0 and (d)

$$C_4 = -C_3 \ln b \tag{j}$$

(3) via $r = c$, φ_1=φ_2 and (c),(d)

$$C_1 \ln c + C_2 = C_3 \ln c + C_4 \tag{k}$$

(4) via $r = c, \varepsilon_1 \dfrac{\partial \varphi_1}{\partial r} = \varepsilon_2 \dfrac{\partial \varphi_2}{\partial r}$ and (c),(d)

$$C_1 \varepsilon_1 = C_3 \varepsilon_2 \tag{l}$$

Using C_1, C_2, C_3, and C_4 we have

$$\varphi_1 = U_0 \left[1 - \frac{\ln(r/a)}{N} \right], \quad a \leqslant r \leqslant c$$

$$\varphi_2 = U_0 \left[\frac{(\varepsilon_1/\varepsilon_2) \ln(b/r)}{N} \right], \quad c \leqslant r \leqslant b$$

The electric field in the two media between the coaxial-line are expressed as

$$\boldsymbol{E}_1 = -\nabla \varphi_1 = \boldsymbol{e}_r \frac{U_0}{rN}, \quad a \leqslant r \leqslant c$$

$$\boldsymbol{E}_2 = -\nabla \varphi_2 = \boldsymbol{e}_r \frac{\left(\dfrac{\varepsilon_1}{\varepsilon_2}\right) U_0}{rN}, \quad c \leqslant r \leqslant b$$

with

$$N = \ln\left(\frac{c}{a}\right) - \frac{\varepsilon_1}{\varepsilon_2} \ln\left(\frac{c}{b}\right)$$

5.4 Using Separation of Variables to Solve Two-Dimension and Three-Dimension Laplace's Equation

In this section we shall introduce the method of separation of variables to solve the two-dimension problem in Cartesian rectangular, cylindrical and spherical coordinates.

5.4.1 Using Separation of Variables to Solve A Rectangular Coordinate Two-Dimension Field Problem

1. Separation of Variables

In rectangular coordinates Laplace's equation can be written as

$$\nabla^2\varphi = \frac{\partial^2\varphi}{\partial x^2} + \frac{\partial^2\varphi}{\partial y^2} + \frac{\partial^2\varphi}{\partial z^2} = 0 \tag{5.17}$$

Suppose that the electric potential φ is a function of x, y and z The boundary surface is given in accordance with the coordinate plane of the chosen axes. If the multi-variate function $\varphi(x, y, z)$ could be written as the product of three single-variate functions, then we would have the product

$$\varphi(x, y, z) = X(x) \cdot Y(y) \cdot Z(z) \tag{5.18}$$

Now, using Equation (5.18) and the product of X, Y, Z to divide Equation (5.18), we can rewrite Equation (5.17) as

$$\frac{1}{X}\frac{\mathrm{d}^2X}{\mathrm{d}x^2} + \frac{1}{Y}\frac{\mathrm{d}^2Y}{\mathrm{d}y^2} + \frac{1}{Z}\frac{\mathrm{d}^2Z}{\mathrm{d}z^2} = 0 \tag{5.19}$$

It appears that in Equation (5.19) the first term only varies with x, the second term only varies with y, and the final term only varies with z. However this is not completely true. Since the other two terms do not vary with x and the sum of three parts is zero, the first part does not vary with x when x changes. Therefore the first part is constant. The same notion holds for the second and the third terms in the above equation. Then the first term, the second term and the third term are constants. Thereofre, all three terms are constant. Suppose the first term, the second term and the third term are k_x^2, k_y^2, k_z^2, then we have

$$\frac{1}{X}\frac{\mathrm{d}^2X}{\mathrm{d}x^2} = k_x^2 \tag{5.20}$$

$$\frac{1}{Y}\frac{\mathrm{d}^2Y}{\mathrm{d}y^2} = k_y^2 \tag{5.21}$$

$$\frac{1}{Z}\frac{\mathrm{d}^2Z}{\mathrm{d}z^2} = k_z^2 \tag{5.22}$$

The above equations represent a system of ordinary differential equations

Using Equation (5.19)~(5.22) we have separated the equation

$$k_x^2 + k_y^2 + k_z^2 = 0 \tag{5.23}$$

After rewriting the multi-variate function φ as the product of three single-variate functions X, Y, Z, and the variable separation, the partial differential equation (Equation

(5.17))becomes the three ordinary differential equations given by Equations(5.20)~(5.22). This technique is referred to as the separation of variables.

2. Two-Dimension Laplace's Equation

We now discuss the solution of two-dimension potential field problem.

We define

$$\varphi(x, y) = X \cdot Y \tag{5.24}$$

then we have

$$\nabla^2\varphi = \frac{\partial^2\varphi}{\partial x^2} + \frac{\partial^2\varphi}{\partial y^2} \tag{5.25}$$

Using Equation(5.24)and the product of $X$and Y; Equation(5.25) can be written as

$$\frac{1}{X}\frac{\mathrm{d}^2X}{\mathrm{d}x^2} + \frac{1}{X}\frac{\mathrm{d}^2Y}{\mathrm{d}y^2} = 0 \tag{5.26}$$

The first term in Equation(5.26)does not vary with y and the sum of the two terms remain zero when y changes. Therefore the two parts are constant and can be expressed as k_x^2 and k_y^2.

We obtain

$$\frac{1}{X}\frac{\mathrm{d}^2X}{\mathrm{d}x^2} = k_x^2 \tag{5.27}$$

$$\frac{1}{Y}\frac{\mathrm{d}^2Y}{\mathrm{d}y^2} = k_y^2 \tag{5.28}$$

and

$$k_x^2 + k_y^2 = 0 \tag{5.29}$$

From Equation(5.29)we have

$$\left.\begin{aligned} |k_x| &= |k_y| = k \\ k_x &= \pm \mathrm{j}k_y \end{aligned}\right\} \tag{5.30}$$

Both k_x and k_y in Equation(5.30)have the same amplitude and one of them ($k_x$and k_y) is real and another is imaginary. Equation(5.27)and Equation(5.28)are linear ordinary differential equations. Their solutions are superposable and the solutions are different functions. In the case of two-dimensions in a rectangular coordinate system the trigonometric function, hyperbolic function and exponential function are rather popular. If we write the solution of Equation(5.27)and Equation(5.28)in hyperbolic function form, then we shall have

$$\varphi = X \cdot Y = (A\mathrm{ch}k_x x + B\mathrm{sh}k_x x) \cdot (C\mathrm{ch}k_y y + D\mathrm{sh}k_y y) \tag{5.31}$$

The exponential function form of the solution is given by

$$\varphi = X \cdot Y = \left(A'\mathrm{e}^{k_x x} + B'\mathrm{e}^{-k_x x}\right) \cdot \left(C'\mathrm{e}^{k_y y} + D'\mathrm{e}^{-k_y y}\right) \tag{5.32}$$

where A, B, C, D andA', B', C', D' in Equations(5.31)and (5.32) are undetermined constants.

Since k_x and k_y in Equations(5.31)and (5.32) have the same amplitude and one of them (k_x and k_y)is real and another is imaginary. We have two states:

(1) When $k_x = k, k_y = \mathrm{j}k$ Equations(5.31)and(5.32)can be written as

$$\varphi_1 = (C_1\mathrm{ch}kx + C_2\mathrm{sh}kx) \cdot (C_3\cos ky + C_4\sin ky) \tag{5.33}$$

$$\varphi_2 = (C_1\mathrm{e}^{kx} + C_2\mathrm{e}^{-kx}) \cdot (C_3\cos ky + C_4\sin ky) \tag{5.34}$$

(2) When $k_x = \mathrm{j}k, k_y = k$ Equations(5.31)and(5.32)can be written as

$$\varphi_3 = (C_1 \cos kx + C_2 \sin kx) \cdot (C_3 \mathrm{ch} ky + C_4 \mathrm{sh} ky) \tag{5.35}$$

$$\varphi_4 = (C_1 \cos kx + C_2 \sin kx) \cdot (C_3 \mathrm{e}^{ky} + C_4 \mathrm{e}^{-ky}) \tag{5.36}$$

If $|k_x| = |k_y| = 0$, from Equations(5.27)and(5.28)we know that the solution of the product is given by $(A_0x+B_0)\cdot(C_0y+D_0)$. Since $\nabla^2\varphi = 0$ is a linear function, Equations(5.33)~(5.36) which are the solutions of function $\nabla^2\varphi = 0$ can add the part $(A_0x + B_0)\cdot(C_0y + D_0)$. Such as

$$\varphi_1 = (A_0x + B_0) \cdot (C_0y + D_0) + (C_1 \mathrm{ch} kx + C_2 \mathrm{sh} kx) \cdot (C_3 \cos ky + C_4 \sin ky) \tag{5.37}$$

When deriving Equations(5.33)~(5.36)we use the following mathematical relationships:

$$\left.\begin{aligned} \mathrm{e}^{\mathrm{j}x} &= \cos x + \mathrm{j} \sin x \\ \mathrm{sh}(\mathrm{j}x) &= \frac{\mathrm{e}^{\mathrm{j}x} - \mathrm{e}^{\mathrm{j}x}}{2} = \mathrm{j} \sin x \\ \mathrm{ch}(\mathrm{j}x) &= \frac{\mathrm{e}^{\mathrm{j}x} + \mathrm{e}^{-\mathrm{j}x}}{2} = \cos x \end{aligned}\right\}$$

Functions hyperbolic sine, sh x and hyperbolic cosine, called ch x are illustrated in Figure 5.2.

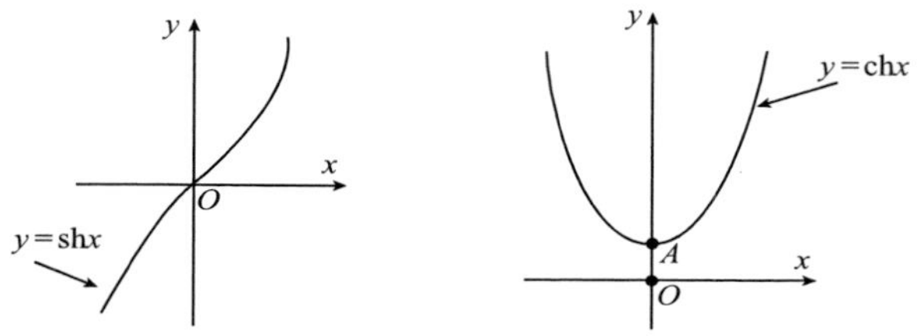

(a) hyperbolic sine (b) hyperbolic cosine

Figure 5.2

The quantities C_1, C_2, C_3, C_4 in Equations (5.33)~(5.36)are undetermined constants and will be determined via the boundary conditions (including boundary value and boundary condition of the field).

Example 5.3 Consider two semi-infinite grounded metal plates which are parallel to each other. The distance between them is given by b, and there are very long strips of metal with an electric potential of U_0 on one end of the two plates. The gap between the plate and the metal strips is small and they are insulated from each other as shown in Figure of Example 5.3(a). Find the electric potential distribution between the two plates.

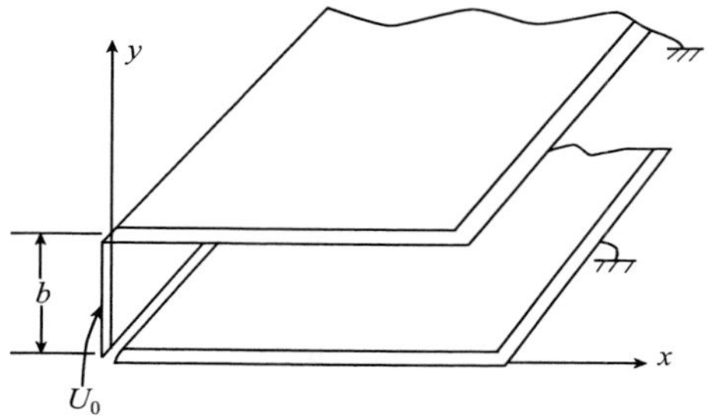

Figure of Example 5.3(a) The parallel metal plate with the side plate electric potential U_0, extends infinitely in the direction of x axis and z axis

Analysis In order to choose the suitable solution form of the electric potential φ, we shall first analyze the characteristics of φ. $\varphi = 0$when y=0, $y = b$. In the direction of y axis φ equals to zero repeatedly. Apparently, φ will vary with y in the trigonometric function form. In the direction of x axis when x=0 $\varphi = U_0$. And when $x \to \infty \varphi = 0$. Then φ varies with x in the exponential function form. Therefore we can write the solution in the following form

$$\varphi = (C_1 \mathrm{e}^{kx} + C_2 \mathrm{e}^{-kx}) \cdot (C_3 \cos ky + C_4 \sin ky)$$

We must now use the given boundary conditions to solve the constants C_1, C_2, C_3, C_4, k and find out the solution of φ. The boundary conditions are given by

(1)when $y = 0, 0 < x < \infty, \varphi = 0$;
(2)when $y = b, 0 < x < \infty, \varphi = 0$;
(3)when $x \to \infty, 0 < y < b, \varphi = 0$;
(4)when $x = 0, 0 < y < b, \varphi = U_0$;

Using the boundary conditions the constants can be represented as follows:
(1) When $y = 0, 0 < x < \infty, \varphi = 0$; using Equation(5.34)we have C_3=0;
(2) When $y = b, 0 < x < \infty, \varphi = 0$; using Equation(5.34)we have

$$k = \frac{n\pi}{b}, \quad n = 1, 2, 3, \cdots \tag{a}$$

(3) When $x \to \infty, 0 < y < b, \varphi = 0$ using Equation(5.34)we have C_1=0 and Equation(5.34)can be written as

$$\varphi = C \sin \frac{n\pi y}{b} \cdot \mathrm{e}^{-\frac{n\pi x}{b}}, \quad n = 1, 2, 3, \cdots \tag{b}$$

where we have the constant$C = C_2 C_4$. The above Equation will satisfy the first three boundary conditions. The last boundary condition is given by $x = 0, 0 < y < b, \varphi = U_0$. Apparently Equation (b) does not satisfy this boundary condition. We use the principle of superposition of the solution of linear differential equations and write the infinite series of Equation (b) as the solution of electric potential φ. Then

$$\varphi = \sum_{n=1}^{\infty} C_n \sin\left(\frac{n\pi y}{b}\right) \cdot \mathrm{e}^{-n\pi x/b} \tag{c}$$

Finally using the last boundary condition we can get the coefficient C_n in Equation (c).
(4) when $x = 0, 0 < y < b, \varphi = U_0$; we have

$$U_0 = \varphi(0, y) = \sum_{n=1}^{\infty} C_n \sin \frac{n\pi y}{b} \tag{d}$$

The next step involves the evaluation of the Fourier series. We can prove mathematically that if the cosine term series are included in Fourier series, Fourier series could be a full group of functions. It shows that any boundary condition could be satisfied by the sum of infinite series. We can deduce the Fourier coefficient C_n.

Both sides of Equation (d) multiple by $\sin \frac{p\pi y}{b}$ where p is integer. Then take integral of y from 0 to b we make

$$\int_0^b U_0 \sin\left(\frac{p\pi y}{b}\right) \mathrm{d}y = \int_0^b \sum_{n=1}^{\infty} C_n \sin\left(\frac{n\pi y}{b}\right) \sin\left(\frac{p\pi y}{b}\right) \mathrm{d}y \tag{e}$$

The left part of Equation(e)can be expressed as

$$\int_0^b U_0 \sin\left(\frac{p\pi y}{b}\right) \mathrm{d}y = \begin{cases} 0, & \text{If } p = 2,4,6,\cdots \\ \dfrac{2bU_0}{p\pi}, & \text{If } p = 1,3,5,\cdots \end{cases} \tag{f}$$

Using the orthogonality of trigonometric functions the right side of Equation (e) is given by

$$\int_0^b C_n \sin\left(\frac{n\pi y}{b}\right) \sin\left(\frac{p\pi y}{b}\right) \mathrm{d}y = \begin{cases} 0, & \text{If } p \neq n \\ C_n \dfrac{b}{2}, & \text{If } p = n \end{cases} \tag{g}$$

We note that equation (f) is equal to Equation (g), and thus we have

$$C_n = \begin{cases} 0, & \text{When } n = 2,4,6,\cdots \\ \dfrac{4U_0}{n\pi}, & \text{When } n = 1,3,5,\cdots \end{cases} \tag{h}$$

Using this value in Equation (c) the solution of the electric potential φ is

$$\varphi = \frac{4U_0}{\pi} \sum_{n=1,3,5,\cdots} \frac{1}{n} \sin\left(\frac{n\pi y}{b}\right) \cdot \mathrm{e}^{-n\pi x/b} \tag{i}$$

We shall use the Fourier series solution of Equation (i) to approximate the boundary condition $x = 0, 0 < y < b, \varphi = U_0$; The results for different values of n are illustrated in Figure of Example 5.3(b).

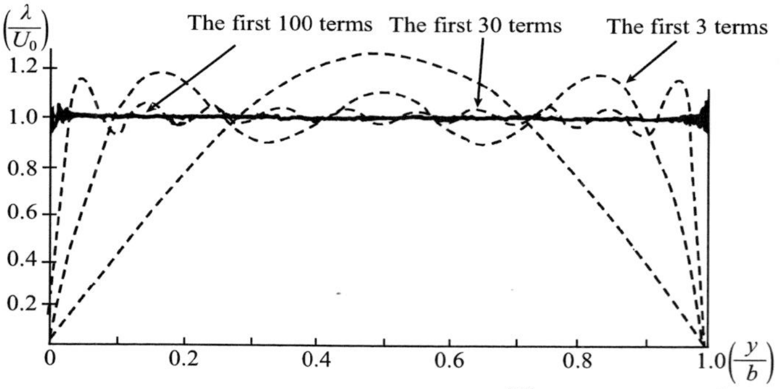

Figure of Example 5.3(b) Approximations of progression $\varphi = \frac{4U_0}{\pi} \sum_{n(\text{Odd number})} \frac{1}{n} \sin\left(\frac{n\pi y}{b}\right)$ to U_0 when taking the first term, first 3 terms, first 30 terms and first 100terms

Example 5.4 Four infinite metal plates which are insulated from each other The distance between the metal plates is small. The four plates constitute a rectangle cavity as shown in Figure of Example 5.4. The length of the rectangle is denoted by a and the width of the rectangle is denoted by b. The electric potential of the top and bottom plates are given by zero. The right plate's electric potential is U_0 while the left plate's normal derivative is zero $\frac{\partial \varphi}{\partial x}$=0. Find the electric potential distribution inside the cavity via two different methods.

Method 1 This is the mixed boundary value problem. Since the metal tube extends infinitely along the z coordinate; the electric field φ does not depend on z coordinate. As shown by the given boundary conditions φ repeatedly is equal to zero along the y coordinate, $\varphi = 0$when y=0, $y = b$ φ varies with y in the trigonometric function form. The distribution of an electric potential is rather complicated along the x coordinate. Specifically, $x = 0, \dfrac{\partial \varphi}{\partial x} = 0; x = a, \varphi = U_0$, and there is no zero point along the x coordinate. Hence, we cannot choose the trigonometric function. While hyperbolic functions such as sh(·) and ch(·) can represent such a characteristic, we shall use the infinite series in Equation(5.33)

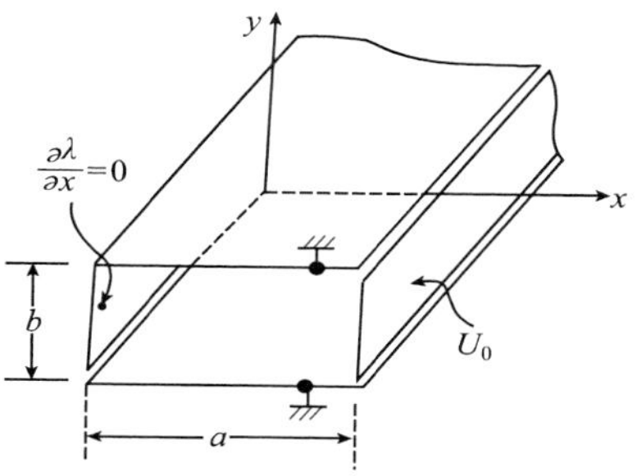

Figure of Example 5.4 Hybrid boundary value problem

$$\varphi = (C_1 \text{ch}kx + C_2 \text{sh}kx) \cdot (C_3 \cos ky + C_4 \sin ky)$$

as the form for the solution. We must now use the given boundary conditions to ensure the constant C_1, C_2, C_3, C_4, k and find a solution for φ. The boundary conditions are

(1)$y = 0, 0 < x < a, \varphi = 0$;

(2)$y = b, 0 < x < a, \varphi = 0$;

(3)$x = 0, 0 < y < b, \dfrac{\partial \varphi}{\partial x} = 0$;

(4)$x = a, 0 < y < b, \varphi = U_0$. $\quad (a)$

By using the boundary conditions the constants will be given by

(1) $y = 0, 0 < x < a, \varphi = 0$;using Equation(5.33)we have C_3=0;

(2) $y = b, 0 < x < a, \varphi = 0$;using Equation(5.33)we have $kb = n\pi$

$$k = \frac{n\pi}{b}, \quad n = 1, 2, 3, \cdots \tag{b}$$

(3) $x = 0, 0 < y < b, \dfrac{\partial \varphi}{\partial x} = 0$ using Equation(5.33)we have

$$\frac{\partial \varphi}{\partial x} |_{x=0} \Rightarrow (C_1 k\text{sh}kx + C_2 k\text{ch}kx) |_{x=0} = 0$$

Therefore

$$C_2 = 0$$

Then Equation(5.33)would be simplified to

$$\varphi = \sum_{n=1}^{\infty} C_n \text{ch}\left(\frac{n\pi x}{b}\right) \sin\left(\frac{n\pi y}{b}\right) \tag{c}$$

Finally through the final boundary condition we can get coefficient C_n in Equation (c).

(4) $x = a, 0 < y < b, \varphi = U_0$;we have

$$U_0 = \sum_{n=1}^{\infty} a_n \sin\left(\frac{n\pi y}{b}\right) \tag{d}$$

where a_n is a Fourier coefficient

$$a_n = C_n \text{ch}\left(\frac{n\pi y}{b}\right) \tag{e}$$

Using the same methodology in Example 5.3 to find a_n

$$a_n = \begin{cases} 0, & \text{When } n = 2,4,6,\cdots \\ \dfrac{4U_0}{n\pi}, & \text{When } n = 1,3,5,\cdots \end{cases} \tag{f}$$

Through Equation (e) and Equation (f) we have

$$C_n = \frac{a_n}{\text{ch}\left(\dfrac{n\pi a}{b}\right)} = \frac{4U_0}{n\pi\text{ch}\left(\dfrac{n\pi a}{b}\right)} \tag{g}$$

The solution of the electric potential φ is obtained as

$$\varphi = \sum_{n=1,3,5,\cdots} \frac{4U_0}{n\pi} \cdot \frac{\text{ch}\left(\dfrac{n\pi x}{b}\right)}{\text{ch}\left(\dfrac{n\pi a}{b}\right)} \sin\left(\frac{n\pi y}{b}\right) \tag{h}$$

Method 2 As mentioned in Method 1, φ does not depend on the z coordinate and varies with y in the trigonometric function form. We have obtained the solution after selecting the hyperbolic function to solve the problem. Subsequently, in the Example 5.4 $x = 0, \dfrac{\partial\varphi}{\partial x} = 0; x = a, \varphi = U_0$ we can choose the exponential function to describe the solution as

$$X = C_1\text{e}^{kx} + C_2\text{e}^{-kx} \tag{i}$$

$\varphi$is represented as

$$\varphi = (C_1\text{e}^{kx} + C_2\text{e}^{-kx}) \cdot (C_3\cos ky + C_4\sin ky)$$

From $x = 0, \dfrac{\partial\varphi}{\partial x}|_{x=0} = 0$we have $C_1 = C_2$ and

$$X = C_1(\text{e}^{kx} + \text{e}^{-kx}) = C\prime\text{ch}kx \tag{j}$$

$C' = 2C_1$is an undetermined constant since

$$\text{ch}kx = \frac{\text{e}^{kx} + \text{e}^{-kx}}{2} \tag{k}$$

Drawing upon Equation (j), Equation(5.34)and considering C_3=0 we have

$$\varphi = \sum_{n=1}^{\infty} C_n\text{ch}(kx)\text{sin}(ky) \tag{l}$$

where $k = \dfrac{n\pi}{b}$ $(n = 1,2,3,\cdots)$.

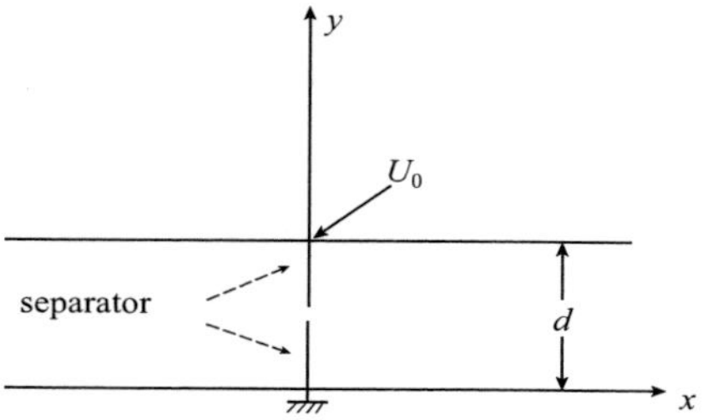

Figure of Example 5.5(a) The metal grooved structure with additional voltage U_0

We note that Equation (l) is equivalent to Equation (c). The solution has been obtained via Method 1

Example 5.5 Along the x, y directions are two very large parallel metal plates. The distance between the two plates is given by d. Along the z direction there exist two metal separators which connect to the top and bottom plate respectively. The two plates constitute a metal tank as shown in Figure of Example5.5. The voltage of the top plate is U_0. Find the electric potential φ in the metal tank.

From Figure of Example 5.5 we know that the potential in the tank varies with x and y and is independent of z. Thus, we have a two-dimension problem. It would facilitate the analysis if we treat this problem as a superposition of two electric fields. We can describe the problem as Figure of Example 5.5(b).

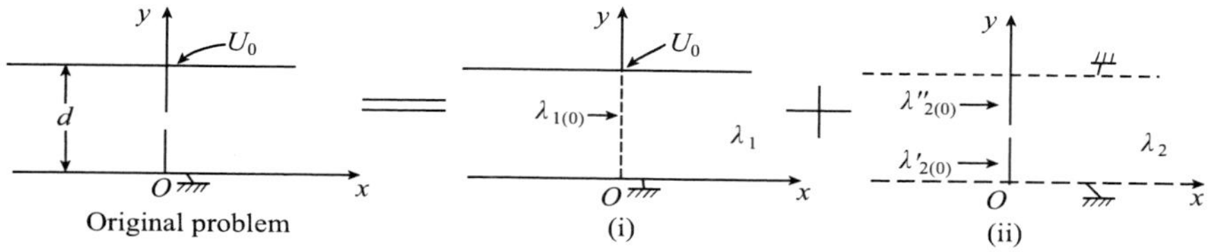

Figure of Example 5.5(b) The decomposition of the original problem

(1) $\varphi_{1(0)}$in Figure (i)

Figure (i) can be treated as a large flat-board capacitor with electric field intensity E_y uniformly distributed as

$$E_y = U_0/d$$

$\varphi_{1(0)}$varies with y. Using the given boundary value we have

$$\varphi_1 = \varphi_{1(0)} = \frac{U_0}{d}y \tag{a}$$

(2) $\varphi'_{2(0)}, \varphi''_{2(0)}$in Figure (ii)

We use the boundary conditions to derive $\varphi'_{2(0)}, \varphi''_{2(0)}$.

$$\begin{aligned} \varphi &= 0, \quad 0 \leqslant y \leqslant \frac{d}{2} \\ \varphi &= U_0, \quad \frac{d}{2} \leqslant y \leqslant d \end{aligned} \tag{b}$$

At x=0 the sum of the two potentials in Figure (i) and Figure(ii) should be equal to the original given boundary value. That is

$$\left.\begin{aligned} \varphi_{1(0)} + \varphi_{2(0)} &= \frac{U_0}{d}y + \varphi_{2(0)} = 0 \\ \varphi_{2(0)} &= -\frac{U_0}{d}y \end{aligned}\right\}, \quad 0 \leqslant y \leqslant \frac{d}{2} \tag{c}$$

and

$$\left.\begin{aligned} \varphi_{1(0)} + \varphi_{2(0)} &= \frac{U_0}{d}y + \varphi_{2(0)} = U_0 \\ \varphi_{2(0)} &= U_0 - \frac{U_0}{d}y \end{aligned}\right\}, \quad \frac{d}{2} \leqslant y \leqslant d \tag{d}$$

On the basis of the uniqueness theorem, the solution to the Laplace's equation that satisfies the boundary conditions of Figure (i) and Figure (ii) will be unique. We have now found that $\varphi_1 = \varphi_{1(0)}$ and can derive φ_2.

Based on the characteristic of the potential distribution in Figure (ii): Along the y axis repeatedly we have zero.In the infinite direction of x when $x \to \infty, \varphi_2 = 0$ infinite along the x and $x \to \infty, \varphi_2 = 0$. φ_2 can be expressed as $\sin(ky){\cdot}\mathrm{e}^{-kx}$. Since we are examining a two-dimensional field we have $|k_x| = |k_y| = k = \dfrac{n\pi}{d}$, and write the solution of φ_2 in the Fourier series form

$$\varphi_2 = \sum_{n=1}^{\infty} A_n \cdot \mathrm{e}^{-\frac{n\pi x}{d}} \cdot \sin\left(\frac{n\pi y}{d}\right) \tag{e}$$

We must now find A_n in the Fourier series.
Drawing upon $\varphi'_{2(0)}, \varphi''_{2(0)}$ and Equation(e) we have

$$\begin{cases} \displaystyle\sum_{n=1}^{\infty} A_n \sin\left(\frac{n\pi y}{d}\right) = -\frac{U_0}{d} y \\ \displaystyle\sum_{n=1}^{\infty} A_n \sin\left(\frac{n\pi y}{d}\right) = U_0 - \frac{U_0}{d} y \end{cases} \tag{f}$$

Using $\sin\left(\dfrac{n\pi y}{d}\right)$ to multiply Equation (f), and evaluating the integral with respect to y we have

$$\int_0^{d/2} A_n \sin^2\left(\frac{n\pi y}{d}\right) \mathrm{d}y + \int_{d/2}^{d} A_n \sin^2\left(\frac{n\pi y}{d}\right) \mathrm{d}y = \int_0^{d} A_n \sin^2\left(\frac{n\pi y}{d}\right) \mathrm{d}y$$
$$= \int_0^{d/2} \left(-\frac{U_0}{d} y\right) \cdot \sin\left(\frac{n\pi y}{d}\right) \mathrm{d}y + \int_{d/2}^{d} \left(U_0 - \left(\frac{U_0}{d}\right) y\right) \cdot \sin\left(\frac{n\pi y}{d}\right) \mathrm{d}y \tag{g}$$

and

$$A_n = \frac{2U_0}{n\pi} \cdot \cos\left(\frac{n\pi}{2}\right) = \begin{cases} 0, & n = 1, 3, 5, \cdots \\ \dfrac{2U_0}{n\pi} (-1)^{n/2}, & n = 2, 4, 6, \cdots \end{cases} \tag{h}$$

$$\varphi_2 = \sum_{n=2,4,\cdots}^{\infty} \frac{2U_0}{n\pi} (-1)^{n/2} \cdot \mathrm{e}^{-\frac{n\pi x}{d}} \cdot \sin\left(\frac{n\pi y}{d}\right) \tag{i}$$

The electric potential in the metal tank is obtained as

$$\varphi = \varphi_1 + \varphi_2 = \frac{U_0}{d} y + \sum_{n=2,4,\cdots}^{\infty} \frac{2U_0}{n\pi} (-1)^{n/2} \cdot \mathrm{e}^{-\frac{n\pi x}{d}} \cdot \sin\left(\frac{n\pi y}{d}\right) \tag{j}$$

5.4.2 Solving the Three-Dimension Field Problem in Cartesian Coordinate By the Method of Separation of Variables

The method for solving problems of a three-dimensional field are similar to those involving a two-dimensional field. The method is also dependent upon the shape of the boundary. First, we must select the appropriate coordinate system. Then, determine the function between the potential and the x, y, z coordinates according to the characteristics of the boundary value. And then we can select the form of the solution of the potential. Finally, determine the constants by means of the given boundary conditions.

Separating the variables of the three-dimensional Laplace's equation yields the three ordinary differential equations

$$\frac{1}{X}\frac{\mathrm{d}^2 X}{\mathrm{d}x^2} = k_x^2$$

$$\frac{1}{Y}\frac{\mathrm{d}^2Y}{\mathrm{d}y^2} = k_y^2 \tag{5.38}$$

$$\frac{1}{Z}\frac{\mathrm{d}^2Y}{\mathrm{d}y^2} = k_z^2$$

as well as the separation equation

$$k_x^2 + k_y^2 + k_z^2 = 0$$

The solution of the three-dimensional potential for φ is

$$\varphi = X(x) \cdot Y(y) \cdot Z(z) \tag{5.39}$$

The relation between φ and x, y, z can be written as an hyperbolic function via

$$\varphi = (C_1 \mathrm{ch} k_x x + C_2 \mathrm{sh} k_x x) \cdot (C_3 \mathrm{ch} k_y y + C_4 \mathrm{sh} k_y y) \cdot (C_5 \mathrm{ch} k_z z + C_6 \mathrm{sh} k_z z) \tag{5.40}$$

or a exponential function as

$$\varphi = (C_1 \mathrm{e}^{k_x^x} + C_2 \mathrm{e}^{-k_x^x}) \cdot (C_3 \mathrm{e}^{k_y^y} + C_4 \mathrm{e}^{-k_y^y}) \cdot (C_5 \mathrm{e}^{k_z^z} + C_6 \mathrm{e}^{-k_z^z}) \tag{5.41}$$

If the quantities k_x, k_y are imaginary numbers, then by Equation(5.23) we will know that $k_z$is real number represented as follows

$$\left.\begin{aligned} k_x &= \mathrm{j}k_x' \\ k_y &= \mathrm{j}k_y' \\ k_z &= k_z' = \sqrt{k_x'^2 + k_y'^2} \end{aligned}\right\} \tag{5.42}$$

Substituting k_x, k_y, k_z given above into Equation(5.40), we arrive at

$$\begin{aligned} \varphi =& (C_1' \cos k_x' x + C_2' \sin k_x' x) \\ & \cdot (C_3' \cos k_x' y + C_4' \sin k_y' y) \cdot (C_5' ch k_z' z + C_6' \mathrm{sh} k_z' z) \end{aligned} \tag{5.43}$$

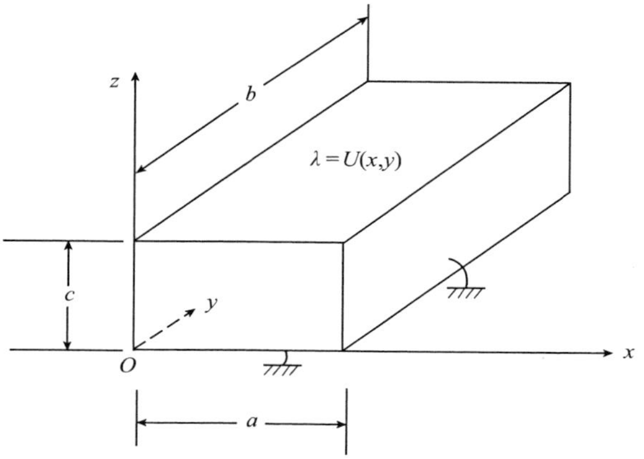

Figure 5.3 Three-dimensional field problems

Consider Figure 5.3 as an example to illustrate the three-dimensional field solution. There is a hexahedron as shown in the Figure 5.3. The potential of the top surface ($z = c$) is

$\varphi = U(x,y)$. The potential of the remaining five surfaces are zero (such as the metal sheet connected with the earth). The size of each surface as shown in Figure 5.3. Find the distribution of the potential at any point within this cube.

Analysis from the given boundary conditions in Figure 5.3, we can find that the potential turn out to be repeatedly zero along the direction of x(x=0, a, φ=0); the potential also turns out to be repeatedly zero along the direction of y ($y = 0, b$, φ=0). Therefore, the relationship between the potential φ and x, y can be described by trigonometric functions. The potential φ=0 when z=0(the bottom surface), and $\varphi = U(x,\ y)$ when $z = c$ (the top surface). We can describe the relationship between z by hyperbolic functions (e.g. sinh, cosh). Based on the above analysis, the potential φ in this case can be expressed in the following form:

$$\begin{aligned}\varphi = &(C_1 \cos k'_x x + C_2 \sin k'_x x) \cdot (C_3 \cos k'_y y + C'_4 \sin k'_y y) \\ &\cdot (C_5 \mathrm{ch} k'_z z + C_6 \mathrm{sh} k'_z z)\end{aligned} \tag{5.44}$$

The above equation is in the same form as Equation (5.43).

We can deduce these constants $C_1 \sim C_6$, and k'_x, k'_y, k'_z. By the given boundary value, the boundary values of the six surfaces are given as follows:

$$\begin{aligned}
&\varphi = 0, && \text{when } x = 0, 0 < y < b, 0 < z < c, \\
&\varphi = 0, && \text{when } x = a, 0 < y < b, 0 < z < c, \\
&\varphi = 0, && \text{when } y = 0, 0 < z < c, 0 < x < a, \\
&\varphi = 0, && \text{when } y = b, 0 < z < c, 0 < x < a, \\
&\varphi = 0, && \text{when } z = 0, 0 < x < a, 0 < y < b, \\
&\varphi = U(x,y), && \text{when } z = c, 0 < x < a, 0 < y < b,
\end{aligned} \tag{5.45}$$

(1) For the three boundary conditions of x=0, φ=0; y=0, φ=0; and z=0, φ=0, from Equation (5.44), we can obtain

$$C_1 = 0, \quad C_3 = 0, \quad C_5 = 0$$

(2) For the second boundary condition $x = a$, φ=0, from Equation (5.44), we can obtain

$$\begin{aligned} &k'_x a = m\pi \\ &k'_x = m\pi/a, \quad m = 1, 2, 3, \cdots \end{aligned} \tag{5.46}$$

(3) For the forth boundary condition $y = b$, φ=0, from Equation(5.44), we can obtain

$$\begin{aligned} &k'_y b = n\pi \\ &k'_y = \frac{n\pi}{b}, \quad n = 1, 2, 3, \cdots \end{aligned} \tag{5.47}$$

(4) From Equations(5.42),(5.46), (5.47), we can obtain

$$k_z = k'_z = \sqrt{k'^2_x + k'^2_y} = \sqrt{\left(\frac{m\pi}{a}\right)^2 + \left(\frac{n\pi}{b}\right)^2} \tag{5.48}$$

Substituting the derived relations $C_1 = 0, C_3 = 0, C_5 = 0, k'_x, k'_y, k'_z$ into Equation(5.44), leads to

$$\varphi_{mn} = C_{mn} \sin\left(\frac{m\pi x}{a}\right) \cdot \sin\left(\frac{n\pi y}{b}\right) \cdot \mathrm{sh}\left[\sqrt{\left(\frac{m\pi}{a}\right)^2 + \left(\frac{n\pi}{b}\right)^2} z\right] \tag{5.49}$$

It should be noted that C_{mn} is a constant in the above equation.

(5) For the sixth boundary condition $z = c$, $\varphi = U(x, y)$, we can deduce the constants and obtain the definite solution through the potential φ taking the form of an infinite series.

We specify

$$\varphi = \sum_{m=1}^{\infty}\sum_{n=1}^{\infty} C_{mn} \sin\left(\frac{m\pi x}{a}\right) \cdot \sin\left(\frac{m\pi y}{b}\right) \cdot \text{sh}\left[\sqrt{\left(\frac{m\pi}{a}\right) + \left(\frac{n\pi}{b}\right)^2} z\right] \tag{5.50}$$

so as to obtain

$$U(x, y) = \sum_{m=1}^{\infty}\sum_{n=1}^{\infty} D_{mn} \sin\left(\frac{m\pi x}{a}\right) \cdot \sin\left(\frac{m\pi y}{b}\right) \tag{5.51}$$

In the above equation we have

$$D_{mn} = C_{mn}\text{sh}\left[\sqrt{\left(\frac{m\pi}{a}\right)^2 + \left(\frac{n\pi}{b}\right)^2} c\right] \tag{5.52}$$

We now find the Fourier series coefficients D_{mn}. The methodology of finding the coefficients is similar to the technique used in the previous example. We multiply both sides of Equation(5.51) by $\sin\left(\frac{p\pi x}{a}\right) \sin\left(\frac{q\pi y}{b}\right)$ and then Integrate with respect to x and y via

$$\begin{aligned}
&\int_0^a \int_0^b U(x, y) \sin\left(\frac{p\pi x}{a}\right) \sin\left(\frac{q\pi y}{b}\right) dxdy \\
&= \int_0^a \int_0^b D_{mn} \sin\left(\frac{m\pi x}{a}\right) \sin\left(\frac{p\pi x}{a}\right) \sin\left(\frac{n\pi y}{b}\right) \sin\left(\frac{q\pi y}{b}\right) dxdy
\end{aligned} \tag{5.53}$$

The right hand side of the above equation is given by

$$\begin{aligned}
&D_{mn} \int_0^a \int_0^b \sin\left(\frac{m\pi x}{a}\right) \sin\left(\frac{p\pi x}{a}\right) \sin\left(\frac{n\pi y}{b}\right) \sin\left(\frac{q\pi y}{b}\right) dxdy \\
&= \begin{cases} 0, & \text{when } p \neq m, q \neq n \\ D_{mn}\dfrac{ab}{4}, & \text{when } p = m, q = n \end{cases}
\end{aligned} \tag{5.54}$$

Applying the above equation and Equation (5.53) allows us to arrive at

$$D_{mn} = \frac{4}{ab} \int_0^a \int_0^b U(x, y) \sin\left(\frac{m\pi x}{a}\right) \sin\left(\frac{m\pi y}{b}\right) dxdy \tag{5.55}$$

Substituting the above expression into Equation(5.52), C_{mn} is given by

$$\begin{aligned}
C_{mn} &= D_{mn} \Big/ \text{sh}\left[\sqrt{\left(\frac{m\pi}{a}\right)^2 + \left(\frac{n\pi}{b}\right)^2} \cdot c\right] \\
&= \left[\frac{4}{ab} \int_0^a \int_0^b U(x, y) \sin\left(\frac{m\pi x}{a}\right) \sin\left(\frac{m\pi y}{b}\right) dxdy\right] \Big/ \text{sh}\left[\sqrt{\left(\frac{m\pi}{a}\right)^2 + \left(\frac{n\pi}{b}\right)^2} \cdot c\right]
\end{aligned} \tag{5.56}$$

Substitution of the expression above into Equation(5.50), allows us arrive at the solution of φ.

Discussions about some special cases of $U(x,y)$.

(1) $U(x, y) = U_0$

From Equation(5.56)we derive

$$C_{mn} = \left(16/nm\pi^2\right) U_0 \Big/ \mathrm{sh}\left[\sqrt{\left(\frac{m\pi}{a}\right)^2 + \left(\frac{n\pi}{b}\right)^2} \cdot c\right], \quad m,n = 1,3,5,\cdots \tag{5.57}$$

Substituting the above equation into Equation(5.50) yields

$$\begin{aligned} \varphi = \frac{16U_0}{mn\pi^2} \sum_{m=1,3,5,\cdots} \sum_{n=1,3,5,\cdots} & \frac{1}{\mathrm{sh}\left[\sqrt{\left(\frac{m\pi}{a}\right)^2 + \left(\frac{m\pi}{b}\right)^2} \cdot c\right]} \\ & \cdot \sin\left(\frac{m\pi x}{a}\right) \sin\left(\frac{n\pi y}{b}\right) \mathrm{sh}\left[\sqrt{\left(\frac{m\pi}{a}\right)^2 + \left(\frac{m\pi}{b}\right)^2} \cdot z\right] \end{aligned} \tag{5.58}$$

(2) $U(x,y) = U_0 \sin\left(\frac{\pi x}{a}\right) \sin\left(\frac{\pi y}{b}\right)$

It is indicated that m=1 and n=1. Besides, since the orthogonality of trigonometric functions we can calculate D_{11} using Equation(5.56). We can also calculate D_{11} by Equation(5.51), as follows

$$U_0 \sin\left(\frac{\pi x}{a}\right) \sin\left(\frac{\pi y}{b}\right) = D_{11} \sin\left(\frac{\pi x}{a}\right) \sin\left(\frac{\pi y}{b}\right)$$

Thus we have

$$D_{11} = U_0 \tag{5.59}$$

and substitute the above equation into Equation(5.52)to obtain C_{11} via the following analysis

$$\begin{aligned} C_{11} &= D_{11}/\mathrm{sh}\left[\sqrt{\left(\frac{\pi}{a}\right)^2 + \left(\frac{\pi}{b}\right)^2} \cdot c\right] \\ &= U_0/\mathrm{sh}\left[\sqrt{\left(\frac{\pi}{a}\right)^2 + \left(\frac{\pi}{b}\right)^2} \cdot c\right] \end{aligned} \tag{5.60}$$

Insertion of the equation above into Equation(5.50)enables us to obtain the following solution for the potential

$$\varphi = \frac{U_0}{\mathrm{sh}\left[\sqrt{\left(\frac{\pi}{a}\right)^2 + \left(\frac{\pi}{b}\right)^2} \cdot c\right]} \sin\left(\frac{\pi x}{a}\right) \sin\left(\frac{\pi y}{b}\right) \cdot \mathrm{sh}\left[\sqrt{\left(\frac{\pi}{a}\right)^2 + \left(\frac{\pi}{b}\right)^2} \cdot z\right] \tag{5.61}$$

(3) $U(x,y) = U_0 \sin\left(\frac{\pi y}{b}\right)$

This equation indicates that n identically equals to 1. From Equation(5.51)we obtain

$$U_0 \sin\left(\frac{\pi y}{b}\right) = \sum_{m=1}^{\infty} C_{m1} \sin\left(\frac{m\pi x}{a}\right) \cdot \sin\left(\frac{\pi y}{b}\right) \mathrm{sh}\left[\sqrt{\left(\frac{m\pi}{a}\right)^2 + \left(\frac{\pi}{b}\right)^2} \cdot c\right]$$

This allows us to obtain

$$U_0 = \sum_{m=1}^{\infty} a_{m1} \sin\left(\frac{m\pi x}{a}\right) \tag{5.62}$$

In the equation

$$a_{m1} = C_{m1}\text{sh}\left[\sqrt{\left(\frac{m\pi}{a}\right)^2 + \left(\frac{\pi}{b}\right)^2} \cdot c\right] \tag{5.63}$$

Using the method of finding the Fourier series coefficients, we obtain a_{m1} in Equation(5.62)as follows

$$a_{m1} = \frac{4U_0}{m\pi}, \quad m = 1, 3, 5, \cdots$$

Insertion of the above equation into Equation(5.63) provides C_{m1}. Subsequently, the substitution of C_{m1} into Equation(5.50) enables us to obtain the potential

$$\varphi = \sum_{m=1,3,5,\cdots} \frac{4U_0}{m\pi\text{sh}\left[\sqrt{\left(\frac{m\pi}{a}\right)^2 + \left(\frac{\pi}{b}\right)^2} \cdot c\right]} \cdot \sin\left(\frac{m\pi x}{a}\right)\sin\left(\frac{\pi y}{b}\right)\text{sh}\left[\sqrt{\left(\frac{m\pi}{a}\right)^2 + \left(\frac{\pi}{b}\right)^2} \cdot z\right] \tag{5.64}$$

5.4.3 Solving the Two-Dimensional Field of Cylindrical-Coordinate Systems by Separation of Variables

Laplace's equation in a cylindrical-coordinate system is expressed as

$$\nabla^2\varphi = \frac{1}{r}\frac{\partial}{\partial r}\left(r\frac{\partial\varphi}{\partial r}\right) + \frac{1}{r^2}\frac{\partial^2\varphi}{\partial\phi^2} + \frac{\partial^2\varphi}{\partial z^2} = 0 \tag{5.65}$$

If the field along the z direction is unchanging and axial symmetric, then we have a one-dimensional field

$$\varphi(r) = C_1\ln r + C_2 \tag{5.66}$$

The solution has been discussed in Part 5.3. We now analyze the solution of a two-dimensional field.

1. The Potential φ is Exclusively A Function of r, z

If the potential is only axisymmetric then it is a two-dimensional field. The Laplace's equation is given by

$$\frac{\partial^2\varphi}{\partial r^2} + \frac{1}{r}\frac{\partial\varphi}{\partial r} + \frac{\partial^2\varphi}{\partial z^2} = 0 \tag{5.67}$$

We define

$$\varphi(r, z) = R(r)Z(z) \tag{5.68}$$

Substituting the above equation into Equation(5.67), and divide by Equation(5.68), leads to an expression for the separation of variables as follows

$$\frac{1}{R}\frac{\mathrm{d}^2R}{\mathrm{d}r^2} + \frac{1}{rR}\frac{\mathrm{d}R}{\mathrm{d}r} + \frac{1}{Z}\frac{\mathrm{d}^2Z}{\mathrm{d}z^2} = 0 \tag{5.69}$$

In the above equation $R(r)$ is solely a function of r, and Z (z) is solely a function of z. The condition that makes the above equations true is

$$\left.\begin{aligned} &\frac{1}{R}\frac{\mathrm{d}^2R}{\mathrm{d}r^2} + \frac{1}{rR}\frac{\mathrm{d}R}{\mathrm{d}r} = -T^2 \\ &\frac{1}{Z}\frac{\mathrm{d}^2Z}{\mathrm{d}z^2} = T^2 \end{aligned}\right\} \tag{5.70}$$

where T is a constant. Solving the two ordinary differential equations, and substitute the solution into Equation(5.68)yields

$$\varphi(r,z) = [C_1\mathrm{J}_0(Tr) + C_2\mathrm{N}_0(Tr)] \cdot [C_3\mathrm{sh}(Tz) + C_4\mathrm{ch}(Tz)] \tag{5.71}$$

When $T = \mathrm{j}\tau$ and τ is real number, the above equation becomes

$$\varphi(r,z) = [C_1' I_0(\tau r) + C_2' k_0(\tau r)] \cdot [C\prime_3 \sin(\tau z) + C_4' \cos(\tau z)] \tag{5.72}$$

The functions $I_0(\tau r)$and$k_0(\tau r)$ denote the first and second variant (virtual variables) Bessel function of zero-order respectively.

2. The Potential φ Is the Function of r, ϕ

If the potential changes with r, ϕ, then it will be a two-dimensional field and Laplace's equation will be given by

$$\nabla^2\varphi = \frac{\partial^2\varphi}{\partial r^2} + \frac{1}{r}\frac{\partial\varphi}{\partial r} + \frac{1}{r^2}\frac{\partial^2\varphi}{\partial\phi^2} = 0 \tag{5.73}$$

Define

$$\varphi(r,\phi) = R(r)\Phi(\phi) \tag{5.74}$$

where $R(r)$ is solely a function of r, and $\Phi(\phi)$is only a function of ϕ. Substituting Equation(5.74) into Equation(5.73), we obtain

$$\Phi\frac{\mathrm{d}^2R}{\mathrm{d}r^2} + \frac{\Phi}{r}\frac{\mathrm{d}R}{\mathrm{d}r} + \frac{R}{r^2}\frac{\mathrm{d}^2\Phi}{\mathrm{d}\phi^2} = 0$$

Multiplying the above equation by $r^2/(R\Phi)$ yields

$$\frac{r^2}{R}\frac{\mathrm{d}^2R}{\mathrm{d}r^2} + \frac{r}{R}\frac{\mathrm{d}R}{\mathrm{d}r} + \frac{1}{\Phi}\frac{\mathrm{d}^2\Phi}{\mathrm{d}\phi^2} = 0 \tag{5.75}$$

It should be noted that the third term in the above equation is not a function of r, In order to ensure that the sum of the three terms is zero, the sum of the first and second terms are also unchanged. Thus we have

$$\frac{r^2}{R}\frac{\mathrm{d}^2R}{\mathrm{d}r^2} + \frac{r}{R}\frac{\mathrm{d}R}{\mathrm{d}r} = k^2 \tag{5.76}$$

and

$$\frac{1}{\Phi}\frac{\mathrm{d}^2\Phi}{\mathrm{d}\phi^2} = -k^2 \tag{5.77}$$

which represent two ordinary differential equations. The quantity k^2 is the separation constant.

The solution of Equation(5.77)can be written as

$$\Phi(\phi) = C_3\sin k\phi + C_4\cos k\phi \tag{5.78}$$

In many practical problems the period of the potential field is 2π. In other words, the potential φ is single-valued when the angle changes 360 degrees, so the potential is given by

$$\varphi(\phi) = \varphi(\phi + 2\pi)$$

Equivalently, we have

$$\varphi(k\phi) = \varphi(k\phi + 2\pi k)$$

Since k must be integer we shall define $k = n$ which transforms Equation(5.78)into

$$\Phi(\phi) = C_3 \sin n\phi + C_4 \cos n\phi \tag{5.79}$$

Set k equal to n, and Equation(5.76)may now be written as:

$$r^2 \frac{\mathrm{d}^2 R}{\mathrm{d}r^2} + r\frac{\mathrm{d}R}{\mathrm{d}r} - n^2 R = 0 \tag{5.80}$$

This is an ordinary differential equation with variable coefficients known as Euler's equation. The solution is known to be

$$R(r) = C_1 r^n + C_2 r^{-n} \tag{5.81}$$

Substituting the above equation and Equation(5.79) into Equation(5.74)yields a general solution of the following form for φ

$$\varphi = (C_1 r^n + C_2 r^{-n}) \cdot (C_3 \sin n\phi + C_4 \cos n\phi) \tag{5.82a}$$

According to the superposition principle we can also express the solution φ in the form of the following series

$$\varphi = \sum_{n=1}^{\infty} (C_{1n} r^n + C_{2n} r^{-n}) \cdot (C_{3n} \sin n\phi + C_{4n} \cos n\phi) \tag{5.82b}$$

Example5.6 Consider an infinite dielectric cylinder along the z direction with radius of a and a dielectric constant of ε_1. The dielectric cylinder is placed in the air with a uniform electric field E_0 along the direction of the x axis. Find the potential and electric fields inside and outside of the dielectric cylinder.

Analysis Since the dielectric cylinder has infinite length along the z axis, bothφ_1(potential inside the cylinder) andφ_2(potential outside the cylinder) are not dependent on the z coordinate. Thus, this is a problem pertaining to a two-dimensional field of cylindrical coordinates. Hence, the basic solution form of the potential φ is can be obtained from Equation(5.82b). We now determine the constants of Equation(5.82b) using the boundary conditions.

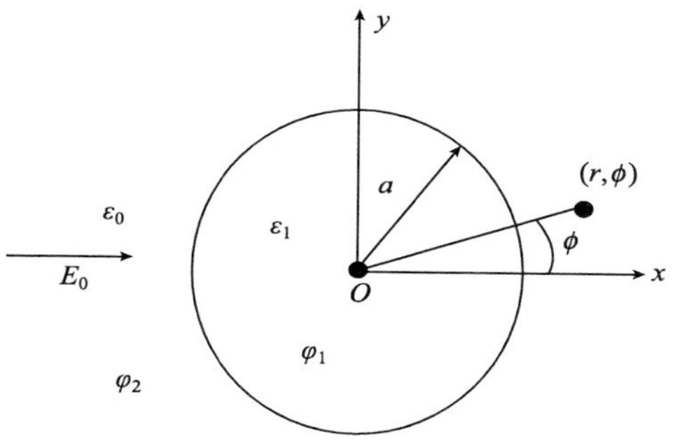

Figure Equation of Example 5.6(a) The cross section of a dielectric cylinder

(1) Here are four boundary conditions.

(i) When $r \to \infty$, $\varphi_2 = -E_0 x = -E_0 r \cos\phi$;

(ii) When $r \to 0$, $\varphi_1$is finite value;

(iii) When $r = a$, $\varphi_1 = \varphi_2$;

(iv) When $r = a$, D_{1n}=D_{2n}, that is $\varepsilon_1 \left.\dfrac{\partial \varphi_1}{\partial r}\right|_{r=a} = \varepsilon_0 \left.\dfrac{\partial \varphi_2}{\partial r}\right|_{r=a}$.

(2) The form of the general solution of the potential inside and outside of the dielectric cylinder.

The form of the general solution of potential inside the dielectric cylinder can be expressed as:

$$\varphi_1 = \sum_{n=1}^{\infty} (C'_{1n} r^n + C'_{2n} r^{-n}) \cdot (C'_{3n} \sin n\phi + C'_{4n} \cos n\phi) \tag{a}$$

The form of the general solution of potential outside the dielectric cylinder can be expressed as:

$$\varphi_2 = \sum_{n=1}^{\infty} (C''_{1n} r^n + C''_{2n} r^{-n}) \cdot (C''_{3n} \sin n\phi + C''_{4n} \cos n\phi) \tag{b}$$

(3) find the determining solution of φ_1, φ_2 by the boundary conditions.

(i) When $r \to \infty$, the impact of the dielectric cylinder on the uniform field can be neglected. Using $\boldsymbol{e}_x E_0 = \boldsymbol{e}_x \left(-\dfrac{\partial \varphi_2}{\partial x}\right)$ we can obtain

$$\varphi_2 = -E_0 x = -E_0 (r \cos\phi) \tag{c}$$

Contrasting the above equation with Equation(b) as $r \to \infty$, we note that $C''_{3n} = 0, n = 1$. Substituting $C''_{3n} = 0, n = 1$ into Equation(b), we obtain

$$\varphi_2 = C''_{11} C''_{41} r \cos\phi + C''_{21} C''_{41} \frac{1}{r} \cos\phi$$

Contrasting the above equation with Equation(c) which was obtained for $r \to \infty$, we note that $C''_{11} C''_{41} = -E_0$ and thus

$$\varphi_2 = -E_0 r \cos\phi + C_2 \frac{1}{r} \cos\phi \tag{d}$$

In the equation$C_2 = C''_{21} C''_{41}$

(ii)When $r \to 0$, φ_1 has a finite value. This allows us to arrive at $C'_{2n} = 0$ from Equation(a), substitute $C'_{2n} = 0$, n=1 and $C'_{3n} = 0$into Equation(a) we can obtain

$$\varphi_1 = C'_{11} C'_{41} r \cos\phi = C_1 r \cos\phi \tag{e}$$

Thus, only the two constants C_1 and C_2 (in Equation(d) and Equation(e), respectively) remain undetermined. Solving C_1 and C_2 will use two boundary conditions given by (iii), and (iv).

(iii)When $r \to a$, $\varphi_1 = \varphi_2$. From Equation(d) and Equation(e), we can obtain

$$C_1 a \cos\phi = -E_0 a \cos\phi + C_2 \frac{1}{a} \cos\phi$$

which simplifies to

$$C_1 a = -E_0 a + \frac{C_2}{a} \tag{f}$$

(iv) When $r \to a, D_{1n} = D_{2n}$. Thus, we have $\varepsilon_1 \left.\dfrac{\partial \varphi_1}{\partial r}\right|_{r=a} = \varepsilon_0 \left.\dfrac{\partial \varphi_2}{\partial r}\right|_{r=a}$, and from Equation(d), Equation(e) we derive

$$C_1 \varepsilon_1 \cos\phi = -\varepsilon_0 E_0 \cos\phi - C_2 \varepsilon_0 \frac{1}{a^2} \cos\phi$$

which simplifies to

$$C_1 \varepsilon_1 = -\varepsilon_0 E_0 - \frac{1}{a^2} C_2 \varepsilon_0 \tag{g}$$

Substituting C_1 and C_2into Equation(e) and Equation(d), respectively. This allows us to obtain the potential of the dielectric cylinder

$$\varphi_1 = \frac{-2\varepsilon_0}{\varepsilon_1+\varepsilon_0}E_0 r\cos\phi \tag{h}$$

$$\varphi_2 = -E_0 r\cos\phi + \frac{(\varepsilon_1-\varepsilon_0)}{r(\varepsilon_1+\varepsilon_0)}a^2 E_0\cos\phi \tag{i}$$

(4) Derive $\boldsymbol{E}_1$ and $\boldsymbol{E}_2$ viaφ_1 andφ_2

After finding the potentialφ_1,$\varphi_2$of the dielectric cylinder we can derive the electrical field strength $\boldsymbol{E}$ inside and outside the dielectric cylinder. Doing so would be more convenient once we take the negative gradient of potential φ_1,φ_2.

The electrical field strength $\boldsymbol{E}_1$ inside the dielectric cylinder is

$$\boldsymbol{E}_1 = -\nabla\varphi_1 = \frac{2\varepsilon_0 E_0}{\varepsilon_1+\varepsilon_0}(\boldsymbol{e}_r\cos\phi - \boldsymbol{e}_\varphi\sin\phi) = \boldsymbol{e}_x\frac{2\varepsilon_0 E_0}{\varepsilon_1+\varepsilon_0} \tag{j}$$

From the above equation, it can be observed that the direction of the electrical field strength inside the dielectric cylinder is the same as the applied electric field $\boldsymbol{E}_0$. Both fields act along the x-axis and are uniform in size. However, the polarized dielectric field will lead to a reversed electric field. Therefore the larger the dielectric constant $\varepsilon_1$of the dielectric cylinder is, the smaller the electric field inside the cylinder is. It's shown as Figure of Example 5.6(b).

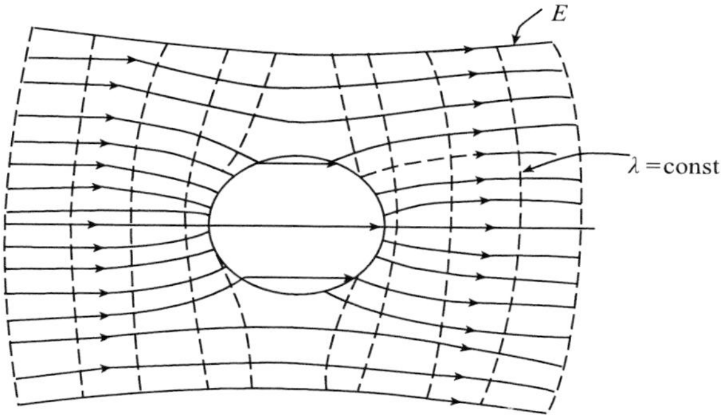

Figure of Example 5.6(b) The dielectric inside and outside of the dielectric cylinder

Since the electric field E is perpendicular to the z coordinate and acts along the x direction, both inside and outside the dielectric cylinder the potentials will satisfy the relation$\varphi(r,\phi) = \varphi(r,-\phi)$ In other words, the potential is an even function of ϕ leading to $C'_{3n} = 0, C''_{3n} = 0$

The electrical field strength $\boldsymbol{E}_2$ outside of the dielectric cylinder is given by

$$\begin{aligned}
\boldsymbol{E}_2 &= -\nabla\varphi_2 \\
&= -\left(\boldsymbol{e}_r\frac{\partial\varphi_2}{\partial r} + \boldsymbol{e}_\phi\frac{1}{r}\frac{\partial\varphi_2}{\partial\phi}\right) \\
&= \boldsymbol{e}_r E_0\cos\phi\cdot\left(1+\frac{\varepsilon_1-\varepsilon_0}{\varepsilon_1+\varepsilon_0}\cdot\frac{a^2}{r^2}\right) + \boldsymbol{e}_\phi E_0\sin\phi\left(\frac{\varepsilon_1-\varepsilon_0}{\varepsilon_1+\varepsilon_0}\cdot\frac{a^2}{r^2}-1\right) \\
&= \boldsymbol{e}_x E_0 + \left[\frac{\varepsilon_1-\varepsilon_0}{\varepsilon_1+\varepsilon_0}\left(\frac{a}{r}\right)^2 E_0(\boldsymbol{e}_r\cos\phi + \boldsymbol{e}_\phi\sin\phi)\right]
\end{aligned} \tag{k}$$

From the above equation , the electric field outside the dielectric cylinder is formed by superposition of the applied uniform electric field $\boldsymbol{E}_0$ (the first term on the right side of the above equation) and the induced electric field (the term in the bracket on the right-hand-side of the above equation) the dielectric cylinder generated after being polarized. We also note that the induced electric field has vanished, and $\boldsymbol{E}_2$ becomes a uniform field when $r \to \infty$. However, the effect of the induction field is notable when r is small. Therefore, the distribution of $\boldsymbol{E}_2$ near the dielectric cylinder is very different from the distribution of the uniform field $\boldsymbol{E}_0$, shown via Figure of Example 5.6(b).

5.4.4 Separation of Variables to Solve A Two-Dimensional Filed with Spherical Coordinates

For a boundary with spherical symmetry, it is more convenient to use spherical coordinates to solve the problem. The Laplace equation in the spherical coordinates is expressed as

$$\nabla^2\varphi = \frac{1}{r^2}\frac{\partial}{\partial r}\left(r^2\frac{\partial\varphi}{\partial r}\right) + \frac{1}{r^2\sin\theta}\frac{\partial}{\partial\theta}\left(\sin\theta\frac{\partial\varphi}{\partial\theta}\right) + \frac{1}{r^2\sin^2\theta}\frac{\partial^2\varphi}{\partial\phi^2} = 0 \tag{5.83}$$

A two-dimensional field is discussed in this session. For instance the potential φ is only a function of $r$and θ.This is an axial symmetric field potential and is independent of the azimuth ϕ. Subsequently, from Equation (5.83), we obtain

$$\frac{\partial}{\partial r}\left(r^2\frac{\partial\varphi}{\partial r}\right) + \frac{1}{\sin\theta}\frac{\partial}{\partial\theta}\left(\sin\theta\frac{\partial\varphi}{\partial\theta}\right) = 0 \tag{5.84}$$

First we use a separation of variables and then solve Equation (5.84). We define

$$\varphi(r,\theta) = R(r)\cdot\Theta(\theta) \tag{5.85}$$

In this equation $R(r)$ is only the function of r, and $\Theta(\theta)$ is only a function of θ. Substituting the above equation into Equation.(5.84) and dividing by the $R\cdot\Theta$, we obtain

$$\frac{1}{R}\frac{\mathrm{d}}{\mathrm{d}r}\left(r^2\frac{\mathrm{d}R}{\mathrm{d}r}\right) + \frac{1}{\Theta\sin\theta}\frac{\mathrm{d}}{\mathrm{d}\theta}\left(\sin\theta\frac{\mathrm{d}\Theta}{\mathrm{d}\theta}\right) = 0 \tag{5.86}$$

Because the first term in the above equation is independent of θ; the equation will not change with θ .The sum of the two terms is zero indicating that the second term does not change with θ either. We conclude that both terms in the equation are constant, thus we have

$$\frac{1}{R}\frac{\mathrm{d}}{\mathrm{d}r}\left(r^2\frac{\mathrm{d}R}{\mathrm{d}r}\right) = k \tag{5.87}$$

and

$$\frac{1}{\Theta\sin\theta}\frac{\mathrm{d}}{\mathrm{d}\theta}\left(\sin\theta\frac{\mathrm{d}\Theta}{\mathrm{d}\theta}\right) = -k \tag{5.88}$$

We shall now discuss the solution of the two ordinary differential equations. Equation(5.87) can be written as

$$r^2\frac{\mathrm{d}^2R}{\mathrm{d}r^2} + 2r\frac{\mathrm{d}R}{\mathrm{d}r} = kR \tag{5.89}$$

The solution of the above equation is given by

$$R = C_1r^n + C_2r^{-(n+1)} \tag{5.90}$$

Substituting the above equation into Equation(5.89), we arrive at

$$k = n(n+1) \tag{5.91}$$

Subsequently, Equation(5.88) can be written as

$$\frac{\mathrm{d}}{\mathrm{d}\theta}\left(\sin\theta\frac{\mathrm{d}\Theta}{\mathrm{d}\theta}\right) + \Theta k\sin\theta = \frac{\mathrm{d}}{\mathrm{d}\theta}\left(\sin\theta\frac{\mathrm{d}\Theta}{\mathrm{d}\theta}\right) + n(n+1)\Theta\sin\theta = 0 \tag{5.92}$$

In order to solve this equation in a convenient manner, we define

$$u = \cos\theta$$

and have

$$\mathrm{d}u = -\sin\theta\mathrm{d}\theta \tag{5.93}$$

$$\mathrm{d}\theta = -\frac{\mathrm{d}u}{\sin\theta}$$

Substituting $\mathrm{d}\theta$ into Equation(5.92) yields

$$\frac{\mathrm{d}}{\mathrm{d}u}\left[\left(1-u^2\right)\frac{\mathrm{d}\Theta}{\mathrm{d}u}\right] + n(n+1)\Theta = 0 \tag{5.94}$$

This equation can be called a Legendre equation. When n is integer, the solution of the Equation(5.49) is polynomial of $\cos\theta$ that is $\Theta = P_n(\cos\theta)+\theta_n(\cos\theta)$. The first and the second terms are called Legendre function of the first kind, and a Legendre function of the second kind, respectively. Drawing upon the relations θ=0, $\cos\theta$=1, $\theta_n(\cos 0°) = \theta_n(1)$ approximate infinity. If the point at which θ=0 is included in the field, the term $\theta_n(\cos\theta)$ in the solution will disappear. As a result Equation(5.94) can be written as

$$\Theta = P_n(u) = P_n(\cos\theta) \tag{5.95}$$

This equation is called the Legendre polynomial.

Substituting Equations(5.95) and(5.90)into Equation(5.85), we can get a series solution of two-dimensional field which exhibits axial symmetry in spherical coordinates

$$\varphi = \sum_{n=0}^{\infty}\left(C_{1n}r^n + C_{2n}r^{-(n+1)}\right)\cdot P_n(\cos\theta) \tag{5.96}$$

For $P_n(\cos\theta)$, n is the order of the Legendre polynomial. When $n = 0, 1, 2\cdots$, Figure 5.4 shows the value of $P_n$and its curve:

$$P_0(\cos\theta) = 1$$
$$P_1(\cos\theta) = \cos\theta$$
$$P_2(\cos\theta) = \frac{1}{2}\left(3\cos^2\theta - 1\right)$$
$$P_3(\cos\theta) = \frac{1}{2}\left(5\cos^3\theta - 3\cos\theta\right)$$
$$P_4(\cos\theta) = \frac{1}{8}\left(35\cos^4\theta - 30\cos^2\theta + 3\right)$$
$$P_5(\cos\theta) = \frac{1}{8}\left(63\cos^5\theta - 70\cos^3\theta + 15\cos\theta\right)$$

Illustration of an important property of Legendre equations: For the Equation(5.91), $n(n+1)=k$. When $n'=-(n+1)$ then:

$$n'(n'+1)=n(n+1)=k$$

This indicates that Equation(5.94) will not change whenn' is replaced by n, So

$$\mathrm{P}_{n'}(\cos\theta)=\mathrm{P}_{-(n+1)}(\cos\theta)=\mathrm{P}_n(\cos\theta)$$

Obviously, when there is a solution to this two-dimensional Laplace's equation

$$\varphi=Cr^n\mathrm{P}_n(\cos\theta)$$

then there will be another solution given by

$$\varphi=C'r^{-(n+1)}\mathrm{P}_{-(n+1)}(\cos\theta)=C'r^{-(n+1)}\mathrm{P}_n(\cos\theta)$$

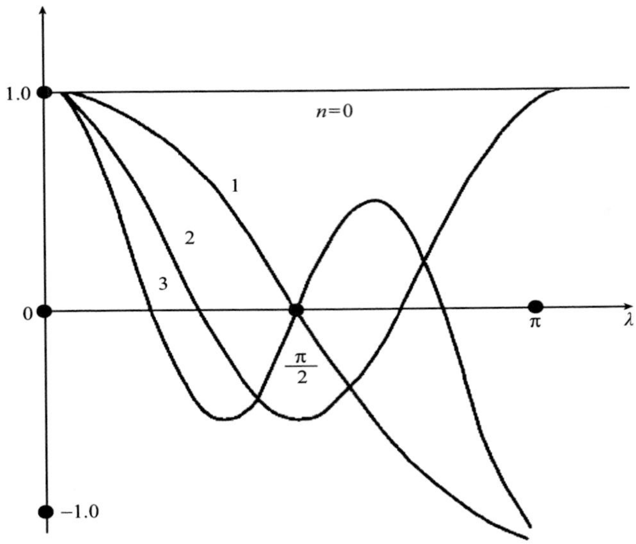

Figure 5.4 Curve of the Legendre polynomial $\mathrm{P}_n(\cos\theta)$

Example 5.7 Media sphere is placed in uniform electric filed. The electric field intensity is $\boldsymbol{E}_0$ and the radius of the sphere is a. The permittivity of the Media sphere is ε and outside the sphere is ε_0. Solve electric field intensity and electric potential of the Media sphere in and outside the sphere respectively.

Solution According to the condition as shown in Figure of Exmaple 5.7(a), electric potential φ doesn't change with ϕ and are only a function of r and θ. Then φ in and outside media sphere can be written according to the Equation(5.96). A series solution of electric potential φ in a media sphere is given by

$$\varphi_1=\sum_{n=0}^{\infty}\left(C_{1n}r^n+C_{2n}r^{-(n+1)}\right)\cdot\mathrm{P}_n(\cos\theta) \tag{a}$$

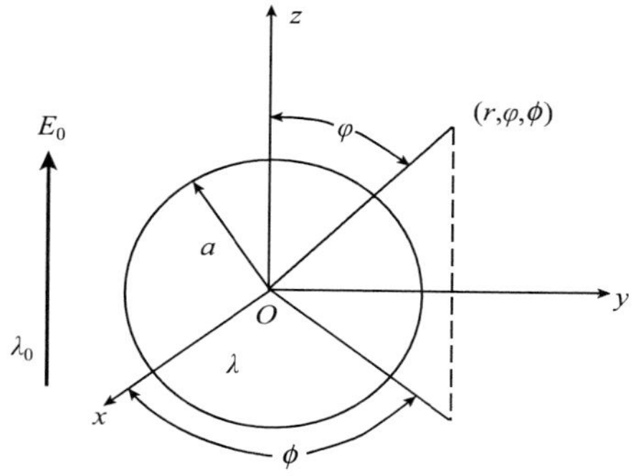

(a) Media sphere in uniform electric field

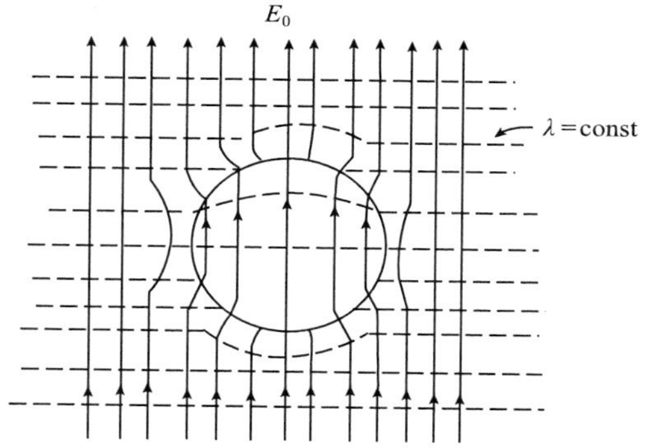

(b) Distribution of the field in and outside media sphere placed in uniform electric field

Figure of Example 5.7

The series solution of electric potential φ outside media sphere is expressed as

$$\varphi_2 = \sum_{n=0}^{\infty} \left(C_{3n} r^n + C_{4n} r^{-(n+1)} \right) \cdot P_n(\cos\theta) \tag{b}$$

Now, all that remains is for us to use the boundary conditions to determine the constants in Equations(a) and (b).

(1) The four boundary conditions are given by:

(i) When $r \to \infty$, $\varphi_2 = -E_0 r \cos\theta$;

(ii) When $r \to 0$, $\varphi_1$is finite;

(iii)When $r = a$, $\varphi_2 = \varphi_1$;

(iv) When $r = a$, $D_{1n} = D_{2n}$, that is $\varepsilon \frac{\partial \varphi_1}{\partial r}|_{r=a} = \varepsilon_0 \frac{\partial \varphi_2}{\partial r}|_{r=a}$. (c)

(2) The constants C_{1n}, C_{2n}, C_{3n} and C_{4n} can be determined using boundary conditions

(i)When $r \to \infty$, the influence of media sphere on uniform electric field can be neglected. So according to $\boldsymbol{E}_0 = \boldsymbol{e}_z E_0 = -\boldsymbol{e}_z \frac{\partial \varphi_2}{\partial z}$, electric potential can be obtained via

$$\varphi_2 = -E_0 z = -E_0 r \cos\theta \tag{d}$$

For Equation(b) when $r \to \infty$, the other terms will be equal to the corresponding terms in Equation(d). $C_{4n} r^{-(n+1)}$ which will disappear too.

$$-E_0 r \cos\theta = \sum_{n=0}^{\infty} C_{3n} r^n \mathrm{P}_n(\cos\theta) \tag{e}$$

From the r in Equation (e) we get n=1. If n=1, the Legendre polynomial is given as $P_1(\cos\,\theta)$=cosθ. Then from Equation(e), there is:

$$C_{3n} = -E_0$$

Substituting n=1 and $C_{3n} = -E_0$ into Equation(b) leads to

$$\varphi_2 = -E_0 r \cos\theta + C_{41} r^{-2} \cos\theta \tag{f}$$

(ii) When $r \to 0$, $\varphi_1$is finite. From Equation(a) we have

$$C_{2n} = 0$$

and thus Equation(a) can be written as

$$\varphi_1 = C_{11} r \cos\theta \tag{g}$$

(iii) When$r = a$, $\varphi_1 = \varphi_2$, we have

$$C_{11} a \cos\theta = -E_0 a \cos\theta + C_{41} a^{-2} \cos\theta$$

and

$$C_{11} a = -E_0 a + C_{41} a^{-2} \tag{h}$$

(iv) When $r = a$, we have $\varepsilon \frac{\partial \varphi_1}{\partial r} = \varepsilon_0 \frac{\partial \varphi_2}{\partial r}$, Substituting φ_1 and φ_2 in Equation(f) and (g) into the above equation, then we can get:

$$\varepsilon C_{11} = -\varepsilon_0 E_0 - 2\varepsilon_0 C_{41} a^{-3} \tag{i}$$

Solving the simultaneous equations given by (h) and (i) we obtain

$$C_{11} = \frac{-3\varepsilon_0}{(\varepsilon + 2\varepsilon_0)} E_0, \quad C_{41} = \frac{(\varepsilon - \varepsilon_0)}{(\varepsilon + 2\varepsilon_0)} E_0 a^3$$

(3) Electric potential φ_1, φ_2 in and outside the media sphere.

Substituting C_{11} and C_{41} into the Equations(g) and (f), the electric potential φ_1 in the media sphere can be obtained

$$\varphi_1 = \frac{-3\varepsilon_0}{(\varepsilon + 2\varepsilon_0)} E_0 r \cos\theta \tag{j}$$

The electric potential φ_2 outside the media sphere can be derived as

$$\varphi_2 = -E_0 r \cos\theta + \frac{(\varepsilon - \varepsilon_0) E_0 a^3}{(\varepsilon + 2\varepsilon_0) r^2} \cos\theta \tag{k}$$

(4) Electric field intensity$\boldsymbol{E}_1$, $\boldsymbol{E}_2$ in and outside the media sphere

The electric field can be derived from the gradient of electric potential after we obtain an expression for φ_1, φ_2. So electric field intensity$\boldsymbol{E}_1$ is given by

$$\begin{aligned} \boldsymbol{E}_1 &= -\nabla\varphi_1 = -\boldsymbol{e}_r \frac{\partial \varphi_1}{\partial r} - \boldsymbol{e}_\theta \frac{1}{r}\frac{\partial \varphi_1}{\partial \theta} \\ &= \frac{3\varepsilon_0}{\varepsilon + 2\varepsilon_0} E_0 (\boldsymbol{e}_r \cos\theta - \boldsymbol{e}_\theta \sin\theta) = \boldsymbol{e}_z \frac{3\varepsilon_0}{\varepsilon + 2\varepsilon_0} E_0 \end{aligned} \tag{l}$$

$$E_1 = E_z = E_0 \frac{3\varepsilon_0}{(\varepsilon + 2\varepsilon_0)}$$

This indicates that the electric field intensity in the media sphere is uniform electric field, and its amplitude is small than the original electric field intensity$\boldsymbol{E}_0$.

The electric field intensity $\boldsymbol{E}_2$ outside the media sphere is expressed as

$$E_2 = e_z E_0 + \left(\frac{\varepsilon - \varepsilon_0}{\varepsilon + 2\varepsilon_0}\right) \cdot \left(\frac{a}{r}\right)^3 \cdot E_0 (e_r 2\cos\theta + e_\theta \sin\theta) \tag{m}$$

There are two portions of the electric field intensity outside the media sphere. One portion is the original electric field intensity $\boldsymbol{E}_0$. The other portion is the electric field generated by the polarization of the media sphere, that is the second term on the right side of the equal sign. Figure of Example 5.7(b) shows the electric potential and the distribution of electric field in and outside the media sphere.

Example 5.8 Metal ball in a uniform electric field. An isolated metal ball with radius a is placed in the uniform electric field whose electric field intensity is given by $\boldsymbol{E}_0$. The metal ball is an equipotential volume and the electric field intensity in the metal ball is zero. The electric field outside the metal ball is a sum of the induction field and original electric field. We shall solve for the electric field intensity and electric potential outside the metal ball.

Analysis Spherical coordinates are chosen due to the spherical boundary involved. Since the metal ball is placed in a uniform electric field, we choose the direction of the polar axis to be the same as the direction of $\boldsymbol{E}_0$. Then, the potential field outside the metal ball is a two-dimensional field $[\varphi(r,\theta)]$that is given by Equation(5.96). The constants C_{1n}, C_{2n}, n can be determined through the boundary conditions.

(1) We have two boundary conditions

(i) When $r \to \infty, \varphi = -E_0 r \cos\theta$

(ii) When $r = a$, $\varphi = 0$ (a)

(2) The constant can be determined through the boundary conditions

(i) When $r \to \infty$, the influence of the metal ball on the uniform electric field disappears, $C_{2n}r^{-(n+1)}$in the Equation(5.96) tend to zero. Thus Equation(5.96) can be written as

$$\varphi = \sum_{n=0}^{\infty} C_{1n} r^n \cdot \mathrm{P}_n(\cos\theta) \tag{b}$$

Compare this equation with $\varphi = -E_0 r \cos\theta$ when $r \to \infty$we get $n = 1$. Furthermore, we have

$$C_{1n} = -E_0 \tag{c}$$

This is due to $\mathrm{P}_1(\cos\theta) = \cos\theta$. Substituting Equation(c) into Equation(5.96) we arrive at

$$\varphi = (-E_0 r + C_{21} r^{-2})\cos\theta \tag{d}$$

(ii) When $r = a$, $\varphi = 0$. From the above equation, we can get

$$0 = \left(-E_0 a + \frac{C_{21}}{a^2}\right)\cos\theta$$

so

$$C_{21} = a^3 E_0 \tag{e}$$

(3) The electric field intensity and electric potential outside the metal ball

Substituting Equation(e) into Equation(d), the electric potential outside the conductive ball is given by

$$\varphi = -E_0 r\cos\theta + E_0\left(\frac{a}{r}\right)^3 r\cos\theta \tag{f}$$

The electric field intensity $\boldsymbol{E}$ can be obtained from the gradient of electric potential

$$\begin{aligned} \boldsymbol{E} &= -\nabla\varphi = e_r E_r + e_\theta E_\theta \\ E_r &= -\frac{\partial\varphi}{\partial r} = E_0\left(1 + \frac{2a^3}{r^3}\right)\cos\theta \\ E_\theta &= -\frac{1}{r}\frac{\partial\varphi}{\partial\theta} = -E_0\left(1 - \frac{a^3}{r^3}\right)\sin\theta \end{aligned} \tag{g}$$

Obviously, if the negative gradient on the two terms in the Equation(f) is used, the first term on the right side of the equality is the original uniform electric field, E_0, that is

$$-\nabla\left(-E_0 r\cos\theta\right)=E_0\left(e_r\cos\theta-e_\theta\sin\theta\right)=e_z E_0 \tag{h}$$

The induction field can be obtained from the negative gradient on the second item on the right side of equal sign in Equation(f). The total electric filed intensity outside the metal ball is a sum of the induction field and uniform electric field. The image of the field distribution in the Figure of Example 5.8

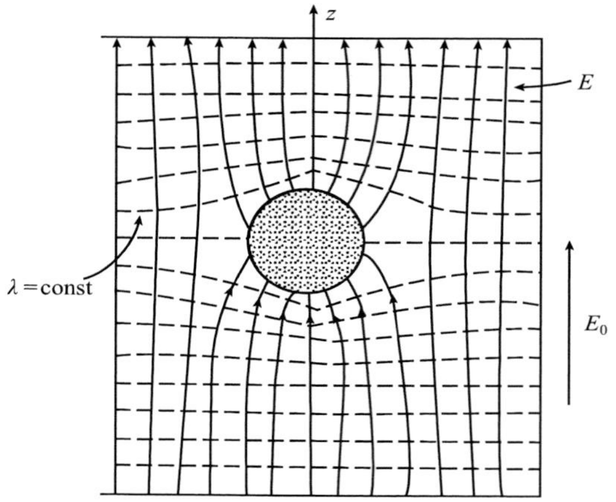

Figure of Example 5.8 Metal ball in the uniform electric field

5.5 Image Method

5.5.1 Image Theory

Solving the Laplace equation is usually difficult for the following situations: ① placing a metal ball near a point charge; ② placing a long and straight live conductor above a huge block of metal plane. If we properly place some charges, referred to as image charges, replace the induced charges that are generated by the original charge on the sphere or plane of the conductor and have an unknown distribution, the electric potential calculated based on the original charge and image charges will not only satisfy the Laplace equation but also satisfy the boundary conditions as long as the original boundary conditions are kept, e.g. the location of the metal sphere still remains equipotential surface. Furthermore, the solution will be unique according to the uniqueness theorem.

As in the case of the mirror image of a charge to a conductor, image theory can be described as substituting the image charges for the induced charges on the surface of the conductor while replacing the conductor by the medium of target field. In other words, *the space consisting of different mediums and boundaries is simplified to a infinite medium space.* Thus, if the boundary value on the surface of the original conductor is guaranteed to be unchanged, we can obtain a unique potential field outside the conductor where *there are no image charges* based on the field source that is made of original charges and image charges.

5.5.2 Flat Image Method

A point charge q is placed in the air above an infinite conductor plane (i.e. in the $y-z$ plane). The conductor plane is connected to the ground, and the distance between the point charge and the plane is given by h as shown in Figure 5.5 (a). In the flat mirror method, the induced charges on the conductor plane surface are replaced by an image charge $-q$ located beneath the conductor plane with a distance h. The conductor plane is then substituted by the air medium as shown in Figure 5.5 (b). Therefore the $y-z$ plane is kept at a zero potential plane, since the original charge q and the image charge $-q$ have the same distance to any point on the $y-z$ plane. This is consistent with the given boundary condition of the infinite conductor plane that connects to the ground. Thus, the image method shown in Figure 5.5 (b) had provided the correct solution. It is worth noting that by using the original charge and the image charge we can only solve a potential field above the conductor plane but not the potential field and electrical field of the region where the conductor is (i.e. below the conductor plane surface). In fact, the conductor is an equal potential body for which the electrical field is always zero.

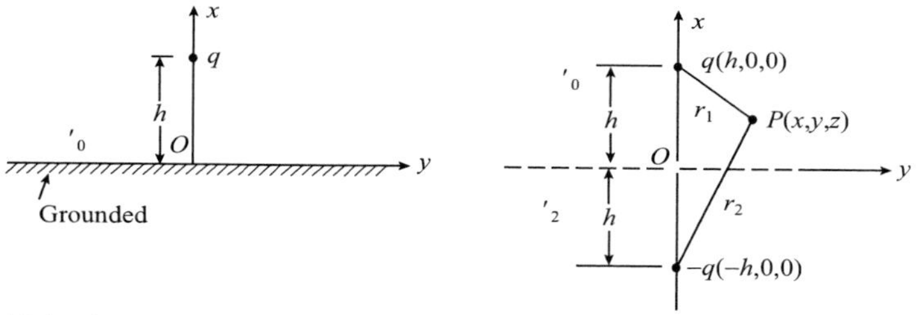

(a) A point charge q is above an infinite grounded conductor plane

(b) Replacing the conductor surface and its corresponding induced charges by an image charge $-q$

Figure 5.5 Flat image of a point charge

Since the original problem is transformed into one in which the space is infinite and the medium is air after the image method, we can write the electric potential φ of point P according to Figure 5.5 (b) as

$$\varphi = \frac{1}{4\pi\varepsilon_0}\left(\frac{q}{r_1} - \frac{q}{r_2}\right) \tag{5.97}$$

where $r_1 = \sqrt{(x-h)^2 + y^2 + z^2}$, $r_2 = \sqrt{(x+h)^2 + y^2 + z^2}$. Substituting r_1 and r_2 into Equation (5.97) to get

$$\varphi = \frac{q}{4\pi\varepsilon_0}\left(\frac{1}{\sqrt{(x-h)^2 + y^2 + z^2}} - \frac{1}{\sqrt{(x+h)^2 + y^2 + z^2}}\right) \tag{5.98}$$

The above equation is an expression for an electric potential of an arbitrary point above the conductor plane except for the location of the charge q where $r_1 = 0$. *It can be proved that the electric potential will satisfy the Laplace equation.*

The electric field can be obtained by taking the gradient of the electric potential. If this requires calculating the induced charges generated by charge q; then we shall only need to obtain the E_x component of the electric field. According to the boundary conditions, the normal electric displacement D_n of the conductor surface is equal to the density of the surface charges ρ_s. That is

$$\rho_s = D_n = \varepsilon_0 E_x \,|_{x=0} \tag{5.99}$$

In fact, the E_x component of the conductor surface (at $x = 0$) can be expressed as

$$E_x|_{x=0} = -\frac{\partial \varphi}{\partial x}|_{x=0} = \frac{-qh}{2\pi\varepsilon_0 (h^2 + y^2 + z^2)^{3/2}} = \frac{-qh}{2\pi\varepsilon_0 (h^2 + R^2)^{3/2}} \tag{5.100}$$

where $R^2 = y^2 + z^2$. Therefore,

$$\rho_s = \varepsilon_0 E_x|_{x=0} = \frac{-qh}{2\pi (h^2 + R^2)^{3/2}} \tag{5.101}$$

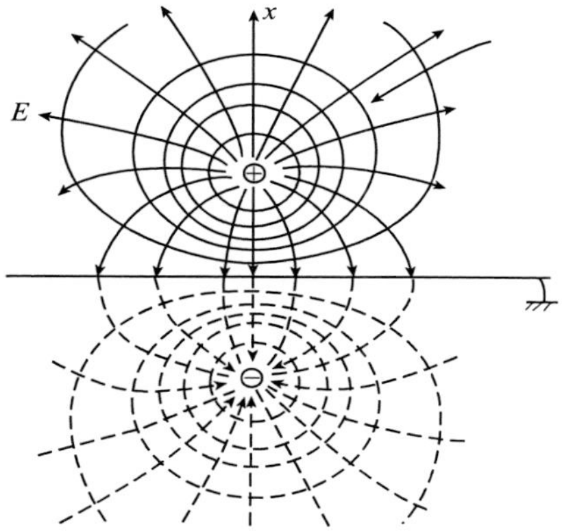

Figure 5.6 The potential field constructed by a point charge and an infinite conductor surface

The total electric charge of the induced charges generated by the original charge q on the surface of the conductor, it can be obtain by the integration of ρ_s to infinite surface on $y - z$ plane. If we let point O be the center of a circle with radius R, the total electric charge (Q) of induced charges can be expressed as

$$Q = \int_S \rho_s \mathrm{d}S = \int_0^{2\pi} \int_0^{\infty} \left[\frac{-qh}{2\pi (h^2 + R^2)^{3/2}}\right] R\mathrm{d}R\mathrm{d}\theta$$

$$= \left.\frac{qh}{(h^2 + R^2)^{1/2}}\right|_0^{\infty} = -q$$

In Figure 5.6, the actual lines above the conductor represent the distributions of the electric fluxline and the equipotential line between the point charge and the infinite conductor surface. The dashed lines below the conductor in the figure are for the virtual image electric field and potential field which do not exist in reality. In fact, the equipotential surface can be obtained by performing a rotation around the x-axis.

In terms of magnitude, the electrostatic attraction between the point charge q and the induced charges on the infinite conductor equals to the force between the charge q and image charge $-q$.

Example 5.9 Consider a long wire which is parallel to the ground and has radius a. The distance from the wire to the ground is h. Find the shunt capacitance of the wire.

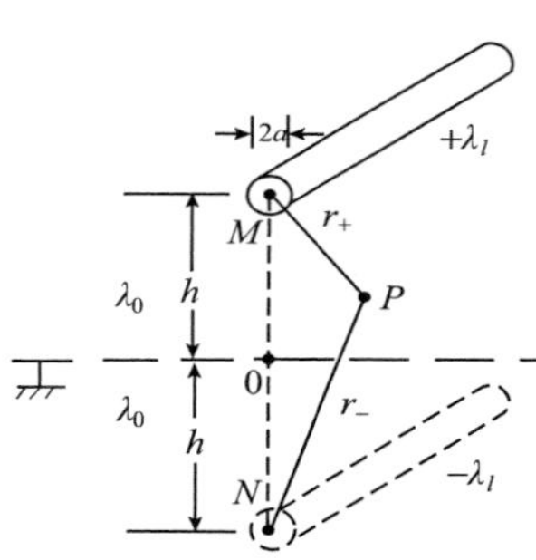

Figure of Example 5.9

Solution The shunt capacitance C_0 of unit length wire can be expressed as

$$C_0 = \frac{\rho_l}{\varphi_{(M)} - \varphi_{(0)}} \tag{a}$$

where $\varphi_{(M)}$ is the electric potential of a single wire, $\varphi_{(0)}$ is the electric potential of the ground ($\varphi_0 = 0$) and ρ_l is the linear electric charge density of the wire. In the following, we calculate $\varphi_{(M)}$.

Let $h \gg a$(so that $+\rho_l$ can be approximately seen as uniformly distributed on the surface of the wire). By using the image method, we remove the ground and induced charges on the ground and add an image single wire ($-\rho_l$) as shown in Figure 5.9. *The electric potential of an arbitrary point above the original ground can be accordingly expressed as*

$$\varphi_P = \frac{\rho_l}{2\pi\varepsilon_0} \ln \frac{r_-}{r_+} \tag{b}$$

The above equation is the expression for a potential of an arbitrary point in space for a parallel leads system. In this system, r_- and r_+ represent the vertical distances from the image wire and the original wire to the point Prespectively. From Equation(b) we have

$$\varphi_{(M)} = \frac{\rho_l}{2\pi\varepsilon_0} \ln \frac{2h-a}{a}$$

Substituting the above equation into Equation(a), we get the shunt capacitance of the wire as follows

$$C_0 = \frac{2\pi\varepsilon_0}{\ln \dfrac{2h-a}{a}} \approx \frac{2\pi\varepsilon_0}{\ln \dfrac{2h}{a}} (\text{F/m}) \tag{c}$$

Accordingly, one can write out an expression for the unit length capacitance C_0' of the parallel leads system as

$$C_0' = \frac{C_0}{2} = \frac{\pi\varepsilon_0}{\ln \dfrac{D-a}{a}} \approx \frac{\pi\varepsilon_0}{\ln \dfrac{D}{a}} (\text{F/m}) \tag{d}$$

where D denotes the distance between the two parallel wires which in this example is $2h$.

For some transmission system such as a single line with ground, double lines, coaxial line, micro-strip line, and strip line, there is a practical value to calculate the capacitance distribution. Knowledge of the capacitance value has practical meaning. More specifically, the unit length capacitance value is involved in the study of an important parameter: the characteristic impedance in TEM transmission systems.

Example 5.10 Find the unit length capacitance C_0 of a pair of homogeneous transmission lines that are parallel with the ground.

Solution In Figure of Example5.10, the cross sections of the pair of homogeneous transmission lines that are parallel with the ground are marked as number 1 and 2. The line radius is a, and the distance between the two lines is D. The quantity h represents the distance from the line to the ground.

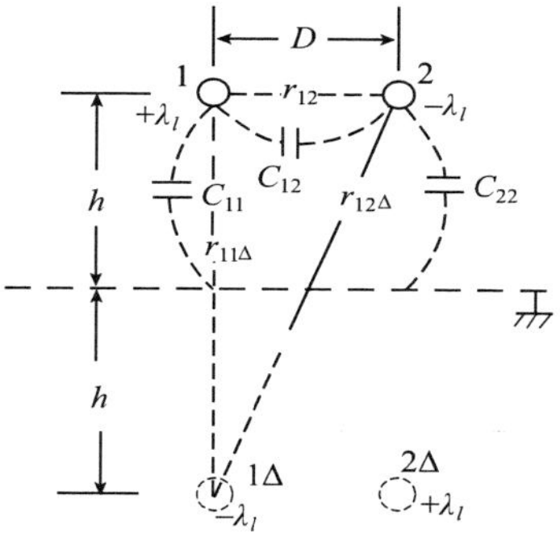

Figure of Example 5.10

Method1 Find C_0 using the partial capacitance concept. Let $1'$ and $2'$ denote the image of lines 1 and 2, respectively. The electric charges of these lines are marked in the figure. Assume that $D \gg a$ and $h \gg a$. C_0 can be seen as the parallel connection of C_{12} and the series capacitor of C_{11} and C_{22} which are the shunt capacitances of line 1 and line 2 respectively. Thus, C_0 can be expressed as

$$C_0 = \frac{C_{11}C_{22}}{C_{11}+C_{22}} + C_{12} \tag{a}$$

In the following analysis we shall derive the value of C_{11}, C_{22} and C_{12}.

Let ρ_{l1} and ρ_{l2} represent the charge density ρ_l of line 1 and the charge density $-\rho_l$ of line 2, respectively. Based on the concept of partial capacitance we have

$$\rho_{l1} = C_{11}\varphi_1 + C_{12}(\varphi_1 - \varphi_2) \tag{b}$$

$$\rho_{l2} = C_{21}(\varphi_2 - \varphi_1) + C_{22}\varphi_2 \tag{c}$$

where φ_1 and φ_2 can be expressed according to Figure of Example 5.10 as

$$\varphi_1 = \frac{1}{2\pi\varepsilon_0}\left[\rho_{l1}\ln\frac{2h}{a} - \rho_{l2}\ln\frac{D}{\sqrt{(2h)^2+D^2}}\right] \tag{d}$$

$$\varphi_2 = \frac{1}{2\pi\varepsilon_0}\left[\rho_{l1}\ln\frac{\sqrt{(2h)^2+D^2}}{D} - \rho_{l2}\ln\frac{a}{2h}\right] \tag{e}$$

From Equations(d) and (e) we can derive new expressions for ρ_{l1} and ρ_{l2} which are similar to Equations(b) and (c). By comparing the two expression of ρ_{l1} and ρ_{l2} we can obtain the following results

$$C_{11} = C_{22}, \quad C_{12} = C_{21} \tag{f}$$

We also have

$$C_{11} = 2\pi\varepsilon_0 \Big/ \ln\left(\frac{2h}{a}\right) - \ln k \tag{g}$$

$$C_{11} = 2\pi\varepsilon_0 \cdot \ln\left(\frac{1}{k}\right) \Big/ \left(\ln\frac{2h}{a}\right)^2 - (\ln k)^2 \tag{h}$$

where $k = D/\sqrt{(2h)^2+D^2}$.

Substituting Equations(g) and (h) into Equation (a) yields

$$C_0 = \frac{\pi\varepsilon_0}{\ln\left(\frac{2h}{a}\right) + \ln k}(\mathrm{F/m}) \tag{i}$$

The above equation is an expression of the unit length capacitance of parallel leads which takes into account the influence of the ground. If the two lines are far from the ground, i.e. $h \gg D$, the influence of the ground can be neglected and the above equation can be reformulated as

$$C_0 \approx \frac{\pi\varepsilon_0}{\ln\left(\frac{2h}{a}\right)}(\mathrm{F/m}) \tag{j}$$

Equation (j) is the unit length capacitance of the parallel-wire transmission lines and is the same as Equation (d) in Example 5.9.

Method2 C_0 can be directly solved by examining the ratio of the *total potential and total charge* ρ_l of the two lines which are worked out after image method. By considering the image we arrive at the relation

$$C_0 = \frac{\rho_l}{\varphi_1 - \varphi_2} \tag{k}$$

where φ_1 andφ_2 are the electric potential of line 1 and line 2, respectively. In fact, φ_1 and ϕ_2 have been already solved for via method 1 of this example. Substituting Equations(d) and (e) into Equation(k) yields

$$C_0 = \frac{\rho_l}{\varphi_1 - \varphi_2} = \frac{\rho_l}{2 \times \dfrac{\rho_l}{2\pi\varepsilon_0}\left[\ln\left(\dfrac{2h}{a}\right) + \ln k\right]} = \frac{\pi\varepsilon_0}{\ln\left(\dfrac{2h}{a}\right) + \ln k} \tag{l}$$

which is consistent with Equation (i).

5.5.3 Sphere Image

1. The Size of the Sphere Image Charge and Its Coordinates

Consider a grounded metal ball with radius a. When there is a point charge q that has a distance D to the center of the ball, what is the electric potential of an arbitrary point outside the ball?

The solution to such a problem involves finding the size of the image charge and its location. Subsequently, it would be straight-forward to derive the potential field outside the ball.

As mentioned above, the key step is to find the image charge. To do so, we first assume an image charge exists inside the ball at point N. The size of the charge is q' as shown in Figure 5.7. Let $ON = d$.

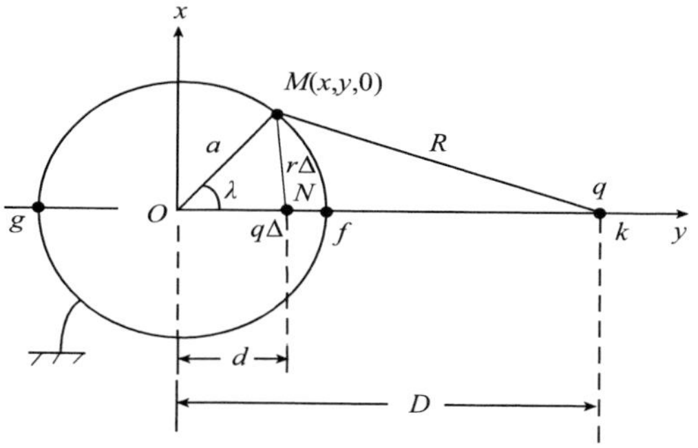

Figure 5.7 Sphere image

There are different approaches for finding the magnitude of q' and its location.

Method 1 Select an arbitrary point M on the surface of the metal ball. We draw upon the fact that the potential of M is always zero regardless of location to obtain q' and d.

Let the distances from the original charge q and image charge q' to M be R and r', respectively. Since the metal ball connects to the ground, the potential of M is equal to zero. That is

$$\varphi_{(M)} = \frac{q}{4\pi\varepsilon_0 R} + \frac{q'}{4\pi\varepsilon_0 r'} = 0 \tag{5.102}$$

with

$$\left.\begin{aligned} R^2 &= a^2 + D^2 - 2aD\cos\theta \\ r'^2 &= a^2 + d^2 - 2ad\cos\theta \end{aligned}\right\} \tag{5.103}$$

Substituting the above equations into Equation (5.102) and after some simplification we have

$$\left[q^2\left(d^2 + a^2\right) - q'^2\left(D^2 + a^2\right)\right] + 2a\cos\theta\left(q'^2 D - q^2 d\right) = 0 \tag{5.104}$$

The above equation will always hold regardless of the location of M. This is because the accuracy of the equation is independent to θ. Thus, the conditions for the validity of Equation (5.104) are

$$\left.\begin{aligned} &q^2\left(d^2 + a^2\right) - q'^2\left(D^2 + a^2\right) = 0 \\ &q'^2 D - q^2 d = 0 \end{aligned}\right\} \tag{5.105}$$

There are two solution for Equations (5.105). One of the solutions is given by

$$d = D, \quad q' = -q \tag{5.106}$$

However, the above solution can not be possible. The other solution is given by

$$d = a^2/D \tag{5.107}$$

$$q' = q\frac{a}{D} \tag{5.108}$$

Method 2 The electric potential at the intersection between the points (f, g) of the sphere and the y axis are the same. The use of this fact allows us to calculate d and q'.

The electric potential at point f is equal to zero

$$\varphi_f = \frac{q}{4\pi\varepsilon_0 (D-a)} + \frac{q'}{4\pi\varepsilon_0 (a-d)} = 0 \tag{5.109}$$

The electric potential at point g is equal to zero as above equation

$$\varphi_g = \frac{q}{4\pi\varepsilon_0 (D+a)} + \frac{q'}{4\pi\varepsilon_0 (a+d)} = 0 \tag{5.110}$$

Combining the two equations above, we have the relation

$$d = \frac{a^2}{D}, \quad q' = -q\frac{a}{D} \tag{5.111}$$

Method 3 Due to the fact that the electrical potential will remain unchanged on the sphere, we can use a rectangular coordinate system to represent R and r' in Figure 5.7 and also calculate the value and position of the image charge.

Suppose that the electrical potential at point M equals zero. We shall have

$$\varphi_{(M)} = \frac{q}{4\pi\varepsilon_0 R} + \frac{q'}{4\pi\varepsilon_0 r'} = 0 \tag{5.112}$$

and then

$$-\frac{q}{q'} = \frac{R}{r'} = \text{const.} \tag{5.113}$$

which leads to

$$\begin{aligned} \left(\frac{R}{r'}\right)^2 &= \frac{x^2+(D-y)^2}{x^2+(d-y)^2} = \frac{(x^2+y^2)-2Dy+D^2}{(x^2+y^2)-2dy+d^2} \\ &= \frac{a^2-2Dy+D^2}{a^2-2dy+d^2} = \text{const.} \end{aligned} \tag{5.114}$$

Since $(R/r')^2 = \text{constant}$, we shall have

$$\frac{\mathrm{d}}{\mathrm{d}y}\left(\frac{R}{r'}\right)^2 = 0 \tag{5.115}$$

Finally, we are ready to arrive at the following result

$$d = \frac{a^2}{D}$$

Substituting the above equation into Equation (5.114) yields

$$\left(\frac{R}{r'}\right)^2 = \frac{D^2-2Dy+a^2}{a^2-2dy+d^2} = \frac{D^2\left(1-2\frac{y}{D}+\frac{a^2}{D^2}\right)}{a^2\left(1-2\frac{y}{D}+\frac{a^2}{D^2}\right)} = \left(\frac{D}{a}\right)^2$$

That is

$$\frac{R}{r'} = \frac{D}{a}$$

We also have

$$-\frac{q}{q'} = \frac{R}{r'}$$

Combining the two above equation, we have

$$q' = -q\frac{a}{D} \tag{5.116}$$

Together with the result formerly calculated

$$d = \frac{a^2}{D} \tag{5.117}$$

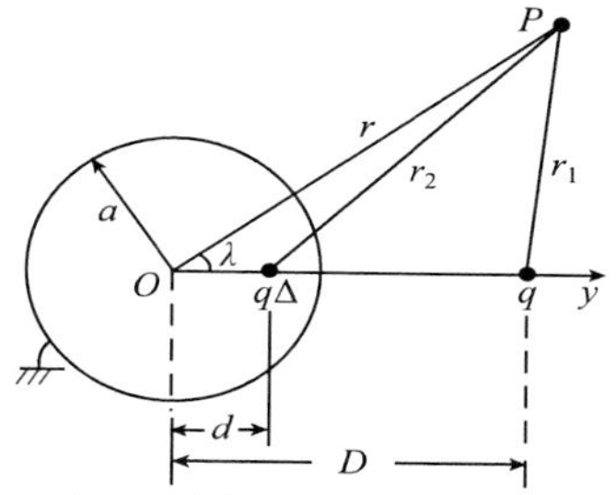

Figure 5.8 Calculation of the electrical potential at any point outside of the sphere

So we have the quantity $q'$and position d of the image charge.

2. The Surface Density and Quantity of Induction Charge on the Sphere

First, we use the above calculated image charge to calculate the electrical potential φ at any point outside of the sphere. On the basis of Figure 5.8, the electrical potential φ at point P is

$$\varphi = \frac{q}{4\pi\varepsilon_0 r_1} + \frac{q'}{4\pi\varepsilon_0 r_2} \tag{5.118}$$

We shall draw upon the law of cosines to get

$$\begin{aligned} r_1 &= \sqrt{D^2 + r^2 - 2Dr\cos\theta} \\ r_2 &= \sqrt{d^2 + r^2 - 2dr\cos\theta} \\ &= \sqrt{\left(\frac{a^2}{D}\right)^2 + r^2 - 2\left(\frac{a^2}{D}\right) r\cos\theta} \end{aligned}$$

Substituting the above two equations and Equation (5.108) into (5.118), we have an expression for the electrical potential at any point outside of the sphere

$$\varphi = \frac{q}{4\pi\varepsilon_0}\left\{\left(D^2 + r^2 - 2Dr\cos\theta\right)^{-\frac{1}{2}} - \frac{a}{D}\left[\left(\frac{a^2}{D}\right)^2 + r^2 - 2r\left(\frac{a^2}{D}\right)\cos\theta\right]^{-\frac{1}{2}}\right\} \tag{5.119}$$

We can now use the electrical potential φ to calculate the induction charge on the metal grounding sphere. First, we calculate the surface density ρ_s of the induction charge on the sphere. On the basis of the boundary conditions, ρ_s will be given by

$$\begin{aligned} \rho_s &= D_n = \varepsilon_0 E_n = \varepsilon_0 E_r|_{r=a} = \varepsilon_0(-)\left.\frac{\partial\varphi}{\partial r}\right|_{r=a} \\ &= \frac{q\left(a^2 - D^2\right)}{4\pi a\left(D^2 + a^2 - 2Da\cos\theta\right)^{3/2}} \end{aligned} \tag{5.120}$$

The total induction charge is the integral of ρ_s on the surface of the metal sphere. If we select the y axis as the dipole axis we have

$$\oint_S \rho_s \mathrm{d}S = \int_0^{\pi}\int_0^{2\pi} \rho_s\left(a^2\sin\theta\mathrm{d}\theta\mathrm{d}\phi\right) = -\frac{a}{D}q \tag{5.121}$$

Obviously, the calculated quantity of induction charge is equal to the image charge q'. Replacing the conducting grounding sphere with the image charge q' will not change the electrical field and potential distribution. Thus, selecting a closed surface outside the conducting sphere, the total electric flux will remain the same whether we replace the conducting sphere with the image charge or not. This indicates that the amount of induction charge on the grounding sphere will be equal to the amount of image charge.

3. Discussing the Sphere Image under Two Different Conditions

(1) The metal sphere is not grounded and the net charge on the sphere is equal to zero. There is a point charge q around the sphere.

Since there is no grounding, the electrical potential will not be equal to zero. We have the prior result of the image charge $q' = -\frac{a}{D}q$ when the electrical potential equals zero on the sphere. However, this condition does not meet the boundary requirement. Thus, it is necessary to keep the electrical potential the same on the sphere and not equal to zero at the same time. A key step is to place a point charge q''at the center of the ball. It remains to be determined to decide upon an appropriate q''. It is known that the electric flux passing through the closed surface around the conducting sphere with no connection with the earth equals zero. This means that the net charge $(q' + q'')$ is equal to zero in the closed surface. We have another boundary condition which states that

$$q' + q'' = 0$$

Thus we have

$$q'' = -q' = q\frac{a}{D} \tag{5.122}$$

where $q''$is the charge needed to be placed at the center of the ball. Deciding the magnitude and position of q' and q'', we can replace the conducting sphere, which is not connected to the earth, with these two charges. As a result, the electrical potential φ at any point outside of the sphere can be easily calculated through the use of q', $q''$and q. At this point it would be simple to calculate the electrical field and the induction charge density on the surface with the result of electrical potential.

(2) The metal sphere is not grounding but has net charge q_0 on the sphere. There is a point charge q around the sphere.

There are two conditions that needs to be satisfied in this problem. The first condition is that the sphere has an equipotential surface and the electrical potential is not zero. The second one is that the conducting sphere has a net charge of q_0 on the sphere. To meet the first requirement, we must place a charge q'' at the ball center beside the image charge q'. According to the second requirement, which is that the conducting sphere has a net charge of q_0 on the sphere, we have

$$q' + q'' = q_0$$

Thus we shall have

$$q'' = q_0 - q' = q_0 + q\frac{a}{D} \tag{5.123}$$

where q'' is the charge that should be placed at the center of the ball. The equation above makes use of the following relation $q' = -q\frac{a}{D}$.

Thus, by substituting the metal sphere with $q'$and q'' and using the values of q', $q''$and q, we can calculate the electrical potential, the electrical field, and the inducting charge density outside the sphere.

Example 5.11 Calculate the grounding capacitance of the metal sphere. This problem will be solved via the infinite image technique. As shown in Figure of Example5.11, we have a metal sphere with a radius of a. The metal sphere has a charge of q and the center of the ball is at a height of h above the ground. Calculate the grounding capacitance of the metal sphere.

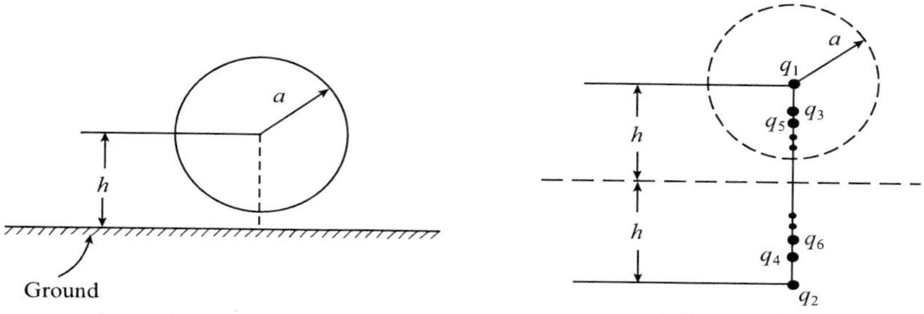

(a) The original problem (b) The ground image of charging metal sphere

Figure of Example 5.11 The infinite image technique

Analysis There are two boundary conditions in this problem. The first boundary condition is that the sphere surface is an equipotential surface and the electrical potential is non-zero. The second boundary condition is that the ground surface is of zero potential. If the sphere is isolated, we can place the charge q at the center of the ball to replace the charge distributed over the surface of the sphere. However, the requirement that the ground surface is of zero potential must also be met. Thus, we place charge q_1 at the center of the ball. Let q_2 be the ground image of q_1 where

$$q_2 = -q_1$$

q_2 meets the second requirement of ground, but at the same time it violates the condition of the equipotential surface of the sphere. To solve this problem, calculate the image charge q_3 for q_2 in the sphere by means of the sphere image. From Equations(5.107)and(5.108), we shall have that

$$q_3 = -q_2\frac{a}{2h} = q_1\frac{a}{2h}, \quad \text{We have} q_2 = -q_1$$

The distance to the ball center is given by $\dfrac{a^2}{2h}$.

The charges q_1, q_2 and q_3 meet the boundary condition of the metal sphere, but at the same time they violate the condition of the zero electrical potential boundary condition of the ground. Thus, a ground image q_4 for q_3 should be introduced. As a result, an infinite series of image charges of the two boundary surfaces (the ground and the metal sphere) should be introduced to describe the interaction between these two boundary surfaces. While the number of the charges is infinite, the method can converge fast. We list the image charges and the distance between the image charges and the ball center.

Image Charge	Distance between the Charge and the Ball Center
$q_2 = -q_1$	$2h$
$q_3 = -q_2\dfrac{a}{2h} = q_1\dfrac{a}{2h} = q_1T$	$a^2/2h$
$q_4 = -q_3$	$2h - a^2/2h$
$q_5 = q_1\dfrac{T^2}{1-T^2}$	$a^2/2h/\left[1-(a/2h)^2\right]$
$q_6 = -q_5$	$\vdots$
$q_7 = q_1\dfrac{T^3}{(1-T^2)\left(1-\dfrac{T^2}{1-T^2}\right)}$	

(a)

In the equation, $T = \dfrac{a}{2h}$.

From Equation (a), we note that the total image point charge in the sphere is given by

$$\begin{aligned} q &= q_1 + q_3 + q_5 + \cdots \\ &= q_1 \left(1 + T + \frac{T^2}{1-T^2} + \frac{T^3}{(1-T^2)\left(1 - \dfrac{T^2}{1-T^2}\right)} + \cdots \right) \end{aligned} \tag{b}$$

The electrical potential on the metal sphere is generated by all of the charges q_1, q_2, $q_3, \cdots$However, the charges pair q_2 and q_3, as well as q_4 and q_5 and so on to maintain the zero electrical potential on the metal sphere. Only the charge q_1 at the ball center makes the metal sphere have an electrical potential of φ given by

$$\varphi = \frac{q_1}{4\pi\varepsilon_0 a} \tag{c}$$

On the basis of Equations (b) and (c), the ground capacitance of the metal sphere is

$$C = \frac{q}{\varphi} = 4\pi\varepsilon_0 a \left[1 + \left(\frac{a}{2h}\right) + \frac{\left(\dfrac{a}{2h}\right)^2}{1 - \left(\dfrac{a}{2h}\right)^2} + \cdots \right] \tag{d}$$

where the first term $4\pi\varepsilon_0 a$ is the capacitance of the isolated metal sphere.

5.5.4 Cylindrical Surface Mirror Image

Consider two parallel cylindrical straight wires. If the wires are close to each other, then the interaction between the charges on the two wires will not be negligible. The charge density is high if the two wires are far away from each other, and lower if they are closer. Here the distribution of the charges is no longer axial symmetry (the geometric axes of the cylindrical wires), which means that unevenly distributed charges cannot be put on the geometric axes. Then where should the mirror image "axis" in place of the uneven charge distribution (which is called the electrical axis) be put is still a question. And this is a problem pertaining to the mirror image of the cylindrical surface. The method used for analyzing such a problem is referred to as the electrical axis method.

Consider the mirror image problem of the parallel double wires with radius of a, and the distance between the geometric axes of $2d$. We wish to find the positions of the mirror image line charges.

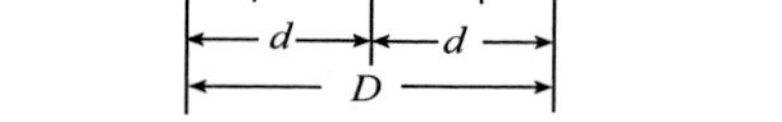

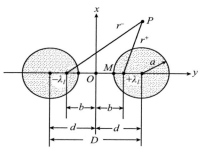

Figure 5.9 Cylindrical surface mirror image

Analysis We can solve this problem by using the electrical potential expression of the parallel double wires at any point in the space, and the boundary conditions of the wire surface which is an equipotential surface.

As shown in Figure 5.9, there are two parallel infinite cylindrical wires. The line charge densities of the left and right wires are $-\rho_l$ and $+\rho_l$, respectively. However, the line charge densities should not be put on the geometric axes of the wires (the cross section coordinates of the axes are $(0,-d)$,$(0,d)$), rather they should be put on the electrical axes (the cross section coordinates of the electrical axes are $(0,-b)$,$(0,b)$). Now we shall derive the position b of the electrical axis.

On the parallel double wires, the potential φ_P at any point P is given by

$$\varphi_P = \frac{\rho_l}{2\pi\varepsilon_0} \ln \frac{r^-}{r^+} \tag{5.124}$$

with

$$\begin{cases} r^- = \sqrt{x^2 + (y+b)^2} \\ r^+ = \sqrt{x^2 + (y-b)^2} \end{cases}$$

Move P to any point on the surface of the wire. Since the surface of the conductor is an equipotential surface, $\frac{r^-}{r_+}$ is a constant which we shall designate as K. In other words

$$\frac{(r^-)}{(r^+)} = K^2 = \frac{x^2 + (y+b)^2}{x^2 + (y-b)^2}$$

From the above formula, we obtain the relation

$$x^2 + \left(y - \frac{K^2+1}{K^2-1} b\right)^2 = \left(\frac{2Kb}{K^2-1}\right)^2 \tag{5.125}$$

which represents the equation for a circle with

$$\text{Center : } \begin{cases} x = 0 \\ y = \dfrac{K^2+1}{K^2-1} b \end{cases} \qquad \text{Radius :} r = \left| \frac{2Kb}{K^2-1} \right| \tag{5.126}$$

Using Figure 5.9, we obtain the value of the center and radius of the circle from the formula above:

$$\text{Center : } \begin{cases} x = 0 \\ y = d = \left| \dfrac{K^2+1}{K^2-1} \right| b \end{cases} \qquad \text{Radius :} r = a = \left| \frac{2Kb}{K^2-1} \right| \tag{5.127}$$

And for the point M on the surface of the wire, K is given by

$$K = \frac{r^-}{r^+} = \frac{b + (d-a)}{a - (d-b)} \tag{5.128}$$

By inserting the above formula into Equation(5.127) , we obtain the relation

$$\frac{d}{a} = \frac{b^2 + (a-d)^2}{b^2 - (a-d)^2}$$

Therefore, we can find an expression for the position b of the electrical axis given by

$$b = \sqrt{d^2 - a^2} \tag{5.129}$$

Figure 5.10 provides the distribution of the electrical axes, the geometric axes, and the equipotential line cluster of the parallel double wires.

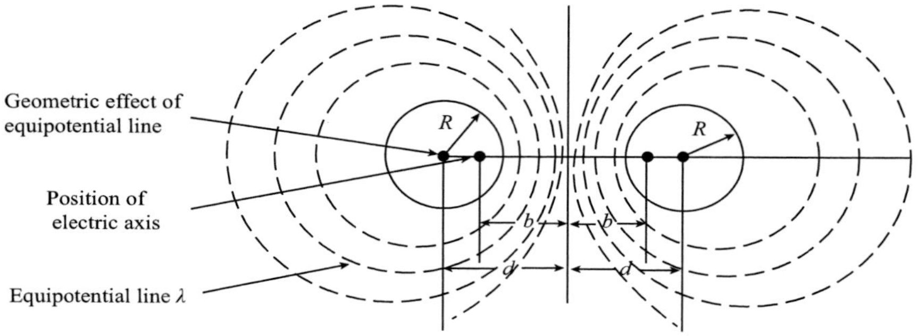

Figure 5.10 The distribution of the electrical axes, the geometric axes and the equipotential line cluster.

When the parallel double wires are fixed, the radius a of the double wires and the distance $2d$ between the wires can be confirmed. Then b can be derived from Formula (5.129). Thus, the electrical axes have been held constant (including the positions of the electrical axes and the charges they carry). The remaining issue is to find the potential field and relevant parameters at any point outside the two parallel wires.

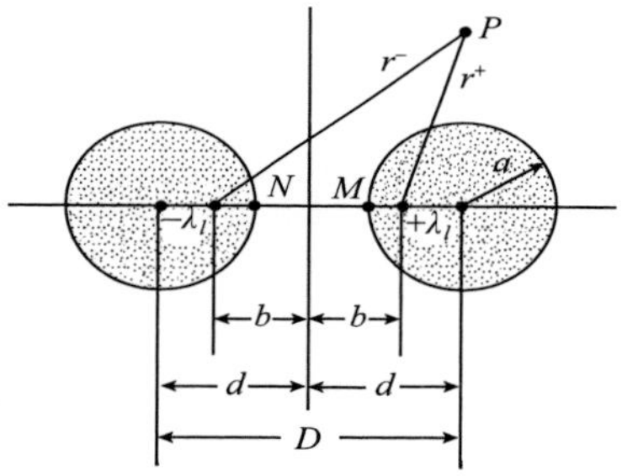

Figure of Example 5.12 Find the capacitance between the parallel double wires

Example 5.12 As shown in Figure of Example 5.12, the radii of the infinite parallel double wires are given by a, and the distance between the two wires is given by D. Find the capacitance per unit length between the parallel double wires.

Solution First, replace the two wires with two electrical axes. The densities of the charges that the two electrical axes carry are given by $+\rho_l$ and $-\rho_l$. In order to find the capacitance between the double wires, we need to find the potential difference between the wires.

The potential $\varphi_{(M)}$ on the right wire is given by(see Formula (5.124))

$$\varphi_M = \frac{\rho_l}{2\pi\varepsilon_0}\ln\frac{r'^-}{r'^+} = \frac{\rho_l}{2\pi\varepsilon_0}\ln\frac{b+(d-a)}{a-(d-b)} \tag{a}$$

The potential $\varphi_{(N)}$ on the right wire is

$$\varphi_N = \frac{\rho_l}{2\pi\varepsilon_0}\ln\frac{r''^-}{r''^+} = \frac{\rho_l}{2\pi\varepsilon_0}\ln\frac{a-(d-b)}{b+(d-a)} \tag{b}$$

thus the potential difference U between M and N is expressed as

$$\begin{aligned} U = \phi_M - \phi_N &= \frac{\rho_l}{\pi\varepsilon_0}\ln\frac{b+(d-a)}{a-(d-b)} \\ &= \frac{\rho_l}{\pi\varepsilon_0}\ln\frac{b+(d-a)}{b-(d-a)} = \frac{\rho_l}{\pi\varepsilon_0}\ln\frac{d+\sqrt{d^2-a^2}}{a} \end{aligned} \tag{c}$$

The formula above is derived using Formula (5.129).

Then the capacitance per unit length of the parallel double wires C_0 is

$$C_0 = \frac{\rho_l}{U} = \frac{\pi\varepsilon_0}{ln\left(\dfrac{d+\sqrt{d^2-a^2}}{a}\right)} \tag{d}$$

When $D \gg a$, the formula above can be reduced to

$$C_0 \approx \frac{\pi\varepsilon_0}{\ln\left(\dfrac{D}{a}\right)} \tag{e}$$

Example 5.13 Consider two infinitely long parallel conductor cylinders. The radii of the cylinders are a_1 and a_2, and the distance between the two axes is D. This is depicted in Figure of Example 5.13. Find the capacitance C_0 per unit length between the two parallel conductor cylinders.

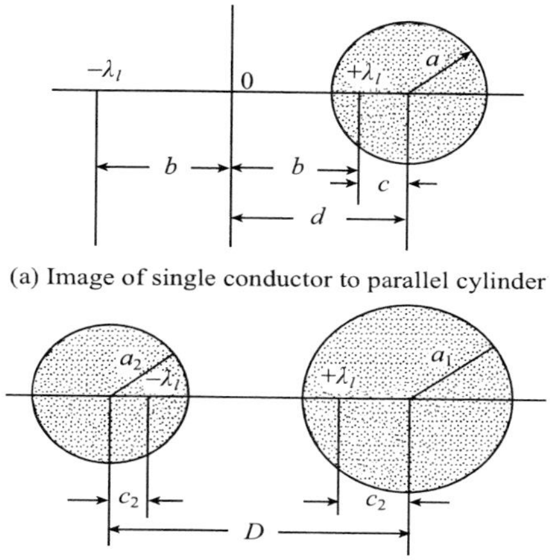

(a) Image of single conductor to parallel cylinder

(b) Image of two parrallel cylinder conductors with different radius

Figure of Example 5.13 (a) is the reference figure of (b)

Solution Let the line charge densities of the two conductor cylinders be $+\rho_l$ and $-\rho_l$, respectively. Assume that the distance D between the two conductor cylinders is not large enough to avoid the interaction between two the wires. As a result, the line charges can not be assumed to be concentrated at the geometric axes. Instead, the line charges will be concentrated at the electrical axes. There are formulas that have already been derived for the mirror image positions (in the cylinders) of the distribution charges on the surface of the cylinders. The problem now is that the radii of the two cylindrical conductors are not equal, and thus we need to determine the positions of the electrical axes respectively. The general formula for the positions of the electrical axes given by $b = \sqrt{d^2 - a^2}$ still needs to be used for this scenario. From Figure of Example 5.13 (a) we can see that the distance c between the electrical axis and the geometric axis of the cylinder is given by

$$\begin{aligned} c &= d - b = \frac{d^2 - b^2}{d + b} \\ &= \frac{d^2 - \left(\sqrt{d^2 - a^2}\right)^2}{d + b} = \frac{a^2}{d + b} \end{aligned}$$

Based on the formula above and according to Figure of Example 5.13, it is easy to find the distances c_1 and c_2 between the two electrical axes and their respective geometric axes. More specifically, we have that

$$c_1 = a_1^2/(D - c_2)$$

$$c_2 = a_2^2/(D - c_1)$$

Solving the two simultaneous equations above , we obtain

$$\left.\begin{aligned} c_1 &= \frac{a_1^2 - a_2^2 + D^2 - N^2}{2D} \\ c_2 &= \frac{a_2^2 - a_1^2 + D^2 - N^2}{2D} \end{aligned}\right\}$$

where $N^2 = \sqrt{a_1^2 - a_2^2 + D^2 - 4a_1^2 D}$.

After finding the positions of the electrical axes, we can find the potential difference between the two conductor cylinders

$$U = \frac{\rho_l}{2\pi\varepsilon_0} \ln \frac{(D - a_1 - c_2)(D - a_2 - c_1)}{(a_1 - c_1)(a_2 - c_2)}$$

The capacitance C_0 per unit length between the two conductor cylinders is given by

$$C_0 = \frac{\rho_l}{U} = \frac{2\pi\varepsilon_0}{\ln \dfrac{(D - a_1 - c_2)\ (D - a_2 - c_1)}{(a_1 - c_1)\ (a_2 - c_2)}}$$

When the radii of the two conductor cylinders are equal, i.e. $a_1 = a_2 = a$, the formula above becomes equivalent to Formula (d) in Example 5.12.

5.5.5 Media Mirror Image

As shown in Figure 5.11(a), division 1 and division 2 represent two different kinds of media. In the case of division 1, q is the original image charge. In order to find the electric field in division 1, we should put a mirror image charge q' in division 2. In order to find the electric field in division 2, we should put a mirror image charge q'' in division 1. Using the boundary conditions of the interface between the two different media, we can obtain the mirror image charges q' and q'', as shown in Firgure 5.11 (b), (c).

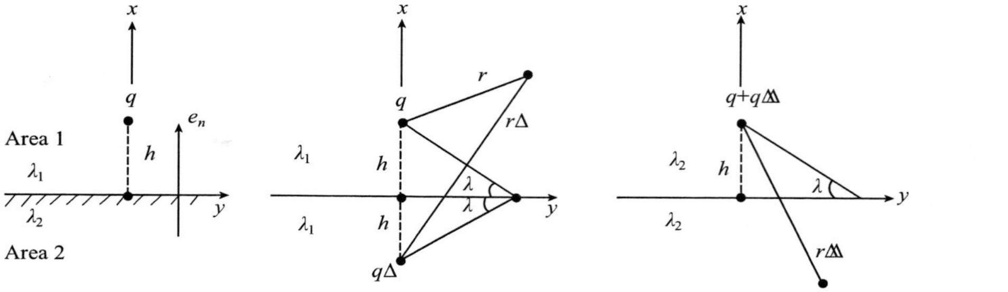

(a) Orginal problem (b) Image charge q′ is in area 2 and field in area 1 is need to be worked out (c) Image charge q″ is in area1 and field in area 2 is need to be worked out

Figure 5.11 Media mirror image

Method 1 Using the boundary conditions of the interface between two different media $E_{1t} = E_{2t}$ and $D_{1n} = D_{2n}$. We can now proceed to find the mirror image charges q' and q''.

Referring to Figure 5.11 (b), (c), and the normal direction e_n (pointing to area ε_1) of the interface between two different media given in Figure 5.11 (a), we would like to derive E_{1t}, E_{2t}, D_{1n}, D_{2n}.

First, we find the tangential electric field

$$E_{1t} = \frac{q}{4\pi\varepsilon_1 r^2}\cos\alpha + \frac{q'}{4\pi\varepsilon_1 r^2}\cos\alpha$$
$$E_{2t} = \frac{q+q''}{4\pi\varepsilon_2 r^2}\cos\alpha \tag{5.130}$$

Second, we find the normal electric field

$$D_{1n} = \varepsilon_1 E_{1n} = \varepsilon_1\left(\frac{-q}{4\pi\varepsilon_1 r^2}\sin\alpha + \frac{q'}{4\pi\varepsilon_1 r^2}\sin\alpha\right)$$
$$D_{2n} = \varepsilon_2 E_{2n} = -\varepsilon_2\frac{(q+q'')}{4\pi\varepsilon_2 r^2}\sin\alpha \tag{5.131}$$

Finally, using the equations $E_{1t} = E_{2t}$ and $D_{1n} = D_{2n}$, we can find the mirror image charges q' and q''

$$q' = -q''$$
$$q'' = q\frac{(\varepsilon_2 - \varepsilon_1)}{(\varepsilon_1 + \varepsilon_2)} \tag{5.132}$$

Method 2 Find the solution by using the electric potential and the boundary conditions of the electric potential. Charge q and q' generate potential φ_1 in region 1. Charge q and q''generate potential φ_2 in region 2. φ_1, φ_2 on the interface of the medium (side x=0) will satisfy the following boundary conditions

$$\varphi_1|_{x=0} = \varphi_2|_{x=0} \tag{5.133}$$

and

$$D_{1n} = D_{2n}$$

We have that

$$\varepsilon_1\left.\frac{\partial\varphi_1}{\partial x}\right|_{x=0} = \varepsilon_2\left.\frac{\partial\varphi_2}{\partial x}\right|_{x=0} \tag{5.134}$$

and given

$$\left.\begin{aligned}\varphi_1 &= \frac{1}{4\pi\varepsilon_1}\left(\frac{q}{r} + \frac{q'}{r'}\right)\\ \varphi_2 &= \frac{1}{4\pi\varepsilon_2}\left(\frac{q+q''}{r''}\right)\end{aligned}\right\} \tag{5.135}$$

where

$$\left.\begin{aligned}r &= (x-h)^2 + y^2 + z^2\\ r' &= (x+h)^2 + y^2 + z^2\end{aligned}\right\}\quad x \geqslant 0$$
$$r'' = (x-h)^2 + y^2 + z^2 \quad x \leqslant 0$$

Substituting r, r', r'' in the Equation (5.135), and then using two boundary conditions in Equations(5.133) and (5.134), we obtain:

$$\left.\begin{aligned}q' &= \frac{\varepsilon_1 - \varepsilon_2}{\varepsilon_1 + \varepsilon_2}q\\ q'' &= -q' = \frac{\varepsilon_2 - \varepsilon_1}{\varepsilon_1 + \varepsilon_2}q\end{aligned}\right\} \tag{5.136}$$

The location of the image charge $q'$and the original charge q are symmetric to the interface of the media. The location of the image charge q'' and the original charge q coincide.

5.6 Conformal Transformation, or Called Conformal Mapping

Conformal transformation is also called complex function method. By means of complex analytic function, it can be used to solve the problem of *two-dimentional stationary field boundary value problems.* It's especially useful for some complex boundary value problems, such as elliptic or hyperbolic metal boundary value problem depicted by Fig. 5.12.

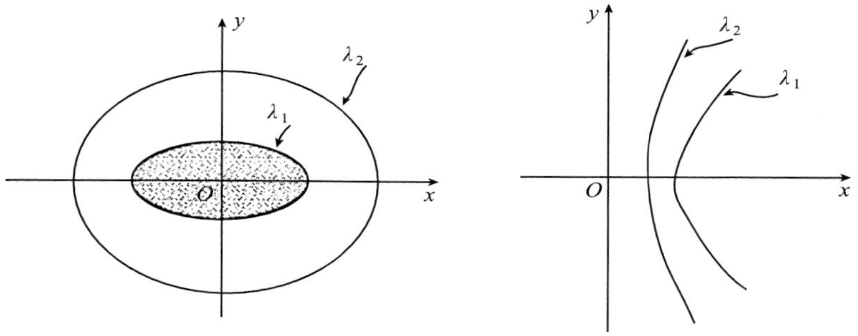

(a) Two elliptic metal plane (b) Two hyperbolic metal plane

Figure 5.12

5.6.1 Complex analytic function $W(z)$

At first dicuss complex analytic function $W(z)$, let

$$W(z) = u(x, y) + \mathrm{j}v(x, y) \tag{5.137}$$

where z is a complex variable,

$$z = x + \mathrm{j}y = r\mathrm{e}^{\mathrm{j}\varphi} \tag{5.138}$$

where x and y are both real number, and r and ϕ are the magnitude and the argument of z respectively.

$W(z)$ is a function of complex variable z, so is called complex function and can be written as $W(z) = u+\mathrm{j}v$.It has all kinds of forms, such as lnz, e^z, chz, cosz, arccosz, arcchz, etc.

The codition of complex analytic function is, at some field, $W(z)$ is to be not only derivable almost everywhere, but also strictly derivable. That is, the $\Delta z(=\Delta x+\mathrm{j}\Delta y)$ can approach zero by different means. Furthermore, the obtainded derivatives must be same, i.e, the only derivative can be obtained. Now we choose two kinds of method to approach zero:

(1) If Δy=0,$\Delta x \to$0, then $\Delta z = \Delta x \to$0.Thus

$$\frac{\mathrm{d}W}{\mathrm{d}z} = \frac{\mathrm{d}W}{\mathrm{d}x} = \frac{\partial u}{\partial x} + \mathrm{j}\frac{\partial v}{\partial x} \tag{5.139}$$

(2) IfΔx=0,$\Delta y \to$0, thenΔz=j$\Delta y \to$0.Thus

$$\frac{\mathrm{d}W}{\mathrm{d}z} = \frac{\mathrm{d}W}{\mathrm{j}\mathrm{d}y} = \frac{\partial u}{\mathrm{j}\partial y} + \mathrm{j}\frac{\partial v}{j\partial y} = \frac{\partial v}{\partial y} - \mathrm{j}\frac{\partial u}{\partial y} \tag{5.140}$$

According to strictly derivable condition, i.e the requirement of having the only derivative, Equations(5.139) and (5.140) must be equal, therefore

$$\frac{\partial u}{\partial x} = \frac{\partial v}{\partial y} \tag{5.141}$$

$$\frac{\partial v}{\partial x} = -\frac{\partial u}{\partial y} \tag{5.142}$$

This is so-called Cauchy-Riemann equation.

At polar coordinates, from Equation(5.138) we can know

$$\begin{aligned} dz &= d(re^{j\phi}) = e^{j\phi}dr + rd(j\phi)e^{j\phi} \\ &= dr \cdot e^{j\phi} + jrd\phi \cdot e^{j\phi} \end{aligned}$$

If let $dr=0$, then $dz=jrd\phi \cdot e^{j\phi} \rightarrow 0$; If $d\phi=0$, then $dz=dre^{j\phi} \rightarrow 0$. Thus we can get the following equation respectively

$$\frac{dW}{dz} = -j\frac{\partial u}{re^{j\phi} \cdot \partial\phi} + \frac{\partial v}{re^{j\phi}\partial\phi}$$

$$\frac{dW}{dz} = \frac{\partial u}{e^{j\phi}\partial r} + j\frac{\partial v}{e^{j\phi} \cdot \partial r}$$

Since the two formulae above are equal, the Cauchy-Riemann Equations in polar coordinate can be obtained as

$$\frac{\partial u}{\partial r} = \frac{\partial v}{r\partial\phi} \tag{5.143}$$

$$\frac{\partial v}{\partial r} = -\frac{\partial u}{r\partial\phi} \tag{5.144}$$

5.6.2 Main Properties of Analytic Functions of Complex Variables

(1) The two real-valued variables u, v in analytic function of complex variables $W(z) = u + jv$ satisfy Laplace's Equation.

Take partial derivative of Equation(5.141)on x, and take partial derivative of Equation(5.142) on y, then sum of these two equations will lead to

$$\frac{\partial^2 u}{\partial x^2} + \frac{\partial^2 u}{\partial y^2} = 0$$

With the similar method we will also have

$$\frac{\partial^2 v}{\partial x^2} + \frac{\partial^2 v}{\partial y^2} = 0$$

The two equations above indicate that, both u and v satisfy the two-dimension Laplace's Equation.

In the same way, with the help of Equations(5.141) and (5.144) we can find that $u$and$v$satisfy the two-dimension Laplace's Equation in cylindrical coordinate.

(2) Analytic functions of complex variables have angle-preserving (conformal) property.

Analytic functions of complex variables are differentiable, or have unique derivatives. Under this condition now it can be proved that: for two curves through point z_0 on Z plane with angle of $\alpha_2 - \alpha_1$ between them, when they are transformed (mapped) to W plane, though their shapes are changed, the angle $\beta_2 - \beta_1$ between the two curves on point W_0 is just equal to $\alpha_2 - \alpha_1$, in other words, the mapping is angle-preserving, as shown in Figure 5.13.

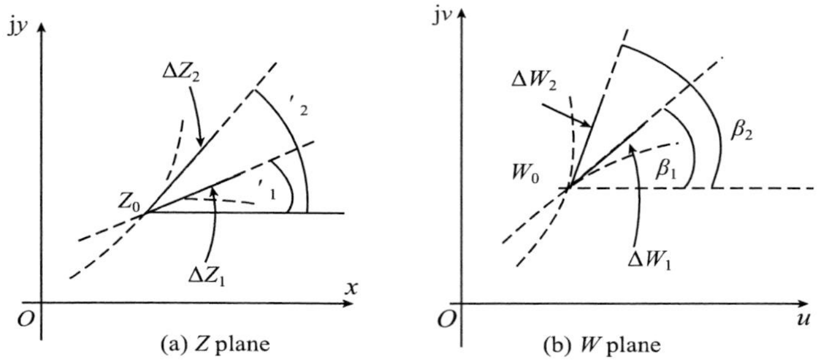

Figure 5.13 Angle-preserving mapping of analytic functions

$\Delta Z_1$and ΔZ_2 are two extremely short sections of lines on Z plane, the angle between them is $\alpha_2 - \alpha_1$, after being mapped to W plane, they become two extremely short sections of lines ΔW_1 and ΔW_2, with angle $\beta_2 - \beta_1$ between them. Because $(\mathrm{d}W/\mathrm{d}z)$ shall have a unique value on point Z_0, we have

$$\lim_{\Delta z_1 \to 0} \frac{\Delta W_1}{\Delta Z_1} = \lim_{\Delta z_2 \to 0} \frac{\Delta W_2}{\Delta Z_2} = \frac{\mathrm{d}W}{\mathrm{d}z} = M\mathrm{e}^{\mathrm{j}\phi} \tag{5.145}$$

where M is the modulus, ϕ is the argument.

From Figure 5.13 it can be known that

$$\begin{cases} \Delta Z_1 = |\Delta Z_1|\, \mathrm{e}^{\mathrm{j}\alpha_1} \\ \Delta Z_2 = |\Delta Z_2|\, \mathrm{e}^{\mathrm{j}\alpha_2} \end{cases}$$

$$\begin{cases} \Delta W_1 = |\Delta W_1|\, \mathrm{e}^{\mathrm{j}\beta_1} \\ \Delta W_2 = |\Delta W_2|\, \mathrm{e}^{\mathrm{j}\beta_2} \end{cases}$$

Substituting the modulus and argument values of ΔZ_1, ΔZ_2, ΔW_1 and ΔW_2 above into Equation(5.145) will result in

$$\frac{|\Delta W_1|}{|\Delta Z_1|} \cdot \mathrm{e}^{\mathrm{j}(\beta_1 - \alpha_1)} = \frac{|\Delta W_2|}{|\Delta Z_2|} \cdot \mathrm{e}^{\mathrm{j}(\beta_2 - \alpha_2)}$$

so

$$(\beta_1 - \alpha_1) = (\beta_2 - \alpha_2)$$

i.e.

$$\alpha_2 - \alpha_1 = \beta_2 - \beta_1$$

which indicates that analytic functions of complex variables have angle-preserving property. Two orthogonal curves on Z plane, after being mapped to W plane, are still orthogonal. The proof is as below.

Method 1 As shown in Figure 5.14, on W plane, two straight lines $u = c_1$ and $v = c_2$ are orthogonal, are they still orthogonal when they are mapped to Z plane? On the curve where u remains constant, the increment of u equals zero, then we have

$$\mathrm{d}u = \frac{\partial u}{\partial x}\mathrm{d}x + \frac{\partial u}{\partial y}\mathrm{d}y = 0$$

so that

$$\left.\frac{\mathrm{d}y}{\mathrm{d}x}\right|_{u=\text{const.}} = -\frac{\partial u}{\partial x} \Big/ \frac{\partial u}{\partial y} \tag{5.146}$$

In the same way, on the curve where v remains constant, the increment of v equals zero, then we have

$$\left.\frac{\mathrm{d}y}{\mathrm{d}x}\right|_{v=\text{const.}} = -\frac{\partial v}{\partial x}\bigg/\frac{\partial v}{\partial y} \tag{5.147}$$

Multiplying (5.146) and (5.147) and applying Cauchy-Riemann Equations, now we have

$$\left.\frac{\mathrm{d}y}{\mathrm{d}x}\right|_{u=\text{const.}} \times \left.\frac{\mathrm{d}y}{\mathrm{d}x}\right|_{v=\text{const.}} = -1 \tag{5.148}$$

which shows that on Z plane, two curves $u(x, y) = c_1$, $v(x, y) = c_2$ are orthogonal.

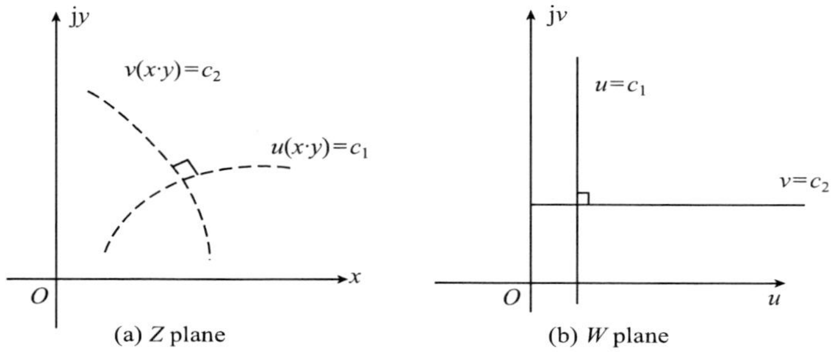

Figure 5.14 The angle-preserving mapping relationship between W plane and Z plane

Method 2 according to the nature which the dot product of orthogonal vector is zero, it can prove that the two curves in Z-plane are orthogonal.

Also using Figure5.14, we can directly give the gradient of u and v expressions

$$\nabla u = e_x\frac{\partial u}{\partial x} + e_y\frac{\partial u}{\partial y}$$
$$\nabla v = e_x\frac{\partial v}{\partial x} + e_y\frac{\partial v}{\partial y}$$

Therefore

$$\begin{aligned}\nabla u \cdot \nabla v &= \frac{\partial u}{\partial x}\frac{\partial v}{\partial x} + \frac{\partial u}{\partial y}\frac{\partial v}{\partial y} \\ &= \frac{\partial u}{\partial x}\left(-\frac{\partial u}{\partial y}\right) + \frac{\partial u}{\partial y}\frac{\partial u}{\partial x} = 0\end{aligned} \tag{5.149}$$

The derivation takes use of the application of the Cauchy - Riemann conditions. Because∇u is absolutely vertical to the curves which $u(x, y)$ equals to the constant c_1 . Thus, $\nabla u \cdot \nabla v = O$can indicate that the curves which u and v equal to the constant zero in z-plane are vertical mutually.

The above discuss two important properties of the analysis of complex function. One is that the two real variables u and v of the analysis of complex function satisfy Laplace equation. It shows u or v can act as a potential function. The other is that the curves which u equals to const and v equals to const are orthogonal mutually. Therefore, when the curves which u means a constant represent the equipotential line, then the curves which v means const are consistent with the power line, and vice versa. Of course, the cluster curves which u and v equal to const also have the nature.

5.6.3 Using $W(z)$ to get electric field strength E directly

Directly take the derivative of $W(z)$ can obtain electric field strength in the z-plane.

According to the nature which analytical complex function is strictly derivative and Cauchy-Riemann equations, it has

$$\begin{aligned}\left|\frac{\mathrm{d}W}{\mathrm{d}z}\right| &= \left|\frac{\mathrm{d}W}{\mathrm{d}x}\right| = \left|\frac{\mathrm{d}W}{\mathrm{d}y}\right| \\ &= \sqrt{\left(\frac{\partial u}{\partial x}\right)^2 + \left(\frac{\partial u}{\partial y}\right)^2} = \sqrt{\left(\frac{\partial v}{\partial x}\right)^2 + \left(\frac{\partial v}{\partial y}\right)^2}\end{aligned} \tag{5.150}$$

Taking use of that electric field strength equals to the negative potential gradient and Cauchy-Riemann equations, when we assume u potential function, there is

$$\begin{aligned}|E| = |-\nabla u| &= \left|-\left(e_x\frac{\partial u}{\partial x} + e_y\frac{\partial u}{\partial y}\right)\right| \\ &= \sqrt{\left(\frac{\partial u}{\partial x}\right)^2 + \left(\frac{\partial u}{\partial y}\right)^2} \\ &= \sqrt{\left(\frac{\partial v}{\partial x}\right)^2 + \left(\frac{\partial v}{\partial y}\right)^2} = |-\nabla v|\end{aligned} \tag{5.151}$$

Comparing Equation (5.150)to(5.151), then

$$|E| = \left|\frac{\mathrm{d}W}{\mathrm{d}z}\right| \tag{5.152}$$

5.6.4 Electric flux function

Assuming u as potential function, then v is the electric flux function. Electric flux through any area can be obtained by the flux function v. There is a simple method of proof: in Figure 5.15, if v denotes potential function, thus, by one side of curve MN and the other side which the unit length along $\boldsymbol{e}_z$ direction, it can form a curved surface, then the electric flux through this curved surface φ_e is

$$\varphi_e = \int_S \varepsilon E \cdot \mathrm{d}s \tag{5.153}$$

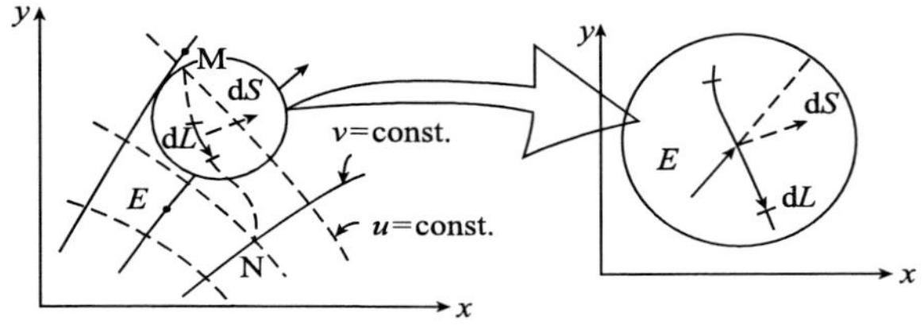

Figure 5.15 The figure of computing electric flux

The figure shows

$$\begin{aligned}E\cdot\mathrm{d}s &= (e_xE_x + e_yE_y)\cdot(e_z\times\mathrm{d}l) \\ &= (e_xE_x + e_yE_y)\cdot[e_z\times(e_x\mathrm{d}x + e_y\mathrm{d}y)] \\ &= -E_x\mathrm{d}y + E_y\mathrm{d}x\end{aligned}$$

Substituting this into Equation(5.153), it can obtain electric flux

$$\begin{aligned}\varphi_e &= \int_M^N \varepsilon\left(-E_x \mathrm{d}y + E_y \mathrm{d}x\right) = \int_M^N \varepsilon\left(\frac{\partial u}{\partial x}\mathrm{d}y - \frac{\partial u}{\partial y}\mathrm{d}x\right) \\ &= \int_M^N \varepsilon\left(\frac{\partial v}{\partial y}\mathrm{d}y + \frac{\partial v}{\partial x}\mathrm{d}x\right) = \int_M^N \varepsilon \mathrm{d}v \\ &= \varepsilon\left(v_N - v_M\right)\end{aligned} \tag{5.154}$$

That is, the electric field flux of surface which through M to N along z direction per unit length, equal to the difference between the flux function of M point and N point.

Another point to note, u and v is the dimensionless number, the potential function and flux function represented by them is usually to be difference of a factor with the potential value and electric flux value of actual problem, they should not be equated directly generally.

***Example* 5.14** Using conformal mapping method to gain the capacitance of per unit length between parallel twin conductor line. It is shown in Figure of Example 5.14. The radius of conductor is a, the distance between lines is $2d$, the electric potentials of two lines are respectively $\frac{U_0}{2}, -\frac{U_0}{2}$(This problem has been solved by axis method in Example 5.12).

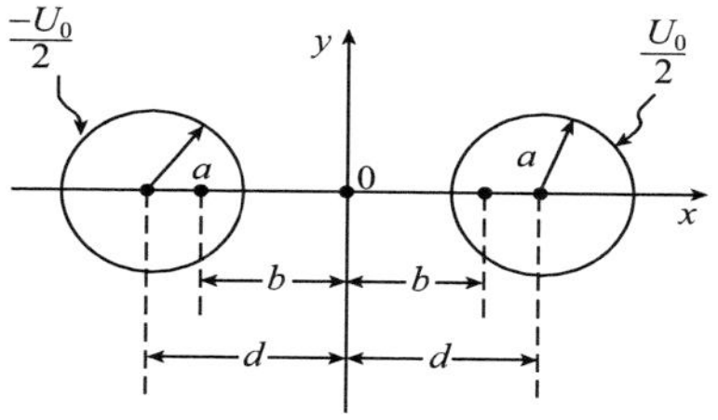

Figure of Example 5.14 Computing the capacity between parallel twin conductor line

Solution By choosing logarithmic function, it can give the same shape of equipotential line as this case. Considering the complex function

$$W = C_1[\ln(z - b) - \ln(z + b)] \tag{a}$$

It is easily to prove that it satisfies Cauchy-Riemann equations, that this is a analysis complex function. According to $W = u+\mathrm{j}v$, $z = x+\mathrm{j}y$ and the equation above, it has

$$u = \frac{C_1}{2}\ln\left[\frac{(x-b)^2 + y_2}{(x+b)^2 + y^2}\right] \tag{b}$$

$$v = C_1\left[\arctan\left(\frac{y}{x-b}\right) - \arctan\left(\frac{y}{x+b}\right)\right] \tag{c}$$

Taking u as the potential function, and set Equation (b) as a const, then

$$\frac{(x-b)^2 + y^2}{(x+b)^2 + y^2} = k^2$$

Where k means a constant. By changing the equation above, it has

$$\left[x - \frac{b\left(1+k^2\right)^2}{1-k^2}\right]^2 + y^2 = \frac{4b^2k^2}{(1-k^2)^2}$$

Obviously, this is an equation of a circle, which is centered at

$$\begin{cases} x = \dfrac{b(1+k^2)}{1-k^2} \\ y = 0 \end{cases} \tag{d}$$

with radius of

$$R = 2bk/(1-k^2) \tag{e}$$

Now let's solve the k and b in the equation above. For the wire on the right side of the figure, the position of the center of the circle is

$$x = d = \frac{b(1+k^2)}{1-k^2}, \quad y = 0$$

with radius of

$$R = a = 2bk/(1-k^2)$$

then we obtain

$$\left.\begin{aligned} b &= \sqrt{d^2 - a^2} \\ k &= \frac{d}{a} + \sqrt{\frac{d^2 - a^2}{a^2}} \end{aligned}\right\} \tag{f}$$

Finally the constant C_1 in Equation (b) can be determined according to the boundary conditions of the wire on the right side with circular cross-section. From the analysis above and the given boundary conditions, equations can be written as

$$\frac{U_0}{2} = \frac{C_1}{2} \ln k^2$$

therefore C_1 is solved as

$$C_1 = \frac{U_0}{2\ln\left(\dfrac{d}{a} + \sqrt{\dfrac{d^2-a^2}{a^2}}\right)} = \frac{U_0}{2\text{arcch}\left(\dfrac{d}{a}\right)} \tag{g}$$

Substitute the equation above into Equations (b) and (c), the functions for electrical potential and flux can be acquired respectively as follows

$$\varphi = u = \frac{U_0}{4\text{arcch}\left(\dfrac{d}{a}\right)} \ln\left[\frac{(x-b)^2 + y^2}{(x+b)^2 + y^2}\right] \tag{h}$$

$$v = \frac{U_0}{2\text{arcch}\left(\dfrac{d}{a}\right)} \left[\arctan\left(\frac{y}{x-b}\right) - \arctan\left(\frac{y}{x+b}\right)\right] \tag{i}$$

To find the capacitance C_0 for the wire of unit length, the amount of charge ρ_l on the wire of unit length is needed. According to Gauss's Law, ρ_l can be obtained through electric flux ψ_e. Let v_∞ denote the flux at $x = 0, y = \infty$, and $v_{-\infty}$ denote the flux at $x = 0, y = -\infty$.

Hence, for a part in $(v_\infty - v_{-\infty})$, or the part in the brackets of Equation (i), its value equals 2π, which leads to

$$\rho_l = \psi_e = \varepsilon_0 (v_\infty - v_{-\infty}) \frac{2\pi\varepsilon_0 U_0}{2\text{arcch}\left(\dfrac{d}{a}\right)}$$

thus

$$\begin{aligned} C_0 &= \frac{\rho_l}{\dfrac{U_0}{2} - \left(-\dfrac{U_0}{2}\right)} = \frac{\pi\varepsilon_0}{\text{arcch}\left(\dfrac{d}{a}\right)} \\ &= \frac{\pi\varepsilon_0}{\ln\left(\dfrac{d + \sqrt{d^2 - a^2}}{a}\right)} \end{aligned} \tag{j}$$

Example 5.15 Find the capacitance C_0 between co-axial metal conductive cylinders in unit length with elliptical cross-section.

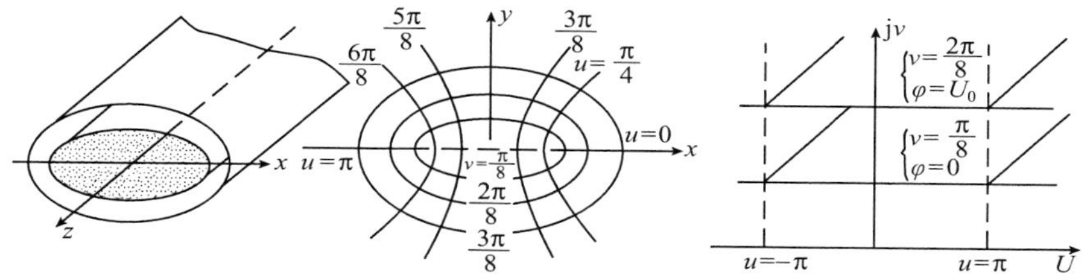

(a) Co-axial metal cylinders with elliptical crosssection
(b) Curves for arccos z
(c) Graph in z plane transformed into W Plane

Figure of Example 5.15

Solution Based on the shape of the surfaces of the conductors shown in Figure of Example 5.15(a), $W(z)$ can be chosen as

$$W(z) = \arccos z = \arccos(x + \mathrm{j}y) \tag{a}$$

It can be easily proved that this function satisfies Cauchy–Riemann equations, namely arccos z is an analytic function of a complex variable. It is known from Equation (a) that

$$W(z) = u + jv = \arccos(x + \mathrm{j}y)$$

in other words

$$x + \mathrm{j}y = \cos(u + \mathrm{j}v) = \cos u \mathrm{ch} v - \mathrm{j} \sin u \mathrm{sh} v$$

Then we have

$$\left.\begin{aligned} x &= \cos u \mathrm{ch} v \\ y &= -\sin u \mathrm{sh} v \end{aligned}\right\} \tag{b}$$

Take square operation on both sides of the equation above, and divide them by $\mathrm{ch}^2 v$ and $\mathrm{sh}^2 v$ respectively, then sum the two equations, we will have

$$\frac{x^2}{\mathrm{ch}^2 v} + \frac{y^2}{\mathrm{sh}^2 v} = \cos^2 u + \sin^2 u = 1 \tag{c}$$

Apparently this is an elliptic equation.

Take square operation on both sides of Equation (b), and divide them by $\cos^2 u$ and $\sin^2 u$ respectively, then subtract one equation from the other, we will have

$$\frac{x^2}{\cos^2 u} - \frac{y^2}{\sin^2 u} = \mathrm{ch}^2 v - \mathrm{sh}^2 v = 1 \tag{d}$$

Apparently this is a hyperbolic equation.

It can be seen that, the ellipse corresponding to $v =$ const. is consistent with the elliptical cross-section of the conductive cylinder in this example. Therefore, $v =$ const. represent equipotentials, while $u =$ const. represent lines with constant flux. In other words, v is a function of potential, and u is a function of flux. The curves corresponding to $W(z) = \arccos(z) = u + jv$ are shown in Figure of Example 5.15(b) .

It is known from Equation(c)that the semimajor axis a and the semiminor axis b of the ellipse are as follows respectively

$$a = \mathrm{ch} v, \quad b = \mathrm{sh} u \tag{e}$$

Using Equation (e) and the given semimajor and semiminor axes of the inner metal cylinder with elliptic cross-section, we can obtain that $v = \frac{\pi}{8}$. Using the given semimajor and semiminor axes of the outer metal cylinder with elliptic cross-section, we can obtain that $v = \frac{2\pi}{8}$, as shown in Figure of Example 5.15(b). For different practical boundary conditions: if the inner conductor is grounded, or when $v = \frac{\pi}{8}$, electric potential $\varphi = 0$; if the electric potential of the outer conductor is U_0, or when $v = \frac{2\pi}{8}$, electric potential $\varphi = U_0$. Thus, the relationship between the function of potential, v, and the electric potential φ, is as follows:

$$\varphi = U_0 \left[\frac{8}{\pi} \left(v - \frac{\pi}{8} \right) \right] \tag{f}$$

The electric potential not only satisfies Laplace Equation, but also the given boundary conditions, so it is the only solution for electric potential.

If the ellipse ($v =$ const) and the hyperbola ($u =$ const) on Z plane are transformed to W plane, they become straight lines, and the equipotentials become planes, as shown in Figure of Example 5.15 (c). Then, it would be very easy to find the capacitance of the unit length C_0 (plate capacitor).

$$C_0 = \frac{\varepsilon S}{d}$$

where d is the distance between the two plates of the capacitor. According to Figure of Example 5.15(c) it can be known that $d = \frac{\pi}{4} - \frac{\pi}{8} = \frac{\pi}{8}$, S is the area of the plate, which equals the width multiplied with unit length. The width is corresponding to the u axis. It can be seen from Figure of Example 5.15(b) that, the distance from $u = 0$ to $u = \pi$ is corresponding to half of the capacitance, so the whole width of the capacitor should be 2π, as shown in Figure of Example 5.15(c). Substituting those values into relevant equations we can have C_0 as

$$C_0 = \frac{\varepsilon 2\pi}{\pi/8} = 16\varepsilon \tag{g}$$

Certainly the end effect of the cylinders with elliptical cross-section is not considered in the equation above, that is, the lengths of them are assumed as infinite, which is a two-dimension field we talked about.

5.7 Finite-Difference Method—Numerical Computation Methods

If analytical method is employed to accurately solve the complicated problems on boundaries, difficulties would be encountered around the math involved, moreover the results are even hard to obtain. Under these circumstances, if numerical computation methods are used, approximate solution with very high accuracy can be found. Finite-difference method is one of the numerical computation methods. The principle of this method is briefly introduced in this section.

In finite-difference method, field region is divided into multiple grids, so that the continuous distribution of the fields in that region can be represented by the discrete values on the nodes of the grids. Essentially, finite-difference method *utilize the relationship among the electric potential at a certain node in the lattices and the electric potentials at the nodes surrounding that node, to establish algebraic equations, that is, finite-difference equations,* in order to solve electric potentials. In that way, *finite-difference method uses the partial differential equations for functions of potentials in solving algebraic equations.* For instance, in a rectangular two-dimension field, the Poisson equation the function of potentials satisfies is

$$\frac{\partial^2 \varphi}{\partial x^2} + \frac{\partial^2 \varphi}{\partial y^2} = -\rho/\varepsilon$$

If there is no source in the field region, namely, $\rho = 0$, the Poisson equation degrades to Laplace equation. If we divide the two-dimension field region into many square grids, each with edges having length of h, generally h is supposed to be very short. It can be seen from Figure 5.16 that each square has four nodes, those nodes on the outer boundaries are called outer nodes, the electric potentials on the outer nodes should be given (that is, the known boundary conditions). Except the outer nodes, all the other nodes are inner nodes. Any inner node is connected with the four surrounding nodes via the edges of the squares, and the finite-difference equations are the equations that link the electric potentials on these nodes together.

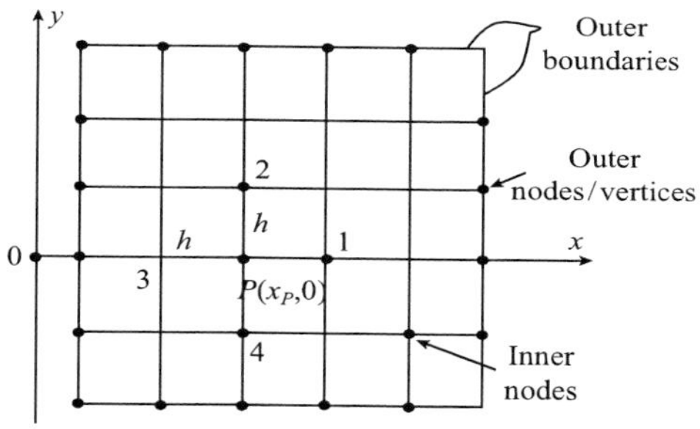

Figure 5.16 Depiction of finite-difference method.

Here we will discuss the establishment of finite-difference equations, which is illustrated by the point P (on x axis) in Figure 5.16 as an example. The coordinates of P are $(x_P, 0)$, let the electric potential on P be φ_P, then on the x axis, any point nearby with coordinates $(x, 0)$ has electric potential φ_x, which can be expanded through Taylor series at the point P as

$$\begin{aligned}\varphi_x = \varphi_P &+ \left(\frac{\partial \varphi}{\partial x}\right)_P (x - x_P) + \frac{1}{2!}\left(\frac{\partial^2 \varphi}{\partial x^2}\right)_P (x - x_P)^2 \\ &+ \frac{1}{3!}\left(\frac{\partial^3 \varphi}{\partial x^3}\right)_P (x - x_P)^3 + \cdots\end{aligned} \tag{5.155}$$

According to which, we can write the expressions for the electric potentials $\varphi_1, \varphi_2, \varphi_3, \varphi_4$ on the points denoted by 1, 2, 3, 4 in Figure 5.16. For example, the electric potential on point 1 is

$$\begin{aligned}\varphi_1 &= \varphi_P + \left(\frac{\partial \varphi}{\partial x}\right)_P (x_P + h - x_P) + \frac{1}{2!}\left(\frac{\partial^2 \varphi}{\partial x^2}\right)_P (x_P + h - x_P)^2 \\ &\quad + \frac{1}{3!}\left(\frac{\partial^3 \varphi}{\partial x^3}\right)_P (x_P + h - x_P)^3 + \cdots \\ &= \varphi_P + \left(\frac{\partial \varphi}{\partial x}\right)_P h + \frac{1}{2!}\left(\frac{\partial^2 \varphi}{\partial x^2}\right)_P h^2 + \frac{1}{3!}\left(\frac{\partial^3 \varphi}{\partial x^3}\right)_P h^3 + \cdots\end{aligned} \tag{5.156}$$

In the similar way the electric potential on point 3 can be obtained as

$$\begin{aligned}\varphi_3 &= \varphi_P + \left(\frac{\partial \varphi}{\partial x}\right)_P (-h) + \frac{1}{2!}\left(\frac{\partial^2 \varphi}{\partial x^2}\right)_P (-h)^2 \\ &\quad + \frac{1}{3!}\left(\frac{\partial^3 \varphi}{\partial x^3}\right)_P (-h)^3 + \cdots \\ &= \varphi_P - \left(\frac{\partial \varphi}{\partial x}\right)_P h + \frac{1}{2!}\left(\frac{\partial^2 \varphi}{\partial x^2}\right)_P h^2 - \frac{1}{3!}\left(\frac{\partial^3 \varphi}{\partial x^3}\right)_P h^3 + \cdots \end{aligned} \tag{5.157}$$

Superpose φ_1 and φ_3, we have

$$\varphi_1 + \varphi_3 = 2\varphi_P + h^2 \left(\frac{\partial^2 \varphi}{\partial x^2}\right)_P + 2 \times \frac{1}{4!}\left(\frac{\partial^4 \varphi}{\partial x^4}\right)_P h^4 + \cdots \tag{5.158}$$

If we draw a straight line in parallel with y-axis through point P, and use the same expansion method as the above, we can have the electric potentials on points 2 and 4. By superposing these two, we have

$$\varphi_2 + \varphi_4 = 2\varphi_P + h^2 \left(\frac{\partial^2 \varphi}{\partial y^2}\right)_P + 2 \times \frac{1}{4!}\left(\frac{\partial^4 \varphi}{\partial y^4}\right)_P h^4 \tag{5.159}$$

Then we add Equations (5.158) and (5.159)together, considering h is extremely small and the fourth order and higher orders of h are omitted (note that approximation is made here), which leads to

$$\varphi_1 + \varphi_2 + \varphi_3 + \varphi_4 = 4\varphi_P + h^2 \left(\frac{\partial^2 \varphi}{\partial x^2} + \frac{\partial^2 \varphi}{\partial y^2}\right)_P$$

If there is some source in the field region under study, we have

$$\left(\frac{\partial^2 \varphi}{\partial x^2} + \frac{\partial^2 \varphi}{\partial y^2}\right)_P = -(\rho/\varepsilon)_P \tag{5.160}$$

Which would equal zero if there is no source in the region under study. Therefore, the finite-difference equation is

$$\varphi_P = \begin{cases} \dfrac{\varphi_1 + \varphi_2 + \varphi_3 + \varphi_4 + h^2 (\rho/\varepsilon)_P}{4}, & \text{for} \quad \rho \neq 0 \\ \dfrac{\varphi_1 + \varphi_2 + \varphi_3 + \varphi_4}{4}, & \text{for} \quad \rho = 0 \end{cases} \tag{5.161}$$

Each inner node has a finite-difference equation, and the number of the inner nodes is the number of the finite-difference equations. By solving these related algebraic equations, we can find the values of the electric potentials on all of the inner nodes, while the electric potentials on the outer nodes are given, thus the electric potentials on all the nodes are determined. Apparently, the smaller the squares are, with more nodes, the more accurate the computational results will be, and the higher the required computational load will be.

Generally speaking, for finding the electric potential on each node, related finite-difference equations are not solved, but iteration method (relaxation method) may be used. That is, *to assume the initial values for the electric potentials on each node as arbitrary values, then, the new value of the electric potential* on a certain inner node will be updated as the average value over the electric potentials on the surrounding four nodes. When we *continue on calculating the electric potentials on the nodes neighboring to that node, the new electric potential value obtained previously will be substituted into the equations.* And this will be iterated over and over again, the electric potential on each node can be achieved with relatively high accuracy (namely, the electric potential on that node is very close to the average electric potential obtained over those of the neighboring four nodes).

Example 5.16 Try to find the distribution of electric potentials on the square-shaped cross-section of a metal tube with infinite length, by using iteration method. The values of electric potentials on the surface of the metal tube are shown in Figure of Example 5.16(a).

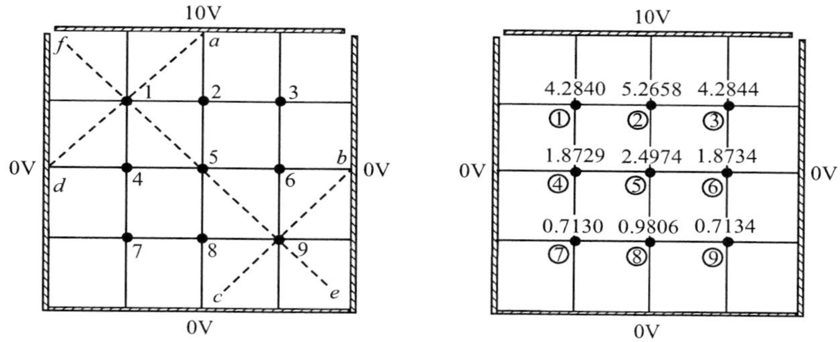

(a) The voltage on the boundary of the metal tube and selection of the initial electric potential values on inner nodes

(b) The electric potential values on inner nodes after 5 iterations

Figure of Example 5.16

(1) Selection of the initial electric potential values on inner nodes (nodes 1, 2, ⋯, 8, 9)

In principle, *the initial electric potential values on inner nodes can be selected arbitrarily.* However, the following selection method may be closer to practical situation. The key point is:

Some of the initial values can be chosen according to the given voltages on the boundary of the metal tube, then these initial values and the given voltages on the boundary of the metal tube will be utilized to find the initial values of other nodes, which has the process as below:

(i) The initial value of the electric potential φ_5 on node 5 can be chosen as

$$\varphi_5 = \frac{1}{4}\left(\varphi_a + \varphi_b + \varphi_c + \varphi_d\right) = \frac{1}{4}\left(10 + 0 + 0 + 0\right) = 2.5(\text{V})$$

(ii) The initial values of the electric potentials φ_1, φ_3 on nodes 1, 3 can be chosen as

$$\varphi_1 = \varphi_3 = \frac{1}{4}\left(\varphi_a + \varphi_5 + \varphi_d + \varphi_f\right) = \frac{1}{4}\left(10 + 2.5 + 0 + \frac{10}{2}\right) = 4.375(\text{V})$$

The approximated value for node f (slot) φ_f can be chosen as

$$\frac{1}{2}(10 + 0) = 5(\text{V})$$

(iii) The initial values of the electric potentials φ_7, φ_9 on nodes 7, 9 can be chosen as

$$\varphi_9 = \varphi_7 = \frac{1}{4}\left(\varphi_b + \varphi_e + \varphi_c + \varphi_5\right) = \frac{1}{4}\left(0 + 0 + 0 + \varphi_5\right) = 0.625(\text{V})$$

(iv) The initial values of the electric potentials φ_4, φ_6 on nodes 4, 6 can be chosen as

$$\varphi_4 = \varphi_6 = \frac{1}{4}(\varphi_1 + \varphi_5 + \varphi_7 + \varphi_d) = \frac{1}{4}(4.375 + 2.5 + 0.625 + 0) = 1.875(\mathrm{V})$$

(v) The initial value of the electric potential φ_2 on node 2 is

$$\varphi_2 = \frac{1}{4}(\varphi_a + \varphi_3 + \varphi_5 + \varphi_1) = \frac{1}{4}(10 + 4.375 + 2.5 + 4.375) = 5.3125(\mathrm{V})$$

(vi)The initial value of the electric potential φ_8 on node 8 is

$$\varphi_8 = \frac{1}{4}(\varphi_5 + \varphi_9 + \varphi_c + \varphi_7) = \frac{1}{4}(2.5 + 0.625 + 0 + 0.625) = 0.9375(\mathrm{V})$$

(2) With the chosen initial values of inner nodes, the electric potentials on the inner nodes can be obtained through iteration method (relaxation method). Computation can be processed based on the sequence of nodes as 1→2→3→ ···. The electric potential of a certain inner node is the sum of the electric potential values of the four neighboring nodes, above, below, on the left, and on the right, of the current node, divided by 4. The electric potential values on inner nodes after 5 iterations are shown in Figure of Example 5.16(b). The Table below gives the electric potential values on inner nodes during the process of iteration.

Initial values and values during iterations	φ_1	ϕ_2	φ_3	φ_4	φ_5	φ_6	φ_7	φ_8	ϕ_9
Initial values	4.3750	5.3125	4.3750	1.8750	2.5000	1.8750	0.6250	0.9375	0.6250
After one iteration	4.2969	5.2930	4.2920	1.8555	2.4903	1.8518	0.6983	0.9534	0.7013
—	—	—	—	—	—	—	—	—	—
After three iterations	4.2841	5.2623	4.2822	1.8688	2.4927	1.8712	0.7105	0.9783	0.7125
—	—	—	—	—	—	—	—	—	—
After five iterations	4.2840	5.2658	4.2844	1.8729	2.4974	1.8734	0.7130	0.9806	0.7134

5.8 Green's Function and Green's First, Second Identities

5.8.1 Green's Function

δ function (Dirac delta function) is described in Section 2.10 of Chapter 2. δ function is the density function of unit point charge. Since it is the density function of charge, its electric potential φ should satisfy Poisson's equation

$$\nabla^2\varphi = -\frac{\delta(r - r')}{\varepsilon_0} \tag{5.162}$$

In homogeneous material at unbounded space, electric potential φ of unit point charge is

$$\varphi = \frac{1}{4\pi\varepsilon_0 |r - r'|} \tag{5.163}$$

It also can be seen from Expression (2.124) that, $\nabla^2\left(\frac{1}{R}\right)$ owns properties of δ function, based on which the relationship between δ function in homogeneous material at unbounded space and $\nabla^2\left(\frac{1}{R}\right)$ can be obtained as

$$\nabla^2\left(\frac{1}{R}\right) = -4\pi\delta\left(r - r'\right)$$

namely

$$\nabla^2\left(\frac{1}{4\pi R}\right) = \nabla^2\left(\frac{1}{4\pi\left|r - r'\right|}\right) = -\delta\left(r - r'\right) \tag{5.164}$$

and $\frac{1}{4\pi\left|r - r'\right|}$ is called Green's function $G(r - r')$, that is

$$G(r - r') = \frac{1}{4\pi\left|r - r'\right|} \tag{5.165}$$

then we have

$$\nabla^2 G(r - r') = -\delta\left(r - r'\right) \tag{5.166}$$

which is the Poisson's equation Green's function satisfies. Comparing Equation(5.163) and Equation(5.165), it can be known that

$$G(r - r') = \varepsilon_0\varphi(r - r') = \varepsilon_0\frac{1}{4\pi\left|r - r'\right|} \tag{5.167}$$

which is the solution for the Poisson's equation of Green's function [Equation (5.166)].

Green's function is suitable for representing the fields where sources are in a distributed manner, and also can be applied to describe the response from unit point source, as shown in Figure 5.17.

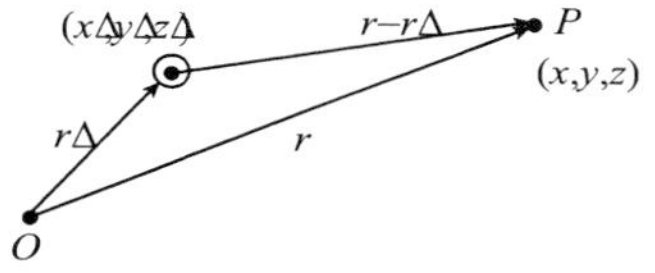

Figure 5.17 Field point (x, y, z), source point (x', y', z'), coordinate origin (O) and r, r'

5.8.2 Green's First and Second Identities

Green's identities are also called Green's theorem, which can be derived from Gauss' divergence theorem and vector identities.

For an arbitrary vector C, its expression for Gauss' divergence theorem is

$$\int_V \nabla \cdot C \mathrm{d}V = \oint_S C \cdot \mathrm{d}S \tag{5.168}$$

if let φ_1, φ_2 be two scalar functions, and let vector C be

$$C = \varphi_1 \nabla \varphi_2 \tag{5.169}$$

With a vector identity we have the following:

$$\nabla \cdot C = \nabla \cdot (\varphi_1 \nabla \varphi_2) = \varphi_1 \nabla^2 \varphi_2 + \nabla\varphi_1 \cdot \nabla\varphi_2 \tag{5.170}$$

Substituting Equation (5.170) into Equation (5.168) we have

$$\int_V (\varphi_1 \nabla^2 \varphi_2 + \nabla\varphi_1 \cdot \nabla\varphi_2)\mathrm{d}V = \oint_S \varphi_1 \nabla\varphi_2 \cdot \mathrm{d}S = \oint_S \varphi_1 \frac{\partial \varphi_2}{\partial n}\mathrm{d}S \tag{5.171}$$

which is Green's first identity, also called Green's first theorem.

When Green's second identity is to be derived, just interchange the positions of φ_1, φ_2 in $C = \varphi_1 \nabla \varphi_2$ and substituting it into Equation(5.168), we will have a formula with the exact form as Green's first identity except that the positions of φ_1, φ_2 are swapped. Subtracting this expression from Green's first identity, leaving

$$\int_V \left(\varphi_1 \nabla^2 \varphi_2 - \varphi_2 \nabla^2 \varphi_1\right) \mathrm{d}V = \oint_S \left(\varphi_1 \frac{\partial \varphi_2}{\partial n} - \varphi_2 \frac{\partial \varphi_1}{\partial n}\right) \mathrm{d}S \tag{5.172}$$

which is Green's second identity, also called Green's second theorem.

Exercises

5.1 Try to derive the Laplace's equation for electric potential φ of electrostatic field, and write its expression in Cartesian coordinates.

5.2 There are two extremely large metal plates in parallel with yOz plane, with the spacing between them is d and filled with air, the voltage on the upper plate (where $x = d$) is U_0, the lower plate (where $x = 0$) is grounded, try to use Laplace's Equation $\nabla^2 \varphi = 0$ to find the electric potential φ and electric field E between the two plates (Figure of Exercise 5.2).

5.3 Under the conditions given by Exercise 5.2, volume charge density ρ_0 is uniformly distributed between the two plates, try to use Poisson equation $\nabla^2 \varphi = -\rho_0/\varepsilon_0$ to find the electric potential φ and electric field E between the two plates.

5.4 There is a point charge $q(h, 0, 0)$ above ground, try to use method of images to find the electric potential ϕ and electric field E at a point $P(x, 0, 0)$ along x axis (Figure of Exercise 5.4).

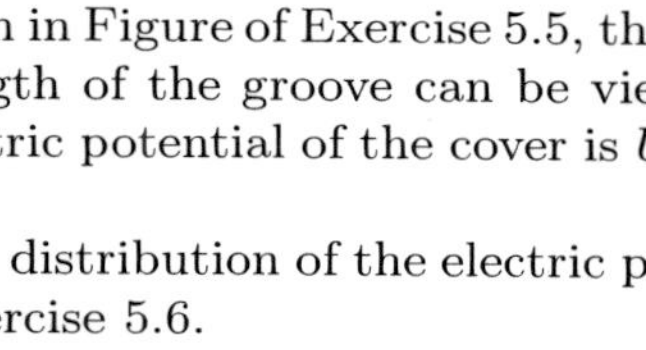

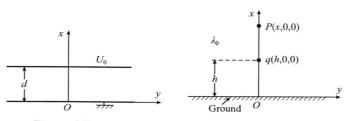

Figure of Exercise 5.2

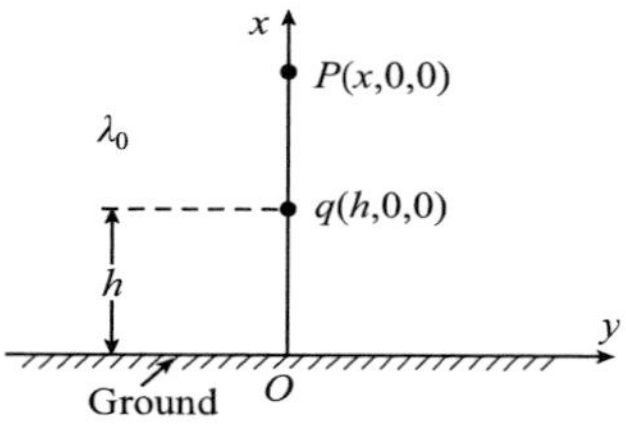

Figure of Exercise 5.4

5.5 As shown in Figure of Exercise 5.5, there is a conductor groove with rectangular cross-section, the length of the groove can be viewed as infinite, and there is a cover insulated with it, the electric potential of the cover is U_0, find the function of electric potentials inside the groove.

5.6 Find the distribution of the electric potential in the two-dimensional region as shown in Figure of Exercise 5.6.

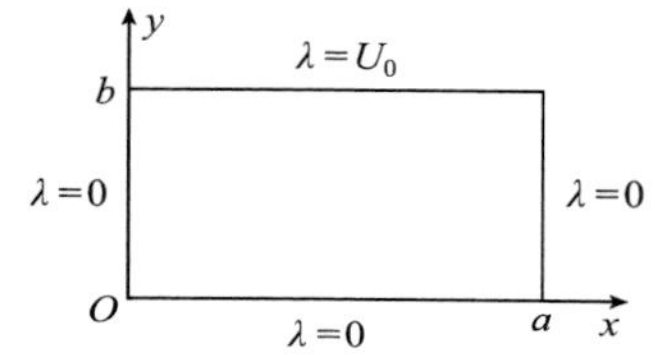

Figure of Exercise 5.5

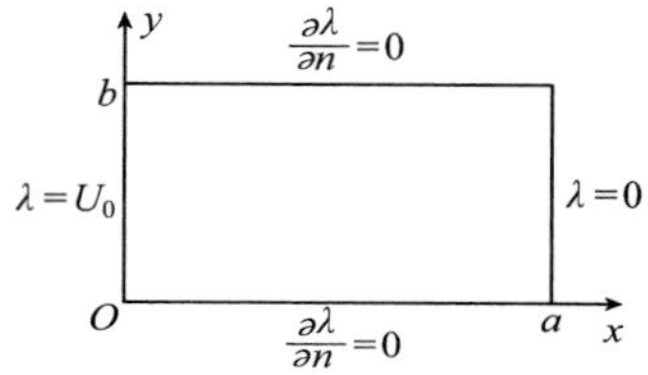

Figure of Exercise 5.6

5.7 There is a region bounded by a groove made of conductor plates, extending infinitely along y and z directions, the electric potentials on the cross-section and the boundaries are shown in Figure of Exercise 5.7, find:

(1) electric potential in that region;

(2) surface charge density on the conductor plates.

5.8 There are two infinitely large conductor plates in parallel with each other, the spacing between them is b, where there is a extremely thin slice of conductor, for which, y falls in $b \geqslant y \geqslant d$, x falls in $-\infty < x < +\infty$, as shown in Figure of Exercise 5.8. The upper plate and the slice has electric potential of U_0, and the lower plate has a potential of zero, find the electric potential between the two plates. Assume that, for the conductor slice, from $y = 0$ through $y = d$, the electric potential is varying linearly $\varphi = \dfrac{U_0}{d} y$.

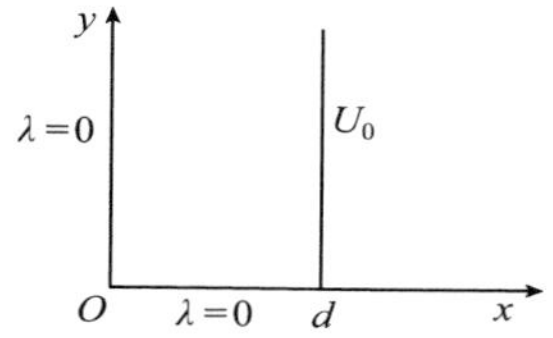

Figure of Exercise 5.7

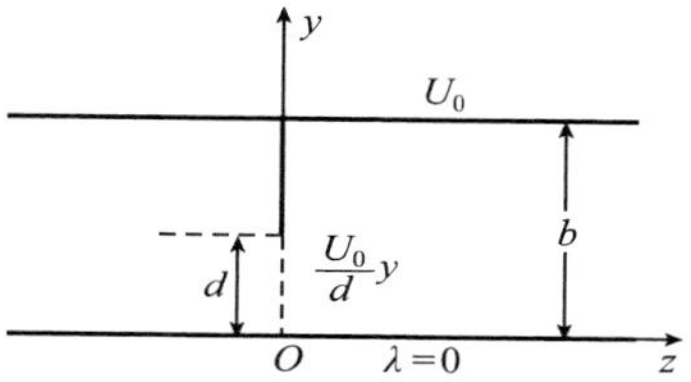

Figure of Exercise5.8

5.9 There is pair of infinitely large conductor plates in parallel with each other and both grounded, between which there is a line charge ρ_l in parallel with z axis, and its position is $(0, d)$. Find the function of electric potential between the plates (Figure of Exercise 5.9).

5.10 The radii of the inner and outer conductors for a co-axial cable are a and b, respectively, and the cable is very long in z direction, the electric potential of the outer conductor is zero (grounded), the electric potential of the inner conductor is U_0, as shown in Figure of Exercise 5.10. Find the distributions of electric potential and field in the co-axial cable.

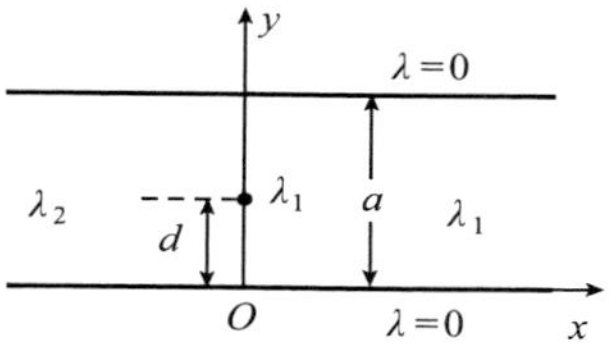

Figure of Exercise 5.9

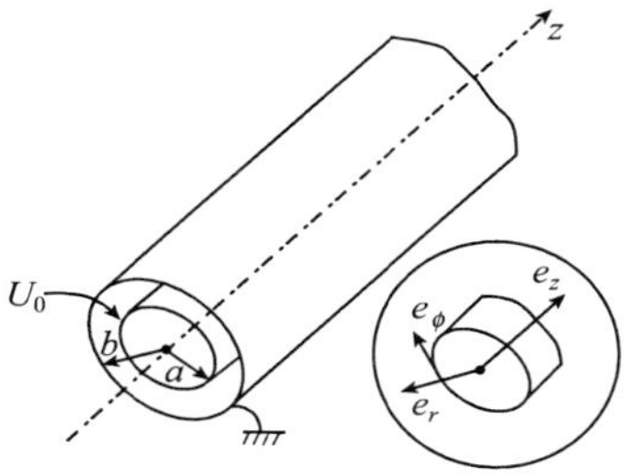

Figure of Exercise 5.10

5.11 There is a cylinder with infinite length and radius being a, the distribution of electric potential on the surface of it is as follows:

$$\varphi(a, \phi) = \begin{cases} -U_0, & 0 < \phi < \pi \\ U_0, & \pi < \phi < 2\pi \end{cases}$$

Find the distributions of electric potential inside of the cylinder.

5.12 In a uniform electric field $E = e_x E_0$, a long conductor cylinder is placed with its axis normal to the direction of the electric field, and its axis coincides with z axis, with radius being a. Find the function of electric potential outside of the conductor cylinder and the induced charge density over the surface of the conductor.

5.13 There is a cylindrical surface of slim conductor, and its radius is b with infinite length, which is divided equally into four parts, as shown in Figure of Exercise 5.13. The cylindrical surface in the second and the fourth quadrants is grounded, and the surface in the first and the third quadrants have electric potentials U_0 and $-U_0$, respectively. Find the distributions of electric potential inside of the cylinder.

5.14 There is a metal cone, whose height is h, and the radius of the base is a, $h \gg a$, the distance between its vertex and ground is r_0 (r_0 is very short), as shown in Figure of Exercise 5.14. Find electric potential φ and electric field E between the cone and ground.

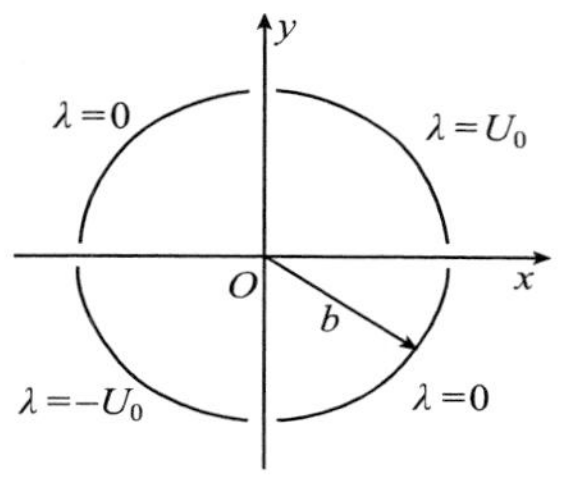

Figure of Exercise 5.13

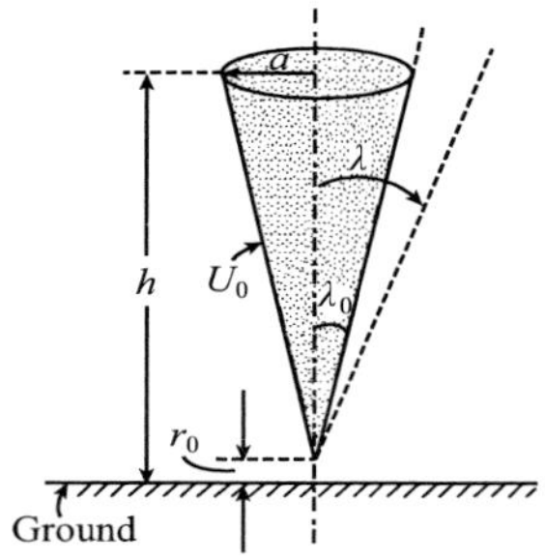

Figure of Exercise 5.14

5.15 A conductor sphere with radius a is placed in a uniform electric field E_0, assume that (1) the voltage on the conductor sphere is U_0; (2) the charge on the conductor sphere is Q. Find the distributions of electric potential around the sphere in the two cases respectively.

5.16 A cylinder with radius a and height l, has zero electric potential on its top and bottom surface but that on the lateral surface is U_0. Find the electric potential inside the cylinder.

5.17 As shown in Figure of Exercise 5.17, there is a point charge q placed at the position where $x = 1, y = 1$, between two conductor plates with the angle between them is $60°$. Find:

(1) the positions and quantities of all the image charges;

(2) the electric potential at the point where $x = 2, y = 1$.

5.18 There is a metal sphere buried under ground, and there is a point charge $+q$ above ground, as shown in Figure of Exercise 5.18. Find the quantity and position of the image charge, as well as the electric potential at point P and the induced surface charge density ρ_s at point O.

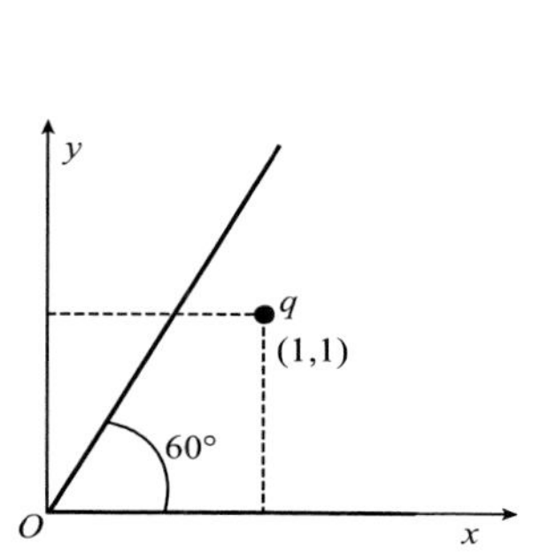

Figure of Exercise 5.17

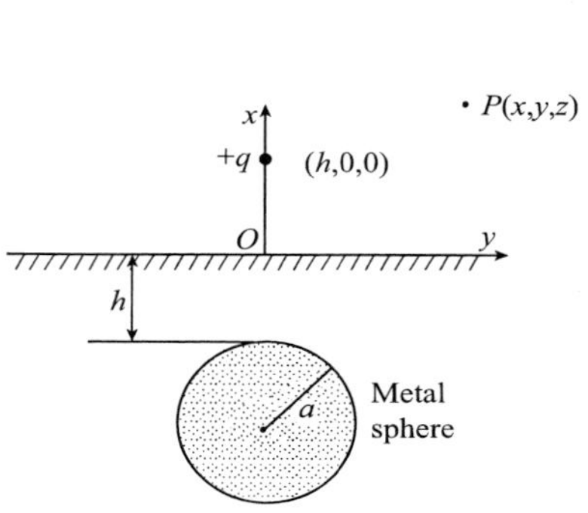

Figure of Exercise 5.18

5.19 There is a long conductor wire with radius a hanging in the air, which is in parallel with conductor planes, and the distance between wire and the two planes are d_1 and d_2, respectively, and $d_1 \gg a, d_2 \gg a$ (Figure of Exercise 5.19). Find the capacitance between the unit length of wire and the conductor planes.

5.20 There is a metal sphere with radius a being grounded, and there is a point charge q placed away from the center of the sphere with distance being d.

(1) prove that the effect of the grounded metal sphere can be represented by an image charge q', $q' = -q\dfrac{a}{d}$, the position of q' is $r' = \dfrac{a^2}{d}$, as shown in Figure of Exercise 5.20;

(2) if the metal sphere has electric potential of U_0, what additional result for (1) should be given?

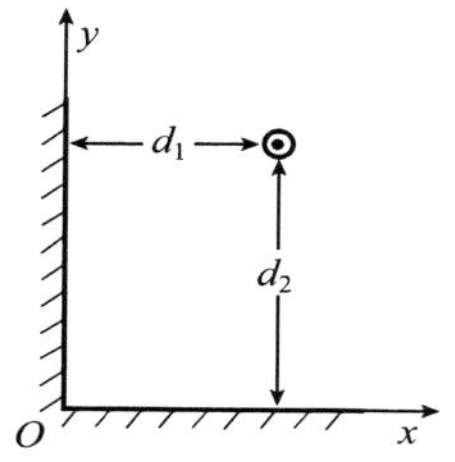

Figure of Exercise 5.19

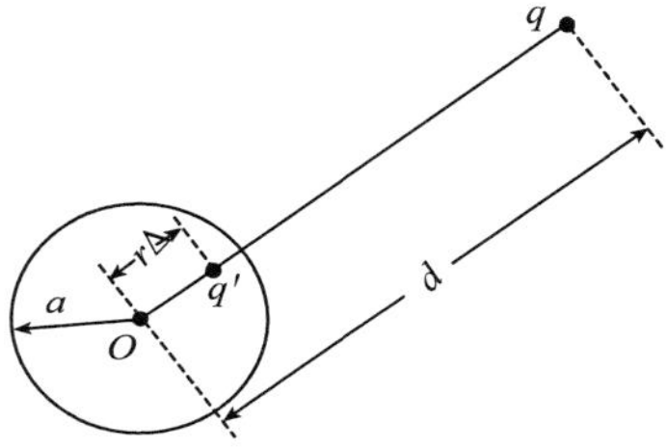

Figure of Exercise 5.20

5.21 As shown in Figure of Exercise 5.21, there are several short antennas (can be viewed as current element) placed in different directions, draw their images against the ground.

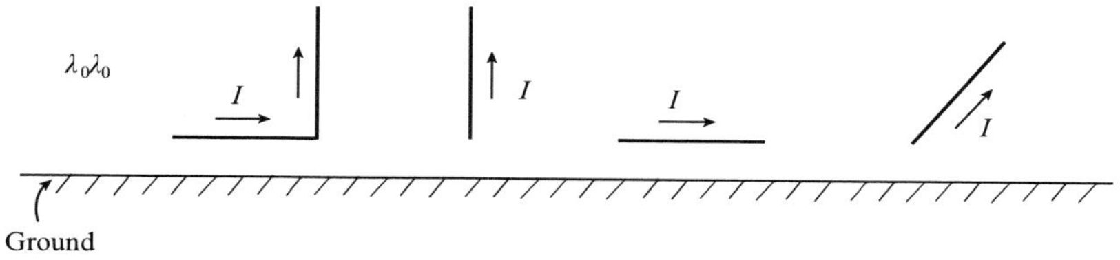

Figure of Exercise 5.21

5.22 In a grounded hollow conductor sphere with radius a, there are point charges $q_1 = q$ and $q_2 = -q$ located at $z = a/3$ and $z = -a/3$, respectively, as shown in Figure of Exercise 5.22. Find the distribution of electric potential in the hollow conductor sphere.

5.23 As shown in Figure of Exercise 5.23, there is a cylindrical grounding device buried in the ground, $h \gg d/2$. Find the grounding resistance.

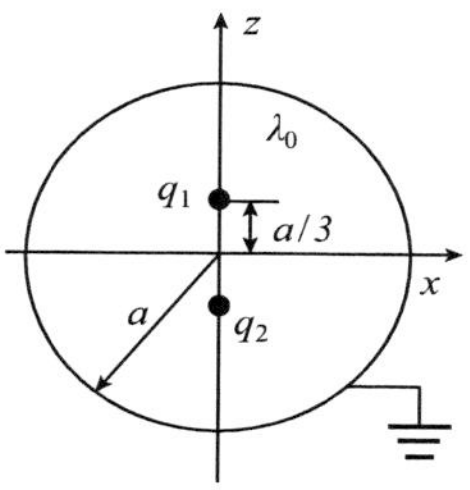

Figure of Exercise 5.22

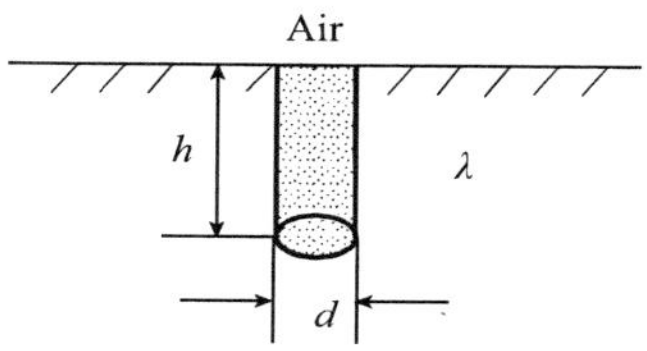

Figure of Exercise 5.23

Chapter 6
Alternating Electromagnetic Fields

This chapter discusses three topics of alternating electromagnetic fields: Firstly, the theory about macroscopic electromagnetic fields, i.e. Maxwell's equations; secondly, the laws that alternating electromagnetic fields follow on the boundaries between different materials, i.e. boundary conditions; lastly, Poynting's theorem which describes the energy conservation of electromagnetic fields.

6.1 Maxwell's Equations

It can be seen from the research on time-invariant fields, electrostatic field is produced only by static charge. Whilst, magnetostatic field is generated only by steady current; there is no interaction between static electric field and static magnetic field. However, when charge and current vary with time, they will generate electric and magnetic fields alternating with time, called alternating electromagnetic fields. Alternating fields are not only functions of position, but also functions of time. Alternating electric field and alternating magnetic field are no longer independent from each other, but they are mutually linked, behave as sources of each other, and closely coupled each other.

Maxwell's equations summarize the basic laws behind macroscopic electromagnetic phenomena in a concise and accurate way. They are a set of fundamental equations for electromagnetism. Firstly it was Faraday who discovered electromagnetic induction phenomenon in 1831, which unveiled the law that time-varying magnetic field can produce (induced) electric field. Then *Maxwell proposed the concept of displacement current while amending Ampere's circuital law* in 1873, which generalized the validity of it and led to the conclusion that time-varying electric field can produce magnetic field. It is indicated that, except for the alternating electric and magnetic fields generated by alternating charges and currents respectively, they can also generate each other and act as the sources of each other. So when alternating fields are analyzed, besides considering the laws for either electric field or magnetic field themselves, more emphasis should be given to the study on the mutual effect between electric and magnetic fields, to the study on the relations between fields and sources as well as the connections between fields and materials. It is the famous Maxwell's equations that build the theoretical architecture that embodies the principles and concepts mentioned above.

Among Maxwell's equations, there is one curl equation with one divergence equation for electric field vector, and there is also one curl equation with one divergence equation for magnetic field vector, altogether four equations. And this conforms to the requirement presented by Helmholtz's theorem that both the curl and divergence of a vector must be given before the vector can be uniquely specified, with its completeness of mathematical description.

Maxwell predicted the propagation of electromagnetic waves based on the concepts that alternating electric field and magnetic field can be mutually excited. After that, Hertz generated electromagnetic waves through experiment in 1888, verifying the correctness of Maxwell's prediction.

6.2 Law of Induction and Maxwell's Second Equation

6.2.1 Generalization of Law of Induction—Integral Form of Maxwell's Second Equation

Faraday's law of induction shows that, if a magnetic flux is varying through an open surface (with its normal direction of $\boldsymbol{n}$) *bounded by a closed loop of circuit*, then it will generate an induced electromotive force (emf) $\mathscr{E}$ in the closed loop, with the expression as

$$\mathscr{E} = -\frac{\mathrm{d}\Phi}{\mathrm{d}t} \tag{6.1}$$

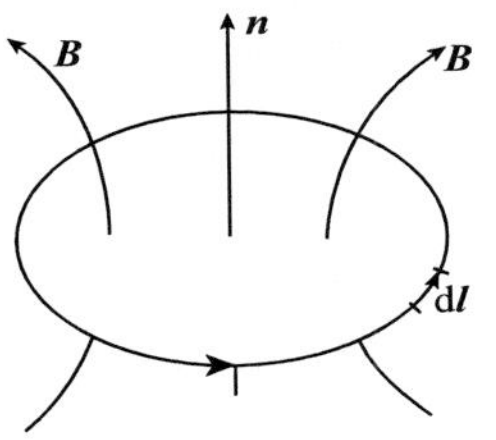

Figure 6.1 The right-handed relationship between d$\boldsymbol{l}$ and $\boldsymbol{B}$

where the negative sign indicates that the emf $\mathscr{E}$ is opposing the changing trends of the magnetic flux. Here it is assumed that the direction of the closed path is the same as the positive direction of $\mathscr{E}$, and it maintains a right-handed relationship with the positive direction of magnetic flux, as shown in Figure 6.1.

Either changing the magnetic field or moving the circuit can alter the magnetic flux through the open surface. What we are interested in is the case where the circuits are stationary but the magnetic fields are varying with time.

The induced emf $\mathscr{E}$ and magnetic flux Φ in Formula (6.1) can be expressed respectively as

$$\mathscr{E} = \oint_l \boldsymbol{E} \cdot \mathrm{d}\boldsymbol{l}$$

$$\Phi = \int_S \boldsymbol{B} \cdot \mathrm{d}\boldsymbol{S}$$

thus Formula (6.1) may be written in the following form

$$\oint_l \boldsymbol{E} \cdot \mathrm{d}\boldsymbol{l} = -\int_S \frac{\partial}{\partial t}\boldsymbol{B} \cdot \mathrm{d}\boldsymbol{S} \tag{6.2}$$

which is the integral form of Maxwell's Second Equation.

Faraday's law of induction was obtained under the presence of conductive circuit loop. While *Maxwell brought up a more generalized condition about the formation of the loop, i.e. the validity of the law of electromagnetic induction is independent from the properties of the materials the loop is made of. The loop can be made of conductive materials, or dielectric materials, or it is just a loop in the abstract sense.*

6.2.2 Differential Form of Maxwell's Second Equation

When $\boldsymbol{E}$ and $\boldsymbol{H}$ are continuous functions of space and time, and continuously differentiate over space and time, by applying Stokes theorem to the left-hand side of Formula (6.2) one can obtain

$$\oint_l \boldsymbol{E} \cdot \mathrm{d}\boldsymbol{l} = \int_S (\nabla \times \boldsymbol{E}) \cdot \mathrm{d}\boldsymbol{S} = -\int_S \frac{\partial \boldsymbol{B}}{\partial t} \cdot \mathrm{d}\boldsymbol{S} \tag{6.3}$$

which is applicable to an arbitrary open surface $\boldsymbol{S}$, yielding the differential form of Maxwell's Second Equation

$$\nabla \times \boldsymbol{E} = -\frac{\partial \boldsymbol{B}}{\partial t} \tag{6.4}$$

When fields do not vary with time, the integral and differential forms of Maxwell's Second Equation degenerates to the basic function for electrostatic field showing its irrotational property.

$$\oint_l \boldsymbol{E} \cdot \mathrm{d}\boldsymbol{l} = 0 \tag{6.5}$$

$$\nabla \times \boldsymbol{E} = 0 \tag{6.6}$$

It can be clearly shown by comparing Equations (6.4) and (6.6) that, static electric field is irrotational, but alternating electric field is not, it is *an electric field with vortex*. Its vortex source (density) is $\partial \boldsymbol{B}/\partial t$.

Maxwell's Second Equation describes the law that alternating magnetic fields generate alternating electric fields.

Example 6.1 In a uniformly distributed alternating magnetic field with $\boldsymbol{B} = \boldsymbol{e}_y B_0 \sin \omega t$, there is a rectangular conductive wire frame with its dimension as $a \times b$, suppose the initial angle between the direction of the wire frame and the y axis is α_0. Find the induced electromotive force (emf) in the wire frame under the following two cases: the wire frame is stationary and it is rotating around x axis at radian frequency ω.

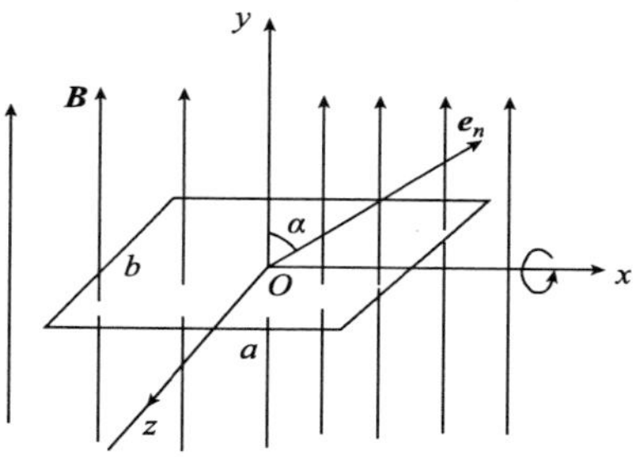

Figure of Example 6.1

Solution To obtain the emf $\mathscr{E}$, the magnetic flux Φ through the wire frame should be found first. Since the magnetic field is distributed uniformly, so Φ can be expressed as

$$\Phi = \int_S \boldsymbol{B} \cdot \mathrm{d}\boldsymbol{S} = \boldsymbol{B} \cdot \boldsymbol{e}_n(ab) = \boldsymbol{e}_y \cdot \boldsymbol{e}_n B(ab) \tag{a}$$

If the wire frame is stationary, then

$$\boldsymbol{e}_y \cdot \boldsymbol{e}_n = \cos \alpha_0 = \text{const.} \tag{b}$$

If the wire frame is rotating around x axis at radian frequency ω, furthermore, suppose $\alpha = 0$ when $t = 0$, or the expression for α is $\alpha = \omega t$, so when the wire frame is rotating

$$\boldsymbol{e}_y \cdot \boldsymbol{e}_n = \cos \alpha = \cos \omega t \tag{c}$$

Substituting Equation (b) into Equation (a) and taking differentiation over time, one can obtain the emf generated when the wire frame is stationary like

$$\begin{aligned} \mathscr{E} &= -\frac{\mathrm{d}\Phi}{\mathrm{d}t} = -\frac{\mathrm{d}}{\mathrm{d}t}\left(B_0 \sin \omega t \cdot \cos \alpha_0\right)(ab) \\ &= -ab\omega B_0 \cos \omega t \cdot \cos \alpha_0 \end{aligned} \tag{d}$$

When substituting Equation (c) into Equation (a) and taking differentiation over time, one can obtain the emf generated during the rotation of the wire frame at radian frequency ω as

$$\begin{aligned} \mathscr{E} &= -\frac{\mathrm{d}\Phi}{\mathrm{d}t} = -\frac{\mathrm{d}}{\mathrm{d}t}\left(B_0 \sin \omega\, t \cdot \cos \omega\, t\right)(ab) \\ &= -B_0 ab\omega \cos 2\omega\, t \end{aligned}$$

6.3 Ampere's Circuital Law and Maxwell's First Equation

Maxwell's key contribution to electromagnetic theory is particularly exhibited by Maxwell's First Equation. He revealed the limitation of simple Ampere's circuital law, then proposed the concept of displacement current to amend simple Ampere's circuital law, leading to the formation of Maxwell's First Equation.

6.3.1 Simple Ampere's Circuital Law and Its Inconsistency for Alternating Electromagnetic Fields

Ampere's circuital law obtained under direct-current condition (also known as simple Ampere's circuital law) is like

$$\left.\begin{aligned} &\oint_l \boldsymbol{H}\cdot \mathrm{d}\boldsymbol{l} = \int_S \boldsymbol{J}\cdot \mathrm{d}\boldsymbol{S} = I \\ &\nabla \times \boldsymbol{H} = \boldsymbol{J} \end{aligned}\right\} \tag{6.7}$$

where I is conduction current, $\boldsymbol{J}$ is conduction current density. Inconsistency appears in this law for alternating fields. Different ways may be used to reveal the inconsistency.

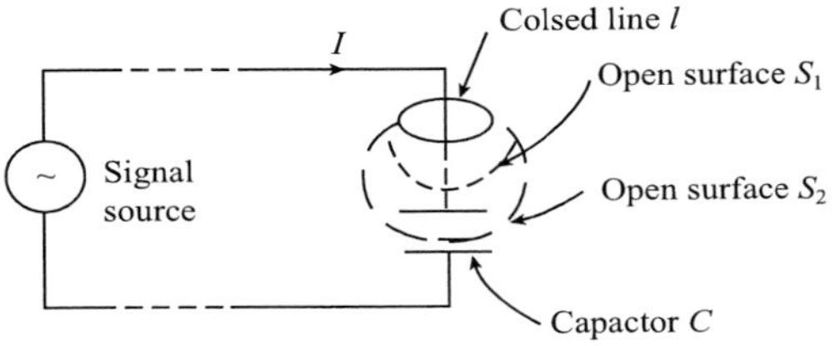

Figure 6.2 AC circuit with capacitor.

Method 1 The integral form of simple Ampere's law $\oint_l \boldsymbol{H}\cdot \mathrm{d}\boldsymbol{l} = I$ can be directly applied to alternating fields to demonstrate its incompleteness.

As shown in Figure 6.2, for a alternating-current circuit with capacitor, there can be many open surfaces with the contour of the closed path l, but the integration of magnetic field along the closed loop l has to be a unique value. In the figure, the surface S_1 is outside the capacitor, part of the surface S_2 is inside the capacitor. Obviously the conduction current through S_1 is I, while the conduction current through S_2 is zero, according to simple Ampere's law, which leaves

$$\oint_l \boldsymbol{H}\cdot \mathrm{d}\boldsymbol{l} = \begin{cases} I, & \text{for surface } S_1 \\ 0, & \text{for surface } S_2 \end{cases}$$

It is evidently contradictory, which indicated that simple Ampere's law would be incomplete when being applied to alternating fields.

Method 2 The differential form of Ampere's law $\nabla \times \boldsymbol{H} = \boldsymbol{J}$, as well as continuity equation, can be directly employed to analyze the contradiction brought up by simple Ampere's law in alternating fields.

With the differential form of simple Ampere's law

$$\nabla \times \boldsymbol{H} = \boldsymbol{J}$$

and a vector identity

$$\nabla \cdot \nabla \times \boldsymbol{H} = 0$$

one can have

$$\nabla \cdot \boldsymbol{J} = 0 \tag{6.8}$$

Apparently, $\nabla \cdot \boldsymbol{J} = 0$ is just a special case for continuity equation $\nabla \cdot \boldsymbol{J} = -\partial\rho/\partial t$, which only applies to DC condition. That is to say, $\nabla \times \boldsymbol{H} = \boldsymbol{J}$ is not universally valid. So a correction to it is needed to make it valid under both AC and DC conditions. That is the issue Maxwell intended to solve.

6.3.2 Maxwell's Concept of Displacement Current and Maxwell's First Equation

To deal with the contradiction encountered by simple Ampere's circuital law under alternating fields, Maxwell proposed the concept of displacement current, that is, *time-varying electric field produces displacement current. Like conduction current, the displacement current can also generate magnetic field.* Based on this assumption, simple Ampere's circuital law was modified leading to Maxwell's First Equation.

Inspiration can be acquired from continuity equation for electric current when solving the inconsistency in simple Ampere's circuital law for alternating fields. Substituting $\nabla \cdot \boldsymbol{D} = \rho$ into continuity equation $\nabla \cdot \boldsymbol{J} = -\partial\rho/\partial t$, one can have

$$\nabla \cdot \boldsymbol{J} + \frac{\partial}{\partial t}\nabla \cdot \boldsymbol{D} = 0$$

Interchanging the order of differentiation over time and over space would not affect the completeness of the equation above, so one can get

$$\nabla \cdot \left(\boldsymbol{J} + \frac{\partial \boldsymbol{D}}{\partial t}\right) = 0 \tag{6.9}$$

where, $\boldsymbol{J}$ is conduction current density, $\boldsymbol{D}$ is electric displacement vector, and $\partial\boldsymbol{D}/\partial t$ is called displacement current density, their vector sum $(\boldsymbol{J} + \partial\boldsymbol{D}/\partial t)$ is called the total current density, and Equation (6.9) is the differential form of the continuity equation for the total current.

Applying divergence theorem to Equation (6.9) one can obtain

$$\int_V \nabla \cdot \left(\boldsymbol{J} + \frac{\partial \boldsymbol{D}}{\partial t}\right) \mathrm{d}V = \oint_S \left(\boldsymbol{J} + \frac{\partial \boldsymbol{D}}{\partial t}\right) \cdot \mathrm{d}\boldsymbol{S} = 0$$

that is

$$\oint_S \left(\boldsymbol{J} + \frac{\partial \boldsymbol{D}}{\partial t}\right) \cdot \mathrm{d}\boldsymbol{S} = 0 \tag{6.10}$$

which is the integral form of continuity equation for total current.

Maxwell believed that, *not only conduction current, but also displacement current, can produce magnetic field.* So displacement current should be introduced to simple Ampere's circuital law, in other words, *conduction current should be replaced with the total current,* which leaves

$$\nabla \times \boldsymbol{H} = \boldsymbol{J} + \frac{\partial \boldsymbol{D}}{\partial t} \tag{6.11}$$

which is the differential form of Maxwell's First Equation.

Using Stoke's theorem to the expression above one can have

$$\int_S (\nabla \times \boldsymbol{H}) \cdot \mathrm{d}S = \oint_l \boldsymbol{H} \cdot \mathrm{d}\boldsymbol{l} = \int_S \left(\boldsymbol{J} + \frac{\partial \boldsymbol{D}}{\partial t} \right) \cdot \mathrm{d}\boldsymbol{S}$$

that is

$$\oint_l \boldsymbol{H} \cdot \mathrm{d}\boldsymbol{l} = \int_S \left(\boldsymbol{J} + \frac{\partial \boldsymbol{D}}{\partial t} \right) \cdot \mathrm{d}\boldsymbol{S} \tag{6.12}$$

which is the integral form of Maxwell's First Equation.

This equation properly resolves the contradiction of *simple Ampere's circuital law* in Figure 6.2, and gives a comprehensive implication of Ampere's circuital law. Hence Maxwell's First Equation is also known as *generalized Ampere's circuital law*, which reveals the law that besides conduction current displacement current also generates magnetic field.

The curl of magnetostatic fields is zero when there is no conduction current. But it does not hold for alternating magnetic fields, which may have nonzero curl if there isn't any conduction current. This is because alternating electric fields are supporting it as 'sources'. *Both alternating currents and alternating electric fields are the vortex sources for alternating magnetic fields.*

6.3.3 Displacement Current and Continuity of Total Current

1. Displacement Current

Both conduction current and displacement current have the dimension of current, and can generate magnetic field. These are their common properties. However, these two kinds of currents are established in quite different ways. There must be free charges moving around for conduction current, but it does not have to be in the case for displacement current, which is corresponding to (or generated by) time-varying electric fields only.

In dielectrics,

$$\boldsymbol{D} = \varepsilon_0 \boldsymbol{E} + \boldsymbol{P}$$

where $\boldsymbol{P}$ is the polarization (dipole moment's vector sum per unit volume) of the medium. Then the displacement current density is

$$\frac{\partial \boldsymbol{D}}{\partial t} = \varepsilon_0 \frac{\partial \boldsymbol{E}}{\partial t} + \frac{\partial \boldsymbol{P}}{\partial t} \tag{6.13}$$

The first component $\varepsilon_0 \frac{\partial \boldsymbol{E}}{\partial t}$ indicates that, displacement current can be produced by time-varying electric fields, and does not require any movement of charged particles. The second component $\frac{\partial \boldsymbol{P}}{\partial t}$ is generated by time-varying electric moment in the polarized dielectrics. However, $\frac{\partial \boldsymbol{P}}{\partial t} = 0$ in vacuum, which means that there is no moving charged particle for displacement current in vacuum.

Displacement current density $\frac{\partial \boldsymbol{D}}{\partial t}$ is a vector. Its magnitude is equal to the changing ratio of $\boldsymbol{D}$ against time, which is corresponding to frequency, the higher the frequency, the greater the displacement current density. The direction of the displacement current density is in accordance with the positive increment for electric displacement $\boldsymbol{D}$ over time t.

2. Total Current's Continuity

The total current continuity equations are (6.9) and (6.10). The total current consists of the conduction current and the displacement current.

(1) Conduction currents are the currents caused by the movement of free charges. Its current density includes the conduction current density $\boldsymbol{J}_{\mathrm{c}}$ and the movement current density $\boldsymbol{J}_{\mathrm{v}}$, that is

$$\boldsymbol{J} = \boldsymbol{J}_{\mathrm{c}} + \boldsymbol{J}_{\mathrm{v}} = \sigma \boldsymbol{E} + \rho \boldsymbol{v} \tag{6.14}$$

(2) The displacement current density $\boldsymbol{J}_{\mathrm{d}}$ is

$$\boldsymbol{J}_{\mathrm{d}} = \frac{\partial \boldsymbol{D}}{\partial t} = \varepsilon_0 \frac{\partial \boldsymbol{E}}{\partial t} + \frac{\partial \boldsymbol{P}}{\partial t} \tag{6.15}$$

$$\boldsymbol{J}_{\mathrm{d}} = \varepsilon_0 \frac{\partial \boldsymbol{E}}{\partial t} \quad (\text{in vacuum}, \boldsymbol{P} = 0) \tag{6.16}$$

(3) If the external (impressed) current source $\boldsymbol{J}_{\mathrm{i}}$ generating fields is not considered, the total current is the sum of $\boldsymbol{J}_{\mathrm{c}} + \boldsymbol{J}_{\mathrm{v}} + \boldsymbol{J}_{\mathrm{d}}$.

(4) The total current preserves continuity. If the conduction current through some closed surface is increasing, the displacement current through it must be decreasing; the amount by which the conduction current increases, is just the amount by which the displacement current decreases, which satisfies $\oint_S \boldsymbol{J} \cdot \mathrm{d}\boldsymbol{S} = -\oint_S \left(\frac{\partial \boldsymbol{D}}{\partial t}\right) \cdot \mathrm{d}\boldsymbol{S}$.

Example 6.2 A closed surface contains one plate of a capacitor, as shown in Figure of Example 6.2. Prove: the conduction current flowing into (or out of) the closed surface is equal to the displacement current flowing out of (or into) the closed surface. To simplify the issue, suppose d is very small for the plate capacitor, which has a relatively large area of S; electric field is uniformly distributed in the capacitor; the voltage on the given capacitor is $U = U_0 \sin \omega t$; the capacitor is filled with air medium.

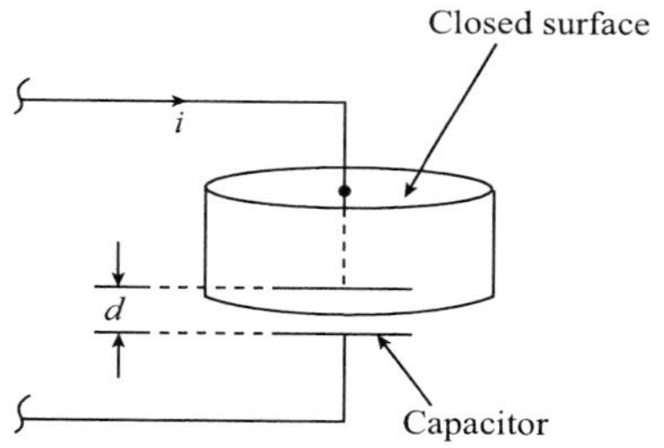

Figure of Example 6.2

Solution (1) the conduction current through the closed surface i is

$$i = \frac{\mathrm{d}q}{\mathrm{d}t} = C\frac{\mathrm{d}U}{\mathrm{d}t} = C\omega U_0 \cos \omega t$$

(2) find the displacement current through the closed surface i_{d}.

The alternating electric field in the capacitor is uniform, and its value is

$$E = \frac{U}{d} = \frac{U_0}{d} \sin \omega t$$

therefore the displacement current in the capacitor is

$$\frac{\partial \boldsymbol{D}}{\partial t} = \varepsilon_0 \frac{\partial \boldsymbol{E}}{\partial t} = \varepsilon_0 \frac{U_0}{d} \omega \cos \omega t$$

Then the total displacement current through the closed surface is identical to the total displacement current in the capacitor:

$$i_{\mathrm{d}} = \left(\frac{\partial \boldsymbol{D}}{\partial t}\right) S = \left(\frac{\varepsilon_0 S}{d}\right) \omega V_0 \cos \omega t = C\omega U_0 \cos \omega t$$

Obviously the conduction current through the closed surface is equal to the displacement current through that closed surface. They are not only equal in magnitude, but if one is flowing into the closed surface, the other must be flowing out of the closed surface, so that the total current continuity Equation (6.10) for alternating fields is satisfied.

6.4 Gauss's Law and Maxwell's Third Equation

The integral and differential forms of Gauss's Law for electrostatic fields are

$$\oint_S \boldsymbol{D} \cdot \mathrm{d}\boldsymbol{S} = \int_V \rho \mathrm{d}V = q \tag{6.17}$$

$$\nabla \cdot \boldsymbol{D} = \rho \tag{6.18}$$

The two equations above also hold for alternating fields, but $\boldsymbol{D}, \rho, q$ are all varying with time. For example, $\boldsymbol{D}$ in Expression (6.18) can be (and should be) viewed as being produced jointly by both alternating charges and alternating magnetic fields, but the electric fields generated by alternating magnetic fields (as sources) have zero divergence, which leads to alternating electric fields activated by alternating charges have non zero divergence, equal to the volume charge density at that point. Accordingly, Gauss's law is still applicable in alternating fields. Then, Equations (6.17) and (6.18) are integral and differential forms of Maxwell's Third Equation, respectively.

6.5 Maxwell's Fourth Equation

Since no magnetic charges or monopoles has been observed, hence magnetic line of force is also closed under alternating fields conditions, namely the magnetic flux through any closed surface is always zero, which can be expressed as

$$\oint_S \boldsymbol{B} \cdot \mathrm{d}\boldsymbol{S} = 0 \tag{6.19}$$

$$\nabla \cdot \boldsymbol{B} = 0 \tag{6.20}$$

$\boldsymbol{B}$ in Equations (6.19) and (6.20) should be viewed as being produced jointly by both conduction currents and alternating electric fields. The magnetic fields generated by alternating electric fields also have closed magnetic line of force, and its magnetic flux through any closed surface is always zero, with its divergence being zero. Hence expressions (6.19) and (6.20) are applicable to magnetostatic fields as well as alternating magnetic fields, and they are known as integral and differential forms of Maxwell's fourth Equation, respectively.

Example 6.3 Under Cartesian coordinates, given that a magnetic field in a source-free region: $H_x = 0, H_y = H_0 \sin k'y \cdot \sin(\omega t - kz)$, where k' and k are constants, find the H_z component of the magnetic field.

Method 1 Find the solution directly from Maxwell's Fourth Equation ($\nabla \cdot \boldsymbol{B} = 0$) .

Based on

$$\nabla \cdot \boldsymbol{B} = \frac{\partial B_x}{\partial x} + \frac{\partial B_y}{\partial y} + \frac{\partial B_z}{\partial z} = \frac{\partial B_y}{\partial y} + \frac{\partial B_z}{\partial z} = 0$$

that is

$$\frac{\partial H_z}{\partial z} = -\frac{\partial H_y}{\partial y} = -H_0 k' \cos k'y \sin(\omega t - kz)$$

$$H_z = -H_0 k' \cos k'y \int \sin(\omega t - kz)\mathrm{d}z$$

$$= -H_0 k' \cdot \frac{1}{k} \cdot \cos k'y \cdot \cos k'y(\omega t - kz) + C$$

Since alternating magnetic field is under study, set the constant C to zero for the indefinite integral, then one can have

$$H_z = -\frac{1}{k} k' H_0 \cos k'y \cos(\omega t - kz)$$

Method 2 Solve the problem based on Maxwell's First and Second Equations
According to Maxwell's First Equation

$$\frac{\partial \boldsymbol{D}}{\partial t} = \nabla \times \boldsymbol{H}$$

$$= \boldsymbol{e}_x \left(\frac{\partial H_z}{\partial y} - \frac{\partial H_y}{\partial z} \right) + \boldsymbol{e}_y (-) \frac{\partial H_z}{\partial x} + \boldsymbol{e}_z \frac{\partial H_y}{\partial x} \quad \text{(because } H_x = 0\text{)}$$

$$= \boldsymbol{e}_x \left(\frac{\partial H_z}{\partial y} - \frac{\partial H_y}{\partial z} \right) \quad \text{(because the field's components are functions of just } y, z, t\text{)}$$

It can be known that the electric field has only the E_x component.
According to Maxwell's Second Equation

$$-\frac{\partial \boldsymbol{B}}{\partial t} = \boldsymbol{e}_x \left(\frac{\partial E_z}{\partial y} - \frac{\partial E_y}{\partial z} \right) + \boldsymbol{e}_y \left(\frac{\partial E_x}{\partial z} - \frac{\partial E_z}{\partial x} \right) + \boldsymbol{e}_z \left(\frac{\partial E_y}{\partial x} - \frac{\partial E_x}{\partial y} \right)$$

$$= \boldsymbol{e}_y \frac{\partial E_x}{\partial z} - \boldsymbol{e}_z \frac{\partial E_x}{\partial y} \quad \text{(because } E_y = 0, E_z = 0\text{)} \tag{a}$$

From Expression (a) one can obtain

$$-\mu \frac{\partial H_y}{\partial t} = \frac{\partial E_x}{\partial z} \tag{b}$$

$$-\mu \frac{\partial H_z}{\partial t} = \frac{\partial E_x}{\partial y} \tag{c}$$

Based on Expression (b) one can have

$$E_x = -\mu \int \frac{\partial H_y}{\partial t} \mathrm{d}z = \frac{kH_0}{\omega\varepsilon} \sin(\omega t - kz) \cdot \sin(k'y)$$

For alternating fields, set the constant to zero for the indefinite integral.
Substituting the obtained E_x into Expression (c) we have

$$H_z = -\frac{k'}{k} H_0 \cos(k'y) \cdot \cos(\omega t - kz)$$

which is derived by using the relationship between parameters like $\omega^2 \mu\varepsilon = k^2$.

6.6 Maxwell's Equations and Auxiliary Equations

There are four Maxwell's Equations, with their integral forms as

$$\left.\begin{aligned}&\oint_l \boldsymbol{H}\cdot \mathrm{d}\boldsymbol{l} = \int_S \left(J + \frac{\partial D}{\partial t}\right)\cdot \mathrm{d}\boldsymbol{S}\\&\oint_l \boldsymbol{E}\cdot \mathrm{d}l = -\int_S \frac{\partial \boldsymbol{B}}{\partial t}\cdot \mathrm{d}\boldsymbol{S}\\&\oint_S \boldsymbol{D}\cdot \mathrm{d}\boldsymbol{S} = \int_V \rho \mathrm{d}V = q\\&\oint_S \boldsymbol{B}\cdot \mathrm{d}\boldsymbol{S} = 0\end{aligned}\right. \tag{6.21}$$

and their differential forms as

$$\left.\begin{aligned}&\nabla\times\boldsymbol{H} = \boldsymbol{J} + \frac{\partial \boldsymbol{D}}{\partial t}\\&\nabla\times\boldsymbol{E} = -\frac{\partial \boldsymbol{B}}{\partial t}\\&\nabla\cdot\boldsymbol{D} = \rho\\&\nabla\cdot\boldsymbol{B} = 0\end{aligned}\right\} \tag{6.22}$$

It should be noted that, the current density $\boldsymbol{J}$ in the equations should include the impressed current $\boldsymbol{J}_{\mathrm{i}}$ from external sources (if they exist), as well as conduction current density $\boldsymbol{J}_{\mathrm{c}} = \sigma\boldsymbol{E}$, and movement current density $\boldsymbol{J}_{\mathrm{v}} = \rho\boldsymbol{v}$.

The Maxwell's Equations above are all obtained under the assumption that the integral region (line, surface, volume) is not varying with time. Total current continuity equation is not included in Maxwell's Equations because it is not an independent one.

There are 4 fundamental field vectors, as $\boldsymbol{E}, \boldsymbol{D}, \boldsymbol{B}, \boldsymbol{H}$, each of which has 3 components, so there are total 12 unknown variables. Accordingly, to determine 12 unknown variables requires 12 scalar equations. *With current continuity equation $\nabla\cdot\boldsymbol{J} = -\partial\rho/\partial t$ taken into account, there are only two independent curl equations*, which can just present 6 scalar equations, while $\boldsymbol{D} = \varepsilon\boldsymbol{E}$ and $\boldsymbol{B} = \mu\boldsymbol{H}$ in the constitutive relations can give 3 equations each, totally 12 scalar equations, which keeps the number of fields equations consistent with the number of unknown variables. Then final solutions can be found.

Constitutive relations of fields indicate the dependencies between the fields and the media, so they are also known as constitutive equations of the media or auxiliary equations, as below

$$\boldsymbol{J} = \sigma\boldsymbol{E}, \quad \boldsymbol{D} = \varepsilon\boldsymbol{E}, \quad B = \mu\boldsymbol{H} \tag{6.23}$$

It can be seen from Maxwell's Equations that, the first equation indicates that not only conduction currents, but also alternating electric fields, generate magnetic fields. Furthermore, the magnetic fields' curl is not zero even in places where the conduction current density is zero, since their 'sources' are alternating electric fields. The second equation shows that alternating magnetic fields can produce electric fields, or electric vortex fields, and their 'sources' are alternating magnetic fields.

Differential forms of the first and the second equations of Maxwell's Equations are two curl equations. *The sources for curls of vector fields are called 'vortex sources'.* Alternating electric fields and magnetic fields are sources for each other.

Differential forms of the third and the fourth Maxwell's Equations are two divergence equations. *The sources for divergences of vector fields are called 'divergence-producing types of sources'.* Obviously, the 'divergence-producing types of sources' for electric field $\boldsymbol{E}$ is charge density ρ, while magnetic flux density is solenoidal (divergence-free).

6.7 Complex Format of Maxwell's Equations

When the source charges or currents are varying in a time-harmonic manner (in sinusoidal or cosine way), the electric fields and magnetic fields they produce will also vary time-harmonically. Time-harmonic variation is a very common case, even for the electromagnetic waves not varying time-harmonically, they can also be decomposed to the sum of the fundamental wave and higher order harmonic waves, then can be handled as harmonic waves individually. Rewriting time-harmonic variables in complex format will be convenient for mathematical operations.

Take the electric field $\boldsymbol{E}$ as an example, assume its instantaneous expression is

$$\begin{aligned}\boldsymbol{E}(x,y,z,t) &= \boldsymbol{E}_0(x,y,z)\cos(\omega t+\phi_0)\\ &= \sqrt{2}E_{\mathrm{e}}(x,y,z)\cos(\omega t+\phi_0)\end{aligned} \tag{6.24}$$

where $\boldsymbol{E}_0\,(x,y,z)$ and ϕ_0 are the amplitude and the initial phase of alternating electric field, and they are functions of space. Let $\dot{\boldsymbol{E}}_0\,(x,y,z)=\sqrt{2}\boldsymbol{E}_{\mathrm{e}}\,(x,y,z)\cdot \mathrm{e}^{\mathrm{j}\phi_0}$, $\dot{\boldsymbol{E}}_0\,(x,y,z)$ is called the complex amplitude vector of the electric field, and *name the following expression*

$$\dot{\boldsymbol{E}}(x,y,z,t)=\sqrt{2}\dot{\boldsymbol{E}}_{\mathrm{e}}(x,y,z)\cdot \mathrm{e}^{\mathrm{j}\omega t} \tag{6.25}$$

as the complex format with $\mathrm{e}^{\mathrm{j}\omega t}$. Taking its real part (denoted by Re), one can have the instantaneous representation of $\boldsymbol{E}$ as

$$\begin{aligned}\boldsymbol{E}(x,y,z,t) &= \mathrm{Re}\left[\sqrt{2}\dot{\boldsymbol{E}}_{\mathrm{e}}(x,y,z)\cdot \mathrm{e}^{\mathrm{j}\omega t}\right]\\ &= \mathrm{Re}\left[\sqrt{2}\boldsymbol{E}_{\mathrm{e}}(x,y,z)\cdot \mathrm{e}^{\mathrm{j}(\omega t+\phi_0)}\right]\\ &= \sqrt{2}\boldsymbol{E}_{\mathrm{e}}(x,y,z)\cos(\omega t+\phi_0)\end{aligned} \tag{6.26}$$

With the help of Equations (6.25) and (6.24), transforming the instantaneous expression as,

$$\nabla\times\boldsymbol{E}(x,y,z,t)=-\frac{\partial}{\partial t}\boldsymbol{B}(x,y,z,t)$$

The left hand side can be represented as

$$\nabla\times\boldsymbol{E}(x,y,z,t)=\mathrm{Re}\left\{\sqrt{2}\mathrm{e}^{\mathrm{j}\omega t}\left[\nabla\times\dot{\boldsymbol{E}}_{\mathrm{e}}(x,y,z)\right]\right\}$$

and the right hand side is

$$-\frac{\partial\boldsymbol{B}(x,y,z,t)}{\partial t}=-\mathrm{Re}\left[\sqrt{2}\mathrm{e}^{\mathrm{j}\omega t}\cdot \mathrm{j}\omega\dot{\boldsymbol{B}}_{\mathrm{e}}(x,y,z)\right]$$

since the two expressions above are equal, then one can obtain the complex format of Maxwell's Second Equation as

$$\nabla\times\dot{\boldsymbol{E}}(x,y,z)=-\mathrm{j}\omega\dot{\boldsymbol{B}}(x,y,z)$$

With similar operation, complex format of Maxwell's Equations can be achieved. For the ease of writing, we usually remove the dot mark from complex field vectors, then the complex format of Maxwell's Equations are

$$\nabla \times \dot{\boldsymbol{H}} = \dot{\boldsymbol{J}} + \mathrm{j}\omega \dot{\boldsymbol{D}} \tag{6.27}$$

$$\nabla \times \dot{\boldsymbol{E}} = -\mathrm{j}\omega \dot{\boldsymbol{B}} \tag{6.28}$$

$$\nabla \cdot \dot{\boldsymbol{D}} = \rho \tag{6.29}$$

$$\nabla \cdot \dot{\boldsymbol{B}} = 0 \tag{6.30}$$

The field vectors in the complex format of Maxwell's Equations do not include time factors. *But in practical usage, to give a complete expression with respect to time and space as well as for convenience in mathematical manipulations, we often introduce time factor* $\mathrm{e}^{\mathrm{j}\omega t}$ *in the complex format of field vectors*, such as Equation (6.25).

Generally speaking, for harmonic waves, for instance, the electric field $E_x = F(z, t)$ can have three types of expression:

(1) Complex expression with $\mathrm{e}^{\mathrm{j}\omega t}$:

$$E_x = E_0 \mathrm{e}^{\mathrm{j}(\omega t + kz + \phi_0)} \tag{6.31}$$

(2) Instantaneous expression [expressed in time-domain, obtained by taking the real (or imaginary) part of Expression (6.31)]

$$E_x = E_0 \cos(\omega t + kz + \phi_0) \tag{6.32}$$

(3) Complex expression [expressed in frequency-domain, obtained by neglecting the time factor $\mathrm{e}^{\mathrm{j}\omega t}$ of Expression (6.31)]

$$E_x = E_0 \mathrm{e}^{\mathrm{j}(kz + \phi_0)} \tag{6.33}$$

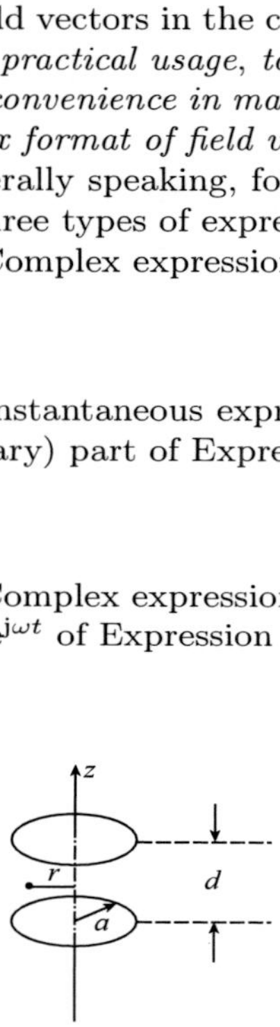

Figure of Example 6.4

Example 6.4 A disc capacitor, with radius $a = 5 \times 10^{-3}$m, the distance between the two discs $d = 10^{-4}$m, the parameters of the filling dielectrics $2.55\varepsilon_0, \mu_0$, and the low frequency voltage on the capacitor $U = U_0 \cos \omega t$, suppose the edge effect is not considered, find inside the capacitor: (1) $\boldsymbol{E}$ and $\boldsymbol{H}$. (2) The Electric field energy W_{e}. (3) The magnetic field energy W_{m}. (4) The ratio between the max. electric field energy $W_{\mathrm{e(m)}}$ and the max. magnetic field energy $W_{\mathrm{m(m)}}$.

Solution (1) Find $\boldsymbol{E}$ and $\boldsymbol{H}$.

Since the edge effect is ignored, the electric field in the disc capacitor is uniformly distributed, as

$$\boldsymbol{E} = \boldsymbol{e}_z \frac{U}{d} = \boldsymbol{e}_z \frac{U_0}{d} \cos \omega t \tag{a}$$

And the displacement current density $\boldsymbol{J}_{\mathrm{d}}$ is also uniformly distributed

$$\boldsymbol{J}_{\mathrm{d}} = \boldsymbol{e}_z \frac{\partial D}{\partial t} = -\boldsymbol{e}_z \omega \varepsilon \left(\frac{U_0}{d}\right) \sin \omega t \tag{b}$$

There are two ways to obtain magnetic field $\boldsymbol{H}$ via $\boldsymbol{J}_{\mathrm{d}}$.

Method 1 The generalized Ampere's circuital law (integral form of the first equation in Maxwell's Equations), can be used to find $\boldsymbol{H}$.

Suppose there is a closed circle with radius r centered on some point on z axis, along which due to the uniform distribution of $\boldsymbol{J}_\mathrm{d}$ inside the capacitor, one can have

$$H_\phi = \frac{\pi r^2 \cdot J_\mathrm{d}}{2\pi r} = -\frac{U_0}{2d}\omega\varepsilon r \sin\omega t \tag{c}$$

Equation (b) is used for deriving the equation above.

Method 2 Utilizing the available $\boldsymbol{J}_\mathrm{d}$ to find $\boldsymbol{H}$ directly via the differential form of the first equation in Maxwell's Equations (under cylindrical coordinates).

$$\nabla \times \boldsymbol{H} = \boldsymbol{e}_z \boldsymbol{J}_\mathrm{d} = \boldsymbol{e}_z \left[\frac{1}{r}\frac{\partial (rH_\phi)}{\partial r} - \frac{1}{r}\frac{\partial H_r}{\partial \phi}\right]$$

Based on the cylindrical symmetry, $\dfrac{\partial}{\partial \phi} = 0$, the equation above is transformed to

$$\frac{1}{r}\frac{\partial (rH_\phi)}{\partial r} = J_\mathrm{d} = -\omega\varepsilon\left(\frac{U_0}{d}\right)\sin\omega t$$

so that H_ϕ is obtained [namely Formula (c)].

(2) The electric field energy W_e.

Different ways can be used to find the electric field energy W_e in the capacitor.

Method 1 The circuit theory is used to obtain W_e.

$$W_\mathrm{e} = \frac{1}{2}CU^2 = \frac{1}{2}\left(\frac{\varepsilon S}{d}\right)U^2 = \frac{1}{2}\left(\frac{\varepsilon\pi a^2}{d}\right)U_0^2\cos^2\omega t \tag{d}$$

Method 2 The electric field's volume energy density w_e is used to find W_e.

Given

$$w_\mathrm{e} = \frac{1}{2}\varepsilon E^2$$

therefore

$$W_\mathrm{e} = \int_v w_\mathrm{e}\mathrm{d}v = \frac{1}{2}\varepsilon\int_0^a\int_0^{2\pi} E^2 (r\mathrm{d}r\mathrm{d}\phi)\, d = \frac{1}{2}\left(\frac{\varepsilon\pi a^2}{d}\right)U_0^2\cos^2\omega t \tag{e}$$

which is derived by using Expression (a).

(3) The magnetic field energy W_m.

The magnetic field's volume energy density w_m is employed to find W_m.

Given

$$w_\mathrm{m} = \frac{1}{2}\mu_0 H^2$$

then

$$W_\mathrm{m} = \int_v w_\mathrm{m}\mathrm{d}V = \frac{1}{2}\mu_0\int_0^a\int_0^{2\pi} H^2 (r\mathrm{d}r\mathrm{d}\phi)\, d = \frac{1}{16d}\left(U_0^2\pi\omega^2\mu_0\varepsilon^2 a^4\right)\sin^2\omega t \tag{f}$$

which is derived by using Expression (c).

(4) When $f = 1$ MHz, 100 MHz, find the maximum ratio between W_e and W_m. Using Expressions (e) and (f) for the maximum value of energy (that is to take $\sin\omega t = 1, \cos\omega t = 1$), which leads to

$$\frac{W_{\mathrm{m(m)}}}{W_{\mathrm{e(m)}}} = \frac{\omega^2\varepsilon_r\varepsilon_0\mu_0 a^2}{8} \tag{g}$$

thus

$$\frac{W_{m(m)}}{W_{e(m)}} = 3.35 \times 10^{-9} \quad (\text{when } f = 1 \text{ MHz})$$

$$\frac{W_{m(m)}}{W_{e(m)}} = 3.35 \times 10^{-5} \quad (\text{when } f = 100 \text{ MHz})$$

It can be seen in the capacitor electric energy is much greater than the magnetic energy.

Example 6.5 On the charge density ρ and the relaxation time τ in conductive media.

Solution In conductive media (linear, isotropic) in compliance with Ohm's law, the differential form of Ohm's law is

$$\boldsymbol{J} = \sigma \boldsymbol{E}$$

If the medium is homogeneous, the following operations can be used for the expression above

$$\nabla \cdot \boldsymbol{J} = \nabla \cdot (\sigma \boldsymbol{E}) = \sigma \nabla \cdot \boldsymbol{E} = \sigma \rho / \varepsilon = -\frac{\partial \rho}{\partial t}$$

which is derived by using

$$\nabla \cdot \boldsymbol{E} = \rho / \varepsilon \quad \text{and} \quad \nabla \cdot \boldsymbol{J} = -\frac{\partial \rho}{\partial t}$$

so an equation about ρ is obtained as

$$-\frac{\partial \rho}{\partial t} = \frac{\sigma}{\varepsilon} \rho$$

for which the solution is

$$\rho = \rho_0 \mathrm{e}^{-(\sigma/\varepsilon)t} = \rho_0 \mathrm{e}^{-(t/\tau)}$$

This expression shows that the free charge density within the conductive media will be exponentially decreased with respect to time.

$\tau = \dfrac{\varepsilon}{\sigma}$ *in the formula is called relaxation time, in unit of second.* τ *is the duration it takes for the charge density from the beginning at* ρ_0 *to be reduced to* $\dfrac{\rho_0}{e}$. *For average conductors, the relaxation time is extremely short.* Take metal copper as an example, $\sigma = 5.8 \times 10^7 \text{S/m}$, $\varepsilon = 10^{-9}/(36\pi)\text{F/m}$, the relaxation time is about 10^{-19}s.

6.8 Boundary Conditions for Alternating Fields

In practical problems of interest, there are often different media. Thus, it is necessary to study the laws that electric fields and magnetic fields follow at the interface between two media (boundary between media), which are called boundary conditions.

The differential form of Maxwell's equations is only applicable at the positions where those field components are differentiable, which represents the relations among various fields in homogeneous media or continuously changing media. However at the boundaries the media properties have abrupt changes and leave discontinuities of electromagnetic fields. Therefore, the integral form of Maxwell's Equations should be employed to study the laws of fields at the places where the media are discontinuous.

6.8.1 Electromagnetic Field in Perfect Conductor

The conductor surface is one of the boundaries we often deal with. A perfect conductor has conductivity σ value of infinity. *One important conclusion is that there is no static electric fields, neither alternating electromagnetic fields, in perfect conductors.*

The static charge exists only on the surface of conductor. This is because if there were an electric field in the conductor at electrostatic state, the electric field would apply force to charges, and the charges in the conductor would have been moved. That is not a static electric field anymore. In addition, if a conductor is placed in electrostatic field, there is still no electrostatic field in the conductor. This is because that the positive/negative charges in the conductor are influenced by the external electric field, and the negative charges will move in the direction with higher electric potential, establishing an inner field that counteracts the external electric field until these two fields are equal. Moreover, the process of counteracting the external electric field to reach a balance is very quick. Hence there is no static electric field inside conductors, and static charges are only distributed along the surface of a conductor.

Are there any alternating electromagnetic fields inside perfect conductors?

It can be seen from the differential form of Ohm's law $\boldsymbol{J} = \sigma \boldsymbol{E}$. Due to the infinity of the perfect conductor's conductivity σ, as long as there is a very small electric field strength $\boldsymbol{E}$, infinite current would be generated inside perfect conductor, then producing infinite magnetic field, which is definitely not possible. Therefore, there is neither alternating electric field, nor the alternating magnetic field, within perfect conductors. Regarding this issue, an in-depth analysis will be given by the section discussing Skin Effect in Chapter 7.

6.8.2 Boundary Conditions for Normal Fields $\boldsymbol{D}_n$ and $\boldsymbol{B}_n$

1. Continuity of D_n at the Interface Between Perfect Dielectrics

Suppose the parameters of the two dielectrics involved are ε_1, μ_1 and ε_2, μ_2, respectively, and the surface charge density on the interface between them is ρ_s. Place a cylinder-shaped closed surface on the interface as shown in Figure 6.3, and the top and bottom faces of this closed surface have extremely small areas of ΔS in parallel with the interface, with the height of the cylinder Δh approaching zero. Integral form of Gauss's law is to be used to discuss the boundary condition for normal electric fields. The direction of $e_{n(S)}$ in the figure is normally pointing outside of the corresponding surface.

Since ΔS is extremely small, the D on ΔS can be viewed as uniformly distributed. Besides, the electrical flux going out through the surrounding side of the cylinder can be neglected (because the Δh approaches zero and the area of the side of the cylinder approaches zero), it can be obtained from Expression (6.17) that

$$D_{1n}\Delta S - D_{2n}\Delta S = \rho_s \cdot \Delta S$$

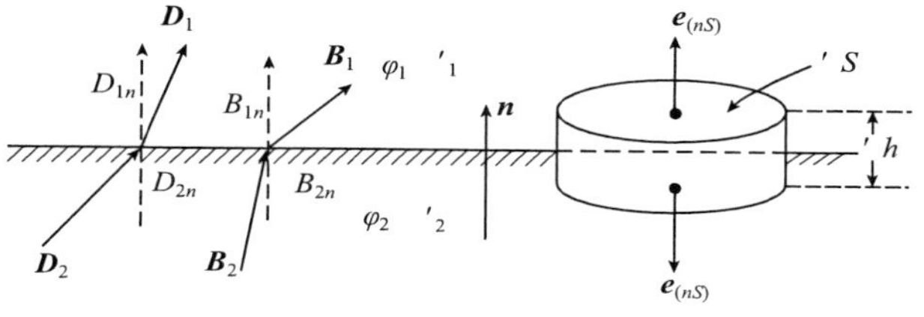

Figure 6.3 On the boundary conditions for surface normal fields

where D_{1n} and D_{2n} are the normal components of $\boldsymbol{D}_1$ and $\boldsymbol{D}_2$, respectively. Therefore

$$D_{1n} - D_{2n} = \rho_s \tag{6.34}$$

The formula above indicates that the difference between the normal components of the electric displacement $\boldsymbol{D}$ along the two sides of the boundary, equals the surface density ρ_s of free charge on the interface.

Generally speaking there is no free charge on the interface between perfect dielectrics, that is $\rho_s = 0$, then

$$D_{1n} = D_{2n} \tag{6.35}$$

which indicates that the normal components of $\boldsymbol{D}$ through the interface between dielectrics are continuous.

2. D_n on the Surface of Perfect Conductor Equals ρ_s

If media 2 is a perfect conductor, since the electric field inside perfect conductor is zero, then $D_{2n} = 0$, and it can be obtained based on Expression (6.34) that

$$D_{1n} = \rho_s \tag{6.36}$$

which indicates that D_n on the surface of perfect conductor equals the surface charge density ρ_s.

3. B_n is Continuous at the Interface Between Dielectrics

Suppose the permeabilities of two dielectrics are μ_1 and μ_2, respectively, by using similar method, we just replace the electric flux density $\boldsymbol{D}$ with the magnetic flux density $\boldsymbol{B}$, with the help of Gauss's law for magnetism, that is $\oint_s \boldsymbol{B} \cdot \mathrm{d}\boldsymbol{S} = 0$, then one can have

$$B_{1n} = B_{2n} \tag{6.37}$$

which indicates that the normal components of $\boldsymbol{B}$ through the interface between dielectrics are continuous.

4. B_n on the Surface of a Perfect Conductor Equals Zero

If media 2 is a perfect conductor, since there is no alternating electromagnetic field inside the perfect conductor, then $B_{2n} = 0$, and based on Expression (6.37) it can be known that

$$B_{1n} = B_{2n} = 0 \tag{6.38}$$

in other words, for alternating fields the normal component of a magnetic field on the surface of a perfect conductor is zero.

6.8.3 Boundary Conditions for Tangential Fields $\boldsymbol{E}_t$ and $\boldsymbol{H}_t$

1. E_t at the Interface Between Perfect Dielectrics is Continuous

As for the interface shown in Figure 6.4, draw a closed rectangular contour, let ab and cd be in parallel with the interface, with their lengths being Δl, and Δl is extremely small. The other two edges ad and bc are normal to the interface, with their heights being Δh, and

Δh approaches zero. Along the path *abcda*, the closed contour line integration is conducted for electric field strength $\boldsymbol{E}$, which leads to

$$\oint_l \boldsymbol{E} \cdot \mathrm{d}\boldsymbol{l} = -\int_S \frac{\partial \boldsymbol{B}}{\partial t} \cdot \mathrm{d}\boldsymbol{S}$$

due to the height Δh approaches zero and Δl is extremely small, the area bounded by the contour approaches zero, besides, as $\dfrac{\partial B}{\partial t}$ has finite value, therefore the surface integral of $\dfrac{\partial B}{\partial t}$ is zero, leaving

$$\oint_l \boldsymbol{E} \cdot \mathrm{d}\boldsymbol{l} = -\int_S \frac{\partial \boldsymbol{B}}{\partial t} \cdot \mathrm{d}\boldsymbol{S} = 0 \tag{6.39}$$

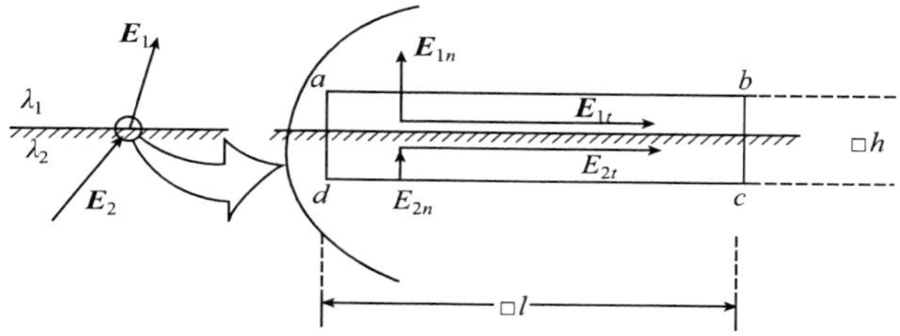

Figure 6.4 On the boundary conditions for tangential fields

For segments $\overline{ab}$ and $\overline{cd}$, tangential electric fields E_{1t} and E_{2t} are applicable for $\oint_l \boldsymbol{E} \cdot \mathrm{d}\boldsymbol{l}$, *as Δl is sufficiently short, on which the tangential field can be viewed as uniformly distributed*; for segments $\overline{bc}$ and $\overline{da}$, normal electric fields E_{1n} and E_{2n} are applicable, due to height $\Delta h \to 0$, however, on which the integral of electric field can be negligible, then one can obtain from the expression above that

$$E_{1t} = E_{2t} \tag{6.40}$$

that is, the tangential components of electric field along the interface between dielectrics are continuous.

2. E_t on the Surface of Perfect Conductor Equals Zero

If media 2 is a perfect conductor, since the electric field inside a perfect conductor is zero, then $E_{2t} = 0$, and it can be obtain according to Expression (6.40) that

$$E_{1t} = 0 \tag{6.41}$$

that is, the tangential component of the electric field E_t on the surface of a perfect conductor is zero.

3. H_t at the Interface Between Perfect Dielectrics is Continuous

The method used here is the same as the one for studying boundary conditions of tangential components of electric fields. Draw a closed rectangular contour *abcda* at the interface between dielectrics, according to Maxwell's First Equation $\oint_l \boldsymbol{H} \cdot dl = I + \int_S \dfrac{\partial \boldsymbol{D}}{\partial t} \cdot \mathrm{d}\boldsymbol{S}$ it can be known that, the displacement current through the area surrounded by the closed contour is $\int_S \frac{\partial \boldsymbol{D}}{\partial t} \cdot \mathrm{d}\boldsymbol{S}$, as the area approaches zero, so its value is negligible. If there exists conduction current I on the interface, and the current is distributed along the surface, or the current is distributed only on the infinitely thin layer of the boundary, the formula above can be written as

$$(H_{1t} - H_{2t})\Delta l = I = J_s \Delta l$$

therefore

$$H_{1t} - H_{2t} = J_s \tag{6.42}$$

where J_s is the surface current density. This expression shows that, as there is surface conduction current density J_s on the interface, the difference between H_{1t} and H_{2t} equals J_s.

Regarding the interface between perfect dielectrics, there is no conduction current on it, that is $J_s = 0$, from Expression (6.42) it can be known that

$$H_{1t} = H_{2t} \tag{6.43}$$

in other words, at the interface between perfect dielectrics H_t is continuous.

4. H_t on the Surface of A Perfect Conductor Equals J_s

If media 2 is a perfect conductor, then $H_{2t} = 0$, it can be obtained from Expression (6.42) that

$$H_{1t} = J_s \tag{6.44}$$

that is, the tangential component of the magnetic field strength H_t on the surface of perfect conductor equals the surface current density J_s.

The vector form of boundary conditions for a perfect conductors is

$$\left.\begin{aligned} \boldsymbol{n} \times \boldsymbol{H} &= \boldsymbol{J}_s \\ \boldsymbol{n} \times \boldsymbol{E} &= 0 \\ \boldsymbol{n} \cdot \boldsymbol{D} &= \rho_s \\ \boldsymbol{n} \cdot \boldsymbol{B} &= 0 \end{aligned}\right\} \tag{6.45}$$

here, $\boldsymbol{n}$ is equivalent to $\boldsymbol{e}_n$, the unit normal vector of the boundary.

It is pointed out again, that when boundary conditions are analyzed, for the Δl and Δh of the closed rectangular contour, as well as for the ΔS and Δh of the closed cylindrical Gaussian surface on the interface between dielectrics, we especially emphasized that $\Delta h \to 0$, Δl and ΔS should be sufficiently small (infinitesimal). *As a matter of fact, if it could be imagined that for boundary conditions, $\Delta l \to 0$, $\Delta S \to 0$ (converging to a "point"), which can provide a more accurate understanding of boundary conditions. Accordingly, the interface between dielectrics, does not have to be a very large planar surface, but the requirement can be relaxed more. For the boundary shown by Figure 6.5, the validity of the conclusions from the previous discussion of boundary conditions still holds.*

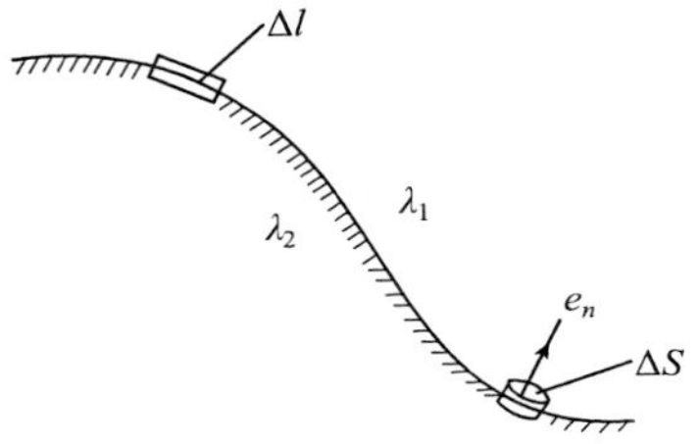

Figure 6.5 Illustration of more generalized boundaries between dielectrics as well as closed contour and Gaussian surface for studying boundary conditions

Example 6.6 There are two types of perfect dielectrics, they have permittivities ε_1 and ε_2 respectively, and no free charges on the boundary. On the interface, the angles between the static electric line of force and the normal of the surface in dielectric 1 and dielectric 2 are α_1 and α_2 respectively, find the relation between α_1 and α_2.

Solution Let $\boldsymbol{D}_1$ and $\boldsymbol{D}_2$ denote the electric flux densities in dielectric 1 and dielectric 2 respectively. Let $\boldsymbol{E}_1$ and $\boldsymbol{E}_2$ denote the electric field strengths in dielectric 1 and dielectric 2 respectively. In isotropic dielectrics $\boldsymbol{D}$ and $\boldsymbol{E}$ have the same direction. Based on boundary conditions $D_{1n} = D_{2n}$ and $E_{1t} = E_{2t}$, one can have

$$\frac{D_{1n}}{E_{1t}} = \frac{D_{2n}}{E_{2t}} \tag{a}$$

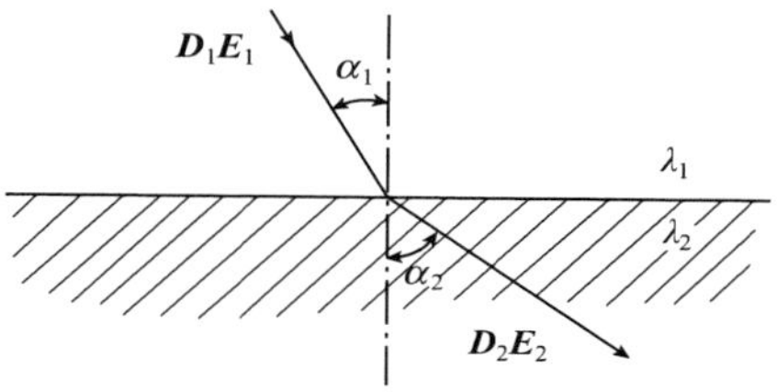

Figure of Example 6.6

It can be known from Figure of Example 6.6 that

$$\left.\begin{aligned} D_{1n} &= D_1 \cos\alpha_1, \quad E_{1t} = E_1 \sin\alpha_1 \\ D_{2n} &= D_2 \cos\alpha_2, \quad E_{2t} = E_2 \sin\alpha_2 \end{aligned}\right\} \tag{b}$$

Substituting Equation (b) into Equation (a) one can have

$$\frac{\tan\alpha_1}{\tan\alpha_2} = \frac{\varepsilon_1}{\varepsilon_2} = \frac{\varepsilon_{r_1}\varepsilon_0}{\varepsilon_{r_2}\varepsilon_0} = \frac{\varepsilon_{r_1}}{\varepsilon_{r_2}} \tag{c}$$

where ε_{r_1} and ε_{r_2} are the relative permittivities of dielectric 1 and dielectric 2 respectively.

It can be obtained from Expression (c) that

$$\alpha_2 = \arctan\left(\frac{\varepsilon_{r_2}}{\varepsilon_{r_1}} \tan\alpha_1\right)$$

6.9 Poynting's Theorem and Poynting Vector

6.9.1 Poynting's Theorem

It is well known that, in circuit theory, voltage and current are used to describe the energy in resistance, inductance and capacitance. Poynting's theorem, however, depicts the conservation of energy in electromagnetic fields, from the point of view of fields.

By employing Maxwell's Equations and the connections with vector operations, one can easily obtain the mathematical expressions of Poynting's theorem. Substituting the differential forms of Maxwell's First and Second Equations into the vector identity below, without considering impressed source $\boldsymbol{J}_i$:

$$\nabla \cdot (\boldsymbol{E} \times \boldsymbol{H}) = \boldsymbol{H} \cdot \nabla \times \boldsymbol{E} - \boldsymbol{E} \cdot \nabla \times \boldsymbol{H}$$

one can have

$$\nabla \cdot (\boldsymbol{E} \times \boldsymbol{H}) = -\boldsymbol{H} \cdot \frac{\partial \boldsymbol{B}}{\partial t} - \boldsymbol{E} \cdot \frac{\partial \boldsymbol{D}}{\partial t} - \boldsymbol{E} \cdot \boldsymbol{J} \tag{6.46}$$

If taking integral of the equation over a volume, one can have

$$-\int_V \nabla \cdot (\boldsymbol{E} \times \boldsymbol{H}) \mathrm{d}V = \int_V \left(\boldsymbol{H} \cdot \frac{\partial \boldsymbol{B}}{\partial t} + \boldsymbol{E} \cdot \frac{\partial \boldsymbol{D}}{\partial t} + \boldsymbol{E} \cdot \boldsymbol{J}\right) \mathrm{d}V \tag{6.47}$$

To have a clearer insight about the Equation on its physical interpretation, further transformations can be conducted. First, applying divergence theorem, we convert the left-hand side of Equation (6.47) from volume integral into surface integral as

$$-\int_V \nabla\cdot(\boldsymbol{E}\times\boldsymbol{H})\mathrm{d}V = -\oint_S (\boldsymbol{E}\times\boldsymbol{H})\cdot\mathrm{d}\boldsymbol{S} \tag{6.48}$$

The right-hand side of Expression (6.46) is also needed to be transformed. Since it is assumed in our scope of study that the media are linear (μ and ε are independent of the magnitude of $\boldsymbol{E}$ and $\boldsymbol{H}$) and isotropic (μ and ε are scalars). Moreover μ, ε and σ do not vary over time, so one can have

$$\left.\begin{aligned} \boldsymbol{H}\cdot\frac{\partial \boldsymbol{B}}{\partial t} &= \mu\boldsymbol{H}\cdot\frac{\partial \boldsymbol{H}}{\partial t} = \frac{1}{2}\mu\frac{\partial \boldsymbol{H}^2}{\partial t} = \frac{\partial}{\partial t}\left(\frac{1}{2}\mu H^2\right) \\ \boldsymbol{E}\cdot\frac{\partial \boldsymbol{D}}{\partial t} &= \varepsilon\boldsymbol{E}\cdot\frac{\partial \boldsymbol{E}}{\partial t} = \frac{1}{2}\varepsilon\frac{\partial \boldsymbol{E}^2}{\partial t^2} = \frac{\partial}{\partial t}\left(\frac{1}{2}\varepsilon E^2\right) \end{aligned}\right\} \tag{6.49}$$

Substituting Expressions (6.48) and (6.49) into (6.47) it can be obtained that

$$-\oint_S (\boldsymbol{E}\times\boldsymbol{H})\cdot\mathrm{d}\boldsymbol{S} = \frac{\partial}{\partial t}\int_V \left(\frac{1}{2}\mu H^2 + \frac{1}{2}\varepsilon E^2\right)\mathrm{d}V + \int_V \boldsymbol{E}\cdot\boldsymbol{J}\mathrm{d}V \tag{6.50}$$

This is the mathematical expression of Poynting's theorem. What is the meaning of it? For the first term on the right-hand side of the equation, $\frac{1}{2}\mu H^2$ and $\frac{1}{2}\varepsilon E^2$ in it represent for the energy densities of electric field and magnetic field, respectively, the volume integral of which equals the total magnetic and electric energy stored in that volume. Taking time differential of the sum of them, we have the time rate of change in the energy stored in the volume, accordingly its value can be large or small, positive or negative, and varying with time. For the second term on the right-hand side of the equation, $\boldsymbol{E}\cdot\boldsymbol{J}$ may be attributed to the conductive material bounded by the closed surface. Since $\boldsymbol{J} = \sigma\boldsymbol{E}$, so $\boldsymbol{E}\cdot\boldsymbol{J} = \sigma E^2$, in other words $\boldsymbol{E}\cdot\boldsymbol{J}$ represents the power dissipation (heating loss) in the unit volume per unit time. The integral of it over the volume will lead to the total power dissipation per unit time, that is the dissipated power, in the whole volume, which always has a positive value, and will never be negative.

The physical interpretation of the two terms on the right-hand side of the Equation (6.50) has been analyzed as above. A situation can happen for the two terms as follows, that is there is not only heating loss in the volume, meanwhile the electromagnetic energy stored in the volume can also be increased somewhat, in other words, the time rate of change for the total stored energy is positive. Then, where do the dissipated energy and the increased stored energy within the volume come from? If there is no impressed source in the volume, the energy must be supplied by the fields' sources outside of the volume, or it comes from the electromagnetic waves generated by the fields' sources outside of the volume which go through the surface bounding the volume, and transported the energy into it.

According to energy conservation law, the energy that comes into the volume from outside should equal to the sum of the two terms on the right-hand side of Equation (6.50).

Obviously, $-\oint_S (\boldsymbol{E}\times\boldsymbol{H})\cdot\mathrm{d}\boldsymbol{S}$ represents the total power flowing into the closed surface. The physical interpretation of vector $\boldsymbol{E}\times\boldsymbol{H}$ is the Poynting Vector is discussed below.

6.9.2 Poynting Vector

Since the integral of $\boldsymbol{E}\times\boldsymbol{H}$ over a closed surface denotes the total power that goes through the closed surface, the vector $\boldsymbol{E}\times\boldsymbol{H}$ represents *the power that flows through unit area at any position of the closed surface, which is also called the power density. It is a vector, generally denoted by the symbol* $\boldsymbol{S}$, *and called the Poynting Vector.*

$$\boldsymbol{S} = \boldsymbol{E} \times \boldsymbol{H} \tag{6.51}$$

$\boldsymbol{S}$ points in the direction of propagation of electromagnetic waves, and determined by the right-hand screw rule for $\boldsymbol{E} \times \boldsymbol{H}$, as depicted in Figure 6.6.

$\boldsymbol{S}$ is the power density, and is the energy passing through unit area per unit time, also known as the energy flux density. Now an illustration of it using a plane wave as an example will be provided. Given the instantaneous values of the electric field and the magnetic field of a plane wave as

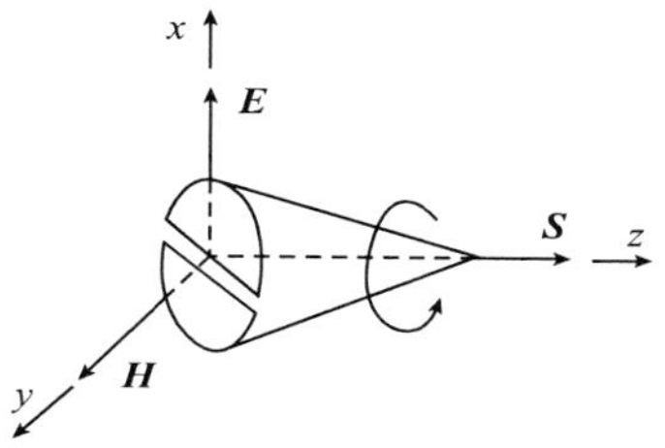

Figure 6.6 Poynting vector

$$\boldsymbol{E} = \boldsymbol{e}_x E_0 \sin(\omega t - kz)$$

$$\boldsymbol{H} = \boldsymbol{e}_y (E_0/\sqrt{\mu/\varepsilon}) \sin(\omega t - kz)$$

Then the instantaneous Poynting vector is

$$\boldsymbol{S}(t) = \boldsymbol{E} \times \boldsymbol{H} = \boldsymbol{e}_z \left(E_0^2 \sqrt{\varepsilon/\mu}\right) \sin^2(\omega t - kz) \tag{6.52}$$

Correspondingly the average Poynting vector (averaged over time) $\boldsymbol{S}_{\text{av}}$ is

$$\begin{aligned} S_{\text{av}} &= \frac{1}{T}\int_0^T S(t)\text{d}t = \frac{1}{T}\int_0^T \left(E_0^2\sqrt{\varepsilon/\mu}\right)\sin^2(\omega t - kz)\,\text{d}t \\ &= \frac{1}{2}\left(\sqrt{\varepsilon/\mu}\right)E_0^2 \end{aligned} \tag{6.53}$$

As the average energy densities of the electric field and the magnetic field, $w_{\text{e(av)}}$ and $w_{\text{m(av)}}$, are as follows respectively

$$w_{\text{e(av)}} = \frac{1}{2}\varepsilon E_{\text{e}}^2 = \frac{1}{4}\varepsilon E_0^2$$

$$w_{\text{m(av)}} = \frac{1}{2}\mu H_{\text{e}}^2 = \frac{1}{4}\mu H_0^2$$

For lossless media

$$w_{\text{m(av)}} = \frac{1}{4}\mu H_0^2 = \frac{1}{4}\mu\left[E_0/\sqrt{\mu/\varepsilon}\right]^2 = \frac{1}{4}\varepsilon E_0^2 = w_{\text{e(av)}}$$

that is

$$w_{\text{e(av)}} = w_{\text{m(av)}} \tag{6.54}$$

With the help of the preceding relation, the connection between $\boldsymbol{S}_{\text{av}}$ and $w_{\text{e(av)}}$, $w_{\text{m(av)}}$, can be obtained. Since

$$S_{\text{av}} = \frac{1}{2}\sqrt{\varepsilon/\mu}E_0^2$$

so

$$\begin{aligned} S_{\text{av}} &= \frac{1}{2}\left(\frac{1}{2}\sqrt{\varepsilon/\mu}E_0^2 + \frac{1}{2}\sqrt{\varepsilon/\mu}E_0^2\right) = \frac{1}{2}\frac{1}{\sqrt{\mu\varepsilon}}\left(\frac{1}{2}\varepsilon E_0^2 + \frac{1}{2}\mu H_0^2\right) \\ &= v\left(\frac{1}{4}\varepsilon E_0^2 + \frac{1}{4}\mu H_0^2\right) = v\left[w_{\text{e(av)}} + w_{\text{m(av)}}\right] \end{aligned} \tag{6.55}$$

where v is the propagation velocity of electromagnetic waves. The average Poynting vector is equal to the product of the electromagnetic energy density and the propagation velocity of electromagnetic waves. In other words, the average Poynting vector is the electromagnetic energy passing through unit area per unit time in surface normal direction.

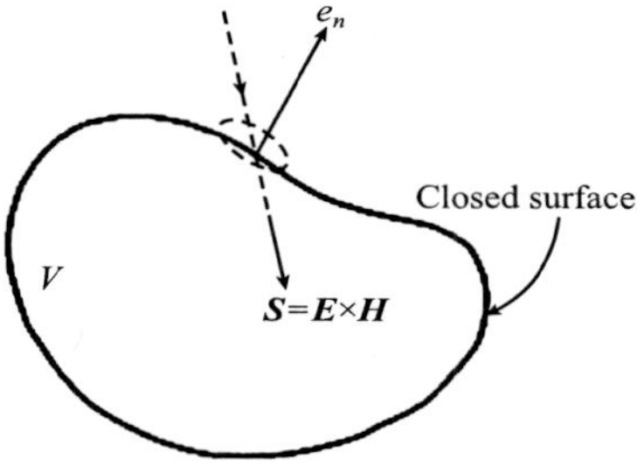

Figure 6.7 The direction of the area element for Poynting's theorem

When the right-hand side of Equation (6.50) is positive-valued, the left-hand side is also positive, which means that the energy flux flows into the closed surface, or the Poynting vector $\boldsymbol{S}$ points in the direction into the closed surface. Therefore, the direction of area element $\mathrm{d}\boldsymbol{S}$, e_n, must be pointing outwards of the closed surface, as shown in Figure 6.7.

If the time rate of change for the energy stored in the volume is negative, or the stored energy is decreasing, in addition, the amount of decreased stored energy is larger than that of dissipation, that is, the right-hand side of Equation (6.50) is negative, thus the excessive energy must go out through the closed surface, namely $(\boldsymbol{E}\times\boldsymbol{H})\cdot\mathrm{d}\boldsymbol{S}$ is positive, which makes the left-hand side of Equation (6.50) also negative.

Since problems in circuit analysis are just special cases of fields, utilizing Poynting vector $\boldsymbol{S}$ to calculate power would have more generalized significance. Now an example will be given below as explanation.

Example 6.7 Suppose a DC current I_z flows through a conductive wire with radius of a, the resistance of the wire in unit length is R, try to calculate the dissipated power for the wire in unit length via Poynting vector.

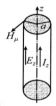

(a) The current,electric field and magnetic field for a single wire

(b) Energy flux density $\boldsymbol{S}$ generated by E_z and H_μ, flows in $-r$ direction inside of the conductor and transformed into heat.

Figure of Example 6.7

Solution Since the cross-section of the wire is round shaped and axisymmetric, cylindrical coordinates are employed.

Because R is the resistance of the wire in unit length, I_zR represents the voltage drop along the wire in unit length, so one can have

$$E_z = I_zR$$

According to Ampere's circuital law, the magnetic field strength on the surface of the wire generated by current I_z will be

$$H_\phi = I_z/(2\pi a)$$

Hence, the Poynting vector $\boldsymbol{S}$ on the surface of the wire produced by E_z and H_ϕ is

$$\boldsymbol{S} = \boldsymbol{e}_zE_z\times\boldsymbol{e}_\phi H_\phi = -\boldsymbol{e}_rE_zH_\phi = -\boldsymbol{e}_rI_z^2R\frac{1}{2\pi a} \tag{a}$$

which shows that *the energy flux density* $\boldsymbol{S}$ *caused by* E_z *and* H_ϕ *on the surface of conductor, will pass the exterior surface of the wire in the surface normal direction, pointing in* $-r$ *direction, namely, the energy is flowing in the surface normal direction into the conductor through its surface. This energy contributes to the heating loss inside of the wire.* The dissipated power of the wire in unit length should be

$$\begin{aligned} P_L &= -\oint_S \boldsymbol{S}\cdot \mathrm{d}\boldsymbol{S} \\ &= -\left[\int_{S_1} \boldsymbol{S}\cdot \boldsymbol{e}_z \mathrm{d}S_1 + \int_{S_2} \boldsymbol{S}\cdot(-\boldsymbol{e}_z)\mathrm{d}S_2 + \int_{S_3} \boldsymbol{S}\cdot \boldsymbol{e}_r \mathrm{d}S_3\right] \end{aligned} \tag{b}$$

where S_1, S_2 and S_3 are the top, bottom and side (cylindrical) surfaces of the wire in unit length, respectively. Due to $\boldsymbol{S}$ pointing in $-\boldsymbol{e}_r$ direction, for the top and bottom surfaces there are $\boldsymbol{S}\cdot \mathrm{d}\boldsymbol{S}_1 = 0$ and $\boldsymbol{S}\cdot \mathrm{d}\boldsymbol{S}_2 = 0$. Thus the expression above will be reduced to the third term, and because the Poynting vector $\boldsymbol{S}$ is uniformly distributed along the cylindrical surface S_3, according to Expression (a) of this example one can have

$$\begin{aligned} P_L &= -\int_{S_3} \boldsymbol{S}\cdot \mathrm{e}_r \mathrm{d}S_3 = \int_{S_3}\left(I_z^2 R\frac{1}{2\pi a}\right)\mathrm{d}S_3 \\ &= \left(I_z^2 R \times \frac{1}{2\pi a}\right)\times 2\pi a \times 1 = I_z^2 R \end{aligned} \tag{c}$$

from which it can be seen that P_L is the dissipated power. It is clearly shown that the dissipated power obtained from Poynting vector is exactly the same as the result based on circuit analysis, both are $I_z^2 R$, which is within our expectation.

Example 6.8 It is known that the electric field and magnetic field of the spherical electromagnetic wave radiated by an antenna are as follows respectively

$$\begin{aligned} E_\theta &= A_0\frac{\sin\theta}{r}\sin(\omega t - kr) \\ H_\phi &= \frac{1}{\eta_0}A_0\frac{\sin\theta}{r}\sin(\omega t - kr) \end{aligned}$$

Find the transmitted power of the antenna.

Solution Based on Poynting vector it can be known that

$$\boldsymbol{S}(t) = \boldsymbol{E}\times\boldsymbol{H} = \boldsymbol{e}_r E_\theta H_\phi = \boldsymbol{e}_r\frac{1}{\eta_0}A_0^2\frac{\sin^2\theta}{r^2}\sin^2(\omega t - kr) \tag{a}$$

which is the instantaneous value of the power density, or the instantaneous power passing through the unit area on the spherical surface, in radial direction.

How to find the total transmit power of the antenna after $\boldsymbol{S}$ is known? Since it is shown by the expression of $\boldsymbol{S}$ that power density $\boldsymbol{S}$ is not uniformly distributed along the spherical surface, but a function of θ, thus integral of power density $\boldsymbol{S}$ over the whole spherical surface should be taken before the instantaneous total power $P(t)$ is achieved, that is, $\oint_S \boldsymbol{S}(t)\cdot\mathrm{d}\boldsymbol{S}$, where $\mathrm{d}\boldsymbol{S}$ is the area element of the spherical surface as below

$$\mathrm{d}\boldsymbol{S} = \boldsymbol{e}_r r^2\sin\theta\mathrm{d}\theta\mathrm{d}\phi \tag{b}$$

hence

$$\begin{aligned} \oint_S \boldsymbol{S}(t)\cdot\mathrm{d}\boldsymbol{S} &= \int_0^{2\pi}\int_0^{\pi}\frac{1}{\eta_0}A_0^2\frac{\sin^2\theta}{r^2}\sin^2(\omega t - kr)\cdot r\mathrm{d}\theta r\sin\theta\mathrm{d}\phi \\ &= \frac{A_0^2}{\eta_0}\sin^2(\omega t - kr)\int_0^{2\pi}\int_0^{\pi}\sin^2\theta\mathrm{d}\theta\mathrm{d}\phi = \frac{8\pi A_0^2}{3\eta_0}\sin^2(\omega t - kr) \end{aligned} \tag{c}$$

which is the instantaneous total power $P(t)$ of the antenna, whose average power should be

$$P_{\text{av}} = \frac{1}{T}\int_0^T \oint_S (\boldsymbol{S}(t)\cdot \mathrm{d}\boldsymbol{S})\mathrm{d}t = \frac{4}{3}\pi\frac{A_0^2}{\eta_0} \tag{d}$$

6.9.3 Active Power, Reactive Power and Complex Form of Poynting's Theorem

1. Active Power and Reactive Power

As we all know that the current and voltage on a resistor have the same phase, and the power consumed by the resistor equals the product of current and voltage on it. It is real-valued, called active power. And this power is converted to the heat over the resistor.

If a load in a circuit is not purely resistive, but contains some reactance components, the current and voltage on the load will show certain phase difference ϕ between them, if the current and voltage are varying in a time-harmonic manner, then their instantaneous complex forms are

$$\left.\begin{aligned} I &= I_0 \mathrm{e}^{\mathrm{j}\omega t} \\ U &= U_0 \mathrm{e}^{\mathrm{j}(\omega t+\phi)} \end{aligned}\right\} \tag{6.56}$$

Under this circumstance, there is not only active power in the circuit but also reactive power. How to calculate them? A very convenient method would be complex number method, that is to take one half of the product of the complex voltage and the conjugate of complex current I^*, so that the average values of active power and reactive power can be obtained, namely,

$$\begin{aligned} \frac{1}{2}UI^* &= \frac{1}{2}U_0\mathrm{e}^{\mathrm{j}(\omega t+\phi)}I_0\mathrm{e}^{-\mathrm{j}\omega t} = \frac{1}{2}U_0 I_0 \mathrm{e}^{\mathrm{j}\phi} \\ &= \frac{1}{2}U_0 I_0\cos\phi + \mathrm{j}\frac{1}{2}U_0 I_0 \sin\phi \end{aligned} \tag{6.57}$$

The real part of the expression above $\left(\frac{1}{2}U_0 I_0\cos\phi\right)$ is the active power; the imaginary part $\left(\frac{1}{2}U_0 I_0\sin\phi\right)$ is the reactive power. When the current and voltage have equal phase (the load is purely resistive), or $\phi = 0$, reactive power is zero, all of the power is real. Otherwise, if the load is purely reactive, then the voltage and current are $\pm 90°$ degrees out of phase, then the real power is zero, all of the power is reactive, and in this condition there is only energy storage or transformation, but no energy transmission or dissipation.

2. Complex Forms of Poynting's Theorem and Poynting Vector

There is an analogy between calculation of power in electromagnetic fields and in ordinary circuit. Due to various reasons like lossy media, electric field and magnetic field can have phase difference between them. Hence the power of electromagnetic fields also consists of two parts as active power and reactive power. Here we will study the relations in electromagnetic fields in homogeneous, non-perfect media (lossy media). The curl equations for harmonic waves $\boldsymbol{E}$ and $\boldsymbol{H}^*$ are

$$\nabla\times\boldsymbol{E} = -\mu\frac{\partial\boldsymbol{H}}{\partial t} = -\mathrm{j}\omega\mu\boldsymbol{H} \tag{6.58}$$

$$\nabla\times\boldsymbol{H}^* = J^* + \frac{\partial\boldsymbol{D}^*}{\partial t} = \sigma\boldsymbol{E}^* - \mathrm{j}\omega\varepsilon\boldsymbol{E}^* \tag{6.59}$$

where the complex fields are

$$\boldsymbol{E} = \boldsymbol{E}_0 \mathrm{e}^{-\mathrm{j}kz}, \quad \boldsymbol{H} = \boldsymbol{H}_0 \mathrm{e}^{-\mathrm{j}kz}$$

$$\boldsymbol{E}^* = \boldsymbol{E}_0^* \mathrm{e}^{\mathrm{j}kz}, \quad \boldsymbol{H}^* = \boldsymbol{H}_0^* \mathrm{e}^{\mathrm{j}kz}$$

where $\boldsymbol{E}^*$, $\boldsymbol{H}^*$ and $\boldsymbol{D}^*$ are the complex conjugates of $\boldsymbol{E}$, $\boldsymbol{H}$ and $\boldsymbol{D}$, respectively; $\boldsymbol{E}_0$ and $\boldsymbol{H}_0$ are the complex amplitudes of the vectors, and they have predetermined magnitudes and early phases.

Now let us examine what the physical interpretation of $\oint_S (\boldsymbol{E} \times \boldsymbol{H}^*) \cdot \mathrm{d}\boldsymbol{S}$ is?

From divergence theorem one can have

$$\oint_S (\boldsymbol{E} \times \boldsymbol{H}^*) \cdot \mathrm{d}\boldsymbol{S} = \int_V \nabla \cdot (\boldsymbol{E} \times \boldsymbol{H}^*) \mathrm{d}V \tag{6.60}$$

then by using vector identity

$$\nabla \cdot (\boldsymbol{E} \times \boldsymbol{H}^*) = \boldsymbol{H}^* \cdot \nabla \times \boldsymbol{E} - \boldsymbol{E} \cdot \nabla \times \boldsymbol{H}^* \tag{6.61}$$

and substituting the curl equations of $\boldsymbol{E}$ and $\boldsymbol{H}^*$ into the preceding expression, one can obtain

$$\begin{aligned} \nabla \cdot (\boldsymbol{E} \times \boldsymbol{H}^*) &= \boldsymbol{H}^* \cdot (-\mathrm{j}\omega\mu \boldsymbol{H}) - E \cdot (\sigma \boldsymbol{E}^* - \mathrm{j}\omega\varepsilon \boldsymbol{E}^*) \\ &= -\mathrm{j}\omega\mu \boldsymbol{H} \cdot \boldsymbol{H}^* + \mathrm{j}\omega\varepsilon \boldsymbol{E} \cdot \boldsymbol{E}^* - \sigma \boldsymbol{E} \cdot \boldsymbol{E}^* \end{aligned}$$

therefore

$$\begin{aligned} \oint_S (\boldsymbol{E} \times \boldsymbol{H}^*) \cdot \mathrm{d}\boldsymbol{S} &= \int_V \nabla \cdot (\boldsymbol{E} \times \boldsymbol{H}^*) \mathrm{d}V \\ &= -\mathrm{j}\omega \int_V [(\mu \boldsymbol{H} \cdot \boldsymbol{H}^*) - (\varepsilon \boldsymbol{E} \cdot \boldsymbol{E}^*)] \, \mathrm{d}V - \int_V (\sigma \boldsymbol{E} \cdot \boldsymbol{E}^*) \mathrm{d}V \end{aligned}$$

with $\boldsymbol{H} \cdot \boldsymbol{H}^* = H_0^2$ and $\boldsymbol{E} \cdot \boldsymbol{E}^* = E_0^2$, the equations above is now transformed into

$$-\oint_S (\boldsymbol{E} \times \boldsymbol{H}^*) \cdot \mathrm{d}\boldsymbol{S} = \int_V \sigma E_0^2 \mathrm{d}V + \mathrm{j}\omega \int_V (\mu H_0^2 - \varepsilon E_0^2) \mathrm{d}V \tag{6.62}$$

which is multiplied by $\frac{1}{2}$ so that the complex representation of Poynting's theorem is obtained like

$$-\frac{1}{2} \oint_S (\boldsymbol{E} \times \boldsymbol{H}^*) \cdot \mathrm{d}\boldsymbol{S} = \int_V \frac{1}{2} \sigma E_0^2 \mathrm{d}V + \mathrm{j}\omega \int_V \frac{1}{2} \left(\mu H_0^2 - \frac{1}{2} \varepsilon E_0^2\right) \mathrm{d}V \tag{6.63}$$

And the right-hand side of the equation above has a real part and an imaginary part. The real part is the average value of the active power, here shown in the form of dissipated power; the imaginary part is reactive power. It can be seen that, *for the complex Poynting vector* $\frac{1}{2}(\boldsymbol{E} \times \boldsymbol{H}^*)$, *its real part just denotes the average active power density* $\boldsymbol{S}_{\text{av}}$ (average over a cycle in time), that is

$$\boldsymbol{S}_{\text{av}} = \frac{1}{2} \mathrm{Re}(\boldsymbol{E} \times \boldsymbol{H}^*) \tag{6.64}$$

where Re denotes "taking the real part of". This equation is quite useful for the calculation of electromagnetic energy, which can not only be applied to the computation of transported power of various transmission lines and transmit power of antennas, but also be employed to calculate the transmission loss incurred in transmission lines, waveguides, optical fibers with different modes, which is of great important significance in practical applications.

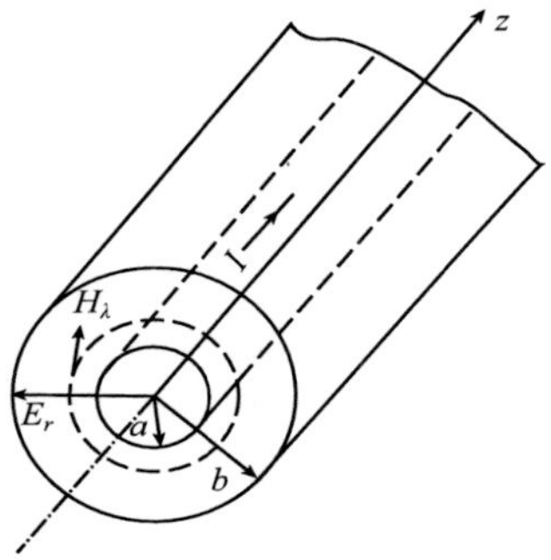

Figure of Example 6.9

Example 6.9 As shown in Figure of Example 6.9, the radii of the inner and outer conductors for the coaxial cable are a and b respectively, and it is filled with air dielectric, with voltage being U and current being I, try to find the transported power via Poynting vector.

Solution TEM waves can be transmitted along a coaxial cable, whose magnetic field has H_ϕ component only, the electric field has E_r component only. The current and voltage in the coaxial cable are $I = I_0 e^{j(\omega t - kz)}$ and $U = U_0 e^{j(\omega t - kz)}$, respectively. As the electromagnetic waves are transmitting along z-axis, it is easy to write the expression of magnetic field H_ϕ according to Ampere's Circuital Law

$$H_\phi = \frac{I_0}{2\pi r} e^{j(\omega t - kz)} \tag{a}$$

then it is convenient to obtain E_r from H_ϕ by using Maxwell's Equations. Here cylindrical coordinates are employed, where electric field has component in r direction only, thus Maxwell's First Equation can be simplified as

$$\boldsymbol{e}_r \left(\frac{1}{r}\frac{\partial H_z}{\partial \phi} - \frac{\partial H_\phi}{\partial z} \right) = \boldsymbol{e}_r \frac{\varepsilon \partial E_r}{\partial t} \tag{b}$$

as $H_z = 0$, based on the preceding expression one can have

$$E_r = \frac{k}{\omega\varepsilon} H_\phi = \sqrt{\frac{\mu_0}{\varepsilon_0}} H_\phi = \eta_0 \frac{I_0}{2\pi r} e^{j(\omega t - kz)} \tag{c}$$

where η_0 is the impedance of air dielectric. In the coaxial cable, the relation between the voltage U and electric field E_r is

$$U = \int_a^b E_r \mathrm{d}r = \int_a^b \eta_0 \frac{I_0}{2\pi r} e^{j(\omega t - kz)} \mathrm{d}r = E_r r \ln\frac{b}{a} \tag{d}$$

that is

$$E_r = \frac{U}{r} \Big/ \ln\frac{b}{a} \tag{e}$$

Therefore the power P transported by coaxial cable is

$$\begin{aligned} P &= \frac{1}{2}\mathrm{Re}\int_S (\boldsymbol{E} \times \boldsymbol{H}^*) \cdot \mathrm{d}S \\ &= \frac{1}{2}\int_a^b \int_0^{2\pi} \frac{I_0 U_0}{2\pi r^2 \ln\dfrac{b}{a}} r \mathrm{d}\phi \mathrm{d}r = \frac{1}{2} I_0 U_0 \end{aligned} \tag{f}$$

which is the same as the average transmitted power in the coaxial cable calculated directly by using voltage and current, as is well known to us.

The method in circuit analysis can be used to calculate the power, and so can be the field approach. However, under most circumstances, there is no current I or voltage U in the scope under study, or even if there is current/voltage, their values could be uncertain, then calculation of power has to be realized by applying points of view from fields, or Poynting vector. For example, calculations of the transmission power in optical fibers and waveguides, and of radiated power of antenna, all fall into this category.

6.10 Potentials and Fields for Alternating Fields

It has been observed during analyzing static fields that the introduction of scalar electric potential φ to solve problems in electrostatic fields, as well as the introduction of vector magnetic potential $\boldsymbol{A}$ to solve problems in magnetostatic fields, would bring great convenience. According to this experience, when alternating electromagnetic fields are under study, scalar electric potential φ and vector magnetic potential $\boldsymbol{A}$ may also be introduced, and the equations in terms of φ and A can be established. After φ and $\boldsymbol{A}$ are solved, it would be easier to find $\boldsymbol{E}$ and $\boldsymbol{H}$.

6.10.1 The Relation between Potentials and Fields for Alternating Fields

First, a brief review of potential functions in static fields is given. For electrostatic fields scalar electric potential φ is introduced to represent electric field strength $\boldsymbol{E}$, namely

$$\boldsymbol{E} = -\nabla\varphi \tag{6.65}$$

For magnetostatic fields vector magnetic potential $\boldsymbol{A}$ is introduced, the connection between $\boldsymbol{A}$ and magnetic induction $\boldsymbol{B}$ is

$$\boldsymbol{B} = \nabla \times \boldsymbol{A} \tag{6.66}$$

For alternating fields, are Expressions (6.65) and (6.66) still valid? The answer would be clear if the differential forms of Maxwell's Equations are evaluated. Rewrite the differential forms of Maxwell's Equations as

$$\nabla \times \boldsymbol{H} = \boldsymbol{J} + \varepsilon\frac{\partial \boldsymbol{E}}{\partial t} \tag{6.67}$$

$$\nabla \times \boldsymbol{E} = -\mu\frac{\partial \boldsymbol{H}}{\partial t} \tag{6.68}$$

$$\nabla \cdot \boldsymbol{D} = \rho \tag{6.69}$$

$$\nabla \cdot \boldsymbol{B} = 0 \tag{6.70}$$

The relation between the current density $\boldsymbol{J}$ and the charge density ρ, or the current continuity equation, is

$$\nabla \cdot \boldsymbol{J} = -\frac{\partial \rho}{\partial t} \tag{6.71}$$

If the vector magnetic potential $\boldsymbol{A}$ in alternating fields is also defined as

$$\boldsymbol{B} = \nabla \times \boldsymbol{A} \tag{6.72}$$

then $\nabla\cdot\boldsymbol{B} = \nabla\cdot\nabla\times\boldsymbol{A} = 0$, which still satisfies Equation (6.70), indicating that $\boldsymbol{B} = \nabla\times\boldsymbol{A}$ still holds valid for alternating fields.

Here we move on to analyze the relation between alternating electric field $\boldsymbol{E}$ and potentials. Substituting Expression (6.72) into Expression (6.68), which yields

$$\nabla \times \boldsymbol{E} = -\frac{\partial}{\partial t}\nabla \times \boldsymbol{A} = -\nabla \times \frac{\partial \boldsymbol{A}}{\partial t}$$

so that

$$\nabla \times \left(\boldsymbol{E} + \frac{\partial \boldsymbol{A}}{\partial t} \right) = 0 \tag{6.73}$$

If suppose $\boldsymbol{E} + \frac{\partial \boldsymbol{A}}{\partial t}$ in the equation above equals the gradient of a certain scalar φ, the equation above still holds. Thus we may suppose

$$-\nabla\varphi = \boldsymbol{E} + \frac{\partial \boldsymbol{A}}{\partial t} \tag{6.74}$$

which leads to the relation between alternating electric field $\boldsymbol{E}$ and potential as

$$\boldsymbol{E} = -\nabla\varphi - \frac{\partial \boldsymbol{A}}{\partial t} \tag{6.75}$$

6.10.2 Differential Equations for Potential Functions

In this section, Maxwell's Equations will be used, as well as Expressions (6.72) and (6.75), to build differential equations for potential functions—inhomogeneous Helmholtz equation.

1. Differential Equation for Vector Magnetic Potential $\boldsymbol{A}$ *(vector's inhomogeneous Helmholtz equation)*

With $\nabla \times \boldsymbol{H} = \varepsilon \frac{\partial \boldsymbol{E}}{\partial t} + \boldsymbol{J}$, and using Expressions (6.72) and (6.75) one can have

$$\nabla \times \nabla \times \boldsymbol{A} = \mu\varepsilon \frac{\partial}{\partial t} \left(-\nabla\varphi - \frac{\partial \boldsymbol{A}}{\partial t} \right) + \mu \boldsymbol{J}$$

The left-hand side and right-hand side of the equation above can also be written as

$$\nabla \times \nabla \times \boldsymbol{A} = \nabla\nabla \cdot \boldsymbol{A} - \nabla^2 \boldsymbol{A}$$

and

$$-\mu\varepsilon \frac{\partial}{\partial t}\nabla\varphi - \mu\varepsilon \frac{\partial^2 \boldsymbol{A}}{\partial t^2} + \mu \boldsymbol{J} = -\mu\varepsilon\nabla \frac{\partial \varphi}{\partial t} - \mu\varepsilon \frac{\partial^2 \boldsymbol{A}}{\partial t^2} + \mu \boldsymbol{J}$$

respectively. Therefore

$$\nabla^2 \boldsymbol{A} - \mu\varepsilon \frac{\partial^2 A}{\partial t^2} = -\mu \boldsymbol{J} + \left(\mu\varepsilon\nabla \frac{\partial \varphi}{\partial t} + \nabla\nabla \cdot \boldsymbol{A} \right) \tag{6.76}$$

Although we intend to give the differential equation of $\boldsymbol{A}$ after all, there is, however, not only $\boldsymbol{A}$ but also scalar potential φ included in the foregoing equation, for which further process should be conducted. Now an apparent question shows up: we only know that $\nabla \times \boldsymbol{A} = \boldsymbol{B}$, but it remains unclear $\nabla \cdot \boldsymbol{A} =$? However, as long as both of the curl and divergence are known, a vector can be determined. So Expression (6.76) would not have a well-defined solution until $\nabla \cdot \boldsymbol{A}$ is known. It can be seen that, if suppose

$$\nabla \cdot \boldsymbol{A} = -\mu\varepsilon \frac{\partial \varphi}{\partial t} \tag{6.77}$$

Expression (6.76) will change into

$$\nabla^2 \boldsymbol{A} - \mu\varepsilon \frac{\partial^2 \boldsymbol{A}}{\partial t^2} = -\mu \boldsymbol{J} \tag{6.78}$$

which is the differential equation for the vector magnetic potential $\boldsymbol{A}$, namely a wave equation of $\boldsymbol{A}$ with sources incorporated.

Equation (6.77) is called Lorentz gauge. Under the condition of alternating fields, as there is continuity Equation (6.71) that connects ρ and $\boldsymbol{J}$ together, there must be a equation that binds $\boldsymbol{A}$ and φ, and this equation is Lorentz gauge, for which further mathematical discussion can be found in Section 6.11.

2. Differential Equation for Scalar Potential φ (scalar's inhomogeneous Helmholtz equation)

With $\nabla \cdot \boldsymbol{D} = \rho$, and using Expressions (6.75) and (6.77), for homogeneous dielectrics one can have

$$\begin{aligned}\nabla \cdot \boldsymbol{E} = \nabla \cdot \left(-\nabla\varphi - \frac{\partial \boldsymbol{A}}{\partial t}\right) &= -\nabla^2\varphi - \frac{\partial}{\partial t}\nabla \cdot \boldsymbol{A} \\ &= -\nabla^2\varphi + \mu\varepsilon\frac{\partial^2\varphi}{\partial t^2} = \rho/\varepsilon\end{aligned}$$

thus

$$\nabla^2\varphi - \mu\varepsilon\frac{\partial^2\varphi}{\partial t^2} = -\rho/\varepsilon \tag{6.79}$$

which is the differential equation for the scalar potential φ, namely a wave equation of φ with sources incorporated.

As for static fields, $\dfrac{\partial^2}{\partial t^2} \Rightarrow 0$, Equations (6.78) and (6.79) will degenerate to

$$\nabla^2\boldsymbol{A} = -\mu\boldsymbol{J} \tag{6.80}$$

and

$$\nabla^2\varphi = -\rho/\varepsilon \tag{6.81}$$

which are the Poisson equations of static $\boldsymbol{A}$ and φ we are familiar with.

For harmonic waves, as $\dfrac{\partial^2}{\partial t^2} \Rightarrow -\omega^2$, Equations (6.78) and (6.79) will be transformed to

$$\nabla^2\boldsymbol{A} + k^2\boldsymbol{A} = -\mu\boldsymbol{J} \tag{6.82}$$

and

$$\nabla^2\varphi + k^2\varphi = -\rho/\varepsilon \tag{6.83}$$

Equation (6.82) is the wave equation with source in terms of harmonically varying $\boldsymbol{A}$, also called the inhomogeneous Helmholtz equation in terms of $\boldsymbol{A}$. And Equation (6.83) is the inhomogeneous Helmholtz equation in terms of φ.

There is one thing that should be pointed out, that the purpose of establishing wave equations for $\boldsymbol{A}$ and φ, is to find $\boldsymbol{A}$ and φ by solving these equations, after which, Equations (6.72) and (6.75) will be utilized to obtain $\boldsymbol{E}$ and $\boldsymbol{H}$. As we look into Lorentz gauge, that is

$$\nabla \cdot \boldsymbol{A} = -\mathrm{j}\omega\mu\varepsilon\varphi$$

one can have

$$\varphi = \frac{-\nabla \cdot \boldsymbol{A}}{\mathrm{j}\omega\mu\varepsilon} \tag{6.84}$$

Substituting Equation (6.84) into Equation (6.75) will lead to

$$\boldsymbol{E} = \frac{\nabla\nabla \cdot \boldsymbol{A}}{\mathrm{j}\omega\mu\varepsilon} - \mathrm{j}\omega\boldsymbol{A} \tag{6.85}$$

Hence, as long as the vector magnetic potential $\boldsymbol{A}$ is acquired, $\boldsymbol{E}$ and $\boldsymbol{H}$ will be found very conveniently through Equations (6.72) and (6.85). And of course, when $\boldsymbol{A}$ is known, $\boldsymbol{H}$ can also be obtained from $\boldsymbol{B} = \nabla \times \boldsymbol{A}$, after which, $\boldsymbol{E}$ can be given via Maxwell's Equations.

6.11 On Lorentz Gauge

The differential equation for the potential's function has been obtained in Section 6.10 as

$$\nabla^2 \boldsymbol{A} - \mu\varepsilon \frac{\partial^2 \boldsymbol{A}}{\partial t^2} = -\mu \boldsymbol{J} + \left(\mu\varepsilon \nabla \frac{\partial \varphi}{\partial t} + \nabla\nabla \cdot \boldsymbol{A}\right) \tag{6.86}$$

If let's take $\mu\varepsilon \nabla \frac{\partial \varphi}{\partial t} + \nabla\nabla \cdot \boldsymbol{A} = 0$ in the equations above, one can have

$$\nabla \cdot \boldsymbol{A} = -\mu\varepsilon \frac{\partial \varphi}{\varphi t} \tag{6.87}$$

Then Equation (6.86) will be changed into

$$\nabla^2 \cdot \boldsymbol{A} - \mu\varepsilon \frac{\partial^2 \boldsymbol{A}}{\partial t^2} = -\mu \boldsymbol{J} \tag{6.88}$$

which is the wave equation of $\boldsymbol{A}$ with sources incorporated, that is an inhomogeneous partial differential equation.

Helmholtz theorem indicates that: as long as the curl and divergence of a vector are known, that vector can be uniquely determined. So far only the curl of $\boldsymbol{A}$ has been known to us, that is, $\nabla \times \boldsymbol{A} = \boldsymbol{B}$, but the divergence is still unknown. That will be the Lorentz gauge to be discussed in this section, namely, Equation (6.87). By using this gauge condition, we can transform Equation (6.86) into Equation (6.88).

Before mathematical discussion is provided, a corollary should be given from some concepts. As already known that in static fields, the scalar electric potential φ is linked to static charges, and the vector magnetic potential $\boldsymbol{A}$ is linked to steady currents, yet the static charges and the steady currents are independent of each other. Therefore in static fields scalar potential φ has nothing to do with vector magnetic potential $\boldsymbol{A}$. However, the situation is totally different for alternating fields, where the alternating charges and currents are related with each other via continuity equation (also called the law of charge conservation). An inspiration rising from this relation (the connection between charge and current) is that, under the conditions for alternating fields, certain connection must exist between scalar potential φ and vector magnetic potential $\boldsymbol{A}$. That connection is Lorentz gauge.

Regarding the inhomogeneous partial differential equation in terms of $\boldsymbol{A}$ (Equation 6.88), if harmonic waves is considered, it will be changed into

$$\nabla^2 \boldsymbol{A} + k^2 \boldsymbol{A} = -\mu \boldsymbol{J} \tag{6.89}$$

for which the solution is

$$\boldsymbol{A} = \frac{\mu}{4\pi} \int_{V'} \frac{\boldsymbol{J} \mathrm{e}^{-\mathrm{j}kr}}{r} \mathrm{d}V' \tag{6.90}$$

Here what is to be explained is that, this solution for $\boldsymbol{A}$ has automatically satisfied Lorentz gauge, or Equation (6.77).

To keep the analysis simple, neglecting the term of $\mathrm{e}^{-\mathrm{j}kr}$, in other words, kr is assumed to approach zero, then taking divergence on Equation (6.90), one can have

$$\begin{aligned}
\nabla\cdot\boldsymbol{A} &= \frac{\mu}{4\pi}\nabla\cdot\int_V \frac{\boldsymbol{J}}{r}\mathrm{d}V' \\
&= \frac{\mu}{4\pi}\int_V \nabla\cdot\left(\frac{\boldsymbol{J}}{r}\right)\mathrm{d}V'^{①} \\
&= \frac{\mu}{4\pi}\int_V \left[\frac{1}{r}\nabla\cdot\boldsymbol{J}+\boldsymbol{J}\cdot\nabla\frac{1}{r}\right]\mathrm{d}V'^{②} \\
&= \frac{\mu}{4\pi}\int_V \left[\boldsymbol{J}\cdot\nabla\frac{1}{r}\right]\mathrm{d}V'^{③} \\
&= \frac{\mu}{4\pi}\int_V -\boldsymbol{J}\cdot\nabla'\frac{1}{r}\mathrm{d}V'^{④} \\
&= \frac{\mu}{4\pi}\int_V \left[\frac{1}{r}\nabla'\cdot\boldsymbol{J}-\nabla'\cdot(\boldsymbol{J}/r)\right]\mathrm{d}V'^{⑤} \\
&= \frac{\mu}{4\pi}\int_V \frac{1}{r}\nabla'\cdot\boldsymbol{J}\mathrm{d}V'^{⑥}
\end{aligned}$$

hence

$$\begin{aligned}
\nabla\cdot\boldsymbol{A} &= \frac{\mu}{4\pi}\int_V \frac{1}{r}(\nabla'\cdot \mathrm{J})\mathrm{d}V' = \frac{\mu}{4\pi}\int_V \frac{1}{r}\left(\frac{\partial\rho}{\partial t}\right)\mathrm{d}V'^{⑦} \\
&= -\mu\varepsilon\frac{\partial}{\partial t}\int_V \frac{\rho}{4\pi\varepsilon r}\mathrm{d}V' = -\mu\varepsilon\frac{\partial\varphi}{\partial t}^{⑧}
\end{aligned}$$

Namely

$$\nabla\cdot\boldsymbol{A} = -\mu\varepsilon\frac{\partial\varphi}{\partial t} \tag{6.91}$$

which is the Lorentz gauge. The continuity equation (law of charge conservation) is utilized during its deduction, which reveals the essential connection existing between it and Lorentz gauge condition.

It is demonstrated by the derivation above that: when Equation (6.90) is taken as the solution to Equation (6.82), Equation (6.91), i.e. the Lorentz gauge, has already been satisfied automatically.

For static fields, Equation (6.91) will degenerate to

---

① Since the integration is performed over the source points, while the divergence is taken for the field points, operator $\nabla\cdot$ can be moved into the integrand.

② Because $\nabla\cdot\left(\frac{\boldsymbol{J}}{r}\right)=\frac{1}{r}\nabla\cdot\boldsymbol{J}+\boldsymbol{J}\cdot\nabla\frac{1}{r}$.

③ Because $\boldsymbol{J}$ is at the source points, and is a function of x', y' and z', but the divergence is applied to field points (x,y,z), thus $\nabla\cdot\boldsymbol{J}=0$.

④ Because $r=\sqrt{(x-x')^2+(y-y')^2+(z-z')^2}$, $\nabla\frac{1}{r}=-\nabla'\frac{1}{r}$.

⑤ Because $\nabla'\cdot\frac{\boldsymbol{J}}{r}=\frac{1}{r}\nabla'\cdot\boldsymbol{J}+\boldsymbol{J}\cdot\nabla'\frac{1}{r}$.

⑥ Because $\int_V \nabla'\cdot\left(\frac{\boldsymbol{J}}{r}\right)\mathrm{d}V'=\oint_S \frac{\boldsymbol{J}}{r}\cdot\mathrm{d}\boldsymbol{S}'$, but the current $\boldsymbol{J}$ either does not equal zero on the surface over which the integral is taken, or is tangential to that surface, then $\oint_S \frac{\boldsymbol{J}}{r}\cdot\mathrm{d}\boldsymbol{S}'=0$.

⑦ Continuity equation, $\nabla\cdot\boldsymbol{J}=-\frac{\partial\rho}{\partial t}$, is employed.

⑧ Because $\varphi=\int_V \frac{\rho}{4\pi\varepsilon r}\mathrm{d}V$.

$$\nabla \cdot \boldsymbol{A} = 0 \tag{6.92}$$

which is just the divergence condition satisfied by the vector magnetic potential $\boldsymbol{A}$ for steady currents, also known as Coulomb gauge.

Exercises

6.1 Alternating electric field and magnetic field are both nonconservative (rotational) fields, try to write down:

(1) the relation between the curl of alternating electric field ($\nabla \times \boldsymbol{E}$) and its vortex sources;

(2) the relation between the curl of alternating magnetic field ($\nabla \times \boldsymbol{H}$) and its vortex sources.

6.2 It is known that the magnetic induction in air is

$$B_y = 10^{-5} \cos(2\pi z) \cos\left(6\pi \times 10^8 t\right)$$

Try to find the displacement current density $\boldsymbol{J}_\mathrm{d}$.

6.3 Try to write down the alternating fields' boundary conditions for the following two interfaces:

(1) boundary conditions for the interface between perfect dielectrics;

(2) vector representation of the boundary conditions for perfect conductors.

6.4 With the help of the current continuity equation: $\nabla \cdot \boldsymbol{J} = -\dfrac{\partial \rho}{\partial t}$, prove that the two curl equations in Maxwell's Equations can be derived from the two divergence equations in it.

6.5 Derive current continuity equation $\nabla \cdot \boldsymbol{J} = -\dfrac{\partial \rho}{\partial t}$ from Maxwell's Equations. (Suggestions: note that the vector identity $\nabla \cdot \nabla \times \boldsymbol{C} = 0$ may be used.)

6.6 A magnetic field strength in free space is $H_y = H_0 \sin(\omega t - kz)$, try to find its electric field strength $\boldsymbol{E}$.

6.7 If an electric field strength in free space is $E_x = E_0 \sin(\omega t - kz)$, find the expression of k in it by using the *two curl equations* in Maxwell's Equations.

6.8 Find what is the sum of conduction current density and displacement current density passing through arbitrary closed surface?

6.9 For linear, lossless, isotropic and heterogeneous media, derive corresponding Maxwell's Equations denoted by electric field strength $\boldsymbol{E}$ and magnetic induction $\boldsymbol{B}$.

6.10 In homogeneous and conductive medium $(\mu, \varepsilon, \sigma)$,

(1) establish the equation of charge density ρ and give its solution.

(2) find relaxation time τ $\left(\text{that is, the time duration for charge density in conductive media from } \rho_0 \text{ when } t = 0, \text{ decreasing to } \dfrac{\rho_0}{e}\right)$.

(3) find the relaxation time τ for copper $\left(\sigma = 5.8 \times 10^7 \mathrm{S/m},\ \varepsilon = \varepsilon_0 = \dfrac{1}{36\pi} \times 10^{-9} \mathrm{F/m}\right)$.

6.11 Given that an alternating magnetic field strength in perfect dielectric as

$$\boldsymbol{H} = \boldsymbol{e}_x A_1 \sin 4x \cos(\omega t - ky) + \boldsymbol{e}_z A_2 \cos 4x \sin(\omega t - ky)$$

find the corresponding displacement current density $\boldsymbol{J}_\mathrm{d}$.

6.12 There is a cylindrical capacitor, the radius of the inner conductor is a, the radius of the outer conductor is b, its length is l, the permittivity of the dielectric filled between the electrodes is ε. When the low frequency voltage applied on it is $u = U_\mathrm{m} \sin \omega t$, find the displacement current density in the dielectric and the displacement current flowing through the cylindrical surface with radius of r $(a < r < b)$; prove this displacement current is equal to the conduction current flowing in the leads of the capacitor.

6.13 Given that an electric field strength in source-free (passive) spatial region is $\boldsymbol{E} = \boldsymbol{e}_y 0.1 \sin(10\pi x) \cos\left(6\pi \times 10^9 t - kz\right) \mathrm{V/m}$, find the corresponding magnetic field strength $\boldsymbol{H}$.

6.14 It is known that the electric field of a spherical wave in free space is $\boldsymbol{E} = \boldsymbol{e}_\theta \left(\frac{E_0}{r}\right) \sin \theta \cos(\omega t - kr)$, find $\boldsymbol{H}$ and k (Clues: obtain two forms of H_ϕ based on the two curl equations in Maxwell's Equations, and they should be equal, thus k is found, which would be relatively simple).

6.15 There is a coaxial cylindrical conductor with its two ends short-circuited, and the radius of the inner conductor is a, the radius of the outer conductor is b, its length is l. The conductivity of the conductor is $\sigma = \infty$, and the media filled in it is air. If the axis of the conductor coincides with the z-axis of cylindrical coordinates, the magnetic field strength between the two conductors will be $\boldsymbol{H} = e_\phi \frac{H_{\mathrm{m}}}{r} \cos kz \cos \omega t$, leaving the surfaces at its two ends located at the positions: $z = 0, l$. Find:

(1) $\boldsymbol{E}$ and k;

(2) $\boldsymbol{J}_s$ and ρ_s on the conductor's surface.

6.16 Given a passive region ($\rho = 0, \boldsymbol{J} = 0$) in free space ($\mu_{\mathrm{r}} = \varepsilon_{\mathrm{r}} = 1, \sigma = 0$), where there exists $\boldsymbol{E} = \boldsymbol{e}_y E_0 \cdot \sin(\omega t - kz)$. Find:

(1) magnetic field strength;

(2) prove that ω/k equals light speed c;

(3) average Poynting vector.

6.17 Given the complex magnitude of magnetic field for sinusoidal electromagnetic fields in vacuum as $\boldsymbol{H}(r) = \boldsymbol{e}_\phi \frac{H_{\mathrm{m}}}{r} \sin \theta \mathrm{e}^{-\mathrm{j}kr}$, where H_{m} and k are constants, it is also known that there is no sources in that region of fields. Find the instantaneous Poynting vector.

6.18 Given that the electric field and magnetic field of a spherical electromagnetic wave radiated by antenna are $\boldsymbol{E} = \boldsymbol{e}_\theta A_0 \frac{\sin\theta}{r} \sin(\omega t - kr)$ and $\boldsymbol{H} = \boldsymbol{e}_\phi \frac{1}{\eta_0} A_0 \frac{\sin\theta}{r} \sin(\omega t - kr)$, respectively. Find the transmit power of the antenna.

6.19 In a region bounded by perfectly conductive walls $0 \leqslant x \leqslant a$ there is an electromagnetic field as follows:

$$E_y = H_0 \mu\omega \left(\frac{a}{\pi}\right) \sin\left(\frac{\pi x}{a}\right) \sin(kz - \omega t)$$
$$H_x = H_0 k \left(\frac{a}{\pi}\right) \sin\left(\frac{\pi x}{a}\right) \sin(kz - \omega t)$$
$$H_z = H_0 \cos\left(\frac{\pi x}{a}\right) \cos(kz - \omega t)$$

Does this electromagnetic field satisfy boundary conditions? If it does, what are the current densities on the conductive walls? Furthermore, find the energy flux density vector and its average value.

6.20 Assume that the instantaneous values of electric field and magnetic field are $\boldsymbol{E} = \boldsymbol{E}_0 \cos(\omega t + \psi_{\mathrm{e}})$ and $\boldsymbol{H} = \boldsymbol{H}_0 \cos(\omega t + \psi_{\mathrm{m}})$, respectively. Prove that the average Poynting vector is $\boldsymbol{S}_{\mathrm{av}} = \frac{1}{2}\boldsymbol{E}_0 \times \boldsymbol{H}_0 \cos(\psi_{\mathrm{e}} - \psi_{\mathrm{m}})$, which could also be derived from $\boldsymbol{S}_{\mathrm{av}} = \mathrm{Re}\left(\frac{1}{2}\boldsymbol{E} \times \boldsymbol{H}^*\right)$.

6.21 Suppose that there is a plate capacitor with its round-shaped plate of radius a, and the distance between the plates is d, filled with homogenous and conductive dielectric with conductivity of σ. Voltage U is exerted on the two plates, and the edge effect is neglected.

(1) find the electric field, magnetic field and energy flux density vector between the two plates;

(2) prove that the dissipated power is just equal to the power coming from outside of the capacitor.

6.22 Given that an electric field strength $\boldsymbol{E}$ in free space is

$$\boldsymbol{E} = \boldsymbol{e}_x E_0 \cos(\omega t - \beta z) + \boldsymbol{e}_y E_0' \cos(\omega t - \beta z)$$

find average Poynting vector $\boldsymbol{S}_{\mathrm{av}}$.

Chapter 7

Propagation of Plane Wave in Infinite Medium

Plane wave is the simplest and the most fundamental electromagnetic wave. This chapter investigates the propagation of plane wave in the infinite medium and some related parameters, including the propagation in perfect dielectric, conducting medium, good conductor, as well as in anisotropic medium(such as ferrite medium). Furthermore, the polarization of electromagnetic wave is discussed.

In a circuit, the establishment of current and voltage needs a process called as "transient process". With this process, the current and voltage can achieve stable states. Likewise, the electromagnetic wave also needs a transient process to reach the stable state.

Here we just focus on the stable-state electromagnetic field. The solutions to the wave equation are also stable. For example, the relationship between the electromagnetic field and time t is described as $\sin \omega t$ or $\cos \omega t$, i.e., $e^{j\omega t}$in the plural form. They represent the harmonic variation of the electromagnetic field with time.

In discussion of the electromagnetic wave, $\boldsymbol{E}$ and $\boldsymbol{H}$ are used, instead of the fundamental quantity $\boldsymbol{E}$ and $\boldsymbol{B}$. $(\boldsymbol{E} \times \boldsymbol{H})$ represents the energy flux density and $\boldsymbol{E}/\boldsymbol{H}$ is related to impedance (wave impedance), both are very important.

7.1 Wave Equations and Solutions

7.1.1 Wave Phenomenon of Electromagnetism and Wave Equation

Time-varying electromagnetic field has wave phenomenon, thus the investigation can start with Maxwell equations. As we all know, the first equation of Maxwell equations implies that time-varying electric fields (represented by displacement current) can usually motivate time-varying magnetic field. However, will this phenomenon stop after the motivation of the magnetic field? It is definitely not. The second equation of Maxwell equations implies that the time-varying magnetic field always motivates a time-varying electric field. In fact, the mutual motivation between time-varying electric field and magnetic field is endless. Because of this interchange process, the time-varying electric field and magnetic field can thus escape from the wave source and then propagate far away. Hence, this kind of electromagnetic fields with wave property is called electromagnetic waves.

The existence of time-varying electric field and magnetic field can support each other, and propagate on the basis of mutual interchange, that is the essence of the propagation of electromagnetic waves.

Wave equation is necessary for the investigation of wave properties for electromagnetic waves. However, how to establish the wave equations?

Method 1 Applying Maxwell's equations and the vector identities, yields wave equations in terms of electric field or magnetic field.

First, consider a simple and general case, saying, the source-free space filled with uniform, ideal and isotropic medium. The conditions are as follows:

$$\mu = \text{const.}, \quad \varepsilon = \text{const.}$$

$$\sigma = 0, \quad J = 0, \quad \rho = 0$$

Then the Maxwell's Equations can be expressed as

$$\left.\begin{array}{l} \nabla \times \boldsymbol{H} = \partial \boldsymbol{D}/\partial t = \varepsilon \partial \boldsymbol{E}/\partial t \\ \nabla \times \boldsymbol{E} = -\partial \boldsymbol{B}/\partial t = -\mu \partial \boldsymbol{H}/\partial t \end{array}\right\} \tag{7.1}$$

$$\left.\begin{array}{ll} \nabla \cdot D = \varepsilon \nabla \cdot E = 0 & (\text{then } \nabla \cdot \boldsymbol{E} = 0) \\ \nabla \cdot \boldsymbol{B} = \mu \nabla \cdot \boldsymbol{H} = 0 & (\text{then } \nabla \cdot \boldsymbol{H} = 0) \end{array}\right\} \tag{7.2}$$

Applying above four Maxwell Equations and the vector identity,

$$\nabla \times \nabla \times \boldsymbol{E} = \nabla \nabla \cdot \boldsymbol{E} - \nabla^2 \boldsymbol{E}$$

Then the source-free wave equation in terms of electric field is

$$\nabla^2 \boldsymbol{E} - \mu\varepsilon \frac{\partial^2 \boldsymbol{E}}{\partial \boldsymbol{t}^2} = 0 \tag{7.3}$$

And the source-free wave equation of magnetic field is

$$\nabla^2 \boldsymbol{H} - \mu\varepsilon \frac{\partial^2 \boldsymbol{H}}{\partial t^2} = 0 \tag{7.4}$$

In general, wave equations include the curl and divergence equations of electric field and magnetic field, completely reflecting the relationship of time-varying electromagnetic fields, as well as the relation between the fields and the sources. Their solutions will reveal the wave properties of time-varying field.

7.1.2 The Solutions of Wave Equations

Now we analyze the solutions of wave equations in source-free region, taking Equation (7.3) as an example.

Generally, electric field intensity is the function of 3D space and time. In Cartesian coordinate system, it can be expressed as

$$E = f(x, y, z, t)$$

Obviosuly, the wave equation is a second-order (for 3D space and time separately) partial differential equation, and it is a vector partial differential equation.

How to solve the wave equation? First, we change vector partial differential equation to scalar equation. In Cartesian coordinate system, the electric field intensity is expressed as

$$\boldsymbol{E} = \boldsymbol{e}_x E_x + \boldsymbol{e}_y E_y + \boldsymbol{e}_z E_z$$

Equation (7.3) can be rewritten as

$$\begin{aligned} \nabla^2 \boldsymbol{E} - \mu\varepsilon \frac{\partial^2 \boldsymbol{E}}{\partial t^2} &= \boldsymbol{e}_x \left[\nabla^2 E_x - \mu\varepsilon \frac{\partial^2 E_x}{\partial t^2}\right] + \boldsymbol{e}_y \left[\nabla^2 E_y - \mu\varepsilon \frac{\partial^2 E_y}{\partial t^2}\right] \\ &+ \boldsymbol{e}_z \left[\nabla^2 E_z - \mu\varepsilon \frac{\partial^2 E_z}{\partial t^2}\right] = 0 \end{aligned} \tag{7.5}$$

From Equation (7.5), we may see that only when three field components of x, y and z direction are all zero, the vector field can thus be zero. In this way, three scalar differential equations with same forms are obtained, only one of them is analyzed in detail as below.

$$\nabla^2 E_z - \mu\varepsilon \frac{\partial^2 E_z}{\partial t^2} = 0 \tag{7.6}$$

It can be rewritten as

$$\frac{\partial^2 E_x}{\partial x^2} + \frac{\partial^2 E_x}{\partial y^2} + \frac{\partial^2 E_x}{\partial z^2} - \mu\varepsilon \frac{\partial^2 E_x}{\partial t^2} = 0 \tag{7.7}$$

This is a second-order partial differential equation, which can be solved with the method of separation of variables.

As said above, only the fundamental electromagnetic wave is investigated in this chapter, in which E_x is one-dimensional variable, for example, just varying with z. With this in mind, Equation (7.7) is simplified as

$$\frac{\partial^2 E_x}{\partial z^2} - \mu\varepsilon \frac{\partial^2 E_x}{\partial t^2} = 0 \tag{7.8}$$

That is the well-known 1-D wave equation, with general solutions expressed as

$$E_x = f\left(t - \frac{z}{v}\right) + f\left(t + \frac{z}{v}\right) \tag{7.9}$$

Where v is a constant

$$v = \frac{1}{\sqrt{\mu\varepsilon}} \tag{7.10}$$

v has the dimension of speed, actually the propagation speed of electromagnetic wave, which will be further discussed in Section 7.2.4.

The first term and second term of Equation (7.9) are two solutions to the wave Equation (7.8), respectively. The first term will be proved to be one solution next.

Let

$$u = t - \frac{z}{v}$$

$$E_x = f(u) = f\left(t - \frac{z}{v}\right)$$

Then we get

$$\frac{\partial f(u)}{\partial z} = \frac{\partial f(u)}{\partial u}\frac{\partial u}{\partial z} = -\frac{1}{v} \cdot f'(u)$$

and

$$\frac{\partial^2 f(u)}{\partial z^2} - \frac{1}{v}\frac{\partial f'(u)}{\partial z} = -\frac{1}{v}\frac{\partial f'(u)}{\partial u}\frac{-\partial u}{\partial z} = \frac{1}{v^2}f''(u)$$

Using same method, we can obtain

$$\frac{\partial^2 f(u)}{\partial t^2} = f''(u)$$

Substituting $\partial^2 f(u)/\partial z^2$ and $\partial^2 f(u)/\partial t^2$ into Equation (7.8), yields

$$\frac{\partial^2 E_x}{\partial z^2} - \mu\varepsilon \frac{\partial^2 E_x}{\partial t^2} = \frac{1}{v^2}f''(u) - \frac{1}{v^2}f''(u) = 0$$

This shows that $f\left(t-\frac{z}{v}\right)$ is the solution of Equation (7.8), Equation (7.10) is used inside.

Similarly, the term $f\left(t+\frac{z}{v}\right)$ can be proved to be another solution of Equation (7.8). Since Equation (7.8) is a linear differential equation, its solutions can be sum up. The sum of $f\left(t-\frac{z}{v}\right)$ and $f\left(t+\frac{z}{v}\right)$ is also the solution of that equation.

7.1.3 The Physical Insight of the Solutions

What do the two terms of the solution of E_x mean? As to$f\left(t-\frac{z}{v}\right)$, it means that E_x is a function of t and z, however, it does not show the function in detail. Now choose an arbitrary function, as shown in Figure 7.1. It shows the distribution of $E_x=\left[f\left(t'-\frac{z'}{v}\right)\right]$ along the space (z direction) at a certain time t'.

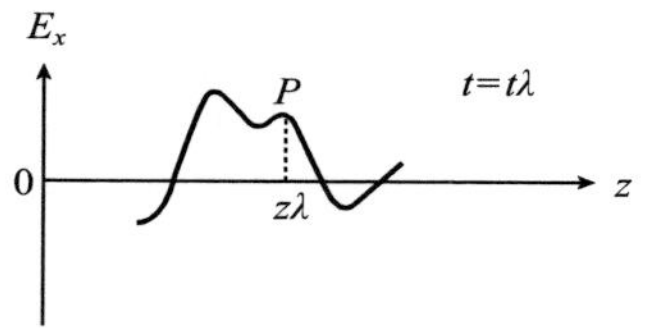

Figure 7.1 Distribution of $E_x=\left[f\left(t'-\frac{z'}{v}\right)\right]$ along the space at a certain moment

In order to clarify the variation of E_x along time t and space z, now observe the movement of an arbitrary point P on the waveform. When point P is fixed, in lossless medium, the function value $\left[f\left(t-\frac{z}{v}\right)\right]$ of point P observed at any time should equal to the function value $\left[f\left(t'-\frac{z'}{v}\right)\right]$ of point P shown in Figure 7.1. That is

$$\left(t_1-\frac{z_1}{v}\right)=\left(t_2-\frac{z_2}{v}\right)=\cdots=\left(t'-\frac{z'}{v}\right)=\text{const.}$$

It shows that when time changes from t_1 to t_2, $t_3\cdots$ the position of point P will change from z_1 to z_2, $z_3\cdots$ accordingly. It can be seen that when time t increases, the position z of point P will also increase, as shown in Figure 7.2. The moving state of point P in the figure represents the moving discipline of electric field$E_x=\left[f\left(t-\frac{z}{v}\right)\right]$. Obviously, point P is traveling in the positive z direction with time, saying, $E_x$is propagating along the positive z direction with velocity v. To summarize, $f\left(t-\frac{z}{v}\right)$ represents that E_x is propagating along the positive

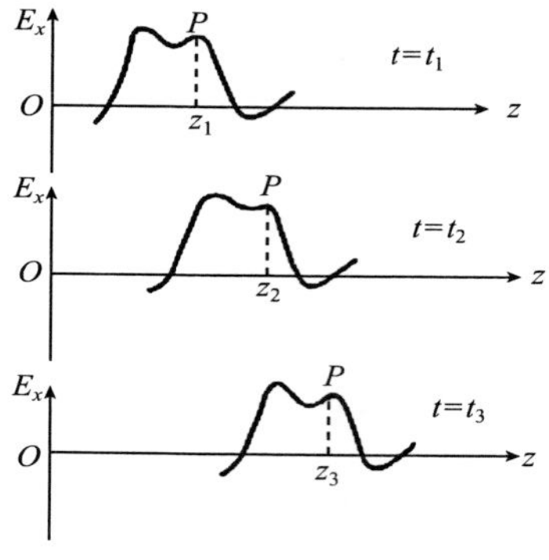

Figure 7.2 $E_x=f\left(t-\frac{z}{v}\right)$ changing with time, the spatial moving state of point P

z direction with velocity v, while $f\left(t+\frac{z}{v}\right)$ represents that E_x is propagating along the negative z direction with velocity v.

7.2 Plane Wave in Perfect Dielectric

Mutually perpendicular electric and magnetic fields make up a plane which is perpendicular to the propagation direction. Furthermore, this plane is an equiphase surface (i.e., phases of both electric field and magnetic field are equal at any point of this plane). This electromagnetic wave is called plane wave.

In free space, the phases of electromagnetic wave radiated by centrally located electric dipole (small antenna element) are the same at each point of the sphere with a constant radius. This electromagnetic wave is called spherical wave.

Since both the electric and magnetic fields are located on a cross section perpendicular to the propagation direction, without any longitudinal field components, it is thus also called transverse electromagnetic wave (TEM wave).

7.2.1 Uniform Plane Wave

The so-called uniform plane wave is defined as that whose equiphase surface is a plane, on which all the amplitude of electric field or magnetic field at any point are equal and the magnitude and phase of electromagnetic field only change along the propagation direction. For example, the uniform plane wave taking $x-y$ plane as a equiphase surface only changes along the z direction, as shown in Figure 7.3(a) and (b).

For the sake of clear understanding, the figure only shows the uniform distribution of electric field and magnetic field on a certain $x-y$ plane at a certain time, rather than showing all the electromagnetic field distributions at different z positions at that time.

Plane wave is the foundation of many more complex wave forms, which can be formed with this simplest plane wave. Therefore, it is necessary and meaningful to make further investigations on plane wave and clarify its parameters and characteristics.

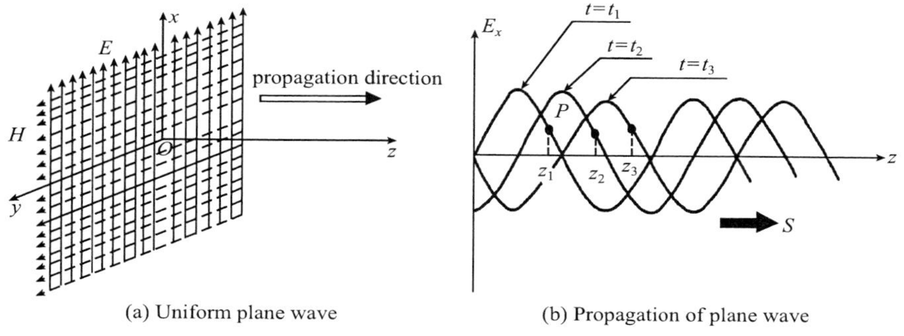

Figure 7.3

7.2.2 Wave Equation of Uniform Plane Wave

As to the uniform plane wave shown in Figure 7.3(a), the electric field only has E_x component and magnetic field only has H_y component. E_x and H_y are not the function of x or y, but the function of z and t. Therefore, Equation (7.8) shall be the wave equation of electric field E_x for uniform plane wave.

Method 2 There is a simpler and more intuitive method to develop the wave equation of uniform plane wave, saying, the wave equation can be established directly from Maxwell's Equations according to the characteristics of uniform plane wave, i.e., there are only $E_x(z, t)$ and $H_y(z, t)$ and E_y, E_z, H_z, H_x are all equal to 0.

For uniform plane wave, the first and second equations in Maxwell's Equations can be simplified as

$$-\frac{\partial H_y}{\partial z} = \varepsilon \frac{\partial E_x}{\partial t} \tag{7.11a}$$

$$\frac{\partial E_x}{\partial z} = \mu \frac{\partial H_y}{\partial t} \tag{7.11b}$$

taking partial differential of Equation (7.11a) with respect to t and taking the partial derivative of Equation (7.11b) with respect to z, we get

$$-\frac{\partial}{\partial t}\left(\frac{\partial}{\partial z} H_y\right) = \varepsilon \frac{\partial^2 E_x}{\partial t^2} \tag{7.12a}$$

$$\frac{\partial^2 E_x}{\partial z^2} = -\mu \frac{\partial}{\partial z}\left(\frac{\partial}{\partial t} H_y\right) \tag{7.12b}$$

In the above equations, the order of $H_y$taking partial differentiation of t and z can be interchanged without affecting the results. Then we can obtain the wave equation of electric field E_x from Equation (7.12),

$$\frac{\partial^2 E_x}{\partial z^2} - \mu\varepsilon \frac{\partial^2 E_x}{\partial t^2} = 0 \tag{7.13}$$

Similarly, we can find the wave equation of magnetic field H_x as below

$$\frac{\partial^2 H_y}{\partial z^2} - \mu\varepsilon \frac{\partial^2 H_y}{\partial t^2} = 0 \tag{7.14}$$

7.2.3 Complex Expression of Uniform Plane Wave

1. Helmholtz Equation

To simplify the computation, the field quantity of harmonic wave is usually expressed in the plural form with time dependence factor $\mathrm{e}^{\mathrm{j}\omega t}$. Therefore, when electric field and magnetic field, in the plural form, are differentiated with respect of time, we can get the following expressions.

$$\frac{\partial}{\partial t} \Rightarrow \mathrm{j}\omega, \quad \frac{\partial^2}{\partial t^2} \Rightarrow -\omega^2$$

Equation (7.13) can be rewritten as

$$\frac{\partial^2 E_x}{\partial z^2} + k^2 E_x = 0 \tag{7.15}$$

where

$$k^2 = \omega^2 \mu\varepsilon \tag{7.16}$$

Equation (7.15) is the wave equation of harmonic wave, which is called Helmholtz Equation.

2. Solution of Helmholtz Equation

The solution of Helmholtz Equation (7.15) can be written in the plural form like following,

$$E_x = E_0 e^{-jkz} + E_0' e^{jkz}$$

However, in order to facilitate application and to provide a complete time and space relation, we always rewrite $e^{j\omega t}$ in the following form.

$$E_x = E_0 e^{j(\omega t - kz)} + E_0' e^{j(\omega t + kz)} \tag{7.17}$$

After finding electric field E_x, it is easy to find H_y by using Maxwell's equation. For example, we can substitute E_x into scalar Maxwell Equation (7.11b). Then,

$$\begin{aligned}\frac{\partial E_x}{\partial z} &= -jkE_0 e^{j(\omega t - kz)} + jkE_0' e^{j(\omega t + kz)} \\ &= -\mu \frac{\partial H_y}{\partial t} = -j\omega\mu H_y\end{aligned}$$

so

$$H_y = \frac{E_0}{\sqrt{\mu/\varepsilon}} e^{j(\omega t - kz)} - \frac{E_0'}{\sqrt{\mu/\varepsilon}} e^{j(\omega t + kz)} \tag{7.18}$$

There are two terms in Equations (7.17) and (7.18). Each one can be the solution of the wave equation separately. These two terms respectively represent two waves propagating along different directions. Supposing the wave propagating along the positive z direction (the first term) is the incident wave, then the one propagating along the negative z direction (the second term) is the reflected wave. Electromagnetic wave propagating in infinite space only contains incident wave but no reflected wave. In that case, Equations (7.17) and (7.18) only have the first term. Therefore, *the complex expression of uniform plane wave propagating along the positive z direction is given as follows.*

$$\left.\begin{aligned}E_x &= E_0 e^{j(\omega t - kz)} \\ H_y &= \left(E_0/\sqrt{\mu/\varepsilon}\right) e^{j(\omega t - kz)}\end{aligned}\right\} \tag{7.19}$$

Its instantaneous fields can be written as

$$\left.\begin{aligned}E_x &= E_0 \sin(\omega t - kz) \\ H_y &= \left(E_0/\sqrt{\mu/\varepsilon}\right) \sin(\omega t - kz)\end{aligned}\right\} \tag{7.20}$$

Example 7.1 The instantaneous electric field intensity $\boldsymbol{E}$ of wave propagating in the air is known as

$$\begin{aligned}\boldsymbol{E} &= \boldsymbol{e}_y E_y + \boldsymbol{e}_z E_z \\ &= \boldsymbol{e}_y E_0 \cos(\omega t - kx) + \boldsymbol{e}_x E_0 \sin(\omega t - kx)\end{aligned}$$

Please find the instantaneous magnetic field intensity $\boldsymbol{H}$ and the Poynting's vector $\boldsymbol{S}$ of this wave.

Solution From the above given electric field, we obtain the magnetic field directly using Maxwell's equation , that is

$$-\mu_0 \frac{\partial H_x}{\partial t} = \frac{\partial E_z}{\partial y} - \frac{\partial E_y}{\partial z} = 0$$

We have

$$H_x = 0$$

from

$$-\mu_0 \frac{\partial H_y}{\partial t} = \frac{\partial E_x}{\partial z} - \frac{\partial E_z}{\partial x} = -\frac{\partial E_z}{\partial x}$$
$$= E_0 k \cos(\omega t - kx)$$

We have

$$H_y = -E_0 \sqrt{\frac{\varepsilon_0}{\mu_0}} \sin(\omega t - kx)$$

Using the same method, we find

$$H_z = E_0 \sqrt{\frac{\varepsilon_0}{\mu_0}} \cos(\omega t - kx)$$

The magnetic field intensity $\boldsymbol{H}$ of the wave is

$$\begin{aligned} \boldsymbol{H} &= \boldsymbol{e}_y H_y + \boldsymbol{e}_z H_z \\ &= [-\boldsymbol{e}_y E_0 \sin(\omega t - kx) + \boldsymbol{e}_z E_0 \cos(\omega t - kx)] \cdot \sqrt{\frac{\varepsilon_0}{\mu_0}} \end{aligned}$$

The poynting's vector $\boldsymbol{S}$ is thus obtained from the known $\boldsymbol{E}$ and $\boldsymbol{H}$

$$\begin{aligned} \boldsymbol{S}(t) &= \boldsymbol{E} \times \boldsymbol{H} = (\boldsymbol{e}_y E_y + \boldsymbol{e}_z E_z) \times (\boldsymbol{e}_y H_y + \boldsymbol{e}_z H_z) \\ &= \boldsymbol{e}_x (E_y H_z - E_z H_y) \\ &= \boldsymbol{e}_x E_0^2 \left\{\cos^2(\omega t - kx) + \sin^2(\omega t - kx)\right\} \cdot \sqrt{\frac{\varepsilon_0}{\mu_0}} \\ &= \boldsymbol{e}_x E_0^2 \cdot \sqrt{\frac{\varepsilon_0}{\mu_0}} \end{aligned}$$

7.2.4 Characteristics and Parameters of Plane Wave

In the following, the characteristics and parameters of uniform plane wave will be clarified from the known solutions.

1. Parameters of Plane Wave

The main parameters of plane wave include propagation velocity v, phase constant k, wave impendence η etc.

(1) Propagation velocity v of plane wave

From Figure 7.3(b) we can see, while observing any point P on the waveform of E_x, the spatial location of P will change over the time t while *the electric field of P keeps unchanged.* In this way, *the phase of point P is always a constant.* That is

$$\omega t_1 - k z_1 = \omega t_2 - k z_2 = \cdots = \omega t_n - k z_n = \text{const.}$$

then

$$\frac{z_2 - z_1}{t_2 - t_1} = \frac{\Delta z}{\Delta t} = \frac{\omega}{k}$$

When $\Delta t \to 0$, the propagation velocity v of electromagnetic wave is

$$v = \frac{\mathrm{d}z}{\mathrm{d}t} = \frac{\omega}{k} \tag{7.21}$$

Substituting $k = \omega\sqrt{\mu\varepsilon}$ into Equation (7.21), we have

$$v = \frac{1}{\sqrt{\mu\varepsilon}} \quad (\text{m/s}) \tag{7.22}$$

v is the well-known propagation velocity of plane wave. To speak strictly, it is the velocity for the wave to propagate along the propagating direction (now the z direction) of plane wave or along the direction perpendicular to the equiphase surface of the wave. It can be easily seen that the propagation velocity of plane wave in perfect dielectrics is only related to the medium parameters μ and ε. The larger permittivity ε usually causes lower propagation velocity of electromagnetic wave. However, the propagation is not related with frequency. The propagation velocity v of electromagnetic wave in vacuum is

$$v = \frac{1}{\sqrt{\mu_0\varepsilon_0}} = 2.997956 \times 10^8 \approx 3 \times 10^8 \quad (\text{m/s})$$

That is to say, the propagation velocity of electromagnetic wave in vacuum (and air) is equal to the speed of light c.

Above is the general discussion about the propagation velocity of plane wave, i.e., the velocity of electromagnetic wave propagating along the direction of plane wave. In practice, it is necessary to further distinguish the difference between the speeds. The velocity is classified into phase velocity and group velocity (energy velocity).

(2) Phase velocity v_{p}

When electromagnetic wave propagates along ξ direction, if the phases of electric field (or magnetic field) in any plane perpendicular to ξ are the same, this plane is called the equiphase surface. *Travelling speed of the equiphase surface is called phase velocity and denoted as v_{p}.*

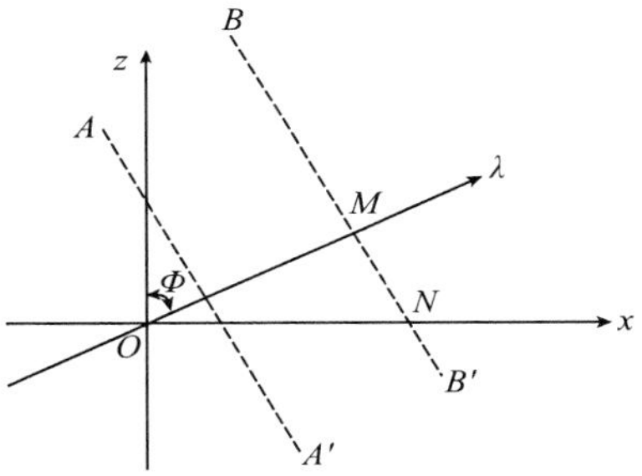

Figure 7.4 Phase-velocity

Method 1 Using the expression of equiphase surface $(\omega t - k_x x) = \text{const}$ to find $v_{\text{p}(x)}$.

Plane wave propagating along ξ direction is related with time t and space ξ as $\text{e}^{\text{j}(\omega t - k\xi)}$. Refer to Figure 7.4 for analysis. From Figure 7.4, ξ is expressed as

$$\xi = \xi(\sin^2\theta + \cos^2\theta) = x\sin\theta + z\cos\theta$$

Substituting the value of ξ

$$\begin{aligned}\text{e}^{\text{j}(\omega t - k\xi)} &= \text{e}^{\text{j}(\omega t - kx\sin\theta - kz\cos\theta)} \\ &= \text{e}^{\text{j}(\omega t - k_x x - k_z z)}\end{aligned}$$

Where

$$\begin{aligned}k_x &= k\sin\theta \\ k_z &= k\cos\theta\end{aligned}$$

k_x and k_z are the phase constants of plane wave propagating along x direction and z direction, respectively.

When the plane wave propagates along ξ direction, the phase velocity of x direction (or z direction) can be known if observing the traveling speed of equiphase surface along x direction (or z direction). When we study the phase variation along x direction, z is a constant, only the remaining $\text{e}^{\text{j}(\omega t - k_x x)}$ is variable. The desired phase velocity in x direction equals to the travelling speed of any equiphase surface along x direction.

Though the chosen equiphase surface locates different positions at different time, however, their phases are always constant. From $(\omega t - k_x x) = \text{const}$, we have

$$x = \left.\frac{\omega t - \text{const.}}{k_x}\right|_{\omega t - k_x x = \text{const.}}$$

Then the phase velocity $v_\mathrm{p}(x)$ in x direction is

$$v_{\mathrm{p}(x)} = \left.\frac{\partial x}{\partial t}\right|_{\omega t - k_x x = \text{const}} = \frac{\omega}{k_x} = \frac{\omega}{k\sin\theta} = \frac{v}{\sin\theta} \tag{7.23a}$$

Similarly, we can get the phase velocity $v_{\mathrm{p}(z)}$ in z direction

$$v_{\mathrm{p}(z)} = \frac{v}{\cos\theta} \tag{7.23b}$$

From Equations (7.23a) and (7.23b), $v_{\mathrm{p}(x)}$ and $v_{\mathrm{p}(z)}$ are both larger than the propagating speed v of plane wave in ξ direction. If the plane wave travels in vacuum, $v = c$, then phase velocity v_p is larger than light speed, which seems to be unreasonable because of being contrary to theory of relativity. However, phase velocity does not represent the travel speed of the true material essentially. It can be seen from Figure 7.4 that the plane AA' perpendiculars to ξ is an equiphase surface on which the phases at any point are the same. plane AA' travels to plane BB' within the time period Δt, and the phases at any point on plane BB' are also the same. Therefore, *while equiphase surface is travelling from AA' to BB', we get different phase velocities if we observe them from different directions (i.e. ξ direction, x direction, z direction etc.).* Furthermore, except that the phase velocity $v_\mathrm{p} = v$, all other phase velocities v_p when observing in any other directions are larger than v.

Method 2 Directly applying the geometric relationship shown in Figure 7.4 to obtain v_p.

$v_{\mathrm{p}(x)}$ and $v_{\mathrm{p}(z)}$ can be obtained directly from the geometric relationship shown in Figure 7.4. From the figure, the phase velocity of plane wave in OM direction is v, while the one in ON direction is $v_{\mathrm{p}(x)}$. Then

$$\frac{\overline{OM}}{\overline{ON}} = \frac{\overline{ON}\times\sin\theta}{ON} = \frac{v}{v_{\mathrm{p}(x)}} = \sin\theta$$

$v_{\mathrm{p}(x)}$is thus available now, $v_{\mathrm{p}(z)}$can be obtained using the same method.

We should mention that, the propagating velocity is the travelling speed of equiphase surface along the propagation direction of electromagnetic wave.

(3) Energy velocity v_g

Energy velocity refers to the propagation velocity of energy, which is the signal group velocity. In this way, energy velocity is also called group velocity. Propagation velocity of any point on envelope of the amplitude-modulated wave is energy velocity. If the AM wave is

$$E_0\cos(\Delta\omega t - \Delta k z)\cdot \mathrm{e}^{\mathrm{j}(\omega t - kz)}$$

Where ω and k are carrier angular frequency and phase constant respectively. $\Delta\omega$ and Δk are the corresponding parameters of AM wave envelope. $\cos(\Delta\omega t - \Delta k z)$ is the envelope of AM wave.

When $\omega \gg \Delta\omega$, then from $(\Delta\omega t - \Delta k z) = \text{const.}$, we can get $z = (\Delta\omega t - \text{const})/\Delta k$. The energy velocity v_g is thus

$$v_\mathrm{g} = \left.\frac{\mathrm{d}z}{\mathrm{d}t}\right|_{\Delta\omega t - \Delta k z = \text{const.}} = \frac{\mathrm{d}\omega}{\mathrm{d}k} = \frac{1}{\dfrac{\partial k}{\partial\omega}} \tag{7.24}$$

(4) Phase Constant (Wave Number) of Plane Wave k

Substituting Equation (7.22) into Equation (7.16), yields

$$k = \omega\sqrt{\mu\varepsilon} = \frac{\omega}{v} = \frac{2\pi f}{f\lambda} = \frac{2\pi}{\lambda}$$

that is

$$k = \frac{2\pi}{\lambda} \tag{7.25}$$

What does the above equation mean? As we know, wave length λ is the transmitting distance of electromagnetic wave during a period of time when phase charges by 360 degree (2π radian). Thus, $k = \frac{2\pi}{\lambda}$ represent the phase change of unit length, and k is thus called phase constant.

Obviously, the higher frequency corresponds to shorter wavelength and larger phase constant k, and vice versa. In addition, the wave travels more slowly in a medium with higher permittivity ε, In such case, the phase constant k will be larger, and vice versa. Since k represents phase variation of unit length, then kl represents the total phase variation within the length of l.

In low-frequency circuit, the wavelength is too much long because of the quite low frequency. For example, when $f = 3\times10^6$Hz, the wavelength is$\lambda = \frac{c}{f} = 100$m. Therefore, in a limited range, i.e. in the circuit with size of several centimeters or several millimeters, the phase variation is negligible. For example, the total phase variation is only 0.036° in a 1cm-long line. In this way, there is very little influence owing to ignorance of this tiny phase variation. That is to say, *the current and voltage can be regarded as functions of time in low-frequency case. Phase variation occuring along the space is not so important. However, as the frequency increases, the phase variation along space is becoming more and more important.* For example, in a 1cm-long line, if wavelength $\lambda = 6$mm (i.e. $f = 5\times10^9$Hz), the phase variation is 60°, a big mistake is definitely caused owing to ignorance of such a large phase variation. Therefore, *phase constant k is an important parameter in high-frequency case.*

(5) Wave Impendence of Plane Wave (also called intrinsic impedance of medium) η

Equation (7.19) shows the expression of uniform plane wave in perfect medium. It can be seen that phases of E_x and H_y are the same. The ratio of their amplitudes is

$$\frac{E_x}{H_y} = \sqrt{\frac{\mu}{\varepsilon}} = \eta \tag{7.26}$$

η which has the impedance dimension, we thus call it wave impendence of plane wave. If the electromagnetic wave propagates in free space, we call η the wave impendence of free space, denoted as η_0. It is

$$\eta_0 = \sqrt{\frac{\mu_0}{\varepsilon_0}} = 120\pi \approx 377(\Omega)$$

It is a real number (pure resistance). *That shows phases of E_x and H_y of plane wave in perfect medium are the same, quite similar to the case of pure resistance.*

2. Wave Characteristic of Plane Wave

Now, we will study how the electromagnetic field changes in space with time t, according to the instantaneous expression of E_x and H_y shown in Equation (7.20).

For the sake of simplicity, only the case of any point (i.e. point P) on a certain equiphase surface of electric field will be studied. As waveform is an integral part, if the variation of that point is clearly understood, then the time and space variation of electric field E_x are thus known.

Since any point P observed is on the equiphase surface, *the field value of point Pis unchanged, and the phase of point P is unchanged too.* Then

$$\omega t_1 - kz_1 = \omega t_2 - kz_2 = \cdots = \omega t_n - kz_n = \text{const}$$

The equation shows that when time changes from $t_1 \to t_2 \to t_3 \to \cdots$, the space location of point P changes from $z_1 \to z_2 \to z_3 \to \cdots$. The corresponding time and space change is shown in Figure 7.5(a). Obviously, the point P travels along the positive z direction, i.e., the whole electric field E_x travels along the positive z direction. The instantaneous electric field E_x can thus be expressed

$$E_x = E_0 \cos(\omega t - kz)$$

Its plural expression is

$$E_x = E_0 \mathrm{e}^{\mathrm{j}(\omega t - kz)}$$

That means electric field E_x propagates in the positive z direction. Obviously, when E_x is

$$E_x = E_0 \mathrm{e}^{\mathrm{j}(\omega t + kz)}$$

It means electric field E_x propagates in the negative z direction.

We call the waves propagating in a certain direction as travelling wave.

Phase variation of travelling wave is continuous. As we all know, the parameter k is a constant representing the phase variation per unit length. kz thus represents the total phase variation in the length of z, as well as an angle $\left(\text{which equals to } \dfrac{2\pi}{\lambda} \cdot z\right)$, while $-kz$ represents a negative angle, which means phase lag. It represents the phase lagging after a distance of z. If z changes continuously, the phase change will also be continuous. According to the phase changing characteristic of travelling wave, we may choose any point z_1 at z axis and then observe the phase variation of the wave on that point. When z_1 is fixed, $-kz$ is thus a constant. Obviously the phase angle $(\omega t - kz_1)$ of that point changes with time t, as shown in Figure 7.5(c).

After choosing a certain time t_1, the phase angle $(\omega t - kz_1)$ changes with z, and have continuous phase without sudden change, as shown in Figure 7.5(b). This is in contrast to the discontinuous change of phase variation of standing wave.

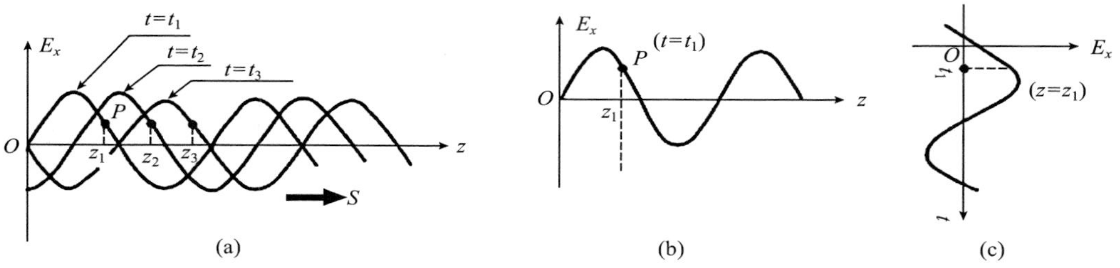

Figure 7.5 Travelling wave

The characteristics of uniform plane wave in homogeneous perfect dielectrics are summarized as follows: the electric field and magnetic field are perpendicular to each other, and they both locate in the cross section without longitudinal (transmitting direction) component, we thus call it transverse electromagnetic wave (TEM wave). Amplitudes of electric and magnetic field of uniform plane wave at any points on the same equiphase surface are the same. The transmitting direction of uniform plane wave is perpendicular to the plane with electric and magnetic fields. The propagation velocity $v = 1/\sqrt{\mu\varepsilon}$ does not change with frequency, we thus call it non-dispersive wave.

Because perfect dielectric is lossless, its amplitudes (E_0, H_0) are constant while propagating.

When the plane wave travels in perfect dielectric, the phases of electric field and magnetic field are the same. That is, electric field and magnetic field can be positive, negative, zero or any other value simultaneously, as shown in Figure 7.6. In this case, the ratio of E_x and H_y (i.e. wave impendence η) is a pure resistance. If we choose air as the medium, the propagation velocity of plane wave equals to light speed c, and wave impendence $\eta_0 = 377\Omega$.

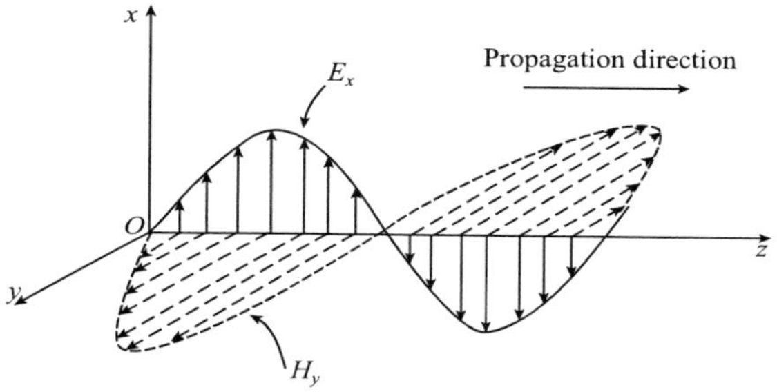

Figure 7.6 Spatial distribution of E_x and H_y of plane wave along the transmitting direction in perfect dielectric

7.3 Polarization of Electromagnetic Wave

Alternating vector fields usually changes with time, i.e., the magnitudes changes with time and even the directions of field vectors at any time usually change with time. The polarization of electromagnetic wave is used to describe this changing mode of field vector. Polarization is usually named according to the moving trajectory of field vector endpoint when it is changing with time. There are three kinds of polarizations for the electromagnetic waves, saying, the linear polarization, circular polarization and elliptical polarization, as shown in Figure 7.7 to Figure 7.9.

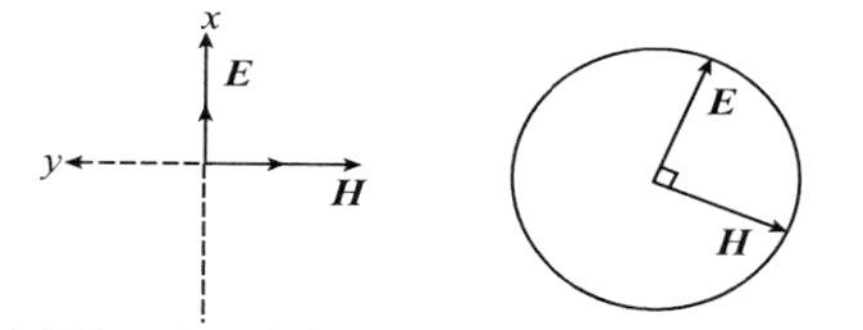

(a) Linearly polarized wave (b) Circularly polarized wave

Figure 7.7

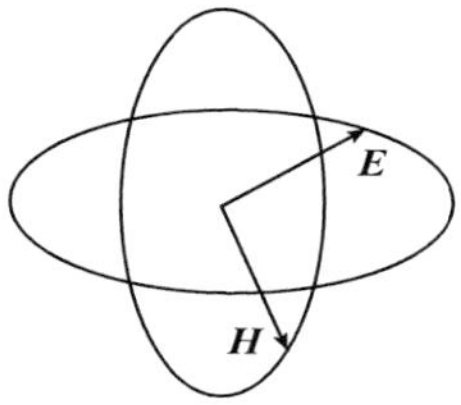

Figure 7.8 Elliptically polarized wave

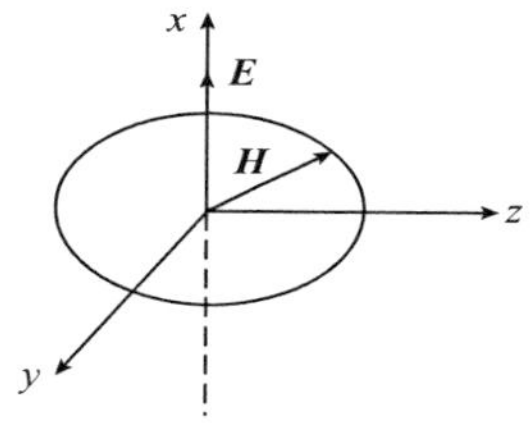

Figure 7.9 Electric field is linear polarized wave and magnetic field is circular (or elliptically) polarized wave

It's important to learn about polarization of electromagnetic wave. It can help us to set the antennas to receive the signals most effectively. Spatial electromagnetic wave of circular polarization is used in modern telecommunication. Magnetic wave with circular polarization is also used in ferrite devices (anisotropic devices).

How to form a variety of polarized waves? As we know, the wave equation is a linear differential equation. According to the superposition principle of linear equation, its solution can be equal to the sum of several solutions. From the physical concept, one kind of polarized wave can be formed by summing of several kinds of polarized waves. Similarly, one kind of wave can also be divided. Of course, their frequency is the same.

7.3.1 Linearly Polarized Wave

As shown in Figure 7.7(a), the electric field is linearly polarized wave along x direction.

$$\boldsymbol{E} = \boldsymbol{e}_x E_0 \sin(\omega t - kz)$$

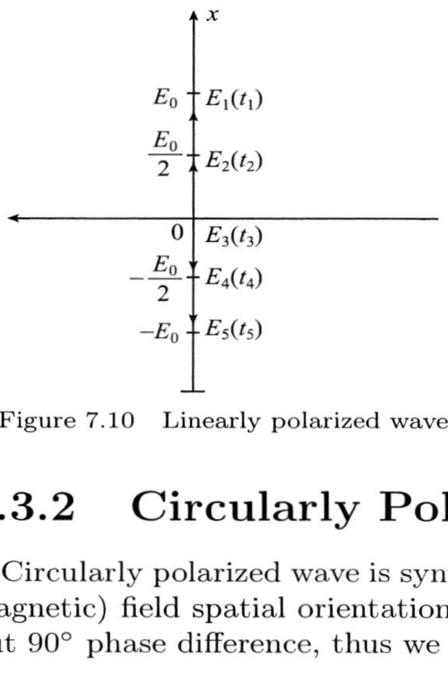

Figure 7.10 Linearly polarized wave

At the fixed position $z = 0$, observing the change with time t, setting

$$t_1 = \frac{3}{12}T, \quad t_2 = \frac{5}{12}T, \quad t_3 = \frac{6}{12}T, \cdots$$

then

$$\omega t_1 = 90°, \quad \omega t_2 = 150°, \quad \omega t_3 = 180°, \cdots$$

The corresponding electric fields E_x are E_0, $\frac{1}{2}E_0$, $0 \cdots$, respectively, as shown in Figure 7.10. The electric field always changing along x-axis with time is called linearly polarized wave.

7.3.2 Circularly Polarized Wave

Circularly polarized wave is synthesized by two linearly polarized waves whose electric (or magnetic) field spatial orientations are perpendicular to each other, with equal amplitudes but 90° phase difference, thus we have

$$\left.\begin{aligned} E_x &= E_0 \cos(\omega t - kz) \\ E_y &= E_0 \cos(\omega t - kz \pm 90°) \end{aligned}\right\} \tag{7.27}$$

Still observing at z=0 position, summing up the square E_x and E_y. Then

$$E_x^2 + E_y^2 = E_0^2 \left[\cos^2(\omega t) + \cos^2(\omega t \pm 90°)\right] = E_0^2$$

that is

$$\left(\frac{E_x}{E_0}\right)^2 + \left(\frac{E_y}{E_0}\right)^2 = 1 \tag{7.28}$$

Obviously, Equation (7.28) is a circular equation. The radius is E_0, the electric field amplitude synthesized by E_x and E_y. The angle between synthesized electric field $\boldsymbol{E}$ and x-axis θ is

$$\theta = \arctan\left(\frac{E_y}{E_x}\right) = \text{arctg}\left[\frac{\mp E_0 \sin(\omega t)}{E_0 \cos(\omega t)}\right] = \mp\omega t$$

Obviously, the synthesized electric field is rotating at a given angular frequency ω. The trajectory of the endpoint of the synthetic electric field vector with time is a circle, as shown in Figure 7.11. *The rotational direction of the synthesized electric field E can be either clockwise or counterclockwise, depending on whether the 90° time phase difference between* E_x *and* E_y *is positive or negative, as well as on the propagating direction.*

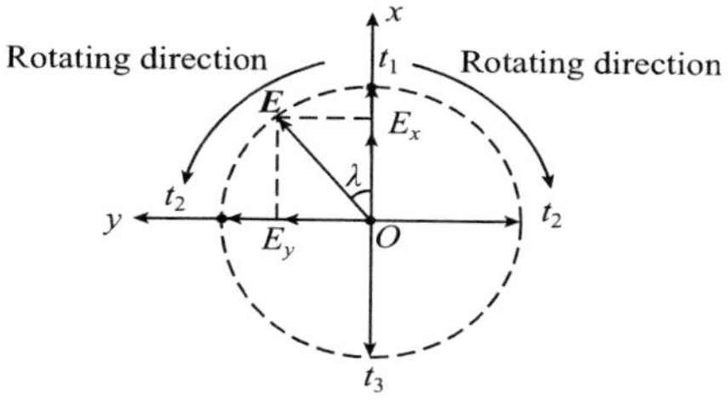

Figure 7.11 Circular polarized wave and its rotating direction

Now study the rotation direction of circularly polarized wave. Still study the position $z = 0$. Let $t_1 = 0$, $t_2 = \dfrac{T}{4}$, $t_3 = \dfrac{2T}{4}$, $\cdots$. From (7.27) we can get the synthesized electric field is

$$\begin{aligned} \boldsymbol{E}_1(t_1) &= \boldsymbol{e}_x \boldsymbol{E}_0 \\ \boldsymbol{E}_2(t_2) &= \mp \boldsymbol{e}_y \boldsymbol{E}_0 \\ \boldsymbol{E}_3(t_3) &= -\boldsymbol{e}_x \boldsymbol{E}_0 \\ &\vdots \end{aligned}$$

Since the offset between $\mp E_y$ at t_2 and E_x at t_1 is $+90°$ and $-90°$, as time changes from $t_1 \to t_2 \to t_3 \cdots$, the rotating direction of synthetic field also has two opposite cases, as shown in Figure 7.11.

In the above analysis, we only observe the variation of synthetic electric field at the position $z = 0$. Actually, the electromagnetic wave propagates along the z direction. It can thus be imagined that at a certain time, the variation of electric field $\boldsymbol{E}$ along z direction of linearly polarized wave is shown in Figure 7.12. The variation of electric field $\boldsymbol{E}$ along z direction of circularly polarized is shown in Figure 7.13.

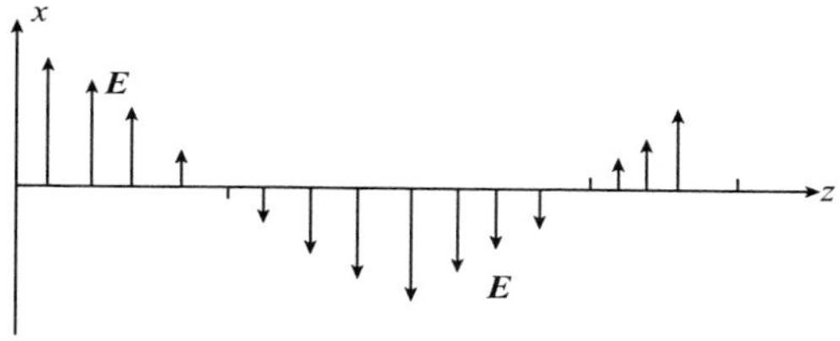

Figure 7.12 Variation of electric field of the linearly polarized wave along z direction at a certain time

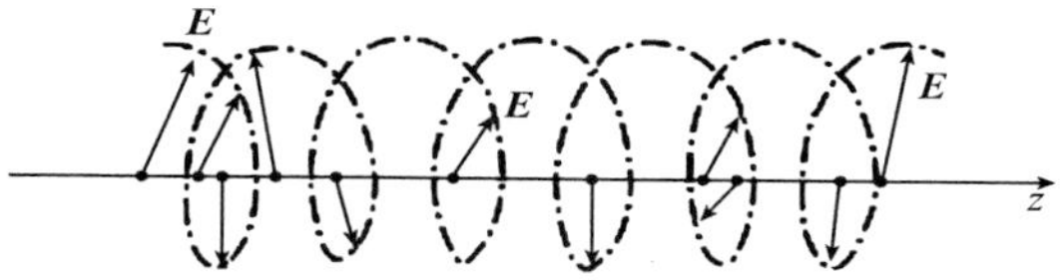

Figure 7.13 Variation of electric field of circularly polarized wave along z direction at a certain time

Actually, the propagation direction must be considered when defining the rotating direction of circularly polarized wave. If observing along the propagating direction, the rotation direction of the electric field (or magnetic field) for the circularly polarized wave is clockwise, we call this wave right-handed circularly polarized wave. On the contrary, we call it left-handed circularly polarized wave, as shown in Figure 7.14.

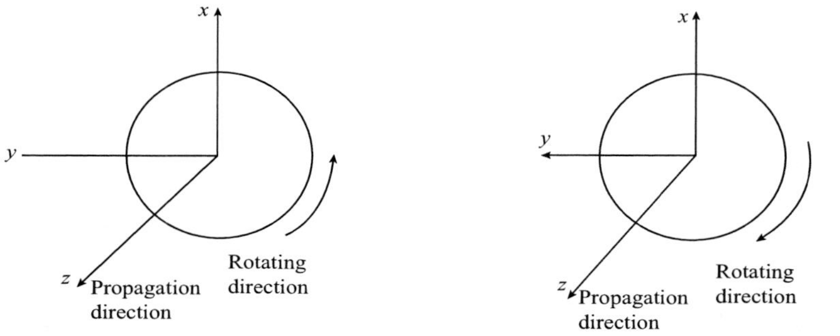

(a) Right-handed citcularly polarized wave (b) Left-handed circularly polarized wave

Figure 7.14

7.3.3 Elliptically Polarized Wave

The elliptically polarized wave can be synthesized by the electric fields (or magnetic fields) of two linear polarized waves, with unequal amplitudes and time phase difference of 90°, spatially perpendicular to each other.

The following two electric fields can constitute the elliptically polarized wave.

$$\left.\begin{aligned} E_x &= E_0' \cos(\omega t - kz) \\ E_y &= E_0 \cos(\omega t - kz \pm 90°) \end{aligned}\right\} \tag{7.29}$$

Still study the position $z = 0$. Divide the E_x above by E_0', and divide E_y by E_0. Then square them and sum up. We get

$$\left(\frac{E_x}{E_0'}\right)^2 + \left(\frac{E_y}{E_0}\right)^2 = \cos^2 \omega t + \cos^2(\omega t + 90°) = 1 \tag{7.30}$$

Obviously, Equation (7.30) is an elliptical equation. In this way, the trajectory of the endpoint of the synthetic electric field vector when changing with time is an ellipse, as shown in Figure 7.15.

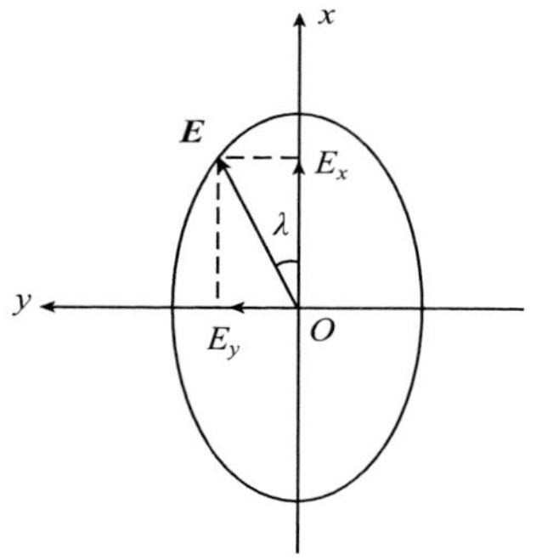

Figure 7.15 Elliptically polarized wave

7.3.4 Decomposition and Synthesis of Polarized Wave

As discussed above, wave equation is a linear equation. According to the superposition principle of linear equation, its solution can be synthesized or be decomposed.

1. The Composition of Linear Polarized Wave

(1) The synthesis of two linear polarized waves perpendicular to each other and with same time phases, is also a linearly polarized wave, as shown in Figure 7.16. *The unit vector*

(*whose amplitude is 1*) $\boldsymbol{E}_{U\cdot L}$ of linearly polarized wave can be written as (omitted the time factor)

$$\boldsymbol{E}_{U\cdot L} = \boldsymbol{e}_x \cos\theta + \boldsymbol{e}_y \sin\theta = \boldsymbol{e}_x E_{x0} + \boldsymbol{e}_y E_{y0} \tag{7.31}$$

(2) The linearly polarized wave may be also formed by two circularly polarized waves with opposite rotation direction, as shown in Figure 7.17. This can be expressed in the following equation

$$\boldsymbol{E}_{U\cdot L} = \frac{1}{\sqrt{2}}\left[\left(\frac{\boldsymbol{e}_x + \boldsymbol{e}_y \mathrm{e}^{\mathrm{j}90^\circ}}{\sqrt{2}}\right) + \left(\frac{\boldsymbol{e}_x + \boldsymbol{e}_y \mathrm{e}^{-\mathrm{j}90^\circ}}{\sqrt{2}}\right)\right] \tag{7.32}$$

The two terms at the right side of the above equation are both circularly polarized waves with opposite rotation directions. Expanding the above equation yields the linearly polarized wave.

$$\boldsymbol{E}_{U\cdot L} = \boldsymbol{e}_x$$

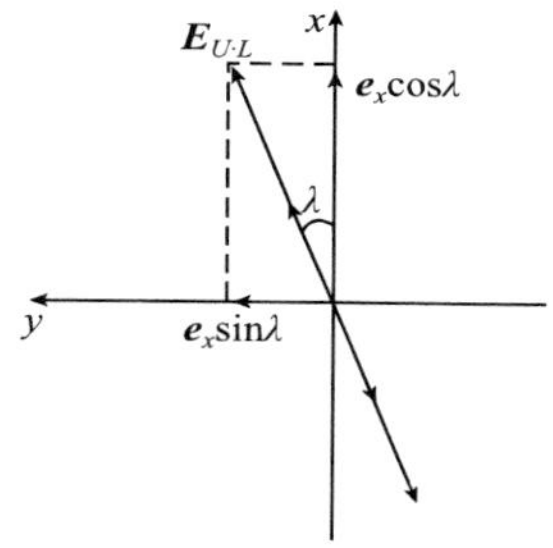

Figure 7.16 Linearly polarized wave and linearly polarized wave

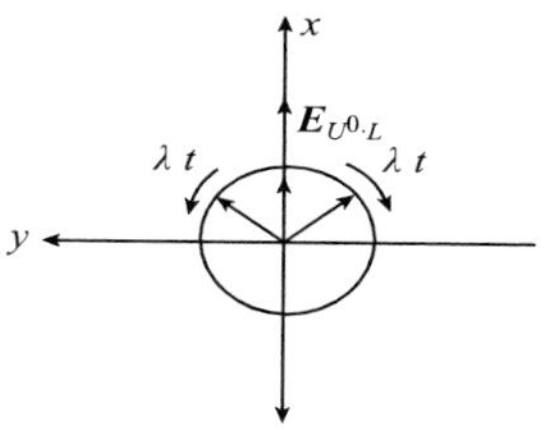

Figure 7.17 Linearly polarized wave and circularly polarized wave

2. The Composition of Circularly Polarized Wave

The circularly polarized wave is consisted of two linearly polarized waves perpendicular to each other, with time phase difference of 90° and equal amplitudes, as shown in Figure 7.18. The unit vector of circularly polarized wave can be written as

$$\boldsymbol{E}_{U\cdot C} = \frac{1}{\sqrt{2}}(\boldsymbol{e}_x + \boldsymbol{e}_y \mathrm{e}^{\mathrm{j}90^\circ}) \tag{7.33}$$

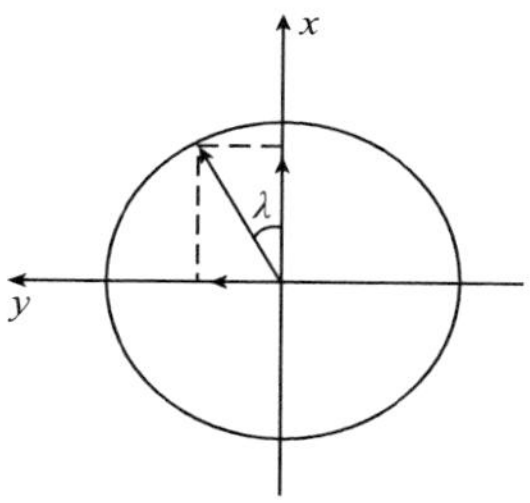

Figure 7.18 The decomposition and synthesis of circular polarized wave

3. The Composition of Elliptically Polarized Wave

(1) The elliptically polarized wave is consisted of two linearly polarized waves perpendicular to each other, having 90° time phase difference and unequal amplitude.

(2) It can also be formed by two circularly polarized waves with unequal amplitudes and opposite rotation directions. If with equal amplitudes ($a = b$), they will form linearly polarized wave, as shown in Figure 7.19.

The elliptically polarized wave can be written as

$$\boldsymbol{E}_{U\cdot e} = \left[\left(\boldsymbol{e}_x \pm \boldsymbol{e}_y \mathrm{e}^{\mathrm{j}90^\circ}\right)\mathrm{e}^{\mp\mathrm{j}\theta} + b\left(\boldsymbol{e}_x \mp \boldsymbol{e}_y \mathrm{e}^{\mathrm{j}90^\circ}\right)\mathrm{e}^{\mp\mathrm{j}\theta}\right] \tag{7.34}$$

The two terms at the right side of the above equation are both circularly polarized waves with unequal amplitudes and opposite rotation directions. They can form elliptically polarized wave or linearly polarized wave, if their amplitudes are equal. Therefore, linearly polarized wave is only the special case of elliptically polarized wave (minor axis equals to zero). θ in the above equation represents the angle between major axis and x axis.

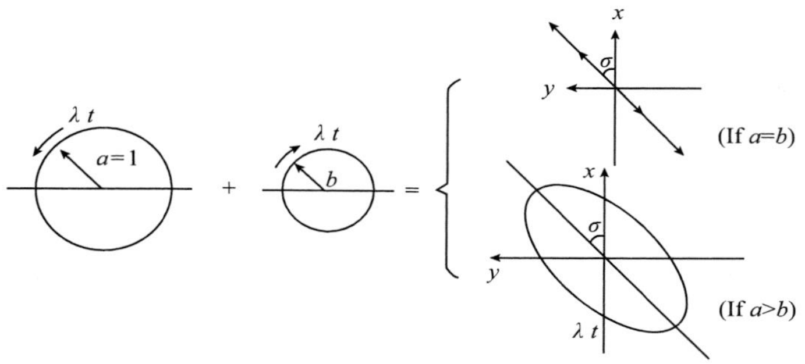

Figure 7.19 The decomposition and synthesis of elliptically polarized wave

Example 7.2 Try to prove that left-handed circularly polarized wave and right-handed circularly polarized wave with equal amplitudes can construct a linearly polarized wave.

Solution The circularly polarized wave consists of two linearly polarized waves which are perpendicular to each other, with 90° time phase difference and equal amplitude. Assume the circularly polarized wave propagates along the positive z direction. Then the amplitudes of the right-handed and left-handed circularly polarized waves are both E_0, written as following:

Right-handed circularly polarized wave

$$(\boldsymbol{e}_x E_0 - \mathrm{j}\boldsymbol{e}_y E_0)\mathrm{e}^{\mathrm{j}(\omega t - kz)} \tag{a}$$

Left-handed circularly polarized wave

$$(\boldsymbol{e}_x E_0 + \mathrm{j}\boldsymbol{e}_y E_0)\mathrm{e}^{\mathrm{j}(\omega t - kz)} \tag{b}$$

Sum up Equations (a) and (b), we have

$$E_0\left[(\boldsymbol{e}_x - \mathrm{j}\boldsymbol{e}_y) + (\boldsymbol{e}_x + \mathrm{j}\boldsymbol{e}_y)\right]\cdot \mathrm{e}^{\mathrm{j}(\omega t - kz)} = \boldsymbol{e}_x(2E_0)\mathrm{e}^{\mathrm{j}(\omega t - kz)}$$

Thus resulting in a linearly polarized wave.

7.4 Plane Wave in A Conducting Medium

A plane wave in a conducting medium usually has some attenuation. The time phase difference between the electric and magnetic field, and other parameters of the wave have also changed.

7.4.1 Wave Equations in A Conducting Medium

Maxwell's equations in a conducting medium are

$$\nabla \times \boldsymbol{H} = \boldsymbol{J} + \varepsilon\frac{\partial \boldsymbol{E}}{\partial t} = \sigma\boldsymbol{E} + \varepsilon\frac{\partial \boldsymbol{E}}{\partial t} \tag{7.35}$$

$$\nabla \times \boldsymbol{E} = -\mu\frac{\partial \boldsymbol{H}}{\partial t} \tag{7.36}$$

$$\nabla \cdot \boldsymbol{D} = 0 \tag{7.37}$$

$$\nabla \cdot \boldsymbol{B} = 0 \tag{7.38}$$

Taking the curl of Equation (7.36) and using Equation (7.35), we obtain

$$\begin{aligned}\nabla \times \nabla \times \boldsymbol{E} &= -\nabla \times \left(\mu \frac{\partial \boldsymbol{H}}{\partial t}\right) = -\mu \frac{\partial}{\partial t}(\nabla \times \boldsymbol{H}) \\ &= -\left[\mu\varepsilon \frac{\partial^2 \boldsymbol{E}}{\partial t^2} + \mu\sigma \frac{\partial \boldsymbol{E}}{\partial t}\right]\end{aligned} \tag{7.39}$$

Using the vector identity and taking into account Equation (7.37), we have

$$\nabla \times \nabla \times \boldsymbol{E} = \nabla\nabla \cdot \boldsymbol{E} - \nabla^2 \boldsymbol{E} = -\nabla^2 \boldsymbol{E} \tag{7.40}$$

Substitution of the above into Equation (7.39) gives

$$\nabla^2 \boldsymbol{E} - \mu\varepsilon \frac{\partial^2 \boldsymbol{E}}{\partial t^2} - \mu\sigma \frac{\partial \boldsymbol{E}}{\partial t} = 0 \tag{7.41}$$

which is known as the wave equation for the electric field in a conducting medium.

Following a similar approach, we can also obtain the wave equation for the magnetic field

$$\nabla^2 \boldsymbol{H} - \mu\varepsilon \frac{\partial^2 \boldsymbol{H}}{\partial t^2} - \mu\sigma \frac{\partial \boldsymbol{H}}{\partial t} = 0 \tag{7.42}$$

If an electromagnetic wave is in the form of a simple harmonic wave, then Equations (7.41) and (7.42) become

$$\nabla^2 \boldsymbol{E} - \gamma^2 \boldsymbol{E} = 0 \tag{7.43}$$

and

$$\nabla^2 \boldsymbol{H} - \gamma^2 \boldsymbol{H} = 0 \tag{7.44}$$

where

$$\gamma^2 = -\left(\omega^2 \mu\varepsilon - \mathrm{j}\omega\mu\sigma\right) \tag{7.45}$$

and γ is called the propagation constant.

Assume that the electric field intensity of a uniform plane wave is E_x, and the magnetic field intensity is H_y. In addition, they are functions of space z and time t only. Thus, Equations (7.43) and (7.44) can be written as

$$\frac{\mathrm{d}^2 E_x}{\mathrm{d}z^2} - \gamma^2 E_x = 0 \tag{7.46}$$

and

$$\frac{\mathrm{d}^2 H_y}{\mathrm{d}z^2} - \gamma^2 H_y = 0 \tag{7.47}$$

The solution of Equation (7.46) can be written as

$$E_x = E_0 \mathrm{e}^{\mathrm{j}\omega t - \gamma z} + E_0' \mathrm{e}^{\mathrm{j}\omega t + \gamma z} \tag{7.48}$$

Thus, the electric field E_x propagating along the positive z direction is

$$E_x = E_0 \mathrm{e}^{\mathrm{j}\omega t - \gamma z} \tag{7.49}$$

Using Maxwell's equations, we get the magnetic field H_y from E_x as

$$H_y = -\frac{1}{\mathrm{j}\omega\mu} \frac{\partial E_x}{\partial z} = \frac{\gamma}{\mathrm{j}\omega\mu} E_x \tag{7.50}$$

Substituting E_x in Equation (7.49) into above equation, we get

$$H_y = \left(\frac{\gamma}{\mathrm{j}\omega\mu}\right) E_0 \mathrm{e}^{\mathrm{j}\omega t - \gamma z} \tag{7.51}$$

Equations (7.49) and (7.51) are known as the expressions for the electric and the magnetic field in a conducting medium.

7.4.2 Electromagnetic Wave Parameters in A Conducting Medium

1. Complex Permittivity ε^e

Maxwell's first equation gives

$$\nabla \times \boldsymbol{H} = \boldsymbol{J} + \varepsilon \frac{\partial \boldsymbol{E}}{\partial t} = \sigma E + \mathrm{j}\omega\varepsilon \boldsymbol{E} = \mathrm{j}\omega \left(\varepsilon - \mathrm{j}\frac{\sigma}{\omega}\right) \boldsymbol{E} = \mathrm{j}\omega\varepsilon^{\mathrm{e}} \boldsymbol{E}$$

Where ε^e is called *the complex permittivity* of the conducting medium. By *denoting its real component ε and imaginary component as ε'* and ε'' respectively, we get

$$\varepsilon^{\mathrm{e}} = \varepsilon' - \mathrm{j}\varepsilon'' = \varepsilon - \mathrm{j}\frac{\sigma}{\omega} \tag{7.52}$$

2. Expressions for γ, α, β

From Equation (7.45) we know that the propagation constant γ is

$$\gamma = \sqrt{\mathrm{j}\omega\mu\sigma - \omega^2\mu\varepsilon} = \sqrt{\mathrm{j}\omega\mu(\sigma + \mathrm{j}\omega\varepsilon)} \tag{7.53}$$

Suppose $\gamma = \alpha + \mathrm{j}\beta$, we find that

$$\gamma^2 = \mathrm{j}\omega\mu\sigma - \omega^2\mu\varepsilon = (\alpha + \mathrm{j}\beta)^2$$

α and β can be found as

$$\alpha = \omega\sqrt{\frac{\mu\varepsilon}{2}\left(\sqrt{1 + \frac{\sigma^2}{\omega^2\varepsilon^2}} - 1\right)} \tag{7.54}$$

$$\beta = \omega\sqrt{\frac{\mu\varepsilon}{2}\left(\sqrt{1 + \frac{\sigma^2}{\omega^2\varepsilon^2}} + 1\right)} \tag{7.55}$$

Here α is the attenuation constant in a conducting medium and has the unit of nepers per meter (Np/m), and β is the phase constant in a conducting medium and has the unit of radians per meter (rad/m).

3. Complex Wave Impedance η^e

From Equation (7.50) we obtain the complex wave impedance η^e in a conducting medium as

$$\left.\begin{aligned} \eta^{\mathrm{e}} &= \frac{E_x}{H_y} = \frac{\mathrm{j}\omega\mu}{\gamma} \\ \text{or} \quad \eta^{\mathrm{e}} &= \sqrt{\mu/\varepsilon^{\mathrm{e}}} \end{aligned}\right\} \tag{7.56}$$

4. Phase Velocity v_p of An Electromagnetic Wave in A Conducting Medium

The phase velocity v_{p} of an electromagnetic wave in a conducting medium is

$$v = \frac{\omega}{\beta} = 1\Bigg/\sqrt{\frac{\mu\varepsilon}{2}\left(\sqrt{1 + \frac{\sigma^2}{\omega^2\varepsilon^2}} + 1\right)} \tag{7.57}$$

From above analysis, we can see that compared with free space (perfect dielectric), a plane wave in a conducting medium has the following features:

(1) An electromagnetic wave in a conducting medium is an attenuated wave. The term $e^{-\alpha z}$ appears in the expressions for the fields. The higher the frequency, or the larger the conductivity σ, the larger the attenuation constant α, the bigger the attenuation.

(2) The wave impedance η^{e} of a plane wave in a conducting medium is a complex quantity, no longer a pure impedance.

(3) That η^{e} is a complex quantity implies that the time phase difference exists between the electric and the magnetic field in a conducting medium.

(4) The propagation velocity of an electromagnetic wave in a conducting medium is no longer a constant.

From Equation (7.57) we can see that the phase velocity v_{p} is a function of frequency. The propagation velocity of electromagnetic wave changes with frequency, caused by the non-zero conductivity σ of the medium. It is usually called the dispersive wave.

(5) In a conducting medium, the magnetic energy density w_{m} is larger than the electric energy density w_{e}.

For a plane wave in a perfect dielectric, the magnetic energy density and the electric energy density are equal

$$\frac{1}{2}\varepsilon E_x^2=\frac{1}{2}\varepsilon(\eta H_y)^2=\frac{1}{2}\varepsilon\left(\sqrt{\frac{\mu}{\varepsilon}}\right)^2 H_y^2=\frac{1}{2}\mu H_y^2$$

For a plane wave in a perfect dielectric, w_e is equal to w_m, and the wave impedance is a pure impedance, and the time phases of the electric and the magnetic field are the same.

In a conducting medium, however, the situation is different, $w_{\mathrm{e}}\neq w_{\mathrm{m}}$. The electric energy density is

$$w_{\mathrm{e}}=\frac{1}{2}\varepsilon E_x^2$$

The magnetic energy density is

$$\begin{aligned}w_{\mathrm{m}}&=\frac{1}{2}\mu H_y^2=\frac{1}{2}\mu\left(\frac{E_x}{|\eta^{\mathrm{e}}|}\right)^2=\frac{1}{2}\mu\left(\left|\sqrt{\varepsilon^{\mathrm{e}}/\mu}\right|\right)^2 E_x^2\\&=\frac{1}{2}E_x^2\left|\varepsilon-\mathrm{j}\frac{\sigma}{\omega}\right|=\frac{1}{2}\varepsilon E_x^2\sqrt{1+\frac{\sigma^2}{\omega^2\varepsilon^2}}=w_e\sqrt{1+\frac{\sigma^2}{\omega^2\varepsilon^2}}\end{aligned}\tag{7.58}$$

Apparently, the magnetic energy is larger than the electric energy in a conducting medium, and its wave impedance is resistive and inductive in nature. Then the electric field of the plane wave leads the magnetic field by an angle ϕ, as given below:

$$\eta^{\mathrm{e}}=\sqrt{\frac{\mu}{\varepsilon^{\mathrm{e}}}}=\sqrt{\frac{\mu}{\varepsilon\left(1-\mathrm{j}\dfrac{\sigma}{\omega\varepsilon}\right)}}=\left\{\sqrt{\frac{\mu}{\varepsilon}}\Big/\left(1+\frac{\sigma^2}{\omega^2\varepsilon^2}\right)^{\frac{1}{4}}\right\}\exp\left(\mathrm{j}\frac{1}{2}\arctan\frac{\sigma}{\omega\varepsilon}\right)$$

thus

$$\phi=\frac{1}{2}\arctan\left(\frac{\sigma}{\omega\varepsilon}\right)\tag{7.59}$$

A diagram of the electromagnetic wave propagating along the positive $z$direction in a conducting medium is shown in Figure 7.20.

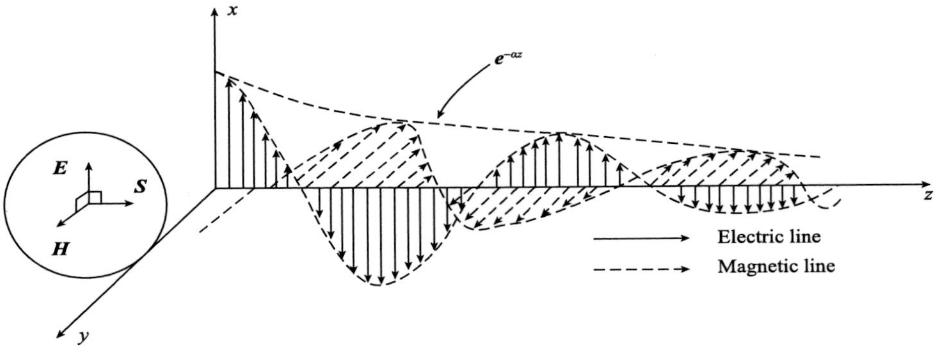

Figure 7.20 The electromagnetic wave propagates in the positive z direction in a conducting medium

7.5 Loss Tangent tan δ and Medium Category

7.5.1 Complex Permittivity ε^{e} of Conductors and Dielectrics

The complex permittivity of a conductor has been given by Equation (7.52). Its loss is caused by σ.

The complex permittivity ε^{e} of a lossy medium can be expressed as

$$\varepsilon^{\mathrm{e}} = \varepsilon' - \mathrm{j}\varepsilon'' \tag{7.60}$$

The dielectric loss is caused by the retarding effects when the dielectric is being polarized, causing non-zero ε''.

With conduct loss and dielectric loss, from Equations (7.52) and (7.60), the complex permittivity ε^{e} of the medium is

$$\varepsilon^{\mathrm{e}} = \varepsilon' - \mathrm{j}\left(\frac{\sigma}{\omega} + \varepsilon''\right) \tag{7.61}$$

7.5.2 Loss Tangent tan δ

The loss tangent tan δ, which can be used to describe the loss of materials, is defined as the ratio of the imaginary part and real part of the complex permittivity ε^{e}

$$\tan\delta = \frac{\dfrac{\sigma}{\omega} + \varepsilon''}{\varepsilon'} \tag{7.62}$$

For conductors, $\dfrac{\sigma}{\omega} \gg \varepsilon''$, thus

$$\tan\delta = \frac{\sigma}{\omega\varepsilon'} \tag{7.63}$$

For dielectrics, $\dfrac{\sigma}{\omega} \ll \varepsilon''$, thus

$$\tan\delta = \frac{\varepsilon''}{\varepsilon'} \tag{7.64}$$

For any material, thus

$$\tan\delta = \frac{\varepsilon''}{\varepsilon'} + \frac{\sigma}{\omega\varepsilon'} \tag{7.65}$$

Table 7.1 lists ε_{r} and tan δ for several dielectric materials.

Table 7.1 Several dielectric materials and their ε_r and tan δ

Dielectric materials \ ε_r and tan δ under different frequencies	ε_r			tan δ		
	60/Hz	1/MHz	10^4/MHz	60/Hz	1/MHz	10^4/MHz
Polystyrene foam	1.03	1.03	1.03	$< 2 \times 10^{-4}$		10^{-4}
Polystyrene	2.55	2.55	2.54	$< 3 \times 10^{-4}$		3×10^{-4}
PTFE	2.10	2.10	2.10	$< 5 \times 10^{-3}$		4×10^{-4}
Polyethylene	2.26	2.26	2.26	$< 2 \times 10^{-4}$		5×10^{-4}
Plexiglass	3.45	2.76	2.50	6.4×10^{-2}	1.4×10^{-2}	5×10^{-3}
Glue boards	4.87	4.74	3.68	8×10^{-2}	2.8×10^{-2}	4.1×10^{-2}

7.5.3 Medium Category

Different medium can be classified by the loss tangent tan δ.

Perfect conductor: $\tan\delta = \dfrac{\sigma}{\omega\varepsilon'} \to \infty$;

Good conductor: $\tan\delta = \dfrac{\sigma}{\omega\varepsilon'} \gg 1$;

Low-loss dielectric: $\tan\delta = \dfrac{\varepsilon''}{\varepsilon'} \ll 1$;

Perfect dielectric: $\tan\delta = \dfrac{\varepsilon''}{\varepsilon'} = 0$.

For different frequencies, the same material may also show different electric conduction performance.

In some of the customary representations, ε' is often written as ε.

7.6 Plane Wave in A Good Dielectric

A good dielectric is a dielectric material which satisfies $\dfrac{\partial}{\omega\varepsilon'} \ll 1$ or $\dfrac{\varepsilon''}{\varepsilon'} \ll 1$. A good dielectric is a low-loss material. A variety of dielectric materials applied in high and ultra-high frequency band should be viewed as good dielectrics.

A loss may be considered to be caused by σ or ε''. Taking into account Equations (6.63) and (6.64), we obtain an alternative relationship between σ and ε'' as

$$\sigma = \omega\varepsilon' \tan\delta = \omega\varepsilon' \frac{\varepsilon''}{\varepsilon'} = \omega\varepsilon'' \tag{7.66}$$

From Equations (7.54)~(7.56), we can get the approximate expressions for α, β and η^{e} in a low-loss material.

(1) For the low-loss case with $\dfrac{\sigma}{\omega\varepsilon} \ll 1$, the approximate expressions are

$$\alpha \approx \frac{1}{2}\sigma\sqrt{\frac{\mu}{\varepsilon}} \tag{7.67}$$

$$\beta \approx \omega\sqrt{\mu\varepsilon}\left[1 + \frac{1}{8}\left(\frac{\sigma^2}{\omega^2\varepsilon^2}\right)\right] = k\left[1 + \frac{1}{8}\left(\frac{\sigma^2}{\omega^2\varepsilon^2}\right)\right] \tag{7.68}$$

$$\eta^{\mathrm{e}} \approx \sqrt{\frac{\mu}{\varepsilon}}\left(1 + \mathrm{j}\frac{\sigma}{2\omega\varepsilon}\right) \tag{7.69}$$

(2) For the low-loss case with $\frac{\varepsilon''}{\varepsilon'} \ll 1$, the approximate expressions, using Equation (7.66), are

$$\alpha \approx \frac{\omega\varepsilon''}{2\varepsilon'}\sqrt{\mu\varepsilon'} \tag{7.70}$$

$$\beta \approx \omega\sqrt{\mu\varepsilon'}\left[1+\frac{1}{8}\left(\frac{\varepsilon''}{\varepsilon'}\right)^2\right] \tag{7.71}$$

$$\eta^{\mathrm{e}} \approx \sqrt{\frac{\mu}{\varepsilon'}}\left(1+\mathrm{j}\frac{\varepsilon''}{2\varepsilon'}\right) \tag{7.72}$$

From Equations (7.67)~(7.72), we can see that when $\frac{\sigma}{\omega\varepsilon}$ or $\frac{\varepsilon''}{\varepsilon'}$ of the high-quality materials approaches zero, the attenuation constant $\alpha \approx 0$, the phase constant $\beta \approx k$, the wave impedance $\eta^{\mathrm{e}} \approx \sqrt{\frac{\mu}{\varepsilon}}$. This case is very similar to case discussed previously, about a plane wave in a perfect dielectric. However, for this case, there is a small transmission loss and dispersion, as well as the time phase difference

Example 7.3 A 550kHz plane wave propagates in a lossy medium with $\tan\delta = 0.02$ and $\varepsilon_{\mathrm{r}} = 2.5$. Find the attenuation constant α, the phase constant $\beta$and the phase velocity v_{p} of the wave.

Method 1 Consider the medium loss caused by $\sigma \neq 0$, and find α, β and v_{p}.

As the loss tangent $\tan\delta = \frac{\sigma}{\omega\varepsilon} = 0.02$, the medium can be regarded as a low-loss dielectric.

$$\frac{\sigma}{\omega\varepsilon} = 0.02 = \frac{\sigma}{2\pi \times 550 \times 10^3 \times \left(2.5 \times \frac{1}{36\pi} \times 10^{-9}\right)}$$

$$\sigma = 1.5278 \times 10^{-6}\mathrm{S}\ /\ \mathrm{m}$$

The attenuation constantα, from Equation (7.67), is

$$\alpha \approx \frac{1}{2}\sigma\sqrt{\frac{\mu}{\varepsilon}} = \frac{\sigma}{2} \cdot \frac{377}{\sqrt{2.5}} = 1.821 \times 10^{-4}\mathrm{Np/m}$$

The phase constant β, from Equation (7.71), is

$$\beta \approx \omega\sqrt{\mu\varepsilon}\left[1+\frac{1}{8}\left(\frac{\sigma}{\omega\varepsilon}\right)^2\right] = 0.01821 \times 1.00005 = 0.018214\mathrm{rad/m}$$

The phase velocity v_{p} is

$$v_{\mathrm{p}} = \frac{\omega}{\beta} = 1.8973 \times 10^8\mathrm{m/s}$$

Method 2 Consider the medium loss caused by ε'', and find α, β and v_{p}.
As $\varepsilon' = \varepsilon_{\mathrm{r}}\varepsilon 0 = 2.5\varepsilon 0$, we have

$$\tan\delta = \frac{\varepsilon''}{\varepsilon'} = 0.02 = \frac{\varepsilon''}{2.5\varepsilon_0}$$

Thus, we get

$$\varepsilon'' = 0.02 \times 2.5\varepsilon_0 = 0.05\varepsilon_0$$

The attenuation constant α, from low-loss Equation (7.70), is given as

$$\alpha = \frac{\omega\varepsilon''}{2\varepsilon'}\sqrt{\mu\varepsilon'} = 1.821 \times 10^{-4}\mathrm{Np/m}$$

The phase constant β, from Equation (7.71), is given as

$$\beta = \omega\sqrt{\mu\varepsilon'}\left[1 + \frac{1}{8}\left(\frac{\varepsilon''}{\varepsilon'}\right)^2\right] = 0.018214\mathrm{rad/m}$$

The phase velocity v_p is

$$v_\mathrm{p} = f\lambda = \frac{\omega}{\beta} = 1.8973 \times 10^8\mathrm{m/s}$$

7.7 Plane Wave in A Good Conductor

In a conducting medium, the expressions for the electromagnetic field of a plane wave are

$$E_x = E_0\mathrm{e}^{-\alpha z}\cdot\mathrm{e}^{\mathrm{j}(\omega t - \beta z)} \tag{7.73}$$

$$H_y = \frac{1}{\eta^\mathrm{e}}E_0\mathrm{e}^{-\alpha z}\cdot\mathrm{e}^{\mathrm{j}(\omega t - \beta z)} \tag{7.74}$$

In the case of a good conductor $\left(\frac{\sigma}{\omega\varepsilon} \gg 1\right)$, the following expressions can be obtained.

(1) Propagation constant γ.

$$\gamma = \sqrt{\mathrm{j}\omega\mu\sigma - \omega^2\mu\varepsilon} = \sqrt{\mathrm{j}\omega\varepsilon(\sigma + \mathrm{j}\omega\varepsilon)} \approx \sqrt{\mathrm{j}\omega\mu\sigma} \tag{7.75}$$

(2) Attenuation constant α and phase constant β.
Write $\sqrt{\mathrm{j}}$ as the following expression:

$$\sqrt{\mathrm{j}} = \sqrt{\mathrm{e}^{\mathrm{j}90^\circ}} = \mathrm{e}^{\mathrm{j}45^\circ} = \frac{1+\mathrm{j}}{\sqrt{2}} \tag{7.76}$$

Substitution of above into Equation (7.75) gives

$$\gamma = \alpha + \mathrm{j}\beta \approx \sqrt{\mathrm{j}\omega\mu\sigma} = \sqrt{\pi f\mu\sigma} + \mathrm{j}\sqrt{\pi f\mu\sigma}$$

thus

$$\alpha \approx \sqrt{\pi f\mu\sigma} \tag{7.77}$$

$$\beta \approx \sqrt{\pi f\mu\sigma} \tag{7.78}$$

(3) Wave impedance η^e.

$$\begin{aligned}\eta^\mathrm{e} &= \sqrt{\frac{\mu}{\varepsilon^\mathrm{e}}} = \sqrt{\frac{\mu}{\varepsilon\left(1-\mathrm{j}\dfrac{\sigma}{\omega\varepsilon}\right)}} \approx \sqrt{\mathrm{j}\frac{\omega\mu}{\sigma}} = \sqrt{\frac{\omega\mu}{\sigma}}\mathrm{e}^{\mathrm{j}45^\circ} \\ &= \sqrt{\frac{\pi f\mu}{\sigma}} + \mathrm{j}\sqrt{\frac{\pi f\mu}{\sigma}}\end{aligned} \tag{7.79}$$

Apparently, the wave impedance in a good conductor is a complex quantity. Moreover, the resistance part and the reactance (inductive) part are equal, that is, the phase angle of η^e is 45°.

(4) Phase velocity v_p.

$$v_p = f\lambda = \frac{\omega}{\beta} = \frac{\omega}{\sqrt{\pi f \mu\sigma}} = \sqrt{\frac{2\omega}{\mu\sigma}} \tag{7.80}$$

The velocity of an electromagnetic wave in a good conductor is a function of frequency and the wave is a dispersive wave. In addition, v_p is inversely proportional to $\sqrt{\sigma}$, that is, the electromagnetic wave travels more slowly in a good conductor with larger σ. For example, when a 465kHz electromagnetic wave propagates in copper, the phase velocity v_p is only 283.15m/s, same order level with the speed of sound in the air.

(5) In a good conductor, the magnetic energy density w_m is much larger than the electric energy density w_e. From Equation (7.58), when $\frac{\sigma}{\omega\varepsilon} \gg 1$, $w_m > w_e$. Obviously, when the conductivity σ is very large, weak electric field intensity will generate very big current density, actually caused by strong magnetic field H.

Example 7.4 An incident plane wave propagates from free space into sea water, the frequency is: (1) $f = 30\text{Hz}$, (2) $f = 10\text{GHz}$, find the penetrated depth of the wave in the sea water when the electric field at that depth drops to 10% of the electric field on the surface of the seawater. The corresponding parameters of the seawater are known as follows:

(1) $f = 30\text{Hz}$, $\mu = \mu_0$, $\varepsilon' = 80\varepsilon_0$, $\sigma = 4\text{S/m}$;

(2) $f = 10\text{GHz}$, $\mu = \mu_0$, $\varepsilon' = 80\varepsilon_0$, $\varepsilon'' = 45\varepsilon_0$;

Analysis When the frequency is very low or very high, we should confirm the characteristics of the seawater (good conductor or dielectric) by computing $\tan\delta$, and then examine the formulas and their parameters and make some specific calculations.

(1) When $f = 30\text{Hz}$

$$\tan\delta = \frac{\sigma}{\omega\varepsilon} = \frac{\sigma}{\omega \times 80\varepsilon_0} \approx 3 \times 10^7$$

Obviously it performs like a good conductor. The expression for the attenuation constant α of a good conductor gives

$$\alpha = \sqrt{\pi f \mu\sigma} = 21.77 \times 10^{-3}\text{Np/m}$$

Suppose the amplitude of the electric field on the surface of the sea water is 1. After penetrating depth h, 10% of its amplitude remains, thus,

$$e^{-\alpha h} = 0.1$$

$$h = \frac{-\ln(0.1)}{21.77 \times 10^{-3}} = 105.8\text{m}$$

(2) When $f = 10\text{GHz}$

$$\tan\delta = \frac{\varepsilon''}{\varepsilon'} = \frac{45\varepsilon_0}{80\varepsilon_0} \approx 0.56$$

Obviously it is a lossy medium. From the given parameters, $\sigma = 0$, and the loss is caused by ε''. Thus, α is

$$\alpha = \text{Re}(\gamma) = \text{Re}\left[\sqrt{j\omega\mu_0(\sigma + j\omega\varepsilon^e)}\right] = \text{Re}\left[\sqrt{j\omega\mu_0(j\omega\varepsilon^e)}\right]$$
$$= \text{Re}\left[\sqrt{j\omega\mu_0 \times j\omega(\varepsilon' - j\varepsilon'')}\right] \approx 508.5(\text{Np/m})$$

Suppose the amplitude of the electric field on the surface of the seawater is 1. After penetrating depth of h, 10% of its amplitude remains, thus,

$$e^{-\alpha h} = e^{-508.5h} = 0.1$$

$$h = \frac{-\ln(0.1)}{\alpha} = 0.0045\text{m}$$

Discussion:

(1) When $f = 10\text{GHz}$ (the wavelength in free space is 3cm), after a deep penetration of 4.5mm, 10% of its amplitude remains, and the attenuation is extremely fast.

(2) When f=30Hz (the wavelength in free space is 10^4km), after a deep penetration of 105.8m, 10% of its amplitude remains, and the attenuation is much smaller than that in the situation about $f = 10\text{GHz}$, that is, in the sea water, the transmission distance of a 30Hz electromagnetic wave is much longer than that of a 10GHz electromagnetic wave.

(3) When $f = 30\text{Hz}$, $\dfrac{\sigma}{\omega\varepsilon} \approx 3 \times 10^7$, the sea water performs like a good conductor. However, the skin depth $\delta = 1/\alpha = 1/\sqrt{\pi f \mu\sigma} = 1/21.77 \times 10^{-3} \approx 45.93(\text{m})$. Since it is a good conductor, why is the skin depth δ so large? This is due to extremely low frequency (30Hz) and small conductivity σ (only 4S/m) of the sea water.

7.8 Skin Effect

One of the characteristics of an electromagnetic wave in a conducting medium is transmission attenuation. The deeper penetration of a wave usually caused larger attenuation, and thus having smaller field amplitude and weaker energy. In other words, the energy is confined to the region near the surface. This phenomenon is called the skin effect.

The skin depth (or penetration depth) is one of the important parameters about the skin effect. It is defined as the distance (depth) travelled by the fields in a conducting medium at which their amplitude falls to $\dfrac{1}{\mathrm{e}}$ of its value on the surface of the conducting medium, denoted as δ. The amplitude of the fields attenuates in accordance with $\mathrm{e}^{-\alpha z}$. From the definition, if we set αz=1, z is the skin depth δ, thus,

$$\delta = \frac{1}{\alpha} \tag{7.81}$$

Substitution of the expression for the attenuation constant α in a conducting medium into above gives

$$\delta = \frac{1}{\omega\sqrt{\dfrac{\mu\varepsilon}{2}\left(\sqrt{1+\left(\dfrac{\sigma}{\omega\varepsilon}\right)^2}-1\right)}} \tag{7.82a}$$

For a good conductor $\left(\dfrac{\sigma}{\omega\varepsilon} \gg 1\right)$, we can approximate the above as

$$\delta = \frac{1}{\alpha} = \frac{1}{\sqrt{\pi f \mu\sigma}} \tag{7.82b}$$

This is the expression for the skin depth in a good conductor, indicating that higher frequency or bigger conductivity of the medium usually cause smaller skin depth δ. Figure 7.21 shows the schematic diagram of the skin-effect phenomenon. In a good conductor, the conductivity σ is so large that the electromagnetic wave attenuates rapidly as soon as it reaches the good conductor, that is, the skin depth δ is very small. That is why a good conductor is opaque to the light unless it is an extremely thin metal film.

For a good conductor, $\alpha = \beta = \sqrt{\pi f \mu\sigma}$, $\beta = 2\pi/\lambda$, and $\delta = 1/\alpha$, the skin depth δ and the wavelength λ thus have the following relationship:

$$\lambda = 2\pi\delta \tag{7.83}$$

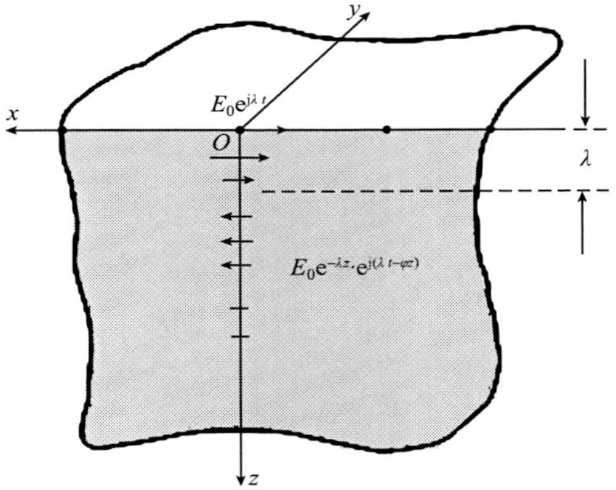

Figure 7.21 Schematic diagram of the skin effect

Table 7.2 lists the skin depth δ for several conductor materials.

Table 7.2 The skin depth δ for several conductors

Materials	Conductivity $\sigma/(\mathrm{S/m})$	Relative permeability	Skin depth δ			
			60Hz/cm	1kHz/mm	1MHz/mm	3GHz/μm
Aluminum	3.54×10^7	1.00	1.1	2.7	0.085	1.6
Brass	1.59×10^7	1.00	1.63	3.98	0.126	2.30
Cr	3.8×10^7	1.00	1.0	2.6	0.081	1.5
Copper	5.8×10^7	1.00	0.85	2.1	0.066	1.2
Gold	4.5×10^7	1.00	0.97	2.38	0.075	1.4
Graphite	1.0×10^5	1.00	20.5	50.3	1.59	20.0
Magnetic iron	1.0×10^7	2×10^2	0.14	0.35	0.011	0.20
Permalloy	0.16×10^7	2×10^4	0.037	0.092	0.0029	0.053
Nickel	1.3×10^7	1×10^2	0.18	4.4	0.014	0.26
Seawater	≈ 5.0	1.00	3×10^3	7×10^3	2×10^3	—
Silver	6.15×10^7	1.00	0.83	2.03	0.064	1.17
Tin	0.87×10^7	1.00	2.21	5.41	0.1714	3.12
Zinc	1.86×10^7	1.00	1.51	3.70	0.117	3.14

For example, the skin depth δ for copper is only 1.2μm at a frequency of 3GHz.

In a conducting medium, the amplitude of the fields attenuates in accordance with $e^{-\alpha z}$, and the power density attenuates in accordance with e^{-2az}. If we set $z = n\delta$, where n is a real number and a multiple of the skin depth δ, then the attenuation of the power density in a conducting medium is

$$e^{-2\alpha z} = e^{-2\alpha n\delta} = e^{-2n}$$

If $S_{\mathrm{av}(0)}$ and $S_{\mathrm{av}(n\delta)}$ are used to express the power density on the surface of a conductor ($z = 0$) and at $z = n\delta$ respectively, then a few sets of data about $n\delta$ and $S_{\mathrm{av}(n\delta)}/S_{\mathrm{av}(0)}$ can be given as follows:

$n\delta$	0	2.3δ	3.45δ	4.6δ	5.75δ	6.9δ
$\dfrac{S_{\mathrm{av}(n\delta)}}{S_{\mathrm{av}(0)}}$	1	10^{-2}	10^{-3}	10^{-4}	10^{-5}	10^{-6}

It can be seen that the power density reduces to 0.01% after penetrating a depth equal to 4.6δ.

7.9 Surface Impedance Z_s of A Good Conductor

The wave equation for current density J in a good conductor can be easily written. Here we rewrite the wave equation [see Equation (7.43)] for the electric field intensity of a plane wave in a conducting medium as

$$\nabla^2 E_x - \gamma^2 E_x = 0$$

Multiplying with the conductivity σ gives

$$\nabla^2 J_x - \gamma^2 J_x = 0 \tag{7.84}$$

which is the wave equation for the current density in a conducting medium.

The magnitude attenuates according to the exponential law when an electromagnetic wave propagates into a conductor, and the solution (after the resumption of $e^{j\omega t}$) for Equation (7.84) can be written as

$$J_x = J_0 e^{-\alpha z} \cdot e^{j(\omega t - \beta z)} \tag{7.85}$$

where J_0 is the magnitude of the current density at $z = 0$ (on the surface of a good conductor). A plane wave propagates along the positive z direction, and the magnitude of the current attenuates according to the exponential law, as shown in Figure 7.22.

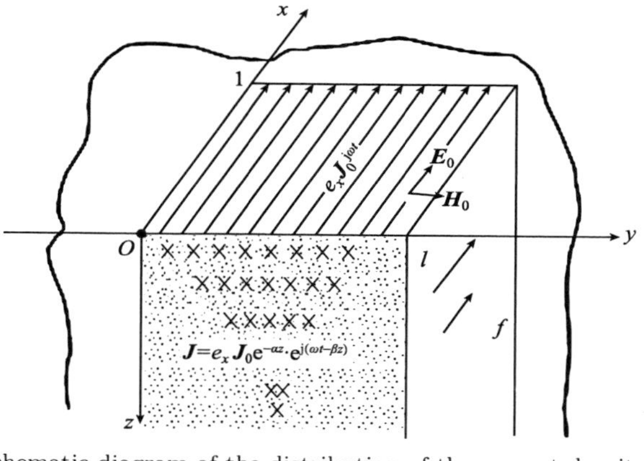

Figure 7.22 Schematic diagram of the distribution of the current density in a conductor

The surface impedance of a conducting medium is defined as

$$Z_s = \frac{E_t}{J_s} \tag{7.86}$$

where E_t is the tangential component of the electric field intensity on the surface of a conducting medium, that is, E_x on the plane $z = 0$ shown in above figure. Therefore, E_t is

$$E_t = E_x|_{z=0} = \frac{1}{\sigma} J_x |_{z=0} = \frac{1}{\sigma} J_0 e^{j\omega t} \tag{7.87}$$

Which is also the voltage per unit length (along the x direction) on the surface of a conducting medium, with unit of V/m.

J_s denotes the total current flowing through such a cross-section (the area is in unit width along the y direction and from zero to infinity along the z direction) along the x direction. Obviously, J_s is

$$J_s = \int_0^{\infty} J_x \mathrm{d}z = \int_0^{\infty} J_0 \mathrm{e}^{-\gamma z} \cdot \mathrm{e}^{\mathrm{j}\omega t} \mathrm{d}z = \frac{J_0}{\gamma} \mathrm{e}^{\mathrm{j}\omega t} \tag{7.88}$$

Substitution of Equations (7.87) and (7.88) into (7.86) gives

$$Z_s = \frac{\gamma}{\sigma} \tag{7.89}$$

which is the general expression for the surface impedance of a conducting medium. For a good conductor, we have

$$\begin{aligned} Z_s &= \frac{\gamma}{\sigma} = \frac{\alpha}{\sigma} + \mathrm{j}\frac{\beta}{\sigma} = R_s + \mathrm{j}X_s \\ &\approx \frac{\sqrt{\pi f \mu \sigma}}{\sigma} + \mathrm{j}\frac{\sqrt{\pi f \mu \sigma}}{\sigma} = \frac{\sqrt{\omega \mu}}{\sigma} \cdot \mathrm{e}^{\mathrm{j}45^\circ} \end{aligned} \tag{7.90}$$

Then the surface resistance (rate) R_s is

$$R_s = \frac{\alpha}{\sigma} = \frac{\sqrt{\pi f \mu \sigma}}{\sigma} = \frac{1}{\sigma \delta} \tag{7.91}$$

where $\alpha\delta = 1$ is used.

From Equations (7.91) and (7.79), we can obtain the relationship between R_s and η^e of a conductor as

$$R_s = R_e(\eta^e) \tag{7.92}$$

The surface reactance X_s is positive and equal to R_s, that is, the reactance is inductive, and the impedance has a 45° phase angle, corresponding to the result previously obtained, saying, the phase of the alternating electric field is 45° in advance of that of the magnetic field. The former is obtained from the concept of the wave impedance, and the latter from the concept of the surface impedance (using the current and the voltage).

The surface resistance (rate) R_s is actually an AC resistance of a conductor having unit width and length but infinite thickness, with flowing current in x direction. From Equation (7.9), it is easy to see that such an AC resistance is equivalent to a DC resistance for a conductor having unit width and length (unit surface area), and z-direction thickness equal to the skin depth δ. *Obviously, for the same conductor, its AC resistance is larger than DC resistance, which is caused by the skin effect.*

The above R_s is obtained under the assumption that the thickness of the conductor is infinite, however, this is not true in practical application. After propagating through the distance of several skin depths, the amplitude of the current is very small, and the skin depth for a high frequency is extremely small in a good conductor. Taking copper as an example, when $f = 100$MHz, the skin depth $\delta \approx 0.00667$mm, and ten times of the skin depth is only 0.0667mm, where the amplitude of the current is only 0.000045 times of that on the surface. Therefore, for a good conductor with a certain thickness, like the waveguide wall, we can still use Equation (7.91) to compute R_s accurately. Even if it is a conductor with a circular cross-section, Equation (7.91) is still valid as long as δ is much smaller than the radius of the conductor. Regarding the circumference as the width, the radius as the thickness, we can find the surface resistance R_s of the round conductor.

The relationship between the conductor resistance R with a surface area $l \times W$ and the surface resistance (rate) R_s is

$$R = R_s \frac{l}{W} \tag{7.93}$$

Where l is the length of a conductor, and W is the width of the conductor. Table 7.3 gives the surface resistance (rate) R_s of several conductors.

Table 7.3 The surface resistance R_s of several conductor materials

Conductor materials \ Conductor parameter	$\sigma/(\text{S/m})$	$\mu/(\text{H/m})$	$\delta/\text{m}(f/\text{Hz})$	$R_s/\Omega(f/\text{Hz})$
Silver	6.15×10^7	$4\pi\times10^{-7}$	$0.06418/\sqrt{f}$	$2.5245\times10^{-7}\sqrt{f}$
Copper	5.80×10^7	$4\pi\times10^{-7}$	$0.06600/\sqrt{f}$	$2.6123\times10^{-7}\sqrt{f}$
Aluminum	3.54×10^7	$4\pi\times10^{-7}$	$0.08459/\sqrt{f}$	$3.3395\times10^{-7}\sqrt{f}$
Brass	1.59×10^7	$4\pi\times10^{-7}$	$0.1262/\sqrt{f}$	$4.9829\times10^{-7}\sqrt{f}$
Solder	0.87×10^7	$4\pi\times10^{-7}$	$0.1706/\sqrt{f}$	$6.7375\times10^{-7}\sqrt{f}$

Example 7.5 A plane wave of frequency $f_1 = 10\text{MHz}$ and $f_2 = 100\text{MHz}$ propagates in copper. The magnetic field intensity on the surface of the metal is known as $H_{y0} = 0.1\text{H/m}$, and the copper is characterized by $\sigma = 5.8\times10^7\text{S/m}$, $\varepsilon_r = 1$, $\mu_r = 1$. Find:

(1) the phase constant β, the attenuation constant α, the phase velocity v_p and the wavelength λ of the wave within the copper;

(2) the wave impedance η^e within the copper and the electric field intensity E_{x0} on the surface of the metal;

(3) the skin depth δ and the surface impedance Z_s;

(4) the average power density S_{av}within the conductor.

Solution As the conductivity σ of the copper is very high, when the frequency is $f_1 = 10\text{MHz}$ and f_2=1000MHz respectively, $[\sigma/(\omega\varepsilon)]$ is very large (about 10^{11} and 10^9 for f_1 and $f_2$respectively). Obviously, whether for f_1 or f_2, the copper is a good conductor. Therefore, the equation for a plane wave in a good conductor can be used for calculation.

(1) Calculations for α, β, v_p and λ.

As $\alpha = \beta = \sqrt{\pi f\mu\sigma} = \sqrt{\pi f \times 4\pi \times 10^{-7} \times 5.8 \times 10^7}$, substitution of f_1 and f_2 into the equation respectively gives

$$\begin{cases} \alpha_1 = 4.785 \times 10^4 \text{ Np/m} \\ \beta_1 = 4.785 \times 10^4 \text{ rad/m} \end{cases}$$

$$\begin{cases} \alpha_2 = 47.85 \times 10^4 \text{ Np/m} \\ \beta_2 = 47.85 \times 10^4 \text{ rad/m} \end{cases}$$

As $v_p = f\lambda = \dfrac{\omega}{\beta} = \dfrac{2\pi f}{\beta}$, substitution of f_1, f_2 and β_1, β_2 into the equation respectively gives

$$v_{p_1} = 1.313 \times 10^3\text{m/s}, \quad v_{p_2} = 13.13 \times 10^3\text{m/s}$$

The propagation velocity of an electromagnetic wave is very small in a good conductor, much less than the speed of the light ($\approx 3 \times 10^8\text{m/s}$). Moreover, the lower the frequency, the slower the speed. As $\lambda = v_p/f$, the corresponding wavelengths for f_1 and f_2 are

$$\lambda_1 = 1.313 \times 10^{-4}\text{m}, \quad \lambda_2 = 1.313 \times 10^{-5}\text{m}$$

The wavelengths for f_1 and f_2 in free space are $\lambda_{1(0)} = 30\text{m}$, $\lambda_{2(0)} = 3\text{m}$, respectively. Obviously, λ_1 and λ_2 are much smaller than their wavelengths in free space.

(2) Wave impedance η^e of a good conductor and E_{x0} on the surface.

$\eta^e = \sqrt{\text{j}\dfrac{\omega\mu}{\sigma}}$, hence the corresponding η^e for f_1 and f_2 are

$$\eta_1^e = 0.00116e^{j45^\circ}\Omega, \quad \eta_2^e = 0.0116e^{j45^\circ}\Omega$$

Obviously, the wave impedance η^e of a metal is a complex quantity. As the conductivity of the copper is very high, its wave impedance is thus very small.

As $\eta^e = E_{x0}/H_{y0}$, that is, $E_{x0} = \eta^e H_{y0}$.

Substitution of η_1^e and η_2^e into above gives

$$E_{x0(1)} = \eta_1^e H_{y0} = 1.16 \times 10^{-4} \cdot e^{j45^\circ} \text{(V/m)}$$

$$E_{x0(2)} = \eta_2^e H_{y0} = 1.16 \times 10^{-3} \cdot e^{j45^\circ} \text{(V/m)}$$

(3) Skin depth δ and surface impedance Z_s.

As $\delta = \dfrac{1}{\alpha}$, substitution of α_1 and α_2 into the equation gives

$$\delta_1 = 2.09 \times 10^{-5}\text{m}$$

$$\delta_2 = 2.09 \times 10^{-6}\text{m}$$

As $Z_s = \dfrac{\sqrt{\pi f \mu \sigma}}{\sigma}(1+j) = \dfrac{\alpha}{\sigma}(1+j) = \dfrac{1}{\sigma\delta}(1+j)$, substitution of δ_1 and δ_2 into the equation gives the corresponding Z_s for f_1 and f_2 as

$$Z_{s(1)} = 8.249 \times 10^{-4}(1+j)(\Omega)$$

$$Z_{s(2)} = 8.249 \times 10^{-3}(1+j)(\Omega)$$

(4) Average power density S_{av} within the conductor.

$$S_{av(1)} = \frac{1}{2}\text{Re}E_{x0(1)}H_{y0}^* = \frac{1}{2}\cos 45^\circ \times 1.16 \times 10^{-4} \times 0.1 = 4.10 \times 10^{-6}(\text{W/m}^2)$$

$$S_{av(2)} = \frac{1}{2}\text{Re}E_{x0(2)}H_{y0}^* = \frac{1}{2}\cos 45^\circ \times 1.16 \times 10^{-3} \times 0.1 = 41 \times 10^{-6}(\text{W/m}^2)$$

Although H_{y0} is the same, the wave impedance η_1^e and η_2^e for f_1 and f_2 in the metal are not equal, and neither are $E_{x0(1)}$ and $E_{x0(2)}$, and nor for $S_{av(1)}$ and $S_{av(2)}$.

Example 7.6 The radius of a round metal wire is a (a is much larger than the skin depth δ). Find the AC resistance R_{ac} per unit length.

Method 1 Write the power loss P_L per unit length in the round conductor with circuit theory and the electromagnetic field respectively. As the results of the two methods should be the same, then R_{ac} can be found.

The circuit expression for P_L is

$$P_L = \frac{1}{2}I^2 R_{ac} \tag{a}$$

The electromagnetic field expression for P_L is

$$P_L = \frac{1}{2}\text{Re}(E \times H^*) \times (2\pi a \times 1)$$

Where $\boldsymbol{E} \times \boldsymbol{H}^*$ is a field quantity on the surface of the round wire with $r = a$. Thus, we have

$$H_\phi = \frac{1}{2\pi a}$$

$$E_z = H_\phi \eta^e = H_\phi \left(\sqrt{\frac{\omega\mu}{\sigma}}\right) e^{j45^\circ}$$

$$P_L = \frac{1}{2}\text{Re}\left[\frac{I^2}{(2\pi a)^2}\sqrt{\frac{\omega\mu}{\sigma}}e^{j45^\circ}\right] \times 2\pi a = \frac{1}{4\sqrt{2}}\cdot\frac{I^2}{\pi a}\sqrt{\frac{\omega\mu}{\sigma}} \tag{b}$$

Equating P_L in Equations (a) and (b) gives

$$R_{ac} = \frac{1}{2\pi a}\sqrt{\frac{\omega\mu}{2\sigma}} = \frac{1}{2\pi a}\sqrt{\frac{\pi f\mu\sigma}{\sigma^2}} = \frac{1}{2\pi a\sigma}\sqrt{\pi f\mu\sigma} = \frac{1}{2\pi a\sigma\delta}$$

Method 2 From Equation (7.91) for the surface resistance R_s, directly find the AC resistance R_{ac} of the round wire in unit length.

The surface resistance (AC resistance) of a conductor with length l, width W and thickness much larger than the skin depth δ, is

$$R_{ac} = R_s\frac{l}{W}$$

Here, we just need to find the AC resistance of the wire in unit length, say $l = 1$; when $a \gg \delta$, the width of the conductor is approximately $W = 2\pi a$, and $R_s = 1/\sigma\delta$. We thus have

$$R_{ac} = \frac{R_s}{2\pi a} = \frac{1}{2\pi a\cdot\sigma\delta}$$

7.10 Power Loss in A Conducting Medium

As stated above, the electromagnetic wave propagating in a conducting medium usually suffers some loss. If the width of medium is much larger than the skin depth δ, then all the entering energy of the electromagnetic wave would dissipate into heat.

Refer to Figure 7.23, the average power density flowing from the surface ($z = 0$) with unit area into the interior of the medium is

$$\begin{aligned}\boldsymbol{S}_{av} &= \frac{1}{2}\text{Re}(\boldsymbol{E}\times\boldsymbol{H}^*) = \frac{1}{2}\text{Re}(|\eta^e|\,e^{j\theta}H_0^2)\boldsymbol{e}_z \\ &= \frac{1}{2}\,|\eta^e|\,H_0^2\cos\theta\,\boldsymbol{e}_z\end{aligned} \tag{7.94}$$

where H_0 is the magnetic field intensity at $z = 0$, and η^e is the complex wave impedance in a conducting medium. Substitution of η^e of a good conductor into the above equation gives

$$S_{av} = \frac{1}{2}H_0^2\sqrt{\frac{\omega\mu}{2\sigma}} \tag{7.95}$$

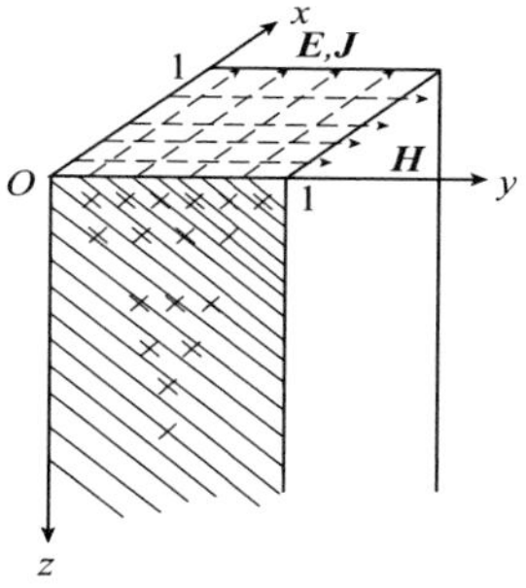

Figure 7.23

The input power is converted into heat loss within the conductor. For a good conductor with large thickness, $H_0 \approx |\boldsymbol{J}_s|$ in the above equation, this can be proved by Ampere's law. In above equation, $\sqrt{\omega\mu/2\sigma}$ is equal to the surface resistance (rate) R_s. Hence, Equation (7.95) can be rewritten as

$$P_L = \frac{1}{2}\,|J_s|^2\,R_s \tag{7.96}$$

This is the power dissipated on the unit surface area of the conductor.

For the surface area S of an actual conductor, the total loss can be obtained by calculating the surface integral of the power loss P_L per unit area

$$P_{\mathrm{L}} = \int_{s} P_{\mathrm{L}} \mathrm{d}S = \int_{s} \left[\frac{1}{2}|J_{\mathrm{s}}|^2 R_{\mathrm{s}}\right] \mathrm{d}S$$

7.11 Dispersive Medium, Dispersive Distortion and Normal Dispersion, Anomalous Dispersion

7.11.1 Dispersive Medium

As mentioned above, for the lossy non-ferromagnetic medium in an alternating electromagnetic field, the corresponding μ remains to be real while the permittivity is a complex ε^{e}, i.e.

$$\varepsilon^{\mathrm{e}} = \varepsilon - \mathrm{j}\frac{\sigma}{\omega} \quad \text{or} \quad \varepsilon^{\mathrm{e}} = \varepsilon' - \mathrm{j}\varepsilon''$$

where the real part is the permittivity of the medium and imaginary part denoted by σ/ω or ε'' represents the loss of the medium.

In fact, under the influence of high frequency electromagnetic field, the movement of charged particles in the medium is slower than the rapid change of high frequency field, hence causing hysteresis. Due to this hysteresis effect, ε, μ and σ all become complex numbers. In addition, as the frequency approaches the intrinsic resonance frequency of the medium, the resonant particle will absorb energy from the alternating electromagnetic field and then scatter monochromatically, causing scattering loss.

Both the heat loss caused by conductivity σ and the loss due to hysteresis effect in high frequency are reflected in the imaginary part of complex permittivity and turn to be the function of frequency. Thus, the propagation velocity of TEM is no longer constant but a function of frequency. We call this kind of waves as dispersive waves and the corresponding lossy medium as dispersive medium.

7.11.2 Dispersive Distortion

Since the attenuation constant α, phase constant β and velocity v have complicated relations with frequency; wave propagation in dispersive medium will cause waveform distortion of group signal. We call the distortion caused by frequency dispersion effect as dispersive distortion. Suppose two adjacent pulses move from $z = 0$ to $z = l$. The two pulses will be broadened and overlap each other and then cause dispersive distortion, as shown in Figure 7.24.

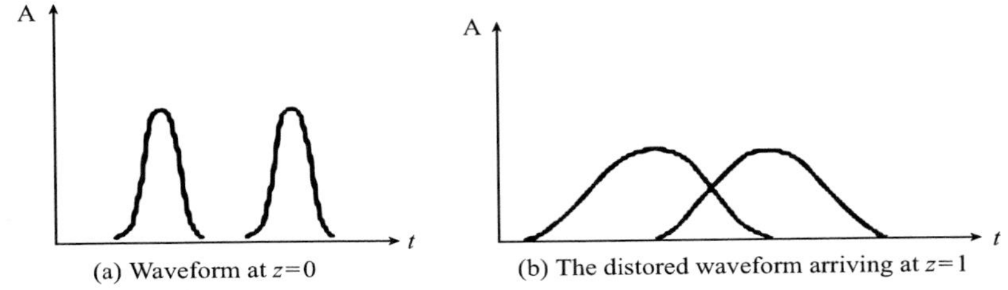

Figure 7.24 The dispersive distortion caused by two adjacent pulses propagating in lossy medium

7.11.3 Normal Dispersion and Anomalous Dispersion

The phase velocity in dispersive medium (lossy medium) can be expressed as

$$v_{\mathrm{p}} = \frac{\omega}{\beta} = \left\{ \frac{\mu\varepsilon}{2} \left[\sqrt{1 + \left(\frac{\sigma}{\omega\varepsilon}\right)^2} + 1 \right] \right\}^{-\frac{1}{2}} \tag{7.97}$$

and the energy velocity is

$$v_{\mathrm{g}} = \frac{\mathrm{d}\omega}{\mathrm{d}\beta} \tag{7.98}$$

Rewriting above equation and applying Equation (7.97) we can obtain

$$v_{\mathrm{g}} = \frac{\mathrm{d}\omega}{\mathrm{d}\beta} = \frac{\mathrm{d}}{\mathrm{d}\beta}(v_{\mathrm{p}}\beta) = v_{\mathrm{p}} + \beta\frac{\mathrm{d}v_{\mathrm{p}}}{\mathrm{d}\omega}\frac{\mathrm{d}\omega}{\mathrm{d}\beta} = v_{\mathrm{p}} + \frac{\omega}{v_{\mathrm{p}}}\frac{\mathrm{d}v_{\mathrm{p}}}{\mathrm{d}\omega}v_{\mathrm{g}}$$

i.e.

$$v_{\mathrm{g}} = v_{\mathrm{p}} + \frac{\omega}{v_{\mathrm{p}}}\frac{\mathrm{d}v_{\mathrm{p}}}{\mathrm{d}\omega}v_{\mathrm{g}} \tag{7.99}$$

From Equations (7.97) and (7.99) we can observe that there is non-linear relation between v_{p}, v_{g} and frequency. The phase velocity varies with the frequency in the following three cases:

$\frac{\mathrm{d}v_{\mathrm{p}}}{\mathrm{d}\omega} < 0$, i.e. the phase velocity decreases as frequency increases. From Equation (7.99), $v_{\mathrm{p}} > v_{\mathrm{g}}$. That is to say the phase velocity is larger than the energy velocity. We call this case as normal dispersion.

$\frac{\mathrm{d}v_{\mathrm{p}}}{\mathrm{d}\omega} > 0$, i.e. the phase velocity increases as frequency increases. Therefore $v_{\mathrm{p}} < v_{\mathrm{g}}$. The phase velocity is smaller than the energy velocity. This is called as anomalous dispersion.

$\frac{\mathrm{d}v_{\mathrm{p}}}{\mathrm{d}\omega} = 0$, the phase velocity is equal to the energy velocity. In this case, the medium is non-dispersive.

The phenomenon of the normal and anomalous dispersion can be used suitably for improving the characteristics of phase against frequency.

Example 7.7 In a lossy medium with relative permittivity $\varepsilon_{\mathrm{r}} = 2.5$ and $\tan\delta = 10^{-3}$. To find the average power loss per cubic meter of the medium when the frequency is 1000MHz and the electric field intensity is 1 V/m.

Method 1 calculate the average power loss of unit volume directly by the expression $\frac{1}{2}\sigma E^2$.

The average power loss of unit volume can be expressed as

$$P_{\mathrm{L}} = \frac{1}{2}JE = \frac{1}{2}\sigma E^2 \tag{a}$$

We can first find the conductivity of the medium. Given that,

$$\tan\delta = \frac{\sigma}{\omega\varepsilon_0\varepsilon_{\mathrm{r}}} = 10^{-3}$$

there is

$$\sigma = 10^{-3} \times (2\pi \times 10^9) \times 2.5 \times \frac{1}{36\pi} \times 10^{-9} = 1.3888 \times 10^{-4}\ (\mathrm{S/m}) \tag{b}$$

Substitute Equation (b) into Equation (a), then we have

$$P_{\mathrm{L}} = \frac{1}{2}\sigma E^2 = 0.6944 \times 10^{-4} \quad (\mathrm{W/m^3})$$

That is to say the average power loss per cubic meter is 0.06944 mW.

Method 2 select two points along the propagation direction and their distance is 1m. Then we can solve the problem by calculating the difference between the S_{av} at these two points.

Since tan $\delta = 10^{-3}$, it can be regarded as a low-loss dielectric. Assume that the wave is propagating along the positive direction of z axis and let $E_0 = 1(\mathrm{V/m})$ at $z = 0$. Thus, the S_{av1} at this point is

$$S_{\mathrm{av1}} = \frac{1}{2}\mathrm{Re}\frac{E_0^2}{\eta^{\mathrm{e}}} = \frac{1}{2}\frac{E_0^2}{|\eta^{\mathrm{e}}|}\cos\theta'$$

The corresponding S_{av2} of the point at $z = 1\mathrm{m}$ can be expressed as

$$S_{\mathrm{av2}} = \frac{1}{2}\frac{E_0^2}{|\eta^{\mathrm{e}}|} \cdot \mathrm{e}^{-2\alpha} \cdot \cos\theta'$$

The average power loss per unit volume (the unit area multiplied by the unit length) P_L should be the difference of the S_{av} at the two points with distance of 1m, i.e.

$$P_L = S_{\mathrm{av1}} - S_{\mathrm{av2}} = \frac{1}{2}\frac{E_0^2}{\eta^{\mathrm{e}}}\cos\theta' \cdot (1 - \mathrm{e}^{-2\alpha}) \tag{c}$$

Next, we need to find out the following parameters: $|\eta^{\mathrm{e}}|$, θ' and α.

$$\tan\delta = \frac{\varepsilon''}{\varepsilon'} = 10^{-3}$$

$$\varepsilon'' = \varepsilon' \times 10^{-3} = 2.5\varepsilon_0 \times 10^{-3}$$

Substituting the above expression into Equation (7.72), hence

$$\eta^{\mathrm{e}} = \sqrt{\frac{\mu}{\varepsilon'}}\left[1 + \mathrm{j}\frac{\varepsilon''}{2\varepsilon'}\right] = |\eta^{\mathrm{e}}|\,\mathrm{e}^{\mathrm{j}\theta'} \approx 238.43\mathrm{e}^{\mathrm{j}0.02865^\circ}$$

From Equation (7.70) we have

$$\alpha = \frac{\omega\varepsilon''}{2\varepsilon'}\sqrt{\mu\varepsilon'} \approx 1.6558 \times 10^{-2} \quad (\mathrm{Np/m})$$

Substitute $|\eta^{\mathrm{e}}|$, θ', α and $E_0 = 1(\mathrm{V/m})$ into c) then we get

$$P_L = \frac{E_0^2}{2|\eta^{\mathrm{e}}|}\cos\theta'(1 - \mathrm{e}^{-2\alpha}) = \frac{1}{2 \times 238.4} \times \cos(0.02865)(1 - 0.967)$$
$$\approx 0.6900 \times 10^{-4} \quad (\mathrm{W/m^3})$$

Example 7.8 Consider the vertical incidence of TEM wave into a good conductor. Show that the entering electromagnetic energy into the conductor will be fully dissipated into heat.

(1) Find out the energy S_{av} entering the conductor from the unit surface area of the conductor.

Assume the surface of the good conductor is located on xOy plane, z axis is orthogonal to the conductor.

The electromagnetic field entering into the conductor can be expressed as

$$\begin{aligned} E_x &= E_0 \mathrm{e}^{-\alpha z} \mathrm{e}^{\mathrm{j}(\omega t - \beta z)} \\ H_y &= E_x / \eta^{\mathrm{e}} \end{aligned}$$

where η^{e}, a complex number, is the wave impedance of the conductor:

$$\eta^{\mathrm{e}} = \sqrt{\frac{\omega\mu}{\sigma}} \cdot \mathrm{e}^{\mathrm{j}45^\circ}$$

The average Poynting vector S_{av} flowing through the unit area of the surface of the good conductor ($z = 0$) can be expressed as

$$\begin{aligned} \boldsymbol{S}_{\mathrm{av}} &= \frac{1}{2}\mathrm{Re}(\boldsymbol{E} \times \boldsymbol{H}^*) = \boldsymbol{e}_z \frac{1}{2}\mathrm{Re}\left(E_0 \mathrm{e}^{\mathrm{j}\omega t} \times E_0 \mathrm{e}^{-\mathrm{j}\omega t} \times \sqrt{\frac{\sigma}{\omega\mu}} \mathrm{e}^{\mathrm{j}45^\circ}\right) \\ &= \boldsymbol{e}_z \frac{\sqrt{2}}{4} E_0^2 \sqrt{\frac{\sigma}{\omega\mu}} \end{aligned} \tag{a}$$

(2) We calculate the power loss of an infinite conductor with unit surface area

Method 1 We can solve the problem by computing the integral of the power loss in unit volume at any point $\left(\text{i.e. } \frac{1}{2}\mathrm{Re}(EJ^*)\right)$, within $[0, \infty]$ with respect to z.

The power loss in unit volume at any point in the conductor can be expressed in the form of electric field as

$$\begin{aligned} \frac{1}{2}\mathrm{Re}(EJ^*) &= \frac{1}{2}\mathrm{Re}\left[E_0 \mathrm{e}^{-\alpha z} \cdot \mathrm{e}^{\mathrm{j}(\omega t - \beta z)} \cdot \sigma E_0 \mathrm{e}^{-\alpha z} \cdot \mathrm{e}^{-\mathrm{j}(\omega t - \beta z)}\right] \\ &= \frac{1}{2}\mathrm{Re}(\sigma E_0^2 \mathrm{e}^{-2\alpha z}) = \frac{1}{2}\sigma E_0^2 \cdot \mathrm{e}^{-2\alpha z} \end{aligned} \tag{b}$$

Then, the total power loss of infinite good conductor with *unit surface area* ($x \times y = 1 \times 1$) can be expressed as

$$\begin{aligned} \int_0^\infty \frac{1}{2}\mathrm{Re}(EJ^*)\mathrm{d}z &= \frac{1}{2}\sigma E_0^2 \cdot \int_0^\infty \mathrm{e}^{-2\alpha z}\mathrm{d}z \\ &= \frac{1}{4}\frac{\sigma E_0^2}{\alpha} = \frac{\sqrt{2}}{4} E_0^2 \sqrt{\frac{\sigma}{\omega\mu}} \end{aligned} \tag{c}$$

From Equations (a) and (c) of this example, we can obtain that the heat loss of power of the conductor is equal to the average power density S_{av} that enters from the surface of the good conductor.

Method 2 By directly applying $\frac{1}{2}|J_{\mathrm{s}}|^2 R_{\mathrm{s}}$, we compute the power loss of the conductor.

For the infinite conductor with unit surface area, the corresponding power loss can be calculated by

$$\frac{1}{2}|J_{\mathrm{s}}|^2 \cdot R_{\mathrm{s}} = \frac{1}{2}\left|\frac{J_0}{\gamma}\right|^2 \cdot \frac{1}{\sigma\delta}$$

where

$$\begin{aligned} J_0^2 &= (\sigma E_0)^2 \\ |\gamma|^2 &= |(\alpha + \mathrm{j}\beta)|^2 \approx |\alpha(1 + \mathrm{j})|^2 = 2\alpha^2 \end{aligned}$$

To derive the above expression we apply $\alpha = \beta$ since it is a good conductor. Thus, the power loss of the conductor with unit surface area can be rewritten as

$$\frac{1}{2}|J_s|^2 R_s = \frac{1}{2}\left(\frac{\sigma^2 E_0^2}{2\alpha^2}\right) \cdot \frac{1}{\sigma\delta} = \frac{1}{4}E_0^2 \cdot \frac{\sigma}{\alpha} = \frac{\sqrt{2}}{4}E_0^2\sqrt{\frac{\sigma}{\omega\mu}} \tag{d}$$

where we apply $\alpha\delta = 1$ to derive the above equations.

In this example, Equations (a), (c) and (d) are all equivalent, indicating that the energy entering into the conductor is completely converted into the heat energy.

7.12 Electromagnetic Waves in Ferrite Medium

7.12.1 Ferrite Materials

The chemical molecules of ferrite materials can be expressed as FeO Fe_2O_3, where ferrum (Fe) can be substituted by manganese (Mn), magnesium (Mg), Aluminum (Al), nickel (Ni) and Zinc (Zn) and so on. One kind of ferrite materials called yttrium iron garnet (YIG) has been widely applied.

What characteristics do ferrite materials have?

First of all, which is of particular importance, ferrite are insulated materials with very high resistivity even up to $10^6 \sim 10^8 \Omega/\text{cm}$. Due to the skin effect the propagation distance of the microwave in the good conductor is very limited. However, microwave is able to go deep into the ferrite materials. Hence it is very necessary to investigate the special features of ferrite materials. If the resistivity is not high enough, microwave cannot enter deeply, the application of ferrite materials will not be of so much importance.

Secondly, the relative permittivity of ferrite materials is also comparatively large, which is between 10~20. As all we know, for an atom of any kind of substance there exist not only the revolutions of its electrons but also the spins of its electrons. These revolutions and rotations of electrons will form electric current and generate magnetic field. For the ordinary substance, nevertheless, the magnetic fields generated by electrons have random directions and can be cancelled out. Therefore, the entire material has no magnetic property. For the ferrite materials, despite the magnetic fields generated by the revolution of electrons are random, they have magnetic property since the magnetic fields caused by the spins cannot be cancelled out completely. In other words, it is the spins of the electrons that generate the magnetic field of ferrite materials. When a ferrite material is exposed to an external steady magnetic field, the directions of magnetic fields generated by the spins will be regulated towards the direction of the steady magnetic field thereby show strong magnetic property.

Thus, studying the interaction between the magnetic field generated by spins and external steady or alternating magnetic field and clarifying the parameters of ferrite materials is prerequisite for further investigation on the electromagnetic fields in ferrite materials.

7.12.2 The Tensor Permeability μ of Ferrite Materials

1. The Precession Movement of the Electron Spin when the Ferrite Material is Biased with An External Steady Magnetic Field $\boldsymbol{B}_0$

What will happen to the movement of the electron and what influence will be caused when a spin electron is biased with an external steady magnetic field $\boldsymbol{B}_0$?

When an electron moves inside the magnetic field (in this case it is the electron spin) it will be subjected to a force. To be specific, a force moment $\boldsymbol{T}$ will be generated when a electron spin with magnetic moment $\boldsymbol{m}$ is subject to an external steady magnetic field $\boldsymbol{B}_0$ and the relation between them can be expressed as

$$\boldsymbol{T} = \boldsymbol{m} \times \boldsymbol{B}_0 \tag{7.100a}$$

This force moment $\boldsymbol{T}$ will change the moment of momentum (or angular momentum) $\boldsymbol{J}$ of the electron spin. Their correlation can be described as the ratio of the moment of momentum with respect to time, i.e.

$$\boldsymbol{T} = \frac{\mathrm{d}\boldsymbol{J}}{\mathrm{d}t} \tag{7.100b}$$

The current caused by an electron spin will generate a magnetic field, defined as a magnetic dipole. The magnetic moment of the magnetic dipole is

$$\boldsymbol{m} = -\frac{e}{m_e}\boldsymbol{J} = r\boldsymbol{J} \tag{7.101}$$

where $r = -\dfrac{e}{m_e}$ is the gyromagnetic ratio, e is the electronic charge and m_e is the electronic mass. The direction of the magnetic moment $\boldsymbol{m}$ is opposite to the direction of $\boldsymbol{J}$, and the direction of current caused by the electron spin are subjected to the right-hand rule. When the direction of $\boldsymbol{m}$ is known, the direction of force moment $\boldsymbol{T}$ can thus be determined according to Equation (7.100a). The direction of the force moment $\boldsymbol{T}$ represents the direction of the increment of $\boldsymbol{J}$, saying, the moment of momentum of the electron spin. Hence, as shown in Figure 7.25, the change of the moment of momentum (or angular momentum) $\Delta\ \boldsymbol{J}$ within time period Δt due to force moment $\boldsymbol{T}$ can be expressed as following

$$\Delta \boldsymbol{J} = (\boldsymbol{J} \sin\theta)\omega_0 \Delta t$$

Therefore

$$\frac{\mathrm{d}\boldsymbol{J}}{\mathrm{d}t} = \boldsymbol{J}\omega_0 \sin\theta \tag{7.102}$$

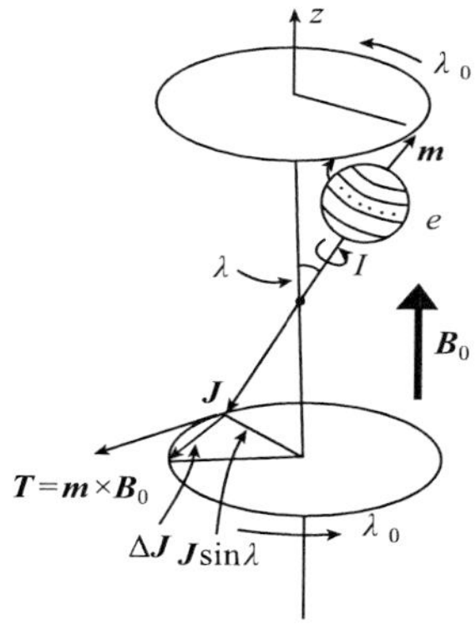

Figure 7.25 The precession movement of electron

Based on Equations (7.100)~(7.102), the angular frequency ω_0 of the precession movement of the electron spin can be derived as

$$\omega_0 = \left(\frac{e}{m_e}\right) B_0 \tag{7.103}$$

Indicating that force moment $\boldsymbol{T}$ of the magnetic moment $\boldsymbol{m}$ of the electron spin biased with an external steady magnetic field $\boldsymbol{B}_0$ can change the moment of momentum $\boldsymbol{J}$ of the electron spin and thereby making the electron spin not only rotate by itself but also revolute around the axis of the external steady magnetic field $\boldsymbol{B}_0$, and thus form the precession movement with angular frequency ω_0.

The precession movement ceases immediately due to the power loss.

It can be also observed from Figure 7.25 that the precession movement of electron is always right-handed under the influence of steady magnetic field. That is to say in the direction of $\boldsymbol{B}_0$ the spin electron always moves around the $\boldsymbol{B}_0$ (i.e. z axis) clockwise.

2. The Tensor Permeability μ

Since that

$$\boldsymbol{m} = r\boldsymbol{J}, \quad \boldsymbol{T} = \frac{\mathrm{d}\boldsymbol{J}}{\mathrm{d}t}, \quad \boldsymbol{T} = \boldsymbol{m} \times \boldsymbol{B}_0$$

we have

$$\frac{\mathrm{d}\boldsymbol{m}}{\mathrm{d}t} = r\frac{\mathrm{d}\boldsymbol{J}}{\mathrm{d}t} = r\boldsymbol{T} = r(\boldsymbol{m} \times \boldsymbol{B}_0)$$

If we let $\boldsymbol{P}_m$ be the magnetization and N represent the atomicity in unit volume, then

$$\boldsymbol{P}_m = N\boldsymbol{m}$$

therefore

$$\frac{\mathrm{d}P_{\mathrm{m}}}{\mathrm{d}t} = r\mu_0(P_{\mathrm{m}} \times H_0) \tag{7.104}$$

Generally, there are steady magnetic field $\boldsymbol{H}_0$ and smaller alternating magnetic field $\boldsymbol{H}_1$ (denoted by $\boldsymbol{h}$ sometimes) simultaneously in the ferrite material. Thus, the total magnetic field is

$$\boldsymbol{H} = \boldsymbol{H}_0 + \boldsymbol{H}_1 = \boldsymbol{e}_x H_{1x} + \boldsymbol{e}_y H_{1y} + \boldsymbol{e}_z(H_0 + H_{1z}) \tag{7.105}$$

where the alternating magnetic field is $\boldsymbol{H}_1 = \boldsymbol{e}_x H_{1x} + \boldsymbol{e}_y H_{1y} + \boldsymbol{e}_z H_{1z}$.

And the total magnetization $\boldsymbol{P}_m$ caused by $\boldsymbol{H}$ can be expressed as

$$\boldsymbol{P}_m = \boldsymbol{e}_x P_{m1x} + \boldsymbol{e}_y P_{m1y} + \boldsymbol{e}_z(P_{\mathrm{m}0} + P_{\mathrm{m}1z}) \tag{7.106}$$

where $P_{\mathrm{m}0}$ represents the saturation magnetization of ferrite material when biased with steady magnetic field $\boldsymbol{H}_0$, and $P_{\mathrm{m}1x}$, $P_{\mathrm{m}1y}$, $P_{\mathrm{m}1z}$ are the magnetization by alternating magnetic field.

Since

$$\frac{\mathrm{d}\boldsymbol{P}_{\mathrm{m}}}{\mathrm{d}t} = r\mu_0\boldsymbol{P}_{\mathrm{m}} \times \boldsymbol{H}$$

Substituting the total magnetic field $\boldsymbol{H}$ in Equation (7.105) and the total magnetization $\boldsymbol{P}_{\mathrm{m}}$ in Equation (7.106) into the above expression and neglecting the product of two alternating magnetic fields (since the strength of the alternating magnetic field is much less than that of the steady magnetic field), meanwhile neglecting subscript '1' of the alternating magnetic field, then we obtain the alternating magnetization

$$\left.\begin{aligned} P_{\mathrm{m}x} &= \frac{(\omega_{\mathrm{M}}\omega_0)H_x + \mathrm{j}\omega\omega_{\mathrm{M}}H_y}{\omega_0^2 - \omega^2} \\ P_{\mathrm{m}y} &= \frac{(\omega_{\mathrm{M}}\omega_0)H_y + \mathrm{j}\omega\omega_{\mathrm{M}}H_x}{\omega_0^2 - \omega^2} \\ P_{\mathrm{m}z} &= 0 \end{aligned}\right\} \tag{7.107}$$

where $\omega_{\mathrm{M}} = r\mu_0 P_{\mathrm{m}0}$ is called eigen-frequency, ω_0 is angular frequency of precession and ω is the angular frequency of external alternating field.

The magnetic induction $\boldsymbol{B}$ corresponding to the alternating magnetic field $\boldsymbol{H}$ in the ferrite material can be expressed as

$$\boldsymbol{B} = \mu_0(\boldsymbol{H} + \boldsymbol{P}_{\mathrm{m}})$$

that is

$$\left.\begin{aligned} B_x &= \mu_0(H_x + P_{\mathrm{m}x}) \\ B_y &= \mu_0(H_y + P_{\mathrm{m}y}) \\ B_z &= \mu_0(H_z + P_{\mathrm{m}z}) \end{aligned}\right\} \tag{7.108}$$

Substituting the Equation (7.107) into Equation (7.108) , we have

$$\left.\begin{aligned} B_x &= \mu_{11}H_x + \mu_{12}H_y \\ B_y &= \mu_{21}H_x + \mu_{22}H_y \\ B_z &= \mu_0 H_z \end{aligned}\right\} \tag{7.109}$$

where

$$\left.\begin{aligned} \mu_{11} &= \mu_{22} = \mu_0\left(1+\frac{\omega_M\omega_0}{\omega_0^2-\omega^2}\right) \\ \mu_{12} &= -\mu_{21} = \mathrm{j}\mu_0\frac{\omega_M\omega}{\omega_0^2-\omega^2} \\ \mu_{33} &= \mu_0 \end{aligned}\right\} \tag{7.110}$$

Therefore, the tensor permeability $\boldsymbol{\mu}$ is

$$\boldsymbol{\mu} = \begin{bmatrix} \mu_{11} & \mu_{12} & \mu_{13} \\ \mu_{21} & \mu_{22} & \mu_{23} \\ \mu_{31} & \mu_{32} & \mu_{33} \end{bmatrix} = \begin{bmatrix} \mu_{11} & \mu_{12} & 0 \\ -\mu_{12} & \mu_{11} & 0 \\ 0 & 0 & \mu_0 \end{bmatrix} \tag{7.111}$$

Let $\mu_{11} = \mu_r\mu_0$, $\dfrac{-\omega_M\omega}{\omega_0^2-\omega^2} = k_r$ and we can get

$$\boldsymbol{\mu} = \mu_0\begin{bmatrix} \mu_r & -\mathrm{j}k_r & 0 \\ \mathrm{j}k_r & \mu_r & 0 \\ 0 & 0 & 1 \end{bmatrix} \tag{7.112}$$

From Equation (7.110), $\omega_M = 0$ if there is no external magnetic field. Hence we have $\mu_{11} = \mu_{22} = \mu_{33} = \mu_0$ and $\mu_{12} = \mu_{21} = \mu_{23} = \cdots = 0$, and

$$\mu = \begin{bmatrix} \mu_0 & 0 & 0 \\ 0 & \mu_0 & 0 \\ 0 & 0 & \mu_0 \end{bmatrix}$$

In this case, μ becomes a scalar.

7.12.3 Propagation of Electromagnetic Wave in Ferrite Medium

When the objective electromagnetic wave propagates in uniform ferrite material, we can just substitute the parameters ε and $\boldsymbol{\mu}$ into the Maxwell equations or the wave equation.

As mentioned above, the precession movement of electron itself is right-handed. If the external alternating magnetic field is also rotating such as right-handed or left-handed circular polarization wave, the ferrite material will demonstrate completely different properties and can be characterized by different parameters. Two cases will be studied below.

(1) The direction of TEM wave (which is the direction of z axis) is coherent with the direction of external steady magnetic field $\boldsymbol{H}_0$. The analysis of parameters of the wave (the magnetic field is circular polarization wave) in the ferrite material can be made as following.

If the magnetic field is right-handed circular polarization wave

$$\boldsymbol{H}_+ = \boldsymbol{e}_x H_+ - \mathrm{j}\boldsymbol{e}_y H_+ = \boldsymbol{e}_x H_x + \boldsymbol{e}_y H_y \tag{7.113}$$

The wave equation of TEM wave in ferrite material can be expressed as

$$\frac{\partial^2}{\partial z^2}\begin{bmatrix} H_x \\ H_y \\ 0 \end{bmatrix} + \omega^2\varepsilon\begin{bmatrix} \mu_{11} & \mu_{12} & 0 \\ \mu_{21} & \mu_{22} & 0 \\ 0 & 0 & \mu_0 \end{bmatrix}\begin{bmatrix} H_x \\ H_y \\ 0 \end{bmatrix} = 0 \tag{7.114}$$

thence

$$\frac{\partial^2 H_x}{\partial z^2} + \omega^2\varepsilon(\mu_{11}H_x + \mu_{12}H_y) = 0 \tag{7.115}$$

$$\frac{\partial^2 H_y}{\partial z^2} + \omega^2\varepsilon(\mu_{21}H_x + \mu_{22}H_y) = 0 \tag{7.116}$$

Substitute H_x, H_y in Equation (7.113) into Equation (7.115)

$$\frac{\partial^2}{\partial z^2}H_+ + \omega^2\varepsilon(\mu_{11} - \mathrm{j}\mu_{12})H_+ = \frac{\partial^2}{\partial z^2}H_+ - r_+^2 H_+ = 0$$

where

$$r_+^2 = -\omega^2\varepsilon(\mu_{11} - \mathrm{j}\mu_{12})$$

Therefore, the propagation constant r_+ of right-handed circular polarization wave is expressed as

$$r_+ = \mathrm{j}\omega\sqrt{\varepsilon}\cdot\sqrt{\mu_{11} - \mathrm{j}\mu_{12}} \tag{7.117}$$

Likewise, the propagation constant r_- of left-handed circular polarization wave can be calculated as

$$r_- = \mathrm{j}\omega\sqrt{\varepsilon}\cdot\sqrt{\mu_{11} + \mathrm{j}\mu_{12}} \tag{7.118}$$

As shown in the expressions of r_+, r_-, right-handed circular polarization wave and left-handed circular polarization wave have different propagation constants.

Denote the permeability of the ferrite material in cases of right-hand polarized wave and left-hand polarized wave by μ_+ and μ_- respectively. From the expressions of r_+, r_-, we can have

$$\mu_+ = \mu_{11} - \mathrm{j}\mu_{12}$$

$$\mu_- = \mu_{11} + \mathrm{j}\mu_{12} \tag{7.119a}$$

Substitute the expressions of μ_+ and μ_- in Equation (7.110) into the above expression

$$\left.\begin{aligned} \mu_+ &= \mu_{11} - \mathrm{j}\mu_{12} = \mu_0\left(1 + \frac{\omega_M}{\omega_0 - \omega}\right) \\ \mu_- &= \mu_{11} + \mathrm{j}\mu_{12} = \mu_0\left(1 + \frac{\omega_M}{\omega_0 + \omega}\right) \end{aligned}\right\} \tag{7.119b}$$

The frequency dependences of μ_+ and μ_- are shown in Figure 7.26.

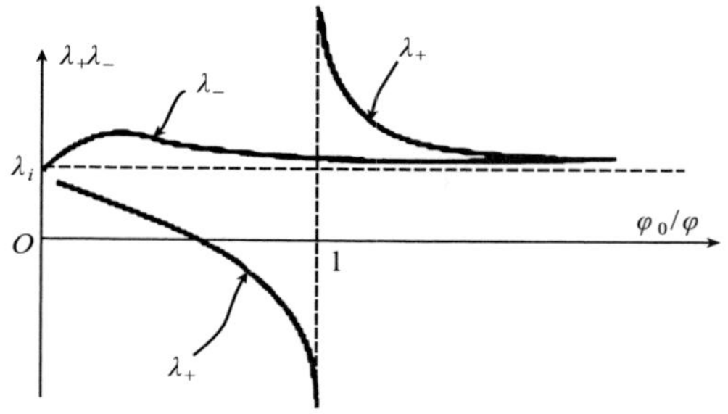

Figure 7.26 The change rules of μ_+ and μ_-

Discussed above is the propagation of TEM wave (of which the magnetic field is circular polarization) in ferrite material while the propagation direction of the wave is coherent with the direction of external magnetic field $\boldsymbol{H}_0$. *Thus, the subsequent formulas are only valid for this case, including the formulas of* μ_+ *and* μ_- *in Equation (7.119) and* r_+, r_- *in Equations (7.117) and (7.118) respectively.*

(2) The direction of TEM wave (which is the direction of x axis) and the direction of external steady magnetic field $\boldsymbol{H}_0$ (which is the direction of z axis) are mutually perpendicular. In this case, the parameters of the wave can be calculated as following.

From the Maxwell equation $\nabla \times \boldsymbol{H} = \mathrm{j}\omega\varepsilon E$we can get

$$\nabla \times \nabla \times \boldsymbol{H} = \mathrm{j}\omega\varepsilon\nabla \times \boldsymbol{E} = \mathrm{j}\omega\varepsilon(-\mathrm{j}\omega\boldsymbol{\mu}\boldsymbol{H}) = \omega^2\varepsilon\boldsymbol{\mu}\boldsymbol{H} \tag{7.120}$$

Assume the wave propagating along the direction of x axis is only the function of variable x, i.e. the relation with respect to space can be expressed as $\mathrm{e}^{-\gamma x}$. Thus the term $\nabla \times \nabla \times \boldsymbol{H}$ in the above expression can be also expressed as

$$\begin{aligned}\nabla \times \nabla \times \boldsymbol{H} &= \nabla(\nabla \cdot \boldsymbol{H}) - \nabla^2\boldsymbol{H} \\ &= \nabla\left(\frac{\partial \boldsymbol{H}_x}{\partial x}\right) - \frac{\partial^2 \boldsymbol{H}}{\partial x^2} = e_x\gamma^2 H_x - \gamma^2\boldsymbol{H}\end{aligned} \tag{7.121}$$

Combining Equations (7.120) and (7.121), we obtain the vector wave equation

$$\boldsymbol{e}_x\gamma^2 H_x - \gamma^2\boldsymbol{H} - \omega^2\varepsilon\mu\boldsymbol{H} = 0 \tag{7.122}$$

Substitute $\boldsymbol{\mu}$ in Equation (7.112) into the above vector formula and then expand it to obtain the following three scalar formulas:

$$\omega^2\varepsilon_\mathrm{r}\varepsilon_0\mu_0(\mu_\mathrm{r}H_x - \mathrm{j}k_\mathrm{r}H_y) = 0 \tag{7.123}$$

$$\gamma^2 H_y + \omega^2\varepsilon_\mathrm{r}\varepsilon_0\mu_0(\mathrm{j}k_\mathrm{r}H_x + \mu_\mathrm{r}H_y) = 0 \tag{7.124}$$

$$\gamma^2 H_z + \omega^2\varepsilon_\mathrm{r}\varepsilon_0\mu_0 H_z = 0 \tag{7.125}$$

There are two possible solutions for the above three formulas.

One solution is $H_z \neq 0$, $H_x = 0$, $H_y = 0$. It can be obtained from Equation (7.126)

$$\gamma^2 = -\omega^2\varepsilon_\mathrm{r}\mu_0\varepsilon_0 = (\mathrm{j}\beta)^2 = -\left(\frac{\omega}{v_\mathrm{p}}\right)^2 \tag{7.126}$$

The only magnetic field component of this wave is H_z and it has the same direction as the external steady magnetic field $\boldsymbol{H}_0$. For the linear polarization magnetic field of this case, the ferrite material does not have any specific properties, just behaving like an ordinary medium. The propagation velocity of the wave can be calculated from Equation (7.126):

$$v_\mathrm{p} = \frac{1}{\sqrt{\varepsilon_\mathrm{r}\mu_0\varepsilon_0}} = \frac{c}{\sqrt{\varepsilon_\mathrm{r}}} \tag{7.127}$$

Another solution is $H_z = 0$ and $H_x \neq 0$, $H_y \neq 0$.

Combining this solution with Equation (7.123) and considering that the direction of the external steady magnetic field is along z axis, we have the following relations:

$$H_z = 0 \tag{7.128}$$

$$H_x = \mathrm{j}\frac{k_\mathrm{r}}{\mu_\mathrm{r}}H_y \tag{7.129}$$

Equation (7.129) indicates that the magnetic field of this wave is elliptic polarization and thus the ferrite material, of which the direction of the magnetic field is along z axis, will react to the wave. Substitute Equation (7.129) into Equation (7.124) we can get that

$$\gamma^2 = -\omega^2 \varepsilon_r \varepsilon_0 \mu_0 \left(\frac{\mu_r^2 - k_r^2}{\mu_r} \right) \tag{7.130}$$

Since the wave propagates along x axis and its field quantity is only the function of x and t, we can find the component of electric field E_z, which is based on the Maxwell equations and Equation (7.130)

$$\boldsymbol{e}_z \varepsilon_r \varepsilon_0 \frac{\partial E_z}{\partial t} = \boldsymbol{e}_z \left(\frac{\partial H_y}{\partial x} - \frac{\partial H_x}{\partial y} \right) = \boldsymbol{e}_z \frac{\partial H_y}{\partial x} = -\boldsymbol{e}_z \gamma H_y$$

$$E_z = -\frac{\eta_0}{\sqrt{\varepsilon_r}} \sqrt{\frac{\mu_r^2 - k_r^2}{\mu_r}} \cdot H_y \tag{7.131}$$

Since there are only two field components H_x and E_z, *the wave is transverse electric (TE) wave instead of TEM wave.* Additionally, since the magnetic field is elliptic polarization, the corresponding μ of the ferrite material can be calculated from Equation (7.130):

$$\mu = \mu_0 \left(\frac{\mu_r^2 - k_r^2}{\mu_r} \right) \tag{7.132}$$

The phase velocity is

$$v_p = \frac{1}{\sqrt{\mu\varepsilon}} = \frac{1}{\sqrt{(\varepsilon_0 \varepsilon_r) \left(\mu_0 \frac{\mu_r^2 - k_r^2}{\mu_r} \right)}} = \frac{c}{\sqrt{\varepsilon_r} \sqrt{\left(\frac{\mu_r^2 - k_r^2}{\mu_r} \right)}} \tag{7.133}$$

Obviously, the electron spin in the ferrite material revolue around the steady magnetic field $\boldsymbol{H}_0$ *dextrorsely; When the external alternating magnetic field is also circular polarization wave, as either TEM or non-TEM electromagnetic waves, the ferrite will demonstrate very strong anisotropy to the left-handed and right-handed circularly polarized waves, provided that the axis of the external circular polarization magnetic field is parallel with that of the axis of the moving electronic spin.* Taking ferrite isolator as an example, the forward transmission loss is very low (for example lower than 0.3dB) while the reverse transmission loss is very large (more than 30dB), which reveals the non-reverse property of attenuation.

Example 7.9 A plane wave with magnetic field intensity $\boldsymbol{H} = (e_x + \mathrm{j} e_y) H_m \mathrm{e}^{\mathrm{j}(\omega t - \beta z)}$ and frequency $f = 3$GHz, propagates in the uniform and infinite ferrite material. The external steady magnetic field is in the direction of z axis. The tensor permeability is expressed as follows

$$[\mu_r] = \begin{bmatrix} 1.2 & -\mathrm{j}0.3 & 0 \\ \mathrm{j}0.3 & 1.2 & 0 \\ 0 & 0 & 1 \end{bmatrix}$$

relative permittivity $\varepsilon_r = 20$. Calculate (1) electric field intensity $\boldsymbol{E}$; (2) propagation velocity of the electromagnetic wave v_p and the wavelength λ; (3) wave impedance η.

Solutions the magnetic field intensity of the given left-handed circular polarization wave is

$$\boldsymbol{h} = (\boldsymbol{e}_x + \mathrm{j}\boldsymbol{e}_y) h_0 \mathrm{e}^{\mathrm{j}(\omega t - \beta z)} = \boldsymbol{e}_x h_0 \mathrm{e}^{\mathrm{j}(\omega t - \beta z)} = \boldsymbol{e}_y h_0 \mathrm{e}^{\mathrm{j}(\omega t - \beta z + 90^\circ)} \tag{a}$$

(1) Calculate the electric field intensity based on Maxwell equations. Since

$$\boldsymbol{e}_y \frac{\partial H_x}{\partial z} = \boldsymbol{e}_y \varepsilon \frac{\partial E_y}{\partial t} = \boldsymbol{e}_y \mathrm{j}\omega\varepsilon E_y$$

there is

$$E_y = -\frac{\beta}{\omega\varepsilon}H_x = -\frac{\beta}{\omega\varepsilon}H_{\mathrm{m}}\mathrm{e}^{\mathrm{j}(\omega t-\beta z)} \tag{b}$$

From the formula

$$-\boldsymbol{e}_x\frac{\partial H_y}{\partial z} = \boldsymbol{e}_x\varepsilon\frac{\partial E_x}{\partial t} = \boldsymbol{e}_x\mathrm{j}\omega\varepsilon E_x$$

we can have

$$E_x = \frac{\beta}{\omega\varepsilon}H_y = \mathrm{j}\frac{\beta}{\omega\varepsilon}H_{\mathrm{m}}\mathrm{e}^{\mathrm{j}(\omega t-\beta z)} \tag{c}$$

Based on Equations (b) and (c) the electric field intensity can be calculated as

$$\boldsymbol{E} = \frac{\beta}{\omega\varepsilon}(\mathrm{j}\boldsymbol{e}_x - \boldsymbol{e}_y)H_{\mathrm{m}}\mathrm{e}^{\mathrm{j}(\omega t-\beta z)} \tag{d}$$

Equations (a) and (d) show that it is a TEM wave and the direction of the wave is along z axis.

(2) The propagation velocity of the electromagnetic wave v_P and the wavelength λ.

Denote the corresponding permeability of the left-handed circular polarization magnetic field of the wave as μ_-. According to the term $\mu_- = \mu_{11} + \mathrm{j}\mu_{12}$ in Equations (7.119a) and (7.112), we can have

$$\mu_- = \mu_0(\mu_{\mathrm{r}} + k_{\mathrm{r}}) \tag{e}$$

The propagation velocity of the electromagnetic wave is thus

$$v_{\mathrm{p}} = \frac{1}{\sqrt{\varepsilon\mu_-}} = \frac{1}{\sqrt{\varepsilon_{\mathrm{r}}\varepsilon_0\mu_0(\mu_{\mathrm{r}} + k_{\mathrm{r}})}} = \frac{c}{\sqrt{\varepsilon_{\mathrm{r}}(\mu_{\mathrm{r}} + k_{\mathrm{r}})}} \tag{f}$$

Substitute $\varepsilon_{\mathrm{r}} = 20$, $\mu_{\mathrm{r}} = 1.2$, $k_{\mathrm{r}} = 0.3$ into the above formula

$$v_{\mathrm{p}} = 5.477 \times 10^7 \quad (\mathrm{m/s})$$

The wavelength λ of the electromagnetic wave in ferrite material is

$$\lambda = \frac{v_{\mathrm{p}}}{f} = 1.826 \quad (\mathrm{cm})$$

(3) The wave impedance of the plane wave in ferrite material

$$\eta = \sqrt{\frac{\mu_-}{\varepsilon}} = \sqrt{\frac{(\mu_{\mathrm{r}} + k_{\mathrm{r}})\mu_0}{\varepsilon_{\mathrm{r}}\varepsilon_0}} = \eta_0\sqrt{\frac{(\mu_{\mathrm{r}} + k_{\mathrm{r}})}{\varepsilon_{\mathrm{r}}}} \tag{g}$$

thus

$$\eta = 120\pi\sqrt{\frac{1.5}{25}} = 103.2 \quad (\Omega)$$

Exercises

7.1 Let the instantaneous electric field intensity $\boldsymbol{E}$ of the electromagnetic wave in free space be

$$\boldsymbol{E} = \boldsymbol{e}_y 37.7\cos(6\pi \times 10^8 t + 2\pi z)\mathrm{V/m}$$

Solve the following problems:
(1) the direction of the wave propagation;
(2) the frequency, wavelength, phase constant and phase velocity;
(3) the instantaneous magnetic field intensity $\boldsymbol{H}$. Is it a uniform plane wave?

7.2 Let the electric field intensity $\boldsymbol{E}$ of the electromagnetic wave in free space be

$$\boldsymbol{E} = \boldsymbol{e}_x E_0 \sin(\omega t - kz) + \boldsymbol{e}_y E_0 \sin(\omega t - kz)$$

Solve the following problems:

(1) the instantaneous magnetic field intensity $\boldsymbol{H}$.

(2) the poynting vector of the wave.

7.3 Let the amplitude of the magnetic field intensity $\boldsymbol{H}$ of a uniform plane wave in the air be $\frac{1}{3\pi}$A/m. Suppose the wave propagates in the direction of $-\boldsymbol{e}_z$ with phase constant $k = 30$rad/m. The direction of $\boldsymbol{H}$ is $(-\boldsymbol{e}_y)$ when $t = 0$ and $z = 0$.

(1) Write the expressions of $\boldsymbol{H}$ and $\boldsymbol{E}$;

(2) Find the frequency and wavelength.

7.4 Let the magnetic field intensity $\boldsymbol{H}$ of a plane wave in vacuum be

$$\boldsymbol{H} = 10^{-6}\left(\frac{3}{2}\boldsymbol{e}_x + \boldsymbol{e}_y + \boldsymbol{e}_z\right)\cos\left[\omega t + \pi\left(x - y - \frac{1}{2}z\right)\right] \text{ A/m}$$

Find:

(1) The propagation direction of the wave;

(2) The wavelength and the frequency;

(3) The electric field intensity $\boldsymbol{E}$;

(4) The average Poynting vector $\boldsymbol{S}_{\text{av}}$.

7.5 Let the electric field intensity $\boldsymbol{E}$ of the plane wave in perfect dielectrics $(\varepsilon_r\varepsilon_0,\ \mu_0)$ be

$$\boldsymbol{E} = \boldsymbol{e}_z 100\text{e}^{\text{j}(2\pi\times10^6 t - 2\pi\times10^{-2}x)}\ \mu\text{V/m}$$

Find:

(1) The magnetic induction $\boldsymbol{B}$;

(2) The relative permittivity ε_r.

7.6 A uniform plane wave $\boldsymbol{E} = \boldsymbol{e}_x 50\text{e}^{\text{j}(10^{10}t - kz)}$ is propagating in the lossless polypropylene $(\mu_r = 1, \varepsilon_r = 2.25)$. Find the following parameters:

(1) The frequency;

(2) The phase constant k;

(3) The instantaneous magnetic field intensity $\boldsymbol{H}(t)$;

(4) The average Poynting vector $\boldsymbol{S}_{\text{av}}$.

7.7 Point out the polarization of the following plane wave.

(1) $\boldsymbol{E} = 3(\boldsymbol{e}_x + \text{j}\boldsymbol{e}_y)\text{e}^{-\text{j}kz}$;

(2) $\boldsymbol{E} = (3\boldsymbol{e}_x + 2\boldsymbol{e}_y)\text{e}^{-\text{j}kz}$;

(3) $\boldsymbol{E} = (3\boldsymbol{e}_x + \boldsymbol{e}_y 4\text{e}^{\text{j}\frac{\pi}{3}})\text{e}^{-\text{j}kz}$;

(4) $\boldsymbol{E} = (-\boldsymbol{e}_x - 2\sqrt{3}\boldsymbol{e}_y + \sqrt{3}\boldsymbol{e}_z)\text{e}^{-\text{j}0.04\pi(\sqrt{3}x - 2y + 3z)}$.

7.8 (1) Prove that an elliptical polarization wave can be divided into two circular polarizations with different rotation directions.

(2) Find the radius of these two circular polarizations respectively.

7.9 A TEM wave is propagating in the direction of z axis. Let the electric field intensity of the wave be

$$\boldsymbol{E} = \boldsymbol{e}_x E_0 \sin(\omega t - kz) + \boldsymbol{e}_y E_0' \cos(\omega t - kz)$$

Find:

(1) the magnetic field intensity $\boldsymbol{H}$;

(2) what kind of polarization the electric field and the magnetic field are, respectively;

(3) the average power density $\boldsymbol{S}_{\text{av}}$.

7.10 The wavelength of the electromagnetic wave be 0.2m. Suppose the wavelength changes to 0.09m when the wave enters into perfect dielectric. Let $\mu_r = 1$. Compute ε_r and the wave velocity in the dielectric.

7.11 Let the wavelength of certain uniform plane wave in the free space is 12cm, and the wavelength changes to 8cm as the wave enters into lossless medium. $|\boldsymbol{E}| = 50$V/m, $|\boldsymbol{H}| = 0.1$A/m. Find out the frequency of the wave and μ_r, ε_r of the lossless medium.

7.12 The far field of the current element located vertically at the origin of spherical coordinate are

$$\boldsymbol{E} = \boldsymbol{e}_\theta \frac{100}{r} \sin\theta \cos(\omega t - kr) \text{ V/m}$$

$$\boldsymbol{H} = \boldsymbol{e}_\varphi \frac{0.265}{r} \sin\theta \cos(\omega t - kr) \text{ A/m}$$

Calculate the average power passing through the hemisphere shell where $r = 1000$m.

7.13 Let the electric field of plane wave in the free space be $\boldsymbol{E} = \boldsymbol{e}_x 150\sin(\omega t - kz)$ V/m. Find the instantaneous and average total power passing through a rectangle surface of 30mm×15mm at $z = 0$.

7.14 Suppose the right-handed circular polarization wave propagating along z axis, consists of two linear polarization waves. For one of the two linear polarization waves, the electric field is along the x axis with the amplitude $\boldsymbol{E}_0$ (V/m) at $z = 0$ and angular frequency ω. Give the expressions of $\boldsymbol{E}$ and $\boldsymbol{H}$ of the circular polarization and prove that the average energy flux density vector with respect to time is the sum of average energy flux density vectors of the two linear polarization waves with respect to time.

7.15 Show that the attenuation of field intensity amplitude is approximately 55dB per wavelength as the electromagnetic propagates in conducting medium.

7.16 Generally, in order to realize effective electromagnetic shielding, the thickness of shielding layer is set to be one wavelength. Find

(1) the thickness of Aluminum ($\sigma = 10^7$S/m, $\varepsilon_r = 1$, $\mu_r = 10^4$) shielding in intermediate frequency transformer of radio (The frequency is 465kHz).

(2) the thickness of iron ($\sigma = 10^7$S/m, $\varepsilon_r = 1$, $\mu_r = 10^4$) shielding in power transformer (The frequency is 50Hz).

7.17 Silver plating ($\sigma = 6.1 \times 10^7$S/m, $\mu_0 = 4\pi \times 10^{-7}$H/m) is used to improve the conductivity of glass material. The operation frequency is 2.4GHz. Find the thickness of silver plating when the thickness of silver layer is one wavelength.

7.18 Let the radius of round copper wire (ε_0, μ_0, $\sigma = 5.8 \times 10^7$S/m) be $a = 1.5$mm, as shown in Figure of Exercise 7.18. For this wire, find

(1) DC resistance R_D per unit length;

(2) Surface resistance R_s at $f = 100$MHz;

(3) AC resistance R_A per unit length at $f = 100$MHz.

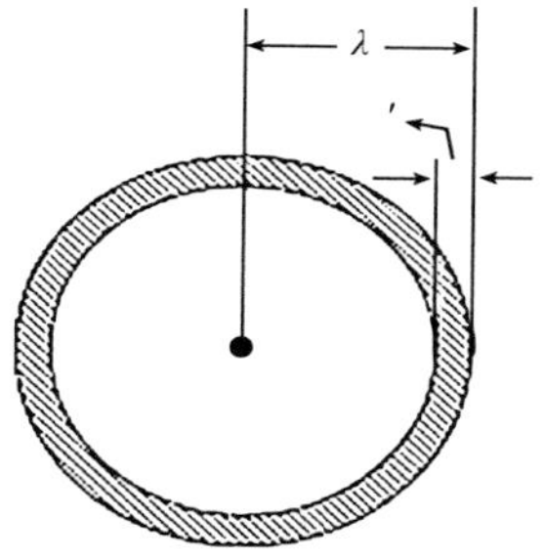

Figure of Exercise 7.18

7.19 The wavelength of the uniform linear polarization plane wave in the air is 60m. $\boldsymbol{E} = \boldsymbol{e}_x \cos\omega t$V/m at the place where it is 1 meters below sea level as the wave enters into the sea along the z axis and propagates down vertically. Find the instantaneous $\boldsymbol{E}$ and $\boldsymbol{H}$, the phase velocity and the wavelength at any point below the sea level. For the sea water, $\sigma = 4$S/m and $\varepsilon_r = 80$, $\mu_r = 1$.

7.20 The conductivity of sea water is $\sigma = 4$S/m and $\varepsilon_r = 8$. Find the attenuation constant, wavelength and wave impedance of the electromagnetic wave in the sea with frequencies of 10kHz, 1MHz, 10MHz and 1GHz.

7.21 A uniform plane wave with frequency 3GHz, its electric field is polarized along the y axis, propagates in a non-magnetic medium with $\varepsilon_r = 2.5$ and loss tangent of 10^{-2}, in the direction of $+x$

(1) What is the propagation distance when half of the amplitude of the wave is attenuated?

(2) Find the intrinsic impedance of the medium, the wavelength and wave velocity.

(3) If $\boldsymbol{E} = \boldsymbol{e}_y 50\sin(6\pi \times 10^9 t + \pi/3)$ at $x = 0$, gives the expression of $\boldsymbol{H}(x, t)$.

7.22 In the free space ($z < 0$), a uniform plane wave propagates along the direction of $+z$ and enters into a conductor with $\sigma = 6.17 \times 10^7$S/m and $\mu_r = 1$ at $z = 0$. The frequency and amplitude of the wave are $f = 1.5$MHz and 1V/m respectively, and $\boldsymbol{E}(0, t) = \boldsymbol{e}_y \sin(2\pi f t)$ at the interface ($z = 0$). Find $\boldsymbol{H}(z, t)$.

Chapter 8

Reflection and Refraction of Electromagnetic Waves

The propagation of electromagnetic waves in an infinite medium is analyzed in previous chapter. Now we are to study the reflection and refraction of waves caused by discontinuity of media.

All the medium discussed in this chapter is still homogeneous, linear and isotropic. Assuming that the permeability μ approximately equals μ_0, and the interfaces between media are planar surfaces.

Owing to the medium discontinuity, the alternating fields usually induce a layer of alternating charges on the interface between the media. These induced time-varying charges can be considered as a new source for fields, and can generate new electromagnetic waves at the two sides of the interface. As a result, there are not only incident wave but also reflected wave in medium 1, and the wave propagating through the interface into medium 2, is refracted wave (or transmitted wave). These three types of waves coexist on the interface between the media.

Investigation on the law of wave reflection and refraction is still based on the boundary conditions on interface.

As we all know, the tangential electric field on the surface of perfect conductors is zero. In other words, on the surface of perfect conductors, when the tangential component of E-fields of the incident wave is totally reflected, the reflected wave is anti-phase, hence satisfying the boundary condition $E_t = 0$. However, on the interface between two arbitrary media, when the electric field parallel to the interface is being reflected, what happened to the magnitude and phase? Will they be in phase, out-of-phase or having phase difference? It depends on the parameters of the dissimilar media, and the situation will be different. The law of wave reflection and refraction propagating through the surface boundary will be given by the law of reflection, law of refraction and Fresnel equations.

In this book, for the sake of clearness and convenience, some assumptions are given below, for the reflection of the tangential electric field of the incident wave, its phase keeps unchanged at the interface. Moreover, the obtained formulas in terms of reflection coefficient are consistent with those widely used in microwave technologies.

8.1 Plane Wave Normally Incident on the Surface of Perfect Conductor

Region 1 is filled with an ideal dielectric (μ_0, ε), while region 2 is filled with perfect conductor, as shown in Figure 8.1. The interface between ideal dielectric and perfect conductor is located on plane xOy, actually the cross section of the interface between media is depicted in the figure.

When the plane wave is normally incident on the surface of perfect conductor from region 1, total wave reflection will occur. As a result, there is not only the incident wave, but also reflected wave in region 1, both are travelling waves. The resulting total wave is another kind of wave — standing wave.

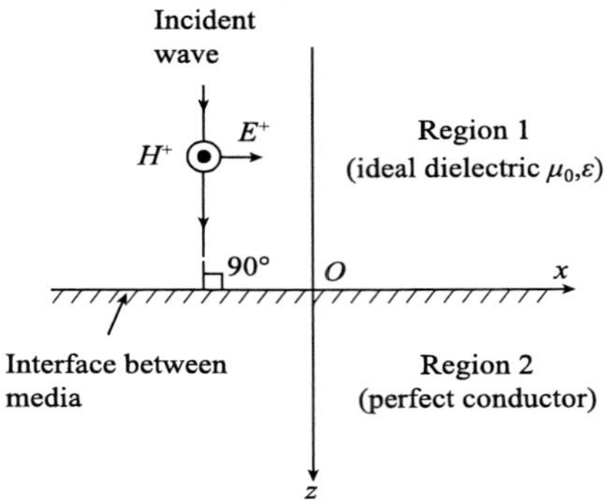

Figure 8.1 Plane wave normally incident upon the surface of perfect conductor

The boundary conditions describe the relationship among the fields and the relationship between the fields and their sources (the sources on the interface between media), on both sides (infinitely close to the interface) of the medium interface. Therefore, *the boundary conditions are the basis for further studying the rules of wave reflection and refraction.*

8.1.1 The Expressions of Electric Field and Magnetic Field for Incident Wave and Reflected Wave

The electric field of normally incident plane wave can be expressed as

$$\boldsymbol{E}^+ = \boldsymbol{e}_x \boldsymbol{E}_x^+ = \boldsymbol{e}_x \boldsymbol{E}_0^+ \mathrm{e}^{\mathrm{j}(\omega t - kz)} \tag{8.1}$$

which is propagating along the positive z direction. The magnitude of the magnetic field of this wave equals to the electric field divided by the wave impedance η. The direction of the magnetic field can be determined by $\boldsymbol{E}^+ \times \boldsymbol{H}^+ = \boldsymbol{S}^+$. As the electric field of incident wave is E_x, the magnetic field is H_y. Then we have

$$\boldsymbol{H}^+ = \boldsymbol{e}_y H_y^+ = \boldsymbol{e}_y \frac{E_0^+}{\eta} \mathrm{e}^{\mathrm{j}(\omega t - kz)} \tag{8.2}$$

where $\boldsymbol{E}^+, \boldsymbol{H}^+$represent the electric and magnetic fields of the forward wave (namely, the incident wave). And the wave propagating in negative z direction, known as the reflected wave, is designated by $\boldsymbol{E}^-, \boldsymbol{H}^-$. *Assuming* that $\boldsymbol{E}^-$still remains in positive x direction, hence the expression for electric field of reflected wave will be

$$\boldsymbol{E}^- = \boldsymbol{e}_x E_0^- \mathrm{e}^{\mathrm{j}(\omega t + kz)} \tag{8.3}$$

The magnetic field of the reflected wave $\boldsymbol{H}^-$ is

$$\boldsymbol{H}^- = \boldsymbol{e}_y H_0^- \mathrm{e}^{\mathrm{j}(\omega t + kz)} = -\boldsymbol{e}_y (E_0^- / \eta) \mathrm{e}^{\mathrm{j}(\omega t + kz)} \tag{8.4}$$

where a minus sign is introduced because it is assumed that the tangential electric field E_x has the same phase when reflected, therefore, the phase of reflected magnetic field has to be reversed to guarantee that the Poynting vector $\boldsymbol{S}^-$ of the reflected wave is oriented along the negative z direction.

8.1.2 The Expressions of Electric Field and Magnetic Field for Total Waves

1. The Expression of Electric Field for Total Waves

In region 1, considering the total electric fields of the normally incident wave and the reflected wave [i.e. Equations (8.1) and (8.3)], we have

$$\boldsymbol{E} = \boldsymbol{E}^+ + \boldsymbol{E}^- = \boldsymbol{e}_x(E_0^+ \mathrm{e}^{-\mathrm{j}kz} + E_0^- \mathrm{e}^{\mathrm{j}kz})\mathrm{e}^{\mathrm{j}\omega t} \tag{8.5}$$

The electric field strength $\boldsymbol{E}$ is in x direction, parallel to the interface between media. According to the boundary conditions of perfect conductor, the tangential electric field is equal to zero, that is $E = 0$ where $z = 0$. Then Equation (8.5) becomes

$$\boldsymbol{E} = \boldsymbol{e}_x(E_0^+ + E_0^-)\mathrm{e}^{\mathrm{j}\omega t} = 0 \tag{8.6}$$

thus

$$E_0^+ = -E_0^- \tag{8.7}$$

which indicates that when tangential electric field is incident on the surface of perfect conductor, total reflection will occur, namely, $|E_0^+| = |E_0^-|$, besides, the phase of the reflected wave should be reversed by 180° to meet the boundary conditions, saying, the tangential electric field should be zero on the surface of perfect conductor. This concept is implied by the minus sign in Equation (8.7).

Substituting Equation (8.7) into Equation (8.5), we have

$$\boldsymbol{E} = \boldsymbol{e}_x E_0^+(\mathrm{e}^{-\mathrm{j}kz} - \mathrm{e}^{\mathrm{j}kz}) \cdot \mathrm{e}^{\mathrm{j}\omega t} = -\boldsymbol{e}_x \mathrm{j}2E_0^+ \sin(kz) \cdot \mathrm{e}^{\mathrm{j}\omega t} \tag{8.8}$$

or

$$E_x = -\mathrm{j}2E_0^+ \sin(kz) \cdot \mathrm{e}^{\mathrm{j}\omega t} \tag{8.9}$$

which is the expression of electric field for the total wave.

2. The Expression of Magnetic Field for the Total Waves

The magnetic fields of incident wave and reflected wave [Equations (8.2) and (8.4)] are summed, and applying the Formula (8.7), we have

$$\boldsymbol{H} = \boldsymbol{H}^+ + \boldsymbol{H}^- = \boldsymbol{e}_y\left(\frac{E_0^+}{\eta}\mathrm{e}^{-\mathrm{j}kz} + \frac{E_0^+}{\eta}\mathrm{e}^{\mathrm{j}kz}\right)\mathrm{e}^{\mathrm{j}\omega t} = \boldsymbol{e}_y 2\frac{E_0^+}{\eta}\cos(kz) \cdot \mathrm{e}^{\mathrm{j}\omega t} \tag{8.10}$$

or

$$H_y = 2\frac{E_0^+}{\eta}\cos(kz)\mathrm{e}^{\mathrm{j}\omega t} = 2H_0^+ \cos(kz)\mathrm{e}^{\mathrm{j}\omega t} \tag{8.11}$$

which is the expression of magnetic field for the total waves.

8.1.3 The Characteristics of Electric Fields and Magnetic Fields for Total Waves

According to the Expressions (8.9) and (8.11) of electric field and magnetic field for total waves, it can be concluded that the normal incidence of plane wave on the surface of perfect conductor results in total wave reflection. The total wave formed by the superposition of two travelling waves, incident wave and the reflected wave, has the following characteristics:

(1) The superposition of electric and magnetic fields are all standing waves, but the distributions of these two standing waves along the z direction are not the same. The distribution of standing wave for electric field along z direction is sin (kz), while that for magnetic field is cos (kz). This distribution can be directly determined based on the boundary conditions of perfect conductor.

(2) *The electric field and magnetic field for the total waves have different spatial distribution (along the z coordinate axis), besides,* E_x *and* H_y *also have a time phase difference of 90°.* Therefore, the average Poynting vector $\boldsymbol{S}_{\text{av}}$along the z direction defined by E_x and H_y must be zero. Applying Equations (8.9) and (8.11), we have

$$\begin{aligned}\boldsymbol{S}_{\text{av}} &= \frac{1}{2}\text{Re}(\boldsymbol{E}\times\boldsymbol{H}^*) = \boldsymbol{e}_z\frac{1}{2}\text{Re}(E_xH_y^*)\\ &= \boldsymbol{e}_z\frac{1}{2}\text{Re}\left[(2E_0^+)(2H_0^+)\sin(kz)\cos(kz)\text{e}^{\text{j}90^\circ}\right] = 0\end{aligned} \tag{8.12}$$

Because the power densities transmitted in the positive and negative z directions are equal, therefore, the average power density is zero. In addition, Equation (8.12) also indicates that, the simple standing wave represented by Equations (8.9) and (8.11) transports no energy, just having exchange of electric energy and magnetic energy.

(3) Due to the time phase difference of 90° between E_x and H_y, when the total electric field of the total waves is maximum, the total magnetic field is equal to zero, and vice versa.

Figure 8.2 shows the standing wave distribution of the electric and magnetic fields for the total waves, which clearly indicate that, when the electric field of standing wave reaches the maximum, the magnetic field of standing wave is zero.

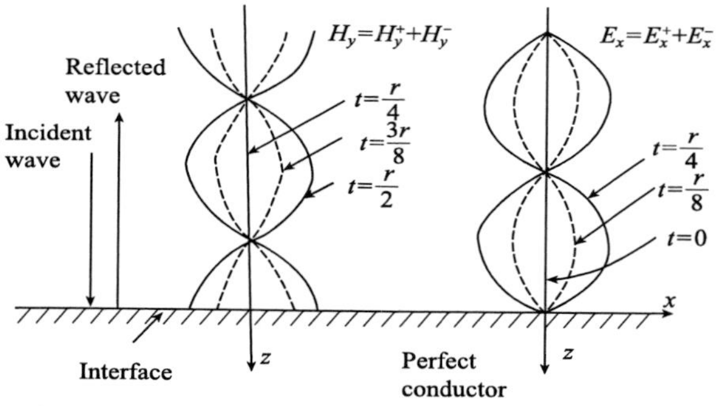

Figure 8.2 The standing wave distribution of the electric and magnetic fields of the total waves

At the positions of interface where the vertical distances from the interface are integer times of the half-wavelength, the total electric field of the total waves is identically zero, and these positions are called wave nodes for electric standing waves. At the positions where the vertical distances from the interface are odd number times of 1/4 wavelength, the amplitude is maximum (twice the amplitude of incident wave), and these positions are called wave peak points of electric field. While the standing wave of magnetic field is different, its wave peak point appears at the positions on the boundary and where the distances from the interface are integer times of the half-wavelength. Its wave node appears at the positions where the distance from interface is odd number times of 1/4 wavelength.

8.2 Plane Wave Normally Incident on the Interface between Perfect Dielectrics

As a plane wave is normally incident on the interface from dielectric 1 (μ_0, ε_1), causing partial wave reflection and partial wave refraction, as shown in Figure 8.3.

Now let's study the laws regarding reflection and refraction.

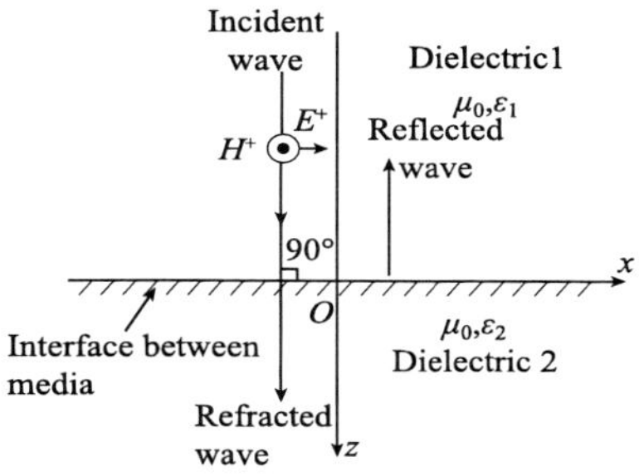

Figure 8.3 Plane wave normally incident upon the interface between perfect dielectrics

8.2.1 The Expressions of Electromagnetic Fields of Incident Wave, Reflected Wave and Refracted Wave

Supposing the expression of electric field for the incident wave as

$$\boldsymbol{E}^+ = \boldsymbol{e}_x E_x^+ = \boldsymbol{e}_x E_0^+ \mathrm{e}^{\mathrm{j}(\omega t - k_1 z)} \tag{8.13}$$

then the magnetic field of the incident wave can be written as

$$\boldsymbol{H}^+ = \boldsymbol{e}_y H_y^+ = \boldsymbol{e}_y H_0^+ \mathrm{e}^{\mathrm{j}(\omega t - k_1 z)} = \boldsymbol{e}_y \frac{E_0^+}{\eta_1} \mathrm{e}^{\mathrm{j}(\omega t - k_1 z)} \tag{8.14}$$

where $\eta_1 = \sqrt{\mu_0/\varepsilon_1}, k_1 = \omega\sqrt{\mu_0\varepsilon_1}$.

1. Expressions of Electric Field and Magnetic Field for the Reflected Wave

The electric field of the reflected wave can be written as

$$\boldsymbol{E}^- = \boldsymbol{e}_x E_x^- = \boldsymbol{e}_x E_0^- \mathrm{e}^{\mathrm{j}(\omega t + k_1 z)} \tag{8.15}$$

thus, the magnetic field of the reflected wave can be written as

$$\boldsymbol{H}^- = \boldsymbol{e}_y H_0^- \mathrm{e}^{\mathrm{j}(\omega t + k_1 z)} = -\boldsymbol{e}_y \frac{E_0^-}{\eta_1} \mathrm{e}^{\mathrm{j}(\omega t + k_1 z)} = -\boldsymbol{e}_y \frac{1}{\eta_1} E_x^- \tag{8.16}$$

2. Expressions of Electric Field and Magnetic Field for the Refracted Wave

The refracted wave is referred to as the wave entering into the dielectric 2 (μ_0, ε_2), and denoted as $\boldsymbol{E}^{\mathrm{T}}, \boldsymbol{H}^{\mathrm{T}}$.

The electric field of the refracted wave can be written as

$$\boldsymbol{E}^{\mathrm{T}} = \boldsymbol{e}_x E_x^{\mathrm{T}} = \boldsymbol{e}_x E_0^{\mathrm{T}} \mathrm{e}^{\mathrm{j}(\omega t - k_2 z)} \tag{8.17}$$

The magnetic field of the refracted wave can be written as

$$\boldsymbol{H}^{\mathrm{T}} = \boldsymbol{e}_y H_y^{\mathrm{T}} = \boldsymbol{e}_y H_y^{\mathrm{T}} \mathrm{e}^{\mathrm{j}(\omega t - k_2 z)} = e_y \frac{1}{\eta_2} E_x^{\mathrm{T}} \tag{8.18}$$

where $\eta_2 = \sqrt{\mu_0/\varepsilon_2}, k_2 = \omega\sqrt{\mu_0\varepsilon_2}$.

8.2.2 The Reflection Coefficient R and Transmission Coefficient T for Electric Fields

The reflection coefficient R of electric field is defined as the ratio of the tangential electric field strength for reflected wave and the tangential electric field strength for incident wave on the interface $(z = 0)$, that is

$$R = \frac{E_0^-}{E_0^+} \tag{8.19}$$

The transmission coefficient T of electric field is defined as the ratio of the tangential electric field strength of refracted wave and the tangential electric field strength of incident wave on the interface ($z = 0$), that is

$$T = \frac{E_0^T}{E_0^+} \tag{8.20}$$

Now let's find R, T of the electric fields in case of normal incidence.

Method 1 To obtain R, T by directly using continuity conditions of tangential electric field and tangential magnetic field on the interface.

(1) Reflection coefficient R of electric fields in case of normal incidence.

From the boundary conditions for tangential electric fields, say $E_{1t} = E_{2t}$, we have

$$E_{1t} = E_0^+ + E_0^- = E_{2t} \tag{8.21}$$

From the boundary conditions for tangential magnetic fields, say $H_{1t} = H_{2t}$, we arrive at the following equation:

$$\begin{aligned} H_{1t} = H_0^+ + H_0^- &= \frac{E_0^+}{\eta_1} - \frac{E_0^-}{\eta_1} \\ = H_{2t} = H_0^T &= \frac{E_0^T}{\eta_2} = \frac{E_{2t}}{\eta_2} \end{aligned}$$

thus

$$E_{2t} = E_0^{\mathrm{T}} = \frac{\eta_2}{\eta_1}(E_0^+ - E_0^-) \tag{8.22}$$

From Equations (8.21) and (8.22), we have

$$E_0^+ + E_0^- = \frac{\eta_2}{\eta_1}(E_0^+ - E_0^-)$$

Therefore, reflection coefficient of normally incident electric fields can be acquired by solving the equation above as

$$R = \frac{E_0^-}{E_0^+} = \frac{\eta_2 - \eta_1}{\eta_2 + \eta_1} \tag{8.23}$$

(2) Transmission coefficient T of electric fields in case of normal incidence.

With the definition of T and the boundary condition $E_{1t} = E_{2t}$, we have

$$T = \frac{E_0^T}{E_0^+} = \frac{E_{2t}}{E_0^+} = \frac{E_{1t}}{E_0^+} = \frac{E_0^+ + E_0^-}{E_0^+} = 1 + R$$

For normal incidence, we have

$$T = 1 + R \tag{8.24}$$

Substituting Equation (8.23) into the preceding equation, the expression of the transmission coefficient of electric field will be

$$T = \frac{2\eta_2}{\eta_2 + \eta_1} \tag{8.25}$$

Method 2 *The tangential electric fields and magnetic fields are continuous on the interface between dielectrics, which means that, the wave impedance Z_z observed along the direction perpendicular to the interface must be continuous. Hence R, T can be obtained by applying impedance continuity.*

At the interface ($z = 0$) we have

$$\left.\begin{aligned} E_{1t} &= E_0^+(1+R) \\ H_{1t} &= \frac{E_0^+}{\eta_1}(1-R) \end{aligned}\right\} \tag{8.26}$$

$$Z_{z1} = \frac{E_{1t}}{H_{1t}} = \eta_1 \frac{1+R}{1-R}$$

and

$$\left.\begin{aligned} E_{2t} &= E_0^+ T \\ H_{2t} &= \frac{E_0^+ T}{\eta_2} \end{aligned}\right\} \tag{8.27}$$

$$Z_{z2} = \frac{E_{2t}}{H_{2t}} = \eta_2$$

Using Z_{z1} and Z_{z2} in above equations, and due to $Z_{z1} = Z_{z2}$, we can obtain the reflection coefficient as

$$R = \frac{\eta_2 - \eta_1}{\eta_2 + \eta_1}$$

Considering the expressions above and Equations (8.26) and (8.27), as well as $E_{1t} = E_{2t}$, the transmission coefficient is given as

$$T = \frac{2\eta_2}{\eta_1 + \eta_2}$$

(3) When the wave is normally incident, the relationship between the power densities of incident wave, reflected wave, refracted wave (S_{av}^+, S_{av}^-, S_{av}^T) and R, T, η_1, η_2.

It can be obtained directly from Expressions (8.23), (8.25) that

$$1 - R^2 = \left(\frac{\eta_1}{\eta_2}\right) T^2 \tag{8.28}$$

which describes the relationship among power densities of incident wave, reflected wave, refracted wave, say S_{av}^+, S_{av}^-, S_{av}^T, on the interface between the dielectrics, *it is actually the relationship of energies*. If the incident power density is S_{av}^+ when it is normally incident, multiplying S_{av}^+ with above equation, we can obtain the following relationship based on law of energy conservation:

$$\left\{\begin{aligned} S_{\text{av}}^- &= R^2 S_{\text{av}}^+ \\ S_{\text{av}}^T &= \left(\frac{\eta_1}{\eta_2}\right) T^2 \cdot S_{\text{av}}^+ \end{aligned}\right. \tag{8.29}$$

Further discussion of this equation will be given in Section 8.2.5.

8.2.3 The Reflection Coefficient R_H and Transmission Coefficient T_H for Magnetic Fields

The reflection coefficient R and transmission coefficient T of electric field are known now, the reflection coefficient R_H and transmission coefficient T_H of magnetic field can be obtained easily.

(1) Reflection coefficient R_H of magnetic field.

$$R_H = \frac{H_0^-}{H_0^+} = \frac{-E_0^-/\eta_1}{E_0^+/\eta_1} = -\frac{E_0^-}{E_0^+} = -R \tag{8.30}$$

which is derived from Equation (8.23).

(2) Transmission coefficient T_H of magnetic field

$$T_H = \frac{H_0^{\mathrm{T}}}{H_0^+} = \frac{E_0^{\mathrm{T}}/\eta_2}{E_0^+/\eta_1} = \frac{\eta_1}{\eta_2}\frac{E_0^{\mathrm{T}}}{E_0^+} = \frac{\eta_1}{\eta_2}T \tag{8.31}$$

Thus

$$R_H = -R, \quad T_H = \frac{\eta_1}{\eta_2}T \tag{8.32}$$

From Equation (8.32), it is known that *the magnitudes of reflection coefficients of electric field and magnetic field are equal*, that is

$$|R| = |R_H| \tag{8.33}$$

which is because that the incident wave and reflected wave are in the same dielectric.

From Equation (8.32), it is known that *the transmission coefficients for electric field and magnetic field are not equal.*

Here comes up with a question, why are the transmission coefficients of E-fields and B-fields not equal in magnitude? Obviously, *though the magnitudes of the reflection coefficients of E-fields and the B-fields* [see Equations (8.23) and (8.30)] *are the same, their phases are reverse.* As a result, for the superposition of the fields on the side of medium 1 at the boundary, if the incident wave and reflected wave of the electric field can be summed due to their same phase, then the total magnetic field has to be reduced due to counteracting, and vice versa. In accordance with the boundary conditions, the composite electric field and magnetic field should be continuous with those in medium 2. Thus, *for the electric field and magnetic field in medium 2, if one of them is the sum of the incident wave and reflected wave, the other one has to be one wave subtracted by another wave.* As a result, transmission coefficients of electric field and magnetic field can not be equal. These unequal transmission coefficients just indicate that in the second medium, the ratio of electric field and magnetic field is changed, which is not η_1 any more, but becomes the wave impedance of the second medium

$$\frac{E_0^T}{H_0^T} = \frac{E_0^+ T}{H_0^+ T_H} = \eta_2 \tag{8.34}$$

which is derived by using Equation (8.31).

8.2.4 For Normally Incident Waves, the Reflection Coefficient and Transmission Coefficient for General Dielectrics in Terms of ε

The permeability of the general dielectrics can be assumed to be μ_0, then the reflection coefficient R [Equation (8.23)] and transmission coefficient T [Equations (8.25)] of E-fields are

$$R = \frac{\sqrt{\varepsilon_1} - \sqrt{\varepsilon_2}}{\sqrt{\varepsilon_1} + \sqrt{\varepsilon_2}} \tag{8.35}$$

$$T = \frac{2\sqrt{\varepsilon_1}}{\sqrt{\varepsilon_1} + \sqrt{\varepsilon_2}} \tag{8.36}$$

The reflection coefficient and transmission coefficient of B-fields can then be written as

$$R_H = \frac{H_0^-}{H_0^+} = \frac{\sqrt{\varepsilon_2} - \sqrt{\varepsilon_1}}{\sqrt{\varepsilon_2} + \sqrt{\varepsilon_1}} \tag{8.37}$$

$$T_H = \frac{H_0^T}{H_0^+} = \frac{2\sqrt{\varepsilon_2}}{\sqrt{\varepsilon_1} + \sqrt{\varepsilon_2}} \tag{8.38}$$

From Expression (8.35) it can be seen that, when $\varepsilon_1 > \varepsilon_2 (\eta_2 > \eta_1)$, *the reflection coefficient R of electric field is positive. And when $\varepsilon_1 < \varepsilon_2 (\eta_2 < \eta_1)$, the reflection coefficient R of electric field is negative.* From Expression (8.36) it can be seen that, *the transmission coefficient T of electric field is always positive.*

8.2.5 Proof of Validity of $S_{\text{av}}^T = \left(\frac{\eta_1}{\eta_2}\right) \cdot T^2 \cdot S_{\text{av}}^+$ [i.e. Equation (8.29)] from Multiple Perspective

First let us take a look at the relationship between S_{av}^+, S_{av}^- and R, because

$$S_{\text{av}}^+ = \frac{1}{2} E_0^+ H_0^+$$

then

$$\begin{aligned} S_{\text{av}}^- &= \frac{1}{2} E_0^- H_0^- = \frac{1}{2} E_0^+ R \cdot (E_0^- / \eta_1) \\ &= \frac{1}{2} E_0^+ R (E_0^+ / \eta_1) = \frac{1}{2} R^2 E_0^+ H_0^+ = R^2 \cdot S_{\text{av}}^+ \end{aligned}$$

that is

$$S_{\text{av}}^- = R^2 \cdot S_{\text{av}}^+ \tag{8.39}$$

How to find S_{av}^T? it is usually to take it for granted that $S_{\text{av}}^T = T^2 S_{\text{av}}^+$, in fact it is wrong. Why? Because it was wrongly considered that the transmission coefficients of the electric field and magnetic field (T and T_H) were equal. Actually, T and T_H are not equal. The right relationship between S_{av}^+, S_{av}^T and the transmission coefficients should be

$$S_{\text{av}}^T = (T \cdot T_H)\, S_{\text{av}}^+$$

The correct method to find S_{av}^T is as follows.

Method 1 Obtain S_{av}^T according to the energy conservation law.

When it is normal incidence, according to the energy conservation law, on the interface of the media

$$S_{\text{av}}^+ - S_{\text{av}}^- = S_{\text{av}}^T \tag{8.40}$$

and because

$$S_{\text{av}}^- = R^2 S_{\text{av}}^+$$

so

$$(1 - R^2) S_{\text{av}}^+ = S_{\text{av}}^T \tag{8.41}$$

Substituting Expression (8.23) into the equation above and considering Equation (8.25) we have

$$S_{\text{av}}^T = \left(\frac{\eta_1}{\eta_2}\right) \cdot T^2 \cdot S_{\text{av}}^+ \tag{8.42}$$

Method 2 By using the relationship among E_0^+, H_0^+, E_0^T and H_0^T, prove $\boldsymbol{S}_{\text{av}}^T = \left(\frac{\eta_1}{\eta_2}\right) \cdot T^2 \cdot \boldsymbol{S}_{\text{av}}^+$ for case of normal incidence.

$$\begin{aligned} S_{\text{av}}^T &= \frac{1}{2} E_0^T H_0^T = \frac{1}{2}(E_0^+ T)(E_0^T/\eta_2) \\ &= \frac{1}{2}(E_0^+ T)(E_0^+ T/\eta_2) = \frac{1}{2} E_0^+ H_0^+ (\eta_1/\eta_2) T^2 \\ &= \left(\frac{\eta_1}{\eta_2}\right) T^2 \cdot S_{\text{av}}^+ \end{aligned} \tag{8.43}$$

Method 3 In case of normal incidence, prove the following equation by using transmission coefficient of electric field T and transmission coefficient of magnetic field T_H:

$$S_{\text{av}}^T = \left(\frac{\eta_1}{\eta_2}\right) T^2 \cdot S_{\text{av}}^+ \tag{8.44}$$

$$S_{\text{av}}^T = \frac{1}{2} E_0^T H_0^T = \frac{1}{2}(E_0^+ T) \cdot (H_0^+ T_H)$$

Substituting $T_H = \left(\frac{\eta_1}{\eta_2}\right) T$ in Equation (8.32) into the equation above, then we have

$$S_{\text{av}}^T = \frac{1}{2} E_0^+ T \left(H_0^+ \frac{\eta_1}{\eta_2} T\right) = \frac{1}{2} E_0^+ H_0^+ \cdot \frac{\eta_1}{\eta_2} \cdot T^2 = \left(\frac{\eta_1}{\eta_2}\right) T^2 \cdot S_{\text{av}}^+ \tag{8.45}$$

8.3 Plane Waves Obliquely Incident upon the Surface of Perfect Conductor

When the issue of oblique incidence is discussed, the concepts of 'parallel' polarized waves and 'perpendicularly' polarized waves will be involved. It is defined in the book as follows: the wave whose electric field is perpendicular to *the incidence plane* (*the plane containing the incidence ray of the wave and the normal of the interface between media is called plane of incidence*) is called the perpendicularly polarized wave; if the electric field is parallel to *the incidence plane*, it is called parallel polarized wave. In addition, there is a provision in engineering: the wave with electric field parallel to *the medium interface* is called the 'horizontal' polarized wave, while the wave with electric field perpendicular to *the medium interface* is known as the 'vertical' polarized wave. This book does not adopt this provision.

8.3.1 The Wave Vector k

Vector form of phase constant $\boldsymbol{k}$ *is also known as wave vector.* The representation, magnitude and direction of $\boldsymbol{k}$ will be analyzed here.

A wave propagates in an arbitrary direction ξ (unit vector is $\boldsymbol{e}_\xi$). It is assumed that the angles between ξ and x, y, z axis are α, β, γ, respectively, so that the arbitrary unit vector $\boldsymbol{e}_\xi$ can be expressed as

$$\boldsymbol{e}_\xi = \boldsymbol{e}_x \cos\alpha + \boldsymbol{e}_y \cos\beta + \boldsymbol{e}_z \cos\gamma$$

so that

$$\cos^2\alpha + \cos^2\beta + \cos^2\gamma = 1 \tag{8.46}$$

The wave vector $\boldsymbol{k}$ can be expressed as

$$\boldsymbol{k} = \boldsymbol{e}_x k_x + \boldsymbol{e}_y k_y + \boldsymbol{e}_z k_z = \boldsymbol{e}_x k \cos\alpha + \boldsymbol{e}_y k \cos\beta + \boldsymbol{e}_z k \cos\gamma \tag{8.47}$$

$$k^2 = k_x^2 + k_y^2 + k_z^2 \tag{8.48}$$

where k_x, k_y and k_z are the phase constant along the x, y and z direction, respectively. $\boldsymbol{k}$ has the maximum value of phase constant, so the direction of $\boldsymbol{k}$ must be perpendicular to the equal-phase plane of the wave. Therefore, the direction of wave vector $\boldsymbol{k}$ coincides with the direction of wave propagation. The wave's spatial phase variation along the direction ξ is

$$\begin{aligned}\boldsymbol{k} \cdot \boldsymbol{\xi} &= (\boldsymbol{e}_x k_x + \boldsymbol{e}_y k_y + \boldsymbol{e}_z k_z) \cdot (\boldsymbol{e}_x x + \boldsymbol{e}_y y + \boldsymbol{e}_z z) \\ &= k_x x + k_y y + k_z z\end{aligned} \tag{8.49}$$

The unit vector of wave vector $\boldsymbol{k}$ is

$$\boldsymbol{e}_k = \frac{\boldsymbol{k}}{|\boldsymbol{k}|} = \frac{\boldsymbol{k}}{\sqrt{k_x^2 + k_y^2 + k_z^2}} \tag{8.50}$$

The magnitude of $\boldsymbol{e}_k$ is 1, and the direction is the same as the propagation direction of TEM wave, say, the direction of $\boldsymbol{E} \times \boldsymbol{H}$, that is to say, the directions of $\boldsymbol{e}_k$, $\boldsymbol{E}$, $\boldsymbol{H}$ are orthogonal to each other. As the wave impedance of TEM is $\eta = \dfrac{E}{H}$, therefore, the following equation can be easily obtained:

$$\left.\begin{aligned}\boldsymbol{E} &= \eta \boldsymbol{H} \times \boldsymbol{e}_k \\ \boldsymbol{H} &= \left(\frac{1}{\eta}\right) \boldsymbol{e}_k \times \boldsymbol{E}\end{aligned}\right\} \tag{8.51}$$

8.3.2 The Electromagnetic Fields of Incident Wave, Reflected Wave and Total Wave

A perpendicularly polarized wave obliquely incident onto the surface of perfect conductor is depicted in Figure 8.4. The incident wave propagates along the positive direction of ξ, and the reflected wave propagates along the negative direction of ξ'.

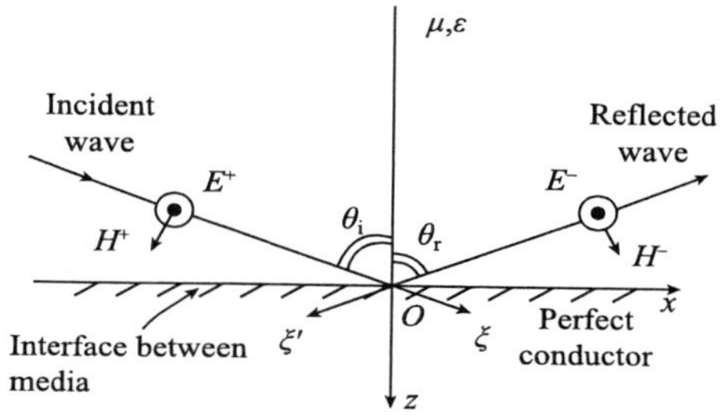

Figure 8.4 Plane wave obliquely incident upon the surface of perfect conductor (Perpendicularly polarized wave)

θ_{i}, θ_{r} are the incidence angle and reflection angle, respectively. Since the electromagnetic wave is incident on a perfect conductor surface, and *both the electric and magnetic fields inside the perfect conductor are zero* (see Section 6.8.1), there is only reflection wave produced but without refraction wave.

Same procedure with that used for normal incident wave will be used for study in this case. First, write the expressions of incident wave and reflected wave, and then determine the relationship between θ_i and θ_r, as well as the amplitude and phase of the reflected wave by using boundary conditions. Finally, the total wave is obtained by combining the incident and reflected waves.

1. The Expressions of Electric Fields of Incident Wave and Reflected Wave

As shown in Figure 8.4, the expression for the electric field of incident wave $\boldsymbol{E}^+$is

$$\boldsymbol{E}^+ = \boldsymbol{e}_y E_y^+ = \boldsymbol{e}_y E_0^+ e^{j(\omega t - k\xi)} \tag{8.52}$$

Now let's find $k\xi$.

Method 1 to find $k\xi$ based on the geometric relations in Figure 8.5.

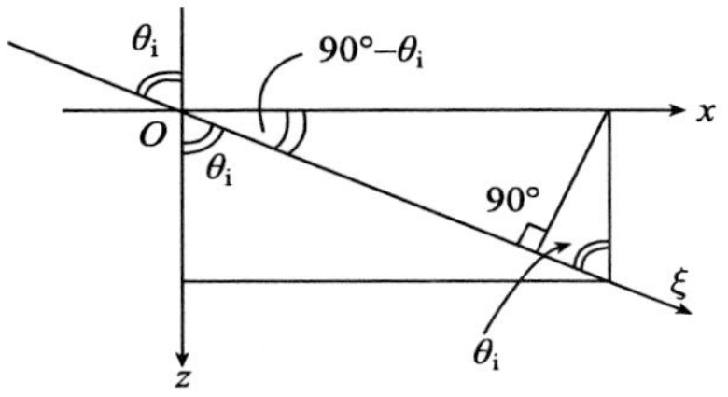

Figure 8.5 The relationship among ξ and x, z, θ_i

Since the direction of ξ is taken as the same with the propagation direction of the wave, then, $\boldsymbol{k} \cdot \boldsymbol{\xi} = k\xi$.

From Figure 8.5, following equations can be obtained

$$\xi = x\cos(90^\circ - \theta_i) + z\cos\theta_i = x\sin\theta_i + z\cos\theta_i \tag{8.53}$$

$$k\xi = k(x\sin\theta_i + z\cos\theta_i) = k_x x + k_z z \tag{8.54}$$

where k_x, k_z represent the phase constants along x, z directions, respectively.

Method 2 to obtain $k\xi$, applying the scalar product of wave vector $\boldsymbol{k}$ and vector ξ. According to Figure 8.5 we have

$$\begin{aligned}\boldsymbol{k} \cdot \boldsymbol{\xi} &= [\boldsymbol{e}_x k\cos(90^\circ - \theta_i) + \boldsymbol{e}_z k\cos\theta_i] \cdot (\boldsymbol{e}_x x + \boldsymbol{e}_z z) \\ &= k\sin\theta_i x + k\cos\theta_i z = k_x x + k_z z\end{aligned} \tag{8.55}$$

Substitute (8.54) into (8.52), which leaves

$$E_y^+ = E_0^+ e^{j(\omega t - kx\sin\theta_i - kz\cos\theta_i)} \tag{8.56}$$

Similarly, from Figure 8.4, it can be known that the expression of electric field of reflected wave is

$$\boldsymbol{E}^- = \boldsymbol{e}_y E_y^- = \boldsymbol{e}_y E_0^- e^{j(\omega t + k\xi')} \tag{8.57}$$

where ξ' is

$$\xi\prime = -x\sin\theta_r + z\cos\theta_r \tag{8.58}$$

Substitute (8.58) into (8.57), yields

$$E_y^- = E_0^- e^{j(\omega t - kx\sin\theta_r + kz\cos\theta_r)} \tag{8.59}$$

2. The Expressions of Magnetic Fields of Incident Wave and Reflected Wave

It can be known from Figure 8.4 that the magnetic field of the incident wave has two components of H_x and H_z, that is

$$\boldsymbol{H}^+ = \boldsymbol{e}_x H_x^+ + \boldsymbol{e}_z H_z^+ = -\boldsymbol{e}_x H^+ \cos\theta_i + \boldsymbol{e}_z H^+ \sin\theta_i \tag{8.60}$$

Based on (8.60) we know that

$$H_x^+ = -H^+ \cos\theta_\mathrm{i} \tag{8.61}$$

where H^+ is the magnetic field of the incident wave,

$$H^+ = H_0^+ \mathrm{e}^{\mathrm{j}(\omega t - k\xi)} = H_0^+ \mathrm{e}^{\mathrm{j}(\omega t - kx\sin\theta_\mathrm{i} - kz\cos\theta_\mathrm{i})}$$

substituting the equation above into (8.61), then H_x^+ of incident wave is

$$H_x^+ = -H_0^+ \mathrm{e}^{\mathrm{j}(\omega t - kx\sin\theta_\mathrm{i} - kz\cos\theta_\mathrm{i})} \cdot \cos\theta_\mathrm{i} \tag{8.62}$$

Similarly, H_z^+ can be obtained as

$$H_z^+ = H_0^+ \mathrm{e}^{\mathrm{j}(\omega t - kx\sin\theta_\mathrm{i} - kz\cos\theta_\mathrm{i})} \cdot \sin\theta_\mathrm{i} \tag{8.63}$$

Following a similar approach, we can find the magnetic fields H_x^-, H_z^- of reflected wave

$$H_x^- = H_0^- \mathrm{e}^{\mathrm{j}(\omega t - kx\sin\theta_\mathrm{r} + kz\cos\theta_\mathrm{r})} \cdot \cos\theta_\mathrm{r} \tag{8.64}$$

and

$$H_z^- = H_0^- \mathrm{e}^{\mathrm{j}(\omega t - kx\sin\theta_\mathrm{r} + kz\cos\theta_\mathrm{r})} \cdot \sin\theta_\mathrm{r} \tag{8.65}$$

3. The Expressions of the Electric Field and Magnetic Field of the Total Wave

By adding the electric field of incident wave E_y^+ [Equation (8.56)] to the electric field of reflected wave E_y^- [Equation (8.59)], we have the electric field of total wave E_y as

$$E_y = E_y^+ + E_y^- = E_0^+ \mathrm{e}^{\mathrm{j}(\omega t - kx\sin\theta_\mathrm{i} - kz\cos\theta_\mathrm{i})} + E_0^- \mathrm{e}^{\mathrm{j}(\omega t - kx\sin\theta_\mathrm{r} + kz\cos\theta_\mathrm{r})} \tag{8.66}$$

Applying the boundary conditions that the tangential electric field is zero at $z = 0$ to the equation above, then we get

$$E_0^+ \mathrm{e}^{-\mathrm{j}kx\sin\theta_\mathrm{i}} + E_0^- \mathrm{e}^{-\mathrm{j}kx\sin\theta_\mathrm{r}} = 0 \tag{8.67}$$

which is the boundary condition ($E_t = 0$) on the surface of perfect conductor, whose validity is independent of x. In addition, E_0^+, E_0^- are known constants, then

$$\mathrm{e}^{-\mathrm{j}kx\sin\theta_\mathrm{i}} = \mathrm{e}^{-\mathrm{j}kx\sin\theta_\mathrm{r}}$$

so

$$\theta_\mathrm{i} = \theta_\mathrm{r} = \theta \tag{8.68}$$

Incidence angle θ_i is equal to the reflection angle θ_r. This is the law of reflection.

Substituting $\theta_\mathrm{i} = \theta_\mathrm{r} = \theta$ into Equation (8.67), we have

$$E_0^+ = -E_0^- \tag{8.69}$$

This expression shows that *when the tangential electric field is incident upon the surface of the perfect conductor there will be total reflection*, that is $|E_0^+| = |E_0^-|$. Meanwhile, the E-fields of reflected wave should have *a reversed phase of* 180°. It is quite reasonable, only with this property the boundary conditions can be satisfied, saying, the tangential electric field on perfect conductor surface is zero.

Substituting $\theta_\mathrm{i} = \theta_\mathrm{r} = \theta$ and $E_0^+ = -E_0^-$ into Equation (8.66), we obtain

$$\begin{aligned} E_y &= E_0^+ (\mathrm{e}^{\mathrm{j}kz\cos\theta} - \mathrm{e}^{\mathrm{j}kz\cos\theta}) \cdot \mathrm{e}^{\mathrm{j}(\omega t - kx\sin\theta)} \\ &= -j2E_0^+ \sin(kz\cos\theta) \cdot \mathrm{e}^{\mathrm{j}(\omega t - kx\sin\theta)} \end{aligned} \tag{8.70}$$

which is the expression of the electric field of total wave.

Regarding the interface of media, the magnetic field of total wave includes the surface normal component H_z and the tangential component H_x. According to the boundary conditions on the surface of perfect conductor, the normal component of alternating magnetic field equals zero, the following equation (where $z = 0$) can be obtained from Equations (8.63) and (8.65)

$$H_z = H_z^+ + H_z^- = H_0^+ \mathrm{e}^{\mathrm{j}(\omega t - kx\sin\theta_\mathrm{i})} \sin\theta_\mathrm{i} + H_0^- \mathrm{e}^{\mathrm{j}(\omega t - kx\sin\theta_\mathrm{r})} \sin\theta_\mathrm{r} = 0 \tag{8.71}$$

As discussed previously, this equation is not related with x, hence the following can be obtained immediately

$$\theta_\mathrm{i} = \theta_\mathrm{r}$$

and

$$H_0^+ = -H_0^- \tag{8.72}$$

Substituting $\theta_\mathrm{i} = \theta_\mathrm{r} = \theta$ and $H_0^+ = -H_0^-$ into the expressions of H_z^+ and H_z^- [i.e. (8.63) and (8.65)], and then taking sum of them, then the normal component H_z of total wave can be obtained

$$\begin{aligned} H_z &= H_z^+ + H_z^- = H_0^+ \sin\theta(\mathrm{e}^{-\mathrm{j}kz\cos\theta} - \mathrm{e}^{\mathrm{j}kz\cos\theta})\mathrm{e}^{\mathrm{j}(\omega t - kx\sin\theta)} \\ &= -\mathrm{j}2H_0^+ \sin\theta \sin(kz\cos\theta) \cdot \mathrm{e}^{\mathrm{j}(\omega t - kx\sin\theta)} \end{aligned} \tag{8.73}$$

Following the similar analysis method, the tangential component H_x of total wave can also be found as

$$H_x = H_x^+ + H_x^- = -2H_0^+ \cos\theta \cos(kz\cos\theta) \cdot \mathrm{e}^{\mathrm{j}(\omega t - k\sin\theta x)} \tag{8.74}$$

Of course, if the electric field E_y of total wave is known, the B-fields can be thus found through Maxwell's equations.

4. The Characteristics of the Electric and Magnetic Fields of Total Wave

When plane wave is obliquely incident upon the surface of perfect conductor, total reflection occurs, then the electric and magnetic fields of total wave, which are the superposition of reflected and incident waves, are as follows

$$E_y = -\mathrm{j}2E_0^+ \sin(kz\cos\theta)\mathrm{e}^{\mathrm{j}(\omega t - kx\sin\theta)} \tag{8.75}$$

$$H_z = -\mathrm{j}\frac{2E_0^+}{\eta} \sin\theta \sin(kz\cos\theta)\mathrm{e}^{\mathrm{j}(\omega t - kx\sin\theta)} \tag{8.76}$$

$$H_x = \frac{-2E_0^+}{\eta} \cos\theta \cos(kz\cos\theta)\mathrm{e}^{\mathrm{j}(\omega t - kx\sin\theta)} \tag{8.77}$$

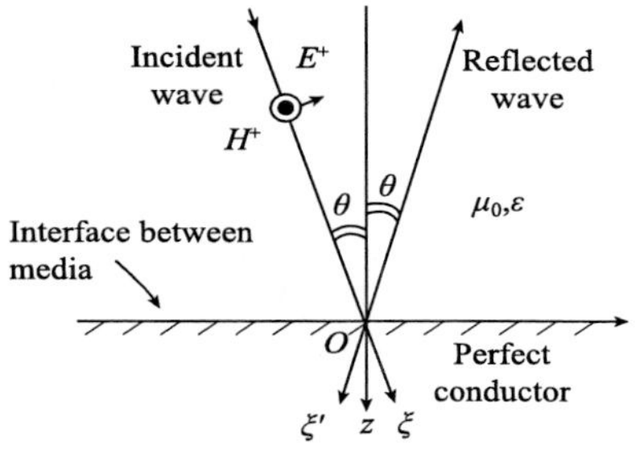

Figure 8.6 Plane wave obliquely incident upon the surface of perfect conductor (parallel polarized wave)

Obviously, the three fields above are all standing waves along z-direction (normal of the interface), but are travelling waves along x-direction.

As indicated by Equations (8.75) and (8.76), E_y and H_z have same time phase, therefore, they form the active power propagation oriented along the positive x-direction.

As observed, E_y and H_x has a 90° time phase difference, thus $-\boldsymbol{e}_z \frac{1}{2}\mathrm{Re}\,(E_y H_x^*) = 0$, implying that the average power density along z-direction is equal to zero. That means, the wave along the z-direction forms a pure standing wave, and there is no power transmission.

The case where the perpendicularly polarized wave is obliquely incident was analyzed above, now the results about the parallel polarized wave obliquely incident upon the surface of perfect conductor will be described briefly, as shown in Figure 8.6.

Because the analysis method is the same as the previous one, it is not repeated here. The readers can derive it as an example by themselves. Now we just give the expressions of electric and magnetic fields of total wave, when parallel polarized plane wave is obliquely incident upon the surface of perfect conductor and results in reflection, as follows.

$$H_y = 2H_0^+ \cos(kz\cos\theta)\mathrm{e}^{\mathrm{j}(\omega t - kx\sin\theta)} \tag{8.78}$$

$$\begin{aligned} E_z &= -2E_0^+ \sin\theta\cos(kz\cos\theta)\mathrm{e}^{\mathrm{j}(\omega t - kx\sin\theta)} \\ &= -2H_0^+\eta\sin\theta\cos(kz\cos\theta)\mathrm{e}^{\mathrm{j}(\omega t - kx\sin\theta)} \end{aligned} \tag{8.79}$$

$$\begin{aligned} E_x &= -\mathrm{j}2E_0^+ \cos\theta\sin(kz\cos\theta)\mathrm{e}^{\mathrm{j}(\omega t - kx\sin\theta)} \\ &= -\mathrm{j}2H_0^+\eta\cos\theta\sin(kz\cos\theta)\mathrm{e}^{\mathrm{j}(\omega t - kx\sin\theta)} \end{aligned} \tag{8.80}$$

For the z coordinate, the incident and reflected waves transmit along opposite directions. Therefore, there must be a standing wave distribution along the z direction. As by the law for this distribution, it all depends on the boundary conditions of perfect conductor surface.

For the x direction, the incident and reflected waves transmit along the same direction, and they both have the same phase velocity along the x direction. Therefore, all the field components of total wave along the x direction are travelling wave.

It is noteworthy that the total wave propagating along the x direction discussed above is no longer a TEM wave. As shown in Figure 8.6, *the total wave contains the longitudinal electric field component E_x, but its magnetic field has only transverse component H_y, which is called transverse magnetic wave, say TM wave. Similarly, as shown in Figure 8.4, the total wave of perpendicularly polarized wave is called transverse electric wave (TE wave).*

5. The Parameters of the Wave

(1) The phase constant k_x, k_z of total wave.

From (8.78), (8.80) and Figure 8.6, it can be known that the phase constant of total wave along the x and z directions are as follows respectively

$$k_x = k\sin\theta \tag{8.81}$$

$$k_z = k\cos\theta \tag{8.82}$$

(2) The wavelength λ_x, λ_z of total wave.

The total wave wavelength λ_z along the z direction is

$$\lambda_z = \frac{2\pi}{k_z} = \frac{2\pi}{\left(\dfrac{2\pi}{\lambda}\right)\cos\theta} = \frac{\lambda}{\cos\theta} \tag{8.83}$$

where λ is the wavelength of plane wave, λ_z is the wavelength of standing wave along the z direction for the total wave, which is equal to the two times of the distance between the adjacent nodes (or antinodes).

The total wave wavelength λ_x along the x direction is

$$\lambda_x = \frac{2\pi}{k_x} = \frac{2\pi}{\left(\dfrac{2\pi}{\lambda}\right)\sin\theta} = \frac{\lambda}{\sin\theta} \tag{8.84}$$

(3) The phase velocity of the total wave $v_{p(x)}$.
The total wave propagates along the x direction, and its phase velocity $v_{p(x)}$ is

$$v_{p(x)} = f\lambda_x = \frac{f\lambda}{\sin\theta} = \frac{v}{\sin\theta} \tag{8.85}$$

$v_{p(x)}$ is greater than the propagation velocity v of incident plane wave (along the ξ direction).

8.4 Plane Wave Obliquely Incident upon the Interface between Perfect Dielectrics

When a plane wave is obliquely incident upon the interface of perfect dielectrics, partial reflection usually occurs while the remaining energy will be transmitted into another dielectric, which, however, deviates from the direction of incident wave. We call it the refracted wave or transmission wave. As shown in Figure 8.7, $\theta_i, \theta_r, \theta_T$ represents the incidence angle, reflection angle and refraction angle, respectively.

8.4.1 The Law of Reflection and the Law of Refraction — Snell's law

The Law of reflection and the law of refraction (also known as Snell's law) indicate the relationship between θ_i and θ_r, as well as the relationship between θ_i, θ_T and refraction index n_1, n_2, respectively. Now we take oblique incidence of parallel polarized wave shown in Figure 8.7 as an example for analysis.

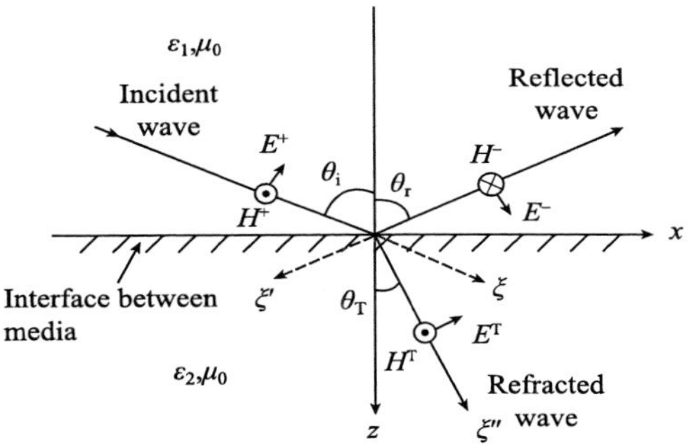

Figure 8.7 Plane wave's incidence, reflection and refraction

Method 1 By using the boundary conditions about the continuity of tangential electric field or tangential magnetic field on the interface of perfect dielectrics, it is easy to reveal the law of reflection and refraction.

For the interface between dielectrics, there exists such field components like tangential electric field E_x, the tangential magnetic field H_y and surface normal electric field E_z, as shown in the figure, only one of the three is analyzed. For example, if the tangential magnetic field H_y is known as the incident wave,

$$H_y^+ = H_0^+ e^{j(\omega t - k_1 z\cos\theta_i - k_1 x\sin\theta_i)} \tag{8.86}$$

where $k_1 = \omega\sqrt{\mu_0\varepsilon_1}$.

The magnetic field of reflected wave is

$$H_y^- = -H_0^- e^{j(\omega t + k_1\xi')} = -H_0^- e^{j(\omega t + k_1 z\cos\theta_r - k_1 x\sin\theta_r)} \tag{8.87}$$

The equation above contains the following relationship:

$$\xi' = z\cos\theta_r - x\sin\theta_r \tag{8.88}$$

The magnetic field of refracted wave

$$H_y^T = H_0^T e^{j(\omega t - k_2\xi'')} = H_0^T e^{j(\omega t - k_2 z\cos\theta_T - k_2 x\sin\theta_T)} \tag{8.89}$$

where $k_2 = \omega\sqrt{\mu_0\varepsilon_2}$,

$$\xi'' = z\cos\theta_T + x\sin\theta_T \tag{8.90}$$

On the interface of perfect dielectrics (where $z = 0$), the tangential component of magnetic field is continuous, saying $H_{1t} = H_{2t}$. According to (8.86), (8.87) and (8.89), this boundary condition can be written specifically as

$$H_y^+ + H_y^- = H_y^T \tag{8.91}$$

that is

$$H_0^+ e^{-jk_1 x \sin\theta_i} - H_0^- e^{-jk_1 x \sin\theta_r} = H_0^T e^{-jk_2 x \sin\theta_T} \tag{8.92}$$

The boundary condition is correct at any point on the interface. In other words, the validity of (8.92) has nothing to do with x.

According to this inference, we can write the following relationship directly from Equation (8.92):

$$k_1\sin\theta_i = k_1\sin\theta_r = k_2\sin\theta_T \tag{8.93}$$

from $k_1\sin\theta_i = k_1\sin\theta_r$ in the previous equation we obtain the law of reflection

$$\theta_i = \theta_r \tag{8.94}$$

From $k_1\sin\theta_i = k_2\sin\theta_T$ in (8.93), the law of refraction can also be given as

$$\frac{\sin\theta_T}{\sin\theta_i} = \frac{k_1}{k_2} = \frac{\sqrt{\varepsilon_1}}{\sqrt{\varepsilon_2}} = \frac{n_1}{n_2} \tag{8.95}$$

The law of reflection and the law of refraction are also known as Snell's law. In the equation above, n_1, n_2 are refractive indices, or indices of refraction (refraction coefficient), of medium 1 and medium 2. $n = \sqrt{\varepsilon_r\mu_r}$, *and for dielectrics*, $n = \sqrt{\varepsilon_r}$. It should be noted that, the law of reflection and the law of refraction are universally applicable, which are valid to the situations regarding any two kinds of media.

Method 2 This method is essentially the same as Method 1, essentially satisfying the boundary conditions. This method deals with the problem from another viewpoint, *the phase constant, phase velocities of incident wave, reflected wave and refracted wave shall be the same along the propagation direction (now x direction) on the interface*, to reveal the laws of reflection and refraction.

At a certain point on the interface, in medium 1 the sum of the incident and reflected wave's tangential magnetic fields, say H_{1t}, must equal the (refractive wave's) tangential magnetic field H_{2t} in medium 2 at the same point. Namely $H_{1t} = H_{2t}$. However, the incident waves and reflected waves on one side of the interface or in medium 1, and refracted wave on the other side of the interface or in medium 2, transmit oriented along the x direction, because that *at any point of the surface between media $H_{1t} = H_{2t}$ should be satisfied, which requires the incident wave, reflected wave and refracted wave should travel along the x direction at the same speed*, or these three waves have the same phase constant along the x direction.

The phase constants of these three waves along the x direction are the same, that is

$$k_1\sin\theta_i = k_1\sin\theta_r = k_2\sin\theta_T \tag{8.96}$$

We thus have

$$\theta_i = \theta_r \tag{8.97}$$

and

$$\frac{\sin\theta_T}{\sin\theta_i} = \frac{k_1}{k_2} = \frac{\sqrt{\varepsilon_1}}{\sqrt{\varepsilon_2}} = \frac{n_1}{n_2} \tag{8.98}$$

Equation (8.97) is the law of reflection, and Equation (8.98) is the law of refraction.

Considering the same phase velocity along the x direction for the incident wave, reflected wave and refracted wave, same results can be found, that is

$$\frac{v_1}{\sin\theta_i} = \frac{v_1}{\sin\theta_r} = \frac{v_2}{\sin\theta_T} \tag{8.99}$$

From the equation above we can have

$$\theta_i = \theta_r \tag{8.100}$$

and

$$\frac{\sin\theta_T}{\sin\theta_i} = \frac{v_2}{v_1} = \frac{\sqrt{\varepsilon_1}}{\sqrt{\varepsilon_2}} = \frac{n_1}{n_2} \tag{8.101}$$

Since $\theta_i = \theta_r$, θ_i, θ_r, they will be represented by θ in the following sections.

8.4.2 Four Vectors, $\boldsymbol{e}_i, \boldsymbol{e}_r, \boldsymbol{e}_T$ and $\boldsymbol{e}_n$ are Coplanar

Followed by the discussion of Snell's law, we briefly describe the relationship among the four vectors, saying, $\boldsymbol{e}_i$, $\boldsymbol{e}_r$, $\boldsymbol{e}_T$ and $\boldsymbol{e}_n$, the unit vector of the incident wave ray, the reflected wave ray, the refracted wave ray, as well as the normal unit vector of the interface. These four vectors are positioned on a same plane, as shown in Figure 8.8.

Since the tangential field on the interface is continuous, then on the interface (along the segment $\overline{OP}$, suppose its vector distance as a unit vector $\boldsymbol{e}_l$), the phase shifts of the incident wave, reflection wave and refracted wave are equal, namely,

$$k_1\boldsymbol{e}_i \cdot \boldsymbol{e}_l = k_1\boldsymbol{e}_r \cdot \boldsymbol{e}_l = k_2\boldsymbol{e}_T \cdot \boldsymbol{e}_l \tag{8.102}$$

where $\boldsymbol{e}_i, \boldsymbol{e}_r, \boldsymbol{e}_T$ are unit vectors for the rays of incident wave, reflected wave, and refracted wave, respectively.

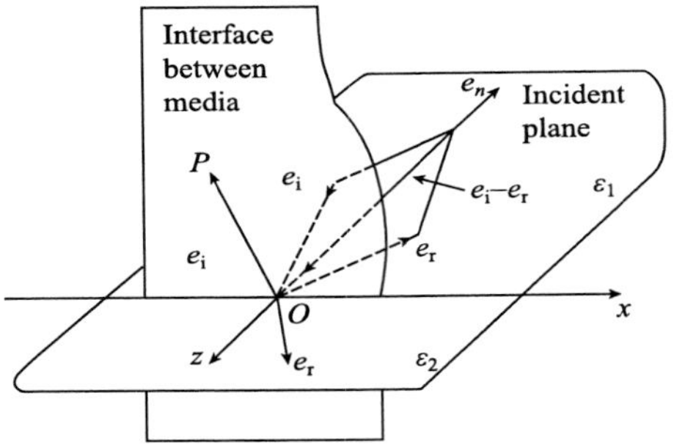

Figure 8.8 A picture for studying $\boldsymbol{e}_i, \boldsymbol{e}_r, \boldsymbol{e}_T$ and $\boldsymbol{e}_n$ as coplanar

For the incident wave and the reflected wave, Equation (8.102) gives that

$$\boldsymbol{e}_i \cdot \boldsymbol{e}_l - \boldsymbol{e}_r \cdot \boldsymbol{e}_l = (\boldsymbol{e}_i - \boldsymbol{e}_r) \cdot \boldsymbol{e}_l = 0 \tag{8.103}$$

Since $\boldsymbol{e}_l$ is oriented along any direction on the surface, the formula above thus indicates that the vector $(\boldsymbol{e}_i - \boldsymbol{e}_r)$ must be perpendicular to the interface, $(\boldsymbol{e}_i - \boldsymbol{e}_r)$ and $\boldsymbol{e}_n$ are in parallel with each other and coplanar. In addition, because vector $(\boldsymbol{e}_i - \boldsymbol{e}_r)$ and $\boldsymbol{e}_i, \boldsymbol{e}_r$ are coplanar, therefore, $\boldsymbol{e}_i, \boldsymbol{e}_r, \boldsymbol{e}_n$ are coplanar.

For the incident wave and the refracted wave, from Equation (8.102) we can also have

$$(k_1\boldsymbol{e}_{\mathrm{i}} - k_2\boldsymbol{e}_{\mathrm{T}}) \cdot \boldsymbol{e}_l = 0 \tag{8.104}$$

With similar analysis then we can conclude that $\boldsymbol{e}_{\mathrm{i}}, \boldsymbol{e}_{\mathrm{T}}, \boldsymbol{e}_n$ are coplanar. Combining the analysis results of Equations (8.103) and (8.104), we can see that the four vectors $\boldsymbol{e}_{\mathrm{i}}, \boldsymbol{e}_{\mathrm{r}}, \boldsymbol{e}_{\mathrm{T}}, \boldsymbol{e}_n$ are coplanar, all on the plane of incidence.

Example 8.1 A TEM wave obliquely incident from the air upon an interface of media at $z = 0$, the electric field strength of incident wave is given as

$$E_y^+ = 5\mathrm{e}^{\mathrm{j}(-3x-2z)}$$

Try to find: (1) The angle of incidence θ_{i};

(2) The unit vector $e_{k_{\mathrm{i}}}$ of the ray of incident wave.

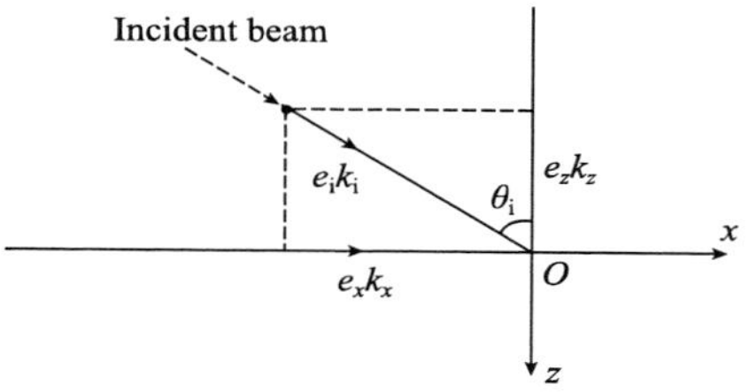

Figure of Example 8.1

Solutions (1) The angle of incidence θ_{i}.

From the figure, we have

$$\begin{cases} k_x = k_{\mathrm{i}} \sin\theta_{\mathrm{i}} = 3 \\ k_z = k_{\mathrm{i}} \cos\theta_{\mathrm{i}} = 2 \end{cases}$$

Thus

$$\tan\theta_{\mathrm{i}} = \frac{3}{2}$$

$$\theta_{\mathrm{i}} = 56.3^\circ$$

(2) The unit vector $e_{k_{\mathrm{i}}}$ of the ray of incident wave

Because

$$\boldsymbol{k}_{\mathrm{i}} = \boldsymbol{e}_x k_x + \boldsymbol{e}_y k_y + \boldsymbol{e}_z k_z = 3\boldsymbol{e}_x + 2\boldsymbol{e}_z$$

then

$$\boldsymbol{e}_{k_{\mathrm{i}}} = \frac{\boldsymbol{k}_{\mathrm{i}}}{|\boldsymbol{k}_{\mathrm{i}}|} = \frac{\boldsymbol{k}_{\mathrm{i}}}{\sqrt{3^2+2^2}} = \frac{3}{\sqrt{13}}\boldsymbol{e}_x + \frac{2}{\sqrt{13}}\boldsymbol{e}_z$$

Example 8.2 Using Snell's law prove the reversibility of wave's rays.

Solution Assume that there are three layers of medium, whose interfaces are infinite plane and parallel to each other. When the plane wave is incident from medium 1, according to Snell's law, on the interface between media 1 and 2,

$$\sqrt{\varepsilon_1}\sin\theta_1 = \sqrt{\varepsilon_2}\sin\theta_2 \tag{a}$$

on the interface of media 2 and 3

$$\sqrt{\varepsilon_2}\sin\theta_2 = \sqrt{\varepsilon_3}\sin\theta_3 \tag{b}$$

Whereas, if the plane wave is incident from the medium 3, then on the interface between medium 3 and 2

$$\sqrt{\varepsilon_3}\sin\theta_3 = \sqrt{\varepsilon_2}\sin\theta_2$$

which is same as (b). While on the interface between medium 2 and 1

$$\sqrt{\varepsilon_2}\sin\theta_2 = \sqrt{\varepsilon_1}\sin\theta_1$$

which is same as (a).

It can be seen that, no matter the plane wave is travelling either from medium 1 to medium 3 through medium 2, or from medium 3 to 1 through 2, the paths of the rays are the same, or the wave rays are reversible.

From Equations (a) and (b), the following equation will be obtained

$$\sqrt{\varepsilon_1}\sin\theta_1 = \sqrt{\varepsilon_2}\sin\theta_2 = \sqrt{\varepsilon_3}\sin\theta_3$$

which can be generalized that, for the case of n layers of media, $n\sin\theta$ *is conserved,* namely

$$\sqrt{\varepsilon_1}\sin\theta_1 = \sqrt{\varepsilon_2}\sin\theta_2 = \cdots = \sqrt{\varepsilon_n}\sin\theta_n$$

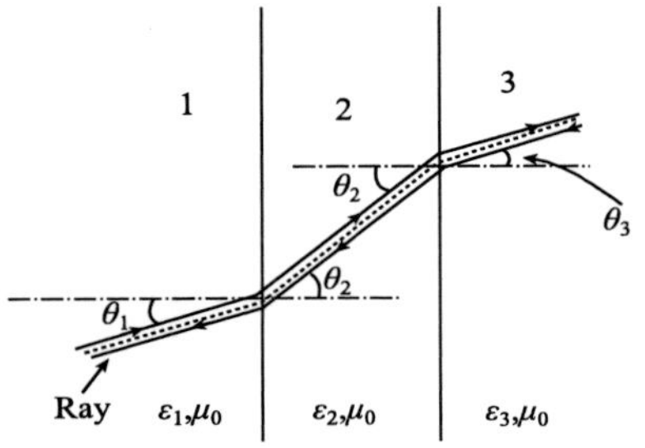

Figure of Example 8.2

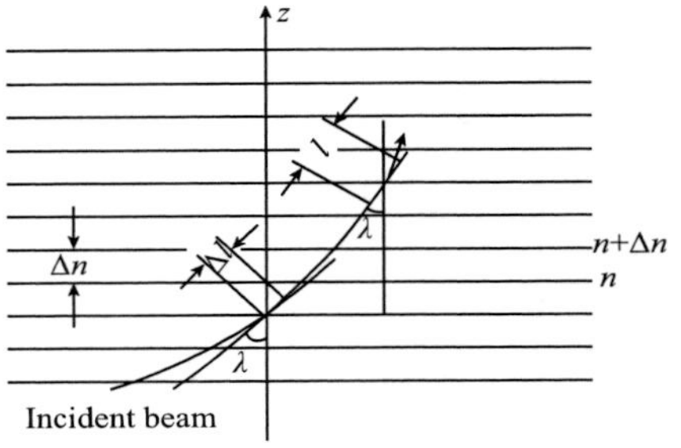

Figure of Example 8.3

Example 8.3 A beam of waves is propagating in layered media, if the refractive index n of the layered media is only a function of coordinate z, and $\frac{\mathrm{d}n}{\mathrm{d}z} > 0$, the angle between incident beam and z-axis isθ, try to prove $\frac{\mathrm{d}\theta}{\mathrm{d}l} < 0$, that is, with the increase of distance l, the angle θ will decrease.

Solution In a continuously varying layered medium, Snell's law can be expressed as

$$n_1\sin\theta_1 = n_2\sin\theta_2 = \cdots = n_n\sin\theta_n \tag{a}$$

As the refractive index n is only a function of coordinate z, Equation (a) is thus valid for any z, hence we have

$$\frac{\mathrm{d}}{\mathrm{d}z}(n\sin\theta) = 0 \tag{b}$$

As stated above, n usually change with z, from Equation (a) we can know that, θ has to change accordingly, in other words, n, θ *are the functions of* z, then based on Equation (b) we have,

$$\frac{\mathrm{d}}{\mathrm{d}z}(n\sin\theta) = n\cos\theta\frac{\mathrm{d}\theta}{\mathrm{d}z} + \sin\theta\frac{\mathrm{d}n}{\mathrm{d}z} = 0 \tag{c}$$

From the figure we have $\frac{\mathrm{d}z}{\mathrm{d}l} = \cos\theta$, so Equation (c) becomes

$$\frac{\mathrm{d}\theta}{\mathrm{d}l} = -\frac{1}{n}\sin\theta\frac{\mathrm{d}n}{\mathrm{d}z} \tag{d}$$

8.4.3 The Reflection Coefficient R and Transmission Coefficient (also Called Refractive Index) T for Plane Waves Obliquely Incident upon the Interface between Perfect Dielectrics — Fresnel Equations

Plane waves obliquely incident and polarized in any direction, always can be decomposed into the obliquely incident parallel polarized wave and perpendicularly polarized wave. Hence we will discuss the issue from the perspectives of the two types of polarizations. *The reflection coefficient R and transmission coefficient T can be defined using tangential electric field (refer to the definition in this book).* As a matter of fact, the total electric field, total magnetic field, tangential magnetic field or power density can also be employed to define R and T. However, the expressions of R, T based on different definitions, are not exactly the same. Anyhow, the electromagnetic fields of the reflected or refracted waves obtained from each definition, are the same.

1. Reflection Coefficient R_P and Transmission Coefficient T_P of Parallel Polarized Waves Obliquely Incident upon the Interface between Dielectrics

Analysis The tangential field components of the incident wave, reflected wave and refracted wave should be found firstly, then the continuity conditions of tangential components can be used to obtain R_P, T_P. The oblique incidence of parallel polarized wave is shown in Figure 8.9.

Next the boundary conditions are used to find R_P, T_P (*defined in terms of tangential electric field in this book*).

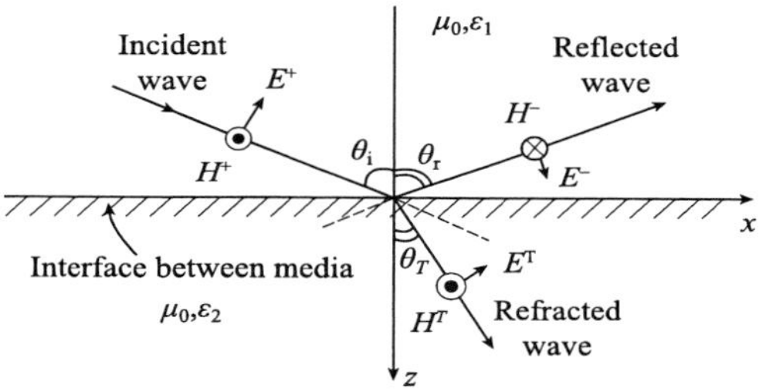

Figure 8.9 Assume that the tangential electric field's phase does not change when being reflected, the directions of electromagnetic fields of incident wave, reflected wave, and refracted wave for parallel polarized waves

(1) Gives the tangential electric and magnetic fields of the incident wave, reflected wave and refracted wave on the interface ($z = 0$).

The tangential electric and magnetic fields of the incident wave on the interface are

$$E_x^+ = E_{x0}^+ e^{j(\omega t - k_1 x \sin\theta)} \tag{8.105}$$

$$H_y^+ = H_{y0}^+ e^{j(\omega t - k_1 x \sin\theta)} \tag{8.106}$$

The tangential fields of the reflected wave on the interface are

$$E_x^- = E_{x0}^- e^{j(\omega t - k_1 x \sin\theta)} \tag{8.107}$$

$$H_y^- = -H_{y0}^- e^{j(\omega t - k_1 x \sin\theta)} \tag{8.108}$$

The tangential fields of the refracted wave on the interface are

$$E_x^T = E_{x0}^T e^{j(\omega t - k_2 x \sin\theta_T)} \tag{8.109}$$

$$H_y^T = H_{y0}^T e^{j(\omega t - k_2 x \sin\theta_T)} \tag{8.110}$$

(2) The boundary conditions $E_{1t} = E_{2t}, H_{1t} = H_{2t}$ are used to obtainR_P, T_P for parallel polarized waves.

Considering the continuity conditions of tangential electric fields on the interface, and Equations (8.105), (8.107), (8.109), we can have

$$E_x^+ + E_x^- = (E_{x0}^+ + E_{x0}^-) \cdot \mathrm{e}^{\mathrm{j}(\omega t - k_1 x \sin\theta)} = E_x^T = E_{x0}^T \mathrm{e}^{\mathrm{j}(\omega t - k_2 x \sin\theta_T)} \tag{8.111}$$

Above equation has no dependence on the spatial location x, hence the equation above can be written as

$$E_{x0}^+ + E_{x0}^- = E_{x0}^T \tag{8.112}$$

With the same analysis method, taking into account that there is anti-phase between H^- and H^+ in Figure 8.9, as well as considering the continuity conditions of tangential magnetic fields on interface, we have

$$H_{y0}^+ - H_{y0}^- = H_{y0}^T \tag{8.113}$$

To find a set of simple expressions of R, T in accordance with the transmission line theory, we then define another two *wave impedances in terms of tangential fields oriented along the z direction, like the wave impedance* Z_{z1} *in medium* 1, *and the wave impedance* Z_{z2} *in medium* 2:

$$Z_{z1} = \frac{E_x^+}{H_y^+} \tag{8.114}$$

$$Z_\mathrm{L} = Z_{z2} = \frac{E_x^T}{H_y^T} \tag{8.115}$$

$Z_{z2} = Z_\mathrm{L}$, namely, the impedance of medium 2 is regarded as the load impedance Z_L.

For the parallel polarized wave in Figure 8.9, the two equations above can also be written as

$$Z_{z1} = \frac{E_x^+}{H_y^+} = \frac{E^+ \cos\theta}{H^+} = \eta_1 \cos\theta \tag{8.116}$$

and

$$Z_\mathrm{L} = \frac{E_x^T}{H_y^T} = \frac{E^T \cos\theta_T}{H^T} = \eta_2 \cos\theta_\mathrm{T} \tag{8.117}$$

By substituting Equations (8.114) and (8.115) into (8.113), the continuity condition of tangential magnetic fields can be expressed as

$$\frac{E_{x0}^+}{Z_{z1}} - \frac{E_{x0}^-}{Z_{z1}} = \frac{E_{x0}^T}{Z_\mathrm{L}} \tag{8.118}$$

From Equations (8.112) and (8.118), *the reflection coefficient of electric field* $R_{(\mathrm{P})}$ can be found as

$$R_{(\mathrm{P})} = \frac{E_x^-}{E_x^+} = \frac{E_{x0}^-}{E_{x0}^+} = \frac{Z_\mathrm{L} - Z_{z1}}{Z_\mathrm{L} + Z_{z1}} \tag{8.119}$$

The transmission coefficient of electric field $T_{(\mathrm{P})}$ is

$$T_{(\mathrm{P})} = \frac{E_x^T}{E_x^+} = \frac{E_{x0}^T}{E_{x0}^+} = \frac{2Z_\mathrm{L}}{Z_\mathrm{L} + Z_{z1}} \tag{8.120a}$$

From Equations (8.119) and (8.120), the relationship between $R_{(\mathrm{P})}$ and $T_{(\mathrm{P})}$ of parallel polarized waves can be achieved as

$$1 + R_{(\mathrm{P})} = T_{(\mathrm{P})} \tag{8.120b}$$

Substituting $Z_{\mathrm{L}} = \eta_2 \cos\theta_T, Z_{z1} = \eta_1 \cos\theta$ and law of refraction into Equations (8.119) and (8.120), for the media with $\mu_1 = \mu_2$ we have

$$R_{(\mathrm{P})} = \frac{-\left(\dfrac{\varepsilon_2}{\varepsilon_1}\right)\cos\theta + \sqrt{\left(\dfrac{\varepsilon_2}{\varepsilon_1}\right) - \sin^2\theta}}{\left(\dfrac{\varepsilon_2}{\varepsilon_1}\right)\cos\theta + \sqrt{\left(\dfrac{\varepsilon_2}{\varepsilon_1}\right) - \sin^2\theta}} \tag{8.121}$$

$$T_{(P)} = \frac{2\sqrt{\left(\dfrac{\varepsilon_2}{\varepsilon_1}\right) - \sin^2\theta}}{\left(\dfrac{\varepsilon_2}{\varepsilon_1}\right)\cos\theta + \sqrt{\left(\dfrac{\varepsilon_2}{\varepsilon_1}\right) - \sin^2\theta}} \tag{8.122}$$

Equations (8.121) and (8.122) are Fresnel Equations of parallel polarized waves. During the derivation, the following equation is employed:

$$\cos\theta_T = \sqrt{1 - \sin^2\theta_T} = \sqrt{1 - \sin^2\theta\left(\frac{\varepsilon_1}{\varepsilon_2}\right)} = \frac{n_1}{n_2}\sqrt{\left(\frac{\varepsilon_2}{\varepsilon_1}\right) - \sin^2\theta}$$

2. Reflection Coefficient $R_{(\mathrm{N})}$ and Transmission Coefficient $T_{(\mathrm{N})}$ of Perpendicularly Polarized waves obliquely incident upon the interface between dielectrics

As to the reflection and refraction of perpendicularly polarized waves, it will not be discussed in detail, because the analysis method is similar to that above. However, it should be noted that, when perpendicularly polarized waves are obliquely incident, the two wave impedances defined previously, namely Z_{z1} and Z_{L} (or Z_{z2}) is changed into the following forms:

$$Z_{z1} = \frac{E_y^+}{H_x^+} = \frac{E^+}{H^+\cos\theta} = \eta_1/\cos\theta \tag{8.123}$$

$$Z_L = Z_{z2} = \frac{E_y^T}{H_x^T} = \frac{E^T}{H^T\cos\theta_T} = \eta_2/\cos\theta_T \tag{8.124}$$

The reflection coefficient $R_{(\mathrm{N})}$ and the transmission coefficient $T_{(\mathrm{N})}$ of electric fields of perpendicularly polarized wave are as below respectively

$$R_{(\mathrm{N})} = \frac{E_y^-}{E_y^+} = \frac{Z_{\mathrm{L}} - Z_{z1}}{Z_{\mathrm{L}} + Z_{z1}} \tag{8.125}$$

$$T_{(\mathrm{N})} = \frac{E_y^T}{E_y^+} = \frac{2Z_{\mathrm{L}}}{Z_{\mathrm{L}} + Z_{z1}} \tag{8.126}$$

And the relationship between $R_{(\mathrm{N})}$ and $T_{(\mathrm{N})}$ is

$$1 + R_{(\mathrm{N})} = T_{(\mathrm{N})} \tag{8.127}$$

Substituting Equation (8.123), (8.124), and the law of refraction into Equation (8.125) and (8.126), when $\mu_1 = \mu_2$, we obtain $R_{(\mathrm{N})}$ and $T_{(\mathrm{N})}$ of perpendicularly polarized waves obliquely incident upon the interface between perfect dielectrics as

$$R_{(\mathrm{N})} = \frac{\cos\theta - \sqrt{\left(\dfrac{\varepsilon_2}{\varepsilon_1}\right) - \sin^2\theta}}{\cos\theta + \sqrt{\left(\dfrac{\varepsilon_2}{\varepsilon_1}\right) - \sin^2\theta}} \tag{8.128}$$

$$T_{(\mathrm{N})} = \frac{2\cos\theta}{\cos\theta + \sqrt{\left(\dfrac{\varepsilon_2}{\varepsilon_1}\right) - \sin^2\theta}} \tag{8.129}$$

Equations (8.128) and (8.129) are Fresnel Equations for perpendicularly polarized waves.

Normal incidence of plane waves is just a special case of oblique incidence. Therefore, either of the Fresnel Equations can be selected from the two cases above (parallel and perpendicularly polarized waves being obliquely incident), to determine the reflection coefficient and transmission coefficient for normal incidence. Here, just notice that $\theta_{\mathrm{i}} = \theta_{\mathrm{r}} = \theta = 0, \theta_{\mathrm{T}} = 0$.

3. Simplified Representation of Fresnel Equations (R, T)

Reflection coefficients $R_{\mathrm{N}}, R_{\mathrm{P}}$, of perpendicularly polarized waves and parallel polarized waves have expressions of same form, similarly, transmission coefficients $T_{\mathrm{N}}, T_{\mathrm{P}}$ have expressions of same form, that is

$$R_{(\mathrm{N}),(\mathrm{P})} = \frac{Z_{\mathrm{L}} - Z_{z1}}{Z_{\mathrm{L}} + Z_{z1}} \tag{8.130}$$

$$T_{(\mathrm{N}),(\mathrm{P})} = \frac{2Z_{\mathrm{L}}}{Z_{\mathrm{L}} + Z_{z1}} \tag{8.131}$$

However, it must be noted that, Z_{L}, Z_{z1} *of parallel polarized waves are different from those of perpendicularly polarized waves.*

For perpendicularly polarized waves

$$\left.\begin{aligned} Z_{\mathrm{L}} &= \frac{\eta_2}{\cos\theta_{\mathrm{T}}} \\ Z_{z1} &= \frac{\eta_1}{\cos\theta} \end{aligned}\right\} \tag{8.132}$$

For the parallel polarized waves

$$\left.\begin{aligned} Z_{\mathrm{L}} &= \eta_2\cos\theta_{\mathrm{T}} \\ Z_{z1} &= \eta_1\cos\theta \end{aligned}\right\} \tag{8.133}$$

This set of simplified Equations (8.130)~(8.133) is applicable to dielectrics and magnetic medium. The wave impedance η in preceding equations is

$$\eta = \sqrt{\frac{\mu_{\mathrm{r}}\mu_0}{\varepsilon_{\mathrm{r}}\varepsilon_0}}$$

4. The Phases of Reflected Waves and Refracted Waves

Incidence of plane wave on the interface between perfect dielectrics will generate reflected wave and refracted wave. What is the relationship between their phases and the incident wave on the interface? Refer to the two sets of Fresnel Equations of parallel and perpendicularly polarized waves in terms of tangential electric field: Equations (8.121) and (8.122), Equations (8.128) and (8.129), the following conclusions can be made:

(1) the phase of refracted wave equals the phase of the incident wave.

When refracted wave propagates in dielectric 2, from the optical theory it is known that θ_{i} is definitely less than the critical angle θ_{c}. That is, the condition $\sin\theta_{\mathrm{i}} < \sin\theta_{\mathrm{c}}$ (and $\sin\theta_{\mathrm{c}} = \sqrt{\varepsilon_2/\varepsilon_1}$) is satisfied. Thus, the transmission coefficients $T_{(\mathrm{P})}, T_{(\mathrm{N})}$ are always positive real numbers. In other words, *the ratio of the tangential component of electric field of refracted wave and that of the incident wave is always positive real number on the interface between dielectrics, hence the two waves have the same phases.*

(2) The tangential components of electric fields of refracted wave and incident wave may be in-phase or out-of-phase.

First the case of perpendicular polarized waves is studied:

For ease of analysis, substituting the following equation into Equation (8.128) yields

$$\cos\theta_{\mathrm{T}} = \sqrt{1 - \sin^2\theta\left(\frac{\varepsilon_1}{\varepsilon_2}\right)} = \frac{n_1}{n_2}\sqrt{\frac{\varepsilon_2}{\varepsilon_1} - \sin^2\theta} \tag{8.134}$$

then the reflection coefficient of perpendicularly polarized wave is given as

$$R_{(\mathrm{N})} = \frac{\dfrac{n_1}{n_2}\cos\theta - \cos\theta_T}{\dfrac{n_1}{n_2}\cos\theta + \cos\theta_T} \tag{8.135}$$

where $n_1 = \sqrt{\varepsilon_{\mathrm{r1}}\mu_{\mathrm{r1}}} = \sqrt{\varepsilon_{\mathrm{r1}}}$, similarly, n_2 equals to the square root of the relative dielectric constant.

When $n_1 > n_2$, from the law of refraction we have $\theta_{\mathrm{T}} > \theta_{\mathrm{i}}$, hence $\cos\theta_{\mathrm{i}} > \cos\theta_{\mathrm{T}}$. Therefore, $R_{(\mathrm{N})}$ has a positive real value, namely, the reflected wave has the same phase as the incident wave. When $n_1 < n_2$, then $\theta_{\mathrm{T}} < \theta_{\mathrm{i}}$, so $\cos\theta_{\mathrm{T}} > \cos\theta_{\mathrm{i}}$. Thus, $R_{(\mathrm{N})}$ is negative, saying the reflected wave and the incident has anti-phase, as shown in Figure 8.10.

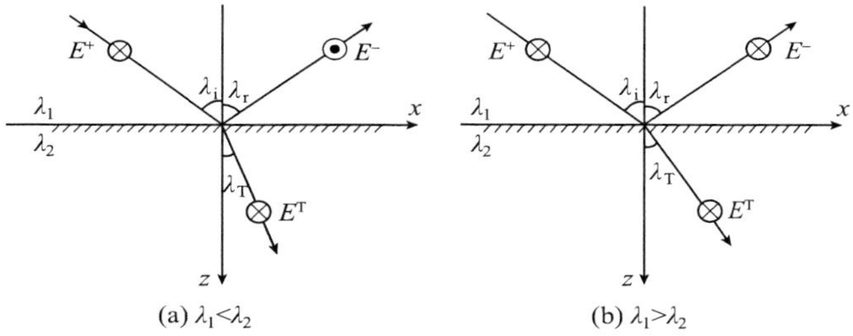

Figure 8.10 The phase relationship among incident wave, reflected wave and refracted wave for perpendicularly polarized wave

Now let us analyze the case of parallel polarized wave:

Substituting Equation (8.134) into (8.121), we have the reflection coefficient of parallel polarized wave

$$R_{(\mathrm{P})} = \frac{-\cos\theta + \left(\dfrac{n_1}{n_2}\right)\cos\theta_{\mathrm{T}}}{\cos\theta + \left(\dfrac{n_1}{n_2}\right)\cos\theta_{\mathrm{T}}} \tag{8.136}$$

Whether $R_{(\mathrm{P})}$ is positive or negative it is independent not only of n_1/n_2, but also of θ (i.e. θ_{i}) and θ_{T}. In other words, for $n_1 > n_2$ or $n_2 > n_1$, $R_{(\mathrm{P})}$ *may be positive or negative.*

Figure 8.11 (a) shows the relationship among the reflection coefficient $R_{(\mathrm{N})}$, transmission coefficient $T_{(\mathrm{N})}$ and θ_{i} of perpendicularly polarized wave, in this case $n_1/n_2 = 1/1.5$, the wave is transmitted from a rare medium to a dense medium. Because $n_1 < n_2$ and it is perpendicularly polarized wave, the reflected wave and incident wave are out-of-phase on the surface. *The transmission coefficient $T_{(\mathrm{N})}$ is positive, that is, the reflected wave and incident wave are in-phase on the surface.*

Figure 8.11 (b) indicates the relationship among the reflection coefficient $R_{(\mathrm{P})}$, transmission coefficient $T_{(\mathrm{P})}$ and θ_{i} of parallel polarized wave, $n_1/n_2 = 1/1.5$ for this case, also means that the wave is transmitted from a rare medium to a dense medium. On the interface, the

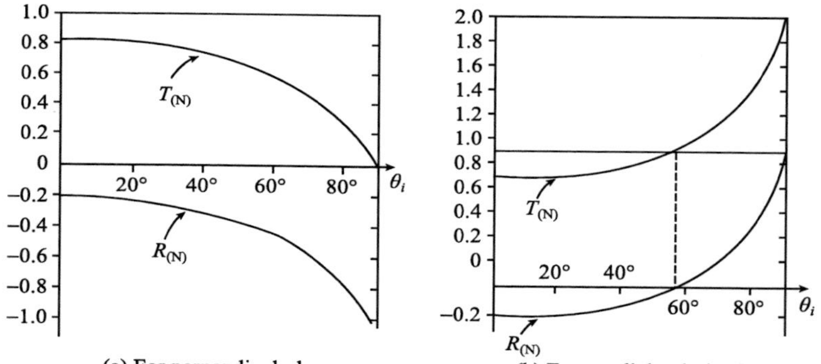

(a) For perpendicularly polarized wave (b) For parallel polarized wave

Figure 8.11 Relationship among transmission coefficient, reflection coefficient in terms of tangential electric field and θ_i (when $n_1/n_2 = 1/1.5$)

phase relationship between the incident wave and reflected wave of parallel polarized wave vary with θ_i when n_1/n_2 is fixed. The range of θ_i is $0° \sim 90°$, $R_{(\mathrm{P})}$ can be either positive or negative. In other words, the reflected wave and the incident wave may be in-phase, or out-of-phase, depending on the value of θ_i. Meanwhile, *the transmission coefficient is always positive, that is, on the interface the refracted is always in-phase with the incident wave.*

8.4.4 Total Reflection of Waves on the Interface between Media — Critical Angle θ_c

Plane waves obliquely incident upon the interface between media will lead to wave reflection and refraction. But under certain circumstances, there is only reflection without refraction, that is, the phenomenon of total waves reflection occurs.

1. Total Reflection of Waves — Critical Angle θ_c

When electromagnetic waves transmit from the optical dense medium (with larger n) to rare medium (with smaller n), and the incidence angle θ_i is greater than or equal to the critical angle θ_c, then the total reflection of the wave occurs.

Now let us find the critical angleθ_c.

Method 1 If $\varepsilon_1 > \varepsilon_2$, when the angle of refraction $\theta_\mathrm{T} = 90°$, then the corresponding θ_i is the critical angle θ_c.

Based on the law of refraction

$$\frac{\sin\theta_\mathrm{T}}{\sin\theta_\mathrm{i}} = \frac{\sqrt{\varepsilon_1}}{\sqrt{\varepsilon_2}} = \frac{n_1}{n_2}$$

if $\varepsilon_1 > \varepsilon_2$, when $\theta_\mathrm{T} = 90°$, that is, there is no refracted wave propagating into the medium 2, the incident angle θ_i at this time is represented by θ_c, called the critical angle. Hence we have

$$\sin\theta_\mathrm{c} = \frac{\sqrt{\varepsilon_2}}{\sqrt{\varepsilon_1}} = \frac{n_2}{n_1} \tag{8.137}$$

Substituting the preceding equation into the reflection coefficient in Equation (8.121), (8.128), it can be found that, when $\theta_\mathrm{i} = \theta_\mathrm{c}$, either for parallel polarized wave or for perpendicularly polarized wave, the absolute reflection coefficient equals 1, that is, the total wave reflection happens on the interface between media.

Method 2 Directly let the reflection coefficients R_{P}, R_{N} in Fresnel Equations be 1, then the critical angle θ_{c} can be obtained.

According to the Equations (8.119) and (8.125) of Fresnel Equations, it is known that the reflection coefficients of parallel polarized and perpendicularly polarized waves can be uniformly expressed as

$$R_{(\mathrm{P}),(\mathrm{N})} = \frac{Z_{\mathrm{L}} - Z_{z1}}{Z_{\mathrm{L}} + Z_{z1}} \tag{8.138}$$

But it should be noted that $Z_{z1(\mathrm{P})}, Z_{\mathrm{L}(\mathrm{P})}$ for the parallel polarized wave and $Z_{z1(\mathrm{N})}, Z_{\mathrm{L}(\mathrm{N})}$ for the perpendicularly polarized wave, are not the same:

For parallel polarized wave

$$Z_{\mathrm{L}(\mathrm{P})} = \eta_2 \cos\theta_{\mathrm{T}}, \quad Z_{z1}\,(\mathrm{P}) = \eta_1 \cos\theta \tag{8.139}$$

For perpendicularly polarized wave

$$Z_{\mathrm{L}(\mathrm{N})} = \eta_2/\cos\theta_{\mathrm{T}}, \quad Z_{z1(\mathrm{N})} = \eta_1/\cos\theta \tag{8.140}$$

when the load $Z_{\mathrm{L}(\mathrm{P})} = 0, Z_{\mathrm{L}(\mathrm{N})} = \infty$, *that is* $\theta_{\mathrm{T}} = 90°$, *then* $R_{(\mathrm{P}),(\mathrm{N})} = \pm 1$, *the total reflection occurs.* θ_{i} *at this time is the critical angle* θ_{c}.

Take parallel polarized wave as an example, from Equations (8.139) and (8.138) we know that when $Z_{\mathrm{L}(\mathrm{P})} = 0$, $\theta_{\mathrm{T}}= 90°$, $R_{\mathrm{P}} = -1$, total reflection occurs, θ_{i} is changed to θ_{c}, as expressed below

$$Z_{\mathrm{L}(\mathrm{P})} = \eta_2 \cos\theta_{\mathrm{T}} = \eta_2\sqrt{1 - \sin^2\theta_{\mathrm{T}}} = \eta_2\sqrt{1 - \left(\frac{\varepsilon_1}{\varepsilon_2}\right)\sin^2\theta_{\mathrm{c}}} = 0$$

$$\sin\theta_{\mathrm{c}} = \frac{\sqrt{\varepsilon_2}}{\sqrt{\varepsilon_1}} \tag{8.141}$$

Method 3 A more efficient approach is *to apply directly the concept of impedance. When* $Z_{\mathrm{L}(\mathrm{P})} = 0$ *or* $Z_{\mathrm{L}(\mathrm{N})} = \infty$, *total reflection happens, then critical angle* θ_{c} *can be found.*

$Z_{\mathrm{L}} = 0, \infty$indicates the load of short circuit or open circuit, which usually cause total wave reflection. Take perpendicularly polarized wave as an example, from $Z_{\mathrm{L}(\mathrm{N})} = \infty$ for (8.140), as well as the law of refraction, it can be obtained that

$$Z_{\mathrm{L}(\mathrm{N})} = \eta_2/\cos\theta_{\mathrm{T}} = \eta_2\Big/\sqrt{1 - \left(\frac{\varepsilon_1}{\varepsilon_2}\right)\sin^2\theta_{\mathrm{c}}} = \infty \tag{8.142}$$

$$\sin\theta_{\mathrm{c}} = \sqrt{\varepsilon_2}/\sqrt{\varepsilon_1}$$

2. The Fields in the Medium 2 (*rare medium*) *when Total Reflection Occurs*

When total wave reflection occurs on the interface, is there any field in the medium 2? Yes, there is. Because there is a total field superpositioned by incident wave and reflected wave infinitely close to the interface on the side of medium 1. This total field should be related to the corresponding field components infinitely close to the interface on side of medium 2, satisfying the boundary conditions. The boundary conditions do not change when total reflection occurs. Therefore, although there is no refracted wave propagating along the z direction in medium 2 when total reflection occurs, there are fields in medium 2. What kinds

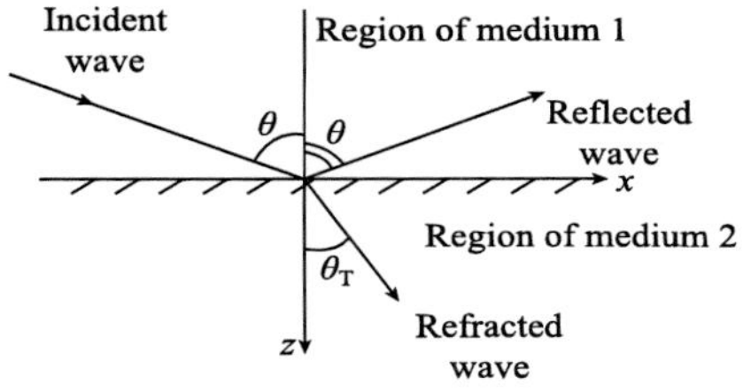

Figure 8.12

of fields are there in medium 2? To clarify this point, we can choose any field component to analyze, such as the magnetic field of refracted wave given by Formula (8.89). The magnetic field of refracted wave is

$$H_y^T = H_0^T \cdot \mathrm{e}^{\mathrm{j}(\omega t - k_2 z \cos\theta_\mathrm{T} - k_2 x \sin\theta_\mathrm{T})}$$

From this equation and Figure 8.12, the wave in medium 2 propagates not only along the x direction, but also along the z direction when no total reflection occurs.

For ease of explanation, the equation above can be transformed, and, without considering the amplitude, we only look into its phase.

The phase variation of the field along the x direction follows

$$\mathrm{e}^{-\mathrm{j}k_2 x \sin\theta_\mathrm{T}} = \exp\left(-\mathrm{j}k_2 x\sqrt{\frac{\varepsilon_1}{\varepsilon_2}}\sin\theta\right) \tag{8.143}$$

The phase variation of the field along the z direction follows

$$\begin{aligned}\mathrm{e}^{-\mathrm{j}k_2 z \cos\theta_\mathrm{T}} &= \exp\left[-\mathrm{j}k_2 z\left(\pm\sqrt{1-\sin^2\theta_\mathrm{T}}\right)\right] \\ &= \exp\left[-\mathrm{j}k_2 z\left(\pm\mathrm{j}\sqrt{\sin^2\theta_\mathrm{T}-1}\right)\right] \\ &= \exp\left(\pm k_2 z\sqrt{\frac{\varepsilon_1}{\varepsilon_2}\sin^2\theta - 1}\right)\end{aligned} \tag{8.144}$$

Here the law of refraction is applied. Let's see the square root operation in Equation (8.144), that is

$$\sqrt{\left(\frac{\varepsilon_1}{\varepsilon_2}\right)\sin^2\theta - 1} = ? \tag{8.145}$$

According to the law of reflection, when $\theta_\mathrm{T} = 90°$, $\theta_\mathrm{i} = \theta_\mathrm{c}$, we have

$$\sin\theta_\mathrm{i} = \sqrt{\frac{\varepsilon_2}{\varepsilon_1}} \text{ when } \theta_\mathrm{i} = \theta_\mathrm{c} \tag{8.146}$$

hence

$$\sin\theta_\mathrm{i} > \sqrt{\frac{\varepsilon_2}{\varepsilon_1}} \text{ when } \theta_\mathrm{i} > \theta_\mathrm{c} \tag{8.147}$$

$$\sin\theta_i < \sqrt{\frac{\varepsilon_2}{\varepsilon_1}} \text{ when } \theta_\mathrm{i} < \theta_\mathrm{c} \tag{8.148}$$

From expressions Equations (8.145) and (8.147), *when* $\theta_\mathrm{i} > \theta_\mathrm{c}$, *or when total reflection occurs, Equation* (8.145) *has a real value, which can be substituted into Equation* (8.144), *we can see that, there is no phase change when waves are travelling along the positive* z *direction, but decayed exponentially* [take negative sign for Equation (8.144)].

The analysis above indicates that there are still fields in medium 2 when total reflection happens. For this field, only its magnitude decreases exponentially along the z direction, but no phase change or propagation take place. From Equation (8.143), the wave propagating along the x direction, or the direction parallel to the interface between media, is called as surface wave.

Although total reflection happens, there is electromagnetic field in medium 2, in this case medium 2 can thus be regarded as pure reactance. For example, from Equation (8.117) we have $Z_{L(P)} = \eta_2 \cos\theta_T$. Using the law of refraction, we can write $Z_{L(P)}$ as the following formula

$$Z_L = \eta_2 \cos\theta_T = \eta_2\sqrt{1-\sin^2\theta_T} = j\eta_2\left(\sqrt{\frac{\varepsilon_1}{\varepsilon_2}\sin^2\theta - 1}\right) \tag{8.149}$$

As described before, when $\theta_i > \theta_c$, total reflection appears, then

$$\sin\theta_i > \sqrt{\frac{\varepsilon_2}{\varepsilon_1}} \tag{8.150}$$

Observing above equations, we say, Z_L *have an imaginary value for total reflection, as it is a pure reactance.* As we all know, for AC circuits, the average power through pure reactance is zero, anyhow, still some current flows through the reactance components. This comparison of circuit and field theory can help to have clearer understanding.

When total reflection of electromagnetic waves occurs, the phase shift incurred, or the sudden change of phase, can be easily obtained through the Fresnel Equations.

Example 8.4 Try to use Snell's law to describe the reflection of electromagnetic waves over ionosphere.

Solution As shown in Figure of Example 8.4, assume that the refractive index n of the ionosphere decreases gradually along the z direction, or the variation of n along a wavelength is negligible.

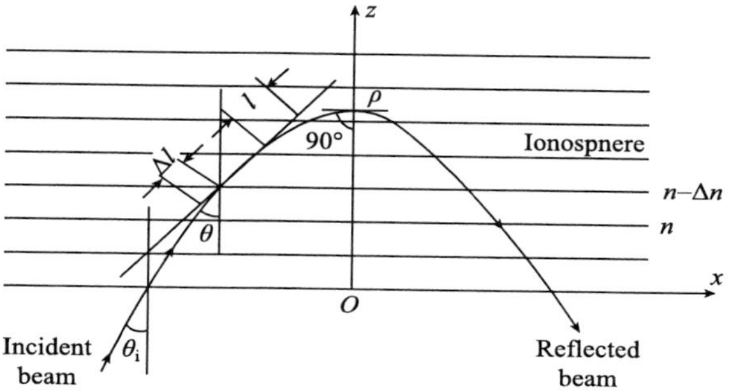

Figure of Example 8.4

The incidence angle for electromagnetic wave transmitting into the ionosphere is θ_i. After entering the ionosphere, the angle θ gradually changes with $n(z)$. Now let us analyze the variation of incident angle θ. From Snell's law it is known that

$$n_1 \sin\theta_i = n\sin\theta$$

that is, *when refraction occurs on the interface between two media,* $n\sin\theta$ *is conserved from one side to another.* If we imagine that the medium whose refractive index n is continuously varying as along many thin layers, then along the ray beam, $n\sin\theta$ is still conserved. If xOy plane is the interface between air and the ionosphere, then $n_i = 1$ at $z = 0$, thus

$$\sin\theta_i = n\sin\theta \tag{a}$$

As $\dfrac{d}{dz}(n\sin\theta) = 0$, and n, θ are functions of z, so

$$\begin{aligned}\frac{\mathrm{d}}{\mathrm{d}z}(n\sin\theta) &= n\frac{\mathrm{d}}{\mathrm{d}z}(\sin\theta)+\sin\theta\frac{\mathrm{d}n}{\mathrm{d}z}\\ &= n\cos\theta\frac{\mathrm{d}\theta}{\mathrm{d}z}+\sin\theta\frac{\mathrm{d}n}{\mathrm{d}z}=0\end{aligned}$$

From the figure, $\mathrm{d}z/\mathrm{d}l=\cos\theta$, then from the equation above we have

$$\frac{\mathrm{d}\theta}{\mathrm{d}l}=-\frac{1}{n}\cdot\frac{\mathrm{d}n}{\mathrm{d}l}\cdot\tan\theta \tag{b}$$

For incident wave, the ion density of the ionosphere grows along $+z$ direction, then the refractive index n decreases z, or $\mathrm{d}n/\mathrm{d}l<0$. Hence from Equation (b) we can see $\frac{\mathrm{d}\theta}{\mathrm{d}l}>0$, indicating that when the ray is going upward, it encounters continuous refraction with continuously increasing l, and θ also increase *until* $\theta=90°$, *eventually reaching the point* P. At that point the ray is oriented along the x direction, then $n_1=n_2$ (because n is a function only of z), based on the formula of the critical angle $\sin\theta_c=\frac{n_2}{n_1}$, hence the critical angle θ_c at point P is just 90°. Therefore, once the incident wave reaches point P, total reflection occurs immediately, and the beam turns into reflected (downward) state, with symmetrical ray to the incident (upward) ray, as shown in figure of Example 8.4.

At the peak of the beam path (point P), $\theta=90°$, it is known from Snell's law that

$$n\sin\theta=n\sin 90°=n_1\sin\theta_i=\sin\theta_i$$

thus

$$n(\theta=90°)=\sin\theta_i \tag{c}$$

where $n(\theta=90°)$ is the refractive index at the points with $\theta=90°$. It is the desirable refractive index, in such case, the beam of electromagnetic wave can be reflected back from the ionosphere for given incident angle θ_i.

Example 8.5 Try to prove that the reflected wave for linearly polarized wave usually becomes elliptically polarized wave after total reflection on the interface between media.

Solution Assume that the angle between the linearly polarized electric field intensity of incident wave $\boldsymbol{E}_1$ and the normal of incident plane, is θ. Firstly, $\boldsymbol{E}_1$ is decomposed into electric field $E_{1(P)}$ in parallel with the incident plane (i.e. the electric field of parallel polarized wave) and electric field $E_{1(N)}$ perpendicular to the incident plane (i.e. the electric field of perpendicularly polarized wave); the time phases of these two components are the same. Then, the amplitudes and phases of the electric fields for parallel and perpendicularly polarized waves after the total reflection should be studied, to determine their polarization states. Given that $\varepsilon_1>\varepsilon_2$, the expression for the reflection coefficient for the electric field is

$$R_{(P)}=\frac{-\frac{\varepsilon_2}{\varepsilon_1}\cos\theta_i+\sqrt{\frac{\varepsilon_2}{\varepsilon_1}-\sin^2\theta_i}}{\frac{\varepsilon_2}{\varepsilon_1}\cos\theta_i+\sqrt{\frac{\varepsilon_2}{\varepsilon_1}-\sin^2\theta_i}}\quad\text{(for parallel polarized wave)}$$

$$R_{(N)}=\frac{\cos\theta_i-\sqrt{\frac{\varepsilon_2}{\varepsilon_1}-\sin^2\theta_i}}{\cos\theta_i+\sqrt{\frac{\varepsilon_2}{\varepsilon_1}-\sin^2\theta_i}}\quad\text{(for perpendicularly polarized wave)}$$

as $\sin\theta_i>\sqrt{\frac{\varepsilon_2}{\varepsilon_1}}$ for total reflection, therefore

$$\sqrt{\frac{\varepsilon_2}{\varepsilon_1}-\sin^2\theta_i}=\mathrm{j}\left(\sqrt{\sin^2\theta_i-\frac{\varepsilon_2}{\varepsilon_1}}\right)$$

The item in the brackets on the right side of the equation above is a real number. So the reflection coefficient can also be written as

$$R_{(P)} = \frac{E_{1(P)}^{-}}{E_{1(P)}^{+}} = \frac{-\dfrac{\varepsilon_2}{\varepsilon_1}\cos\theta_i + j\sqrt{\sin^2\theta_i - \dfrac{\varepsilon_2}{\varepsilon_1}}}{\dfrac{\varepsilon_2}{\varepsilon_1}\cos\theta_i + j\sqrt{\sin^2\theta_i - \dfrac{\varepsilon_2}{\varepsilon_1}}} = e^{j2\phi_{(P)}}$$

$$R_{(N)} = \frac{E_{1(N)}^{-}}{E_{1(N)}^{+}} = \frac{\cos\theta_i - j\sqrt{\sin^2\theta_i - \dfrac{\varepsilon_2}{\varepsilon_1}}}{\cos\theta_i + j\sqrt{\sin^2\theta_i - \dfrac{\varepsilon_2}{\varepsilon_1}}} = e^{j2\phi_{(N)}}$$

Since it is total reflection, so the magnitudes of R_P, R_N are equal to 1, however, in general, for incident wave, $\left|E_{1(P)}^{+}\right| \neq \left|E_{1(N)}^{+}\right|$, so

$$\left|E_{1(P)}^{-}\right| \neq \left|E_{1(N)}^{-}\right|$$

Obviously, the amplitudes of electric fields of parallel and perpendicularly polarized waves are not equal after reflection, and their time phases $2\phi_{(P)}, 2\phi_{(N)}$ are not the same:

$$\phi_{(P)} = -\arctan\frac{\sqrt{\sin^2\theta_i - \dfrac{\varepsilon_2}{\varepsilon_1}}}{\dfrac{\varepsilon_2}{\varepsilon_1}\cos\theta_i} \quad \text{(for parallel polarized wave)}$$

$$\phi_{(N)} = -\arctan\frac{\sqrt{\sin^2\theta_i - \dfrac{\varepsilon_2}{\varepsilon_1}}}{\cos\theta_i} \quad \text{(for perpendicularly polarized wave)}$$

Therefore, the reflected wave usually becomes elliptically polarized wave.

One special case is that, when the angle between the electric field intensity of incident wave $\boldsymbol{E}_1$ and the normal of the incident plane, $\theta = 45°$, $E_{1(P)}^{+} = E_{1(N)}^{+}$, now if $E_{1(P)}^{-}$ and $E_{1(N)}^{-}$ could have a time phase difference of 90° , namely, $2\phi_{(P)} - 2\phi_{(N)} = 90°$. Then the electric fields of reflected wave $E_{1(P)}^{-}$ and $E_{1(N)}^{-}$ are spatially orthogonal, with equal amplitude and time phase difference of 90°, hence the electric field of reflected wave turns into a circularly polarized wave.

Example 8.6 A beam of waves in air is incident upon one end of dielectric rod at arbitrary angle, it is required that all the waves must be constrained in the dielectric rod and propagate to the other end. Try to find the minimum value of the permittivity of this dielectric rod with electromagnetic waves transmission.

Solution It can be seen from the Figure of Example 8.6, in the dielectric rod the conditions for total reflection is

$$\sin\theta_1 \geqslant \sin\theta_c \tag{a}$$

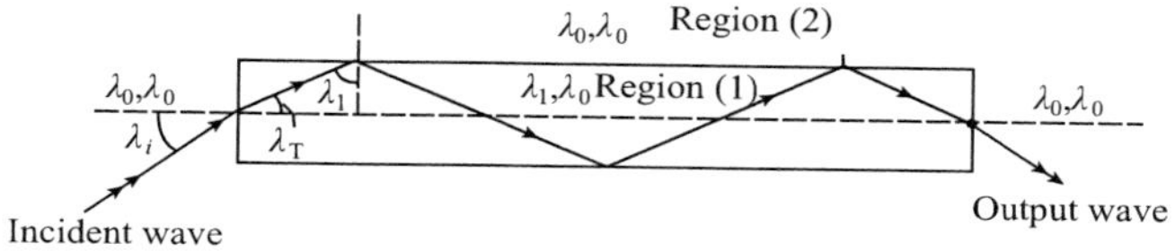

Figure of Example 8.6

The dielectric rod that guides electromagnetic waves all by internal reflection and

$$\theta_i = 90° - \theta_T$$

Thus expression Equation (a) can be written as

$$\cos\theta_T \geqslant \sin\theta_c \tag{b}$$

In the dielectric rod, when the waves transmit from dielectric rod (ε_1) to air (ε_0), then

$$\sin\theta_c = \frac{\sqrt{\varepsilon_0}}{\sqrt{\varepsilon_1}} = 1/\sqrt{\varepsilon_{r1}} \tag{c}$$

When it is incident from air to dielectric rod, we have

$$\sin\theta_T = \sin\theta_i \cdot \frac{1}{\sqrt{\varepsilon_{r1}}} \tag{d}$$

From expression Equation (b), it is known that

$$\sqrt{1-\sin^2\theta_T} \geqslant \sin\theta_c \tag{e}$$

Substituting Equation (d) into the expression above, we get

$$\sqrt{1-\frac{1}{\varepsilon_{r1}}\sin^2\theta_i} \geqslant \frac{1}{\sqrt{\varepsilon_{r1}}}$$

Namely

$$\varepsilon_{r1} \geqslant 1+\sin^2\theta_i \tag{f}$$

Because the incident angle is arbitrary, when θ_i=90°, the right side of formula above reaches the maximum, and its value is 2. Therefore, the relative permittivity $\varepsilon_{r1}$of the dielectric rod in this case is atleast equal to 2. Actually, the permittivity of ordinary glass and quartz materials usually take this value.

8.4.5 Total Refraction of Waves on the Interface between Media — Brewster Angle θ_B

When electromagnetic waves are obliquely incident upon interface between media without any reflection, the corresponding incidence angle is known as Brewster angle θ_B. Just let the reflection coefficient R equal zero, θ_B can thus be found. The following discussion are still made for the two cases of perpendicularly polarization and parallel polarization, and under the condition $\mu_r = 1$.

For perpendicularly polarized waves, with, let the numerator of the given reflection coefficient $R_{(N)}$ [Equation (8.128)] be zero, saying, $R_{(N)} = 0$, then we have

$$\cos\theta = \sqrt{\frac{\varepsilon_2}{\varepsilon_1} - \sin^2\theta} \tag{8.151}$$

which is only valid when$\varepsilon_1 = \varepsilon_2$. Therefore, for perpendicularly polarized waves, the incidence of any angle can result in some reflection, implying that *Brewster's angle does not exist for perpendicularly polarized waves. Only parallel polarized waves have Brewster's angle* θ_B.

1. Find Brewster Angle θ_B

Method 1 Directly let the reflection coefficient of parallel polarized wave equal to zero, namely $R_{(P)}=0$, and thus find θ_B.

According to the reflection coefficient of parallel polarized wave in (8.121), let its numerator be zero, that is, $R_{(P)}=0$, we have

$$\frac{\varepsilon_2}{\varepsilon_1}\cos\theta_B = \sqrt{\frac{\varepsilon_2}{\varepsilon_1}-\sin^2\theta_B} \tag{8.152}$$

where θ_B is used instead of θ_i. Solving the equation above, we obtain that

$$\sin\theta_B = \sqrt{\frac{\varepsilon_2}{\varepsilon_1+\varepsilon_2}} \tag{8.153}$$

or

$$\tan\theta_B = \sqrt{\frac{\varepsilon_2}{\varepsilon_1}} \tag{8.154}$$

Equation (8.153) or (8.154) is the expression of Brewster angle.

Method 2 From the concept of impedance, if $Z_L = Z_{z1}$, namely impedance matching, reflection will not occur [$R_{(P)}=0$], then we can find θ_B.

For parallel polarized waves

$$\left.\begin{array}{l} R_{(P)} = (Z_L - Z_{z1})/(Z_L + Z_{z1}) \\ Z_L = \eta_2\cos\theta_T \\ Z_{z1} = \eta_1\cos\theta_i \end{array}\right\} \tag{8.155}$$

When $Z_L = Z_{z1}$, $R_{(P)}=0$, then the corresponding θ_i is θ_B. Because

$$Z_L = \eta_2\cos\theta_T = \eta_2\sqrt{1-\sin^2\theta_T} = \eta_2\sqrt{1-\frac{\varepsilon_1}{\varepsilon_2}\sin^2\theta_B}$$
$$Z_{z1} = \eta_1\cos\theta_B = \eta_1\sqrt{1-\sin^2\theta_B}$$

Using $Z_L = Z_{z1}$ we obtain

$$\sin\theta_B = \sqrt{\varepsilon_2/(\varepsilon_1+\varepsilon_2)} \tag{8.156}$$

or

$$\tan\theta_B = \sqrt{\varepsilon_2/\varepsilon_1} \tag{8.157}$$

Noting that, the law of refraction is applied in above derivation.

When the angle of incidence equals to Brewster's angle, no reflection of the parallel polarized wave occurs on the interface between media, all the energy is transported into medium 2. Therefore, for the incident wave polarized along arbitrary direction, if $\theta_i = \theta_B$, when it reaches the interface between media, its parallel polarized component will enter into the medium 2 completely, and only part of the perpendicularly polarized component of the wave is reflected, which can be used for polarization filting. Observing Figure 8.13, the critical angle θ_c is slightly larger than the Brewster angle θ_B, due to

$$\sin\theta_c = \sqrt{\frac{\varepsilon_2}{\varepsilon_1}},\quad \sin\theta_B = \sqrt{\frac{\varepsilon_2}{\varepsilon_1+\varepsilon_2}}$$

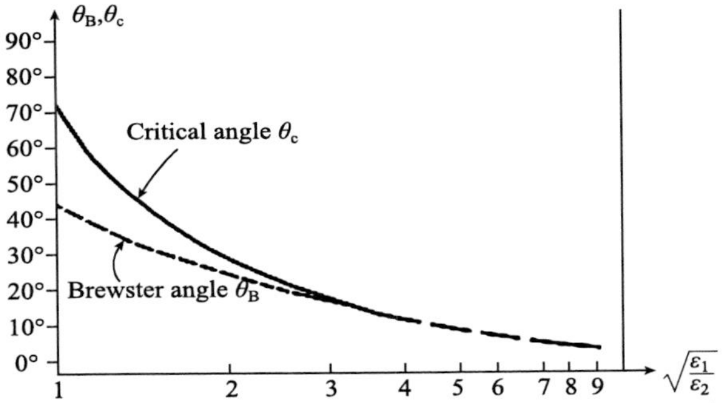

Figure 8.13 Comparison between θ_c and θ_B (for parallel polarized wave) when a wave is transmitted from medium 1 to medium 2

2. When $\theta_i = \theta_B$, Why Do Parallel Polarized Waves not Reflect on the Interface between Media?

Different concepts can be applied to analyze the reasons why no reflection happens.

Method 1 For parallel polarized waves, when the incident angle θ_i equals to θ_B, $Z_{z1} = Z_L$, that is *impedance matching, hence no reflection occurs.* Readers can continue to prove it by themselves.

Method 2 The reason for 'no reflection' can also be found in the characteristics of radiated electromagnetic waves by electric dipole after medium being polarized.

Given that

$$\sin\theta_B = \sqrt{\frac{\varepsilon_2}{\varepsilon_1 + \varepsilon_2}} \tag{8.158}$$

When $\theta_i = \theta_B$

$$\sin\theta_T = \sin\theta_B \cdot \frac{\sqrt{\varepsilon_1}}{\sqrt{\varepsilon_2}} = \sqrt{\frac{\varepsilon_1}{\varepsilon_1 + \varepsilon_2}} \tag{8.159}$$

From Equations (8.158) and (8.159) it is known that

$$\sin^2\theta_B + \sin^2\theta_T = 1$$

$$\theta_B + \theta_T = 90° \tag{8.160}$$

that is, the reflected ray and refracted ray are perpendicular to each other, as shown in Figure 8.14.

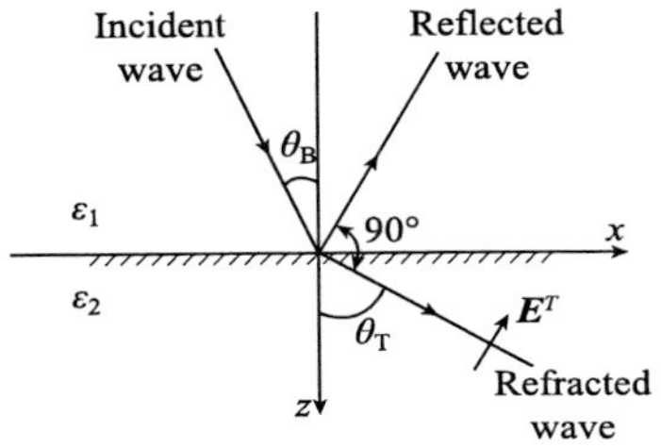

Figure 8.14

After the dielectric is polarized by the electric field of refracted wave $\boldsymbol{E}^T$, electric dipoles appear and begin to radiate at the same frequency. The direction of the resulting electric dipoles after the polarization is the same as the external electric field ($\boldsymbol{E}^T$). The electric dipole has no radiation in its axis direction (the direction of maximum radiation is perpendicular to the axis of the electric dipole. Refer to Section 12.2.4 for details). The electric field of refracted wave $\boldsymbol{E}^T$ for parallel polarized wave is in parallel with the incidence plane, moreover, it is in parallel with the rays of reflected wave, that is, *the axis of electric dipole is in parallel with the reflected rays, as there is no electromagnetic radiation along the axis of dipole, so when $\theta_i = \theta_B$, no reflection occurs for the parallel polarized wave.*

Example 8.7 Plane wave is being incident from air to the surface of pure water (the relative permittivity of pure water is 80), try to find:

(1) The Brewster angle θ_B of parallel polarized wave and the refraction angle $\theta_T$when $\theta_i = \theta_B$;

(2) When the incident angle of perpendicularly polarized wave is the same as the angle of θ_B in the case of (1), find its reflection coefficient and transmission coefficient.

Solutions (1) for parallel polarized wave, from Equation (8.157) we have

$$\theta_B = \arctan\sqrt{\frac{\varepsilon_2}{\varepsilon_1}} = \arctan\sqrt{\frac{80}{1}} = 83.62°$$

Based on Snell's law we get the refraction angle θ_T

$$\sin\theta_T = \frac{\sqrt{\varepsilon_1}}{\sqrt{\varepsilon_2}}\sin\theta_i = \frac{\sqrt{\varepsilon_1}}{\sqrt{\varepsilon_2}}\sin\theta_B = \frac{\sin 83.62°}{\sqrt{80}}$$

$$\theta_T = \arcsin\left(\frac{\sin 83.62°}{\sqrt{80}}\right) \approx 6.379°$$

(2) $R_{(N)}$ and $T_{(N)}$ of perpendicularly polarized wave.

The known incident angle is $\theta_i = 83.62°$ from Equation (8.128) we have

$$R_{(N)} = \frac{\cos\theta_i - \sqrt{\dfrac{\varepsilon_2}{\varepsilon_1} - \sin^2\theta_i}}{\cos\theta_i + \sqrt{\dfrac{\varepsilon_2}{\varepsilon_1} - \sin^2\theta_i}} \approx -0.97528$$

From Equation (8.129) it can be obtained that

$$T_{(N)} = \frac{2\cos\theta_i}{\cos\theta_i + \sqrt{\dfrac{\varepsilon_2}{\varepsilon_1} - \sin^2\theta_i}} \approx 0.0247$$

Satisfying

$$1 + R_N = T_N$$

Example 8.8 A circularly polarized wave is propagating in air, the transmission power density is $S_{av} = 1\text{W/m}^2$, the operating frequency of the electromagnetic wave is very high. Try to extract the power density of linearly polarized wave out of this circularly polarized wave by using a large triangular prism made of glass ($\varepsilon_r = 4$), and calculate the output power density of the linearly polarized wave.

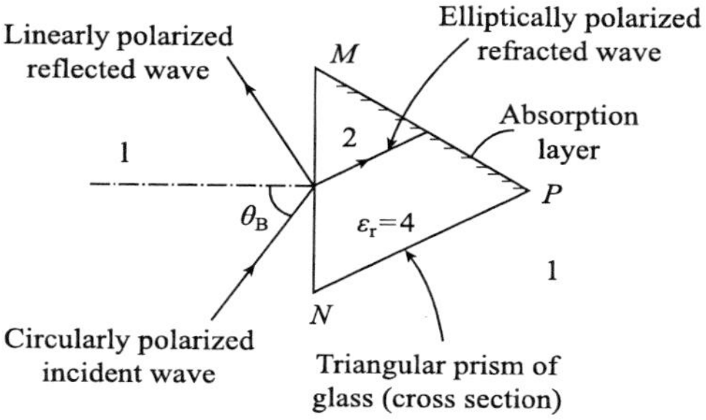

Figure of Example 8.8

Solution Firstly, the electric field of circularly polarized wave is decomposed into two spatially orthogonal electric fields of linearly polarized wave of equal amplitude and time phase difference of 90°, If one electric field is perpendicular to the plane of incidence (called perpendicularly polarized wave), another field must be in parallel with the plane of incidence (called parallel polarized wave).

If rotating the prism to make the incident angle θ_i on its surface MN equal to Brewster angle θ_B, that is

$$\theta_i = \theta_B = \arctan\sqrt{\frac{\varepsilon_2}{\varepsilon_1}} = \arctan\sqrt{4} \approx 63.43^\circ$$

It is known that, when $\theta_i = \theta_B$, for a dielectric like glass ($\mu_r = 1$), only perpendicularly polarized wave (linearly polarized wave) can result in reflection, and the electric field refracted into the glass prism consists of two parts, namely, the electric fields of all parallel polarized waves and part of the perpendicularly polarized waves, which become elliptically polarized waves. They will be incident upon surface MP of the prism covered with a resistive film as absorption layer, so that the incident elliptically polarized wave will no longer be reflected.

Applying the analysis above, the output power of linearly polarized wave can be obtained from the circularly polarized wave.

Let the amplitude of electric field of perpendicularly polarized wave be $E_{0i(N)}$, one of the two linearly polarized waves constituting the circularly polarized waves. Because

$$\sin\theta_i = \sin\theta_B = \sin 63.43^\circ \approx 0.894$$

$$n_2 = 2n_1 \text{ (since the glass}' \ \varepsilon_r = 4)$$

Hence the reflection coefficient $R_{(N)}$ of electric field of perpendicularly polarized wave is

$$R_{(N)} = \frac{\cos\theta - \sqrt{\dfrac{\varepsilon_2}{\varepsilon_1} - \sin^2\theta}}{\cos\theta + \sqrt{\dfrac{\varepsilon_2}{\varepsilon_1} - \sin^2\theta}} \approx -0.6$$

Then the amplitude of electric field of perpendicularly polarized reflected wave E_{0r} is

$$E_{0r} = R_{(N)} E_{0i(N)}$$

Therefore, the average power density of perpendicularly polarized reflected wave $S_{r(av)}$ is

$$\begin{aligned} S_{r(av)} &= \frac{1}{2} E_{0r} H_{0r} = \frac{1}{2}\left(\frac{E_{0r}^2}{\eta_1}\right) \\ &= \frac{1}{2} R_{(N)}^2 E_{0i(N)}^2 \frac{1}{\eta_1} = R_{(N)}^2 \left(\frac{1}{2}\frac{1}{\eta_1} E_{0i(N)}^2\right) \end{aligned}$$

As the power density of circularly polarized wave is the sum of the power densities of two linearly polarized waves, for circularly polarized wave ($E_{0i(N)} = E_{0i(P)}$),$S_{av} = 1\text{W/m}^2$, therefore,

$$2 \times \left(\frac{1}{2}\frac{1}{\eta_1} E_{0i(N)}^2\right) = S_{av} = 1$$

that is

$$\frac{1}{2}\frac{1}{\eta_1} E_{0i(N)}^2 = \frac{1}{2}$$

Hence the average power density of perpendicularly polarized reflected wave is obtained as

$$S_{r(av)} = R_{(N)}^2 \left(\frac{1}{2}\frac{1}{\eta_1} E_{0i(N)}^2\right) \approx 0.18 \left(\text{W/m}^2\right)$$

8.5 Reflection and Refraction of Waves on the Interface between Conductive Media

On the interface between conductive media, the boundary conditions still holds that the tangential components of electric and magnetic fields are continuous, thus, what have been obtained previously, like the law of reflection, the law of refraction (Snell's law) and Fresnel Equations, are still valid. However, in this case, the permittivity is complex. Next we will analyze the characteristics of reflected and refracted waves when the waves are incident from perfect dielectrics into the good conductors.

For conductor (medium 2, with parameters as ε_2, μ_0, σ), its permittivity is equivalent to a complex number $\varepsilon_2^{\mathrm{e}}$

$$\varepsilon_2^{\mathrm{e}} = \varepsilon_2 + \frac{\sigma}{\mathrm{j}\omega} = \varepsilon_2 - \mathrm{j}\frac{\sigma}{\omega} \tag{8.161}$$

Then the wave impedance of the conductor η_2 is also a complex number

$$\eta_2 = \sqrt{\frac{\mu_0}{\varepsilon_2^{\mathrm{e}}}} = \sqrt{\frac{\mathrm{j}\omega\mu_0}{\sigma + \mathrm{j}\omega\varepsilon_2}} \tag{8.162}$$

For a good conductor, there are

$$\eta_2 \approx \sqrt{\frac{\mathrm{j}\omega\mu_0}{\sigma}} = \sqrt{\frac{\pi f \mu_0}{\sigma}}(1+\mathrm{j}) \tag{8.163}$$

and

$$k_2 = \omega\sqrt{\mu_0 \varepsilon_2^{\mathrm{e}}} \approx \sqrt{\pi f \mu \sigma}(1-\mathrm{j}) = \frac{1}{\delta}(1-\mathrm{j}) \tag{8.164}$$

where δ is the skin depth.

When electromagnetic waves are incident to a surface of good conductor, the characteristics of reflected and refracted waves are as follows:

(1) The angle of refraction $\theta_{\mathrm{T}} \approx 0°$.

According to the law of refraction, for $|\varepsilon_2^{\mathrm{e}}| \gg \varepsilon_1$, it can be known that

$$\sin\theta_{\mathrm{T}} = \sin\theta_{\mathrm{i}}\left(\frac{\sqrt{\varepsilon_1}}{\sqrt{\varepsilon_2^{\mathrm{e}}}}\right) \approx 0$$

thus

$$\theta_{\mathrm{T}} \approx 0° \tag{8.165}$$

In this case, *for any incidence upon the surface of a good conductor, the direction of the refracted wave into the conductor is always perpendicular to the interface.*

(2) The reflection coefficients $R_{(\mathrm{N})}$, $R_{(\mathrm{P})} \approx 1$, the transmission coefficients $T_{(\mathrm{N})}, T_{(\mathrm{P})} \ll 1$.

Noting that $|\varepsilon_2^{\mathrm{e}}| \gg \varepsilon_1$, namely $\eta_1 \gg \eta_2$, the following results can be easily obtained from Fresnel Equations:

From Equations (8.121) and (8.122) we know that, $R_{(\mathrm{P})}$, $T_{(\mathrm{P})}$ of parallel polarized wave are

$$R_{(\mathrm{P})} \approx \frac{-\left(\dfrac{\varepsilon_2^{\mathrm{e}}}{\varepsilon_1}\right)\cos\theta}{\left(\dfrac{\varepsilon_2^{\mathrm{e}}}{\varepsilon_1}\right)\cos\theta} = -1 \tag{8.166}$$

$$T_{(\mathrm{P})} = \frac{2\sqrt{\left(\dfrac{\varepsilon_2^{\mathrm{e}}}{\varepsilon_1}\right) - \sin^2\theta}}{\dfrac{\varepsilon_2^{\mathrm{e}}}{\varepsilon_1}\cos\theta + \sqrt{\left(\dfrac{\varepsilon_2^{\mathrm{e}}}{\varepsilon_1}\right) - \sin^2\theta}} \ll 1 \tag{8.167}$$

From Equations (8.128) and (8.129), $R_{(\mathrm{N})}$, $T_{(\mathrm{N})}$ of perpendicularly polarized wave are

$$R_{(\mathrm{N})} \approx \frac{-\sqrt{\dfrac{\varepsilon_2^{\mathrm{e}}}{\varepsilon_1} - \sin^2\theta}}{\sqrt{\dfrac{\varepsilon_2^{\mathrm{e}}}{\varepsilon_1} - \sin^2\theta}} = -1 \tag{8.168}$$

$$T_{(\mathrm{N})} = \frac{2\cos\theta}{\cos\theta + \sqrt{\dfrac{\varepsilon_2^{\mathrm{e}}}{\varepsilon_1} - \sin^2\theta}} \ll 1 \tag{8.169}$$

Obviously, either for parallel polarized wave or perpendicularly polarized wave, the tangential electric field of the reflected wave will reverse its phase on the surface of good conductor, almost total reflection with little refraction occurs. However, the tangential E-fields of refraction wave is not inversed.

(3) When the refracted wave propagates in good conductor, there is not only phase shift along the propagation direction, but also the amplitude of the wave decays exponentially.

Example 8.9 When a plane wave is normally incident upon the surface of copper (medium 2) from air (medium 1), given that the parameters for copper are: $\mu_2 = \mu_0, \varepsilon_2 = \varepsilon_0$, $\sigma_2 = 5.8 \times 10^7\mathrm{S/m}$, find the reflection coefficient and transmission coefficient of the wave when $f = 10^6\mathrm{Hz}$.

Solution According to Equations (8.119) and (8.120), or Equations (8.125) and (8.126), which give the expressions of reflection coefficient R and transmission coefficient T of the electric fields when the plane wave is normally incident upon the interface between media, as follows,

$$R = \frac{\eta_2 - \eta_1}{\eta_2 + \eta_1}$$

$$T = \frac{2\eta_2}{\eta_2 + \eta_1}$$

Since $\eta_1 = 377\Omega$, η_2 can be given based on (8.162) as

$$\eta_2 = \sqrt{\frac{\mu_0}{\varepsilon_2^{\mathrm{e}}}} = \sqrt{\frac{\mathrm{j}\omega\mu_0}{\sigma_2 + \mathrm{j}\omega\varepsilon_2}}$$

Substituting the parameters of medium 2 into the equation above, then we have

$$\eta_2 = 3.69 \times 10^{-4}\angle 45°\Omega$$

Substituting η_1, η_2 into the expression of reflection coefficient R, yields

$$R = \frac{E^-}{E^+} = \frac{\eta_2 - \eta_1}{\eta_2 + \eta_1} = -0.9999986\angle 0.000079°$$

If it is a perfect conductor, the reflection coefficient of electric field is -1, saying, the electric field is totally reflected, and, the phase of electric field is inverted during reflection. From the known reflection coefficient in this case, little difference between magnitude or phase for this case and for perfect conductor can be observed. In other words, copper can be regarded as totally reflecting material for the radio frequencies.

Substituting η_1, η_2 into the expression of transmission coefficient T, we obtain

$$T = \frac{E^T}{E^+} = \frac{2\eta_2}{\eta_2 + \eta_1} = 0.00000196\angle 45^\circ$$

That is, at the copper side of the interface , the electric field is about 2 millionths of the incident E-fields.

As we see, the wave impedance of medium 2 (copper) is $\eta_2 = \dfrac{E^T}{H^T} = 3.69 \times 10^{-4}\angle 45^\circ\Omega$ when $f_0 = 10^6$ Hz, actually almost zero. Therefore, the copper surface can be viewed as a surface with zero impedance.

8.6 Plane Waves Normally Incident upon the Interfaces among Multi-layered Media

Previously the problems of reflection and refraction of waves were discussed when plane waves being normally incident upon the interface between two media. In this section we will study the wave reflection and refraction for the cases of multi-layer-media, as shown in Figure 8.15.

In this section, we will find the impedance observed from $z = -l$ towards positive z direction from field theory. This impedance can be discussed as below, at $z = -l$, the ratio of the total electric field and total magnetic field on the side of medium 1, is denoted by $Z_L(l)$; at $z = -l$, the ratio of the total electric field and total magnetic field on the side of medium 2, is denoted by $Z_{in}(l)$. Therefore, $Z_L(l)$ is the load impedance for the first medium at $z = -l$, and $Z_{in}(l)$ is the input impedance for the second medium at $z = -l$. Because of the continuity of tangential fields at $z = -l$, saying $H_{1t} = H_{2t}$, $E_{1t} = E_{2t}$, then $Z_L(l) = Z_{in}(l)$. Sometimes this impedance is generally referred as the equivalent wave impedance η_{eff}, hence $\eta_{eff} = Z_L(l) = Z_{in}(l)$.

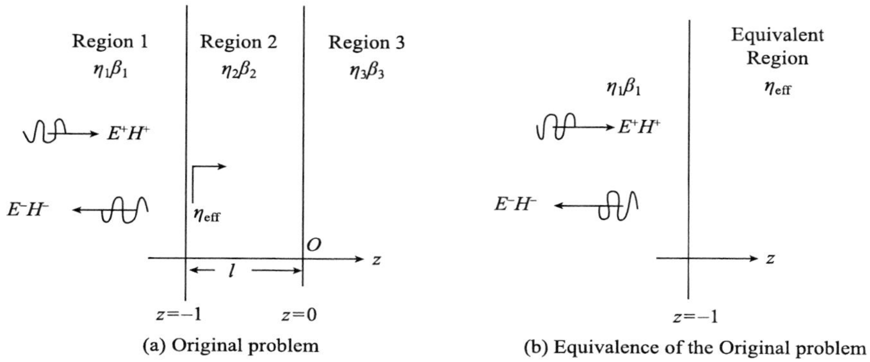

Figure 8.15 The normal incidence for multi-layered media

Now find the equivalent wave impedance η_{eff}.

Method 1 Following the expression of the reflection coefficient (8.23), find η_{eff}according to the field in region (2).

It is assumed that the incident field in the region 2 is

$$E_2^+ = E_{20}^+ e^{-j\beta_2 z}$$
$$H_2^+ = \frac{E_{20}^+}{\eta_2} e^{-j\beta_2 z} \tag{8.170}$$

Then the total electromagnetic field in the region 2 can be expressed as

$$\begin{aligned} E_2(z) &= E_{20}^+ \left[e^{-j\beta_2 z} + R(0)e^{j\beta_2 z}\right] \\ H_2(z) &= \frac{E_{20}^+}{\eta_2} \left[e^{-j\beta_2 z} + R(0)e^{j\beta_2 z}\right] \end{aligned} \qquad (-l \leqslant z \leqslant 0) \tag{8.171}$$

where $R(0)$ is the reflection coefficient of electric field at $z = 0$, whose magnitude and phase depend on η_2 and η_3. Following Equation (8.23), then $R(0)$ can be written as

$$R(0) = (\eta_3 - \eta_2)/(\eta_3 + \eta_2) \quad (\text{for } z = 0) \tag{8.172}$$

If the equivalent wave impedance η_{eff} is defined at $z = -l^+$ (that is, on the side of medium 2 at the first interface between media), the corresponding ratio of the E-field and B-fields is

$$\eta_{\text{eff}} = \left.\frac{E_2}{H_2}\right|_{z=-l^+} = \eta_2 \frac{e^{j\beta_2 l} + R(0)e^{-j\beta_2 l}}{e^{j\beta_2 l} - R(0)e^{-j\beta_2 l}} \tag{8.173}$$

Substituting $R(0)$ of Equation (8.172) into the equation above and representing $e^{\pm j\beta_2 l}$ with Euler's formula, after some simplification, we can obtain the equivalent impedance at $z = -l$ as

$$\eta_{\text{eff}} = \eta_2 \frac{\eta_3 + j\eta_2 \tan \beta_2 l}{\eta_2 + j\eta_3 \tan \beta_2 l} \tag{8.174}$$

According to Equation (8.23) and Figure 8.15 (b), it's easy to write the reflection coefficient $R(-l)$ at $z = -l$ as

$$R(-l) = \left.\frac{E_1^-}{E_1^+}\right|_{z=-l} = \frac{\eta_{\text{eff}} - \eta_1}{\eta_{\text{eff}} + \eta_1} \tag{8.175}$$

Method 2 Applying the boundary conditions of *two interfaces* to find the equivalent impedance η_{eff}.

There are five kinds of waves in the three media: their own incident and reflected waves of region (1) and region (2), incident wave of region (3). If the incident wave of region (1) is given, and the remaining four are unknown, as there are two boundary conditions ($E_{1t} = E_{2t}, H_{1t} = H_{2t}$) for each interface between media, two interfaces will give four boundary conditions, then all the fields can be solved and η_{eff} can be found.

Based on Figure 8.15, the waves of three regions can be easily written as follows.

Region (1)

$$E_{1t} = E_1^+ + E_1^- = E_{10}^+ e^{-j\beta_1(z+l)} + E_{10}^- e^{j\beta_1(z+l)}$$
$$H_{1t} = \frac{1}{\eta_1}\left[E_{10}^+ e^{-j\beta_1(z+l)} - E_{10}^- e^{j\beta_1(z+l)}\right] \tag{8.176}$$

Region (2)

$$E_{2t} = E_2^+ + E_2^- = E_{20}^+ e^{-j\beta_2 z} + E_{20}^- e^{j\beta_2 z}$$
$$H_{2t} = \frac{1}{\eta_2}\left(E_{20}^+ e^{-j\beta_2 z} - E_{20}^- e^{j\beta_2 z}\right) \tag{8.177}$$

Region (3)

$$\begin{aligned} E_{3t} &= E_{30}^{+}\mathrm{e}^{-\mathrm{j}\beta_3 z} \\ H_{3t} &= \frac{1}{\eta_3}E_{30}^{+}\mathrm{e}^{-\mathrm{j}\beta_3 z} \end{aligned} \tag{8.178}$$

On the first interface ($z = -l$), from the boundary conditions $E_{1t} = E_{2t}$ and $H_{1t} = H_{2t}$, the following can be obtained respectively as

$$E_{10}^{+} + E_{10}^{-} = E_{20}^{+}\left[\mathrm{e}^{\mathrm{j}\beta_2 l} + R(0)\mathrm{e}^{-\mathrm{j}\beta_2 l}\right] \tag{8.179}$$

$$\frac{1}{\eta_1}(E_{10}^{+} - E_{10}^{-}) = \frac{E_{20}^{+}}{\eta_2}\left[\mathrm{e}^{\mathrm{j}\beta_2 l} - R(0)\mathrm{e}^{-\mathrm{j}\beta_2 l}\right] \tag{8.180}$$

where $R\,(0)$ is the reflection coefficient on the second interface at $z = 0$.

The ratio of the total electric and magnetic fields on the first interface ($z = -l$), is the equivalent wave impedance η_{eff}.

Dividing Equation (8.179) by Equation (8.180), we have

$$\eta_{\mathrm{eff}} = \eta_2\frac{\mathrm{e}^{\mathrm{j}\beta_2 l} + R(0)\mathrm{e}^{-\mathrm{j}\beta_2 l}}{\mathrm{e}^{\mathrm{j}\beta_2 l} - R(0)\mathrm{e}^{-\mathrm{j}\beta_2 l}} \tag{8.181}$$

On the second interface ($z = 0$), considering the boundary conditions $E_{2t} = E_{3t}$ and $H_{2t} = H_{3t}$, with Equations (8.177) and (8.178), we get

$$E_{20}^{+} + E_{20}^{-} = E_{303}^{++} \tag{8.182}$$

$$\frac{1}{\eta_2}(E_{20}^{+} - E_{20}^{-}) = \frac{1}{\eta_3}E_{30}^{+} \tag{8.183}$$

From the two equations above, it is found that

$$R(0) = \left.\frac{E_{20}^{-}}{E_{20}^{+}}\right|_{z=0} = \frac{\eta_3 - \eta_2}{\eta_3 + \eta_2} \tag{8.184}$$

Substituting the equation above into Equation (8.181), yields

$$\eta_{\mathrm{eff}} = \eta_2\frac{\eta_3 + \mathrm{j}\eta_2\tan\beta_2 l}{\eta_2 + \mathrm{j}\eta_3\tan\beta_2 l} \tag{8.185}$$

8.7 On the Multiformity of the Definitions of Fresnel Equations (R, T)

Generally, different definitions can be used for studying Fresnel Equations (R, T) in the many electromagnetic textbooks, resulting in different expressions and confusing the new beginners.

In this section, several different expressions of reflection coefficient $R_{(\mathrm{P})}$, refractive index $T_{(\mathrm{P})}$ of parallel polarized waves, derived through various definitions, as well as their relationships, will be introduced. Some importance points can concluded as follows:

(1) The expressions of $R_{(\mathrm{P})}$, $T_{(\mathrm{P})}$ from each definition are all correct, since all of them are obtained based on wave equations and boundary conditions.

(2) The expressions and related figures of $R_{(\mathrm{P})}$, $T_{(\mathrm{P})}$ for each definition are self-contained, and should not be mixed up with other formulas and corresponding figures of $R_{(\mathrm{P})}$, $T_{(\mathrm{P})}$

for other definitions. In other words, a set of definitions should be used separately and completely.

(3) For specific issues, R, T from a certain definition may be more convenient to use.

Several different definitions of Fresnel Equations are introduced as follows and several examples will also be given.

Method 1 R, T are defined in terms of tangential electric fields. Assuming that when the parallel polarized wave is reflected, the phases of tangential electric fields do not change (as defined in the previous sections of this book).

The results in this book are directly used here [obtained from Equations (8.119), (8.120), (8.116) and (8.117)]

$$R_{(\mathrm{P})} = \frac{Z_{\mathrm{L}} - Z_{z1}}{Z_{\mathrm{L}} + Z_{z1}} = \frac{\eta_2 \cos\theta_{\mathrm{T}} - \eta_1 \cos\theta_{\mathrm{i}}}{\eta_2 \cos\theta_{\mathrm{T}} + \eta_1 \cos\theta_{\mathrm{i}}} \tag{8.186}$$

$$T_{(\mathrm{P})} = \frac{2Z_{\mathrm{L}}}{Z_{\mathrm{L}} + Z_{z1}} = \frac{2\eta_2 \cos\theta_{\mathrm{T}}}{\eta_2 \cos\theta_{\mathrm{T}} + \eta_1 \cos\theta_{\mathrm{i}}} \tag{8.187}$$

and

$$1 + R_{(\mathrm{P})} = T_{(\mathrm{P})} \tag{8.188}$$

Method 2 $R_{(\mathrm{P})}, T_{(\mathrm{P})}$ are defined in terms of total electric fields. Assuming that when the parallel polarized wave is reflected, the phases of tangential electric fields do not change.

As stated above, when the parallel polarized wave is reflected, the phases of tangential electric fields do not change, saying, the directions of electromagnetic fields of incident, reflected and refracted waves under parallel polarization, defined in Figure 8.9, are employed. (The process is omitted)

Applying the boundary condition $E_{1t} = E_{2t}$ at $z = 0$, we get

$$E^+ \cos\theta_{\mathrm{i}} \mathrm{e}^{-\mathrm{j}k_1 x \sin\theta_{\mathrm{i}}} + E^- \cos\theta_{\mathrm{r}} \mathrm{e}^{-\mathrm{j}k_1 x \sin\theta_{\mathrm{r}}} = E^T \cos\theta_{\mathrm{T}} \mathrm{e}^{-\mathrm{j}k_2 x \sin\theta_{\mathrm{T}}}$$

The reflection coefficient R and the transmission coefficient T are defined as

$$R_{(\mathrm{P})} = \frac{E^-}{E^+}, \quad T_{(\mathrm{P})} = \frac{E^T}{E^+}$$

Considering the laws of reflection and refraction, we thus have

$$E^+ \cos\theta_{\mathrm{i}} + R_{(\mathrm{P})} E^+ \cos\theta_{\mathrm{i}} = T_{(\mathrm{P})} E^+ \cos\theta_{\mathrm{T}} \tag{8.189}$$

Using the boundary condition $H_{1t} = H_{2t}$, it is obtained that

$$\frac{E^+}{\eta_1} - R_{(\mathrm{P})} \frac{E^+}{\eta_1} = T_{(\mathrm{P})} \frac{E^+}{\eta_2} \tag{8.190}$$

thus we find

$$R_{(\mathrm{P})} = \frac{\eta_2 \cos\theta_{\mathrm{T}} - \eta_1 \cos\theta_{\mathrm{i}}}{\eta_2 \cos\theta_{\mathrm{T}} + \eta_1 \cos\theta_{\mathrm{i}}} \tag{8.191}$$

$$T_{(\mathrm{P})} = \frac{2\eta_2 \cos\theta_{\mathrm{i}}}{\eta_2 \cos\theta_{\mathrm{T}} + \eta_1 \cos\theta_{\mathrm{i}}} \tag{8.192}$$

and

$$1 + R_{(\mathrm{P})} = T_{(\mathrm{P})} \cdot \frac{\cos\theta_{\mathrm{T}}}{\cos\theta_{\mathrm{i}}} \tag{8.193}$$

Method 3 $R_{(\mathrm{P})}, T_{(\mathrm{P})}$ are defined in terms of total electric fields. However, here assuming that when the parallel polarized wave is reflected, the phases of the tangential electric fields are reversed, saying, having change of 180°, hence both the E-fields and B-fields of the reflected wave shown in Figure 8.9 should be reversed.

The deriving process is similar to those of Method 1 and Method 2, which is omitted here, and only the results are given:

$$R_{(\mathrm{P})} = \frac{\eta_1 \cos\theta_\mathrm{i} - \eta_2 \cos\theta_\mathrm{T}}{\eta_1 \cos\theta_\mathrm{i} + \eta_2 \cos\theta_\mathrm{T}} \tag{8.194}$$

$$T_{(\mathrm{P})} = \frac{2\eta_2 \cos\theta_\mathrm{i}}{\eta_1 \cos\theta_\mathrm{i} + \eta_2 \cos\theta_\mathrm{T}} \tag{8.195}$$

and

$$1 + R_{(\mathrm{P})} = T_{(\mathrm{P})} \cdot \frac{\eta_1}{\eta_2} \tag{8.196}$$

Method 4 R, T are defined in terms of power density $\boldsymbol{S}_\mathrm{av}$ (omitted).

Example 8.10 A parallel polarized wave is obliquely incident from air upon the surface of a perfect dielectric. The incidence angle $\theta_\mathrm{i} = 30°$, dielectric parameters are $4\,\varepsilon_0$, μ_0. The amplitude of incident electric field $E_0^+ = 1\mathrm{V/m}$, try to use different definitions of Fresnel Equations to find the following items *on the interface* between dielectrics:

(1) The amplitude of electric field E_0^- of reflected wave and $S^-_{\mathrm{av}(0)}$ of the reflected wave;

(2) The amplitude of electric field E_0^T of refracted wave and $S^T_{\mathrm{av}(0)}$ of the reflected wave.

Solution (1) First a few basic parameters should be solved:

$$\eta_1 = \sqrt{\mu_0/\varepsilon_0} = 120\pi\,(\Omega)\,, \quad \eta_2 = \sqrt{\mu_0/4\varepsilon_0} = 60\pi\,(\Omega)$$

From the law of refraction we have

$$\sin\theta_\mathrm{T} = \sin\theta_\mathrm{i}\frac{\sqrt{\varepsilon_1}}{\sqrt{\varepsilon_2}} = \frac{1}{2}\sin\theta_\mathrm{i} = \frac{1}{4}$$

$$\theta_\mathrm{T} = 14.48°, \quad \theta_\mathrm{i} = 30°$$

$$\cos\theta_\mathrm{T} = 0.968$$

$$\cos\theta_\mathrm{i} = \cos 30° = 0.866$$

On the interface, the power density of incident wave is

$$\begin{aligned} S^+_{\mathrm{av}(0)} &= S^+_{\mathrm{av}} \cos\theta_i = \frac{1}{2}E_0^+ H_0^+ \cos\theta_\mathrm{i} \\ &= \frac{1}{2}(E_0^+)^2 \frac{1}{120\pi}\cos 30° \\ &= 1.326 \times 10^{-3} \times 0.866 = 1.1483 \ \ (\mathrm{mW/m^2}) \end{aligned}$$

(2) Find E_0^-, $S^-_{\mathrm{av}(0)}$ and E_0^T, $S^T_{\mathrm{av}(0)}$ on *the interface between dielectrics.*

Making use of the formula system in Method 1, i.e., Equations (8.186)~(8.188) ($R_{(\mathrm{P})}, T_{(\mathrm{P})}$ defined in terms of tangential electric fields, and assuming no phase change for E_t^+ after reflection), to solve the problem

(a) Find E_0^-, $S^-_{\mathrm{av}(0)}$.

First $R_{(\mathrm{P})}$ is obtained before, E_0^-, $S^-_{\mathrm{av}(0)}$ are thus obtained.

$$R_{(\mathrm{P})} = \frac{\eta_2 \cos\theta_\mathrm{T} - \eta_1 \cos\theta_\mathrm{i}}{\eta_2 \cos\theta_\mathrm{T} + \eta_1 \cos\theta_\mathrm{i}} = -0.2829$$

$$E_0^- = R_{(\mathrm{P})} E_0^+ = -0.2829(\mathrm{V/m})$$

$$S^-_{\mathrm{av}(0)} = S^-_{\mathrm{av}} \cdot \cos\theta_{\mathrm{r}} = S^-_{\mathrm{av}} \cdot \cos\theta_{\mathrm{i}} = \frac{1}{2}E^-_0 H^-_0 \cdot \cos\theta_{\mathrm{i}}$$
$$= \left[\frac{1}{2}(E^-_0)^2 \frac{1}{120\pi}\right] \cdot \cos 30^\circ = 0.09188 \quad (\mathrm{mW/m^2})$$

(b) Find E^T_0, $S^T_{\mathrm{av}(0)}$.
The tangential component of electric field of refracted wave on the interface is

$$E^T_{0(t)} = T_{(\mathrm{P})} E^+_{0(t)}$$
$$E^T_0 \cos\theta_{\mathrm{T}} = T_{(\mathrm{P})} E^+_0 \cos\theta_{\mathrm{i}} = T_{(\mathrm{P})} \cos\theta_{\mathrm{i}}$$
$$E^T_0 = T_{(\mathrm{P})} \frac{\cos\theta_{\mathrm{i}}}{\cos\theta_{\mathrm{T}}}$$

And Equation (8.188) in Method 1 shows that

$$T_{(\mathrm{P})} = 1 + R_{(\mathrm{P})} = 0.7171$$
$$E^T_0 = 0.7171 \cdot \left(\frac{\cos\theta_{\mathrm{i}}}{\cos\theta_{\mathrm{T}}}\right) = 0.64145$$
$$S^T_{\mathrm{av}(0)} = S^T_{\mathrm{av}} \cdot \cos\theta_{\mathrm{T}} = \frac{1}{2}E^T_0 \left(\frac{E^T_0}{\eta_2}\right) \cos\theta_{\mathrm{T}}$$
$$= \frac{1}{2}\frac{(0.64145)^2}{60\pi} \times 0.968 = 1.056(\mathrm{mW/m^2})$$

Another method to obtain $S^T_{\mathrm{av}(0)}$ is

$$S^T_{\mathrm{av}} = \frac{1}{2}E^T_0 H^T_0 = \frac{1}{2}(E^T_0)^2 \frac{1}{\eta_2}$$
$$= \frac{1}{2}\left(\frac{E^T_{0(t)}}{\cos\theta_T}\right)^2 \frac{1}{\eta_2} \qquad (E^T_{0(t)} = E^T_0 \cos\theta_{\mathrm{T}})$$
$$= \frac{1}{2}\left(\frac{T_{(\mathrm{P})} E^+_{0(t)}}{\cos\theta_{\mathrm{T}}}\right)^2 \frac{1}{\eta_2} \qquad (E^T_{0(t)} = E^+_{0(T)} T_{(\mathrm{P})})$$
$$= \frac{1}{2}\left(\frac{T_{(\mathrm{P})} E^+_0 \cos\theta_1}{\cos\theta_{\mathrm{T}}}\right)^2 \cdot \frac{1}{\eta_2} \qquad (E^+_{0(t)} = E^+_0 \cos\theta_{\mathrm{i}})$$
$$= \frac{1}{2}T^2_{(\mathrm{P})}(E^+_0)^2 \cdot \frac{1}{\eta_2}\left(\frac{\cos\theta_{\mathrm{i}}}{\cos\theta_{\mathrm{T}}}\right)^2 = 1.0912 \quad (\mathrm{mW/m^2})$$

$$S^T_{\mathrm{av}(0)} = S^T_{\mathrm{av}} \cos\theta_{\mathrm{T}} = 1.0912 \times \cos 14.48^\circ \approx 1.056(\mathrm{mW/m^2})$$

The results above are obviously accurate, because they satisfy the relationship of energy conservation, saying, on the interface between dielectrics (the plane at $z = 0$) the following is satisfied

$$S^+_{\mathrm{av}(0)} - S^-_{\mathrm{av}(0)} = S^T_{\mathrm{av}(0)}$$

that is

$$1.1483 - 0.09188 = 1.056(\mathrm{mW/m^2})$$

Using the formula in Method 2, saying Equations (8.191)~(8.193) ($R_{(\mathrm{P})}, T_{(\mathrm{P})}$ defined in terms of total electric fields, and assuming no phase change for E^+_t), to solve the problem

It is noted that the $R_{(\mathrm{P})}$ in Method 1 and Method 2 [Equations (8.186) and (8.191)] are exactly the same, hence the results above can be used directly, that is

$$R_{(\mathrm{P})} = -0.2829, \quad S^{+}_{\mathrm{av}(0)} = 1.1483 \quad (\mathrm{mW/m^2})$$

However, $T_{(\mathrm{P})}$ in Method 2 is

$$T_{(\mathrm{P})} = \frac{2\eta_2 \cos\theta_{\mathrm{i}}}{\eta_2 \cos\theta_{\mathrm{T}} + \eta_1 \cos\theta_{\mathrm{i}}} = 0.64145$$

(a) Find E_0^-, $S^{-}_{\mathrm{av}(0)}$.

$$E_0^- = R_{(\mathrm{P})} E_0^+ = R_{(\mathrm{P})} = -0.2829$$

$$\begin{aligned} S^{-}_{\mathrm{av}(0)} &= S^{-}_{\mathrm{av}} \cdot \cos\theta_{\mathrm{r}} \\ &= \left[\frac{1}{2}(E_0^-)^2 \cdot \frac{1}{120\pi}\right] \cdot \cos 30^\circ \\ &= 0.09188 \quad (\mathrm{mW/m^2}) \end{aligned}$$

(b) Find E_0^T, $S^{T}_{\mathrm{av}(0)}$.

$$\begin{aligned} E_0^T &= T_{(\mathrm{P})} E_0^+ = T_{(\mathrm{P})} = 0.64145 \quad (\mathrm{V/m}) \\ S^{T}_{\mathrm{av}(0)} &= S^{T}_{\mathrm{av}} \cdot \cos\theta_{\mathrm{T}} \end{aligned}$$

And

$$\begin{aligned} S^{T}_{\mathrm{av}} &= \frac{1}{2} E_0^T H_0^T = \frac{1}{2}(E_0^T)^2 \frac{1}{\eta^2} = \frac{1}{2}(T_{(\mathrm{P})} E_0^+)^2 \cdot \frac{1}{\eta^2} \\ &= \frac{1}{2}(0.64145)^2 \cdot \frac{1}{\eta_2} \approx 1.0914 \quad (\mathrm{mW/m^2}) \\ S^{T}_{\mathrm{av}(0)} &= S^{T}_{\mathrm{av}} \cdot \cos\theta_{\mathrm{T}} \approx 1.056 \quad (\mathrm{mW/m^2}) \end{aligned}$$

It is obvious that the results satisfy the relationship of energy conservation on the interface between dielectrics, i.e.

$$S^{+}_{\mathrm{av}(0)} - S^{(-)}_{\mathrm{av}(0)} = S^{T}_{\mathrm{av}(0)}$$

that is

$$1.1483 - 0.09188 \approx 1.056 \quad (\mathrm{mW/m^2})$$

Using the formula in Method 3, say Equations (8.194)~(8.196) ($R_{(\mathrm{P})}, T_{(\mathrm{P})}$ defined in terms of total electric fields, still assuming no phase change for E_t^+ after reflection), to solve the problem

Substituting the known $\eta_1, \eta_2, \theta_{\mathrm{i}}, \theta_{\mathrm{r}}$ into Equations (8.194)~(8.196) of Method 3, then we have

$$R_{(\mathrm{P})} = 0.2829, \quad T_{(\mathrm{P})} = 0.64145$$

And they satisfy the relationship in Method 3

$$1 + R_{(\mathrm{P})} = T_{(\mathrm{P})} \frac{\eta_1}{\eta_2} = 1.2829$$

given that $S^{+}_{\mathrm{av}(0)} = 1, \theta_{\mathrm{i}} = 30^\circ, \theta_{\mathrm{T}} = 14.48^\circ, E_0^+ = 1\mathrm{V/m}$.

(a) Find E_0^-, $S^-_{\text{av}(0)}$

$$E_0^- = R_{(\text{P})}E_0^+ = R_{(\text{P})} = 0.2829$$

$$S^-_{\text{av}(0)} = S^-_{\text{av}} \cdot \cos\theta_\text{r} = \left[\frac{1}{2}(E_0^-)^2 \cdot \frac{1}{120\pi}\right]\cos 30^\circ = 0.09188(\text{mW/m}^2)$$

(b) Find E_0^T, $S^T_{\text{av}(0)}$

$$E_0^T = T_{(\text{P})}E_0^+ = T_{(\text{P})} = 0.64145$$

$$\begin{aligned} S^T_{\text{av}(0)} &= S^T_{\text{av}} \cdot \cos\theta_\text{T} = \left(\frac{1}{2}E_0^T H_0^T\right) \cdot \cos\theta_\text{T} \\ &= \left[\frac{1}{2}(E_0^T)^2 \frac{1}{\eta^2}\right] \cdot \cos\theta_\text{T} \\ &= \left[\frac{1}{2}(T_{(\text{P})}E_0^+)^2 \frac{1}{\eta^2}\right] \cdot \cos\theta_\text{T} \\ &= 1056 \quad (\text{mW/m}^2) \end{aligned}$$

Apparently, on the interface between dielectrics, the relationship of energy conservation is satisfied

$$S^+_{\text{av}(0)} - S^-_{\text{av}(0)} = S^T_{\text{av}(0)}$$

namely

$$1.1483 - 0.09188 \approx 1.056 \quad (\text{mW/m}^2)$$

Although $R_{(\text{P})}, T_{(\text{P})}$ under various definitions are different, E_0^-, $S^-_{\text{av}(0)}$, E_0^T, $S^T_{\text{av}(0)}$ obtained from different definitions are still same.

Exercises

8.1 A uniform plane wave is incident from air normally upon the surface of perfect conductor at $z = 0$ (xOy plane), given the electric field of incident wave $E_x^+ = E_0^+\text{e}^{\text{j}(\omega t - kz)}$, try to find:

(1) magnetic field of incident wave H_y^+;

(2) magnetic field of reflected wave H_y^-;

(3) magnetic field of the total wave $H_y = H_y^+ + H_y^- = ?$

8.2 In Cartesian coordinates, a TEM wave is propagating along arbitrary direction, with electric field $\boldsymbol{E} = \boldsymbol{E}_0\text{e}^{\text{j}(\omega t - k_x x + k_y y - k_z z)}$, try to solve the following problems:

(1) What do k_x, k_y, k_z represent, respectively?

(2) Write the relationship between k_x, k_y, k_z and the phase constant k of TEM wave. ?

8.3 As shown in Figure of Exercises 8.3, the electric field of a TEM wave in air is $E_y = E_0\text{e}^{\text{j}(\omega t - 2x - 3z)}$, try to find:

(1) k_x, k_z;

(2) θ (the angle between the incident ray and z axis);

(3) phase constant k.

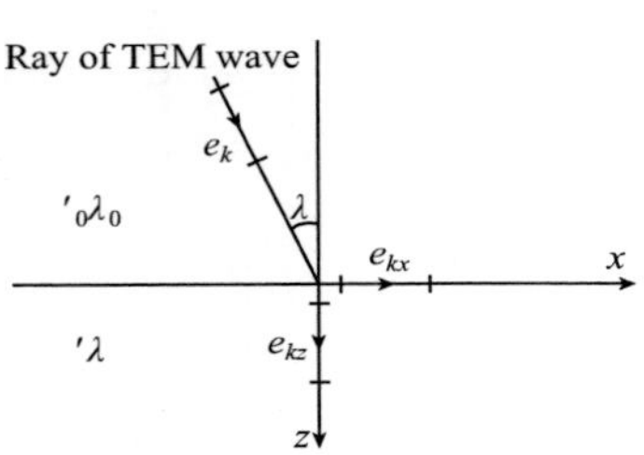

Figure of Exercises 8.3

8.4 To satisfy the boundary conditions on the interface between media (such as $E_{1t} = E_{2t}$, and etc.), the propagation velocity of incident wave, reflected wave, refracted wave along certain direction on the interface (such as the x direction) should be equal ($v^+_{\text{p}(x)} = v^-_{\text{p}(x)} = v^T_{\text{p}(x)}$), that is, $\dfrac{v_1}{\sin\theta_\text{i}} = \dfrac{v_1}{\sin\theta_\text{r}} = \dfrac{v_2}{\sin\theta_\text{T}}$($v_1, v_2$ are the propagation velocity of TEM wave in medium 1 and medium 2, respectively), try to find:

(1) the expression of law of reflection;

(2) the expression of law of refraction.

8.5 A harmonically varying uniform plane wave is incident from air upon the surface of perfect conductor at $z = 0$ (xOy plane), given that the incident electric field is: $\boldsymbol{E}^+ = \boldsymbol{e}_y 10\mathrm{e}^{\mathrm{j}(\omega t - 6x - 8z)}\mathrm{V/m}$. Find:

(1) the incident angle θ_{i} of the wave;

(2) the frequency f and the wavelength λ;

(3) write the complex form of electric field for reflected wave;

(4) write the expression of the electric field of the total wave.

8.6 A uniform plane wave, with frequency of 1GHz and the amplitude of electric field of 1V/m, is normally incident from air upon the surface of copper conductor. Find the average power absorbed by the surface of copper per square meter.

8.7 A TEM wave is normally incident from air upon the surface of a dielectric $(\varepsilon_0\varepsilon_{\mathrm{r}}, \mu_0)$, 20% of the power is being reflected. Find the relative permittivity ε_{r} of the dielectric.

8.8 A uniform plane wave is incident from air to the surface of a perfect conductor at $x = 0$, where the electric field of incident wave is $\boldsymbol{E}^+ = \boldsymbol{e}_y 20\mathrm{e}^{\mathrm{j}(4x-2z)}\mathrm{V/m}$. Find:

(1) the direction of incident wave $\boldsymbol{e}_{k(\mathrm{i})}$;

(2) the incidence angle;

(3) the expression of magnetic field of reflected wave;

(4) the average power density along x direction.

8.9 A uniform plane electromagnetic wave is normally incident from a medium with impedance η upon the surface of a good conductor with conductivity σ and permeability μ, try to prove that the ratio of power flux density penetrated into the conductor and the power density of incident wave is approximately equal to $4R_{\mathrm{s}}/\eta$, where R_{s} is the surface resistance.

8.10 Assuming that a plane wave is normally incident upon the metallic surface, prove that: all the electromagnetic energy penetrated into the mental is dissipated as the Joule heat.

8.11 In free space, a plane wave with the amplitude of electric field of 100V/m penetrates through a silver film with thickness of 5μm , as shown in Figure of Exercises 8.11, the conductivity $\sigma = 6.17 \times 10^7$ S/m, $f = 200$MHz, find E_2, E_3, E_4.

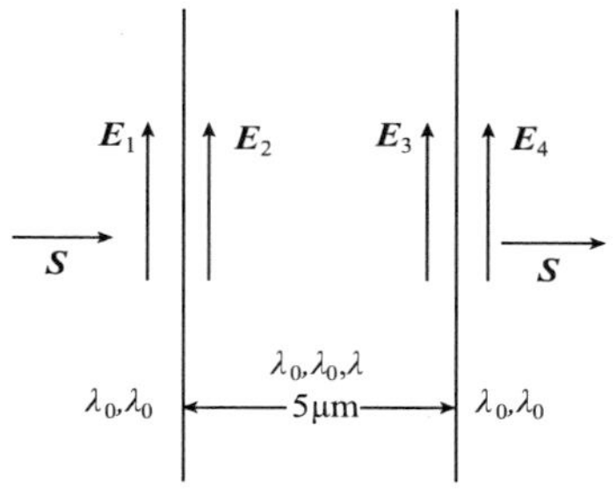

Figure of Exercises 8.11

8.12 Try to derive the formula of group velocity for uniform plane wave in good conductor.

8.13 A plane wave with electric field intensity $\boldsymbol{E}^+ = \boldsymbol{e}_x E_0 \sin\omega\left(t - \dfrac{z}{v_1}\right)$, is normally incident from air onto the interface (z=0) of glass ($\varepsilon_r = 4$, μ_r= 1). Find:

(1) electric field and magnetic field of reflected wave: $\boldsymbol{E}^-, \boldsymbol{H}^-$;

(2) electric field and magnetic field of refracted wave: $\boldsymbol{E}^T, \boldsymbol{H}^T$.

8.14 A uniform plane wave with electric field intensity$\boldsymbol{E}^+ = E_0(\boldsymbol{e}_x + \mathrm{j}\boldsymbol{e}_y)\mathrm{e}^{-\mathrm{j}\beta z}$, is normally incident from air to a lossless dielectric board (it is dielectric for $z \geqslant 0$, its $\mu_{\mathrm{r}} = 1$, $\varepsilon_{\mathrm{r}} = 4$). Find:

(1) the electric field intensity of reflected wave;

(2) the magnetic field intensity of transmission wave;

(3) the respective polarization situation of incident wave, reflected wave and transmission wave.

8.15 A plane wave is normally incident from air upon the surface of a certain medium $(\varepsilon_0\varepsilon_{\mathrm{r}}, \mu_0)$, if it is assumed that the reflection coefficient and transmission coefficient (refractive index) are equal in magnitude. Find:

(1) ε_{r};

(2) For incident wave $S_{\mathrm{av}}^+ = 1\mathrm{mW/m^2}$, find the S_{av}^- and S_{av}^T corresponding to the reflected and refracted waves.

8.16 A uniform plane wave (the electric and magnetic fields are oriented along x-, y- directions respectively, and transmitting in z direction) is normally incident from air upon the surface plane of a perfect medium (the plane where $z = 0$), it is known that the electric field of incident wave on the interface is $E_0^+ = 1.5 \times 10^{-3}\mathrm{V/m}$, the magnetic field of reflected wave is $H_0^- = 1.326 \times 10^{-6}\mathrm{A/m}$.

(1) $\mu_r = 1$ for the medium, find the ε_r;

(2) if $\omega = 3 \times 10^8$ rad/s, find the $\boldsymbol{E}$ and $\boldsymbol{H}$ at any point in the medium.

8.17 A uniform plane wave is normally incident from perfect dielectric 1 to the interface of perfect dielectric 2, if for the two media there is $\mu_{r1} = \mu_{r2} = 1$. Find:

(1) the ratio between the relative permittivity of the two dielectrics is $\varepsilon_{r2}/\varepsilon_{r1}$, in which 10% of the power of incident wave is reflected;

(2) the ratio between the relative permittivity of the two dielectrics $\varepsilon_{r2}/\varepsilon_{r1}$, in which 10% of the power of incident wave is transmitted into the dielectric 2.

8.18 The amplitude of electric field of a uniform plane wave is $E_{0x}^+ = 100\mathrm{e}^{\mathrm{j}0^\circ}$ V/m, which is normally incident from air upon the surface of a lossless dielectric ($z = 0$, for the dielectric, $\mu_2 = \mu_0, \varepsilon_2 = 4\varepsilon_0, \sigma_2 = 0$). Find the amplitudes of electric field of reflected wave and transmission wave.

8.19 The electric field intensity of a uniform plane electromagnetic wave is $\boldsymbol{E} = \boldsymbol{e}_x 10\mathrm{e}^{-\mathrm{j}6z}$, and the wave is normally incident from air upon the surface ($z = 0$) of a lossy medium (ε_r= 2.5, loss tangent is 0.5). Find:

(1) the instantaneous electric field $\boldsymbol{E}^-$ and magnetic field $\boldsymbol{H}^-$ of reflected wave, and the instantaneous electric field $\boldsymbol{E}^T$ and magnetic field $\boldsymbol{H}^T$ of transmission wave;

(2) the average Poynting vector $\boldsymbol{S}_{\mathrm{av}}$ in air and the lossy medium.

8.20 A TEM wave is obliquely incident from air to the surface of a perfect conductor ($z = 0$), given that the electric field of incident wave is $E_y^+ = E_0\mathrm{e}^{\mathrm{j}(\omega t - k_x x - k_z z)}$, if $k_x = 3, k_z = 4$, find:

(1) the operating frequency f;

(2) the incidence angle θ_i;

(3) the expression of electric field of reflected wave $\boldsymbol{E}^-$;

(4) the expression of electric field of total wave $\boldsymbol{E}$.

8.21 A uniform plane wave is obliquely incident from air to the surface ($z = 0$) of a perfect dielectric with $\mu = \mu_0, \varepsilon = 3\varepsilon_0$, if given that the magnetic field of incident wave is

$$\boldsymbol{H}^+ = \left(\sqrt{3}\boldsymbol{e}_x - \boldsymbol{e}_y + \boldsymbol{e}_z\right) \sin\left(\omega t - Ax - 2\sqrt{3}z\right) \text{ A/m},$$

to find:

(1) the constants ω and A in the expression of $\boldsymbol{H}^+$;

(2) the electric field of incident wave $\boldsymbol{E}^+$;

(3) the incidence angleθ_i.

8.22 A lunar satellite transmits radio waves to the moon, the measured Brewster angle is 60°, find the relative permittivity ε_r of the moon surface.

8.23 A linearly polarized plane wave is incident from free space upon the interface of a medium with $\varepsilon_r = 4$, $\mu_r = 1$, if the angle between the electric field of incident wave and the plane of incidence is 45°. Try to find:

(1) the incidence angle θ_i if the reflected wave is only perpendicularly polarized wave;

(2) the ratio of the average power flux of incident wave and the reflected wave.

8.24 A linearly polarized wave is incident upon the interface between two homogeneous non-conductive and non-magnetic media ($\sigma_1 = \sigma_2 = 0, \mu_1 = \mu_2 = \mu_0$) , suppose $\varepsilon_1 > \varepsilon_2$, and the incident angle satisfies $\sin\theta_i > \sqrt{\dfrac{\varepsilon_2}{\varepsilon_1}}$. Prove that reflection coefficient of power is equal to 1 when total reflection occurs.

8.25 Find the critical angle θ_c and Brewster angle θ_B when light rays is incident from the glass ($n = 1.5$) into air; and prove that under normal circumstances, the critical angle is always greater than the Brewster angle.

8.26 A perpendicularly polarized plane electromagnetic wave is incident from under water upon its interface with air, the incidence angle $\theta_i = 20°$, for fresh water, $\varepsilon_r = 81$, $\mu_r = 1$. Find:

(1) the critical angle θ_c;

(2) the reflection coefficient R_N;

(3) the refractive index T_N.

8.27 As shown in Figure of Exercises 8.27, region $z > 0$ is filled with medium of permittivity ε_2. A dielectric board is placed in front of the medium with thickness d and permittivity ε_1. For a TEM wave normally incident from the left side, prove that: when $\varepsilon_{1r} = \sqrt{\varepsilon_{2r}}$ and $d = \lambda/(4\sqrt{\varepsilon_{1r}})$ (λ is the wavelength in the free space), no reflection occurs at $z = -d$.

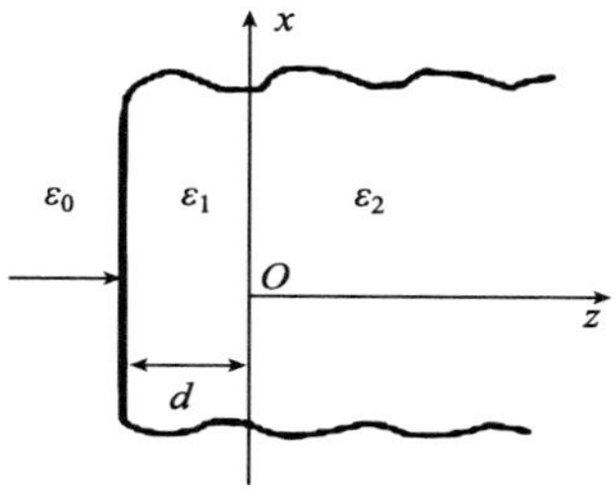

Figure of Exercises 8.27

Chapter 9

Two-Conductor Transmission Lines — Transverse Electromagnetic Wave Guiding System

9.1 Introduction

In the previous chapters, electromagnetic wave propagation was discussed. It was found that the propagation direction of the wave, consisting of the obliquely incident wave and the reflected wave, is parallel to the surface between two dielectric mediums. This interface between dielectric mediums serves to guide electromagnetic waves. The surface of the metal can also guide the electromagnetic waves.

What is a transmission line? A transmission line is a system that can guide electromagnetic waves. Generally, a connecting line can be taken as a transmission line when its signal transmission characteristics must be considered. One of the basic characteristics of transmission lines is that it takes time when the signal transmits through transmission lines. The main characteristics of transmission lines are as follows:

Electrical parameters are distributed in the whole space occupied by the transmission line.

Signal transmission takes time, and the length of the transmission line affects the characteristics of the signal directly.

Signal is the function of both time t and its position.

Several kinds of common guided-wave devices are shown in Figure 9.1. They could be made of metal, such as two-wire lines, or dielectric, such as optical fibers. They could be two-conductor, such as coaxial cables, or single-conductor, such as rectangular waveguides.

Figure 9.1(a), (b), (c) and (f) show the two-conductor transmission lines that serve to guide TEM waves; Figure 9.1(d) and (e) illustrate the single conductor waveguides carrying on TE or TM modes; and Figure 9.1(g) shows the dielectric transmission lines — optical fiber, in which the surface waves are usually guided.

This chapter introduces TEM waves guided by "transmission lines".

Generally, there are three types of guided waves: TE wave, TM wave and TEM wave, representing the transverse electric wave, the transverse magnetic wave and the transverse electromagnetic wave respectively.

When the magnetic field of a guided wave has a component in the direction of its propagation, while the electric field is only in the transverse direction, the guided wave is called a transverse electric wave. The guided wave is referred as a transverse magnetic wave when the magnetic field is only in the transverse direction and the electric field has a component in its direction of propagation. TE and TM waves can exist within a single hollow conductor. The transverse electromagnetic wave requires two or more conductors for its existence. As we know, both the electric field and the magnetic field of the transverse electromagnetic wave are orthogonal to the propagation direction of the wave. This kind of TEM wave, very similar to a plane wave, can be excited at any frequency.

For the TEM waves, the electric field and magnetic field are orthogonal to each other, and both of them are orthogonal to the propagation direction.

In this chapter, to simplify the analysis, the materials of transmission lines, the cross-sections of transmission lines and the characteristic of dielectrics are regarded as the same along the transmission line; such kind of transmission line is called a uniform transmission line. The transmission lines referred in the following text are all uniform transmission lines.

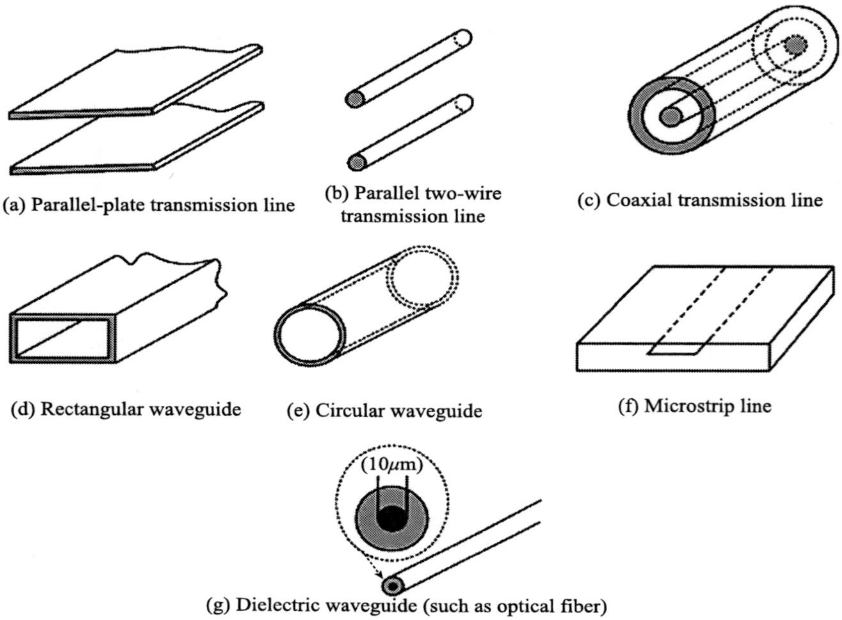

Figure 9.1 Several kinds of transmission lines

9.2 Properties of Wave Equations for TEM Waves

In this chapter, the medium of transmission lines is linear, homogeneous and isotropic which is called L.H. I medium. "Linear" means μ and ε are constant for different field strength. "Homogeneous" means the parameters at any points in the medium are the same. "Isotropic" means neither μ nor ε is affected by the propagation direction.

In Chapter 7, wave equations for $\boldsymbol{E}$ and $\boldsymbol{H}$ were obtained as follows:

$$\begin{cases} \nabla^2 \boldsymbol{E} - \mu\varepsilon \dfrac{\partial^2 \boldsymbol{E}}{\partial t^2} = 0 & (9.1) \\ \nabla^2 \boldsymbol{H} - \mu\varepsilon \dfrac{\partial^2 \boldsymbol{H}}{\partial t^2} = 0 & (9.2) \end{cases}$$

For a simple harmonic wave, wave equations become

$$\begin{cases} \nabla^2 \boldsymbol{E} + k^2 \boldsymbol{E} = 0 & (9.3) \\ \nabla^2 \boldsymbol{H} + k^2 \boldsymbol{H} = 0 & (9.4) \end{cases}$$

where

$$k^2 = \omega^2 \mu \varepsilon$$

For the TEM waves in longitudinal direction ($\boldsymbol{e}_z$ direction), the phase factor included in $\boldsymbol{E}$ and $\boldsymbol{H}$ can be written as $\mathrm{e}^{\mathrm{j}(\omega t - kz)}$. Considering the symmetry of the equations, the electric field $\boldsymbol{E}$ is solved only. Here is

$$\nabla^2 \boldsymbol{E} = \frac{\partial^2 \boldsymbol{E}}{\partial x^2} + \frac{\partial^2 \boldsymbol{E}}{\partial y^2} + \frac{\partial^2 \boldsymbol{E}}{\partial z^2} = \nabla_T^2 \boldsymbol{E} + \frac{\partial^2 \boldsymbol{E}}{\partial z^2} = \nabla_T^2 \boldsymbol{E} - k^2 \boldsymbol{E}$$

where ∇_T is the partial differentiation with respect to the transverse component. Substituting the above equation into Equation (9.3), we obtain

$$\nabla_T^2 \boldsymbol{E} = 0 \tag{9.5}$$

Considering the static field in uniform medium, two basic equations for electrostatic fields in a source-free region are

$$\nabla \cdot \boldsymbol{E} = 0$$

$$\nabla \times \boldsymbol{E} = 0$$

Due to the above equations and vector identities, we obtain

$$\nabla \times \nabla \times \boldsymbol{E} = \nabla(\nabla \cdot \boldsymbol{E}) - \nabla^2 \boldsymbol{E} = -\nabla^2 \boldsymbol{E} = 0$$

Thus there is

$$\nabla^2 \boldsymbol{E} = 0 \tag{9.6}$$

The static electric field along a two-wire line with infinite length is constant in axial direction ($\boldsymbol{e}_z$). Thus we can obtain the following equation from Equation (9.6):

$$\nabla^2 \boldsymbol{E} = \nabla_T^2 \boldsymbol{E} + \frac{\partial^2 \boldsymbol{E}}{\partial z^2} = \nabla_T^2 \boldsymbol{E} + 0 = \nabla_T^2 \boldsymbol{E} = 0$$

or

$$\nabla_T^2 \boldsymbol{E} = 0 \tag{9.7}$$

which is the differential equation that the static electric field in two-conductor system should satisfy. Comparing Equation (9.5) and (9.7), we can find the equations that the static electric field and the electric field of TEM waves satisfy are the same. In addition to the same boundary conditions, it can be concluded that the distribution of the electric field of TEM waves in a cross-section and that of the static electric field in the same cross-section are the same. Similarly, the distribution of magnetic field of TEM waves in a cross-section is identical to that of the static electric current in the same cross-section.

Therefore, an important conclusion is obtained: any uniform transmission system which can establish electrostatic fields can transmit TEM waves. According to this conclusion, the voltage and current can be related with the electric field and the magnetic field for analysis. With this relation, the analysis can be simplified.

9.3 Parallel-Plate Transmission System

In this section, we will discuss TEM waves in parallel-plate transmission lines, as well as basic equations for voltage and current, or transmission-line equations. The parameters of transmission lines, such as the characteristic impedance Z_{c}, will be obtained.

9.3.1 Parallel-Plate Transmission Systems and the Propagating TEM Waves

A parallel-plate transmission system, as shown in Figure. 9.2, is referred to two metal plates parallel to each other. The width of the metal plates and the distance between them are b and a respectively, and $b \gg a$. The plates are infinitely long in z direction. The energy is transmitted along z direction.

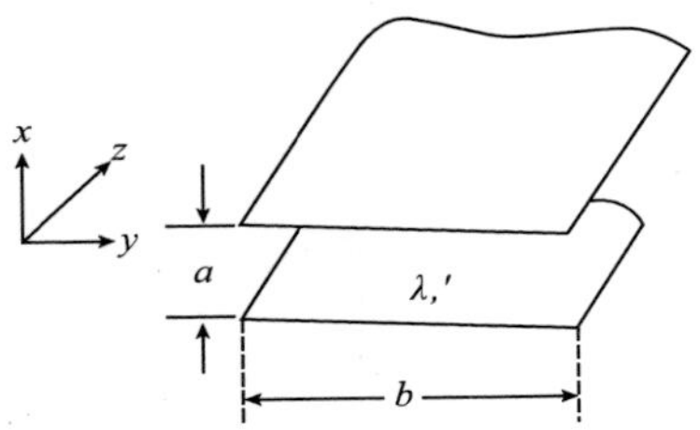

Figure 9.2 A metal parallel-plate transmission system

According to the analysis in section 9.2, the static field could be established and TEM waves could propagate in this system. Meanwhile, the field distributions of E and H of TEM waves in the cross-section are identical to those of a static field. Under the condition of $b \gg a$ (fringe effect can be neglected), the distributions of the electric field component E_x and the magnetic field component H_y are almost uniform in the cross section. As a result, the TEM wave consisting of E_x and H_y is a uniform plane wave. However, the propagation of the uniform plane wave is restricted between the plates and only transmits along the direction of z. Since it is a TEM wave, according to the characteristic of plane waves, the expressions for the fields can be written as follows:

$$E_x = E_0 \mathrm{e}^{\mathrm{j}(\omega t - kz)}$$

$$H_y = E_x/\eta = \frac{1}{\eta} E_0 \mathrm{e}^{\mathrm{j}(\omega t - kz)}$$

9.3.2 Transmission-Line Equations (or telegraph equations) in Parallel-Plate Transmission Systems

Transmission-line equations are the differential equations that the voltage and current satisfies.

As discussed previously, the wave in the parallel-plate transmission system is a uniform plane wave. Thus, we can write the relation between E_x and the voltage applied to the parallel-plates:

$$E_x = \frac{U}{a} \tag{9.8}$$

and the relation between H_y and the current on the parallel-plate transmission system:

$$H_y = J = \frac{I}{b} \tag{9.9}$$

Note that the directions of the surface current density J on the upper and lower plates are opposite, and so are the normal directions of the two plates. Consequently, the direction of the magnetic field on the upper and lower inner surface and that between plates are identical. The value is identical to surface current density J, which is shown in Equation (9.9).

If E_x and H_y are given, telegraph equations can be obtained according to Maxwell's curl equations. Two curl equations become

$$-\frac{\partial H_y}{\partial z} = \varepsilon \frac{\partial E_x}{\partial t} \tag{9.10}$$

and

$$\frac{\partial E_x}{\partial z} = -\mu \frac{\partial H_y}{\partial t} \tag{9.11}$$

Substitute Equations (9.8) and (9.9) into Equations (9.10) and (9.11) respectively, we can obtain

$$\frac{\partial I}{\partial z} = -\varepsilon \frac{b}{a}\left(\frac{\partial U}{\partial t}\right) \tag{9.12}$$

and

$$\frac{\partial U}{\partial z} = -\mu \frac{a}{b}\left(\frac{\partial I}{\partial t}\right) \tag{9.13}$$

Then we can easily determine the capacitance and inductance per-unit-length of the parallel-plate transmission lines as

$$C_0 = \frac{\varepsilon b}{a}(\mathrm{F/m}) \tag{9.14}$$

and

$$L_0 = \frac{\mu a}{b}(\mathrm{H/m}) \tag{9.15}$$

From the above equations, we can obtain

$$L_0 C_0 = \mu\varepsilon \tag{9.16}$$

This is the common relationships between the dielectric parameters and the distribution parameters in uniform two-conductor transmission system when TEM waves propagate. We will demonstrate it further in Example 9.1.

Substituting Equations (9.14) and (9.15) into Equations (9.12) and (9.13) respectively, we can obtain

$$\frac{\partial I}{\partial z} = -C_0 \frac{\partial U}{\partial t} \tag{9.17}$$

and

$$\frac{\partial U}{\partial z} = -L_0 \frac{\partial I}{\partial t} \tag{9.18}$$

Equations (9.17) and (9.18) are known as the transmission-line equations.

From Equation (9.13) and the partial differentiation of Equation (9.12) with respect to z, we can obtain

$$\frac{\partial^2 I}{\partial z^2} = \mu\varepsilon\left(\frac{\partial^2 I}{\partial t^2}\right) \tag{9.19}$$

From Equation (9.12) and the partial differentiation of Equation (9.13) with respect to z, we can obtain

$$\frac{\partial^2 U}{\partial z^2} = \mu\varepsilon\left(\frac{\partial^2 U}{\partial t^2}\right) \tag{9.20}$$

Assume that the voltage and current are both simple harmonic waves, we can obtain $\frac{\partial^2}{\partial t^2} = -\omega^2$. Thus Equation (9.19) and (9.20) can be manipulated to yield:

$$\frac{\mathrm{d}^2 I}{\mathrm{d}z^2} + k^2 I = 0 \tag{9.21}$$

and

$$\frac{\mathrm{d}^2 U}{\mathrm{d}z^2} + k^2 U = 0 \tag{9.22}$$

where

$$k^2 = \omega^2\mu\varepsilon = \omega^2 L_0 C_0 \tag{9.23}$$

Equations (9.21) and (9.22) are called current and voltage wave equations, respectively.

If time element is considered, the solutions of Equations (9.21) and (9.22) can be rearranged as

$$U = U_0 e^{j(\omega t - kz)} \tag{9.24}$$

and

$$I = I_0 e^{j(\omega t - kz)} \tag{9.25}$$

We considered only the forward waves propagating along z direction in the above equations. It is valid for the cases of infinite long or matched transmission systems.

9.3.3 Characteristic Impedance of Parallel-Plate Transmission Systems

The characteristic impedance is one of the most important parameters in TEM wave transmission systems. The characteristic impedance is defined as the ratio of voltage to current in the infinite transmission system (the system contains only incident waves without reflected waves, or the voltage and current waves are travelling waves), that is:

$$Z_c = \frac{U}{I} \tag{9.26}$$

or else

$$Z_c = \frac{U^+}{I^+} = -\frac{U^-}{I^-} \tag{9.27}$$

where U^+, I^+ and U^-, I^- are the voltage and current of incident waves and reflected waves, respectively. Undoubtedly both incident waves and reflected waves are travelling waves.

To determine the specific expression of the characteristic impedance, let the voltage of the incident wave be

$$U = U^+ = U_0 e^{j(\omega t - kz)}$$

Substituting the above equation into Equation (9.13), we obtain

$$\frac{\partial U^+}{\partial z} = -\mu \frac{a}{b} \frac{\partial I^+}{\partial t}$$

From $\dfrac{\partial U^+}{\partial z} = -jkU_0 e^{j(\omega t - kz)}$ and $-\mu \dfrac{a}{b} \dfrac{\partial I^+}{\partial t} = -j\omega\mu \dfrac{a}{b} I^+$, we obtain

$$\begin{aligned} I^+ &= \frac{k}{\omega\mu} \frac{b}{a} U_0 e^{j(\omega t - kz)} \\ &= \sqrt{\frac{\varepsilon}{\mu}} \cdot \frac{b}{a} \cdot U_0 e^{j(\omega t - kz)} \\ &= \sqrt{\frac{\varepsilon}{\mu}} \frac{b}{a} \cdot U^+ \end{aligned} \tag{9.28}$$

Thus, the characteristic impedance of a parallel-plate transmission line is as:

$$Z_c = \frac{U^+}{I^+} = \sqrt{\frac{\mu}{\varepsilon}} \cdot \frac{a}{b} = \eta \frac{a}{b} \tag{9.29}$$

The capacitance and inductance per-unit-length of the parallel-plate transmission lines are given in Equations (9.14) and (9.15), and here is

$$\sqrt{\frac{L_0}{C_0}} = \sqrt{\frac{\left(\dfrac{\mu a}{b}\right)}{\left(\dfrac{\varepsilon b}{a}\right)}} = \sqrt{\frac{\mu}{\varepsilon}} \cdot \frac{a}{b} = \eta \frac{a}{b} \tag{9.30}$$

Comparing Equations (9.29) and (9.30), we obtain

$$Z_c = \sqrt{\frac{L_0}{C_0}} \tag{9.31}$$

This equation is the characteristic impedance of uniform ideal TEM wave transmission systems. It can be used not only in a parallel-plate transmission system but also in other uniform ideal TEM wave transmission systems.

Considering Equation (9.16), the characteristic impedance can be written as

$$Z_c = \sqrt{\frac{L_0}{C_0}} = \frac{\sqrt{\mu\varepsilon}}{C_0} = \frac{L_0}{\sqrt{\mu\varepsilon}} \tag{9.32}$$

Note that the characteristic impedance can be determined from Equation (9.32) as long as C_0 or L_0 of the transmission system is known.

The characteristic impedance is an important parameter of transmission systems, and it is determined by the shape, dimension of the transmission lines and the parameters of the medium around the transmission lines.

9.3.4 Propagation Velocity of the TEM Wave in A Parallel-Plate Transmission System

The velocity of propagation of the TEM wave in a parallel-plate transmission system is:

$$v = \frac{\omega}{k} = \frac{1}{\sqrt{\mu\varepsilon}} = \frac{1}{\sqrt{L_0 C_0}} \tag{9.33}$$

If the medium between plates is air, the velocity of propagation of TEM wave is identical to the velocity of light, and the wavelength is the same with the wavelength of TEM wave in free space.

9.3.5 Higher Order Modes in A Parallel-Plate Transmission System

TEM wave is a dominant wave (or dominant mode) in a parallel-plate transmission system, and it can propagate at any frequency. Besides, the parallel-plate transmission system can transmit higher order modes, such as TE modes and TM modes. However, TE modes and TM modes are restricted by the cutoff frequency, and they cannot propagate when the operating frequency is below their cutoff frequencies.

The cutoff frequency of TE_{10} and TM_{10} waves is the lowest in all the TE and TM waves in parallel-plate transmission systems. Assuming that the width of parallel-plates is infinite and the distance between plates is a, the cutoff frequency f_c can be given as:

$$f_c = \frac{v}{2a} \tag{9.34}$$

where v is the velocity. If the operating frequency is lower than the cutoff frequency, all the TE and TM waves are cut off, and only the TEM wave can propagate.

9.4 Two-Wire Transmission Lines

A two-wire transmission line is shown in Figure 9.3. It is also one kind of uniform transmission systems that can support static fields. Consequently, it can transmit TEM waves. Two kinds of two-wire transmission lines – ideal and low loss transmission lines are investigated in this section.

9.4.1 TEM Waves Propagating along A Two-Wire Transmission Line

The cross-section of a uniform two-wire transmission line is shown is Figure 9.3. Two wires M and N are parallel to each other. TEM waves propagate in it, so there is no longitudinal component of electromagnetic field, saying, $E_z = 0, H_z = 0$, and only E_x, E_y, H_x and H_y exist.

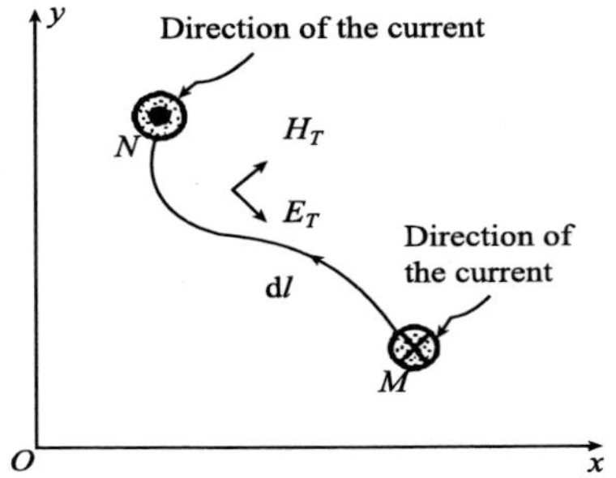

Figure 9.3 The cross-section of two-wire transmission line

The relationship for these components of field in two-wire transmission systems can be written directly from Maxwell Equations. In the Cartesian coordinate system, there is

$$\nabla \times \boldsymbol{H} = \frac{\partial \boldsymbol{D}}{\partial t} \tag{9.35}$$

so we can obtain

$$-\frac{\partial H_y}{\partial z} = \frac{\partial D_x}{\partial t} \tag{9.36}$$

and

$$\frac{\partial H_x}{\partial z} = \frac{\partial D_y}{\partial t} \tag{9.37}$$

From the following equation

$$\nabla \times \boldsymbol{E} = -\frac{\partial \boldsymbol{B}}{\partial t}$$

we can obtain

$$\frac{\partial E_y}{\partial z} = \frac{\partial B_x}{\partial t} \tag{9.38}$$

and

$$\frac{\partial E_x}{\partial z} = -\frac{\partial B_y}{\partial t} \tag{9.39}$$

The relationship for the current and the magnetic field as well as the voltage and the electric field can be written as follows respectively:

$$I = \oint_l \boldsymbol{H}_\mathrm{T} \cdot \mathrm{d}\boldsymbol{l} \tag{9.40}$$

and

$$U = -\int_M^N \boldsymbol{E}_\mathrm{T} \cdot \mathrm{d}\boldsymbol{l} \tag{9.41}$$

where $\boldsymbol{E}_\mathrm{T}$ and $\boldsymbol{H}_\mathrm{T}$ are the transverse components of field, and U is the electric potential difference (from M to N) between two wires.

The field distribution of TEM waves in a two-wire transmission line is shown in Figure 9.4.

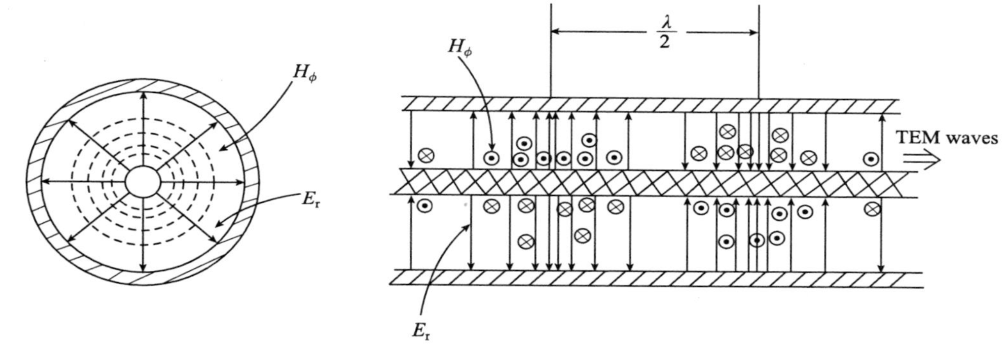

(a) Distribution of the field on the cross-section

(b) Distribution of the field along the longitudinal direction

Figure 9.4 The TEM wave propagating along a two-wire transmission line system

Figure 9.4(a) shows the distribution of E_T and H_T on the cross-section. The electric field lines start from one wire and end at another, supported by positive and negative charges respectively. Magnetic field lines are closed curves surrounding the wires. Figure 9.4(a) shows the distribution of longitudinal electromagnetic field at a certain time.

9.4.2 Method One: Establishing Transmission Line Equations (or telegraph equations) According to Electromagnetic Field Theory

Electromagnetic field theory is more general than electric circuit theory. Electric circuit theory is an approximation of electromagnetic field theory under the special assumption. In this section, two-wire transmission line equations are established on the basis of field theory by the method of combining the "field" and "circuit" theory.

In Cartesian coordinate system, $\boldsymbol{E}_\mathrm{T}, \boldsymbol{H}_\mathrm{T}$, and $d\boldsymbol{l}$ in Equations (9.40) and (9.41) can be written as:

$$\left.\begin{array}{l} \boldsymbol{E}_\mathrm{T} = \boldsymbol{e}_x E_x + \boldsymbol{e}_y E_y \\ \boldsymbol{E}_\mathrm{T} = \boldsymbol{e}_x H_x + \boldsymbol{e}_y H_y \\ \mathrm{d}\boldsymbol{l} = \boldsymbol{e}_x \mathrm{d}x + \boldsymbol{e}_y \mathrm{d}y \end{array}\right\} \tag{9.42}$$

Substituting Equation (9.42) into (9.40), we can obtain:

$$I = \oint_l (e_x H_x + e_y H_y) \cdot (e_x \mathrm{d}x + e_y \mathrm{d}y)$$
$$= \oint_l (H_x \mathrm{d}x + H_y \mathrm{d}y) \tag{9.43}$$

From the partial differentiation of the above equation with respect to z, and applying Equations (9.36) and (9.37), there is:

$$\frac{\partial I}{\partial z} = \oint_l \left(\frac{\partial H_x}{\partial z} \mathrm{d}x + \frac{\partial H_y}{\partial z} \mathrm{d}y \right)$$
$$= \oint_l \left(\frac{\partial D_y}{\partial t} \mathrm{d}x - \frac{\partial D_x}{\partial t} \mathrm{d}y \right)$$
$$= -\frac{\partial}{\partial t} \oint_l (D_x \mathrm{d}y - D_y \mathrm{d}x) \tag{9.44}$$

According to Figure 9.5, we obtain:

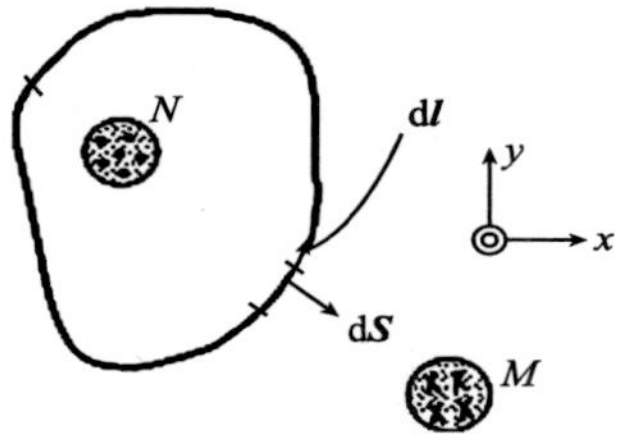

Figure 9.5

$$\boldsymbol{D} \cdot \mathrm{d}\boldsymbol{S} = \boldsymbol{D} \cdot (-e_z \times \mathrm{d}\boldsymbol{l})$$
$$= -(\boldsymbol{e}_x D_x + \boldsymbol{e}_y D_y) \cdot (\boldsymbol{e}_y \mathrm{d}x - \boldsymbol{e}_x \mathrm{d}y)$$
$$= D_x \mathrm{d}y - D_y \mathrm{d}x \tag{9.45}$$

It is noted that $(D_x \mathrm{d}y - D_y \mathrm{d}x)$ is the electric flux through the surface element $\mathrm{d}\boldsymbol{S}$ perpendicularly. Thus, the integral part in Equation (9.44) (the integral path is a closed curve on the cross-section surrounding one wire, shown in Figure 9.5 is the whole electric flux excited by per-unit-length of the wire. This is the quantity of charges ρ_t per-unit-length of wire. And the quantity of charges is the product of the voltage of the transmission line U and the capacitance per-unit-length of the wire C_0. Then Equation (9.44) becomes:

$$\frac{\partial I}{\partial z} = -\frac{\partial}{\partial t} q$$
$$= -\frac{\partial}{\partial t}(C_0 U) = -C_0 \frac{\partial U}{\partial t} \tag{9.46}$$

It is a basic equation expressing the relationship of the voltage and the current in two-wire transmission lines. Similarly, another basic equation can be obtained by rearranging Equation (9.41). The procedure is as follows:

Substituting Equation (9.42) into (9.41), we can obtain

$$U = -\int_M^N \boldsymbol{E}_\mathrm{T} \cdot \mathrm{d}\boldsymbol{l} = -\int_M^N (\boldsymbol{e}_x E_x + \boldsymbol{e}_y E_y) \cdot (\boldsymbol{e}_x \mathrm{d}x + \boldsymbol{e}_y \mathrm{d}y)$$
$$= -\int_M^N (E_x \mathrm{d}x + E_y \mathrm{d}y) \tag{9.47}$$

From partial differentiation of the above equation with respect to z, and applying Equations (9.38) and (9.39), there is

$$\frac{\partial U}{\partial z} = -\frac{\partial}{\partial t} \int_M^N (-B_y \mathrm{d}x + B_x \mathrm{d}y) \tag{9.48}$$

Similar to Equation (9.45), $(-B_y\mathrm{d}x + B_x\mathrm{d}y)$ is electric flux through the surface element $\mathrm{d}\boldsymbol{S}$ perpendicularly. And the integral part in Equation (9.48) is the whole magnetic flux through the surface that consists of curve M-N and longitudinal per unit length shown in Figure 9.3. The magnetic flux should be identical to the product of the inductance per-unit-length L_0 and the current of the transmission line, that is L_0I. Thus Equation (9.48) becomes

$$\frac{\partial U}{\partial z} = -\frac{\partial}{\partial t}(L_0 I) = -L_0\frac{\partial I}{\partial t} \tag{9.49}$$

Consequently, two basic equations are obtained

$$\frac{\partial I}{\partial z} = -C_0\frac{\partial U}{\partial t}$$

$$\frac{\partial U}{\partial z} = -L_0\frac{\partial I}{\partial t}$$

which are called transmission line equations.

From partial differentiation of Equation (9.46) with respect to z and Equation (9.49), there is

$$\begin{aligned}\frac{\partial^2 I}{\partial z^2} &= -C_0\frac{\partial}{\partial z}\left(\frac{\partial U}{\partial t}\right)\\ &= -C_0\frac{\partial}{\partial t}\frac{\partial U}{\partial z} = L_0C_0\frac{\partial^2 I}{\partial t^2}\end{aligned} \tag{9.50}$$

From partial differentiation of Equation (9.49) with respect to z and Equation (9.46), there is

$$\begin{aligned}\frac{\partial^2 U}{\partial z^2} &= -L_0\frac{\partial}{\partial z}\left(\frac{\partial I}{\partial t}\right)\\ &= -L_0\frac{\partial}{\partial t}\frac{\partial I}{\partial z} = L_0C_0\frac{\partial^2 U}{\partial t^2}\end{aligned} \tag{9.51}$$

If the voltage and current wave are harmonic waves, there is $\dfrac{\partial^2}{\partial t^2} = -\omega^2$, and then Equations (9.50) and (9.51) become

$$\frac{\mathrm{d}^2 I}{\mathrm{d}z^2} + k^2 I = 0 \tag{9.52}$$

and

$$\frac{\mathrm{d}^2 U}{\mathrm{d}z^2} + k^2 U = 0 \tag{9.53}$$

The two equations are wave equations for current and voltage wave respectively.

9.4.3 Method Two: Establishing Transmission Line Equations (or telegraph equations) According to the Concept of Distribution Parameters

In this section, we will establish transmission line equations from circuit theory directly. There are resistance and leakage conductance in a real transmission line, while the electric field energy and the magnetic field energy both exist around the transmission line, meaning that there are distribution capacitance and distribution inductance. If the transmission line under investigation is ideal, the resistance and leakage conductance are both zero, and the equivalent circuit for a two-wire line is shown in Figure 9.6. L_0 and C_0 indicate the inductance and capacitance per-unit-length of transmission lines, respectively. Thus, the inductance and capacitance of a very short section of length $\mathrm{d}z$ of a transmission line are $L_0\mathrm{d}z$ and $C_0\mathrm{d}z$, respectively.

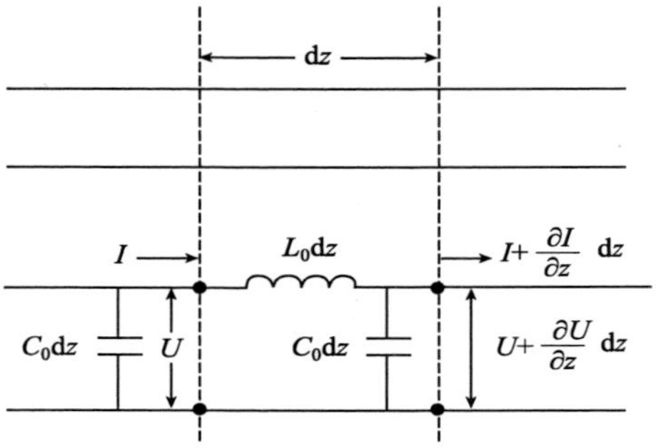

Figure 9.6 The equivalent circuit for an ideal transmission line

The voltage drop along the section of the transmission line of length dz is the product of the equivalent series inductance and the partial differentiation of current with respect to time. Then, there is

$$-\frac{\partial U}{\partial z}\mathrm{d}z = (L_0\mathrm{d}z)\frac{\partial I}{\partial t}$$

so

$$\frac{\partial U}{\partial z} = -L_0\frac{\partial I}{\partial t} \tag{9.54}$$

Similarly, the reduction of current on the section of the transmission line of length $\mathrm{d}z$ is the product of the equivalent parallel conductance and the partial differentiation of voltage with respect to time. That is

$$-\frac{\partial I}{\partial z}\mathrm{d}z = (C_0\mathrm{d}z)\frac{\partial U}{\partial t}$$

so

$$\frac{\partial I}{\partial z} = -C_0\frac{\partial U}{\partial t} \tag{9.55}$$

Equations (9.54) and (9.55) are called the transmission-line equations, from which wave equations for harmonic voltage and current wave can be easily obtained.

9.4.4 Characteristic Impedance of An Ideal Two-Wire Transmission Line

As the investigated transmission lines are ideal, that means there is neither conductor loss nor dielectric loss, we can conclude that the characteristic impedance Z_c should be real. It means Z_c is a pure resistance.

The solution of wave equation for voltage is given by Equation (9.53):

$$\begin{aligned} U &= U^+ + U^- \\ &= U_0^+\mathrm{e}^{\mathrm{j}(\omega t-kz)} + U_0^-\mathrm{e}^{\mathrm{j}(\omega t+kz)} \end{aligned} \tag{9.56}$$

where U^+ and U^- indicate the incident and the reflected waves of voltages, respectively.

Substituting Equation (9.56) into telegraph equations Equation (9.49) yields:

$$-\mathrm{j}kU_0^+\mathrm{e}^{\mathrm{j}(\omega t-kz)}+\mathrm{j}kU_0^-\mathrm{e}^{\mathrm{j}(\omega t+kz)}=-\mathrm{j}\omega L_0 I$$

Then there is

$$\begin{aligned} I &= I^+ + I^- \\ &= \frac{k}{\omega L_0}U_0^+\mathrm{e}^{\mathrm{j}(\omega t-kz)}-\frac{k}{\omega L_0}U_0^-\mathrm{e}^{\mathrm{j}(\omega t+kz)} \end{aligned} \tag{9.57}$$

where, I^+ and I^- indicate the incident and the reflected waves of currents, respectively.

According to the definition of the characteristic impedance, we obtain

$$Z_\mathrm{c}=\frac{U^+}{I^+}=-\frac{U^-}{I^-}=\frac{\omega L_0}{k}=\sqrt{\frac{L_0}{C_0}}$$

that is

$$Z_\mathrm{c}=\sqrt{\frac{L_0}{C_0}} \tag{9.58}$$

Now, we shall demonstrate that the parameter relation $L_0C_0=\mu\varepsilon$ is valid in all the transmission system with uniform cross-section (propagating TEM waves) in Example 9.1

Example 9.1 Prove that $L_0C_0=\mu\varepsilon$ is valid when uniform ideal transmission system propagates TEM waves, where L_0 and C_0 are the capacitance and inductance per-unit-length of the transmission system, respectively.

Solution The shaded portion is the cross-section of uniform transmission system, shown in Figure for Example 9.1. TEM waves propagate along $+z$ direction in the system.

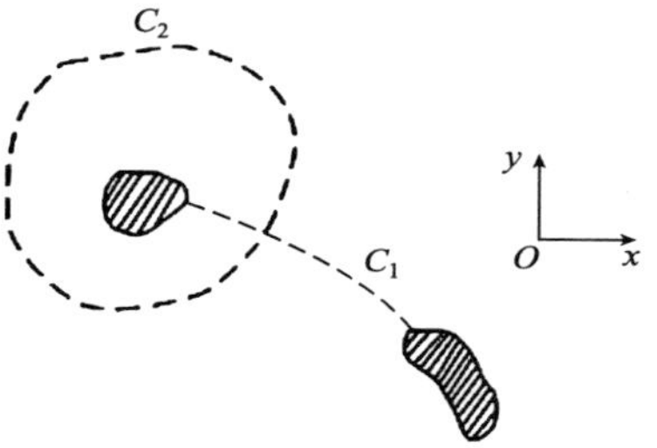

Figure for Example 9.1

Considering equations:

$$U^+=\int_{\mathrm{c}_1}\boldsymbol{E}^+\cdot\mathrm{d}\boldsymbol{l}$$

$$I^+=\oint_{\mathrm{c}_2}\boldsymbol{H}^+\cdot\mathrm{d}\boldsymbol{l}$$

From the above equations we can derive the relationship of L_0, C_0 and the characteristic impedance Z_c, and then prove $L_0C_0=\mu\varepsilon$. Integral paths c_1 and c_2 are shown in Figure for Example 9.1. Assuming that c_2 overlaps with a closed magnetic line, we obtain $\boldsymbol{H}_T^+\cdot\mathrm{d}\boldsymbol{l}=H_T^+\mathrm{d}l$, thus there is

$$\begin{aligned} I^+ &= \oint_{\mathrm{c}_2}H^+\cdot\mathrm{d}\boldsymbol{l}=\oint_{\mathrm{c}_2}H_T^+\mathrm{d}\boldsymbol{l}=\oint_{\mathrm{c}_2}(\boldsymbol{H}_T^+/\eta)\mathrm{d}\boldsymbol{l}=\oint_{\mathrm{c}_2}(\boldsymbol{E}_T^+/\eta)\mathrm{d}\boldsymbol{l} \\ &= \frac{1}{\eta\varepsilon}\oint_{\mathrm{c}_2}D_T\mathrm{d}l \end{aligned}$$

There is $D_T=\rho_s$ when the closed curve is on the surface of the transmission line, then:

$$I^+=\frac{1}{\eta\varepsilon}\oint_{\mathrm{c}_2}\rho_s\mathrm{d}l=\frac{1}{\eta\varepsilon}C_0U^+$$

so

$$Z_c = \frac{U^+}{I^+} = \eta\varepsilon/C_0$$

According to the following equations

$$U^+ = \int_{c_1} \boldsymbol{E}^+ \cdot \mathrm{d}\boldsymbol{l} = \int_{c_1} E_T^+ \mathrm{d}\boldsymbol{l} = \int_{c_1} (\eta H_T^+)\mathrm{d}l = \frac{\eta}{\mu}\int_{c_1} B_T^+ \mathrm{d}l = \left(\frac{\eta}{\mu}\right) L_0 I^+$$

The characteristic impedance can be written as

$$Z_c = \frac{U^+}{I^+} = \eta L_0/\mu$$

Comparing the above equation and $Z_c = \eta\varepsilon/C_0$, we obtain

$$L_0 C_0 = \mu\varepsilon$$

The above equation is derived for the case of uniform ideal transmission systems with arbitrary cross-sections that propagate TEM waves. It is valid in uniform transmission systems with various cross-sections that propagate TEM waves.

Example 9.2 Determine the characteristic impedance of uniform transmission systems according to the definition of the characteristic impedance and the relationship of "field" and "circuit".

Solution The definition of the characteristic impedance is

$$Z_c = \frac{U^+}{I^+}$$

TEM waves (the electric field and magnetic field are represented by E_T^+ and H_T^+, respectively) propagate in the system, so the relationship for current and magnetic field is

$$\begin{aligned} I^+ &= \oint_l H_T^+ \mathrm{d}l = \frac{1}{\eta}\oint_l E_T^+ \mathrm{d}l \\ &= \frac{1}{\eta\varepsilon}\oint_l D_T^+ \mathrm{d}l = \frac{1}{\eta\varepsilon}\oint_l \rho s \mathrm{d}l = \frac{1}{\eta\varepsilon} C_0 U^+ \end{aligned}$$

In above equation, we have applied the boundary condition for perfect conductors $D_n = \rho_s$, where ρ_s is the surface current density on the surface of transmission system. So we have

$$Z_c = \frac{U^+}{I^+} = \frac{\eta\varepsilon}{C_0} = \frac{\sqrt{\mu\varepsilon}}{C_0} = \sqrt{\frac{L_0}{C_0}}$$

The equation $L_0 C_0 = \mu\varepsilon$ is used in the derivation of the above equation. In fact, the expression for the characteristic impedance Z_c has been obtained using another method in the section 9.3.

Example 9.3 Determine the characteristic impedance of a two-wire transmission line with radius a and distance D between wires, where $D \gg a$.

Solution Due to $D \gg a$, the variety of the distribution of the charge and current on the lines due to the influence between lines could be ignored. Thus, we can use the expression for capacitance per-unit-length C_0 directly (the space between two wires is filled with air).

$$C_0 = \frac{\pi\varepsilon_0}{\ln\dfrac{D-a}{a}} \tag{a}$$

In addition there is

$$Z_c = \sqrt{\frac{L_0}{C_0}} = \frac{\sqrt{\mu_0 \varepsilon_0}}{C_0}$$

Substituting Equation (a) into the above equation yields

$$Z_c = 120 \ln \frac{D-a}{a} \approx 120 \ln \left(\frac{D}{a}\right) \tag{b}$$

Assuming that the space between two wires is filled with the medium with $\mu = \mu_0$ and $\varepsilon = \varepsilon_r \varepsilon_0$, Z_c can be expressed as

$$\begin{aligned} Z_c &= \sqrt{\frac{\mu_0}{\varepsilon_r \varepsilon_0}} \cdot \frac{1}{\pi} \ln \left(\frac{D-a}{a}\right) \\ &= \frac{120}{\sqrt{\varepsilon_r}} \cdot \ln \left(\frac{D-a}{a}\right) \approx \frac{120}{\sqrt{\varepsilon_r}} \ln \left(\frac{D}{a}\right) \end{aligned} \tag{c}$$

Example 9.4 Determine the characteristic impedance of a two-wire transmission line with radius a and distance D between wires, where $D \gg a$ is not satisfied.

Solution As $D \gg a$ is not satisfied, we should determine the capacitance per-unit-length by the electric axis method, and then determine the characteristic impedance Z_c of the two-wire transmission line using the capacitance per-unit-length achieved above.

In Example 5.12, C_0 is determined by the electric axis method

$$C_0 = \frac{\pi \varepsilon}{\ln \left[\dfrac{(D/2) + \sqrt{(D/2)^2 - a^2}}{a}\right]} \tag{a}$$

on the other hand

$$Z_c = \frac{\sqrt{\mu_0 \varepsilon}}{C_0}$$

Substituting Equation (a) into the above equation yields

$$Z_c = \frac{120}{\sqrt{\varepsilon_r}} \cdot \ln \left[\frac{D}{2a} + \sqrt{\left(\frac{D}{2a}\right)^2 - 1}\right] \tag{b}$$

Assuming that the space between two wires of the transmission line is filled with air and $D \gg a$ is not satisfied, the characteristic impedance is

$$Z_c = 120 \ln \left[\frac{D}{2a} + \sqrt{\left(\frac{D}{2a}\right)^2 - 1}\right] \tag{c}$$

9.4.5 Input Impedance of An Ideal Two-Wire Transmission Line

The input impedance Z_{in} on one point of a two-wire transmission line is defined as the ratio of the voltage wave to the current wave on that point.

$$Z_{in} = \frac{U}{I} \tag{9.59}$$

where U and I indicate the actual voltage wave and current wave on an arbitrary point, respectively. Both the voltage wave and current wave are the sum of the incident wave and the reflected wave.

Assume that the transmission line is loaded by an arbitrary load Z_{L}, and let Z_{L} be located at the origin of the coordinate, as shown in Figure 9.7.

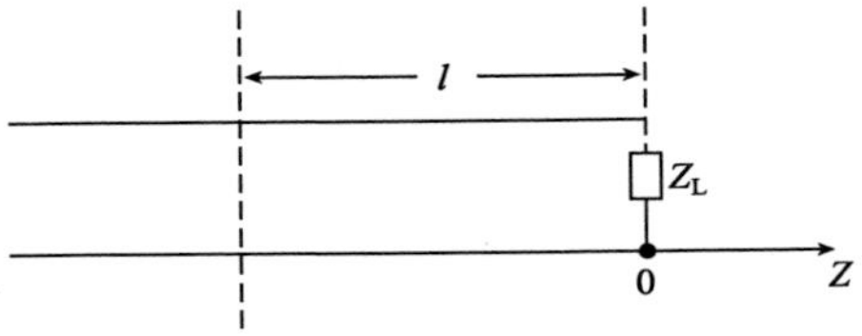

Figure 9.7 Two-wire transmission line terminated in a load Z_{L}

To determine the input characteristic impedance Z_{in}, voltage reflection coefficient at $z = 0$ should be calculated firstly

$$R_U(0) = \frac{U^-(0)}{U^+(0)} \tag{9.60}$$

From Equation (9.56), there is

$$U^+(0) + U^-(0) = U_{\mathrm{L}} \tag{9.61}$$

where U_{L} is the voltage across the load Z_{L}. From Equations (9.57) and (9.58), we can derive the current through the load (at $z = 0$) is

$$I(0) = \frac{U_{\mathrm{L}}}{Z_{\mathrm{L}}} = I^+ + I^- = \frac{U^+(0)}{Z_{\mathrm{c}}} - \frac{U^-(0)}{Z_{\mathrm{c}}}$$

that is

$$U^+(0) - U^-(0) = \frac{Z_{\mathrm{c}}}{Z_{\mathrm{L}}} U_{\mathrm{L}} \tag{9.62}$$

$U^+(0)$ and $U^-(0)$ can be calculated from Equations (9.61) and (9.62), and then substituting them into Equation (9.60) yields the voltage reflection coefficient:

$$R_U(0) = \frac{Z_{\mathrm{L}} - Z_{\mathrm{c}}}{Z_{\mathrm{L}} + Z_{\mathrm{c}}} \tag{9.63}$$

The above equation indicates that $R_U(0)$ is a complex number if Z_{L} is arbitrary and Z_{c} is a pure resistance. However, the phase of $R_U(0)$ is easily determined if Z_{L} is a pure resistance. If $Z_{\mathrm{L}} > Z_{\mathrm{c}}$, the phase of $R_U(0)$ is 0°, or else $Z_{\mathrm{L}} < Z_{\mathrm{c}}$, the phase of $R_U(0)$ is 180°. These are similar to the phase characteristics of the voltage reflection coefficient when the terminal is open-circuited or short-circuited, respectively.

The input impedance at $-z = l$ could be written as follow according to the definition:

$$\begin{aligned} Z_{\mathrm{in}} &= \frac{U}{I} = \frac{U^+(l) + U^-(l)}{I^+(l) + I^-(l)} \\ &= \frac{U_0^+ \mathrm{e}^{-\mathrm{j}kz} + U_0^- \mathrm{e}^{\mathrm{j}kz}}{\dfrac{1}{Z_{\mathrm{c}}}(U_0^+ \mathrm{e}^{-\mathrm{j}kz} - U_0^- \mathrm{e}^{\mathrm{j}kz})} \\ &= Z_{\mathrm{c}} \frac{\mathrm{e}^{\mathrm{j}kl} + R_U(0)\mathrm{e}^{-\mathrm{j}kl}}{\mathrm{e}^{\mathrm{j}kl} - R_U(0)\mathrm{e}^{-\mathrm{j}kl}} \end{aligned} \tag{9.64}$$

Substituting Equation (9.63) into the above equation yields:

$$Z_{\rm in} = Z_{\rm c}\frac{Z_{\rm L} + {\rm j}Z_{\rm c}\tan kl}{Z_{\rm c} + {\rm j}Z_L \tan kl} \tag{9.65}$$

Equation (9.65) is the expression of the input impedance of ideal transmission lines. The input impedance $Z_{\rm in}$ is a periodic function of l. It is obvious that when the load impedance $Z_{\rm L}$ equals to the characteristic impedance $Z_{\rm c}$, $Z_{\rm in} = Z_{\rm c}$, the circuit is said to be matched, and only the incident waves can exist, without any reflected waves.

Similarly, the input admittance $Y_{\rm in}$ of a transmission line can be obtained

$$Y_{\rm in} = 1/Z_{\rm in} = Y_{\rm c}\frac{Y_{\rm L}\cos kl + {\rm j}Y_{\rm c}\sin kl}{Y_{\rm c}\cos kl + {\rm j}Y_{\rm L}\sin kl} \tag{9.66}$$

or

$$Y_{\rm in} = Y_{\rm c}\frac{Y_{\rm L} + {\rm j}Y_{\rm c}\tan kl}{Y_{\rm c} + {\rm j}Y_{\rm L}\tan kl} \tag{9.67}$$

where $Y_{\rm L}$ and $Y_{\rm c}$ are the load admittance and the characteristic admittance respectively.

9.4.6 Velocity and Wavelength of TEM Waves in Two-Wire Transmission Lines

The velocity of propagation is

$$v = \frac{1}{\sqrt{\mu\varepsilon}} \tag{9.68}$$

The velocity of TEM waves in two-wire transmission lines filled with air is identical to that in the free space.

The wavelength of TEM waves depends on the filled medium

$$\lambda = \frac{v}{f} = \frac{1}{f\sqrt{\mu\varepsilon}} \tag{9.69}$$

9.4.7 Characteristics and Applications of Transmission Lines with Various Loads

The input impedance of a transmission line with fixed length is different when it is terminated in different loads. Furthermore, the input impedance varies with the length when the terminal load is fixed. Observing Equation (9.65), we may see that the input impedance depends upon the length of the line, the characteristic impedance, the load impedance, and the operating frequency. It is very useful to summarize the variation rule of the input impedance. The input impedance in case of ideal transmission lines will be considered.

1. The Dependence of $Z_{\rm in}$ on the Length of A Short-Circuited Transmission Line

When a circuit is shorted, that is $Z_{\rm L} = 0$. The input impedance $Z_{\rm in}$ is determined by substituting $Z_{\rm L} = 0$ into Equation (9.65) as

$$Z_{\rm in} = {\rm j}Z_{\rm c}\tan kl \tag{9.70}$$

Therefore, the input impedance $Z_{\rm in}$ along the line can be illustrated in Figure 9.8. In Figure 9.8(a), a section of a line is shown, and the length l from the load is marked in the figure. In Figure 9.8(b), the distribution of the composite voltage wave and the composite current wave along the line is shown. The composite wave is a pure standing wave, as the load is zero, thus the total reflection is caused. The phase difference of the composite voltage wave and the composite current wave is 90 degree. In Figure 9.8(c), the variation of the impedance along the line is shown. Z_{in} varies with the length l, and it can be inductive or capacitive. So it could be a series resonance or a parallel resonance. Z_{in} is proportional to $\tan kl$, thus the characteristic of Z_{in} repeats at every one-half wavelength. In Figure 9.8(d), the equivalent circuit of Z_{in} at different l is shown.

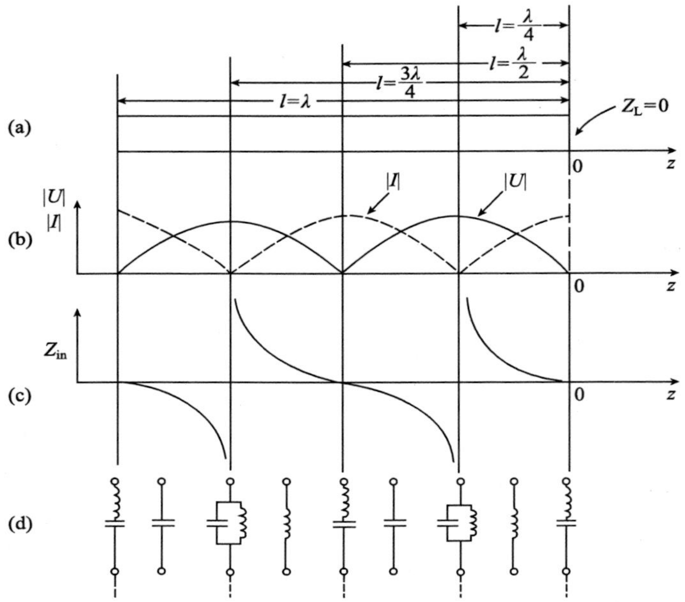

Figure 9.8 The variation and the equivalent circuit of Z_{in} along the transmission line with the terminal short circuited

2. The Dependence of Z_{in} on the Length of An Open-Circuited Transmission Line

When a circuit is opened, that is $Z_{\text{L}} = \infty$. The input impedance is determined from substituting $Z_{\text{L}} = \infty$ into Equation (9.65) as

$$Z_{\text{in}} = -\text{j}Z_{\text{c}} \cot kl \tag{9.71}$$

According to the above equation, the value of Z_{in} along the line and the equivalent circuit can be obtained.

3. Implementation of An Inductor or A Capacitor Using A Stub Line

1) Implementation of an inductor

In practical applications, a stub line terminated into a short-circuit is usually used to implement an inductor at ultrahigh frequencies. The input impedance of a transmission line terminated into a short-circuit is as

$$Z_{\text{in}} = \text{j}Z_{\text{c}} \tan kl$$

Obviously, as long as the following equation is satisfied

$$0 < kl < 90°$$

or

$$0 < l < \lambda/4 \tag{9.72}$$

The input impedance is inductive. When the length of a line is $n\lambda/2$ longer, where n is an integer, that is

$$n\frac{\lambda}{2} < l < \frac{\lambda}{4} + n\frac{\lambda}{2}, \quad n = 0, 1, 2, \cdots \tag{9.73}$$

The impedance is still inductive.

2) Implementation of a capacitor

In practical application, a stub line terminated into an open-circuit is usually used to implement a capacitor at ultrahigh frequency. The input impedance of a transmission line terminated into an open-circuit is as

$$Z_{\text{in}} = -\mathrm{j}Z_{\text{c}} \cot kl$$

Obviously, if the following equation is satisfied

$$0 < kl < 90°$$

or

$$0 < l < \lambda/4 \tag{9.74}$$

The input impedance is thus capacitive. When the following equation is satisfied

$$n\frac{\lambda}{2} < l < \frac{\lambda}{4} + n\frac{\lambda}{2}, \quad n = 0, 1, 2, \cdots \tag{9.75}$$

The impedance is still capacitive.

4. Implementation of Series and Parallel Resonance Circuit Using Stub Line

1) Implementation of series resonance circuits

When an ideal transmission line is terminated into an open-circuit, and the length of the line is $(2n-1)\lambda/4$, where n is an integer, that is

$$l = (2n-1)\frac{\lambda}{4}, \quad n = 1, 2, 3, \cdots$$

The input impedance is zero, that is

$$\begin{aligned} Z_{\text{in}} &= -\mathrm{j}Z_{\text{c}} \cot kl \\ &= -\mathrm{j}Z_{\text{c}} \cot\left[(2n-1) \times 90°\right] = 0 \end{aligned} \tag{9.76}$$

The input impedance being zero indicates a series resonance circuit.

If an ideal transmission line is terminated into an short-circuit and the length of the line is $n\lambda/2$, where n is an integer, the input impedance is zero, that is

$$Z_{\text{in}} = \mathrm{j}Z_{\text{c}} \tan kl = \mathrm{j}Z_{\text{c}} \tan(n\pi) = 0 \tag{9.77}$$

This kind of ideal transmission lines is also corresponding to a series resonance circuit.

2) Implementation of parallel resonance circuits

The input impedance of an ideal parallel resonance circuit is infinite, and the input impedance of an ideal transmission line of length $(2n-1)\lambda/4$ terminated into short-circuit is also infinite. Meanwhile, an ideal transmission line of length $n\lambda/2$ terminated into an open-circuit is infinite. Both of them are equivalent to a parallel resonance circuit.

5. Quarter-Wavelength Impedance Converters

The input impedance is not equal to the original load impedance after a section of a transmission line of length$l$. This section of the line is an impedance converter. When the length of the line is $l = \lambda/4$, the converter has a special application.

When the length of the line is $l = \lambda/4$, or $kl = 90°$, Equation (9.65) becomes

$$Z_{\text{in}} = \frac{Z_{\text{c}}^2}{Z_{\text{L}}} \tag{9.78}$$

After a section of a transmission line of length $\lambda/4$, the input impedance Z_{in} and the load impedance Z_{L} are reciprocal to each other. Consequently, a section of a transmission line with length $\lambda/4$ is also called an impedance inverter. The characteristics are as follows:

When the load impedance Z_{L} increases, the input impedance Z_{in} decreases.

When the load impedance Z_{L} decreases, the input impedance Z_{in} increases.

If the load impedance Z_{L} is equivalent to a series circuit of a resistor R_{L} and an inductive reactance X_{L}, the input impedance is equivalent to a parallel circuit of a resistor R_{in} and a capacitive reactance X_{in}, that is

$$Y_{\text{in}} = \frac{1}{Z_{\text{in}}} = \frac{Z_{\text{L}}}{Z_{\text{c}}^2} = \frac{R_{\text{L}}}{Z_{\text{c}}^2} + \text{j}\frac{X_{\text{L}}}{Z_{\text{c}}^2} \tag{9.79}$$

More importantly, a quarter-wavelength line of the characteristic impedance Z_{c} can be used as an impedance converter. When the load impedance Z_{L} is a pure resistance, the input impedance after a quarter-wavelength line is also a pure resistance, and it is

$$Z_{\text{in}} = \frac{Z_{\text{c}}^2}{Z_{\text{L}}} \tag{9.80}$$

or

$$Z_{\text{c}} = \sqrt{Z_{\text{in}} Z_{\text{L}}} \tag{9.81}$$

If the required load is Z_{02} (pure resistance) and the actual load is Z_{01} (pure resistance), we can use a quarter-wavelength line with the characteristic impedance $Z_{\text{c}} = \sqrt{Z_{01} \cdot Z_{02}}$. Thus, the actual load is connected to the terminal, and the impedance becomes exactly Z_{02} after the conversion of the section line, the circuit is thus matched, as shown in Figure 9.9.

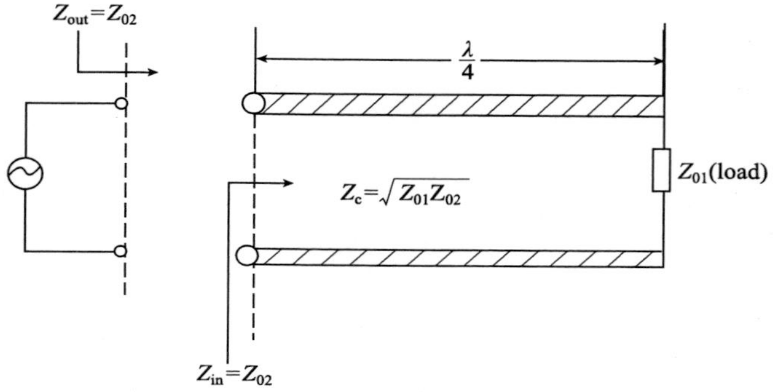

Figure 9.9 A quarter-wavelength impedance converter

For the implementation of a reactance component using a section of a transmission line, it should be noted that both the inductance L and capacitance C vary with frequency. This is different from the lumped inductance and capacitance at low frequencies.

In addition, the expressions such as Equations (9.65), (9.70) and (9.71) for the input impedance of the ideal transmission line in case of a short-circuit terminal or an open-circuit terminal from transmission line theory are valid not only for ideal two-wire transmission lines but also for all the uniform ideal transmission line systems propagating TEM waves, such as parallel-plate transmission systems, strip lines, microstrip lines, coaxial cables and so on. However, different uniform transmission systems propagating TEM waves have different expressions of the characteristic impedance. For example:

Parallel-plate transmission systems

$$Z_{\mathrm{c}} = \eta \frac{a}{b} \quad (\text{when } b \gg a)$$

where a and b are the distance between the plates and the width of the plates, respectively.

Coaxial cable

$$Z_{\mathrm{c}} = \frac{\eta}{2\pi} \ln\left(\frac{b}{a}\right)$$

where a and b are the inner and outer radius of a coaxial cable, respectively.

Two-wire transmission line

$$Z_{\mathrm{c}} = \frac{\eta}{\pi} \ln\left[\frac{D}{2a} + \sqrt{\left(\frac{D}{2a}\right)^2 - 1}\right]$$

where a and D are the radius of the line and the distance between lines, respectively.

From the above three kinds of expressions for the characteristic impedance, it could be concluded that the characteristic impedance has different geometry factors in different wave guiding systems. For instance, the factors in the parallel-plate system, the coaxial cable and the two-wire line are $\left(\frac{a}{b}\right)$,$\ln\left(\frac{a}{b}\right)$, and $\ln\left[\frac{D}{2a} + \sqrt{\left(\frac{D}{2a}\right)^2 - 1}\right]$, respectively. However, the velocity of TEM waves is $v = 1/\sqrt{\mu\varepsilon}$ for all geometries.

9.4.8 Low Loss Transmission Line

1. Transmission Line Equations of A Lossy Transmission Line

The equivalent circuit of a lossy transmission line is shown in Figure 9.10. L_0, C_0, R_0 and G_0 are the inductance, capacitance, resistance and conductance per-unit-length, respectively. They are all distributed parameters.

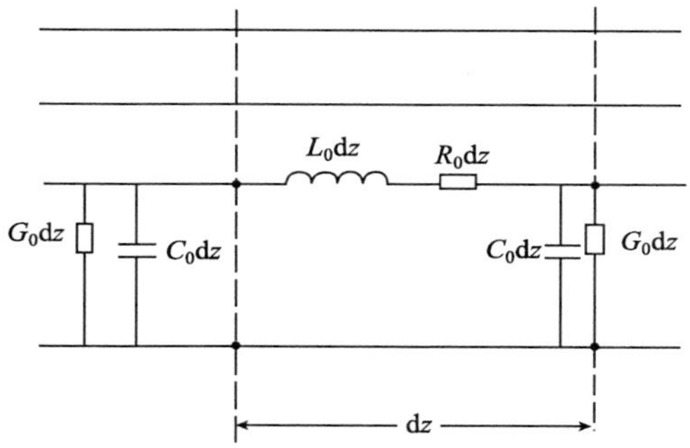

Figure 9.10 The equivalent circuit of a low lossy transmission line

Comparing with the ideal transmission line, the difference is that the series impedance and parallel admittance per-unit-length change from $j\omega L_0$ and $j\omega C_0$ to $(j\omega L_0 + R_0)$ and $j\omega C_0 + G_0$, respectively. According to Equations (9.54) and (9.55), transmission line equations for lossy systems can be written as

$$\frac{dU}{dz} = -(R_0 + j\omega L_0)I \tag{9.82}$$

$$\frac{dI}{dz} = -(G_0 + j\omega C_0)U \tag{9.83}$$

Substituting Equation (9.83) into the derivative of Equation (9.82) with respect to z yields the wave equation for the voltage

$$\frac{d^2U}{dz^2} = \gamma^2 U \tag{9.84}$$

Substituting Equation (9.82) into the partial differentiation of Equation (9.83) with respect to z yields the wave equation for the current

$$\frac{d^2I}{dz^2} = \gamma^2 I \tag{9.85}$$

where γ is the propagation constant:

$$\gamma^2 = (R_0 + j\omega L_0) \cdot (G_0 + j\omega C_0) \tag{9.86}$$

Writing the solutions of Equations (9.84) and (9.85) in complex format as

$$U = U^+ + U^- = U_0^+ e^{j\omega t - rz} + U_0^- e^{j\omega t + \gamma z} \tag{9.87}$$

$$I = I^+ + I^- = I_0^+ e^{j\omega t - rz} + I_0^- e^{j\omega t + \gamma z} \tag{9.88}$$

In the above equations, the first term indicates the incident wave, and the second term indicates the reflected wave. If there is only the incident wave, rearranging Equation (9.87) and Equation (9.82) yields

$$\frac{dU^+}{dz} = \frac{d}{dz}(U_0^+ e^{j\omega t - rz}) = -\gamma U^+ = -(R_0 + j\omega L_0)I^+$$

or

$$U^+\gamma = I^+(R_0 + j\omega L_0)$$

According to the definition of the characteristic impedance Z_c of a transmission line, we can obtain the characteristic impedance from the above equations as

$$Z_c = \frac{U^+}{I^+} = \frac{R_0 + j\omega L_0}{\gamma}$$

$$= \sqrt{\frac{(R_0 + j\omega L_0)^2}{(R_0 + j\omega L_0)(G_0 + j\omega C_0)}} = \sqrt{\frac{R_0 + j\omega L_0}{G_0 + j\omega C_0}} \tag{9.89}$$

The above equation indicates that the characteristic impedance of a lossy transmission line is not a pure resistance. However, the characteristic impedance can be treated as a pure resistance when the loss is low and both R_0 and G_0 tend to be zero.

2. Input Impedance of Lossy Transmission Lines

Method 1 Write the solutions of U and I in the format of hyperbolic functions, and then derive Z_{in}.

The solution of Equations (9.84) and (9.85) can be written in the format of hyperbolic function as

$$U = (C_1 \text{ch}\gamma z + C_2 \text{sh}\gamma z) \cdot e^{j\omega t} \tag{9.90}$$

$$I = (C_3 \text{ch}\gamma z + C_4 \text{sh}\gamma z) \cdot e^{j\omega t} \tag{9.91}$$

where $C_1 \sim C_4$ are unknown constants which could be determined from the terminal condition of the transmission line. Assume that

$$\text{At} \quad z = 0, \qquad U = U_{\text{L}}, \quad I = I_{\text{L}};$$

$$\text{At} \quad -z = l, \qquad U = U(l), \quad I = I(l);$$

Substitute the terminal conditions into Equations (9.90) and (9.91), and differentiate the result with respect to z, then $C_1 \sim C_4$ can be found from Equations (9.82) and (9.83). Substituting them into Equations (9.90) and (9.91) yields:

$$U(l) = (U_{\text{L}} \text{ch}\gamma l + Z_{\text{c}} I_{\text{L}} \text{sh}\gamma l) \cdot e^{j\omega t} \tag{9.92}$$

$$I(l) = \left(I_{\text{L}} \text{ch}\gamma l + \frac{U_{\text{L}}}{Z_{\text{c}}} \text{sh}\gamma l\right) \cdot e^{j\omega t} \tag{9.93}$$

From the above equations, the input impedance of a lossy transmission line can be obtained:

$$Z_{\text{in}} = \frac{U(l)}{I(l)} = \frac{U_{\text{L}} \text{ch}\gamma l + Z_{\text{c}} I_{\text{L}} \text{sh}\gamma l}{I_{\text{L}} \text{ch}\gamma l + \dfrac{U_{\text{L}}}{Z_{\text{c}}} \text{sh}\gamma l}$$

$$= Z_{\text{c}} \frac{Z_{\text{L}} \text{ch}\gamma l + Z_{\text{c}} \text{sh}\gamma l}{Z_{\text{c}} \text{ch}\gamma l + Z_{\text{L}} \text{sh}\gamma l} \tag{9.94}$$

Method 2 By replacing jkl by γl in Equations (9.64) and (9.94) could be derived.

If a transmission line is terminated into a short-circuit ($Z_{\text{L}} = 0$), the input impedance could be obtained from Equation (9.94):

$$Z_{\text{in}} = Z_{\text{c}} \text{th}(\gamma l) = Z_{\text{c}} \text{th}\left[(\alpha + j\beta)l\right]$$

$$= Z_{\text{c}} \frac{\text{sh}\alpha l \cos\beta l + j\text{ch}\alpha l \sin\beta l}{\text{ch}\alpha l \cos\beta l + j\text{sh}\alpha l \sin\beta l} \tag{9.95}$$

If $Z_{\text{L}} = 0$ and the length of a transmission line is $l = (2n-1)\lambda/4$, $n = 1, 2, \cdots$, the input impedance could be obtained from Equation (9.95):

$$Z_{\text{in}} = Z_{\text{c}} \frac{\text{ch}\alpha l}{\text{sh}\alpha l} \tag{9.96}$$

This is the input impedance of a transmission line with length $l = (2n-1)\lambda/4$ and terminated into a short-circuit.

If $Z_{\text{L}} = \infty$ and the length is $l = (2n-1)\lambda/4$, $n = 1, 2, \cdots$, the input impedance could be obtained from Equation (9.94):

$$Z_{\text{in}} = Z_{\text{c}} \frac{\text{ch}\gamma l}{\text{sh}\gamma l} = Z_{\text{c}} \frac{\text{sh}\alpha l}{\text{ch}\alpha l} \tag{9.97}$$

This is the input impedance of a transmission line with length $l = (2n-1)\lambda/4$ and terminated into an open-circuit.

3. Attenuation Constant α and Phase Constant β of Low Loss Transmission Lines

Since the low loss transmission lines satisfy the following conditions:

$$\omega L_0 \gg R_0$$

$$\omega C_0 \gg G_0$$

γ in Equation (9.86) can be expanded with the binomial expression as

$$\begin{aligned}\gamma = \alpha + \mathrm{j}\beta &= \sqrt{(R_0 + \mathrm{j}\omega L_0)(G_0 + \mathrm{j}\omega C_0)} \\ &= \mathrm{j}\omega\sqrt{L_0 C_0\left[\left(\frac{R_0 + \mathrm{j}\omega L_0}{\mathrm{j}\omega L_0}\right)\left(\frac{G_0 + \mathrm{j}\omega C_0}{\mathrm{j}\omega C_0}\right)\right]} \\ &= \mathrm{j}\omega\sqrt{L_0 C_0}\left[1 + \frac{R_0}{2\mathrm{j}\omega L_0} + \frac{R_0^2}{8\omega^2 L_0^2} + \cdots\right]\left[1 + \frac{G_0}{2\mathrm{j}\omega C_0} + \frac{G_0^2}{8\omega^2 C_0^2} + \cdots\right] \\ &\approx \frac{1}{2}\left[\mathrm{R}_0\sqrt{\frac{C_0}{L_0}} + G_0\sqrt{\frac{L_0}{C_0}}\right]\left[1 + \frac{1}{8\omega^2}\left(\frac{R_0}{L_0} - \frac{G_0}{C_0}\right)^2\right] \\ &\quad + \mathrm{j}\omega\sqrt{L_0 C_0}\left[1 + \frac{1}{8\omega^2}\left(\frac{R_0}{L_0} - \frac{G_0}{C_0}\right)^2\right]\end{aligned} \tag{9.98}$$

Neglecting the second order term in the binomial expression, there are

$$\beta \approx \omega\sqrt{L_0 C_0} \tag{9.99}$$

$$\alpha \approx \frac{R_0}{2Z_\mathrm{c}} + \frac{G_0 Z_0}{2} \tag{9.100}$$

For a low loss transmission line, α and β from the above equations are preferable approximations.

The attenuation constant α in Equation (9.100) consists of two parts: one is $\alpha_\mathrm{c} = \dfrac{R_0}{2Z_\mathrm{c}}$ caused by conductor loss; the other is $\alpha_\mathrm{d} = \dfrac{G_0 Z_\mathrm{c}}{2}$ caused by dielectric loss. The dielectric loss could be neglected when the dielectric is air. Hence the attenuation constant is determined by only conductor loss, and it is

$$\alpha = \frac{R_0}{2Z_\mathrm{c}} \tag{9.101}$$

Substituting the above equation into Equation (9.96) and αl tending to zero yield

$$Z_\mathrm{in} = Z_\mathrm{c}\frac{\mathrm{ch}\alpha l}{\mathrm{sh}\alpha l} \approx \frac{Z_\mathrm{c}}{\alpha l} = \frac{2Z_\mathrm{c}^2}{R_0 l} \tag{9.102}$$

This is the input impedance of a low loss transmission line of length $l = (2n-1)\lambda/4$ and terminated into a short-circuit.

The phase constant β in Equation (9.98) is

$$\beta = \omega\sqrt{L_0 C_0}\left[1 + \frac{1}{8\omega^2}\left(\frac{R_0}{L_0} - \frac{G_0}{C_0}\right)^2\right]$$

Thus, the phase velocity v_p is

$$v_\mathrm{p} = \frac{\omega}{\beta} = \frac{1}{\left\{\sqrt{L_0 C_0}\left[1 + \dfrac{1}{8\omega^2}\left(\dfrac{R_0}{L_0} - \dfrac{G_0}{C_0}\right)^2\right]\right\}} \tag{9.103}$$

It is observed that the phase velocity is not a constant but the function of frequency, which means that it is a dispersion wave. In the previous text, it is denoted that dispersion could cause dispersive distortion. Dispersion may widen the pulse when a digital signal is transmittedresulting in inter-symbol interference.

From Equation (9.103), there is

$$\left(\frac{R_0}{L_0} - \frac{G_0}{C_0}\right)^2 \Rightarrow 0 \tag{9.104}$$

The dependence of phase velocity v_p on frequency will be weakened.

4. Input Impedance Z_{in} at Any Point of A Low Loss Resonant Line

In this subsection, the input impedance at any point of a resonance stub will be discussed.

The resonant line of length $\frac{\lambda}{4}$ with the characteristic impedance Z_c is shown in Figure 9.11, which is terminated in a short-circuit. If the transmission line has low loss and high Q, the distributions of current and voltage on the transmission line are in accordance with the sinusoidal or cosinusoidal function. The input impedance looking from an arbitrary point $a - a'$ to the short-circuit terminal is inductive, which is $jZ_c \tan \beta l'$. Here the distance of the point $a - a'$ to the terminal is l'. Similarly, the input impedance looking from the point to the open-circuit terminal is capacitive, and the capacitance is

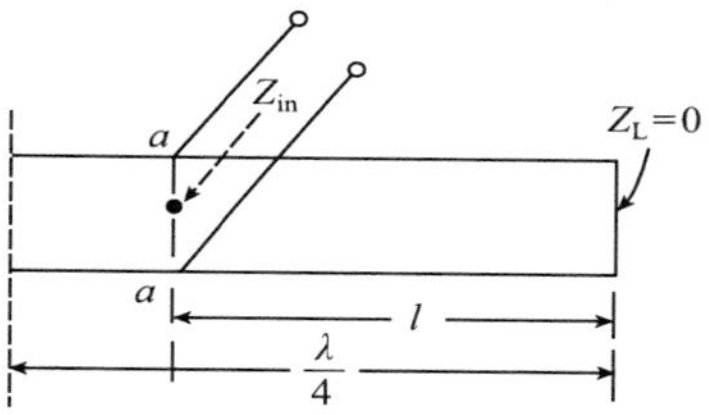

Figure 9.11 Quarter-wavelength resonant line

$$\begin{aligned} -jZ_c \cot\left[\beta\left(\frac{\lambda}{4} - l'\right)\right] &= -jZ_c \cot(90^\circ - \beta l') \\ &= -jZ_c \tan \beta l' \end{aligned}$$

It could be concluded that the reactances from any point to the two directions have the same value but opposite characteristic, so the susceptance of both in parallel is zero. The input impedance is purely resistive. How to determine the input impedance? First, due to the loss characteristics of the transmission line, the distribution of voltage or current along the line can be still taken as sinusoidal or cosinusoidal function. Then the power dissipation in this resonant line is known after the voltage and current are determined. The power dissipation is unrelated with the location of input point. In this way, the input impedance at any point could be easily calculated.

If the input point is located at the open-circuit terminal of a quarter-wavelength line, the input power is

$$\frac{U_0^2}{Z_{in}\left(\frac{\lambda}{4}\right)} = \frac{U_0^2 R_0 l}{2Z_c^2} \tag{9.105}$$

where U_0 and $Z_{in}\left(\frac{\lambda}{4}\right)$ are the voltage and the input impedance of a quarter-wavelength line terminated by a terminal open-circuit respectively. Equation (9.102) is used in the derivation of above equation.

If the input point is located ata $- a'$, the input power is

$$\frac{U^2(l')}{Z_{in}(l')} = \frac{(U_0 \sin \beta l')^2}{Z_{in}(l')} \tag{9.106}$$

The input power expressed in (9.105) is identical to that expressed in Equation (9.106), so the input impedance at any point is

$$Z_{\text{in}}(l') = R_{\text{in}}(l') = \left(\frac{2Z_{\text{c}}^2}{R_0 l}\right) \sin^2 \beta l' \tag{9.107}$$

It can be observed that the input impedance at any point of a low loss quarter-wavelength line terminated in a short-circuit is a pure resistance, and it equals to the product of the resistance at the open-circuit terminal $2Z_{\text{c}}^2/R_0 l$ and $\sin^2 \beta l'$, where l' is the distance between the point and the short-circuit terminal.

Example 9.5 A lossy transmission line of length l has the characteristic impedance Z_{c}. The input impedance is $Z_{\text{in(o)}}$ when the line is terminated in an open-circuit, and the input impedance is $Z_{\text{in(s)}}$when the line is terminated in a short-circuit. Demonstrate that:

(1) $Z_{\text{c}} = \sqrt{Z_{\text{in(o)}} \cdot Z_{\text{in(s)}}}$;

(2) $\gamma = \dfrac{1}{l}\text{arcth}\dfrac{Z_{\text{in(s)}}}{Z_{\text{in(o)}}}$.

Solution According to Equation (9.94), when $Z_{\text{L}} \to \infty$, the input impedance $Z_{\text{in(o)}}$ is

$$Z_{\text{in(o)}} = Z_{\text{c}}\frac{\text{ch}\gamma l}{\text{sh}\gamma l} = Z_{\text{c}} \cdot \text{cth}\gamma l \tag{a}$$

When $Z_{\text{L}} = 0$, the input impedance $Z_{\text{in(s)}}$ is

$$Z_{\text{in(s)}} = Z_{\text{c}}\frac{\text{sh}\gamma l}{\text{ch}\gamma l} = Z_{\text{c}} \cdot \text{th}\gamma l \tag{b}$$

Thus, the characteristic impedance can be obtained from Equations (a) and (b) as

$$Z_{\text{c}} = \sqrt{Z_{\text{in(o)}} \cdot Z_{\text{in(s)}}} \tag{c}$$

And the propagation constant γ is

$$\gamma = \frac{1}{l}\text{arcth}\sqrt{\frac{Z_{\text{in(s)}}}{Z_{\text{in(o)}}}} \tag{d}$$

Example 9.6 A 15-cm-long (less than $\dfrac{\lambda}{4}$) low loss transmission line has the input impedance $Z_{\text{in(o)}} = -\text{j}54.6\Omega$ and $Z_{\text{in(s)}} = \text{j}103\Omega$ when it is terminated in an open-circuit and a short-circuit respectively. Determine the characteristic impedance Z_{c} and the propagation constant γ.

Solution The input impedances when the line is terminated in an open-circuit and a short-circuit are known, hence, we can calculate directly by using Equation (c) and (d) in Example 9.5. Applying Equation (c) in Example 9.5 yields

$$Z_{\text{c}} = \sqrt{Z_{\text{in(o)}} \cdot Z_{\text{in(s)}}} = \sqrt{-\text{j}54.6 \times \text{j}103} = 74.99\Omega$$

And applying Equation (d) in the above example yields

$$\gamma = \frac{1}{l}\text{arcth}\sqrt{\frac{Z_{\text{in(s)}}}{Z_{\text{in(o)}}}} = \frac{1}{0.15}\text{arcth}\sqrt{\frac{\text{j}103}{-\text{j}54.6}} = \text{j}6.28\text{rad/m}$$

9.5 Coaxial Cable

A coaxial cable is also one kind of system which can support static fields. Hence, it can transmit TEM waves.

9.5.1 Coaxial Cables and TEM Waves Propagating on It

Let the voltage and current across the coaxial cable be U and I respectively. The distributions of the electric field and magnetic field of TEM waves in a coaxial cable are identical to that of the static field, so we can directly write the following equations

$$H_\phi = \frac{1}{2\pi r} \tag{9.108}$$

$$E_r = \frac{q}{2\pi\varepsilon r} = \frac{C_0 U}{2\pi\varepsilon r} \tag{9.109}$$

where C_0 is the capacitance per-unit-length of a coaxial cable and C_0U is the quantity per-unit-length of charges on the surface of the inner conductor of the coaxial cable (the charges on the inner and outer conductor have the same value and opposite polarity). Assuming that TEM waves are simple harmonic waves and propagate along $+z$ direction, the expressions for E_r and H_ϕ are

$$H_\phi = \frac{I}{2\pi r} = \frac{1}{2\pi r} I_0 \mathrm{e}^{\mathrm{j}(\omega t - kz)} \tag{9.110}$$

$$E_r = \frac{C_0 U}{2\pi\varepsilon r} = \frac{C_0}{2\pi\varepsilon r} U_0 \mathrm{e}^{\mathrm{j}(\omega t - kz)} \tag{9.111}$$

The distribution of E_r and H_ϕ in a coaxial cable is shown in Figure 9.12.

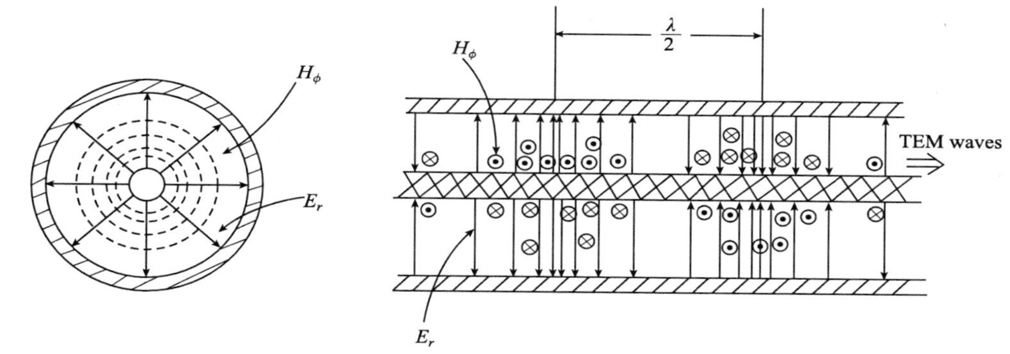

(a) Distribution of the field on the cross-section

(b) Distribution of the field along the longitudinal direction

Figure 9.12 TEM waves in the coaxial cables

9.5.2 Transmission Line Equations for Voltage and Current in Coaxial Cables

The relationship of E_r and H_ϕ in the coaxial cable (in the cylindrical coordinate) can be written from Maxwell's Equations directly.

$$\frac{\partial E_r}{\partial z} = -\mu \frac{\partial H_\phi}{\partial t} \tag{9.112}$$

and

$$-\frac{\partial H_\phi}{\partial z} = \varepsilon \frac{\partial E_r}{\partial t} \tag{9.113}$$

Substitute Equations (9.108) and (9.109) into Equation (9.112) and apply $L_0C_0 = \mu\varepsilon$, we obtain

$$\frac{\partial U}{\partial z} = -L_0\frac{\partial I}{\partial t} \tag{9.114}$$

Substitute Equation (9.108) and (9.109) into Equation (9.113) and apply $L_0C_0 = \mu\varepsilon$, we have

$$\frac{\partial I}{\partial z} = -C_0\frac{\partial U}{\partial t} \tag{9.115}$$

Equations (9.114) and (9.115) are the transmission line equations.

From the partial differentiation of Equation (9.114) with respect to z and Equation (9.115), wave equations for voltage are obtained

$$\frac{\partial^2 U}{\partial z^2} = L_0C_0\frac{\partial^2 U}{\partial t^2} \tag{9.116}$$

From the partial differentiation of Equation (9.115) with respect to z and Equation (9.114), wave equations for current are found

$$\frac{\partial^2 I}{\partial z^2} = L_0C_0\frac{\partial^2 I}{\partial t^2} \tag{9.117}$$

Assume that the voltage and current are both simple harmonic waves, then the above two equations can be written as

$$\left.\begin{aligned}\frac{\mathrm{d}^2 U}{\mathrm{d}z^2} + k^2U = 0\\ \frac{\mathrm{d}^2 I}{\mathrm{d}z^2} + k^2I = 0\end{aligned}\right\} \tag{9.118}$$

Wave equations for the simple harmonic voltage and current wave are given in Equation (9.118) respectively.

If the coaxial cable is infinitely long or terminated in a matched load, there is no reflected wave in the coaxial cable. Thus the solution of voltage U in Equation (9.118) can be written as

$$U = U_0\mathrm{e}^{\mathrm{j}(\omega t - kz)} \tag{9.119}$$

Substituting the above equation into Equation (9.114) or (9.115) yields

$$I = \frac{k}{\omega L_0}U_0\mathrm{e}^{\mathrm{j}(\omega t - kz)} = \sqrt{\frac{C_0}{L_0}}U_0\mathrm{e}^{\mathrm{j}(\omega t - kz)} \tag{9.120}$$

In above derivation, we have applied $k = \omega\sqrt{\mu\varepsilon} = \omega\sqrt{L_0C_0}$.

9.5.3 Characteristic Impedance of Coaxial Cables

The characteristic impedance is defined as the ratio of voltage to current of an incident wave (or reflected wave) in coaxial cables, that is

$$Z_c = \frac{U^+}{I^+} = -\frac{U^-}{I^-} \tag{9.121}$$

From Equations (9.119)~(9.121) we obtain

$$Z_c = \frac{U^+}{I^+} = \sqrt{\frac{L_0}{C_0}} = \frac{\sqrt{\mu\varepsilon}}{C_0} \tag{9.122}$$

The specific expression for the characteristic impedance of a coaxial cable will be determined in the following text. The capacitance per-unit-length C_0 of a coaxial cable is known as

$$C_0 = \frac{2\pi\varepsilon}{\ln\left(\dfrac{b}{a}\right)} \tag{9.123}$$

Substitute the above equation into Equation (9.122), and we obtain the characteristic impedance of a coaxial cable as

$$Z_c = \frac{1}{2\pi}\eta \ln\left(\frac{b}{a}\right)$$

When the medium in a coaxial cable is air, there is $\eta = \sqrt{\dfrac{\mu_0}{\varepsilon_0}} \approx 120\pi$. Thus, the characteristic impedance of the coaxial cable is

$$Z_c = 60 \ln\left(\frac{b}{a}\right) \tag{9.124}$$

If the parameters of the dielectric filled in the coaxial cable are $\mu = \mu_0$, $\varepsilon = \varepsilon_r\varepsilon_0$, the characteristic impedance of the coaxial cable is then

$$Z_c = \frac{\eta}{2\pi} \ln\left(\frac{b}{a}\right) = \frac{60}{\sqrt{\varepsilon_r}} \ln\left(\frac{b}{a}\right) \tag{9.125}$$

It is noted that the operational frequency of TEM wave in a coaxial cable can range from very low to very high frequencies. However, when the frequency is high up to some extent, the higher modes will occur. The higher the frequency, the more modes exist. The highest operating frequency (or the shortest operating wavelength) of a coaxial cable is limited by the cutoff frequency of the first higher mode. If λ_{min} indicates the shortest operating wavelength, in order to restrain the higher modes the following expression must be satisfied.

$$\lambda_{min} > \pi(a + b) \tag{9.126}$$

where a and b are the radius of inner and outer conductors, and $\pi(a + b)$ is the cutoff frequency of TE_{11}wave in coaxial cables. Moreover, the TE_{11} wave has the longest cutoff wavelength in all TE and TM waves. Other TE and TM modes cannot propagate as long as the TE_{11} wave cannot propagate at a certain operating frequency. The concept of TE and TM modes in coaxial cables will be elaborately discussed in section 10.3.

Example 9.7 A coaxial cable filled with air transmits TEM wave. The radius of the inner and outer conductors are a and b respectively. The conductor attenuation constant of the coaxial cable is

$$\alpha_c = R_0/2Z_c = \frac{R_s}{2\pi b} \cdot \frac{1 + \dfrac{b}{a}}{\left[120 \ln\left(\dfrac{b}{a}\right)\right]}$$

where R_s is the surface resistance. Determine the condition of the minimum propagation attenuation of the coaxial cable.

Solution The condition of the minimum propagation attenuation of the coaxial cable can be determined from the equation below

$$\frac{d\alpha_c}{d\left(\dfrac{b}{a}\right)} = \frac{d}{d\left(\dfrac{b}{a}\right)}\left\{\frac{R_s}{2\pi b} \cdot \frac{1 + \dfrac{b}{a}}{\left[120 \ln\left(\dfrac{b}{a}\right)\right]}\right\} = 0$$

The solution is

$$\frac{b}{a} \approx 3.59$$

It means that the conductor loss of the coaxial cable has the minimum value when $\frac{b}{a} \approx 3.59$. However, the change of the attenuation constant is not obvious when $\frac{b}{a}$ varies in a wide range. Comparing to the minimum value, the change of the attenuation is less than 0.5% when $\frac{b}{a}$ varies from 3.2 to 4.1. The attenuation increase 5% when $\frac{b}{a}$ is from 2.6 to 5.2.

When $\frac{b}{a} = 3.59$ for the coaxial cable filled with air, the characteristic impedance is $Z_c = 76.7\Omega$, and the attenuation constant is minimum. When $\frac{b}{a} = 1.649$, the coaxial cable has the maximum power handling capacity and the characteristic impedance is $Z_c = 30\Omega$. Considering the maximum power handling capacity and the minimum attenuation, we usually choose the coaxial cable of $\frac{b}{a} = 2.303$ with $Z_c = 50\Omega$.

Example 9.8 The radius of inner and outer conductors of a coaxial cable filled with air are a and b respectively. The characteristic impedance of the coaxial cable is $Z_c = 75\Omega$. Determine the inductance and capacitance per-unit-length of the coaxial cable.

Solution The values of L_0 and C_0 are both related with the radius ratio of the inner conductor and the outer conductor. The expression for inductance per-unit-length is known as

$$L_0 = \frac{\mu_0}{2\pi} \ln \frac{b}{a} \tag{a}$$

The capacitance per-unit-length could be determined from $L_0 C_0 = \mu\varepsilon$

$$C_0 = \frac{\mu_0 \varepsilon_0}{L_0} = \frac{2\pi\varepsilon_0}{\ln \frac{b}{a}} \tag{b}$$

The value of Z_c is known and $\frac{b}{a}$ can be determined from the expression for the characteristic impedance

$$Z_c = 60 \ln \frac{b}{a} = 75\Omega$$

hence

$$\frac{b}{a} = e^{\left(\frac{75}{60}\right)} = 3.4903 \tag{c}$$

Substituting the value of $\frac{b}{a}$ into Equations (a) and (b) yields

$$L_0 = \frac{\mu_0}{2\pi} \ln(3.4903) = 2.5 \times 10^{-7}(\text{H/m})$$

$$C_0 = \frac{2\pi\varepsilon_0}{\ln(3.4903)} = 44.51 \times 10^{-12}(\text{F/m})$$

9.6 Quasi-TEM Waves in Lossy Transmission Lines

9.6.1 Theoretical Analysis

In practical situation, there are two loss factors, one is the non-ideal dielectric filled between the conductors, the loss tangent is not zero; another is the imperfect conductor, the resistance exists.

To investigate the performances of the real transmission lines, two parameters need to be defined: the leakage conductance per-unit-length G_0 between the two conductors and the resistance per-unit-length R_0.

Take parallel plates as an example, and assume that the permittivity and conductivity of the dielectric between plates are ε and σ, respectively. From "electrostatic analogue method" in Chapter 4, we obtain

$$G_0 = C_0\frac{\sigma}{\varepsilon} = \sigma \cdot \frac{b}{a} \tag{9.127}$$

where C_0 is the capacitance between plates, and b and a are the width of the conductor plates and the distance between plates shown in Fig. 9.2.

In general, the conductors will be assumed as good conductors. Let the conductivity and permeability of the conductor plates are σ_c and μ (usually it is identical to μ_0) respectively. The definition of the surface wave impedance of good conductors can be used to analyze the real conductor plates.

The surface impedance of a good conductor is

$$Z_s = R_s + jX_s = (1+j)\sqrt{\frac{\pi f \mu}{\sigma}} \tag{9.128}$$

The series surface resistance of the per-unit-length upper and lower conductor plates is

$$R = 2\frac{R_s \cdot 1}{b} = \frac{2}{b}\sqrt{\frac{\pi f \mu}{\sigma}} \tag{9.129}$$

It should be noted that the longitudinal electric field component and loss will be induced when real conductor plates are used as transmission lines. If the longitudinal electric field component does not exist, there is no x component of average Poynting's vector on the surface of the conductor, or it can be said that the energy can propagate forward and will not spread. Thus there is no loss in the transmission system. However, it is contradictory with the given condition.

Just due to the existence of the field component along the propagation direction, strictly speaking, the wave in the lossy transmission line is no longer TEM wave. However, for case of the good conductor, the component of the electric field along z direction is much less than the transverse component E_x. This kind of wave is usually called "Quasi-TEM wave" which can be taken as TEM wave to simplify the analysis.

9.6.2 Examples

The conductivity and permeability of good conductor copper are $\sigma = 5.8 \times 10^7$S/m and $\mu = \mu_0$ respectively. When the frequency $f = 3$GHz, determine the ratio of the longitudinal electric field to the transverse electric field in the parallel-plate transmission line filled with air.

$$\frac{|E_z|}{|E_x|} = \frac{|H_y| \cdot |Z_s|}{|H_y \cdot \eta|} = \frac{|Z_s|}{\eta} = \sqrt{\frac{\varepsilon}{\mu}}\sqrt{\frac{2\pi f \mu}{\sigma}} = \sqrt{\frac{2\pi f \varepsilon}{\sigma}} \tag{9.130}$$

$$\frac{|E_z|}{|E_x|} = \sqrt{\frac{2\pi \times 3 \times 10^9 \times 10^{-9}}{36\pi \times 5.8 \times 10^7}} = 5.4 \times 10^{-5}$$

From the result of the example, we can easily see that, if "quasi-TEM wave" is viewed as TEM wave, in low loss transmission systems the effect of the component of electromagnetic field along propagation direction to the analysis results is extremely small.

Exercises

9.1 An ideal two-wire transmission line of length l has a characteristic impedance of Z_c. Determine the input impedance Z_{in} under one of the following conditions.

(1) $Z_L = Z_c$; (2) $Z_L = 0$; (3) $Z_L = \infty$.

9.2 An ideal two-wire transmission line of length $l = \frac{\lambda}{4}$ with the characteristic impedance Z_c is terminated in an inductive load $Z_L = j\omega L$. Solve the following problems:

(1) Is the input impedance Z_{in} inductive or capacitive?

(2) Calculate the resistance component R_{in} and the reactance component X_{in} of the input impedance.

9.3 A 2-m-long lossless transmission line with a characteristic impedance 50Ω is terminated in a load $Z_L = (40 + j30)\Omega$. The operating frequency is 200MHz. Calculate the input impedance.

9.4 Now the polypropylene ($\varepsilon_r = 2.25$) is used as the medium of uniform transmission lines. Assume that the loss can be neglected.

(1) The radius of the conductor in a 300Ω two–wire transmission line is $d = 0.6$mm. Determine the distance$D$between lines;

(2) The radius of the inner conductor of a 75Ω coaxial cable is $a = 0.6$mm. Determine the radius b of the outer conductor.

9.5 A lossless transmission line has a characteristic impedance of $Z_L = 500\Omega$. The input impedance at the start end is an inductance of 250Ω when the line is terminated in a short-circuit. Determine the minimum length of the transmission line. If the line is terminated in an open-circuit, determine the length.

9.6 A transmission line consists of two same metal plates, and the cross-section is shown in Figure of Exercise 9.6. Assume $b \gg a$, and a uniform and lossless medium is filled between plates, and the conductivity of the metal plates is σ. Derive the expressions for power P, attenuation constant α and characteristic impedance Z_c when TEM wave propagates.

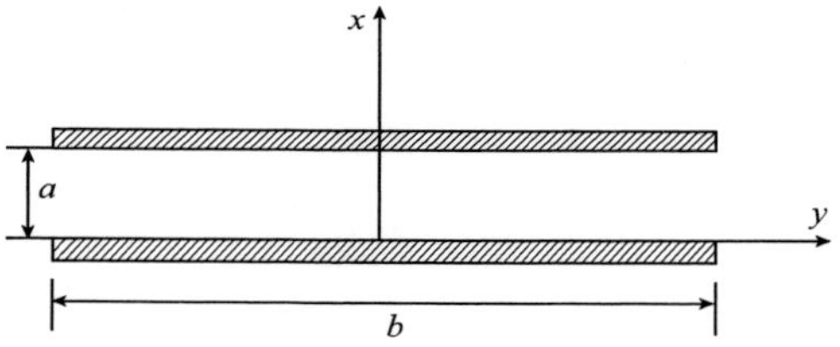

Figure of Exercise 9.6

9.7 Suppose the conductivity of the metal plate of the parallel-plates transmission system is σ_c. And the parameters of the uniform low loss medium between plates are μ, ε and σ_d. Prove that the expression for the attenuation constant can be written as follows when TEM wave propagates.

$$\alpha = \alpha_c + \alpha_d = \frac{1}{2}\frac{R_0}{Z_c} + \frac{1}{2}G_0 Z_c$$

where Z_c, R_0 and G_0 are the characteristic impedance, resistance and leakage conductance per-unit-length respectively.

9.8 The length of a low loss transmission line is 15cm (less than $\frac{\lambda}{4}$). The input impedances are $Z_{in(o)} = -j54.6\Omega$ and $Z_{in(s)} = j103\Omega$ when $Z_L \to \infty$ (open-circuit) and $Z_L = 0$ (short-circuit) respectively. Find the characteristic impedance Z_c and the propagation constant γ of the line.

9.9 Determine the characteristic impedance of a two-wire transmission line filled with air. Let the radius of the line and the distance between lines be a and D, where $D \gg a$.

9.10 Find the characteristic impedance of a two-wire transmission line filled with air. Let the radius of the line and the distance between lines be a and D, where $D \gg a$ is not satisfied.

9.11 Considering a lossless line:

(1) Calculate the characteristic impedance of the line with a load impedance $Z_L = (40 - j30)\Omega$ to achieve minimum VSWR.

(2) Calculate the minimum VSWR and the voltage reflection coefficient.

9.12 The distribution parameters of a transmission line are $R_0 = 10.4\Omega/m$, $C_0 = 8.35 \times 10^{-12}F/m$, $L_0 = 1.33 \times 10^{-6}H/m$ and $G_0 = 0.8 \times 10^{-6}S/m$ respectively at frequency 1GHz. Calculate the characteristic impedance, attenuation constant, phase constant, wavelength and the phase velocity on the line.

9.13 When a uniform lossless transmission line is terminated in different loads, standing waves will occur. Assume that the first wave node be located respectively: (1) at the load; (2) on the point $\lambda/4$ away from the load; (3) between the load and the point $\lambda/4$ away from the load; (4) between the point $\lambda/4$ away from the load and the point $\lambda/2$ away from the load. Specify the characteristics (such as the inductance, capacitance, pure resistance and so on) of the load under different conditions.

9.14 In the high-frequency circuit of operating length λ, a 75Ω transmission line with a short-circuit terminal has an inductive input impedance of 300Ω. Determine the length of the line. To obtain a capacitive input impedance of 300Ω, please determine the length of the line.

9.15 A 75-m-long uniform lossless transmission line has a characteristic impedance of $Z_c = 300\Omega$. The parameters of the medium are $\varepsilon_r = 1$ and $\mu_r = 1$ respectively. The voltage applied to the start end is $U = 100\sin(6\pi \times 10^6 t)$V. The VSWR is $S = 1.8$ and the voltage at the terminal is maximum when the line is terminated in a load. Try to determine

(1) The phase velocity v_p and wavelength λ of the voltage wave on the line;

(2) The load resistance R_L and the power consumption P on the load;

(3) The effective value of the current at the start end.

9.16 The VSWR on a lossless uniform transmission line with a characteristic impedance $Z_c = 300\Omega$ is $S = 2.0$. The nearest minimum voltage value node is 0.3λ away from the load. Please determine:

(1) The reflection coefficient at the terminal;

(2) The load impedance;

(3) The positions where the input impedances are pure resistances and the value of the resistance, respectively.

9.17 A uniform lossless transmission line with a characteristic impedance $Z_{01} = 50\Omega$ is terminated in a semi-infinite length uniform lossless transmission line with a characteristic impedance Z_{02}. A sine-wave power supply at an unknown frequency is applied. The VSWR is $S = 3.0$, and the velocity of wave is $v = 10^8$m/s. Two adjacent nodes where the voltage has the minimum value are located at 15cm and 25cm away from the connection respectively. Determine the characteristic impedance of the semi-infinite length line and the frequency of the power supply.

9.18 A coaxial cable with an inner radius a and an outer radius b is filled with air. Prove the expression $L_0 C_0 = \mu_0 \varepsilon_0$ is satisfied when TEM wave propagates. L_0 and C_0 are the inductance and the capacitance per-unit-length of the coaxial cable respectively.

9.19 A coaxial cable filled with air with an inner radius a and an outer radius b has a characteristic impedance of $Z_c = 75\Omega$. Determine the inductance per-unit-length L_0 and the capacitance per-unit-length C_0 of the coaxial cable.

9.20 A coaxial cable with an inner diameter 12.7mm and an outer diameter 31.75mm is filled with air. The breakdown strength of air is 30kV/cm. The frequency of TEM wave is $f = 9.375$GHz. Determine the maximum power handling capacity of the transmission line.

Chapter 10

TE and TM Modes Transmission System—Waveguide

The regular waveguides which can transmit TE and TM mode are discussed in this chapter. Waveguides are actually metal pipes with some specific cross sections, which takes on a special shape, such as rectangular, circular, or elliptic, etc. "Regular" means the shape of cross section, the size of cross section and the filling medium are not uniform and do not change along the longitudinal direction. At the same time, assuming the filling medium in waveguide is ideal, uniform, linear and isotropic.

Can TEM mode transmit in waveguide alone axial direction (z direction)? If it does, then its electric field component and magnetic field component should be parallel to the cross section (i.e., xOy plane). Owing to the restriction of the waveguide boundary condition, the magnetic line must be closed and forms a plane on the cross section However, there is no source in the waveguide, and the electric field is parallel with the cross section, therefore, no electric current or displacement current flow through the cross section, and then no magnetic field would exist. In this way, the imaginable closed magnetic line would not exist without the desirable current. That is to say, TEM mode cannot transmit in waveguide along the axial direction.

In other words, the electric field parallel to cross section of waveguide can generate displacement current of same direction. According to Maxwell's equations, it is reasonable to have magnetic line around this displacement current. However, in this case, the closed magnetic field must contain the longitudinal component H_z. If so, the field is not TEM mode any more.

Actually, the waveguide can propagate TE waves (transverse electric waves, also referred to as H-waves, having longitudinal magnetic field component, but no longitudinal electric field component) and TM waves (transverse magnetic waves, also referred to as E-waves, having longitudinal electric field component, but no longitudinal magnetic field component)

The electromagnetic wave in waveguide is the function of 3-D space and time. The method of separation of variables can be used to solve the wave function characterized by longitudinal (z direction) field component, and to obtain E_z and H_z. Then all the field components could be extracted by using the relation of transverse field components and the longitudinal field components, saying, $E_z$and H_z.

Solving the electromagnetic field using Maxwell's equations subject to the boundary conditions of metal waveguide pipe is mainly discussed in this chapter. Their characteristics and parameters are also discussed. The analysis of circular waveguide method is similar, then few of the analysis of circular waveguide will be introduced. The higher modes in coaxial lines are discussed in the end.

10.1 Rectangular Waveguide

Rectangular Waveguide is one of the mostly used waveguides. It is a hollow metal pipe with rectangular cross section shown in Figure 10.1. The width of the broad side is a, and the width of the narrow side is b.

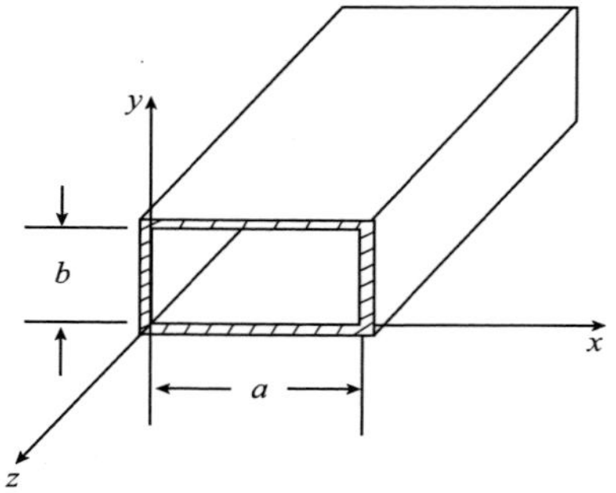

Figure 10.1 Rectangular waveguide

10.1.1 The Wave Equation in Waveguide

The source-free Maxwell's equations of simple harmonic wave in uniform and ideal medium are

$$\nabla \times \boldsymbol{H} = \varepsilon \frac{\partial \boldsymbol{E}}{\partial t} = \mathrm{j}\omega\varepsilon \boldsymbol{E} \tag{10.1}$$

$$\nabla \times \boldsymbol{E} = -\mu \frac{\partial \boldsymbol{H}}{\partial t} = -\mathrm{j}\omega\mu \boldsymbol{E} \tag{10.2}$$

$$\nabla \cdot \boldsymbol{D} = \varepsilon \nabla \cdot \boldsymbol{E} = 0 \tag{10.3}$$

$$\nabla \cdot \boldsymbol{B} = \mu \nabla \cdot \boldsymbol{H} = 0 \tag{10.4}$$

The wave equations can be derived by using Maxwell's equations and vector identities

Take curl of Equation (10.2), and using Equation (10.1), (10.3), and vector identities, we can get

$$\nabla \times \nabla \times \boldsymbol{F} = \nabla\nabla \cdot \boldsymbol{F} - \nabla^2 \boldsymbol{F}$$

so

$$\nabla^2 \boldsymbol{E} + k^2 \boldsymbol{E} = 0 \tag{10.5}$$

The formula which has been already given in the plane-wave chapter is the well-known Helmholtz equation, also known as the wave equation.

Considering Equation (10.5) in waveguide systems, and assuming the simple harmonic waves in the waveguide propagate along the z direction, that is, each field components include the factor of $\mathrm{e}^{\mathrm{j}\omega t - \gamma z}$, the wave equations in the waveguide can thus be found.

$$\begin{aligned}\nabla^2 \boldsymbol{E} + k^2 \boldsymbol{E} &= \nabla_T^2 \boldsymbol{E} + \frac{\partial^2 \boldsymbol{E}}{\partial z^2} + k^2 \boldsymbol{E} \\ &= \nabla_T^2 \boldsymbol{E} + \gamma^2 \boldsymbol{E} + k^2 \boldsymbol{E} \\ &= \nabla_T^2 \boldsymbol{E} + (\gamma^2 + k^2) \boldsymbol{E} \\ &= \nabla_T^2 \boldsymbol{E} + k_{\mathrm{c}}^2 \boldsymbol{E} = 0\end{aligned}$$

that is

$$\nabla_T^2 \boldsymbol{E} + k_{\mathrm{c}}^2 \boldsymbol{E} = 0 \tag{10.6}$$

In the equation

$$k_{\mathrm{c}}^2 = \gamma^2 + k^2 \tag{10.7}$$

Equation (10.6) is the wave equation of the waveguide characterized by the electric field. Following a similar approach we can get the wave equation of the waveguide in terms of the magnetic field.

$$\nabla_T^2 \boldsymbol{H} + k_c^2 \boldsymbol{H} = 0 \tag{10.8}$$

In the equations, $\boldsymbol{E}$ and $\boldsymbol{H}$ are not the horizontal components, but the longitudinal (z direction) component. In addition, $\boldsymbol{E}$ and $\boldsymbol{H}$ are not only the functions of z in form of ($e^{-\gamma z}$), but also the functions of x, y.

10.1.2 The Wave Equations of Longitudinal Field Component (z component) and Their Solutions

Equations (10.6) and (10.8) are the wave equations of vector electric and magnetic fields. In Cartesian coordinates, the wave equation for each field component (scalar) can be given. For example, the electric field equations can be written as

$$\nabla_T^2 E_x + k_c^2 E_x = 0 \tag{10.9}$$

$$\nabla_T^2 E_y + k_c^2 E_y = 0 \tag{10.10}$$

$$\nabla_T^2 E_z + k_c^2 E_z = 0 \tag{10.11a}$$

If E_x, E_y, E_z are replaced by H_x, H_y, H_z, we can get the scalar wave equation in terms of the magnetic field. The following are the solutions to the wave equation of E_z and H_z.

1. The Solutions to $\boldsymbol{E}_z$ of the TM Wave

The wave Equation (10.11a) of E_z can be written as

$$\frac{\partial^2 E_z}{\partial x^2} + \frac{\partial^2 E_z}{\partial y^2} + k_c^2 E_z = 0 \tag{10.11b}$$

1) Use the method of separation of variables to solve Equation (10.11b)
Suppose

$$E_z(x, y, z, t) = X \cdot Y \cdot e^{j\omega t - \gamma z} \tag{10.12}$$

In the equation, X, Y are the function of x, y, respectively.
Substituting Equation (10.12) into (10.11b) to obtain

$$\frac{1}{X}\frac{\partial^2 X}{\partial x^2} + \frac{1}{Y}\frac{\partial^2 Y}{\partial y^2} + k_c^2 = 0 \tag{10.13}$$

As X and Y are only the function of x, y, respectively, x and y is not related to each other. Therefore, in the waveguide, by the usual separation of variables argument, each of the terms in Equation (10.13) must be equal to a constant, so we define separation constants k_x and k_y, such that

$$\frac{1}{X}\frac{\partial^2 X}{\partial x^2} = -k_x^2 \tag{10.14}$$

$$\frac{1}{Y}\frac{\partial^2 Y}{\partial y^2} = -k_y^2 \tag{10.15}$$

so we get

$$\frac{d^2 X}{dx^2} + k_x^2 X = 0 \tag{10.16}$$

$$\frac{\mathrm{d}^2Y}{\mathrm{d}y^2} + k_y^2 y = 0 \tag{10.17}$$

From Equation (10.13)~(10.15), we have

$$k_c^2 = k_x^2 + k_y^2 \tag{10.18}$$

Equations (10.16) and (10.17) are two ordinary differential equations, the general solutions in the rectangular waveguide can be represented by optional trigonometric form, that is

$$X = A\cos(k_x x) + B\sin(k_x x) \tag{10.19}$$

$$Y = C\cos(k_y y) + D\sin(k_y y) \tag{10.20}$$

Substituting Equations (10.19) and (10.20) into (10.12), we then have

$$E_z = [A\cos(k_x x) + B\sin(k_x x)]\cdot[C\cos(k_y y) + D\sin(k_y y)]\mathrm{e}^{\mathrm{j}\omega t-\gamma z} \tag{10.21}$$

Equation (10.21) is the general form (general solution) of the TM-wave electric field component E_z in the rectangular waveguide. If we hope to find the specific solution of E_z, it is necessary to use boundary conditions to determine the various constants, A, B, C, D, k_x, k_y in Equation (10.21).

2) The boundary conditions satisfied by E_z

As we know, E_z is parallel to the four walls of rectangular waveguide,it is the tangential electric field component. In an ideal conductor surface the tangential electric field is zero. Accordingly, the above several constants can be evaluated.

The specific boundary conditions of the tangential electric field component E_z in Rectangular waveguide is:

When $x = 0, x = a, y$ is a arbitrary value between $0 \sim b$, $E_z = 0$; when $y = 0, y = b, x$ is a arbitrary value between $0 \sim a$, $E_z = 0$

3) use the boundary conditions to evaluate unknown constants

(1) When $x = 0$, y is an arbitrary value between $0 \sim b$, $E_z = 0$; and from Equation (10.21) we can see $A = 0$.

(2) When $y = 0$, x is an arbitrary value between $0 \sim a$, $E_z = 0$; and from Equation (10.21) we can get $C = 0$.

Substituting $A = 0, C = 0$ into Equation (10.21), and replace BD with another constant B_0, then get

$$E_z = B_0\sin(k_x x)\sin(k_y y)\mathrm{e}^{\mathrm{j}\omega t-\gamma z} \tag{10.22}$$

(3) When $x = a, y$ is a arbitrary value between $0 \sim b$, $E_z = 0$, and from Equation (10.22) we can get $\sin(k_x a) = 0$, so

$$k_x = \frac{m\pi}{a}, \quad m = 0, 1, 2, 3, \cdots \tag{10.23}$$

(4) When $y = b, x$ is a arbitrary value between $0 \sim a$, $E_z = 0$, and from Equation (10.22) we can get sin $(k_y b) = 0$, so

$$k_y = \frac{n\pi}{b}, \quad n = 0, 1, 2, 3, \cdots \tag{10.24}$$

Substituting the value of k_x, k_y into Equation (10.22), then get the expression of E_z of TM wave.

$$E_z = B_0\sin\left(\frac{m\pi}{a}\dot{x}\right)\sin\left(\frac{n\pi}{b}y\right)\mathrm{e}^{\mathrm{j}\omega t-\gamma z} \tag{10.25}$$

The constant B_0 should be determined by the field source.

2. The Solution to H_z of the TE Wave

The method to seek solutions of H_z for TE wave is similar to that for E_z of TM, except for that a little physical concept should be applied for judgment when we use the boundary conditions to determine the unknown constants, it is summarized as below.

The wave equation of H_z is

$$\frac{\partial^2 H_z}{\partial x^2}+\frac{\partial^2 H_z}{\partial y^2}+k_c^2 H_z=0 \tag{10.26}$$

If setting

$$H_z=X_1\cdot Y_1 e^{j\omega t-\lambda z} \tag{10.27}$$

In the equation, X_1 and Y_1 are respectively the function of x, y. Substituting this equation into Equation (10.26), after some procession, two ordinary differential equations same with Equations (10.16) and (10.17) can be obtained. Substituting their solutions into Equation (10.27), then we have

$$H_z=[A_1\cos(k_x x)+B_1\sin(k_x x)].[C_1\cos(k_y y)+D_1\sin(k_y y)]e^{j\omega t-\lambda z} \tag{10.28}$$

Now applying the boundary conditions to determine the constants of A_1, B_1, C_1, D_1, k_x, k_y.

1) The boundary conditions satisfied by H_z

H_z is the tangential component of magnetic field, from the reflection law of the waves in an ideal conductor surface, we know that on the waveguide walls, it should be the maximum point of standing wave of H_z and should satisfy the boundary conditions of $\frac{\partial}{\partial x}=0$ and $\frac{\partial}{\partial y}=0$. Applying this concept we can determine the unknown constants.

2) use the boundary conditions to determine the unknown constants

(1) Observe along the x axis, at positions of $x=0$, a, H_z get the maximum value. From Equation (10.28) we can see, for all the terms related to x, only the term with $\cos k_x x$ exist, and the term with $\sin k_x x$ should disappear, that is $B_1=0$.

(2) Follow a similar analysis, in Equation (10.28), $D_1=0$. Substituting $B_1=0$, $D_1=0$ into (10.28), and assuming $(A_1\cdot C_1)=H_0, H_0$ is a constant, then get

$$H_z=H_0\cos(k_x x)\cos(k_y y)\cdot e^{j\omega t-\lambda z} \tag{10.29}$$

(3) When $x=a$, y is a arbitrary value between 0~ b, H_z is the maximum value. From Equation (10.29), assuming $|\cos(k_x a)|=1$, so

$$k_x=\frac{m\pi}{a},\quad m=0,1,2,3,\cdots \tag{10.30}$$

(4) When $y=b$, x is a arbitrary value between 0~ a, H_z is the maximum value. From Equation (10.29) and assuming $|\cos(k_y b)|=1$, so

$$k_y=\frac{n\pi}{b},\quad n=0,1,2,3,\cdots \tag{10.31}$$

Substituting the value of k_x, k_y into Equation (10.29), then get

$$H_z=A_0\cos\left(\frac{m\pi}{a}x\right)\cos\left(\frac{n\pi}{b}y\right)e^{j\omega t-\lambda z} \tag{10.32}$$

In fact, the expression of k_x in the field component of TE waves is the same as that in TM wave. The difference is that, for TM waves, m, n in k_x, k_y cannot be zero. From Equation

(10.25) we can see, if either m or n is zero, E_z will be zero, thus all the electromagnetic fields will be zero. However, for the TE wave, either m or n in k_x, k_y can be zero, but not both at the same time. Thus, from Equation (10.32) we can see, $H_z \neq 0$, therefore, the other field components are not all zero.

Form Equation (10.25) and Equation (10.32) we can see, m is the number of the half-wave of the standing wave of electric field (or magnetic field) along the broad-side of rectangular waveguide (x direction), and n is the number of the half-wave of the standing wave of electric field (or magnetic field) along the narrow-side (y direction). Different m and n represent different distribution of the electromagnetic field with different field structures. TE_{mn} or H_{mn} are used to represent the transverse electric waves, while TM_{mn} or E_{mn} are used to stand represent the transverse magnetic wave.

10.1.3 The Horizontal Field E_x, E_y, H_x, H_y Characterized by E_z, H_z

The H_z of TE wave and E_z of TM wave are derived above, now we will characterize the components E_x, E_y, H_x, H_y, by H_z and E_z. In this way, all the field components of TE, TM wave will be known. This is a very brief way, and now we need to find out the relationship. In the Cartesian coordinate system, write the three components in the form of two curl equations of Maxwell's equation.

$$\left.\begin{aligned} &\frac{\partial H_z}{\partial y} + \gamma H_y = \mathrm{j}\omega\varepsilon E_x \\ &-\gamma H_x - \frac{\partial H_z}{\partial x} = \mathrm{j}\omega\varepsilon E_y \\ &\frac{\partial H_y}{\partial x} - \frac{\partial H_x}{\partial y} = \mathrm{j}\omega\varepsilon E_z \end{aligned}\right\} \tag{10.33}$$

and

$$\left.\begin{aligned} &\frac{\partial E_z}{\partial y} + \gamma E_y = -\mathrm{j}\omega\mu H_x \\ &-\gamma E_x - \frac{\partial E_z}{\partial x} = \mathrm{j}\omega\mu H_y \\ &\frac{\partial E_y}{\partial x} - \frac{\partial E_x}{\partial y} = \mathrm{j}\omega\mu H_z \end{aligned}\right\} \tag{10.34}$$

in the equations above, substituting

$$\frac{\partial}{\partial z} = -\gamma$$

From Equations (10.33) and (10.34), we can get the expression of E_x, E_y, H_x and H_y which are expressed by E_z, H_z.

$$\left.\begin{aligned} E_x &= -\frac{1}{k_\mathrm{c}^2}\left(\gamma\frac{\partial E_z}{\partial x} + \mathrm{j}\omega\mu\frac{\partial H_z}{\partial y}\right) \\ E_y &= \frac{1}{k_\mathrm{c}^2}\left(-\gamma\frac{\partial E_z}{\partial y} + \mathrm{j}\omega\mu\frac{\partial H_z}{\partial x}\right) \\ H_x &= \frac{1}{k_\mathrm{c}^2}\left(\mathrm{j}\omega\varepsilon\frac{\partial E_z}{\partial y} - \gamma\frac{\partial H_z}{\partial x}\right) \\ H_y &= -\frac{1}{k_\mathrm{c}^2}\left(\mathrm{j}\omega\varepsilon\frac{\partial E_z}{\partial x} + \gamma\frac{\partial H_z}{\partial y}\right) \end{aligned}\right\} \tag{10.35}$$

in the equation, $k_\mathrm{c}^2 = \gamma^2 + k^2$, while $k^2 = \omega^2\mu\varepsilon$.

10.1.4 The Field Components of TE Wave and TM Wave in Rectangular Waveguide

1. The Field Components of TE Wave

TE waves only have the longitudinal magnetic field component H_z, and $E_z = 0$, then substituting $E_z = 0$ and H_z of Equation (10.32) into (10.35), and assume that the waveguide is lossless, that is, $\alpha = 0$, and $\gamma = \alpha + \mathrm{j}\beta = \mathrm{j}\beta$, so we can get

$$\left.\begin{aligned}
H_z &= H_0 \cos\left(\frac{m\pi}{a}x\right)\cos\left(\frac{n\pi}{b}y\right)\mathrm{e}^{\mathrm{j}(\omega t-\beta z)}\\
H_x &= \mathrm{j}\frac{\beta}{k_\mathrm{c}^2}\frac{m\pi}{a}H_0 \sin\left(\frac{m\pi}{a}x\right)\cos\left(\frac{n\pi}{b}y\right)\mathrm{e}^{\mathrm{j}(\omega t-\beta z)}\\
H_y &= \mathrm{j}\frac{\beta}{k_\mathrm{c}^2}\frac{n\pi}{b}H_0 \cos\left(\frac{m\pi}{a}x\right)\sin\left(\frac{n\pi}{b}y\right)\mathrm{e}^{\mathrm{j}(\omega t-\beta z)}\\
E_x &= \mathrm{j}\frac{\omega\mu}{k_\mathrm{c}^2}\frac{n\pi}{b}H_0 \cos\left(\frac{m\pi}{a}x\right)\sin\left(\frac{n\pi}{b}y\right)\mathrm{e}^{\mathrm{j}(\omega t-\beta z)}\\
E_y &= -\mathrm{j}\frac{\omega\mu}{k_\mathrm{c}^2}\frac{m\pi}{a}H_0 \sin\left(\frac{m\pi}{a}x\right)\cos\left(\frac{n\pi}{b}y\right)\mathrm{e}^{\mathrm{j}(\omega t-\beta z)}\\
E_z &= 0
\end{aligned}\right\} \tag{10.36}$$

In the equations

$$k_\mathrm{c}^2 = k_x^2 + k_y^2 = \left(\frac{m\pi}{a}\right)^2 + \left(\frac{n\pi}{b}\right)^2, \quad m = 0, 1, 2, \cdots; n = 0, 1, 2, \cdots$$

2. The Field Components of TM Wave

TM waves only have the longitudinal electric field component E_z, and $H_z = 0$, then substituting $H_z = 0$ and E_z of Equation (10.25) into Equation (10.35), and assume that the waveguide is lossless, so we can get

$$\left.\begin{aligned}
E_z &= E_0 \sin\left(\frac{m\pi}{a}x\right)\sin\left(\frac{n\pi}{b}y\right)\mathrm{e}^{\mathrm{j}(\omega t-\beta z)}\\
E_x &= -\mathrm{j}\frac{\beta}{k_\mathrm{c}^2}\frac{m\pi}{a}E_0 \cos\left(\frac{m\pi}{a}x\right)\sin\left(\frac{n\pi}{b}y\right)\mathrm{e}^{\mathrm{j}(\omega t-\beta z)}\\
E_y &= -\mathrm{j}\frac{\beta}{k_\mathrm{c}^2}\frac{n\pi}{b}E_0 \sin\left(\frac{m\pi}{a}x\right)\cos\left(\frac{n\pi}{b}y\right)\mathrm{e}^{\mathrm{j}(\omega t-\beta z)}\\
H_x &= \mathrm{j}\frac{\omega\varepsilon}{k_\mathrm{c}^2}\frac{n\pi}{b}E_0 \sin\left(\frac{m\pi}{a}x\right)\cos\left(\frac{n\pi}{b}y\right)\mathrm{e}^{\mathrm{j}(\omega t-\beta z)}\\
H_y &= -\mathrm{j}\frac{\omega\varepsilon}{k_\mathrm{c}^2}\frac{m\pi}{a}E_0 \cos\left(\frac{m\pi}{a}x\right)\sin\left(\frac{n\pi}{b}y\right)\mathrm{e}^{\mathrm{j}(\omega t-\beta z)}\\
H_z &= 0
\end{aligned}\right\} \tag{10.37}$$

In the equations $m = 1, 2, \cdots$; $n = 1, 2, \cdots$.

10.1.5 Parameters of TE Wave, TM-Wave in Rectangular Waveguide

The important electromagnetic parameters in the waveguide are mainly phase constant β, waveguide wavelength λ_g, cutoff wavelength λ_c, phase velocity v_p, energy (group) velocity

v_g, wave impedance Z_w and so on. For the sake of convenience, rewrite the two important formulas.

$$k_c^2 = \gamma^2 + k^2$$

$$k_c^2 = k_x^2 + k_y^2 = \left(\frac{m\pi}{a}\right)^2 + \left(\frac{n\pi}{b}\right)^2$$

1. Cutoff Wavelength λ_c

As the electromagnetic wave propagating in the waveguide is subject to the restriction of waveguide boundary condition, therefore, not all the electromagnetic waves of any frequency can transmit in the waveguide, there is a lowest limit of the transmission frequency. At a given operating frequency, only those $f_c < f$ will propagate; modes with $f_c > f$ will lead to an imaginary β, meaning that all field components will decay exponentially away from the source of excitation. Such modes are referred to as cutoff, or evanescent, modes. Now two parameters can be referred, saying, the cutoff frequency f_c, and the corresponding cutoff wavelength λ_c.

When the electromagnetic wave is just subject to cutoff, the propagation constant γ of the ideal waveguide is equal to zero, that is, the phase shift constant β is equal to zero. For this case, from Equations (10.7) and (10.18), we can see

$$k_c^2 = \gamma^2 + k^2 = k^2 = \left(\frac{2\pi}{\lambda_c}\right)^2$$

and

$$k_c^2 = k_x^2 + k_y^2 = \left(\frac{m\pi}{a}\right)^2 + \left(\frac{n\pi}{b}\right)^2$$

The cut-off wavelength λ_c is

$$\lambda_c = \frac{2\pi}{\sqrt{k_x^2 + k_y^2}} = \frac{2}{\sqrt{\left(\frac{m}{a}\right)^2 + \left(\frac{n}{b}\right)^2}} \tag{10.38}$$

2. Phase Constant β

Assuming the waveguide is lossless, then substituting $\gamma = \mathrm{j}\beta$ into Equation (10.7), we get

$$\gamma^2 = (\mathrm{j}\beta)^2 = -\beta^2 = k_c^2 - k^2$$

so

$$\beta = k\sqrt{1 - (\lambda/\lambda_c)^2} \tag{10.39}$$

In the equation, λ is the operating wavelength of TEM wave.

3. Guided Wavelength λ_g

The phase constant β is the phase shift per unit length of TE, TM waves propagating along the longitudinal direction (z direction) of waveguide. Correspondingly, the wavelength is known as the guided wavelength λ_g. So from Equation (10.39)

$$\lambda_g = \frac{2\pi}{\beta} = \frac{2\pi}{k\sqrt{1 - (\lambda/\lambda_c)^2}} = \frac{\lambda}{\sqrt{1 - (\lambda/\lambda_c)^2}} \tag{10.40}$$

4. Phase Velocity v_p

The velocity of equal-phase plane of TE waves, TM waves travelling longitudinally along the waveguide is called the phase velocity v_p.

Method 1 directly from $\beta = 2\pi/\lambda_g$, we can get the phase velocity v_p.

$$\beta = \frac{2\pi}{\lambda_g} = \frac{2\pi f}{\lambda_g f} = \frac{\omega}{v_p}$$

$$v_p = \frac{\omega}{\beta} = \frac{\omega}{k\sqrt{1-(\lambda/\lambda_c)^2}} = \frac{v}{\sqrt{1-(\lambda/\lambda_c)^2}}$$

that is

$$v_p = \frac{v}{\sqrt{1-(\lambda/\lambda_c)^2}} \tag{10.41}$$

in the equation, v is the propagation velocity of TEM wave.

Method 2 From the definition of v_p (the phase speed is the velocity of equal-phase plane) find v_p.

Because the phase factor of TE wave, TM wave is $e^{j(\omega t-\beta z)}$, waves propagate along the positive z direction. If you choose $(\omega t_1 - \beta z_1)$ as the phase values of the equal-phase plane which you observe, this equal-phase plane moves towards the location of $z = z_2$ at $t = t_2$, so $(\omega t_1 - \beta z_1) = (\omega t_2 - \beta z_2)$, we can get

$$\omega t_1 - \beta z_1 = \omega t_2 - \beta z_2 = \cdots = \omega t_n - \beta z_n = \text{const.}$$

so we have

$$\frac{z_2 - z_1}{t_2 - t_1} = \cdots = \frac{z_n - z_{n-1}}{t_n - t_{n-1}} = \frac{\Delta z}{\Delta t} = \frac{\omega}{\beta}$$

When Δt approached zero, we then get

$$v_p = \frac{dz}{dt} = \frac{\omega}{\beta} = \frac{v}{\sqrt{1-(\lambda/\lambda_c)^2}}$$

Obviously, the phase velocity v_p is greater than the propagation speed of TEM-wave.

If the waveguide is filled with air medium, then the phase velocity v_p is greater than the speed of light. In fact, it is caused by interference between the two obliquely incident TEM waves to the waveguide walls and the reflected waves; it does not represent the velocity of electromagnetic energy.

5. Group Velocity v_g

Group velocity refers to the velocity of group signals which is composed of many frequency signals. The velocity of wave package (envelope) of AM wave is the group velocity. As the propagation velocity of TE, TM wave in the waveguide is related to frequency, and therefore, the group velocity of concern must subject to the condition, saying, the signal bandwidth of the group signal is very small compared with the carrier frequency.

Method 1 Using the expression of AM wave to find v_g.

Consider there is a AM wave

$$A_0 \cos(\Delta\omega t - \Delta\beta z) e^{j(\omega t - \beta z)} \tag{10.42}$$

In the equation, A_0 is a constant. ω, β are the angular frequency and phase constant of the carrier frequency, $\Delta\omega$ (angular frequency variation) and $\Delta\beta$ (phase-shift variation) are the corresponding parameters of the group signals.

The velocity of the envelope $A_0 \cos(\Delta\omega t - \Delta\beta z)$ along the z direction can be derived from the following formula

By

$$\Delta\omega t - \Delta\beta z = \text{const.}$$

that is

$$z = \frac{\Delta\omega t - \text{const.}}{\Delta\beta}$$

Then we can get v_g

$$v_g = \frac{dz}{dt} = \frac{\Delta\omega}{\Delta\beta} \tag{10.43}$$

When $\omega \gg \Delta\omega$, the equation above can also be written as $v_g = \dfrac{d\omega}{d\beta} = 1\Big/\dfrac{d\beta}{d\omega}$

while

$$\beta = \sqrt{k^2 - k_c^2} = \sqrt{\left(\frac{\omega}{v}\right)^2 - \left(\frac{2\pi}{\lambda_c}\right)^2}$$

Substituting this expression into $v_g = 1\Big/\dfrac{d\beta}{d\omega}$, to obtain the group velocity

$$v_g = v\sqrt{1 - (\lambda/\lambda_c)^2} \tag{10.44}$$

Obviously, v_g is less than the speed v of the TEM wave.

Method 2 use the transmit power and the total energy per unit length of waveguide to find v_g.

With regard to power speed, we can analyze it in this way: the transmission power P of the electromagnetic wave (TE, TM waves) in the waveguide should be equal to the integral of S_{av} with respect to the cross section of the waveguide. It is the total energy flowing through the waveguide cross-section during unit time. This total energy accounts for the space of the entire waveguide with the length of l, if the total energy (that is, the energy through the waveguide cross-section in a second) is divided by the average electromagnetic energy density $w_{av(UL)}$ stored in the waveguide of unit length, the result is distance of the energy propagation in one second, saying, the energy speed. Therefore, the energy velocity can be written as $P/w_{av(UL)}$ and get the same results as Equation (10.44). That is, the group velocity represents the propagation velocity of energy. Form Equations (10.41) and (10.44), we can get

$$v_p \cdot v_g = v^2 \tag{10.45}$$

6. The Wave Impedance $Z_{W(TE)}$, $Z_{W(TM)}$ of TE, TM Wave

The wave impedance is defined as the ratio of transverse electric field E_T and the transverse magnetic field H_T in waveguide. With the fields of the modes known, it is thus easy to get the wave impedance for TE and TM modes.

For the TM wave ($H_z = 0$), the following two equations can be written

$$\left.\begin{aligned} j\omega\varepsilon E_x &= \gamma H_y \\ j\omega\varepsilon E_y &= -\gamma H_x \end{aligned}\right\} \tag{10.46}$$

From Equation (10.46) we can get

$$\boldsymbol{E}_T = \boldsymbol{e}_x E_x + \boldsymbol{e}_y E_y = \boldsymbol{e}_x\left(\frac{\gamma}{j\omega\varepsilon}H_y\right) + \boldsymbol{e}_y\left(\frac{-\gamma}{j\omega\varepsilon}H_x\right) = \frac{\gamma}{j\omega\varepsilon}(\boldsymbol{e}_x H - \boldsymbol{e}_y H_x)$$

Applying vector product with $\boldsymbol{e}_z$, and note $\boldsymbol{e}_z \times \boldsymbol{e}_x = \boldsymbol{e}_y, \boldsymbol{e}_z \times \boldsymbol{e}_y = -\boldsymbol{e}_x$, we can get

$$\boldsymbol{e}_z \times \boldsymbol{E}_T = \frac{\gamma}{j\omega\varepsilon}(\boldsymbol{e}_x H_x + \boldsymbol{e}_y H_y) = \frac{\gamma}{j\omega\varepsilon}\boldsymbol{H}_T$$

so

$$Z_{W(TM)} = \frac{E_T}{H_T} = \frac{\gamma}{j\omega\varepsilon} \tag{10.47}$$

From Equations (10.46) and (10.47), we can see, $Z_{W(TM)}$ can also be written as

$$Z_{W(TM)} = \frac{E_T}{H_T} = \frac{E_x}{H_y} = \frac{-E_y}{H_x} = \frac{\gamma}{j\omega\varepsilon} \tag{10.48}$$

The wave impedance $Z_{W(TE)}$ of TE can be obtained in the same way as mentioned above, it is not discussed in detail here. Below is the result

$$Z_{W(TE)} = \frac{E_T}{H_T} = \frac{E_x}{H_y} = \frac{-E_y}{H_x} = \frac{j\omega\mu}{\gamma} \tag{10.49}$$

If the electromagnetic waves has no transmission loss, that is, $\lambda = j\beta$. Substituting this into Equations (10.47) and (10.49), and apply the following formula

$$\beta = k\sqrt{1 - \left(\frac{\lambda}{\lambda_c}\right)^2}, \quad \eta = \sqrt{\frac{\mu}{\varepsilon}} = \frac{k}{\omega\varepsilon} = \frac{\omega\mu}{k}$$

Then we can get

$$Z_{W(TM)} = \eta\sqrt{1 - (\lambda/\lambda_c)^2} \tag{10.50a}$$

and

$$Z_{W(TE)} = \frac{\eta}{\sqrt{1 - (\lambda/\lambda_c)^2}} \tag{10.50b}$$

10.1.6 The Dispersion Characteristic and the Cut-Off Characteristics of Waveguide

1. The Dispersion Characteristic of Waveguide

In waveguide, the phenomenon in which the propagation velocity of the TE and TM wave changes with frequency is known as dispersion. This can be clearly seen from the expression of v_p, v_g Equation (10.41) and (10.44).

We still remember, the velocity v of TEM wave in an ideal medium is a constant. Therefore, *there is no dispersion phenomenon when TEM-wave propagates in an ideal medium.*

Writing the complex permittivity in the conductive medium, that is, $\varepsilon^e = \varepsilon - j\dfrac{\sigma}{\omega}$, as a result, the phase velocity v_p is

$$v_p = \omega/\beta = \left[\frac{\mu\varepsilon}{2}\left(\sqrt{1 + \left(\frac{\sigma}{\omega\varepsilon}\right)^2} + 1\right)\right]^{-\frac{1}{2}}$$

It can be seen, when *TEM waves propagate in conductive media, its speed changes with the frequency, it is also dispersive wave, which is caused by the dispersion medium.*

2. The Cutoff Characteristic of Waveguide

Observe the expressions of phase velocity v_p, group velocity v_g, as well as the phase constant β, we can clearly see the cutoff properties of waveguide. In the text above, we have

$$v_p = \frac{v}{\sqrt{1 - (\lambda/\lambda_c)^2}}$$

$$v_g = v\sqrt{1-(\lambda/\lambda_c)^2}, \quad \beta = k\sqrt{1-(\lambda/\lambda_c)^2}$$

When $\lambda < \lambda_c$, that is, when the operating wavelength is less than the cutoff wavelength, TE, TM waves can transmit properly. v_p, v_g values are meaningful, β is a real number.

When $\lambda = \lambda_c$, that is ,when the operating wavelength is equal to the cutoff wavelength, the phase velocity v_p is infinite, group velocity v_g is equal to zero, phase constant β is also equal to zero. It shows that *when $\lambda = \lambda_c$, the waves are cutoff- no energy transfer along the z direction, no phase shift.*

When $\lambda > \lambda_c$, v_p, v_g become imaginary number, it is meaningless; phase constant β has become an imaginary number [see Equation (10.39)], at this time, the common factor $e^{j(\omega t-\beta z)}$ in the electromagnetic field components becomes the following form.

$$e^{j(\omega t-\beta z)} = e^{j\omega t} \cdot e^{-|\beta| z} \tag{10.51}$$

That is, the phase factor $e^{-j\beta z}$ under the transmission condition has become a "decay factor" $e^{-|\beta| z}$ under the cutoff conditions. In this case, TE, TM waves decay exponentially along the z direction. This is the cutoff phenomenon of waveguide. However, it should be noted that, "decay" does not really mean the loss of energy; it refers to a reactive loss, the energy is stored in the form of electric or magnetic field, and they can exchange each other.

From the wave impedance of TE, TM expressed by Equation (10.50), we can also see that when $\lambda = \lambda_c$, $Z_{W(TM)}$ is equal to zero, while $Z_{W(TE)}$ is infinite. When $\lambda < \lambda_c$ (transmission wave), then $Z_{W(TM)}$ and $Z_{W(TE)}$ are real numbers. If the wave is cutoff, that is, $\lambda > \lambda_c$, then the wave impedance becomes imaginary number, actually a reactance. For the TE waves, imaginary wave impedance represents of inductance in case of cutoff, for the TM wave, represent the capacitance in case of cutoff. The nature of this reactance will cause waves "attenuation". However, this "decay" is not resistive, as mentioned earlier, it does not cause the loss of energy.

To conclude, only when the operating wavelength is smaller than the cutoff wavelength, TE and TM waves can transmit in the waveguide, it is cutoff when the operational wavelength is equal to or greater than the cutoff wavelength, and this is the cutoff properties of waveguides.

10.1.7 The Field Components and Parameters of TE_{10} Wave

From Equation (10.36), we com get the parameters of TE_{10} wave.

$$\left.\begin{aligned} E_y &= E_0 \sin\left(\frac{\pi x}{a}\right) e^{j(\omega t-\beta z)} \\ H_x &= -\left(\frac{\beta}{\omega\mu}\right) E_0 \sin\left(\frac{\pi x}{a}\right) e^{j(\omega t-\beta z)} \\ H_z &= j\left(\frac{\pi}{\omega\mu a}\right) E_0 \cos\left(\frac{\pi x}{a}\right) e^{j(\omega t-\beta z)} \end{aligned}\right\} \tag{10.52}$$

Parameters of TE_{10} wave:

Considering the case $m = 1, n = 0$, from Equation (10.38), we can get the cutoff wavelength λ_c,

$$\lambda_c = 2a \tag{10.53}$$

In case of $m = 1, n = 0$, and $\lambda_c = 2a$, from Equation (10.39), we can get the phase constant β of TE_{10} wave,

$$\beta = k\sqrt{1-(\lambda/2a)^2} \tag{10.54}$$

From Equations (10.40), (10.41), (10.44) and (10.50), we can get the guided wavelength λ_g of TE_{10} wave,

$$\lambda_g = \frac{\lambda}{\sqrt{1-(\lambda/2a)^2}} \tag{10.55}$$

The phase velocity v_p of TE$_{10}$ wave is

$$v_p = \frac{v}{\sqrt{1-(\lambda/2a)^2}} \tag{10.56}$$

The group (energy) velocity v_g of TE$_{10}$ wave is

$$v_g = v\sqrt{1-(\lambda/2a)^2} \tag{10.57}$$

The wave impedance $Z_{W(TE_{10})}$ of TE$_{10}$ wave is

$$Z_{W(TE_{10})} = \frac{\eta}{\sqrt{1-(\lambda/2a)^2}} \tag{10.58}$$

10.1.8 The Field Distribution of TE$_{10}$ Wave

The characteristics of electric and magnetic field of TE$_{10}$wave are discussed below:

(1) There is a time phase difference of 90° between E_y and H_z. In Equation (10.52), the difference between E_y and H_z is "j". This result is reasonable, because when the energy transmits along the z direction in the waveguide, then the average poynting vector along the x direction is zero.

(2) E_y and H_x has no time phase difference, which results in the energy transmission along the z direction.

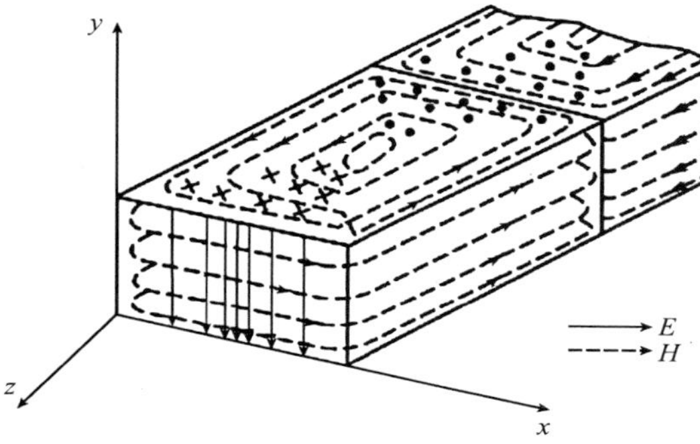

(a) TE$_{10}$ wave field structure (full lines for electric,dotted lines for magnetic)

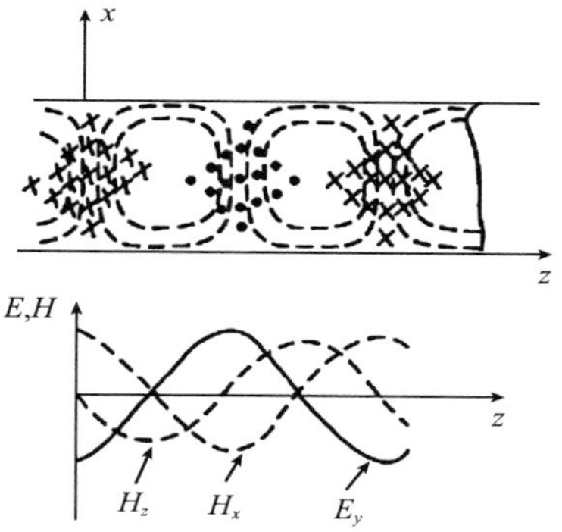

(b) Distribution of the electric and magnetic field for TE$_{10}$ wave in the xOz plane

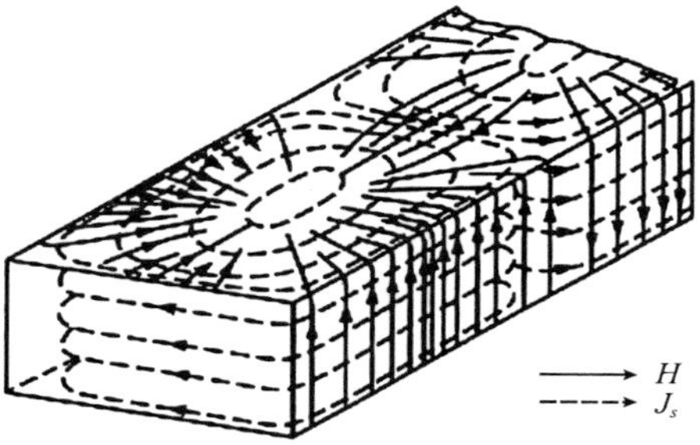

(c) Magneic field of TE$_{10}$ wave and the current distribution on the surface of the waveguide

Figure 10.2 TE$_{10}$ wave

(3) The electric field of TE_{10} wave has only $E_y$component, and its electric field lines are perpendicular to the upper and lower broad-side of the waveguide. E_y distributes sinusoidally along the x direction. At the center of the broad-side $\left(x=\dfrac{a}{2}\right)$, E_y is the biggest, and decrease gradually towards two narrow sides. At $x=0$ and $a, E_y=0$.

(4) The magnetic field of TE_{10} wave has two components of H_x, H_z, forming a closed magnetic field lines parallel to the broad-side of waveguide.

The field structure of TE_{10}wave:

According to Equation (10.52), the field structure of TE_{10} wave is illustrated in Figure 10.2. It is consistent with the characteristics of TE_{10} wave.

Applying the boundary conditions $\boldsymbol{n}\times\boldsymbol{H}=\boldsymbol{J}_\mathrm{s}$ on perfect conducting surface, according to the distribution of the magnetic field on the four walls of waveguide, we can draw the current distribution of TE_{10}-wave, as shown in Figure 10.2 (c), where the dotted lines are for the magnetic line while the solid lines are for the surface currents.

10.1.9 The Transmission Power of TE_{10} Wave

There are two ways to evaluate the transmission power P of the TE_{10}wave in rectangular waveguide, first, applying the concept of power density, second, using the definition of "energy density".

Method 1 Using the power density—that is, using Poynting vector to find power P.

The average transmit power of TE_{10} wave in rectangular waveguide can be expressed as

$$P=\int_s S_\mathrm{av}\cdot \mathrm{d}s=\int_0^a\int_0^b S_\mathrm{av}\mathrm{d}x\mathrm{d}y \tag{10.59}$$

where $\mathrm{d}\boldsymbol{S}$ is the surface element of waveguide cross section, $\boldsymbol{S}_\mathrm{av}$ is the average power density, which transmit along the z direction.

From E_y, H_x in Equation (10.52), we have

$$\boldsymbol{S}_\mathrm{av}=\frac{1}{2}\mathrm{Re}(\boldsymbol{E}\times\boldsymbol{H}^*)=\boldsymbol{e}_z\frac{1}{2}E_0^2\left(\frac{\beta}{\omega\mu}\right)\sin^2\left(\frac{\pi x}{a}\right) \tag{10.60}$$

Substituting this into Equation (10.59), yields

$$P=\frac{1}{2}\left(\frac{\beta}{\omega\mu}\right)E_0^2\int_0^a\int_0^b\sin^2\left(\frac{\pi x}{a}\right)\mathrm{d}x\mathrm{d}y=\frac{1}{4}ab\left(\frac{\beta}{\omega\mu}\right)E_0^2 \tag{10.61}$$

Method 2 Using the "energy density" to determine power P.

"Energy density" refers to the average electromagnetic field energy per unit length of the waveguide, expressed by $w_{\mathrm{av}(UL)}$. The relationship between P and $w_{\mathrm{av}(UL)}$, v_g is

$$P=w_{\mathrm{av}(UL)}v_\mathrm{g} \tag{10.62}$$

or

$$v_\mathrm{g}=\frac{P}{w_{\mathrm{av}(UL)}} \tag{10.63}$$

while

$$w_{\mathrm{av}(UL)}=\int_0^a\int_0^b\left(w_{\mathrm{av}(E)}+w_{\mathrm{av}(M)}\right)\mathrm{d}x\mathrm{d}y \tag{10.64}$$

In the equation, $w_{\mathrm{av(E)}}$, $w_{\mathrm{av(M)}}$ represent the average energy density of electric fields and magnetic fields, respectively. So

$$w_{\mathrm{av(E)}}=\frac{1}{2}\varepsilon E_{y(\mathrm{e})}^2 \tag{10.65}$$

$$w_{\mathrm{av(M)}} = \frac{1}{2}\mu H_{x(\mathrm{e})}^2 + \frac{1}{2}\mu H_{z(\mathrm{e})}^2 \tag{10.66}$$

where $E_{y(\mathrm{e})}, H_{x(\mathrm{e})}, H_{z(\mathrm{e})}$ is the valid value of field. Solve $w_{\mathrm{av(E)}}, w_{\mathrm{av(M)}}$ and substitute them into Equation (10.64), to obtain

$$w_{\mathrm{av}(UL)} = \frac{1}{4}ab\varepsilon E_0^2 \tag{10.67}$$

By using Equation (10.45), (10.62), (10.67) we obtain,

$$v_{\mathrm{g}} = v^2/v_{\mathrm{p}} = \frac{1}{\mu\varepsilon}\cdot\frac{1}{v_{\mathrm{p}}} = \frac{1}{\mu\varepsilon}\cdot\frac{1}{f\lambda_{\mathrm{g}}} = \frac{\beta}{\omega\mu\varepsilon} \tag{10.68}$$

so

$$P = w_{\mathrm{av}(UL)}v_{\mathrm{g}} = \left(\frac{1}{4}abE_0^2\varepsilon\right)\left(\frac{\beta}{\omega\mu\varepsilon}\right) = \frac{1}{4}ab\left(\frac{\beta}{\omega\mu}\right)E_0^2 \tag{10.69}$$

From Equation (10.69), we can have more specific insight of the group (energy) velocity.

10.1.10 The Attenuation of Rectangular Waveguide

The above analysis is based on the assumption that the waveguide is ideal, saying, the conductor is perfect and the medium is lossless. Therefore, there is no loss during the transmission of electromagnetic wave in the waveguide, that is, $\alpha = 0$, the propagation constant $\gamma = \mathrm{j}\beta$. However, even though the waveguide is filled with a medium with very low loss (such as air), the actual metal waveguide wall always produce loss. Now, let's study how to calculate the energy loss caused by the waveguide walls.

Because of the existence of loss, the attenuation factor $\mathrm{e}^{-\alpha z}$ should be introduced into the expression for electric and magnetic fields. Therefore, the transmission power in waveguide can be written as

$$\boldsymbol{P} = \boldsymbol{P}_0\mathrm{e}^{-2\alpha z} \tag{10.70}$$

where $P = P_0$ when $z = 0$. In Equation (10.70), take the derivative of z, then we can get the power loss P_{L} of per unit length in the waveguide, that is

$$P_{\mathrm{L}} = -\frac{\mathrm{d}P}{\mathrm{d}z} = 2\alpha P \tag{10.71}$$

so

$$\alpha = \frac{P_{\mathrm{L}}}{2P} = \frac{\text{the loss of power of per unit length}}{2(\text{transmit power})} \tag{10.72}$$

The expression of transmit power P is

$$P = \frac{1}{2}\mathrm{Re}\int_0^a\int_0^b E_T H_T^*\mathrm{d}x\mathrm{d}y = \frac{1}{2}Z_{\mathrm{W}}\int_0^a\int_0^b |H_T|^2\,\mathrm{d}x\mathrm{d}y \tag{10.73a}$$

where Z_{W} is the wave impedance of TM wave or TE wave, E_{T}, H_{T} are the transverse field components. The power loss P_{L} per unit length of waveguide can be written as

$$P_{\mathrm{L}} = 2\left[\int_0^a \frac{1}{2}|J_1|^2 R_{\mathrm{s}}\mathrm{d}x\right] + 2\left(\int_0^b \frac{1}{2}|J_2|^2 R_{\mathrm{s}}\mathrm{d}y\right) \tag{10.73b}$$

where J_1 represent the current density on the upper and lower broad surface, and J_2 is the current density on the left and right narrow surface of the waveguide. R_{s} is the waveguide surface resistance. The value of J_1, J_2 can be solved using the boundary condition $\boldsymbol{n}\times\boldsymbol{H} = \boldsymbol{J}_{\mathrm{s}}$ of an ideal conductor. Of course, for good conductors, this is not very accurate, but it is a good approximation.

According to Equation (10.72) (10.73a), (10.73b),we can obtain the attenuation constant α of TE_{10} wave in rectangular waveguide

$$\alpha = \frac{R_s}{b\eta\sqrt{1-(\lambda/2a)^2}}\left[1 + 2\frac{b}{a}(\lambda/2a)^2\right] \text{ Np/m} \tag{10.74}$$

The attenuation constant α of other modes can be found from Equation (10.72) and (10.73).

Figure 10.3 shows the frequency dependence of attenuation constant α for TE_{10} and TM_{11} mode, in such case, the width a is fixed, while $\frac{b}{a}$ changes. In the vicinity of cutoff frequency, the attenuation of modes goes up quickly. Moreover, when $\frac{b}{a}$ is fixed, the attenuation of TE_{10} mode is smaller than TM_{11}. For either TE_{10} or TM_{11} mode, the attenuation decrease subsequently with the increase of $\frac{b}{a}$.

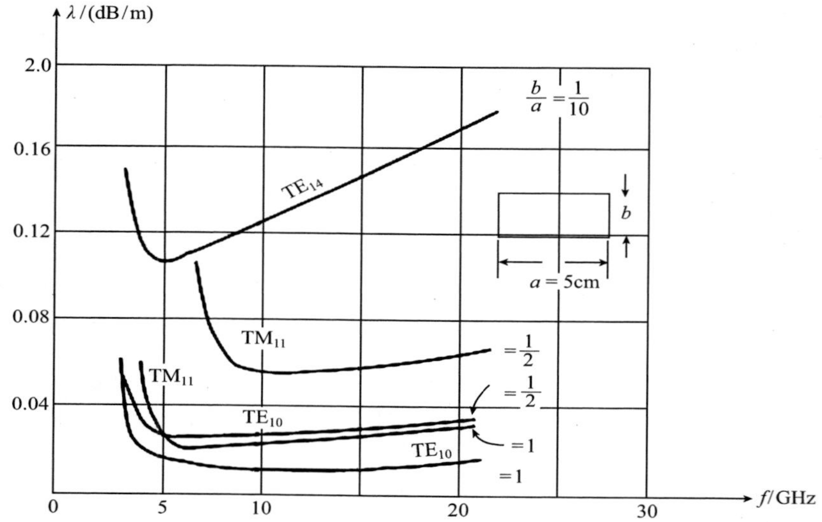

Figure 10.3 The loss caused by metal wall (copper) of the main mode of TE_{10}, TM_{11} in rectangular waveguide when a is 5cm, $\frac{b}{a}$ varies

Example 10.1 Try to prove that the conductor attenuation constant α of TE_{10} mode in rectangular waveguide is

$$\alpha = \frac{R_s}{b\eta\sqrt{1-(\lambda/2a)^2}}\left[1 + 2\frac{b}{a}(\lambda/2a)^2\right]$$

Solution Since $\alpha = P_L/2P$, it is necessary to determine firstly the conductor power loss P_L per unit length for TE_{10} mode. From Equation (10.73b), we know

$$\begin{aligned} P_L &= 2\left\{\left[\int_0^a \frac{1}{2}|J_1|^2 R_s \mathrm{d}x\right] + \left[\int_0^b \frac{1}{2}|J_2|^2 R_s \mathrm{d}y\right]\right\} \\ &= R_s\left\{\left[\int_0^a |J_1|^2 \mathrm{d}x\right] + \left[\int_0^b |J_2| \mathrm{d}y\right]\right\} \end{aligned} \tag{a}$$

From the field components of TE_{10} wave given in Equation (10.52), we can see

$$|J_1|^2_{y=0} = \left[H_x^2 + H_z^2\right]_{y=0} = \left(\frac{\beta}{\omega\mu}\right)^2 \cdot E_0^2 \cdot \sin^2\left(\frac{\pi x}{a}\right) + \left(\frac{\pi}{\omega\mu a}\right)^2 \cdot E_0^2 \cdot \cos^2\left(\frac{\pi x}{a}\right)$$

$$|J_2|^2_{x=0} = \left[H_z^2\right]_{x=0} = \left[\left(\frac{\pi}{\omega\mu a}\right)^2 \cdot E_0^2 \cdot \cos^2\left(\frac{\pi x}{a}\right)\right]_{x=0} = \left(\frac{\pi}{\omega\mu a}\right)^2 \cdot E_0^2$$

Substituting $|J_1|^2$ and $|J_2|^2$ into the expression (a) of P_L, yields

$$\begin{aligned} P_\text{L} &= R_\text{s}\left\{\left[\left(\frac{\beta}{\omega\mu}\right)^2 E_0^2 + \left(\frac{\pi}{\omega\mu a}\right)^2 E_0^2\right] \cdot \frac{a}{2} + \left[\left(\frac{\pi}{\omega\mu a}\right)^2 \cdot E_0^2\right] \cdot b\right\} \\ &= R_\text{s}\left\{\left(\frac{\beta}{\omega\mu}\right)^2 E_0^2 \cdot \frac{a}{2} + \left(\frac{\pi}{\omega\mu a}\right)^2 E_0^2 \cdot \left(\frac{a}{2} + b\right)\right\} \end{aligned} \quad \text{(b)}$$

From Equations (10.73a) and (10.52), we can get the transmission power of TE_{10} mode

$$\begin{aligned} P &= \frac{1}{2} Z_{\text{W}(\text{TE}_{10})} \int_0^a \int_0^b |H_T|^2 \, \text{d}x\text{d}y \\ &= \frac{1}{2} Z_{\text{W}(\text{TE}_{10})} \int_0^a \int_0^b \left(\frac{\beta}{\omega\mu}\right)^2 \cdot E_0^2 \cdot \sin^2\left(\frac{\pi x}{a}\right) \text{d}x\text{d}y \\ &= Z_{\text{W}(\text{TE}_{10})} \left(\frac{\beta}{\omega\mu}\right)^2 \cdot E_0^2 \cdot \frac{ab}{4} = \left(\frac{\beta}{\omega\mu}\right) E_0^2 \cdot \frac{ab}{4} \end{aligned} \quad \text{(c)}$$

Substituting Equations (b) and (c) into the formula of α, then get

$$\begin{aligned} \alpha &= \frac{P_\text{L}}{2P} \\ &= \frac{R_\text{s}}{\eta\sqrt{1-(\lambda/2a)^2}} \left[1 - (\lambda/2a)^2\right] \times \left[1 + (1 + 2\frac{b}{a})(\lambda/2a)^2 \frac{1}{1-(\lambda/2a)^2}\right] \\ &= \frac{R_\text{s}}{b\eta\sqrt{1-(\lambda/2a)^2}} \left[1 + 2\frac{b}{a}(\lambda/2a)^2\right] \end{aligned}$$

Example 10.2 A TE_{10} mode of 10GHz transmit in the rectangular waveguide ,waveguide dimensions are $a = 15\text{cm}$, $b = 0.6\text{cm}$, if the waveguide is made of brass, its conductivity is $\sigma = 1.57 \times 10^7 (\text{S/m})$, the waveguide is filled with polyethylene medium with $\varepsilon_\text{r} = 2.25$ and $\mu_\text{r} = 1$, try to determine the parameters of TE_{10} mode: (1) guided wavelength λ_g; (2) the phase velocity v_p; (3) phase constant β; (4) wave impedance $Z_{\text{W}(\text{TE}_{10})}$; (5) Attenuation constant α of waveguide wall.

Solution The wavelength λ of TEM wave in Polyethylene is

$$\lambda = \frac{c}{\sqrt{\varepsilon_\text{r}} f} = \frac{3 \times 10^8}{\sqrt{2.25} \times 10^{10}} = 0.02 \quad (\text{m})$$

(1) Guided wavelength

$$\lambda_\text{g} = \frac{\lambda}{\sqrt{1-(\lambda/2a)^2}} = \frac{0.02}{\sqrt{1-\left(\dfrac{0.02}{0.03}\right)^2}} = 0.0268 \quad (\text{m})$$

(2) The phase velocity v_p

$$v_p = f\lambda_g = 10^{10} \times 0.0268 = 2.68 \times 10^8 \quad \text{(m/s)}$$

or

$$v_p = \left(\frac{c}{\sqrt{\varepsilon_r}}\right) \cdot \frac{1}{\sqrt{1-\left(\dfrac{\lambda}{2a}\right)^2}} = 2 \times 10^8 \times \frac{1}{0.7454} = 2.68 \times 10^8 \text{(m/s)}$$

(3) phase constant β

$$\beta = \frac{2\pi}{\lambda_g} = 234 \quad \text{(rad/m)}$$

or

$$\beta = k\sqrt{1-\left(\frac{\lambda}{2a}\right)^2} = \frac{2\pi}{\lambda}\sqrt{1-\left(\frac{\lambda}{2a}\right)^2} = 234 \quad \text{(rad/m)}$$

(4) The wave impedance $Z_{W(TE_{10})}$

$$Z_{W(TE_{10})} = \frac{\eta}{\sqrt{1-\left(\dfrac{\lambda}{2a}\right)^2}} = \frac{377}{\sqrt{2.25}} \frac{1}{\sqrt{1-\left(\dfrac{\lambda}{2a}\right)^2}}$$

$$= \frac{251.3}{0.745} = 337.36(\Omega)$$

(5) The attenuation constant α of waveguide wall is [see Equation (10.74)]

$$\alpha = \frac{R_s}{\eta b\sqrt{1-\left(\dfrac{\lambda}{2a}\right)^2}}\left[1 + 2\frac{b}{a}\left(\frac{\lambda}{2a}\right)^2\right]$$

in the equation

$$R_s = \sqrt{\frac{\pi f \mu}{\sigma}} = \sqrt{\frac{\pi \times 10^{10} \times (4\pi \times 10^{-7})}{1.5 \times 10^7}} = 0.0501 \quad (\Omega)$$

so

$$\alpha = 0.0526\text{Np/m}(= 0.4569\text{dB/m})$$

10.1.11 The Single-Mode Transmission of Rectangular Waveguide

Various TE_{mn} and TM_{mn} modes and their linear combinations can occur in rectangular waveguide. When the operating wavelength is less than the cutoff wavelength, these modes are transmission modes; in this case, the waveguide is multi-mode waveguide.

When the waveguide dimensions a and b are given, from Equation (10.38), it is easy to find the cutoff wavelength λ_c of various modes.

When $a = 2b$, λ_c of several main modes are

$\lambda_{c(10)} = 2a$ for TE_{10} mode

$\lambda_{c(20)} = a$ for TE_{20} mode

$\lambda_{c(01)} = 2b = a$ for TE_{01} mode

$\lambda_{c(11)} = 2a/\sqrt{1+(a/b)^2} = 2a/\sqrt{5}$ for TE_{11}, TM_{11} mode

$\lambda_{c(30)} = \dfrac{2}{3}a$ for TE_{30} mode

$\lambda_{c(02)} = b = \dfrac{1}{2}a$ for TE_{02} mode

It can be seen from Figure 10.4, in all modes, the cutoff wavelength of TE_{10} is the longest; it is thus called the lowest mode, or the dominant mode. In addition, TE_{11} and TM_{11} mode have same cutoff wavelength. Different modes with same cutoff wavelength can be named as degenerate modes. For instance, when $a = 2b$, TE_{20} and TE_{01} are degenerate modes.

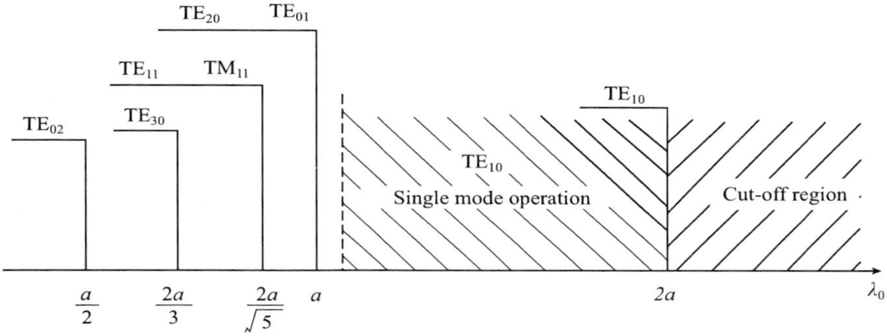

Figure 10.4 The cutoff wavelength of the main modes in rectangular waveguide when $a = 2b$

As the multi-mode transmission will not only cause inconvenience to stimulate and coupling, at the same time, the signal distortion caused by the mode coupling usually occur in the discontinuity in the waveguide. Because of the practical importance of the TE_{10} mode, single mode of TE_{10} mode is thus critically preferable. How to achieve single-mode transmission? In a rectangular waveguide, the cutoff wavelength of TE_{10} mode is the longest ($\lambda_c = 2a$), and it is the lowest mode. *Its transmission conditions should be* $\lambda < 2a$. To achieve TE_{10} single-mode transmission, all other higher modes should be cutoff. For the numerous higher modes, the longest cutoff wavelength is a, and *therefore it should be chosen as* $\lambda > a$. In this way, *the single-mode conditions for* TE_{10} *mode in waveguide is given as*

$$\left.\begin{array}{c} 2a > \lambda > a \\ \text{or} \quad \lambda/2 < a < \lambda \end{array}\right\} \qquad (10.75)$$

and

$$b \leqslant a/2$$

Example 10.3 Consider the figure of Example 10.3, TE_{10} wave transmit in rectangular waveguide. If an ideal metal is used for short-circuit of the waveguide at $z = l$, then total reflection of TE_{10} wave would occur. Again an ideal metal is used for short-circuit the waveguide at $z = 0$, and assuming the length of l be $\lambda_g/2$ (for the TE_{10} wave), try to determine the electromagnetic field components within the metal cavity of the rectangular (known as the rectangular resonant cavity).

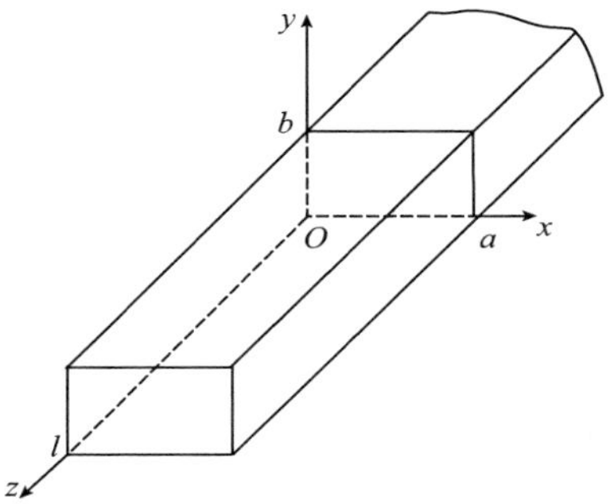

Figure of Example 10.3

Solution Suppose TE_{10} wave propagate along the positive z direction, and the field component of the incident wave is

$$\left.\begin{aligned} E_y^+ &= E_0 \sin\left(\frac{\pi x}{a}\right) \cdot e^{j(\omega t-\beta z)} \\ H_x^+ &= -\left(\frac{\beta}{\omega\mu}\right) E_0 \sin\left(\frac{\pi x}{a}\right) \cdot e^{j(\omega t-\beta z)} \\ H_z^+ &= j\left(\frac{\pi}{\omega\mu a}\right) E_0 \cos\left(\frac{\pi x}{a}\right) \cdot e^{j(\omega t-\beta z)} \end{aligned}\right\} \quad \text{(a)}$$

The incident TE_{10} wave encounters an ideal metal plate at $z = l$ and causes total reflection of wave. It is easy to write the field components of the reflected wave using boundary conditions of an ideal conductor

$$\left.\begin{aligned} E_y^- &= -E_0 \sin\left(\frac{\pi x}{a}\right) \cdot e^{j(\omega t+\beta z)} \\ H_x^- &= -\left(\frac{\beta}{\omega\mu}\right) E_0 \sin\left(\frac{\pi x}{a}\right) \cdot e^{j(\omega t+\beta z)} \\ H_z^- &= -j\left(\frac{\pi}{\omega\mu a}\right) E_0 \cos\left(\frac{\pi \dot{x}}{a}\right) \cdot e^{j(\omega t+\beta z)} \end{aligned}\right\} \quad \text{(b)}$$

Adding the incident waves of the TE_{10} wave with the reflected wave, we thus obtain the total wave, that is,

$$\left.\begin{aligned} E_y = E_y^+ + E_y^- &= -j2E_0 \sin\left(\frac{\pi x}{a}\right) \cdot \sin(\beta z) \cdot e^{j\omega t} \\ H_x = H_x^+ + H_x^- &= -2\left(\frac{\beta}{\omega\mu}\right) E_0 \sin\left(\frac{\pi x}{a}\right) \cdot \cos(\beta z) \cdot e^{j\omega t} \\ H_z = H_z^+ + H_z^- &= 2\left(\frac{\pi}{\omega\mu a}\right) E_0 \cos\left(\frac{\pi x}{a}\right) \cdot \sin(\beta z) \cdot e^{j\omega t} \end{aligned}\right\} \quad \text{(c)}$$

We may note that, at $z = l = \dfrac{\lambda_g}{2}$ (that is, the short-circuit metal plate), $\beta z = \beta l = 180°$, $E_y = 0$, $H_z = 0$, H_x is the peak point of the standing wave along the z direction, and it completely satisfy the ideal conductor boundary conditions here. $\beta z = 0, E_y = 0, H_z = 0, H_x$ are also the peak point of the standing wave at $z = 0$. Therefore, *placing an ideal metal short-circuit plates at $z = 0$ has no effect on the distribution of synthetic waves. After placing this piece of metal plate, it thus becomes a rectangular cavity.* As $\beta l = \pi$, therefore, $\beta = \dfrac{\pi}{l}$. Substitute this value into Formula (c), we then obtain the electromagnetic field in the rectangular resonant cavity (dimensions $a \times b \times l$)

$$\left.\begin{aligned} E_y &= -j2E_0 \sin\left(\frac{\pi x}{a}\right) \cdot \sin\left(\frac{\pi z}{l}\right) \cdot e^{j\omega t} \\ H_x &= -2\left(\frac{\beta}{\omega\mu}\right) E_0 \sin\left(\frac{\pi x}{a}\right) \cdot \cos\left(\frac{\pi z}{l}\right) \cdot e^{j\omega t} \\ H_z &= 2\left(\frac{\pi}{\omega\mu a}\right) E_0 \cos\left(\frac{\pi x}{a}\right) \cdot \sin\left(\frac{\pi z}{l}\right) \cdot e^{j\omega t} \end{aligned}\right. \quad \text{(d)}$$

As the resonant cavity is a closed metal cavity, therefore, the spatial distribution of the electric field and the magnetic field must be in the form of standing wave. Moreover, there is a time phase difference of 90° between electric field E_y and the two magnetic field components H_x, H_z. That is, the electric field reaches its maximum when the magnetic field is zero, and vice versa. This is the basic features between the electric and magnetic fields in resonance state. Formula (d) represents the field component of TE_{101} mode in the rectangular resonant cavity , the superpose 101 stands for $m = 1, n = 0, p = 1$. p standing for the number of half-sine wave along the z-distribution.

Example 10.4 Try to use Formula (d) of example 10.3, that is, try to use the expression of electromagnetic field in a rectangular resonant cavity (dimensions $a \times b \times l$) to evaluate that the maximum electric energy storage is equal to the maximum magnetic energy storage when the cavity is resonant.

Solution From example 10.3, the instantaneous field components of TE_{101} mode in rectangular resonant cavity are

$$\left.\begin{aligned} E_y &= E_{101} \sin\left(\frac{\pi x}{a}\right) \cdot \sin\left(\frac{\pi z}{l}\right) \cdot \sin \omega t \\ H_x &= -\left(\frac{\beta}{\omega\mu}\right) E_{101} \sin\left(\frac{\pi x}{a}\right) \cdot \cos\left(\frac{\pi z}{l}\right) \cdot \cos \omega t \\ H_z &= \left(\frac{\pi}{\omega\mu a}\right) E_{101} \cos\left(\frac{\pi x}{a}\right) \cdot \sin\left(\frac{\pi z}{l}\right) \cdot \cos \omega t \end{aligned}\right\} \tag{a}$$

First, find the instantaneous electric field energy storage $\bar{w}_e(t)$ in the volume of the cavity.

$$\begin{aligned} \bar{w}_e(t) &= \int_v \frac{1}{2}\varepsilon E_y^2(t)\mathrm{d}V \\ &= \frac{1}{2}\varepsilon \int_0^a \int_0^b \int_0^l E_{101}^2 \sin^2\left(\frac{\pi x}{a}\right) \cdot \sin^2\left(\frac{\pi z}{l}\right) \cdot \sin^2 \omega t \cdot \mathrm{d}x\mathrm{d}y\mathrm{d}z \\ &= \frac{1}{8}\varepsilon E_{101}^2 (a \cdot b \cdot l) \sin^2 \omega t \end{aligned} \tag{b}$$

The instantaneous magnetic energy storage in the cavity $\bar{W}_m(t)$ is

$$\begin{aligned} \bar{W}_m(t) &= \int_v \frac{1}{2}\mu H^2(t)\mathrm{d}V = \int_V \frac{1}{2}\mu \left[H_x^2(t) + H_z^2(t)\right] \mathrm{d}V \\ &= \frac{1}{2}\mu \int_0^a \int_0^b \int_0^l \frac{\beta^2}{\omega^2\mu^2} E_{101}^2 \cdot \sin^2\left(\frac{\pi x}{a}\right) \cdot \cos^2\left(\frac{\pi x}{l}\right) \cos^2 \omega t \cdot \mathrm{d}x\mathrm{d}y\mathrm{d}z \\ &+ \frac{1}{2}\mu \int_0^a \int_0^b \int_0^l \frac{\pi^2}{\omega^2\mu^2 a^2} E_{101}^2 \cdot \cos^2\left(\frac{\pi x}{a}\right) \cdot \sin^2\left(\frac{\pi z}{l}\right) \cdot \cos^2 \omega t \cdot \mathrm{d}x\mathrm{d}y\mathrm{d}z \\ &= \frac{1}{8} E_{101}^2 (a \cdot b \cdot l) \cdot \left(\frac{\beta^2}{\omega^2\mu} + \frac{\pi^2}{\omega^2\mu a^2}\right) \cdot \cos^2 \omega t \\ &= \frac{1}{8}\beta\varepsilon E_{101}^2 (a \cdot b \cdot l) \cdot \cos^2 \omega t \end{aligned} \tag{c}$$

From Formulas (b) and (c), we can clearly see:

(1) When the electric field energy storage $\bar{w}_e(t)$ reaches its maximum, the magnetic field energy storage $\bar{w}_m(t)$ is zero, and vice versa.

(2) As $\frac{1}{T}\int_0^T \sin^2 \omega t\mathrm{d}t = \frac{1}{T}\int_0^T \cos^2 \omega t\mathrm{d}t$, therefore, the time for average electric field energy storage in cavity is equal to the time for average magnetic energy storage, that is

$$\frac{1}{T}\int_0^T \bar{w}_e(t)\mathrm{d}t = \frac{1}{T}\int_0^T \bar{w}_m(t)\mathrm{d}t$$

(3) As no loss is considered, so the sum of the electromagnetic energy storage in the cavity is a constant. Moreover, the electromagnetic energy storage is equal to the maximum

electric field energy storage, or the maximum magnetic energy storage. that is

$$\bar{w}_e(t) + \bar{w}_m(t) = \frac{1}{8}\varepsilon E_{101}^2(a \cdot b \cdot l) \cdot \left(\sin^2 \omega t + \cos^2 \omega t\right) = \frac{1}{8}\varepsilon E_{101}^2(a \cdot b \cdot l)$$
$$= [\bar{w}_e(t)]_{\max} = [\bar{w}_m(t)]_{\max} = \text{const.}$$

10.2 Circular Waveguide

Figure 10.5 shows the circular waveguide with inner radius a. Since a cylindrical geometry is involved, it is appropriate to employ cylindrical coordinates. Like the rectangular waveguide, the circular waveguide can propagate TE and TM modes, but not TEM waves. As the case in Cartesian coordinate system, E_z or H_z in cylindrical coordinates can be derived from the wave equations of the circular waveguide, subject to some boundary conditions. The transverse field component of the cylindrical waveguide can thus be derived from E_z and H_z.

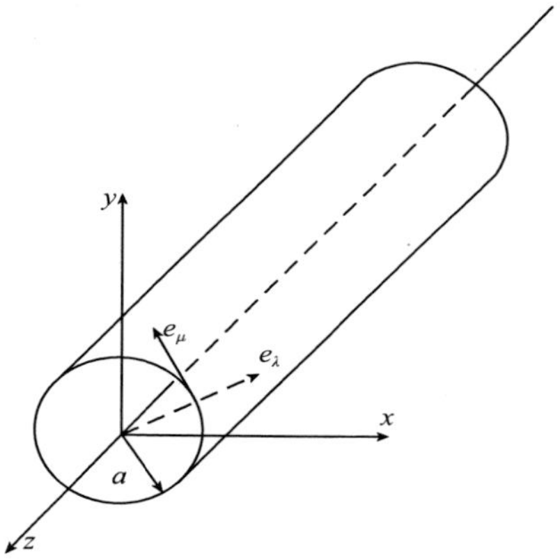

Figure 10.5 Circular waveguide

10.2.1 Wave Equation in Terms of E_z, H_z and the Relation between Transverse Field Components and E_z, H_z

Assumed the harmonic wave propagates along z directions in circular waveguide. Hence, the factor $e^{j\omega t-\gamma z}$ is involved in all field components. The wave equations (i.e. electric field) are also

$$\nabla^2 \boldsymbol{E} + k^2 \boldsymbol{E} = 0$$

or

$$\nabla_T^2 \boldsymbol{E} + \frac{\partial^2 \boldsymbol{E}}{\partial z^2} + k^2 \boldsymbol{E} = \nabla_T^2 \boldsymbol{E} + (\gamma^2 + k^2)\boldsymbol{E} = 0$$

hence

$$\nabla_T^2 \boldsymbol{E} + k_c^2 \boldsymbol{E} = 0 \tag{10.76}$$

where $k_c^2 = \gamma^2 + k^2$.

From Equation (10.76), the wave equation of E_z in cylindrical coordinates is

$$\frac{1}{r}\frac{\partial}{\partial r}\left(r\frac{\partial H_z}{\partial r}\right)+\frac{1}{r^2}\frac{\partial^2 E_z}{\partial \phi^2}+k_{\mathrm{c}}^2 E_z=0 \tag{10.77}$$

Also for the wave equation of H_z

$$\frac{1}{r}\frac{\partial}{\partial r}\left(r\frac{\partial H_z}{\partial r}\right)+\frac{1}{r^2}\frac{\partial^2 H_z}{\partial \phi^2}+k_{\mathrm{c}}^2 H_z=0 \tag{10.78}$$

The relations among E_r, E_ϕ, H_r, H_ϕ and E_z, H_z:

As the case in the rectangular coordinates, the transverse field components can be derived from E_z and H_z by applying Maxwell's curl equations.

$$E_r=-\frac{1}{k_{\mathrm{c}}^2}\left(\gamma\frac{\partial E_z}{\partial r}+\frac{\mathrm{j}\omega\mu}{r}\frac{\partial H_z}{\partial \phi}\right) \tag{10.79}$$

$$E_\phi=\frac{1}{k_{\mathrm{c}}^2}\left(-\gamma\frac{\partial E_z}{\partial \phi}+\mathrm{j}\omega\mu\frac{\partial H_z}{\partial r}\right) \tag{10.80}$$

$$H_r=\frac{1}{k_{\mathrm{c}}^2}\left(\frac{\mathrm{j}\omega\varepsilon}{r}\frac{\partial E_z}{\partial \phi}-\gamma\frac{\partial H_z}{\partial r}\right) \tag{10.81}$$

$$H_\phi=-\frac{1}{k_{\mathrm{c}}^2}\left(\mathrm{j}\omega\varepsilon\frac{\partial E_z}{\partial r}+\frac{\gamma}{r}\frac{\partial H_z}{\partial \phi}\right) \tag{10.82}$$

As a consequence of the factor $\mathrm{e}^{\mathrm{j}\omega t-\gamma z}$ involved in the harmonic wave field, in the derivation of the equations above, the partial differentiation with respects to z and t are used as below,

$$\frac{\partial}{\partial z}=-\gamma,\quad \frac{\partial}{\partial t}=\mathrm{j}\omega$$

10.2.2 TM Mode of Circular Waveguide

TM waves are characterized by $E_z\neq 0$ and $H_z=0$. The field components can be derived by solving E_z and Equation (10.72)~(10.82). Firstly solve E_z.

By wave equation of E_z (10.77), the following equation can be derived

$$\frac{\partial^2 E_z}{\partial r^2}+\frac{1}{r}\frac{\partial E_z}{\partial r}+\frac{1}{r^2}\frac{\partial^2 E_z}{\partial \phi^2}+k_{\mathrm{c}}^2 E_z=0 \tag{10.83}$$

E_z is a function of r, ϕ, z and t. However, as we know, the field component is harmonic wave and propagates along the z direction. Hence, the only problem is to solve the relation between E_z and (r,ϕ). So let

$$E_z=R\Phi\mathrm{e}^{\mathrm{j}\omega t-\gamma z} \tag{10.84}$$

where R is the function of r, Φ and ϕ. Substitute Equation (10.84) into (10.83):

$$\Phi\frac{\partial^2 R}{\partial r^2}+\frac{\Phi}{r}\frac{\partial R}{\partial r}+\frac{R}{r^2}\frac{\partial^2 \Phi}{\partial \phi^2}+k_{\mathrm{c}}^2 R\Phi=0$$

Multiply $r^2/R\Phi$ and apply the method of separation of variables, we have

$$\left(\frac{r^2}{R}\frac{\partial^2 R}{\partial r^2}+\frac{r}{R}\frac{\partial R}{\partial r}+k_{\mathrm{c}}^2 r^2\right)+\frac{1}{\Phi}\frac{\partial^2 \Phi}{\partial \phi^2}=0 \tag{10.85}$$

where the term in bracket is a variable of r and the term out of bracket of ϕ. Obviously, for all r and ϕ, they are constant. And the sum of these two constants is zero, which means they have the same value but different sign. Assuming it be m^2, we can get

$$\frac{r^2}{R}\frac{\mathrm{d}^2 R}{\mathrm{d}r^2} + \frac{r}{R}\frac{\mathrm{d}R}{\mathrm{d}r} + k_{\mathrm{c}}^2 r^2 = m^2 \tag{10.86}$$

and

$$\frac{1}{\Phi}\frac{\mathrm{d}^2\Phi}{\mathrm{d}\phi^2} = -m^2$$

or

$$\frac{\mathrm{d}^2\Phi}{\mathrm{d}\phi^2} + m^2\Phi = 0 \tag{10.87}$$

After using the separation of variables, Equation (10.86) and (10.87) can be found. Now discuss their solution. The solution of Equation (10.87) is

$$\Phi = c\begin{cases}\sin m\phi \\ \cos m\phi\end{cases} \tag{10.88}$$

According to the circular symmetry of the circular waveguide, the field is the variable of ϕ in period 2π.

$$\Phi(\phi) = \Phi(\phi \pm 2\pi)$$

To satisfy this equation, m can only be an integer, and then guarantee $\Phi(m\phi) = \Phi[m(\phi \pm 2\pi)]$.

Now solve Equation (10.86), we can have

$$r^2\frac{\mathrm{d}^2 R}{\mathrm{d}r^2} + r\frac{\mathrm{d}R}{\mathrm{d}r} + (k_{\mathrm{c}}^2 r^2 - m^2)R = 0 \tag{10.89}$$

It is a Bessel equation and the solution is m order Bessel function

$$R = C_1\mathrm{J}_m(k_{\mathrm{c}}r) + C_2\mathrm{N}_m(k_{\mathrm{c}}r) \tag{10.90}$$

where J_m is m order Bessel function of the first kind and N_m is of second kind-Neumann function. Figure 10.6 shows the curve of $\mathrm{J}_m(k_{\mathrm{c}}r)$, $\mathrm{J}_m'(k_{\mathrm{c}}r)$ and $\mathrm{N}_m(k_{\mathrm{c}}r)$. $\mathrm{J}_m'(k_{\mathrm{c}}r)$ is the differential Bessel function of the first kind. From Figure 10.6(c), we can see that the $\mathrm{N}_m(k_{\mathrm{c}}r)$ is infinity for$r$=0. To avoid this case, we take $C_2 = 0$ in Equation (10.90). Then from Equation (10.90), R is

$$R = C_1\mathrm{J}_m(k_{\mathrm{c}}r) \tag{10.91}$$

Substitute Equations (10.88) and (10.91) into Equation (10.84), E_z is now

$$E_z = E_0\mathrm{J}_m(k_{\mathrm{c}}r)\begin{cases}\sin m\phi \\ \cos m\phi\end{cases} \cdot \mathrm{e}^{\mathrm{j}\omega t - \gamma z} \tag{10.92}$$

Assumed the electric magnetic wave propagate in the circular waveguide without any loss, then $\alpha = 0$ and $\gamma = \mathrm{j}\beta$. Substitute $\gamma = \mathrm{j}\beta$ into Equation (10.92) and then use the result for Equations (10.79)~(10.82), yields

$$E_r = -\mathrm{j}\frac{\beta}{k_{\mathrm{c}}}E_0\mathrm{J}_m'(k_{\mathrm{c}}r)\begin{cases}\sin m\phi \\ \cos m\phi\end{cases} \cdot \mathrm{e}^{\mathrm{j}(\omega t - \beta z)} \tag{10.93}$$

$$E_\phi = \mathrm{j}\frac{m\beta}{k_{\mathrm{c}}^2 r}E_0\mathrm{J}_m(k_{\mathrm{c}}r)\begin{cases}-\cos m\phi \\ \sin m\phi\end{cases} \cdot \mathrm{e}^{\mathrm{j}(\omega t - \beta z)} \tag{10.94}$$

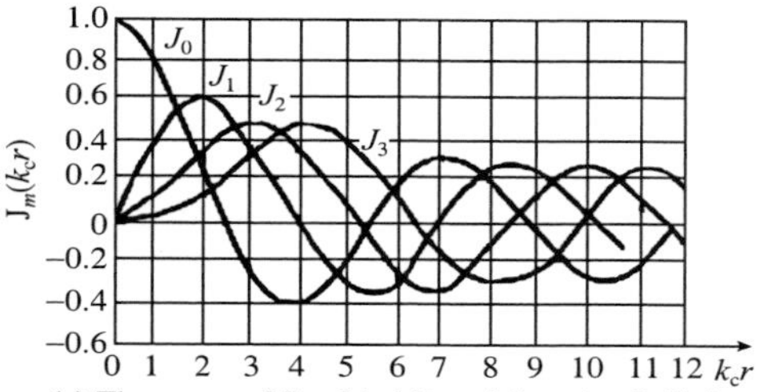

(a) The curve of first kind Bessel function $J_m(k_cr)$

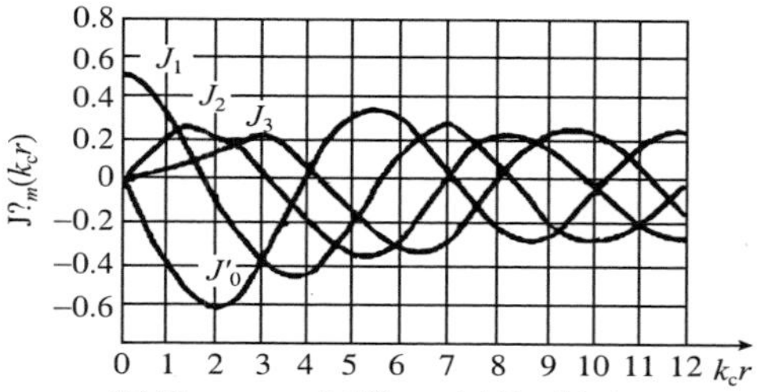

(b) The curve of differential first kind bassel function $J'_m(k_cr)$

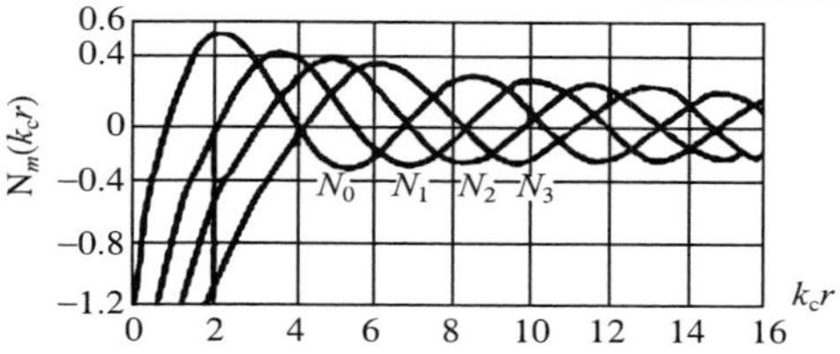

(c) The curve of second kind Bessel fuction $N_m(k_cr)$

Figure 10.6

$$E_z = E_0 \mathrm{J}_m(k_c r) \begin{cases} \sin m\phi \\ \cos m\phi \end{cases} \cdot \mathrm{e}^{\mathrm{j}(\omega t - \beta z)} \tag{10.95}$$

$$H_r = -\mathrm{j}\frac{\omega\varepsilon m}{k_c^2 r} E_0 \mathrm{J}_m(k_c r) \begin{cases} -\cos m\phi \\ \sin m\phi \end{cases} \cdot \mathrm{e}^{\mathrm{j}(\omega t - \beta z)} \tag{10.96}$$

$$H_\phi = -\mathrm{j}\frac{\omega\varepsilon}{k_c} E_0 \mathrm{J}'_m(k_c r) \begin{cases} \cos m\phi \\ \sin m\phi \cdot \mathrm{e}^{\mathrm{j}(\omega t - \beta z)} \end{cases} \tag{10.97}$$

$$H_z = 0 \tag{10.98}$$

The equations above are the field components for TM mode in circular waveguide. Each component can be expressed in the format of sine and cosine factors. This means when the azimuthal is fixed, for a same mode, there are two possible field distribution with 90° spatial phase difference. These modes are named as degenerate mode. If one is the format of sine, then another must be in the format of cosine. Figure 10.7 shows the degenerate wave TE_{11}. Because of the non-dependence on ϕ, the wave having axial symmetry ($m = 0$) do not have the degenerate phenomenon.

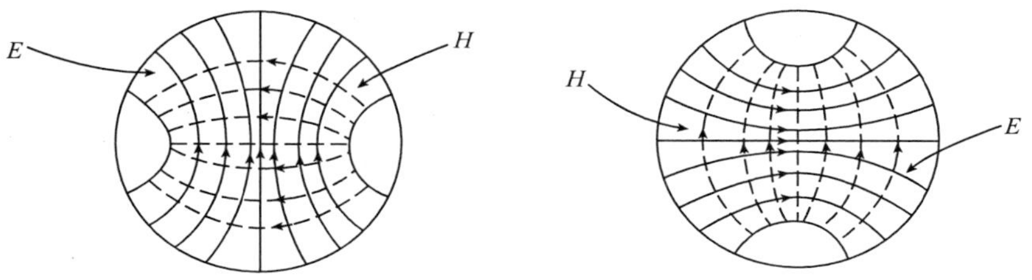

(a) (b)

Figure 10.7 Polarization direction with 90° degree rotation of wave TE_{11}

Apply the boundary condition to determine the constant k_c.

E_z is parallel to the waveguide wall and assume the waveguide as an ideal conductor. Hence $E_z = 0$ for $r = a$. From Equation (10.95), we have

$$J_m(k_c a) = 0 \tag{10.99}$$

We can see from k_c. $(k_c a)$ is the roots of m order Bessel function of first kind. However, there are multiple roots, as shown in Figure 10.6(a). u_{mn} is used to represent the nth root of m order Bessel function of first kind. The first few values of u_{mn} are listed in Table 10.1.

Table 10.1 The roots u_{mm} of first kind Bessel function

m \ n (u_{mm})	1	2	3	4
0	2.405	5.520	8.654	11.792
1	3.832	7.016	10.174	13.324
2	5.136	8.417	11.620	14.796
3	6.379	9.761	13.015	16.220
4	7.586	11.065	14.372	—

Since

$$k_c a = u_{mn}$$

Hence

$$k_c = \frac{u_{mn}}{a} \quad \text{(for TM wave)} \tag{10.100}$$

10.2.3 TE Mode of Circular Waveguide

TE waves are characterized by $E_z = 0$ and $H_z \neq 0$. As in the TM mode case, we can solve H_z

$$H_z = H_0 J_m(k_c r) \begin{Bmatrix} \sin m\phi \\ \cos m\phi \end{Bmatrix} \cdot e^{j(\omega t - \beta z)} \tag{10.101}$$

Substitute Equation (10.101) into Equations (10.79)~(10.82), we have

$$E_r = j\frac{\omega\mu m}{k_c^2 r} H_0 J_m(k_c r) \begin{Bmatrix} -\cos m\phi \\ \sin m\phi \end{Bmatrix} \cdot e^{j(\omega t - \beta z)} \tag{10.102}$$

$$E_\phi = j\frac{\omega\mu}{k_c} H_0 J'_m(k_c r) \begin{Bmatrix} \sin m\phi \\ \cos m\phi \end{Bmatrix} \cdot e^{j(\omega t - \beta z)} \tag{10.103}$$

$$E_z = 0 \tag{10.104}$$

$$H_r = -j\frac{\beta}{k_c} H_0 J'_m(k_c r) \begin{Bmatrix} \sin m\phi \\ \cos m\phi \end{Bmatrix} \cdot e^{j(\omega t - \beta z)} \tag{10.105}$$

$$H_\phi = j\frac{\beta m}{k_c^2 r} H_0 J_m(k_c r) \begin{Bmatrix} -\cos m\phi \\ \sin m\phi \end{Bmatrix} \cdot e^{j(\omega t - \beta z)} \tag{10.106}$$

$$H_z = H_0 J_m(k_c r) \begin{Bmatrix} \cos m\phi \\ \sin m\phi \end{Bmatrix} \cdot e^{j(\omega t - \beta z)} \tag{10.107}$$

This is the field component for TE mode in circular waveguide.

Apply boundary condition to obtain k_c. For $r = a$, on the internal surface of circular waveguide, the tangential electric field is zero. From Equation (10.103), we have

$$J'_m(k_c a) = 0$$

We can get k_c of TE mode. $(k_c a)$ is the roots of m order Bessel function of first kind. There are also multiple roots for above equation, as shown in Figure 10.6(b). Similarly, u'_{mn} is used to represent the nth root of differential m order Bessel function of first kind. The first few values of u'_{mn} are listed in Table 10.2.

Table 10.2 Roots of differential Bessel function of first kind

u_{mm} \ n / m	1	2	3	4
0	3.832	7.016	10.174	13.324
1	1.841	5.332	8.536	11.706
2	3.054	6.705	9.965	13.170
3	4.201	8.015	11.344	
4	5.317	9.282	12.682	

since

$$k_c a = u'_{mn}$$

hence

$$k_c = u\prime_{mn}/a \quad \text{(TE Mode)} \tag{10.108}$$

10.2.4 The Parameters of TE and TM Mode in Circular Waveguide

From Equations (10.93)~(10.98) or Equations (10.102)~ (10.107), the meaning of subscript mn for TE_{mn} and TM_{mn} can be observed, m means the numbers of the maximum field distribution occurring along the circumference of the waveguide, and n means numbers of the maximum field distribution occurring along radius (r from $0 \sim a$).

As a result of the reflection of the metal surface of the circular waveguide, the electric magnetic distribution on the cross section (ϕ direction) is characterized by standing wave. The standing wave distribution of the field is trigonometric function along circumference whereas Bessel function is along radius.

Now discuss the parameters of the circular waveguide.

1. Wave Number k_c

$$k_c = \frac{u_{mn}}{a} \quad \text{(For TM mode)}$$

$$k_c = \frac{u'_{mn}}{a} \quad \text{(For TE mode)}$$

$$k_c^2 = \gamma^2 + k^2 \tag{10.109}$$

2. Cutoff Wavelength λ_c

The cutoff wavelength is the wavelength when propagation constant $\gamma = 0$. From Equation (10.109), it must have

$$k_c = k = \frac{2\pi}{\lambda_c} \tag{10.110}$$

So from Equations (10.100) and (10.108) and the equation above, we can find

$$\lambda_c = \frac{2\pi a}{u_{mn}} \quad \text{(For TM mode)} \tag{10.111}$$

and

$$\lambda_c = \frac{2\pi a}{u'_{mn}} \quad \text{(For TE mode)} \tag{10.112}$$

λ_c of several ordinary modes are listed in Table 10.3.

Figure 10.8 shows the cutoff wavelength λ_c of TE and TM mode, ranging from the longest to the lowest. The maximum of λ_c is TE_{11}, which is $3.41a$. Operating at this mode, the circular waveguide can propagate the single mode wave (except for polarization degeneracy). So TE_{11} is the fundamental mode of the circular waveguide.

Table 10.3 λ_c of several ordinary TE and TM mode

Mode	TE_{01}	TE_{02}	TE_{03}	TE_{11}	TE_{12}	TE_{13}	TE_{21}	TE_{22}
λ_c	$1.64a$	$0.90a$	$0.62a$	$3.41a$	$1.18a$	$0.74a$	$2.06a$	$0.94a$
Mode	TE_{01}	TE_{02}	TE_{03}	TE_{11}	TE_{12}	TE_{13}	TE_{21}	TE_{22}
λ_c	$2.62a$	$1.14a$	$0.72a$	$1.64a$	$0.90a$	$0.62a$	$1.22a$	$0.75a$

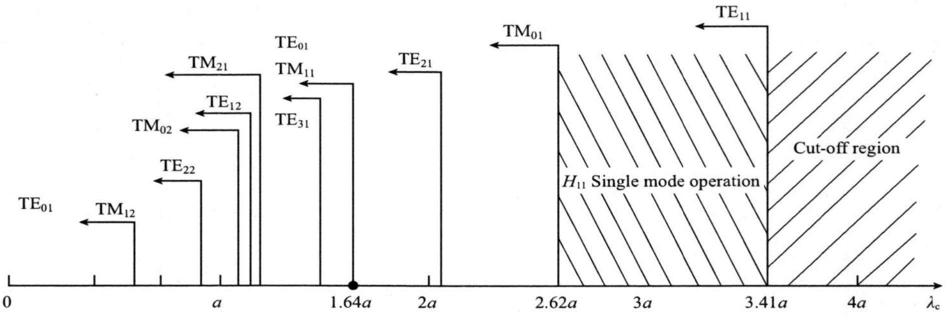

Figure 10.8 The ranged λ_c of the circular waveguide

From the figure above, it can be clearly found *the single-mode condition for TE_{11} is*

$$2.62a < \lambda < 3.41a \tag{10.113}$$

3. Phase Constant β

For the lossless case, $\gamma = j\beta$ and $k_c^2 = \gamma^2 + k^2$, so

$$\beta = \sqrt{k^2 - k_c^2} = k\sqrt{1 - \left(\frac{\lambda}{\lambda_c}\right)^2} \tag{10.114}$$

4. Guide Wavelength λ_g

$$\lambda_g = \frac{2\pi}{\beta} = \frac{\lambda}{\sqrt{1 - \left(\frac{\lambda}{\lambda_c}\right)^2}} \tag{10.115}$$

5. Phase Velocity v_p

$$v_p = \frac{\omega}{\beta} = \frac{\omega}{k\sqrt{1 - \left(\frac{\lambda}{\lambda_c}\right)^2}} = \frac{v}{\sqrt{1 - \left(\frac{\lambda}{\lambda_c}\right)^2}} \tag{10.116}$$

6. Group Velocity v_g

Since

$$v_g = \frac{1}{d\beta/d\omega}$$

and

$$\beta = \sqrt{k^2 - k_c^2} = \sqrt{\left(\frac{\omega}{v}\right) - \left(\frac{2\pi}{\lambda_c}\right)^2}$$

hence

$$v_g = v\sqrt{1 - \left(\frac{\lambda}{\lambda_c}\right)^2} \tag{10.117}$$

7. Wave Impedance Of TE and TM Mode $Z_{\mathrm{W(TE)}}$, $Z_{\mathrm{W(TM)}}$

From Equations (10.102)~(10.107), the wave impedance of TE mode is

$$Z_{\mathrm{W(TE)}} = \frac{E_r}{H_\phi} = -\frac{E_\phi}{H_r} = \frac{\omega\mu}{\beta} = \eta\frac{1}{\sqrt{1-\left(\frac{\lambda}{\lambda_c}\right)^2}} \tag{10.118}$$

From Equations (10.93)~(10.98), the wave impedance of TM mode is

$$Z_{\mathrm{W(TM)}} = \frac{E_r}{H_\phi} = -\frac{E_\phi}{H_r} = \frac{\beta}{\omega\varepsilon} = \eta\frac{1}{\sqrt{1-\left(\frac{\lambda}{\lambda_c}\right)^2}} \tag{10.119}$$

10.2.5 Field Structure and Attenuation of Various Modes in Circular Waveguide

Figure 10.9 shows the field structure of $TE_{11}(H_{11})$, $TE_{01}(H_{01})$, $TM_{01}(E_{01})$ in the circular waveguide. Observe that the TE_{11} is the lowest mode which can take single-mode propagation. And TE_{01} is an important mode owing to its loss and decreases with increasing frequency.

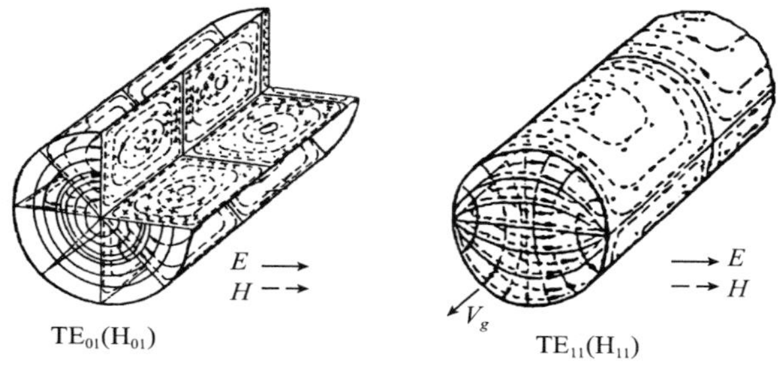

Figure 10.9 Field structure of TE_{01}, TE_{11} mode in a circular waveguide

Figure 10.10 shows the attenuation of TE_{11}, TE_{01}, TM_{01} in a circular waveguide with $a = 2.5$cm. Observe the attenuation of the mode TE_{11} is lower than TM_{01}. However, the attenuation of TE_{01} decreases further with increasing frequency while the attenuation of TE_{11} and TM_{01} increase with the frequency. This is because H_z of TE_{01} is the only magnetic component tangential to the circular waveguide wall. It produces the current along ϕ direction. When the energy of the propagation is finite, H_z decreases with increasing frequency. Thus the current on the wall decreases, resulting in

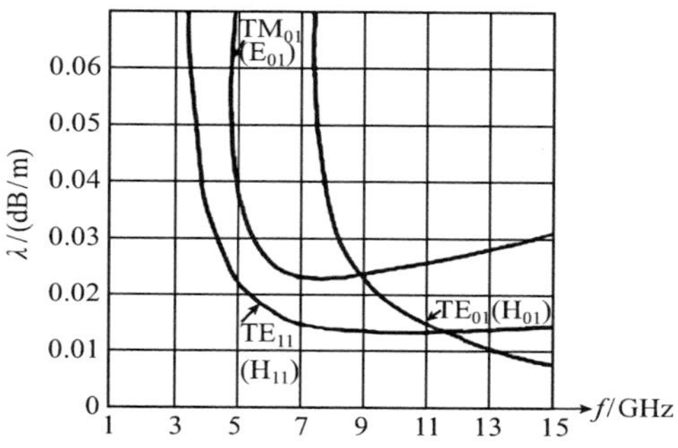

Figure 10.10 Attenuation of TE_{01}, TE_{11}, TM_{01} modes in a circular waveguide with $2a = 5$cm

the attenuation reduction. In fact, all $TE_{0n}(n = 1, 2, 3, \cdots)$ take this feature. With this good feature, TE_{01} mode can be used to manufacture high Q circular cavity.

10.2.6 Comparison between the Attenuation in Circular and Rectangular Waveguide

In general, the attenuation in circular waveguide is a little smaller than rectangular waveguide.

The energy in a waveguide propagates through the cross section whereas the attenuation is caused by the metal wall of the waveguide. Obviously, when the perimeter is same, the cross section of the circular is larger than the rectangular, which means the attenuation in circular waveguide is smaller than rectangular. Figure 10.11 shows the attenuation curve of rectangular and circular waveguide with same perimeter. From the figure, we can see the attenuation for $TE_{11}^{\square}$ in circular waveguide is smaller than that for $TE_{10}^{\square}$ in rectangular waveguide.

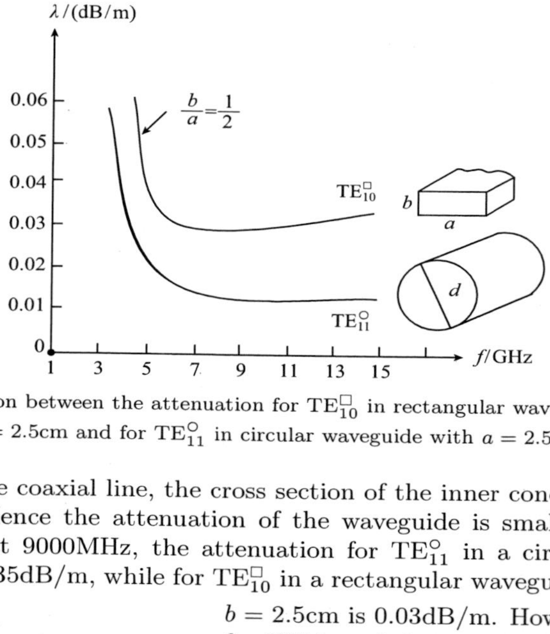

Figure 10.11 Comparison between the attenuation for $TE_{10}^{\square}$ in rectangular waveguide with $a = 5$cm and $b = 2.5$cm and for TE_{11}° in circular waveguide with $a = 2.5$cm

Comparing with the coaxial line, the cross section of the inner conductor in a waveguide is relatively large. Hence the attenuation of the waveguide is smaller than coaxial line. From Figure 10.11, at 9000MHz, the attenuation for TE_{11}° in a circular waveguide with $2a = 5$cm is only 0.0135dB/m, while for $TE_{10}^{\square}$ in a rectangular waveguide with $a = 5$cm and $b = 2.5$cm is 0.03dB/m. However, the attenuation for TEM mode in a 50-7 coaxial line (air medium) is 0.32dB/m at the same frequency. This value is more than 10 times larger than in the waveguide. Furthermore the attenuation of the radio coaxial line filled with dielectric material (such as polyvinyl chloride) is larger than the coaxial line filled with air.

As we know, the propagation attenuation in the optical fiber is very small and even smaller than 0.3dB/km. This is comparable with the waveguides above and the coaxial line.

Example 10.5 Find the resonant frequency of the rectangular resonant cavity

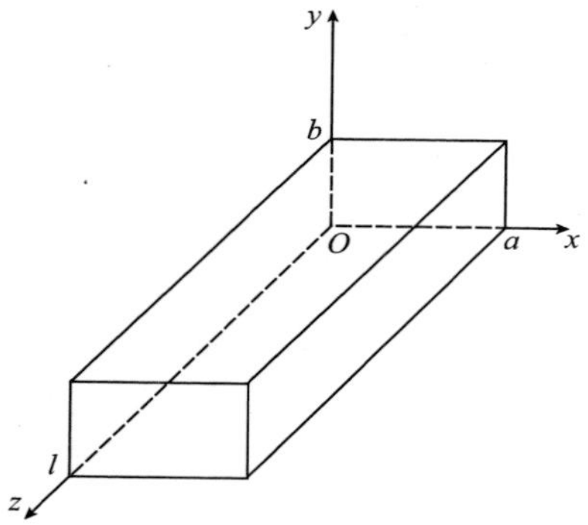

Figure of Example 10.5

Solution Assumed the TE and TM mode propagate along z direction. For rectangular waveguide, from Equations (10.7) and (10.8), we have

$$k_c^2 = \gamma^2 + k^2 \tag{a}$$

and

$$k_c^2 = k_x^2 + k_y^2 = \left(\frac{m\pi}{a}\right)^2 + \left(\frac{n\pi}{b}\right)^2 \tag{b}$$

For lossless waveguide, it has

$$\gamma^2 = (j\beta)^2 = -\beta^2 \tag{c}$$

For the waveguide, phase constant β is along z direction. But for resonant cavity, the wave along z direction inside the cavity is standing wave, which is in the same case for x and y direction. Therefore, β^2 should be replaced by k_z^2 correspondingly.

$$k_z = \frac{p\pi}{l} \tag{d}$$

Substitute Formula (b), (c) and (d) to (a), we find

$$k^2 = k_x^2 + k_y^2 + k_z^2 = \left(\frac{m\pi}{a}\right)^2 + \left(\frac{n\pi}{b}\right)^2 + \left(\frac{p\pi}{l}\right)^2 \tag{e}$$

Since $k^2 = \omega^2\mu\varepsilon$, the resonant angular frequency of the rectangular cavity is

$$\omega_{mnp} = \frac{1}{\sqrt{\mu\varepsilon}}\sqrt{\left(\frac{m\pi}{a}\right)^2 + \left(\frac{n\pi}{b}\right)^2 + \left(\frac{p\pi}{l}\right)^2} \tag{f}$$

where the subscript m, n and p means the number of the half sine standing-wave along x, y and z directions, respectively.

For TM_{mnp} mode, neither m nor n can be zero whereas p can be zero. For TE_{mnp}, either m or n can be zero whereas p cannot be zero.

The degenerate resonant wave can propagate in rectangular resonant cavity i.e. at the same frequency it has more than one resonant mode. For instance, it has three lowest resonant modes at the same time for $a = b = l$. These degenerated modes are TM_{110}, TE_{011}, TE_{101}, and their resonant frequencies are

$$f = \frac{1}{a\sqrt{2\mu\varepsilon}}$$

Example 10.6 Find the resonant frequency of the circular resonant cavity

Solution From Equations (10.1000), (10.108) and (10.109)

$$k_c = \frac{u_{mn}}{a} \quad \text{(For TM mode)} \tag{a}$$

$$k_c = \frac{u'_{mn}}{a} \quad \text{(For TE mode)} \tag{b}$$

$$k_c^2 = \gamma^2 + k^2 \tag{c}$$

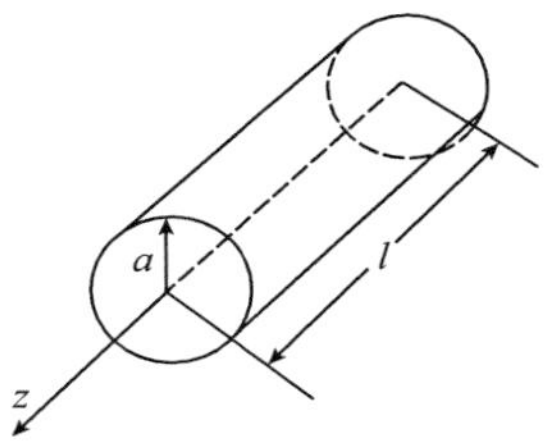

Figure of Example 10.6

Take l as the length of the circular waveguide and a as the radius. Placing a metal plate at the end can form a circular resonant cavity. The electromagnetic wave thus becomes standing wave (sine and cosine) along z direction. Hence, $\gamma = j\beta$ for the lossless condition and replace β by k_z.

$$k_z = \frac{p\pi}{l} \tag{d}$$

Substitute Formula (a), (b) and (d) into (c), it must have

$$k^2 = k_c^2 + k_z^2 = k_c^2 + \left(\frac{p\pi}{l}\right)^2 \tag{e}$$

So the resonant frequency of the circular cavity is

$$\left.\begin{aligned} \omega &= \frac{1}{\sqrt{\mu\varepsilon}}\sqrt{\left(\frac{u'_{mn}}{a}\right)^2 + \left(\frac{p\pi}{l}\right)^2} \quad \mathrm{TE}_{mnp} \quad \text{mode} \\ \omega &= \frac{1}{\sqrt{\mu\varepsilon}}\sqrt{\left(\frac{u_{mn}}{a}\right)^2 + \left(\frac{p\pi}{l}\right)^2} \quad \mathrm{TM}_{mnp} \quad \text{mode} \end{aligned}\right\} \tag{f}$$

10.3 Higher Modes in Coaxial Line

Coaxial line commonly propagate TEM wave. The operational frequency range is very wide from DC to microwave. However, there is a frequency limitation for the the coaxial line to work on TEM mode. If the operation frequency goes beyond the limitation, higher modes occur. This section is to discuss the higher modes in coaxial line.

For the circular waveguide, the higher modes (TE and TM) of coaxial line will be analyzed by using the method of separation of variables to solve the wave equation in the circular coordinate, subject to some certain boundary conditions.

1. TM Mode in the Coaxial Line

From the analysis of the circular waveguide, the solution of E_z for TM wave is known as

$$\begin{aligned} E_z &= R\Phi e^{j(\omega t - \beta z)} \\ &= E_0 \left[C_1 \mathrm{J}_m(k_c r) + C_2 \mathrm{N}_m(k_c r)\right] \cdot \left\{ \begin{array}{l} \sin m\phi \\ \cos m\phi \end{array} \right. \cdot e^{j(\omega t - \beta z)} \end{aligned} \tag{10.120}$$

In the circular waveguide, to avoid E_z from infinite for $r = 0$, let $C_2 = 0$ in above equation, saying, the field of the circular waveguide is only dependent with the Bessel function of the first kind. However, r does not range from zero but from a to b ($a$and b are the inner and outer radius of coaxial line, respectively) in the coaxial line. Therefore the solution of E_z includes the first kind and the second kind Bessel function. The mode and the parameters of mode in coaxial line is different from those in circular waveguide. As mentioned above, only TEM mode is used, not the higher modes (TE and TM mode) of coaxial line. Here only the cutoff wavelength for TE and TM modes will be evaluated, which can be in usage for the cutoff case.

2. Determine k_c of TM Mode by the Boundary Condition

According to the boundary condition that the tangential electronic field is zero on the ideal conductor surface, we know $E_z = 0$ for $r = a$ and $r = b$. From Equation (10.120), it must have

$$C_1 \mathrm{J}_m(k_c a) + C_2 \mathrm{N}_m(k_c a) = 0 \quad (\text{for } r = a) \tag{10.121}$$

and

$$C_1 \mathrm{J}_m(k_c b) + C_2 \mathrm{N}_m(k_c b) = 0 \quad (\text{for } r = b) \tag{10.122}$$

From the two equations above, we have

$$\frac{\mathrm{J}_m(k_c a)}{\mathrm{J}_m(k_c b)} = \frac{\mathrm{N}_m(k_c a)}{\mathrm{N}_m(k_c b)} \tag{10.123}$$

From this equation we can find k_c. However, above equation is a transcendental function, it is quite difficult to solve. Here we just get its approximate solutions. The number of the solution k_c is infinite. Each k_c corresponds to a mode and a cutoff wavelength.

The general way is to use the asymptotic expression of the Bessel function of first and second kind. Then take the appropriate expression for the condition $(k_c a)$ and $(k_c b)$ which are much larger than 1. Substituting them into Equation (10.123) to obtain the appropriate solution of k_c for TM mode as

$$k_c \approx \frac{n\pi}{b-a} \quad (n = 1, 2, 3, \cdots) \tag{10.124}$$

The cutoff wavelength for TM_{mn} is

$$\lambda_c \approx \frac{2(b-a)}{n} \quad (n = 1, 2, 3, \cdots) \tag{10.125}$$

The two equations above are independent of m.

3. k_c for TE Mode in the Coaxial Line

Similar with the TM mode, the way to find k_c for TE mode is using the boundary condition E_ϕ is zero for $r = a$ and $r = b$ to get the transcendental function.

$$\frac{\mathrm{J}'_m(k_c a)}{\mathrm{J}'_m(k_c b)} = \frac{\mathrm{N}'_m(k_c a)}{\mathrm{N}'_m(k_c b)} \tag{10.126}$$

Then we get

$$\lambda_c \approx \frac{\pi(a+b)}{m} \quad (n = 1, m = 1, 2, 3, \cdots) \tag{10.127}$$

and

$$\lambda_c \approx \frac{2(b-a)}{n} \quad (n = 1, m = 1, 2, 3, \cdots) \tag{10.128}$$

From Equations (10.125), (10.127) and (10.128), TE_{11} is the lowest mode in the coaxial line and its cut-off wavelength is largest, it is

$$\lambda_c \approx \quad \pi(a+b) \quad (\text{the lowest mode } \mathrm{TE}_{11}) \tag{10.129}$$

Obviously, to *assure only the fundamental mode TEM propagates in the coaxial line, the operation wavelength must be larger than the cutoff wavelength λ_c of TE_{11}*, say

$$\lambda > \pi(a+b) \tag{10.130}$$

Equation (10.130) can be referred when we design or use coaxial lines. Figure 10.12 and 10.13 show the cutoff wavelength $\lambda_c$of several higher modes, as well as their field distributions.

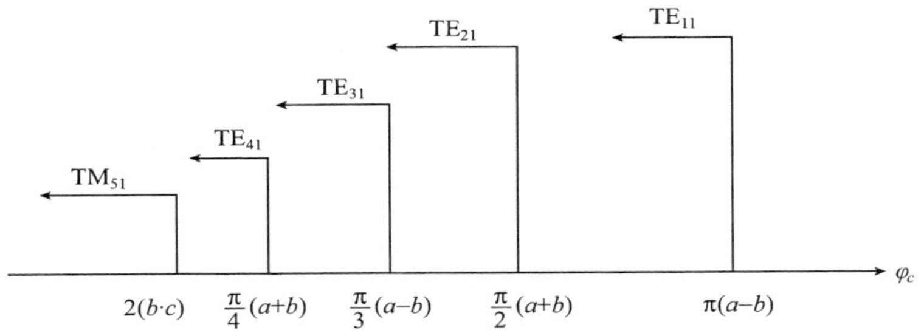

Figure 10.12 Cut-off wavelength λ_c of the higher order mode in coaxial line

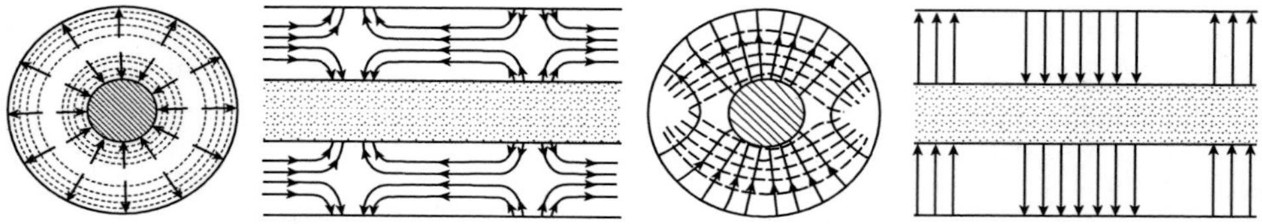

TM$_{01}$ TE$_{11}$

Figure 10.13 Field distribution of some higher order mode in coaxial line

Exercises

10.1 Consider a rectangular waveguide with cross section $a \times b$, propagate TE$_{01}$ mode, and the electronic field is

$$E_x = E_0 \sin(\pi y/b) e^{j(\omega t - \beta z)}$$

To determine, (1) H_y; (2) the power density $\boldsymbol{S}_{av}$ along z direction; (3) The propagating power P for TE$_{01}$ mode.

10.2 The following two rectangular waveguides operate at the same frequency, try to compare their cutoff frequencies for TM$_{11}$ mode.

(1) $a \times b = 23\text{mm} \times 10\text{mm}$;

(2) $a \times b = 16.5\text{mm} \times 16.5\text{mm}$.

10.3 Rectangular waveguide with $a = 2\text{cm}$ $b = 2.5\text{cm}$, determine the length of the waveguide if the phase difference between the upper and lower side-band frequency is 180 degree, and the modulated wave is $(1 + \cos \omega_m t) \cos \omega t$ with $f_m = 20\text{kHz}$, $f = 10\text{GHz}$.

10.4 The longitudinal electric field for TM mode in rectangular waveguide is

$$E_z = E_0 \sin \frac{\pi}{3} x \sin \frac{\pi}{3} y \cos \left(\omega t - \frac{\sqrt{2}}{3} \pi z \right)$$

where the unit of x, y, z is in cm

(1) Find cutoff wavelength and guided wavelength;

(2) If TM$_{32}$ can propagate, determine the size of the waveguide.

10.5 Rectangular waveguide with $a = 7\text{cm}$, $b = 3\text{cm}$, which is filled with air or lossless medium with $\varepsilon_r = 4$, $\mu_r = 1$.

(1) Find the cutoff frequency and guided wavelength for TE$_{10}$, TE$_{20}$, TE$_{01}$, TE$_{11}$, TM$_{11}$ modes.

(2) Which modes can propagate at $f = 3 \times 10^9\text{Hz}$.

10.6 Rectangular waveguide with cross section $23\text{mm} \times 10\text{mm}$, which modes can propagate inside if $\lambda = 10\text{mm}$? and if $\lambda = 30\text{mm}$?

10.7 Find the range of the waveguide width a if TE$_{10}$, TE$_{01}$, TE$_{11}$ and TM$_{11}$can propagate in the vacuum rectangular waveguide, at frequency 15GHz.

10.8 Rectangular waveguide with cross section $22.86\text{mm} \times 10.16\text{mm}$, can it propagate the wave with free space wavelength 2cm, 3cm, 5cm? if yes, which modes?

10.9 Rectangular waveguide with size $2\text{cm} \times 2\text{cm}$ and propagate TE$_{10}$ wave with 10GHz, please find (1) Wave vector $\boldsymbol{k}$; (2) Expression of the electromagnetic field components; (3) Guided wavelength of the waveguide; (4) The phase velocity and group velocity.

10.10 Rectangular waveguide is made of ideal conductor. The width of the cross section is a and height is b. Assume the waveguide axis is parallel to the z axis. Proof (1) The waveguide cannot propagate the monochromatic wave $\boldsymbol{E} = \boldsymbol{e}_x E_0 e^{j(\omega t - \beta z)}$ where E_0, ω, β are constant; (2) On the waveguide wall, the magnetic intensity $\boldsymbol{B}$ satisfy

$$\begin{cases} \dfrac{\partial B_y}{\partial x} = \dfrac{\partial B_z}{\partial x} = 0 \quad (x = 0, a) \\[2mm] \dfrac{\partial B_z}{\partial y} = \dfrac{\partial B_x}{\partial y} = 0 \quad (y = 0, b) \end{cases}$$

10.11 Rectangular waveguide with cross section $7.2\text{mm} \times 3.4\text{mm}$, which is filled with the air dielectric.

(1) Which mode is available if the wavelength is 16cm, 8cm and 6.5cm?

(2) If demanding the lowest frequency is more than 5% of the cutoff frequency of TE_{10} mode and the highest frequency is less than 5% of the cutoff frequency of the higher modes. Try to determine the frequency range appropriate for TE_{10} propagation.

10.12 Rectangular waveguide with cross section 2.29cm×1.02cm, which is filled with the air dielectric. If TE_{10} mode propagates with operation frequency $f = 10\text{GHz}$, find

(1) Phase constant β and wave impedance η_{TE};

(2) β and η_{TE} for TE_{10} mode if filled with ideal dielectric $\mu_r = 1, \varepsilon_r = 4$

(3) If the operation frequency is below 5GHz (lower than the cutoff frequency), try to find the attenuation constant α of TE_{10} mode and wave impedance η_{TE}. Also calculate the distance when the field is attenuated to e^{-1}.

10.13 A copper rectangular waveguide propagate TE_{10} mode and the operational frequency $f = 20\text{GHz}$. If the size of the waveguide is 0.5cm×0.6cm, the copper conductivity $\sigma = 1.57\times10^7\text{S/m}$, $\mu_r = 1, \varepsilon_r = 2.25$ and the loss tangent is 4×10^{-4}. Find:

(1) Phase constant β, guided wavelength λ_g, phase velocity v_p and wave impedance η_{TE};

(2) Attenuation constant α caused by medium and waveguide wall.

10.14 Air-filled rectangular waveguide. The electric field of the fundamental mode TE_{10} is

$$E_y = E_0 \sin \frac{\pi}{a} x e^{-j\beta z}$$

(1) To prove that the propagation power is $P_{TE10} = \dfrac{ab}{4\eta_0}\sqrt{1-\left(\dfrac{\lambda}{2a}\right)^2}E_0^2$;

(2) The waveguide with size of 2.25cm×1.00cm, find the maximum power (not breakdown) if operating at 10GHz.

10.15 For TEM mode, if the medium conductivity $\sigma \neq 0$, the dielectric attenuation is then $\alpha_d = \dfrac{\omega\sqrt{\mu\varepsilon}}{2}\left(\dfrac{\sigma}{\omega\varepsilon}\right)$.

10.16 A red copper rectangular waveguide with cross section of 23mm×10mm and operation frequency $f = 10\text{GHz}$. Find (1) the attenuation for TE_{10} mode (dB/m) if the waveguide is filled with air; (2) the attenuation for TE_{10} mode (dB/m) if the waveguide is filled with dielectric $\varepsilon_r = 2.54$ and the loss tangent 3×10^{-4}.

10.17 Find the propagation power of TE_{0n} wave in circular waveguide.

10.18 In circular waveguide, if the conductor attenuation constant$\alpha = \dfrac{R_s}{a\eta\sqrt{1-(f_c/f)^2}}$, please determine that the attenuation constant achieves minimum at $f = \sqrt{3}f_c$.

10.19 Propagate TE_{10} wave in circular waveguide. The conductor attenuation constant is $\alpha = \dfrac{R_s(f_c/f)^2}{a\eta\sqrt{1-(f_c/f)^2}}$. The diameter of the circular waveguide is 12cm and made of the red copper, also propagate TE_{01} wave and the operation frequency is 87.5GHz. If the attenuation for long-distance communication between two relay stations is not allowed to go beyond 30dB, what is the maximum distance between these two stations? (Surface resistance of the red copper is 0.0505Ω)

10.20 Coaxial line with outer radius $b = 23\text{mm}$ and inner radius $a = 10\text{mm}$. If filled with air or lossless medium with $\varepsilon = 2.25$, determine the characteristic impedance respectively.

Chapter 11

Electromagnetic Radiation

In previous chapters, the electromagnetic waves in unbounded space, the propagation of electromagnetic waves in different media boundary surfaces, i.e, the reflection and refraction problem, as well as the electromagnetic wave in the waveguide transmission system, have been discussed. However, all these discussions haven't considered the wave source. This chapter will discuss the contents of the relevant wave source theory. The first is the lag potential of time-varying field, and then deal with the simplest radiation element — electric dipole and magnetic dipole and their electromagnetic field, as well as some related parameters; finally, the introduction of the duality theorem and, on such basis, further discuss the radiation field of the dual magnetic dipole.

The source of electromagnetic waves, using the terminology of electromagnetics, is actually the time-varying charge and current. In order to generate an effective radiation, the distribution of the charge and current must follow a special way. Antenna is actually an energy conversion device, which is designed in a prescribed manner and can produce effective radiation. Therefore, the antenna, known as the wave source, can produce electromagnetic radiation. The radiation field strength of the source, the spatial distribution of the radiation field, as well as the radiated intensity and energy conversion efficiency etc are all our concerns.

Antenna radiation problem, actually an EM problem with complex boundary conditions, is quite difficult to solve. Even though the antenna structure of concern is very simple, the accurate distribution of charge and current in case of a certain excitation source is still very difficult to obtain. In this case, only some approximate methods can be used for practical application.

According to different structures, the actual antenna can be divided into wire antennas and aperture antennas. A wire antenna can be regarded to consist of infinite line elements carrying with it time-varying electric (or magnetic) current, the basic line element is called electric dipole (electric resonator, electric current element) or magnetic dipole (magnetic resonator, magnetic current element). Similarly, the aperture antenna is considered to consist of infinite facet element carrying with it time-varying electric and magnetic current, the facet element is also known as Huygens element. Therefore, if the radiation of above-mentioned three basic elements are known and understood, the radiation characteristics of all other antennas can be obtained from the superposition principle of electromagnetic fields, taking into account the direction of each element, their spatial distribution of amplitude and phase etc.

11.1 Lag Potential of Alternating Field

The time-varying current and charge on antennas would produce electromagnetic waves. This space field can be obtained by solving Maxwell's equations when the boundary conditions of the antenna are satisfied. The current and the produced electromagnetic field on the antenna are interactive, the current can produce electromagnetic which, in turn, affect the current distribution on the antenna. However, rigorous mathematical analysis method often encounters great difficulties, and sometimes it is even impossible to solve the problems. Therefore, some approximate methods are usually utilized for such purpose, saying, firstly, the source distribution on antenna can be found approximately, consequently, the spatial field around the antenna can be obtained from the known source. Now we will consider the method to calculate the spatial field from the vector potential $\boldsymbol{A}$ and scalar potential φ.

In Chapter 6, the non-homogeneous Helmholtz equations in terms of potential function A and φ have been derived, as shown below:

$$\nabla^2 A - \mu\varepsilon\frac{\partial^2 A}{\partial t^2} = -\mu J \tag{11.1}$$

$$\nabla^2 \varphi - \mu\varepsilon\frac{\partial^2 \varphi}{\partial t^2} = -\rho/\varepsilon \tag{11.2}$$

For the time-harmonic field, the non-homogeneous Helmholtz equation is simplified as

$$\nabla^2 A + k^2 A = -\mu J \tag{11.3}$$

$$\nabla^2 \varphi + k^2 \varphi = -\rho/\varepsilon \tag{11.4}$$

Different methods can be used to solve Equation (11.1) and (11.2). First, the solutions for $\boldsymbol{A}$ and φ can be deduced from the physical concept, the current attention is focused on the physical meaning of the solutions. Later, the rigorous methods can be used to search for the exact solutions of $\boldsymbol{A}$ and φ.

11.1.1 Method 1: Deduction of Solution of A and φ — Non-Homogeneous Helmholtz Equation

The scalar φ is generated by time-varying point charge q in the space. It is assumed that the point charge only exists at the position of q, Equation (11.2) thus becomes homogeneous Helmholtz equation.

$$\nabla^2 \varphi - \mu\varepsilon\frac{\partial^2 \varphi}{\partial t^2} = 0 \tag{11.5}$$

In the spherical coordinate system where q is located at the origin, owing to the spherical symmetry, the scalar φ is only related to r, that is $\varphi = \varphi(r,t)$, then Equation (11.5) becomes

$$\frac{1}{r^2}\frac{\partial}{\partial r}\left(r^2\frac{\partial \varphi}{\partial r}\right) - \mu\varepsilon\frac{\partial^2 \varphi}{\partial t^2} = 0 \tag{11.6}$$

Set $\varphi(r,t) = \dfrac{1}{r}U(r,t)$, Equation (11.6) changes to

$$\frac{\partial^2 U}{\partial r^2} - \frac{1}{v^2}\frac{\partial^2 U}{\partial t^2} = 0 \tag{11.7}$$

where $v = \dfrac{1}{\sqrt{\mu\varepsilon}}$ is wave propagation velocity. Equation (11.7) is one-dimensional wave equation, and the general solution is

$$U(r,t) = f_1\left(t - \frac{r}{v}\right) + f_2\left(t + \frac{r}{v}\right) \tag{11.8}$$

When $U(r,t)$ takes $f_1\left(t - \dfrac{r}{v}\right)$ as its solution, then

$$\varphi(r,t) = \frac{1}{r}f_1\left(t - \frac{r}{v}\right) \tag{11.9}$$

As we know, electrostatic field is a special case of time-varying field, it is then possible to obtain the solution of time-varying scalar potential φ by comparing $\varphi(r, t)$ with $\varphi(r)$ of the electrostatic field. Scalar potential produced by the static charge at the origin is known as

$$\varphi(r) = \frac{q}{4\pi\varepsilon r} \tag{11.10}$$

Comparing Equation (11.9) and (11.10), the scalar potential of time-varying field can be obtained

$$\varphi(r, t) = \frac{q\left(t - \dfrac{r}{v}\right)}{4\pi\varepsilon r} \tag{11.11}$$

If put the point charge to the position (x', y', z'), then Equation (11.1) changes to

$$\varphi(r, t) = \frac{q\left(t - \dfrac{R}{v}\right)}{4\pi\varepsilon R} \tag{11.12}$$

where $R = |r - r'| = \sqrt{(x - x')^2 + (y - y')^2 + (z - z')^2}$ is the distance between the source and the field point.

The scalar potential φ of static field is only a function of space, however, for the time-varying field, the scalar potential φ is not only the function of space, but also the function of time. The value of the scalar potential will change with the source intensity. However, according to the wave theory, it takes time for the microwave to propagate from source point (x', y', z') to the field point $P(x, y, z)$ of distance R with a velocity v, the propagation time is R/v. It means, at time t, the function φ in point P is not produced by the source at the same time. It is produced by the source at time $\left(t - \dfrac{R}{v}\right)$. Equation (11.12) is the reflection of this feature of electromagnetic waves. It shows that the change of potential function for the field is slower than that of the currents and the charges. Because of the delay of the potential function in the propagation, it is called the lag potential of the alternating field. Lag phenomenon is widespread in nature; it is very typical in acoustic and astronomy. For example, the sound that we heard now must be produced before. One obvious astronomy phenomena, saying, we can only watch a star in such a state a few thousand years ago, not its current status.

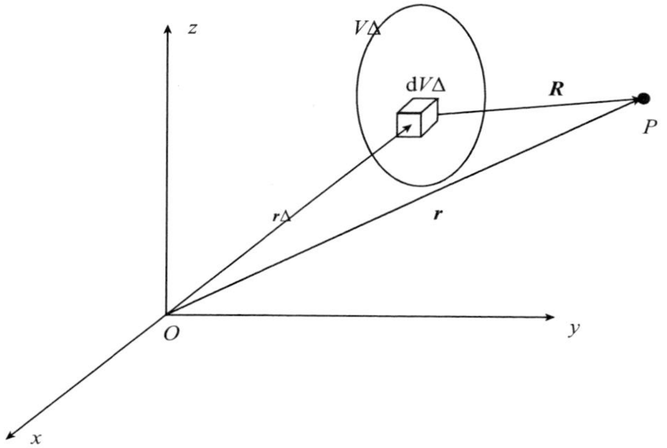

Figure 11.1 $\boldsymbol{r}', \boldsymbol{r}$ and $\boldsymbol{R}$

Above discussed is the situation for the time-varying point charge. If the charges distribute in volume V' as the form of volume charge density ρ, as shown in Figure 11.1, the scalar potential produced by charges in the space is

$$\varphi(r,t) = \int_{V'} \frac{q\left(t - \frac{R}{v}\right)}{4\pi\varepsilon R} \mathrm{d}V' \tag{11.13}$$

Corresponding magnetic vector potential $\boldsymbol{A}$ in terms of volume current density $\boldsymbol{J}$ can be expressed as

$$\boldsymbol{A}(r,t) = \int_{V'} \frac{\mu \boldsymbol{J}\left(t - \frac{R}{v}\right)}{4\pi R} \mathrm{d}V' \tag{11.14}$$

For the time-harmonic field, the plurals of Equations (11.12)~(11.14) are

$$\varphi(r) = \frac{q\mathrm{e}^{-\mathrm{j}kR}}{4\pi\varepsilon R} \tag{11.15}$$

$$\varphi(r) = \int_{V'} \frac{q\mathrm{e}^{-\mathrm{j}kR}}{4\pi\varepsilon R} \mathrm{d}V' \tag{11.16}$$

$$A(r) = \int_{V'} \frac{\mu J\mathrm{e}^{-\mathrm{j}kR}}{4\pi R} \mathrm{d}V' \tag{11.17}$$

If $\varphi(r)$ and $A(r)$ are known, the magnetic and electric fields can be obtained from $\boldsymbol{B} = \nabla \times \boldsymbol{A}$ and $\boldsymbol{E} = -\nabla\varphi - \mathrm{j}\omega\boldsymbol{A}$.

11.1.2 Method 2: The Solutions of Non-Homogeneous Helmholtz Equation in Terms of $\boldsymbol{A}$ and φ

Analysis: firstly, let's see the physical meaning of $\frac{1}{R}\mathrm{e}^{-\mathrm{j}kR}$, and then determine that the source function of $\boldsymbol{A}$ and φ is $\left(\nabla^2 \frac{1}{R}\right)$. After that, the special features of δ function of this source function is to determine the relative source function $\nabla^2 \frac{\rho}{4\pi\varepsilon R}$ of the actual point source. Finally, the relative solution of $\boldsymbol{A}$ and φ of the actual point source can be determined.

Following analysis is an example, aiming to find the electric potential φ of Equation (11.4).

1. $\frac{1}{R}\mathrm{e}^{-\mathrm{j}kR}$*'s Meaning*

Expand $\nabla^2\left(\frac{1}{R}\mathrm{e}^{-\mathrm{j}kR}\right)$ directly to

$$\nabla^2\left(\frac{1}{R}\mathrm{e}^{-\mathrm{j}kR}\right) = \nabla^2\left(\frac{1}{R}\right) \cdot \mathrm{e}^{-\mathrm{j}kR} - k^2\left(\frac{1}{R}\mathrm{e}^{-\mathrm{j}kR}\right) \tag{11.18}$$

then

$$\nabla^2\left(\frac{1}{R}\mathrm{e}^{-\mathrm{j}kR}\right) + k^2\left(\frac{1}{R}\mathrm{e}^{-\mathrm{j}kR}\right) = \nabla^2\left(\frac{1}{R}\right) \cdot \mathrm{e}^{-\mathrm{j}kR} \tag{11.19}$$

compared with Equation (11.4), we can see

$$\varphi = \frac{1}{R}\mathrm{e}^{-\mathrm{j}kR} \tag{11.20}$$

That, $\frac{1}{R}\mathrm{e}^{-\mathrm{j}kR}$ is the solution of the Non-homogeneous Helmholtz equation, moreover, the source function to this solution is

$$\nabla^2\left(\frac{1}{R}\right)\mathrm{e}^{-\mathrm{j}kR} \tag{11.21}$$

2. Source Function $\nabla^2\left(\frac{1}{R}\right)$

There may be a problem: the source function of (11.21) is $\nabla^2\left(\frac{1}{R}\right)\cdot\mathrm{e}^{-\mathrm{j}kR}$, but right now, why $\nabla^2\left(\frac{1}{R}\right)$ is called the source function, not $\mathrm{e}^{-\mathrm{j}kR}$?

When the features of the δ function was studied in section 2.10, an important formula has been obtained [Equation (2.125)]

$$\nabla^2\left(\frac{1}{R}\right) = -4\pi\delta(r - r')$$

Now use this formula to explain why $\nabla^2\left(\frac{1}{R}\right)$ is source function.

We can see, the source function is the product of $\nabla^2\left(\frac{1}{R}\right)$ and $\mathrm{e}^{-\mathrm{j}kR}$, however, because $\nabla^2\left(\frac{1}{R}\right)$ has the special properties of δ function [refer to Equation (2.119) and (2.125)], when $R \neq 0 (r \neq r')$, $\nabla^2\left(\frac{1}{R}\right) = 0$, and when $R = 0$, $\nabla^2\left(\frac{1}{R}\right) = \infty$.

In this way, when $R = 0$ is satisfied, $\nabla^2\left(\frac{1}{R}\right)$ is not equal to zero. However, at this time, $\mathrm{e}^{-\mathrm{j}kR} = 1$, then the source function can be written as

$$\nabla^2\left(\frac{1}{R}\right) \tag{11.22}$$

Conclusion: $\frac{1}{R}\mathrm{e}^{-\mathrm{j}kR}$ represents the solution of the non-homogeneous Helmholtz equation, the source function of this solution is $\nabla^2\left(\frac{1}{R}\right)$. Therefore Equation (11.19) should be written as

$$\nabla^2\left(\frac{1}{R}\mathrm{e}^{-\mathrm{j}kR}\right) + k^2\left(\frac{1}{R}\mathrm{e}^{-\mathrm{j}kR}\right) = \nabla^2\left(\frac{1}{R}\right) \tag{11.23}$$

3. The Solution of Non-Homogeneous Helmholtz Equation When the Actual Point Source is $-\frac{\rho}{\varepsilon}$

According to the selective character of the δ function [Equation (2.123)]:

$$\int_V f(r)\delta(r - r')\mathrm{d}V = f(r')$$

and

$$\nabla^2\left(\frac{1}{R}\right) = -4\pi\delta(r - r')$$

When the actual point source is $-\dfrac{\rho}{\varepsilon}$,

$$\int_V f(r)\nabla^2\left(\frac{1}{R}\right)\mathrm{d}V = -\int_V f(r)\left[4\pi\delta(r-r^1)\right]\mathrm{d}V = -4\pi f(r') = -\frac{\rho}{\varepsilon} \tag{11.24}$$

so

$$f(r') = \frac{\rho}{4\pi\varepsilon} \tag{11.25}$$

We can see from Equation (11.24), the point source function related to the actual point source $-\dfrac{\rho}{\varepsilon}$ is

$$f(r')\nabla^2\left(\frac{1}{R}\right) = \frac{\rho}{4\pi\varepsilon}\nabla^2\left(\frac{1}{R}\right) = \nabla^2\left(\frac{\rho}{4\pi\varepsilon R}\right) \tag{11.26}$$

It can be seen from Equation (11.23), the non-homogeneous Helmholtz equation related to the source function $\nabla^2\left(\dfrac{\rho}{4\pi\varepsilon R}\right)$ is

$$\nabla^2\left(\frac{\rho}{4\pi\varepsilon R}\mathrm{e}^{-\mathrm{j}kR}\right) + k^2\left(\frac{\rho}{4\pi\varepsilon R}\mathrm{e}^{-\mathrm{j}kR}\right) = \nabla^2\left(\frac{\rho}{4\pi\varepsilon R}\right) \tag{11.27}$$

So the solution of φ is

$$\varphi = \frac{\rho}{4\pi\varepsilon R}\mathrm{e}^{-\mathrm{j}kR} \tag{11.28}$$

4. The Solution of φ When the Point Source is Volume Distribution

If the charge $\dfrac{\rho}{\varepsilon}$ is arbitrary volume distribution, superposition theory can be used to obtain the solution of φ

$$\varphi = \int_{V'} \frac{\rho}{4\pi\varepsilon R}\mathrm{e}^{-\mathrm{j}kR}\mathrm{d}V' \tag{11.29}$$

According to the same method, we can obtain the solution of the vector potential of non-homogeneous Helmholtz equation Equation (11.3).

$$\boldsymbol{A} = \int_{V'} \frac{\mu \boldsymbol{J}}{4\pi R}\mathrm{e}^{-\mathrm{j}kR}\mathrm{d}V' \tag{11.30}$$

If the current source is line distribution or surface distribution, the equation above is changed to line integral or surface integral.

It is easy to prove that Equation (11.30) is the solution of Equation (11.3). Now we have

$$\nabla^2\boldsymbol{A} + k^2 A = \frac{\mu}{4\pi}\left[\int_{V'} J(x',y',z')\nabla^2\left(\frac{\mathrm{e}^{-\mathrm{j}kR}}{R}\right)\mathrm{d}V' + k^2\int_{V'} J(x',y',z')\frac{\mathrm{e}^{-\mathrm{j}kR}}{R}\mathrm{d}V'\right] \tag{11.31}$$

Because ∇^2 is the partial derivative of (x, y, z), but integral is with respects to (x', y', z'), so taking the factor ∇^2 inside the integral part does not affect the results.

Compared with Equations (11.31) and (11.23), the following equation is equal to Equation (11.31):

$$\int_{V'} \mu J(x',y',z')\nabla^2\left(\frac{1}{4\pi R}\right)\mathrm{d}V'$$

From Equation (2.125) we can see $\nabla^2\left(\dfrac{1}{4\pi R}\right)$ is point source (it has the character of the point source function), clearly, the above equation is equal to μJ, this proves that Equation (11.30) is the solution of Equation (11.3).

11.2 Electric Dipole

Electric dipole, also known as Hertz dipole, is a very short current element Idl. Idl is much less than the wavelength, the current at each point can be regarded to have equal amplitude and phase. Because it is a short current element in isolation, the charges with equal quality but reverse polarity can appear at the two end, one is $+q$ while the other is $-q$, just like two time-varying "electric pole" , so it is called the "electric dipole", whose direction is from $-q$ to $+q$.

In fact, no electric dipole can be seen in Figure 11.2, however, the actual antenna can be viewed as the connection of numerous such kind of electric dipoles. If the electric and magnetic field of an electric dipole are known, then total field radiated by the antenna can be obtained by integral method on the basis of superposition theory.

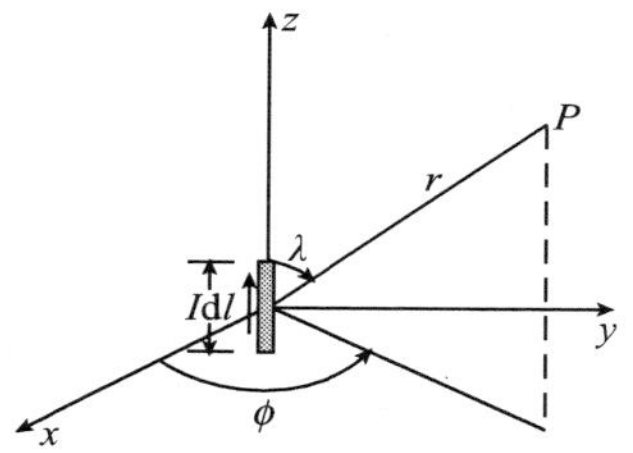

Figure 11.2 Electric dipole

The simplest method to solve the electromagnetic field is to use the vector magnetic potential $\boldsymbol{A}$, and then obtain the magnetic field strength $\boldsymbol{H}$ by equation $\nabla \times \boldsymbol{A} = B$, finally, solve the electric field strength E by using Maxwell's equation.

11.2.1 The Electromagnetic Field of Electric Dipole

The expression of magnetic vector potential A was given in Section 11.1.1, it is

$$\boldsymbol{A} = \frac{\mu}{4\pi}\int_{V'} \frac{\boldsymbol{J}\left(x', y', z', t - \dfrac{R}{v}\right)}{R} \mathrm{d}V'$$

Above volume integral is for the electric dipole itself. It is equal to the multiply product Idl. Considering the situation that the current is harmonic, the magnetic vector potential A of the electric dipole at any point in space can be written as

$$\boldsymbol{A} = \boldsymbol{e}_z \frac{\mu I \mathrm{d}l}{4\pi r} \mathrm{e}^{\mathrm{j}(\omega t - kr)} \text{①} \tag{11.32}$$

1. Solve the Magnetic ield of Electric Dipole through Magnetic Vector Potential $\boldsymbol{A}$

It is most convenient to solve the field of the electric dipole in spherical coordinates system, and put the dipole in the origin.

Writing the expression of the vector magnetic potential $\boldsymbol{A} = e_z A_z$ with respect to the spherical coordinates system, as shown in Figure 11.2

$$A = e_r A_r + e_\theta A_\theta + e_\phi A_\phi = e_r A_z \cos\theta - e_\theta A_z \sin\theta$$

so

$$A_r = A_z \cos\theta = \frac{\mu I \mathrm{d}l}{4\pi r} \cos\theta \mathrm{e}^{\mathrm{j}(\omega t - kr)} \tag{11.33}$$

$$A_\theta = -A_z \sin\theta = -\frac{\mu I \mathrm{d}l}{4\pi r} \sin\theta \mathrm{e}^{\mathrm{j}(\omega t - kr)} \tag{11.34}$$

$$A_\phi = 0 \tag{11.35}$$

① In this equation, $I = I_0 \mathrm{e}^{j\phi}$ is a phasor current, ϕ_0 is the phase angle of the cosine function. For example, if $I(t) = I_0 \sin(\omega t + \phi') = I_0 \cos(\omega t + p' - 90°)$, Then $\phi_0 = \phi' - 90°$. If $I(t) = I_0 \cos(\omega t)$, the $\phi_0 = 0$ and $I = I_0$.

Applying $\boldsymbol{B} = \nabla \times \boldsymbol{A}$, and substituting Equation (11.33)~(11.35) into the curl form of $\boldsymbol{A}$, we can obtain

$$\nabla \times \boldsymbol{A} = e_\phi \frac{1}{r}\left[\frac{\partial}{\partial r}(rA_0) - \frac{\partial A_r}{\partial \theta}\right] = e_\phi \mu H_\phi$$

so

$$H_\phi = \frac{Idl}{4\pi r}\left(\mathrm{j}k + \frac{1}{r}\right)\sin\theta \mathrm{e}^{\mathrm{j}(\omega t - kr)} = \frac{Idlk^2}{4\pi}\left[\frac{\mathrm{j}}{kr} + \frac{1}{(kr)^2}\right]\sin\theta \mathrm{e}^{\mathrm{j}(\omega t - kr)} \tag{11.36}$$

Obviously, the magnetic field of the electric dipole has only one magnetic field component H_ϕ.

2. Solving the Electric Field of the Electric Dipole

Using Maxwell's Equations and (11.36), the electric field strength $\boldsymbol{E}$ can be obtained.

$$E_r = -\mathrm{j}\frac{Idl}{2\pi\omega\varepsilon}\frac{1}{r^2}\left(\mathrm{j}k + \frac{1}{r}\right)\cos\theta \mathrm{e}^{\mathrm{j}(\omega t - kr)} = \frac{Idlk^3}{2\pi\omega\varepsilon}\left[\frac{1}{(kr)^2} - \frac{1}{(kr)^3}\right]\cos\theta \mathrm{e}^{\mathrm{j}(\omega t - kr)} \tag{11.37}$$

$$\begin{aligned} E_\theta &= -\mathrm{j}\frac{Idl}{4\pi\omega\varepsilon}\frac{1}{r}\left(-k^2 + \frac{\mathrm{j}k}{r} + \frac{1}{r^2}\right)\sin\theta \mathrm{e}^{\mathrm{j}(\omega t - kr)} \\ &= \frac{Idlk^3}{4\pi\omega\varepsilon}\left[\frac{\mathrm{j}}{kr} + \frac{1}{(kr)^2} - \frac{\mathrm{j}}{(kr)^3}\right]\sin\theta \mathrm{e}^{\mathrm{j}(\omega t - kr)} \end{aligned} \tag{11.38}$$

$$E_\phi = 0 \tag{11.39}$$

All electromagnetic field produced by the electric dipole are now given in Equations (11.36)~(11.38).

11.2.2 The Near Field and the Far Field of the Electric Dipole

It is too complicated to clarify the relationship between these fields in Equations (11.36)~(11.38). If the electric dipole field is divided into the near-field and far-field, the field characteristics and the relationship will be very clear. Next we will discuss the near- and far-field of the electric dipole.

1. The Near Field of the Electric Dipole

The near field means that kr is much smaller than 1, i.e., the distance between the source point and the field point r is much smaller than $\frac{\lambda}{2\pi}$. Because kr is much smaller than 1, only three field components have a little big amplitude [refer to Equations (11.36)~(11.38)], they are

$$H_\phi \approx \frac{Idl}{4\pi r^2}\sin\theta \mathrm{e}^{\mathrm{j}(\omega t - kr)} \tag{11.40}$$

$$E_r \approx -j\frac{Idl}{2\pi\omega\varepsilon}\cdot\frac{1}{r^3}\cos\theta \mathrm{e}^{\mathrm{j}(\omega t - kr)} \tag{11.41}$$

$$E_\theta \approx -j\frac{Idl}{4\pi\omega\varepsilon}\cdot\frac{1}{r^3}\sin\theta \mathrm{e}^{\mathrm{j}(\omega t - kr)} \tag{11.42}$$

The significant feature of the near field is that there has a 90° phase difference between the electric field strength and the magnetic field strength. E_r and E_θ have a factor j different from H_ϕ. It means the average power density produced by $E_r H_\phi$ and $E_\theta H_\phi$ is equal to zero. It belongs to the induction field, in other words, for induction field, there is only energy exchange, but no energy spread. Therefore, the near field is normally called induction field.

2. The Far Field of the Electric Dipole

The far field means that kr is much larger than 1, the distance between the source point and the field point r is much bigger than $\frac{\lambda}{2\pi}$. Because kr is much bigger than 1, only the following two field components can exist [refer to Equations (11.36)~(11.38)].

$$H_\phi \approx j\frac{Idl}{2\lambda r}\sin\theta e^{j(\omega t-kr)} \tag{11.43}$$

$$E_\theta \approx j\frac{Idl}{2\lambda r}\left(\frac{k}{\omega\varepsilon}\right)\sin\theta e^{j(\omega t-kr)} \tag{11.44}$$

It is thus called far field. *The significant feature of the far field is that E_θ and H_ϕ are in-phase.* Therefore, active power density is produced, and outward spreading of energy is formed. Therefore, it is normally known as radiation field.

It should be mentioned that, *both induction field and radiation field can exist in the near field or far field, however, the induction field is dominant in near field, while the radiation field is dominant in the far field.*

The ratio of the electric field and magnetic field in far field is defined as wave impedance, $\eta_0 = 120\pi\Omega$ in free space.

11.2.3 The Relationship between the Induction Field and the Charge Current of Electric Dipole

The magnetic field H_ϕ of the electric dipole consists of two part, one part contains $\frac{1}{r}$, indicating the radiation field; another part contains $\frac{1}{r^2}$, which represents the induction field.

The region for near field is $r \ll \frac{\lambda}{2\pi}$, so the phase delays $(-kr)$ is still small, which can be approximated as$(\omega t - kr) \approx \omega t$, therefore, near field magnetic field strength H_ϕ can be written as

$$H_\phi \approx \frac{Idl}{4\pi r^2}\sin\theta e^{j\omega t}$$

This equation can be derived directly from Biot-Savart Law, produced by time-varying current element $Idle^{j\omega t}$. In other words, the induction part of H_ϕ is excited directly by current element $Idle^{j\omega t}$.

Since kr is quite small, the electric field strength of the near field can be approximated as

$$E_r \approx -j\frac{Idl}{2\pi\omega\varepsilon}\cdot\frac{1}{r^3}\cos\theta e^{j\omega t} \tag{11.45}$$

$$E_\theta \approx -j\frac{Idl}{4\pi\omega\varepsilon}\cdot\frac{1}{r^3}\sin\theta e^{j\omega t} \tag{11.46}$$

In the previous study to the static field, the electric field strength produced by the electric dipole $+q$ and $-q$ has been obtained,

$$E_r \approx \frac{qdl}{2\pi\varepsilon}\cdot\frac{1}{r^3}\cos\theta \tag{11.47}$$

$$E_\theta \approx \frac{qdl}{4\pi\varepsilon}\cdot\frac{1}{r^3}\sin\theta \tag{11.48}$$

For harmonic cases, $e^{j\omega t}$ should be multiplied to above two equations.

Due to the relationship between current I and charge q is

$$I = \frac{\mathrm{d}q}{\mathrm{d}t} = \mathrm{j}\omega q \tag{11.49}$$

According to Equations (11.45) and (11.46), we can obtain the same expressions of Equations (11.47) and (11.48). indicating that the induction parts of E_r, E_θ are excited by the time-varying charge of the electric dipole.

Due to the 90° phase difference between the current and the charge, the electromagnetic fields excited by the electric dipole also have a phase difference of 90°. Therefore, the corresponding average Poynting vector is zero.

11.2.4 Parameters of Electric Dipole

1. Average Power Density S_{av} of the Electric Dipole

In order to calculate the radiation power of the electric dipole, it is necessary to find average power density S_{av} of electric dipole firstly. Later, the radiation power can be calculated from surface integration. For clear understanding the complicated features of the electromagnetic energy produced by the electric dipole, we only consider the total electric field and magnetic field, and forget the near- or far-field at this moment. From Equations (11.6)~(11.38), S_{av} can be expressed as

$$\begin{aligned}\boldsymbol{S}_{\mathrm{av}} &= \frac{1}{2}\mathrm{Re}(E \times H^*) = \frac{1}{2}\mathrm{Re}\left[(\boldsymbol{e}_r E_r + \boldsymbol{e}_\theta E_\theta) \times \boldsymbol{e}_\phi H_\phi^*\right] \\ &= -\boldsymbol{e}_\theta \frac{1}{2}\mathrm{Re}(E_r H_\phi^*) + \boldsymbol{e}_r \frac{1}{2}\mathrm{Re}(E_\theta H_\phi^*) \\ &= -\boldsymbol{e}_\theta \boldsymbol{S}_{\mathrm{av}(\theta)} + \boldsymbol{e}_r \boldsymbol{S}_{\mathrm{av}(r)}\end{aligned} \tag{11.50}$$

From Equations (11.36)~(11.37),

$$\begin{aligned}\boldsymbol{S}_{\mathrm{av}(\theta)} &= -\frac{1}{2}\mathrm{Re}(E_r H_\phi^*) \\ &= \frac{I^2 \mathrm{d}l^2}{16\pi^2 \omega \varepsilon r^3} \sin\theta \cos\theta \mathrm{Re}\left(\mathrm{j}k^2 + \frac{k}{r} - \frac{k}{r} + \mathrm{j}\frac{1}{r^2}\right) \\ &= 0\end{aligned} \tag{11.51}$$

Obviously, the average Poynting vector produced by E_{r} and H_ϕ in the direction θ is zero, which means there is only imaginary energy in the θ direction.

Now, considering $S_{\mathrm{av}(r)}$ in case of medium $(\varepsilon_{\mathrm{r}}\varepsilon_0, \mu_0)$,

$$\begin{aligned}S_{\mathrm{av}(r)} &= -\frac{1}{2}\mathrm{Re}(E_\theta H_\phi^*) \\ &= \frac{I^2 \mathrm{d}l^2}{32\pi^2 \omega \varepsilon r^2} \sin^2\theta \cos\theta \mathrm{Re}\left(k^3 - \mathrm{j}\frac{k^2}{r} - \frac{k}{r^2} + \mathrm{j}\frac{k^2}{r} + \frac{k}{r^2} - \frac{j}{r^3}\right) \\ &= \frac{I^2 dl^2}{32\pi^2 \omega \varepsilon r^2} \sin^2\theta \cdot (k^3) \\ &= \frac{15\pi}{\sqrt{\varepsilon_{\mathrm{r}}}} \left(\frac{I\mathrm{d}l}{\lambda r}\right)^2 \sin^2\theta\end{aligned} \tag{11.52}$$

It can be seen that $S_{\mathrm{av}(r)}$ is related to six components $\left(k^3 - \mathrm{j}\frac{k^2}{r} - \frac{k}{r^2} + \mathrm{j}\frac{k^2}{r} + \frac{k}{r^2} - \frac{\mathrm{j}}{k^3}\right)$, however, if only taking the real part, only k^3 exists. The other five components have no contribution to the real part. The $S_{av(r)}$ related to k^3 is produced by far field E_θ and H_ϕ of Equations (11.44) and (11.43). It means that the outward spread of energy of the electric dipole depends on the far field.

For the electromagnetic waves in the free space, the average Poynting vector S_{av} is

$$S_{av} = e_r \frac{1}{2} \mathrm{Re}(E_\theta H_\phi^*) = e_r 15\pi \left(\frac{I\mathrm{d}l}{\lambda r}\right)^2 \sin^2\theta \tag{11.53}$$

2. The Radiation Power of the Electric Dipole

The calculation of the radiation power P of the electric dipole is that

$$P = \oint_s S_{av} \cdot \mathrm{d}S = \oint_s S_{av} e_r \cdot \mathrm{d}S e_r = \oint_s S_{av} \mathrm{d}S \tag{11.54}$$

Where $\mathrm{d}S$ is the surface element of the spherical surface, as shown in Figure 11.3,

$$\begin{aligned} \mathrm{d}S &= (r\mathrm{d}\theta) \cdot (r d\beta) = (r\mathrm{d}\theta) \cdot (r\mathrm{d}\phi \sin\theta) \\ &= r^2 \sin\theta \mathrm{d}\theta \mathrm{d}\phi \end{aligned} \tag{11.55}$$

then

$$\begin{aligned} P &= 15\pi \frac{I^2 \mathrm{d}l^2}{\lambda^2} \int_0^{2\pi} \int_0^{\pi} \sin^3\theta \mathrm{d}\phi \mathrm{d}\theta \\ &= 30\pi^2 \frac{I^2 \mathrm{d}l^2}{\lambda^2} \left[\frac{-\cos\theta}{3} (\sin^2\theta + 2)\right]_0^{\pi} \\ &= 40\pi^2 I^2 \left(\frac{\mathrm{d}l}{\lambda}\right)^2 \end{aligned} \tag{11.56}$$

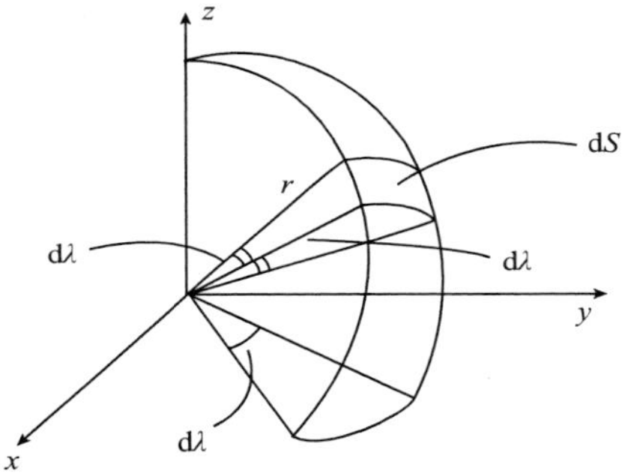

Figure 11.3 The spherical surface element dS

It is the expression of the radiation power of the electric dipole in the air.

3. The Radiation Resistance of the Electric Dipole

Rewriting Equation (11.56), we can obtain

$$P = I^2 \times 40\pi^2 \left(\frac{\mathrm{d}l}{\lambda}\right)^2 = I_e^2 \times 80\pi^2 \left(\frac{\mathrm{d}l}{\lambda}\right)^2 \tag{11.57}$$

where I_e is the effective value of the current, and the power can be expressed as the product of I_e^2 and the resistance, that is

$$P = I_e^2 R_r \tag{11.58}$$

Comparing Equations (11.57) and (11.58), the R_r of electric dipole in air is

$$R_r = 80\pi^2 \left(\frac{\mathrm{d}l}{\lambda}\right)^2 \tag{11.59}$$

where R_r is the radiation resistance of the electric dipole. We may see that, when the length dl increases or the wavelength λ decrease, the radiation resistance increases, the radiation capability thus increases.

4. The Directional Factor and Radiation Pattern of the Electric Dipole

The radiation pattern is used to describe the variation of the electromagnetic field with θ

and φ in spherical coordinate system. From Equations (11.43) and (11.44), E_θ and H_ϕ have no relationship with ϕ, but are the sinusoidal functions of θ. We call the function relationship between the fields and θ, φ as directional factor $F(\theta,\varphi)$, that is

$$F(\theta,\phi) = \sin\theta \tag{11.60}$$

The curve of the directional factor in the spherical coordinate system is called the radiation pattern of the electric dipole, as shown in Figure 11.4. Figure 11.4(a) expresses the planar plot of the radiation field varying with θ. The azimuthal radiation field of the electric dipole is zero, and it achieves maximal value in the direction vertical to the dipole. Figure 11.4 (b) shows that radiation field has no relationship with ϕ.

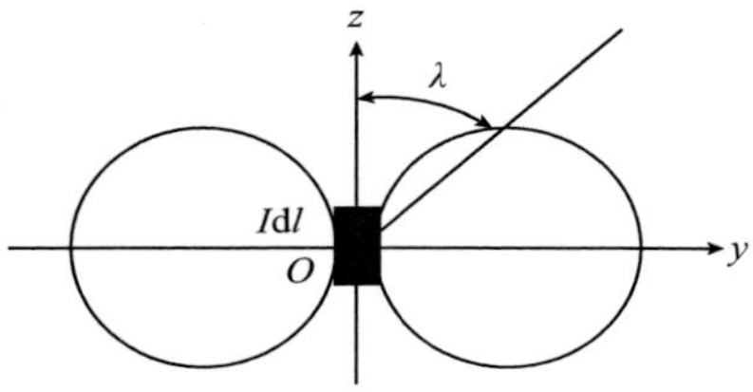

(a) The radiation pattern in Equatorial plane

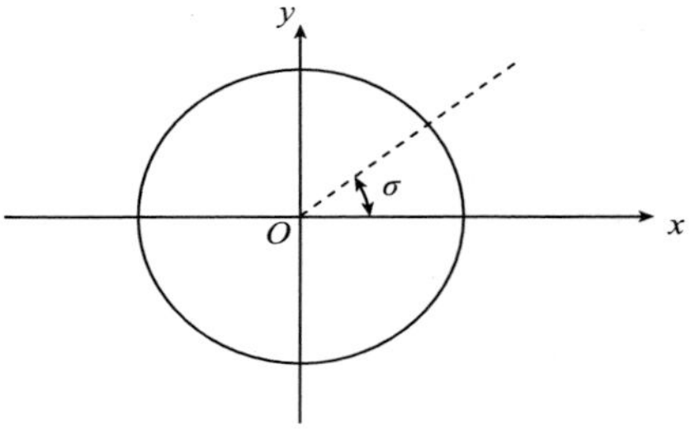

(b) The radiation pattern in Meridian plane

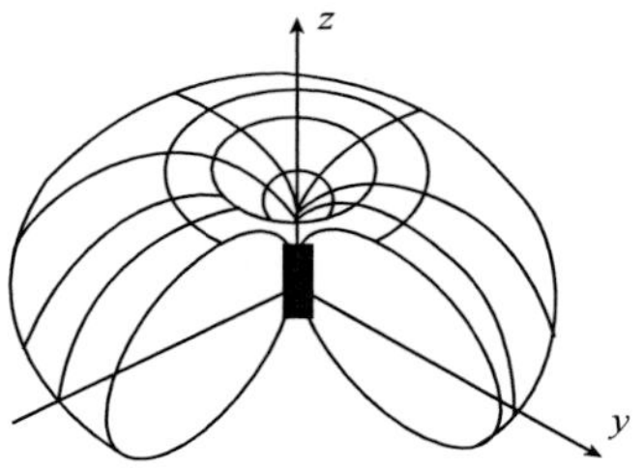

(c) 3-D radiation pattern

Figure 11.4 Radiation pattern of the electric dipole

In general, the plane containing the antenna axis is called meridian plane, as plane yOx, and the plane crossing the origin of the antenna and vertical to the axis is called equatorial plane, like plane xOy.

The actual radiation field is space distribution, so the radiation pattern should be three-dimensional. Because the intensity of the radiation field has no relationship with ϕ, rotating Figure 11.4(a) around the axis, the 3-D radiation pattern of Figure 11.4(c) will not change.

Example 11.1 The electric field of linear polarization planar waves $E_z = E_0 e^{j(\omega t - kr)}$ is incident to an ideal spherical medium, which is polarized. Its dipole moment p varies with time, therefore, the radiation is produced. Assume that the wavelength of plane waves is much bigger than the spherical radius a of the medium, now solving the radiation field and the energy current density S.

Solution The electric potential outside the spherical medium is

$$\varphi_2 = -E_0 r\cos\theta + \left(\frac{\varepsilon - \varepsilon_0}{\varepsilon + 2\varepsilon_0}\right)\frac{E_0 a^3}{r^2}\cos\theta \tag{a}$$

The second term of above equation is the electric potential outside the spherical medium after polarization. The spherical polarization can be expressed by an electric dipole element p in the origin.

The electric potential of the dipole is

$$\varphi = p\frac{\cos\theta}{4\pi\varepsilon_0 r^2}\left(=\frac{p.r}{4\pi\varepsilon_0 r^2}\right) \tag{b}$$

From Formulas (a) and (b), we can get the equivalent electric dipole element p as

$$p = \frac{\varepsilon - \varepsilon_0}{\varepsilon + 2\varepsilon_0}4\pi\varepsilon_0 a^3 E_0 \tag{c}$$

The relationship between electric dipole $I\mathrm{d}l$ and dipole element p is

$$I\mathrm{d}l = \frac{\mathrm{d}q}{\mathrm{d}t}\mathrm{d}l = \mathrm{j}\omega q\mathrm{d}l = \mathrm{j}\omega p \tag{d}$$

Substituding this equation into Equation (11.44), we have

$$\begin{aligned} E_\theta &\approx \mathrm{j}\frac{I\mathrm{d}l}{2\lambda r}\left(\frac{k}{\omega\varepsilon_0}\right)\sin\theta \mathrm{e}^{\mathrm{j}(\omega t - kr)} \\ &= -\frac{pk}{2\lambda\varepsilon_0 r}\sin\theta\mathrm{e}^{\mathrm{j}(\omega t-kr)} \end{aligned} \tag{e}$$

So the far field magnetic field H_ϕ is

$$H_\phi = E_\theta/\eta = -\frac{pk}{2\lambda\varepsilon_0 r}\sin\theta\mathrm{e}^{\mathrm{j}(\omega t-kr)}$$

The average radiation power density $\boldsymbol{S}_{\mathrm{av}}$ is

$$\boldsymbol{S}_{\mathrm{av}} = \frac{1}{2}\mathrm{Re}\left(\boldsymbol{E}\times\boldsymbol{H}^*\right) = \boldsymbol{e}_r\frac{1}{2}\left(\frac{pk}{2\lambda\varepsilon_0 r}\right)^2\sin^2\theta \tag{g}$$

Example 11.2 An *electric small antenna*, which is centrally fed, and the central current is maximum, and linear decrease at the two ends. The current at the end point is zero, as shown in Figure of Example 11.2. If there is another antenna with same size, which has a uniform current distribution and the uniform current is equal to the maximal current I_0 shown in example 11.2, please try to prove the radiation power P' and radiation resistance R'_{r} in the former case are only a quarter of those in latter case.

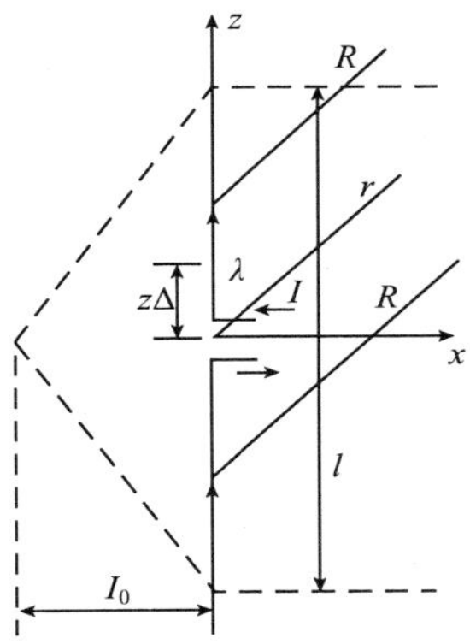

Figure of Example 11.2 Electric small antenna

Solution Take a small segment $\mathrm{d}z'(\mathrm{d}z' \ll \lambda)$, the current above is $I(z')$, which is equivalent to an electric dipole. According to the expression of the field produced by the electric dipole in the space, the far-field produced by $I(z')\mathrm{d}z'$ can be expressed as

$$\mathrm{d}E_\theta = \mathrm{j}\eta_0\frac{I(z')\mathrm{d}z'}{2\lambda R}\sin\theta\mathrm{e}^{-\mathrm{j}kR} \tag{a}$$

The total field of the small antenna is

$$E_\theta = \mathrm{j}\eta_0\int_{-l/2}^{l/2} I(z')\mathrm{d}z'$$

$$\bar{I} = \frac{\mathrm{j}\eta_0 l}{2\lambda r}\sin\theta \mathrm{e}^{-\mathrm{j}kr}\frac{1}{l}\int_{-l/2}^{l/2} I(z')\mathrm{d}z' = \frac{\mathrm{j}\eta_0 l}{2\lambda r}\sin\theta \mathrm{e}^{-\mathrm{j}kr}\bar{I} \tag{c}$$

$\bar{I}$ is the average current flowing through the small antenna.

According to the Fig in example 11.2, the average current is only half of its maximal value, therefore, the amplitude of E_θ and H_ϕ is decreased to 1/2 compared with the dipole with uniform current distribution. So the power density is decreased to 1/4. Thus, when there are same input port currents, the radiation resistance of the former antenna is decreased to 1/4 of the latter.

$$R'' = \frac{1}{4}R_\mathrm{r} = 20\pi^2\left(\frac{l}{\lambda}\right)^2 \ (\Omega) \tag{d}$$

Example 11.3 A vertical antenna is put in the ground, which is much smaller than a wavelength, the height is $l/2$, and its current distribution is zero at the end point and is linearly increasing. It achieves its maximal value when close to the ground, as shown in Figure of Example 11.3. Evaluate the relationship between radiation power and resistance R''_r with the radiation resistance R'_r of last example.

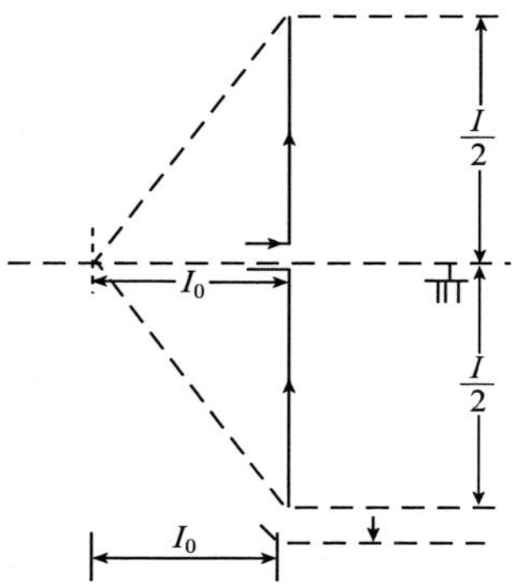

Figure of Example 11.3 Vertical antenna and its mirror

Solution Taking the mirror for the vertical antenna, so we get the same current distribution as Figure of Example 11.2. However, its radiation power is only half of that in above example. In this way, the radiation resistance is only a half, saying,

$$R''_\mathrm{r} = \frac{1}{2}R'_\mathrm{r} = \frac{1}{8}R_\mathrm{r} = 10\pi^2\left(\frac{l}{\lambda}\right)^2$$

11.3 The Magnetic Dipole

11.3.1 The Magnetic Vector Potential *A* of the Magnetic Dipole

A small ring with radius r is shown in Figure 11.5. If the circumference of the ring is quite

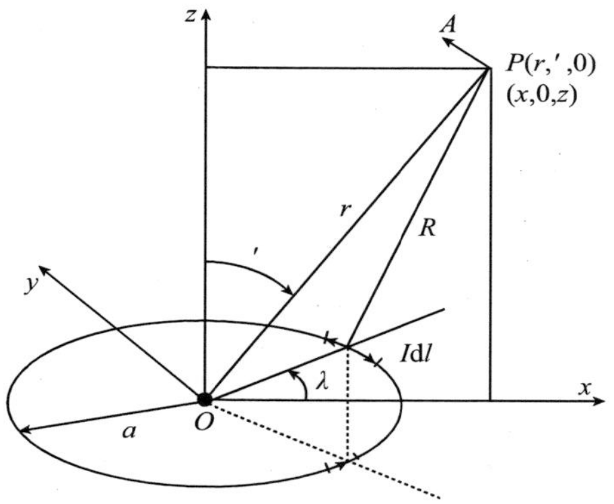

Figure 11.5 The spherical coordinate system of magnetic dipole

smaller than the wavelength, assuming the amplitude and phase of the current are constant in the ring, this small current ring can be then regarded as a magnetic dipole.

Put the ring at the origin of the spherical coordination system, the distance from the origin to the source point is r, and $r \gg a$, calculate the magnetic vector potential $\boldsymbol{A}$, which is expressed as

$$\boldsymbol{A} = \frac{\mu}{4\pi} \oint_l \frac{I\mathrm{d}\boldsymbol{l}}{R} \tag{11.61}$$

Because each current element $I\mathrm{d}\boldsymbol{l}$ in the ring only had $x\,y$ components. It can be seen from Figure 11.5., for any field point with R, there is always a couple of small current segments $I\mathrm{d}l$, which are symmetrical. In this way, the x component of $\boldsymbol{A}$ produced by the current element couple are canceled out, while the y components are superpositioned. Therefore, in Figure 11.5, $\boldsymbol{A}$ of the field point only has y components, so $\boldsymbol{A}$ in the spherical coordination system only has ϕ component, expressed as

$$\boldsymbol{A} = \boldsymbol{e}_\phi A_\phi = \boldsymbol{e}_\phi \frac{\mu I}{4\pi} \int_0^{2\pi} \frac{(a\mathrm{d}\phi)}{R} \cos\phi \tag{11.62}$$

where $a\mathrm{d}\phi = \mathrm{d}l$.

There is a phase lag in the magnetic vector potential of the time-varying magnetic dipole, therefore, the term $\mathrm{e}^{\mathrm{j}(\omega t - kR)}$ should be added, that is

$$\boldsymbol{A} = \boldsymbol{e}_\phi A_\phi = \boldsymbol{e}_\phi \frac{\mu I}{4\pi} \int_0^{2\pi} \frac{(a\mathrm{d}\phi)}{R} \cos\phi \cdot \mathrm{e}^{\mathrm{j}(\omega t - kR)} \tag{11.63}$$

$I\mathrm{e}^{\mathrm{j}\omega t}$ is the current of the magnetic dipole. Above equation is the magnetic vector potential $\boldsymbol{A}$ produced by all the currents elements in the ring. Expressing R in the spherical coordinate system, above equation can be written as following form:

$$\boldsymbol{A} = \boldsymbol{e}_\phi \frac{\mu I}{4\pi} \int_0^{2\pi} \frac{a}{R} \cos\phi \cdot \mathrm{e}^{\mathrm{j}(\omega t - kR)} \cdot \mathrm{e}^{\mathrm{j}k(r-R)} \cdot \mathrm{d}\phi \tag{11.64}$$

Because the size of the ring is much less than wavelength, so

$$k(r - R) \ll 1 \tag{11.65}$$

$\mathrm{e}^{\mathrm{j}k(r-R)}$ in Equation (11.64) can thus be written as

$$\mathrm{e}^{\mathrm{j}k(r-R)} = \cos[k(r-R) + \mathrm{j}\sin[k(r-R)]] \approx 1 + \mathrm{j}[k(r-R)] \tag{11.66}$$

Substituting Equation (11.66) into (11.64), yields

$$\boldsymbol{A} = \boldsymbol{e}_\phi \frac{\mu a I \mathrm{e}^{\mathrm{j}(\omega t - kr)}}{4\pi r} \int_0^{2\pi} \left[\frac{r}{R} + \mathrm{j}kr\left(\frac{r}{R} - 1\right)\right] \cos\phi \mathrm{d}\phi \tag{11.67}$$

From Equation (3.48), we can see

$$\frac{r}{R} \approx 1 + \frac{a}{r} \sin\theta \cos\phi \tag{11.68}$$

Then we have

$$\boldsymbol{A} = \boldsymbol{e}_\phi \frac{\mu(I\pi a^2)k}{4\pi r} \left(\frac{1}{kr} + \mathrm{j}\right) \sin\theta \mathrm{e}^{\mathrm{j}(\omega t - kr)} = \boldsymbol{e}_\phi \frac{\mu m k}{4\pi r} \left(\frac{1}{kr} + \mathrm{j}\right) \sin\theta \mathrm{e}^{\mathrm{j}(\omega t - kr)} \tag{11.69}$$

Rewriting this equation as the following:

$$\boldsymbol{A} = \frac{\mu \boldsymbol{m} \times \boldsymbol{e}_r}{4\pi r} k \left(\frac{1}{kr} + \mathrm{j}\right) \mathrm{e}^{\mathrm{j}(\omega t - kr)} \tag{11.70}$$

Where magnetic dipole moment is $\boldsymbol{m}$

$$\boldsymbol{m} = (I\pi a^2)\boldsymbol{e}_z \tag{11.71}$$

Obviously, the direction of m is vertical to the ring plane, and satisfies the right-hand relationship with the current direction.

Equations (11.69) and (11.70) are the expression of time-varying magnetic vector potential $\boldsymbol{A}$, however, it must be noted that, this expression of $\boldsymbol{A}$ is valid only if $r \gg a$.

11.3.2 The Electro-Magnetic Field of the Magnetic Dipole

Different method can be used to solve the electromagnetic field of the magnetic dipole. Firstly, the magnetic vector potential can be used, and then the duality principle can be used also.

Use the magnetic vector potential to solve the electromagnetic field of the magnetic dipole.

The electromagnetic field of the time-varying magnetic dipole, can be obtained from the lag magnetic vector potential $\boldsymbol{A}$, i.e., from Equation (11.69) and $\boldsymbol{B} = \nabla \times \boldsymbol{A}$.

$$\begin{aligned}\boldsymbol{B} = \nabla \times \boldsymbol{A} = &\boldsymbol{e}_r \frac{\mu m k^2}{2\pi r}\left(\frac{1}{k^2r^2} + \mathrm{j}\frac{1}{kr}\right)\cos\theta \cdot \mathrm{e}^{\mathrm{j}(\omega t - kr)} \\ &+ \boldsymbol{e}_\theta \frac{\mu m k^2}{4\pi r}\left(\frac{1}{k^2r^2} + \mathrm{j}\frac{1}{kr} - 1\right)\sin\theta \cdot \mathrm{e}^{\mathrm{j}(\omega t - kr)}\end{aligned} \tag{11.72}$$

When $kr \gg 1$, from Equation (11.72), we have

$$\boldsymbol{H} = \boldsymbol{e}_\theta \frac{-mk^2}{4\pi r}\sin\theta \mathrm{e}^{\mathrm{j}(\omega t - kr)} \tag{11.73}$$

This is the expression of the far field magnetic field strength of the time-varying magnetic dipole. In far field, only taking component H_θ, for the component H_r is very small. Now we have the expression of the magnetic field, the electric field can be solved conveniently by utilizing Maxwell's equations.

$$\boldsymbol{E} = \boldsymbol{e}_\phi E_\phi = \boldsymbol{e}_\phi \frac{\eta m k^2}{4\pi r}\sin\theta \mathrm{e}^{\mathrm{j}(\omega t - kr)} \tag{11.74}$$

Thus, we obtain the far field electric and magnetic field of time-varying magnetic dipole.

$$E_\phi = \frac{\eta m k^2}{4\pi r}\sin\theta \mathrm{e}^{\mathrm{j}(\omega t - kr)}$$

$$H_\theta = -\frac{mk^2}{4\pi r}\sin\theta \mathrm{e}^{\mathrm{j}(\omega t - kr)}$$

The direction of E_ϕ, H_θ and $\boldsymbol{S}$ are illustrated in Figure 11.6.

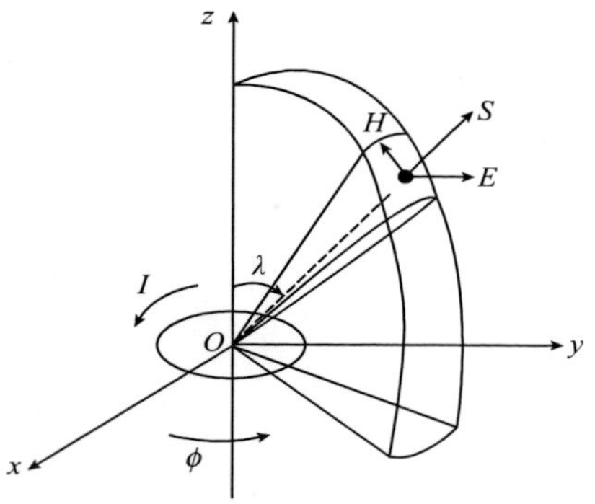

Figure 11.6 The electromagnetic field and the poynting vector of the magnetic dipole

The electric field, magnetic field of the magnetic dipole is quite similar with the electric dipole. (Here means the far field in case of $r \gg \lambda$). The difference is that $\boldsymbol{E}$ and $\boldsymbol{H}$ exchange their space position. Additionally, the polarities of the field are different, and this difference ensure that the Poynting vector is always along the positive r direction. The dualities of these two dipoles agree with the duality law of the field, which will be discussed in the next section.

11.3.3 The Radiation Power and Radiation Resistance of the Magnetic dipole

The average Poynting vector $\boldsymbol{S}_{\mathrm{av}}$ of the magnetic dipole is

$$\boldsymbol{S}_{\mathrm{av}} = \frac{1}{2}\mathrm{Re}(\boldsymbol{E} \times \boldsymbol{H}^*) = -\frac{1}{2}\mathrm{Re}(E_\phi H_\theta^*)\boldsymbol{e}_r$$

Substituting $\boldsymbol{H}, \boldsymbol{E}$ of Equations (11.73) and (11.74) into above equation, and consider the case of air medium, that is

$$\boldsymbol{S}_{\mathrm{av}} = \boldsymbol{e}_r \frac{\eta m^2 k^4}{32\pi^2 r^2}\sin^2\theta = \boldsymbol{e}_r 1.1937\frac{m^2k^4}{r^2}\sin^2\theta \tag{11.75}$$

The radiation power of the magnetic dipole is

$$P = \oint_{\mathrm{s}} \boldsymbol{S}_{\mathrm{av}} \cdot \mathrm{d}\boldsymbol{S} = 1.1937 m^2 k^4 \int_0^{2\pi}\int_0^{\pi} \frac{\sin^2\theta}{r^2}(r^2\sin\theta\mathrm{d}\theta\mathrm{d}\phi) = 10m^2k^4 \tag{11.76}$$

or

$$P = 10\left(\pi a^2 I\right)^2 k^4 \tag{11.77}$$

The radiation resistance R_{r} can be solved by following equation:
Because

$$P = \frac{1}{2}I^2 R_{\mathrm{r}} = I_{\mathrm{e}}^2 R_{\mathrm{r}} = 20(\pi a^2)^2 k^4 I_{\mathrm{e}}^2$$

so

$$R_{\mathrm{r}} = 20(\pi^2 a^4)\left(\frac{2\pi}{\lambda}\right)^4 = 320\pi^6\left(\frac{a}{\lambda}\right)^4 \tag{11.78}$$

Equation (11.78) is the radiation resistance of the magnetic dipole in air medium.

From Equation (11.78) and (11.59), we can see that the radiation resistance of the magnetic dipole is inversely proportional to the square four of the wavelength, but for electric dipole, it is inversely proportional to the square of the wavelength. Comparing these two cases, we can see that the frequency sensitivity of the radiation resistance of magnetic dipole is higher than that of the electric dipole.

Example 11.4 Making a line antenna and a ring antenna with a metal wire of 0.2m long, please find the radiation resistance of each antenna at frequency of 30MHz.

Solution We can regard the line antenna as an electric dipole and the ring antenna as a magnetic dipole.

The radiation resistance of the electric dipole is

$$R_{\mathrm{r(e)}} = 80\pi^2\left(\frac{l}{\lambda}\right)^2 = 80\pi^2\left(\frac{0.2}{10}\right)^2 = 0.316(\Omega) \tag{a}$$

The radiation resistance of the magnetic dipole is

$$R_{\mathrm{r(m)}} = 320\pi^6\left(\frac{a}{\lambda}\right)^4 = 320\pi^6\left(\frac{0.2}{2\pi}\times\frac{1}{10}\right)^4 = 0.316\times 10^{-4} \tag{b}$$

From Formula (a) and (b), we can see that the radiation resistance of the magnetic dipole is too small, so they are weak radiator of the electromagnetic power.

The ratio of these two resistances is

$$\frac{R_{r(e)}}{R_{r(m)}} = 4\left(\frac{\lambda}{l}\right)^2 \tag{c}$$

Because the size of dipole is much smaller than the wavelength, so the radiation resistance of the electric dipole is much bigger than that of the magnetic dipole. In this example, $R_{r(e)}/R_{r(m)} = 10^4$, if the wire length is still 0.2m, but the frequency changes from 30MHz to 3MHz, so the value of $R_{r(e)}/R_{r(m)}$ changes from 10^4 to 10^6. We can conclude, in the condition of same current amplitude, the radiation power of the electric dipole is much bigger than that of the magnetic dipole, and the lower frequency can cause bigger difference between them.

11.4 Dipole Antenna and the Concept of Antenna Array

11.4.1 Dipole Antenna

Because the electric length (dl/λ) of the electric dipole is very small, therefore, its radiation ability is weak and directivity is not strong. For improving the radiation and direction, the length of the electric dipole should be increased to become a practical symmetrical dipole antenna.

1. The Current Distribution of the Dipole Antenna

The dipole antenna can be regarded to transform from a two-wire transmission line terminated with open-circuit, as shown in Figure 11.7(a).

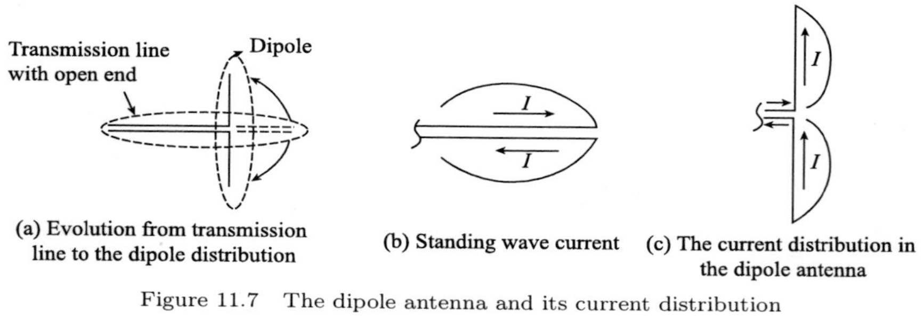

Figure 11.7 The dipole antenna and its current distribution

It is difficult to solve the exact solution of the current distribution in the dipole antenna. Because for a symmetry oscillator with certain thickness and length, its boundary shape does not match well with the spherical coordinates, besides, the coupled oscillators is not ideal conductor, therefore, it is very difficult to find the rigorous solutions.

In engineering application, some approximations are usually assumed: it is assumed that the symmetry oscillator and the uniform transmission line have similar current distribution, saying, sinusoidal standing wave distribution, as shown in Figure 11.7(b) and (c). Clearly, this is an approximation. Because the symmetrical dipole is not a uniform transmission line, and actual oscillator is not ideal conductor. However, the significance is that the dipole radius a is much smaller than the wavelength (that is, a/λ is very small), and the metal conductivity σ of the dipole is very high (for purple copper, $\sigma = 5.8 \times 10^7 \mathrm{S/m}$). Thus, the above approximation to the current distribution can help to achieve enough accuracy in engineering.

2. The Far-Field

A central feeding dipole with length l is shown in Figure 11.8. Its current expression is given:

$$I(z') = I_0 \sin\left[k\left(\frac{l}{2} - |z'|\right)\right] \tag{11.79}$$

According to that current distribution above, the space far-field electric field strength E_θ can be solved.

Method 1 Using Equation (11.44) directly. $I(z')\mathrm{d}z'$ is used to replace $I\mathrm{d}l$, and consider the phase difference caused by the distance difference., then take integral with respect to $\mathrm{d}z'$ from $-\frac{l}{2}$ to $\frac{l}{2}$, thus far-field electric field E_θ of the dipole is obtained.

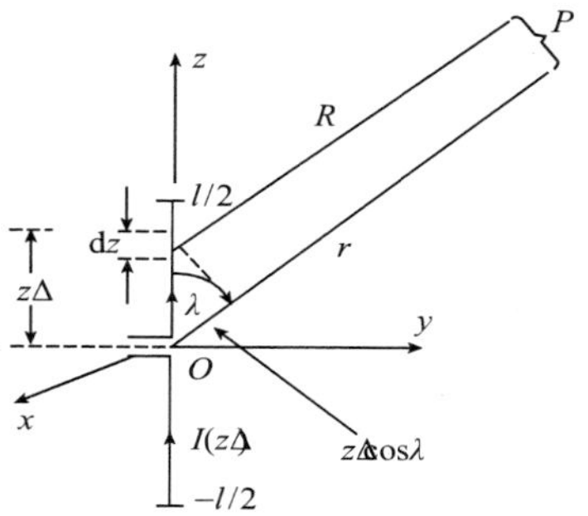

Figure 11.8 The auxiliary graph for analysis of the dipole antenna

Take a small segment $\mathrm{d}z'(\mathrm{d}z' \ll \lambda)$ in the dipole, which is equivalent to an electric dipole with a current $I(z')$. According to Equation (11.44) of the far-field for the electric dipole, we can obtain the far-field $\mathrm{d}E_\theta$ produced by $I(z')\mathrm{d}z'$

$$\mathrm{d}E_0 = \mathrm{j}\eta_0 \frac{I(z')\mathrm{d}z'}{2\lambda R} \sin\theta \mathrm{e}^{-\mathrm{j}kR}$$

Thus, the total E_θ is

$$E_\theta = \int \mathrm{d}E_\theta = \mathrm{j}\eta_0 \int_{-l/2}^{l/2} \frac{I(z')\mathrm{d}z'}{2\lambda R} \sin\theta \mathrm{e}^{-\mathrm{j}kR} \tag{11.80}$$

For the far-field, as shown in Figure 11.8, R and r are nearly parallel, thus

$$R \approx r - z' \cos\theta$$

In the denominator of Equation (11.80), making $R \approx r$ and ignoring the impact of $z' \cos\theta$ on the amplitude of the field. *However, for the phase issues, because the phase shift caused by distance difference is* $kz' \cos\theta = \frac{2\pi}{\lambda} z' \cos\theta$, *and the total length* l *of the dipole is comparable to the wavelength, this impact can thus not be ignored, but should be kept in the phase factor.* Take the expression Equation (11.79) into (11.80), and we have

$$E_\theta = \mathrm{j}\frac{60\pi I_0}{\lambda r} \sin\theta \mathrm{e}^{-\mathrm{j}kr} \int_{-l/2}^{l/2} \sin k\left(\frac{l}{2} - |z'|\right) \mathrm{e}^{\mathrm{j}kz' \cos\theta} \mathrm{d}z'$$

Let the exponential form in the integral equation be written as the trigonometric form,

$$\mathrm{e}^{\mathrm{j}kz' \cos\theta} = \cos(kz' \cos\theta) + \mathrm{j} \sin(kz' \cos\theta)$$

For its imaginary part $\sin(kz' \cos\theta)$ is the odd function of z', and the integral is thus zero. Then, let the remaining part $\sin\left(k\frac{l}{2} - k|z'|\right) \cdot \cos(kz' \cos\theta)$ be a trigonometric form. After integral,

$$E_\theta = \frac{\mathrm{j}60I_0}{r}\mathrm{e}^{-\mathrm{j}kr}\frac{\cos\left(k\frac{l}{2}\cdot\cos\theta\right)-\cos\left(k\frac{l}{2}\right)}{\sin\theta} \tag{11.81}$$

This formula is the far-field electric field of the symmetrical oscillator. The far-field magnetic field is

$$H_\phi = E_\theta/\sqrt{\mu_0/\varepsilon_0} \tag{11.82}$$

Method 2 Solve the far-field magnetic vector potential $\boldsymbol{A}$ according to the current $I(z')$, then solve the far-field electric field E_θ.

Because there is only z component current, the magnetic vector potential at point P is $\boldsymbol{A} = \boldsymbol{e}_z A_z$, and

$$A_z = \frac{1}{4\pi}\int_{-l/2}^{l/2}\frac{\mu I(z')\mathrm{e}^{-\mathrm{j}kR}}{R}\mathrm{d}z' \tag{11.83}$$

For the situation $r \gg l$, we can see from Figure 11.8

$$R \approx r - z'\cos\theta$$

A_z can thus be expressed as

$$\begin{aligned} A_z &= \frac{1}{4\pi}\int_{-l/2}^{l/2}\frac{\mu I(z')\mathrm{e}^{-\mathrm{j}k(r-z'\cos\theta)}}{r-z'\cos\theta}\mathrm{d}z' \\ &\approx \frac{\mu\mathrm{e}^{-\mathrm{j}kr}}{4\pi r}\int_{-l/2}^{l/2} I(z')\mathrm{e}^{\mathrm{j}kz'\cos\theta}\mathrm{d}z' \end{aligned} \tag{11.84}$$

When dealing with this equation, the term $r - z'\cos\theta$ is kept in the phase factor, but in the amplitude term, the $z'\cos\theta$ in $r - z'\cos\theta$ is deleted.

Applying Equation (11.84) and following equations:

$$\boldsymbol{E} = -\mathrm{j}\omega\boldsymbol{A} + \frac{1}{\mathrm{j}\omega\varepsilon\mu}\nabla(\nabla\cdot\boldsymbol{A}) \tag{11.85}$$

$$\boldsymbol{B} = \nabla\times\boldsymbol{A} \tag{11.86}$$

We can obtain the radiation field. For example, substituting A_z in Equation (11.84) into (11.85), we can obtain the far-field [only term $(1/r)$ is remained]

$$\begin{aligned} E_\theta &= -E_z\sin\theta = \mathrm{j}\omega A_z\sin\theta \\ H_\phi &= E_\theta/\sqrt{\mu/\varepsilon} \end{aligned} \tag{11.87}$$

Substituting current expression $I(z')$ into Equation (11.84), then yields

$$\begin{aligned} A_z &= \frac{\mu\mathrm{e}^{-\mathrm{j}kr}}{4\pi r}\int_{-l/2}^{l/2} I_0\sin\left[k(l/2-|z'|)\right]\mathrm{e}^{\mathrm{j}kz'\cos\theta}\cdot\mathrm{d}z' \\ &= \frac{\mu I_0\mathrm{e}^{-\mathrm{j}kr}}{4\pi r}\cdot\frac{2\left[\cos\left(k\frac{l}{2}\cos\theta\right)-\cos\left(k\frac{l}{2}\right)\right]}{k\sin^2\theta} \end{aligned} \tag{11.88}$$

And substituting A_z into Equation (11.87), the far-field is

$$E_\theta = \frac{\mathrm{j}60I_0\mathrm{e}^{-\mathrm{j}kr}}{r}\left[\frac{\cos\left(k\frac{l}{2}\cos\theta\right)-\cos\left(k\frac{l}{2}\right)}{\sin\theta}\right] \tag{11.89}$$

The magnetic field is

$$H_\phi = E_\theta/\sqrt{\mu_0/\varepsilon_0}$$

3. Direction Factor and Radiation Pattern

From the far-field electro-magnetic field Equation (11.81), it is known that, for same radius r, the electromagnetic field changes only with θ but not with ϕ, the directivity factor $F(\theta)$ is

$$F(\theta) = \left[\frac{\cos\left(k\frac{l}{2}\cos\theta\right) - \cos\left(k\frac{l}{2}\right)}{\sin\theta}\right] \tag{11.90}$$

For half-wave dipole antenna (total length $l = \lambda/2$), the directivity factor is

$$F(\theta) = \frac{\cos(90° \cdot \cos\theta)}{\sin\theta} \tag{11.91}$$

Applying Equation(11.90), we can obtain the radiation pattern as Figure 11.9. As shown in Figure 11.9(a), when the dipole total length $l \leqslant \lambda$, the directivity increases and only has main lobe but no side lobe when l/λ increase. In addition, the direction of the main lobe is always vertical to the axis of the dipole.

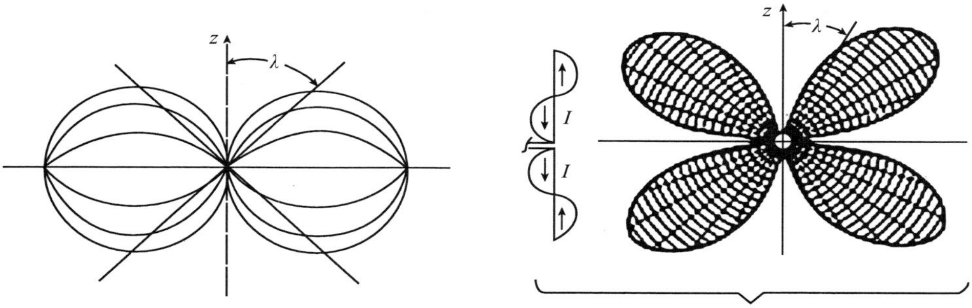

(a) The radiation pattern of the electric dipole, half-wave dipole and full-wave dipole

(b) The current distribution and radiation pattern of the dipole antenna when $l = 2\lambda$

Figure 11.9 The radiation pattern of the dipole antenna

When total length $l > \lambda$, there is not only $+z$ direction current, but also $-z$ direction current. Thus, there are more waves that have different phases and amplitudes in space, they interfere with each other. As a result, there is not only main lobe but also side lobe in the radiation pattern. If l/λ increases, the phenomenon of the transformation of main lobe and side lobe will occur. When $l = 2\lambda$, the current distributions in the two arms of the dipole antenna are symmetrical. Thus, four same wave beams will appear, as shown in Figure 11.9(b).

4. Radiation Power and Radiation Resistance

1) The average power density S_{av} of the dipole

From the radiation expression Equations (11.81) and (11.82), we can obtain

$$\begin{aligned}\boldsymbol{S}_{\text{av}} &= \frac{1}{2}\text{Re}\,(\boldsymbol{E}\times\boldsymbol{H}^*) = \frac{1}{2}\boldsymbol{e}_r\frac{1}{2}\text{Re}E_\theta H_\phi^* \\ &= \boldsymbol{e}_r\frac{30I_{\text{e}}^2}{\pi r^2}\frac{\left[\cos\left(\frac{kl}{2}\cos\theta\right) - \cos\frac{kl}{2}\right]^2}{\sin^2\theta}\end{aligned} \tag{11.92}$$

2) The radiation power P

By calculating the curved surface integration of $\boldsymbol{S}_{\text{av}}$ on dipole-centered spherical surface, radiation power will be

$$\begin{aligned} P &= \oint_{s} \boldsymbol{S}_{\text{av}} \cdot \mathrm{d}s = \oint_{s} S_{\text{av}} \mathrm{d}s = \int_0^{\pi} \int_0^{2\pi} S_{\text{av}} \left(r^2 \sin\theta \mathrm{d}\theta \mathrm{d}\phi\right) \\ &= 60 I_{\text{e}}^2 \int_0^{\pi} \frac{\left[\cos\left(\frac{kl}{2}\cos\theta\right) - \cos\frac{kl}{2}\right]^2}{\sin\theta} \mathrm{d}\theta \end{aligned} \tag{11.93}$$

3) The radiation resistance R_{r}

The relationship between radiation power P and radiation resistance R_{r} is

$$P = \frac{1}{2} I_0^2 R_{\text{r}} = I_{\text{e}}^2 R_{\text{r}}$$

Where I_{e} is the effective value of the current, and I_0 is the amplitude of the current. According to above two formulas, we can obtain the radiation resistance of dipole antenna with any length

$$R_{\text{r}} = 60 \int_0^{\pi} \frac{\left[\cos\left(\frac{kl}{2}\cos\theta\right) - \cos\frac{kl}{2}\right]^2}{\sin\theta} \mathrm{d}\theta^{①} \tag{11.94}$$

According to Equation (11.59) and above equation, the radiation resistance of electric dipole, full and half-wave dipole antenna can be found as following table:

Antenna type	Electric dipole (for $\mathrm{d}l = \lambda/100$)	Half-wave dipole	Full-wave dipole
Radiation resistance	0.079 Ω	73.1 Ω	199 Ω

11.4.2 Antenna Array

The system formed by many antenna elements and excited in some way are called antenna array.

From the discussion above, we may see that, the antenna directivity increase with the increases of the total length l of the dipole, but this increasing is restricted. When l/λ, the side lobe starts to appear in the radiation pattern, when $l/\lambda = 1.27$, the directivity reaches maximal value, but the directivity becomes worse with further length increase. So we need to find new method to increase antenna directivity, and antenna array is one of those effective methods.

In this section, we take a binary array of wire antenna (as dipole) as an example to illustrate the directivity enhancement by the antenna array.

1. Enhance Directivity by Adding Antenna Elements

Equation (11.90) gives the directivity factor $F(\theta)$ of single dipole with total length l

$$F(\theta) = \frac{\cos\left(k\frac{l}{2}\cos\theta\right) - \cos\left(k\frac{l}{2}\right)}{\sin\theta}$$

By computing integral to expression ①, we can get the expression below:

$R_{\text{r}} = 30\{2[r + \ln(kl)] - \text{Ci}(kl)] + \sin(kl)[\text{Si}(2kl) - 2\text{Si}(kl)] + \cos(kl)[r + \ln(kl/2) + \text{Ci}(2kl) - 2\text{Ci}(kl)]\}$

In addition, the expression $\text{Ci}(x) = \int_{\infty}^{x} (\cos u/u)\mathrm{d}u$ represents cosine integral, and $\text{Si}(x) = \int_0^{x} (\sin u/u)\mathrm{d}u$ the sine integral. The constant r=0.5772 is known as Euler-Mascheroni constant.

Now analyzing the binary array as shown in Figure 11.10, which consists of two same dipole elements (with length l), both are vertical to the paper plane, and the distance between the two elements is d.

When only feeding one of the elements with power P_0, the current amplitude in the antenna is I_0', the radiation electric field in far-field at point M is E'. When the same input power is fed and equally divided by the two elements of the array, the current is $I_{0(1)} = I_{0(2)} = I_0 (I_{0(1)}\ I_{0(2)}$ have same amplitudes and phases). The electric fields at point M produced by them are $E_1\,E_2$. For the distance between the point M and the two elements of the antenna is the same, thus, $E = E_1 + E_2 = 2E_1$.

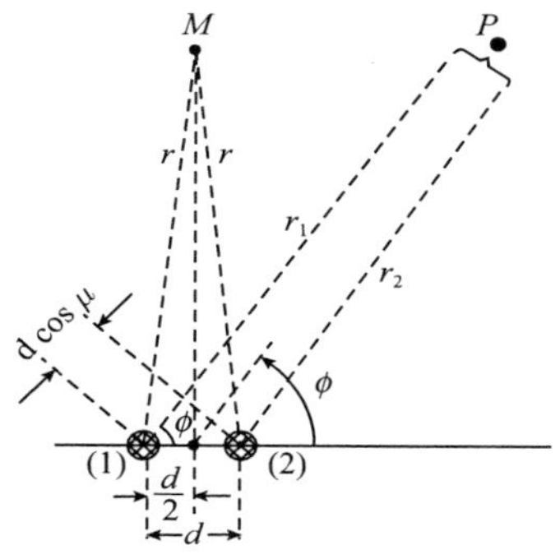

Figure 11.10 The binary array

Now, let's see $E/E' =$? Because the two elements are all the same, their electric parameters are all the same too. Thus, in the condition of same total input power, we have $I_0 = I_0'/\sqrt{2}$. In addition, because the electric field of the dipole is proportional with the current, we have

$$\frac{E}{E'} = \frac{2E_1}{E'} = \frac{2I_0}{I_0'} = \sqrt{2} \tag{11.95}$$

Conclusion: if the input power for a single antenna is equal to the input power of a binary antenna array, the far-zone electric field of the binary array is $\sqrt{2}$ times of that for the single antenna. The directivity of antenna is enhanced.

2. Directivity Function $F(\theta, \phi)$ and Radiation Pattern of Binary Array

1) Directivity function $F(\phi)$ in case of $\theta = 90°$

As shown in Figure 11.10 is the binary array. $I_{0(1)}$ is the current of dipole (1), $I_{0(2)}$ is the current of dipole (2),

$$I_{0(2)} = m\mathrm{e}^{\mathrm{j}\alpha_0} \cdot I_{0(1)}$$

Where m is constant. α_0 is the phase difference of the antenna current. Now we will solve the total field E in far-field point P produced by the binary array. For the P point in far-zone, r_1 and r_2 are parallel, therefore, when dealing with amplitude of the field, we can ignore the distance difference of r_1 and r_2, take $r_1 = r_2$ approximately; however, when dealing with the phase, we must consider the distance difference $d\cos\phi$.

Considering these factors, we can obtain the far-field in case of $\theta = 90°$,

$$E_1 = \mathrm{j}\frac{60I_{0(1)}}{r_1}\left(1 - \cos k\frac{l}{2}\right)\mathrm{e}^{-\mathrm{j}kr_1}$$

$$E_2 = \mathrm{j}\frac{60I_{0(1)}}{r_1}m\left(1 - \cos k\frac{l}{2}\right)\mathrm{e}^{-\mathrm{j}kr_1} \cdot \mathrm{e}^{\mathrm{j}(kd\cos\phi + \alpha_0)}$$

The total E at point P is

$$\begin{aligned} E &= E_1 + E_2 \\ &= \mathrm{j}\frac{60I_{0(1)}}{r_1}\mathrm{e}^{-\mathrm{j}kr_1} \cdot \left\{\left(1 - \cos k\frac{l}{2}\right)\left[1 + m\mathrm{e}^{\mathrm{j}(kd\cos\phi+\alpha_0)}\right]\right\} \end{aligned} \tag{11.96}$$

Therefore the directivity function $F(\phi)$ in case of $\theta = 90°$ is

$$F(\phi) = \left(1 - \cos k\frac{l}{2}\right) \cdot \left[1 + m\mathrm{e}^{\mathrm{j}(kd\cos\phi+\alpha_0)}\right] \quad (\text{when } \theta = 90°) \tag{11.97}$$

Module of $F(\phi)$ is

$$|F(\phi)| = \left(1 - \cos k\frac{l}{2}\right) \cdot \sqrt{1 + m^2 + 2m\cos(kd\cos\phi + \alpha)} = F_{\mathrm{e}}(\phi) \cdot f_{\mathrm{a}}(\phi) \qquad (11.98\ )$$

In this equation, the first term $F_{\mathrm{e}}(\phi) = \left(1 - \cos k\frac{l}{2}\right)$ is called element factor, which is a directivity factor of single element, and can be obtained from Equation (11.90). The second term is called array factor, which is decided by those relative parameters: m, α_0, d. Thus, the directivity function $F(\phi)$ can be obtained by the product of the two terms.

Conclusion: the directivity function of the antenna is equal to the product of element factor and the array factor.

In case of $\theta = 90°$, the element factor is

$$f_(\phi) = 1 - \cos k\frac{l}{2} \quad (\text{when } \theta = 90°) \qquad (11.99)$$

And the array factor is

$$f_(\phi) = \sqrt{1 + m^2 + 2m\cos(kd\cos\phi + \alpha_0)} \qquad (11.100)$$

If $m = 1$, $\alpha_0 = 1$, the array factor $f_{\mathrm{a}}(\phi)$ in Equation (11.100) can be expressed as

$$f_{\mathrm{a}}(\phi) = 2\cos\left(\frac{kd}{2}\cos\phi\right) \quad (m = 1, \alpha_0 = 0) \qquad (11.101)$$

So for case of $\theta = 90°$, and $m = 1$, $\alpha_0 = 1$, the directivity function of the binary array is

$$F(\phi) = \left(1 - \cos k\frac{l}{2}\right) \times 2\cos\left(\frac{kd}{2}\cos\phi\right) \quad (m = 1, \alpha_0 = 0, \theta = 90°) \qquad (11.102)$$

As shown in Equation (11.101), in case of $\phi = \pm 90°$, for any d, the array factor has the maximal value, that is $f_a(\phi = \pm 90°) = 2$. Thus, the directivity function $F(\theta = 90°, \phi = \pm 90°)$ is two times of element factor $f_{\mathrm{e}}(\phi)$. That is

$$\begin{aligned} F(\theta, \phi) &= 2f_{\mathrm{e}}(\phi) \\ &= 2\left(1 - \cos k\frac{l}{2}\right) \quad (m = 1, \alpha_0 = 0, \theta = 90°, \phi = \pm 90°) \end{aligned} \qquad (11.103)$$

So at point M, the electric field amplitude of binary array is two times of that of the single antenna, and the directional radiation energy is enhanced significantly.

For binary array, when $m = 1, \alpha_0 = 0, \theta = 90°, \phi = \pm 90°$, and $L = \lambda$ (full-wave dipole), we can know from Equation (11.102) that the directivity function $F(\theta, \phi)$ is equal to 4, significant directivity improvement is achieved.

2) The radiation pattern in the equatorial plane of binary array

If using dipole as antenna element, and the current in the two elements has the same amplitudes and phases, that is $l = \lambda/2, m = 1, \alpha_0 = 0°$, set the distance between elements be $d = \lambda/2$, from Equation (11.102), we know that the directivity function of the binary array in the equatorial plane is

$$\begin{aligned} F(\theta = 90°, \phi) &= 2\cos\left(\frac{kd}{2}\cos\phi\right) \\ &= 2\cos(90° \cos\phi) \end{aligned} \qquad (11.104)$$

Applying above equation, we can obtain the radiation pattern of the binary array in equatorial plane, as shown in Figure 11.11(a).

3) The radiation pattern in the meridian plane

According to the similar method, we can obtain the directivity function of the binary array in the meridian plane

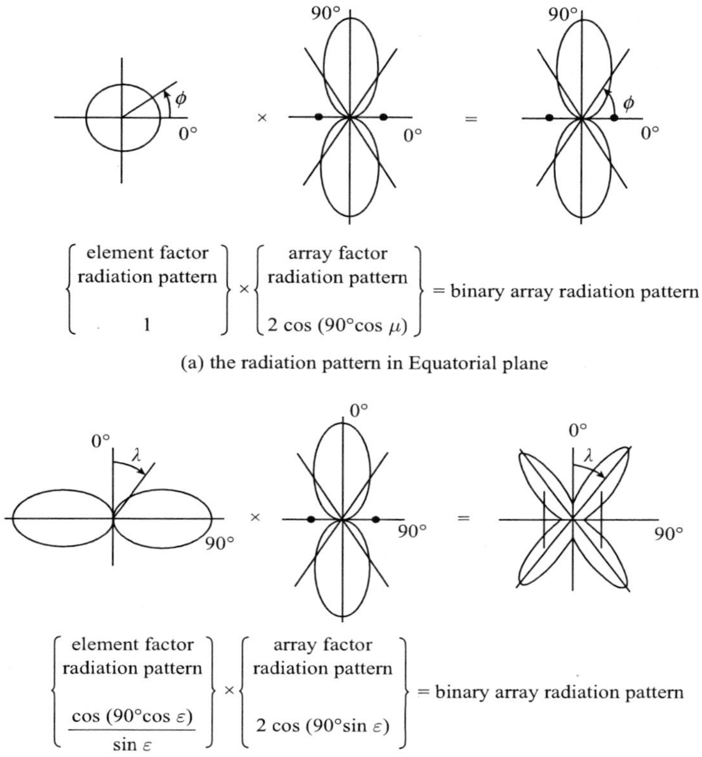

(a) the radiation pattern in Equatorial plane

(b) The radiation pattern in Meridian plane

Figure 11.11

$$F(\theta) = \frac{\cos\left(k\dfrac{l}{2}\cos\theta\right) - \cos\left(k\dfrac{l}{2}\right)}{\sin\theta} \times 2\cos\left(k\frac{\mathrm{d}}{2}\sin\theta\right) \tag{11.105}$$

When $l = \lambda/2$, $m = 1$, $\alpha_0 = 0$, $d = \lambda/2$, $F(\theta)$ changes to

$$F(\theta) = \frac{\cos(90^\circ \cdot \cos\theta)}{\sin\theta} \times 2\cos(90^\circ \cdot \sin\theta) \tag{11.106}$$

Applying above equation, we can obtain the radiation pattern of the binary array in meridian plane, as shown in Figure 11.11(b).

4) N element array

We can use more antennas to build up antenna array. Due to the increasing of the elements, when the electromagnetic waves arrive at a certain point in space, more different distance differences occur, and then cause more complex wave interference, and lead to the change of the radiation pattern. Moreover, we can change the distance between two elements by selecting different types of antennas, or change the amplitude and phase difference in the antennas to change the radiation pattern. No further discussion will be made here. Figure 11.12 shows the radiation pattern in equatorial plane of an eight-element array.

Eight pair antennas have the same current amplitudes and phases, all vertical to the paper plane, and the distance between two elements is $d = \lambda/2$. Obviously, when $\phi = 90^\circ$, the directivity in equatorial plane of the eight-element antenna array has been enhanced heavily.

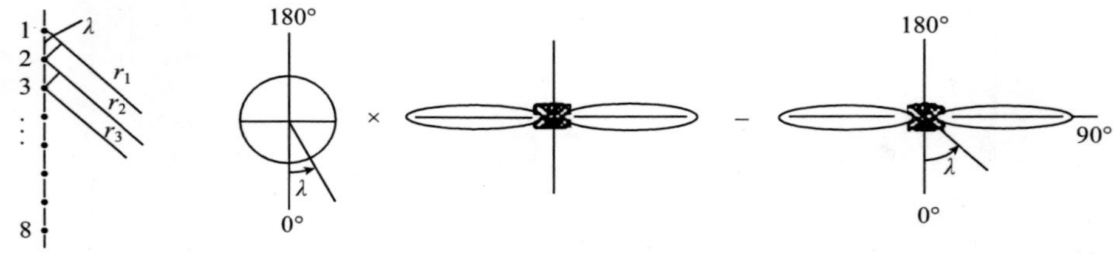

(a) Eight-element array (b) Element factor radiation pattern (c) Array factor radiation pattern (d) Antenna array radiation pattern

Figure 11.12 The radiation pattern of the eight element array in Equatorial plane

11.5 Duality Theory

11.5.1 Duality Law

There exist electric currents and charges in nature, but no magnetic currents and charges; thus, the Maxwell's equations are not symmetrical.

If we introduce magnetic currents and charges factitiously, the Maxwell's equations thus become symmetrical:

$$\nabla \times \boldsymbol{H} = \boldsymbol{J} + \mathrm{j}\omega \boldsymbol{D} \tag{11.107a}$$

$$\nabla \times \boldsymbol{E} = -\boldsymbol{J}_\mathrm{m} - \mathrm{j}\omega \boldsymbol{B} \tag{11.107b}$$

$$\nabla \cdot \boldsymbol{B} = \rho_\mathrm{m} \tag{11.107c}$$

$$\nabla \cdot \boldsymbol{D} = \rho \tag{11.107d}$$

where ρ_m is the magnetic charge density, $\boldsymbol{J}_m$ is the magnetic current density.

According to the superposition law of the electromagnetic in linear medium, the total $\boldsymbol{E}$ and $\boldsymbol{H}$ produced by electric currents/charges and magnetic currents/charges are the superposition of $\boldsymbol{E}_\mathrm{e}$, $\boldsymbol{H}_\mathrm{e}$ and $\boldsymbol{E}_\mathrm{m}$, $\boldsymbol{H}_\mathrm{m}$, produced by separate electric and magnetic currents and charges. That

$$\begin{aligned} \boldsymbol{E} &= \boldsymbol{E}_\mathrm{e} + \boldsymbol{E}_\mathrm{m} \\ \boldsymbol{H} &= \boldsymbol{H}_\mathrm{e} + \boldsymbol{H}_\mathrm{m} \end{aligned} \tag{11.108}$$

When $q_\mathrm{m} = 0$, $\boldsymbol{J}_\mathrm{m} = 0$ and $q_\mathrm{e} \neq 0$, $\boldsymbol{J}_\mathrm{e} \neq 0$, space field is $\boldsymbol{E}_\mathrm{e}$, $\boldsymbol{H}_\mathrm{e}$, Equation (11.107) change to

$$\begin{aligned} &\nabla \times \boldsymbol{H}_\mathrm{e} = \boldsymbol{J} + \mathrm{j}\omega\varepsilon \boldsymbol{E}_\mathrm{e} \\ &\nabla \times \boldsymbol{E}_\mathrm{e} = -\mathrm{j}\omega\mu \boldsymbol{H}_\mathrm{e} \\ &\nabla \cdot \boldsymbol{B}_\mathrm{e} = 0 \\ &\nabla \cdot \boldsymbol{D}_\mathrm{e} = \rho \end{aligned} \tag{11.109}$$

When $q_\mathrm{e} = 0, \boldsymbol{J}_\mathrm{e} = 0$ and $q_\mathrm{m} \neq 0$, $\boldsymbol{J}_\mathrm{m} \neq 0$, space field is $\boldsymbol{E}_\mathrm{m}$, $\boldsymbol{H}_\mathrm{m}$, Equation (11.107) changes to

$$\begin{aligned} &\nabla \times \boldsymbol{H}_\mathrm{m} = \mathrm{j}\omega\varepsilon \boldsymbol{E}_\mathrm{m} \\ &\nabla \times \boldsymbol{E}_\mathrm{m} = -\boldsymbol{J}_\mathrm{m} - \mathrm{j}\omega\mu \boldsymbol{H}_\mathrm{m} \\ &\nabla \cdot \boldsymbol{B}_\mathrm{m} = \rho_\mathrm{m} \\ &\nabla \cdot \boldsymbol{D}_\mathrm{m} = 0 \end{aligned} \tag{11.110}$$

The fields $\boldsymbol{E}_\mathrm{e}, \boldsymbol{H}_\mathrm{e}$ produced by electric currents and charges can be solved by using magnetic vector potential, many typical problems have been solved with this method. Similar method can be used to solve $\boldsymbol{E}_{\rm m}, \boldsymbol{H}_{\rm m}$ produced by magnetic currents and charges, however, they are not so easy. Anyway, if considering the duality of Equations (11.109) and (11.110), we can obtain the solution of Equation (11.110) through Equation (11.109) directly, and make the process much easier.

Duality theory: if the equations which describe different phenomenon have the similar mathematical forms, their solutions will have same form too. Equations (11.109) and (11.110) are dual equations. The parameters located at the same important positions in dual equations are called dual parameters. Such as $\boldsymbol{E}_{\rm e} \rightarrow \boldsymbol{H}_{\rm m}$, $\boldsymbol{H}_{\rm e} \rightarrow -\boldsymbol{E}_{\rm m}$, $\boldsymbol{J} \rightarrow \boldsymbol{J}_{\rm m}$, $\rho \rightarrow \rho_{\rm m}$, $\varepsilon \rightarrow \mu$, $\mu \rightarrow \varepsilon$. We can use dual relationship to substitute the dual parameters, and obtain the solution of Equation (11.110) directly through Equation (11.109).

According to duality law, we can get the boundary condition of magnetic currents and charges through the boundary condition of electric currents and charges. They are dual also. The boundary condition of electric currents and charges are

$$\begin{aligned} \boldsymbol{n} \times (\boldsymbol{E}_{2\rm e} - \boldsymbol{E}_{1\rm e}) &= 0 \\ \boldsymbol{n} \times (\boldsymbol{H}_{2\rm e} - \boldsymbol{H}_{1\rm e}) &= \boldsymbol{J}_{\rm s} \\ \boldsymbol{n} \cdot (\boldsymbol{D}_{2\rm e} - \boldsymbol{D}_{1\rm e}) &= \rho_{\rm s} \\ \boldsymbol{n} \cdot (\boldsymbol{B}_{2\rm e} - \boldsymbol{B}_{1\rm e}) &= 0 \end{aligned} \tag{11.111}$$

So the boundary condition of magnetic currents and charges are

$$\begin{aligned} \boldsymbol{n} \times (\boldsymbol{H}_{2\rm m} - \boldsymbol{H}_{1\rm m}) &= 0 \\ \boldsymbol{n} \times (\boldsymbol{E}_{2\rm m} - \boldsymbol{E}_{1\rm m}) &= -\boldsymbol{J}_{\rm ms} \\ \boldsymbol{n} \cdot (\boldsymbol{B}_{2\rm m} - \boldsymbol{B}_{1\rm m}) &= \rho_{\rm ms} \\ \boldsymbol{n} \cdot (\boldsymbol{D}_{2\rm m} - \boldsymbol{D}_{1\rm m}) &= 0 \end{aligned} \tag{11.112}$$

11.5.2 The Application of the Duality Theory

1. Solving the Field Produced by Magnetic Field

According to the duality theory and replacing the dual parameter, we can obtain the field of the magnetic dipole with a current $I_{\rm m}$ and length l

$$\begin{aligned} H_r &= \frac{I_{\rm m}{\rm d}l}{4\pi}\frac{2}{\omega\mu}\cos\theta\left(\frac{k}{r^2} - \frac{\rm j}{r^3}\right){\rm e}^{-{\rm j}kr} \\ H_\theta &= \frac{I_{\rm m}{\rm d}l}{4\pi}\frac{1}{\omega\mu}\sin\theta\left({\rm j}\frac{k^2}{r} + \frac{k}{r^2} - {\rm j}\frac{1}{r^3}\right){\rm e}^{-{\rm j}kr} \\ H_\phi &= 0 \\ E_r &= E_\theta = 0 \\ E_\phi &= -\frac{I_{\rm m}{\rm d}l}{4\pi}\sin\theta\left(\frac{{\rm j}k}{r} + \frac{1}{r^2}\right){\rm e}^{-{\rm j}kr} \end{aligned} \tag{11.113}$$

Similarly, we can get the far-field of the magnetic dipole from the far-field of the electric dipole [refer to Equations (11.43) and (11.44)]

$$\begin{aligned} H_\theta &= {\rm j}\frac{1}{\eta_0}\frac{I_{\rm m}{\rm d}l}{2\lambda r}\sin\theta{\rm e}^{-{\rm j}kr} \\ E_\phi &= -{\rm j}\frac{I_{\rm m}{\rm d}l}{2\lambda r}\sin\theta{\rm e}^{-{\rm j}kr} \end{aligned} \tag{11.114}$$

If we not only consider the dual relationship but also the substitution relationship $I_{\rm m}{\rm d}l = {\rm j}\omega\mu(I\pi a^2) = {\rm j}\omega\mu m$, we can change above equation to the totally similar forms of Equations (11.73) and (11.74).

2. Electromagnetic Field of Basic Slot Dipole

Basic slot dipole is a narrow slot on a ideal conductor plane with infinite size and zero thickness, which has length l and width d, $l \ll \lambda$ and $d \ll l$, as shown in Figure 11.13.

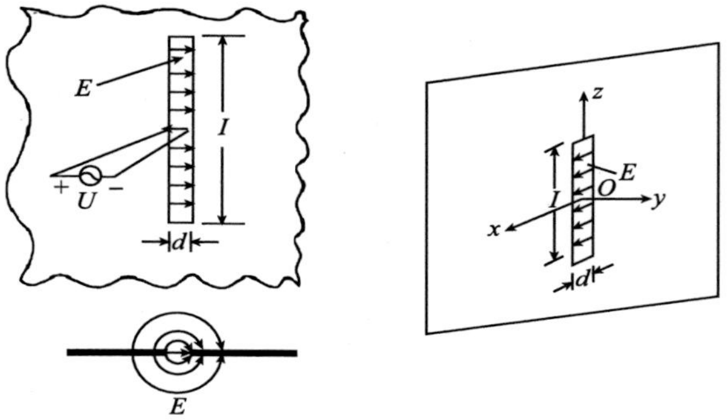

(a) Baic slot **(b) Basic slot in the cartesian coordinate**

Figure 11.13 The graph of the basic slot

Assuming adding a high frequency electric source on the slot, the electric field in the slot will be excited, as shown in Figure 11.13(b). Due to $d \ll l$, if we ignore the margin effect on the slot terminal, we can know that the electric field distribution in the slot is uniform along x direction. The electric field near the slot is approximately stable, which has similar electric field distribution as electrostatic field.

The center of the slot is set to be the origin of the Cartesian coordination system, and the axis of the slot is along the z axis. The metal plane is in plane xOz. According to the boundary condition of the ideal conductor, in plane xOz, the tangential component of electric field should be zero outside the slot zone. Only in the slot, the tangential component of the electric field E_x is non-zero

$$E_x = U/d$$

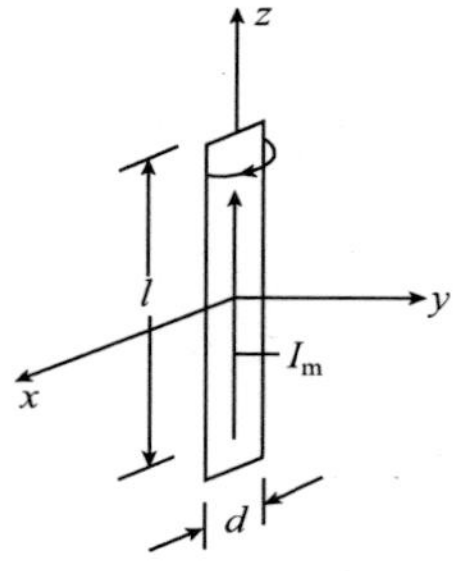

Figure 11.14 Magnetic current element

We will solve the excited electromagnetic field of slot dipole in the right half space ($y > 0$). The boundary condition of magnetic current is $\boldsymbol{n} \times (\boldsymbol{E}_{2\mathrm{m}} - \boldsymbol{E}_{1\mathrm{m}}) = -\boldsymbol{J}_{\mathrm{ms}}$, in the right half space slot, $\boldsymbol{n} = \boldsymbol{e}_y$, $\boldsymbol{E}_{2\mathrm{m}} = \boldsymbol{e}_x \boldsymbol{E}_x$, set $\boldsymbol{E}_{1\mathrm{m}} = 0$, according to equivalence principle, the space field is produced by the slot field. Then in the slot position, the equivalent surface density is $\boldsymbol{J}_{\mathrm{ms}} = (-\boldsymbol{e}_y) \times \boldsymbol{e}_x E_x = \boldsymbol{e}_z E_x$, and the equivalent magnetic current is $\boldsymbol{I}_{\mathrm{m}} = \boldsymbol{J}_{\mathrm{ms}} d = -\boldsymbol{e}_z E_x d$. As shown in Figure 11.14, the slot can be equivalent to a magnetic dipole in an infinite ideal conductor plane, which has a length $l \ll \lambda$, width $d \ll l$. I_{m} is the magnetic current flowing through the dipole. The effect of the infinite metal plane on the right space field of the magnetic dipole can be replaced by the mirror of the magnetic dipole. Therefore, the right space field produced by infinite metal plane is the summation of fields by the magnetic dipole and its mirror. Regarding the mirror of the infinite metal plane to the magnetic dipole, there is a magnetic dipole having same amplitude and direction. In this way, the field in the infinite right half space, is a field produced by a magnetic current $2\boldsymbol{I}_{\mathrm{m}} = \boldsymbol{e}_z 2E_x d$, and a magnetic dipole of length$l$in free space. From Equation (11.113), we can obtain the field of basic slot dipole in the right half space:

$$
\begin{aligned}
H_r &= \frac{E_x dl}{\pi}\frac{1}{\omega\mu}\cos\theta\left(\frac{k}{r^2}-\frac{j}{r^3}\right)\mathrm{e}^{-\mathrm{j}kr} \\
H_\theta &= \frac{E_x dl}{2\pi}\frac{1}{\omega\mu}\sin\theta\left(\mathrm{j}\frac{k^2}{r}+\frac{k}{r^2}-\mathrm{j}\frac{1}{r^3}\right)\mathrm{e}^{-\mathrm{j}kr} \\
H_\phi &= 0 \\
E_r &= E_\theta = 0 \\
E_\phi &= -\frac{E_x dl}{2\pi}\sin\theta\left(\frac{\mathrm{j}k}{r}+\frac{1}{r^2}\right)\mathrm{e}^{-\mathrm{j}kr}
\end{aligned}
\tag{11.115}
$$

The far-field expression of the basic slot element in the right half space is

$$
\begin{aligned}
H_\theta &= \mathrm{j}\frac{1}{\eta_0}\frac{E_x ld}{\lambda r}\sin\theta\mathrm{e}^{-\mathrm{j}kr} \\
E_\phi &= -\mathrm{j}\frac{E_x dl}{\lambda r}\sin\theta\mathrm{e}^{-\mathrm{j}kr}
\end{aligned}
\tag{11.116}
$$

Similar method can be used to solve the field produced by the basic slot element in the left half space ($y < 0$), the boundary condition is $\boldsymbol{n} \times (\boldsymbol{E}_{2\mathrm{m}} - \boldsymbol{E}_{1\mathrm{m}}) = -\boldsymbol{J}_{\mathrm{ms}}$. In the left half slot, $\boldsymbol{n} = -\boldsymbol{e}_y$, $\boldsymbol{E}_{2\mathrm{m}} = \boldsymbol{e}_x E_x$, set $\boldsymbol{E}_{1\mathrm{m}} = 0$, so the equivalence magnetic current density is $\boldsymbol{J}_{\mathrm{ms}} = -(-\boldsymbol{e}_y)\times\boldsymbol{e}_x E_x = -\boldsymbol{e}_z E_x$, and the equivalent magnetic current is $\boldsymbol{I}_{\mathrm{m}} = \boldsymbol{J}_{\mathrm{ms}} d = -\boldsymbol{e}_z E_x d$. Similarly, the fields produced by the slot in the left half space can be found, which almost have the same form expression with those in the right half space, only a negative sign is added.

The field expressions in the left half space are

$$
\begin{aligned}
H_r &= -\frac{E_x dl}{\pi}\frac{1}{\omega\mu}\cos\theta\left(\frac{k}{r^2}-\frac{\mathrm{j}}{r^3}\right)\mathrm{e}^{-\mathrm{j}kr} \\
H_\theta &= -\frac{E_x dl}{2\pi}\frac{1}{\omega\mu}\sin\theta\left(\mathrm{j}\frac{k^2}{r}+\frac{k}{r^2}-\mathrm{j}\frac{1}{r^3}\right)\mathrm{e}^{-\mathrm{j}kr} \\
H_\phi &= 0 \\
E_r &= E_\theta = 0 \\
E_\phi &= \frac{E_x dl}{2\pi}\sin\theta\left(\frac{\mathrm{j}k}{r}+\frac{1}{r^2}\right)\mathrm{e}^{-\mathrm{j}kr}
\end{aligned}
\tag{11.117}
$$

The far-field expressions produced by the basic slot element in the left half space are

$$
\begin{aligned}
H_\theta &= -\mathrm{j}\frac{1}{\eta_0}\frac{E_x ld}{\lambda r}\sin\theta\mathrm{e}^{-\mathrm{j}kr} \\
E_\phi &= \mathrm{j}\frac{E_x dl}{\lambda r}\sin\theta\mathrm{e}^{-\mathrm{j}kr}
\end{aligned}
\tag{11.118}
$$

We can see from the expression that the radiation fields of the right slot are the same with the magnetic dipole, however, the radiation field of the left slot are opposite to the magnetic dipole.

Exercises

11.1 If the magnetic vector potential is known as $\boldsymbol{A} = \frac{\mu_0 I\mathrm{d}l}{4\pi r}\left[\boldsymbol{e}_r\cos\theta - \boldsymbol{e}_\theta\sin\theta\right]\cdot\mathrm{e}^{\mathrm{j}(\omega t-kr)}$, find:

(1) $\boldsymbol{H}$;

(2) Far-field magnetic field H_ϕ, electric field E_θ;

(3) The average Poynting vector $\boldsymbol{S}_{\mathrm{av}}$.

11.2 A 0.01λ length electric dipole, which has an effective current value: $I_{\mathrm{e}} = 300\mathrm{mA}$, find:

(1) radiation resistance R_{r};

(2) radiation power P.

11.3 Find the radiation resistance R_{r} of a 0.01λ electric dipole in infinite space filled with medium of $\varepsilon_r = 9$.

11.4 Assume that an electric dipole produces an electric field with amplitude 100V/m at the position 100km vertical to its axis direction, find its radiation power.

11.5 The radiation power $P_{\mathrm{r}} = 100\mathrm{W}$, find the electric field strength at $r = 10\mathrm{km}$, $\theta = 0°, 45°, 90°$.

11.6 Derive the far-field of the electric dipole antenna.

11.7 As shown in Figure of Exercise 11.7, there is an Electric dipole $I\boldsymbol{dl}$ very close to the ground plane, when its angle to the ground is 30°, the measured transmit power is 10mW. If this antenna is vertical to the ground, find the new transmit power?

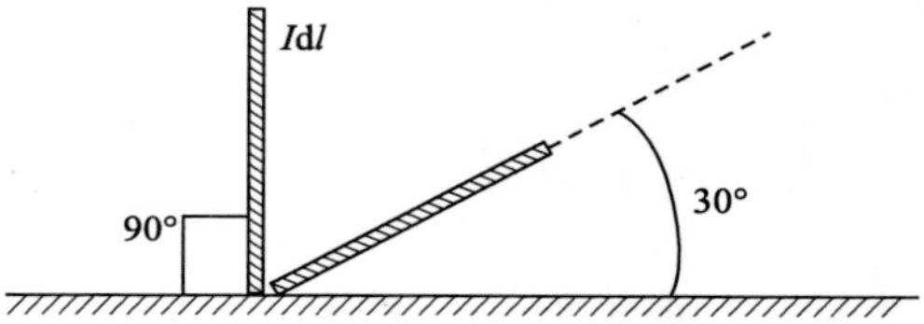

Figure of Exercise 11.17

11.8 In a $2h(h \ll \lambda)$ short central- fed antenna, its current distribution is

$$I(z) = I_0 \left(1 - \frac{|z|}{h}\right)$$

Find:

Radiation field $\boldsymbol{E}, \boldsymbol{H}$;

Radiation resistance R_{r}.

11.9 One current element in $\boldsymbol{e}_z$ direction is located at the origin of the spherical; please prove that its far-field magnetic field can be simplified as

$$\boldsymbol{H} = \boldsymbol{e}_\phi H_\phi = -\boldsymbol{e}_\phi \frac{1}{\mu_0} \sin\theta \frac{\partial A_z}{\partial r}$$

11.10 Assume the current distribution of the half-wave antenna is

$$I = I_{\mathrm{m}} \cos kz \quad \left(-\frac{l}{2} \leqslant z \leqslant \frac{l}{2}\right)$$

(1) Please prove: when $r_0 \gg l$, the vector potential is

$$A_z = \frac{\mu_0 I_{\mathrm{m}} \mathrm{e}^{-\mathrm{j}kr_0}}{2\pi k r_0} \cdot \frac{\cos\left(\frac{\pi}{2}\cos\theta\right)}{\sin^2\theta}$$

(2) Find the far-field magnetic field and electric field;

(3) Find the Poynting vector.

11.11 The current amplitude of the half-wave antenna is 1A, find the maximal electric field at the position 1km away from the antenna position.

11.12 There is an electric dipole antenna in free space, with length $l = 0.1\mathrm{m}$, the frequency is $f = 30\mathrm{MHz}$. Find the electric field, magnetic field and the radiation power, radiation resistance when the antenna is at the plane vertical to the axis of the electric dipole 1km away.

11.13 Find the radiation resistance of a wire segment of length $l = 1\mathrm{m}$ at $f = 1\mathrm{MHz}$:

(1) Assume the wire is straight;

(2) Assume the wire is a ring.

11.14 Refer to Figure of Exercise 11.14, assume that the axis of the element antenna is from the east to the west, find the maximal electric field strength of a receiver at the south direction. When the receiver is moving towards the circumference in which the antenna is positioned at the origin, the electric field is gradually decreasing. Now when the electric field decreases to $1/\sqrt{2}$ times of the maximal value, what is the deviation angle of the receiver to the south position? If the receiver is not moving, what is the change of the received electric field when the element antenna is moving?

If the receiver is an element antenna too, what is the effect of the relative direction of the transmit and receive antenna on the test results?

11.15 As shown in Figure of Exercise 11.15, assume that there are two electric dipoles operating at same frequencies, but the phase difference is $\pi/2$, the dipole elements of these two dipoles are p_0, the distance between the dipoles is d, and d is quite smaller than the wavelength of the radiation electromagnetic waves, find the far-field electromagnetic field, the average energy density and the total radiation power.

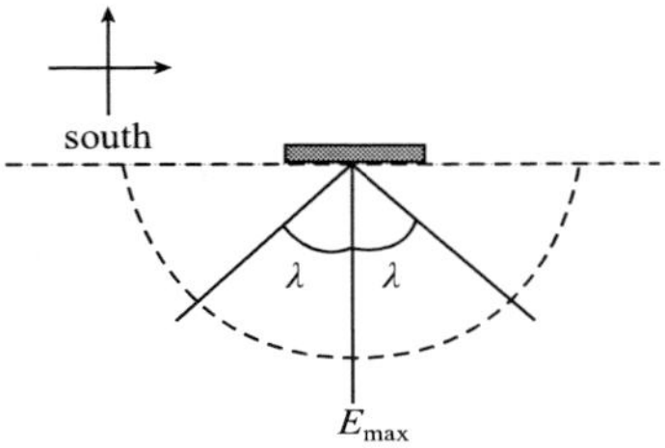

Figure of Exercise 11.14

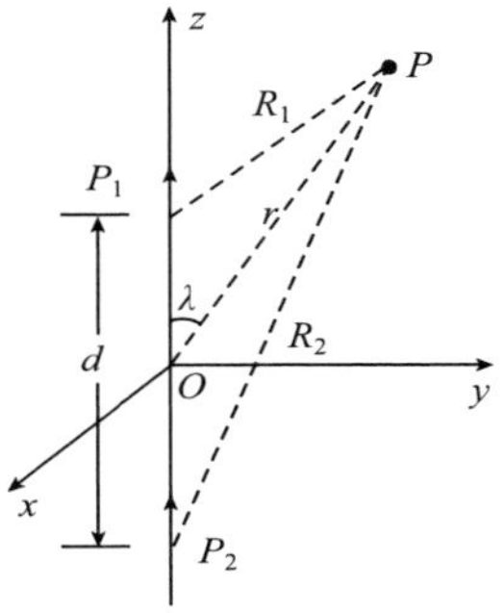

Figure of Exercise 11.15

11.16 For producing an electric field with effective value 100μV/m at the position 100km away from the direction vertical to the axis, what is the radiation power of the electric dipole?

11.17 An electric dipole and a small current ring are set in the origin of the coordinate system, as seen in Figure of Exercise 11.17. If the condition $I_1 \mathrm{d}l = kI_2\pi a^2$ is satisfied, where $k = \omega\sqrt{\mu_0\varepsilon_0}$, prove that in any point of the far-field zone, the electromagnetic is circular polarization, and is it a left circular polarization wave or a right circular polarization?

11.18 As shown in Figure of Exercise 11.18, an electric dipole with a dipole element p_0 is located in the origin of the xOy plane, and rotates towards the z axis with a velocity ω, find the far-field radiation field, the average energy density and the power.

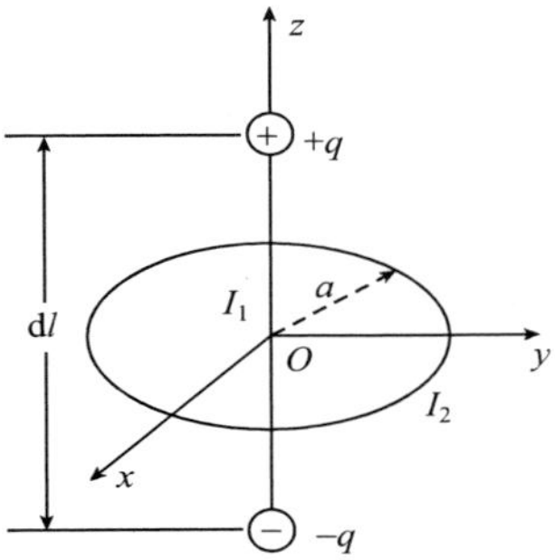

Figure of Exercise 11.17

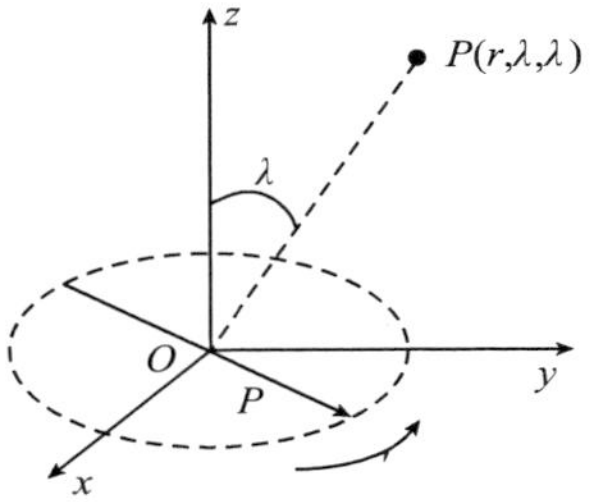

Figure of Exercise 11.18

11.19 As shown in Figure of Exercise 11.19, prove the array directivity factor of the binary array is

$$F(\theta, \phi) = \sqrt{1 + m^2 + 2m\cos(a + kd\sin\theta\cos\phi)}$$

where, m is the ratio of the amplitude of I_1 and I_2; α is the phase difference of I_1 and I_2, d is the distance of the two antenna.

11.20 The distance between the adjacent elements of the uniform linear antenna array is $d = \lambda/2$, if it is required that the maximal radiation direction is at the $\pm 60^\circ$ deviation angle to the axis, what is the phase difference between each element?

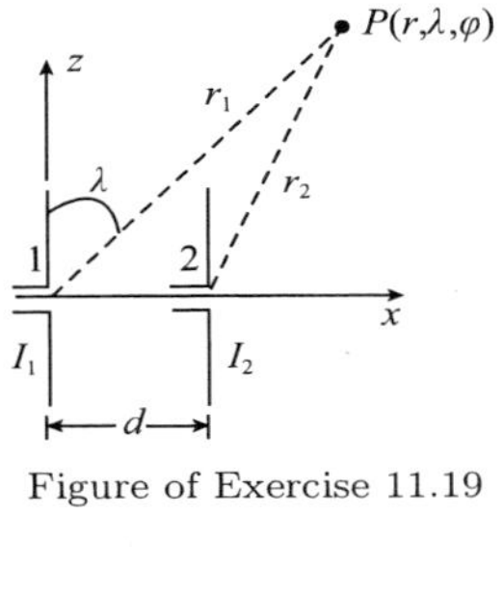

Figure of Exercise 11.19

References

Bi D X. 1985. Electromagnetic Theory. Beijing: Publishing House of Electrics Industry

Cheng D K. 2007. Field and Wave Electromagnetics, Addisonwesley Pub, Co.

Cheng D. 1984. Translated by Zhao Yaotong, Li Binhong. Electromagnetic Fields and Waves. Shanghai: Shanghai Jiaotong University Press

Collin R E. 1961. Principles and Applications of Electromagnetic Fields. New York: McGraw Hill

Collin R E. 1966. Translated by Hou Yuanqing, Field Theory of Guided Waves. Shanghai: Shanghai Scientific & Technical Publishers

Demarest K R. 1983. Engineering Electromagnetics. English reprint copyright 2003 by Science Press

Harrington R F. 1961. Time Harmonic Electromagnetic Fields. New York: McGraw-Hill

Jiao Q X, Wang D D. 1994. Electromagnetic Theory. Beijing: Beijing University of Posts and Telecommunications Press

Jin A K. 2003. Translated by Wu Ji. Electromagnetic Wave Theory. Beijing: Publishing House of Electrics Industry

Jordan E C, Balman K G. 1968. Electromagnetic Waves and Radiating Systems. 2nd ed. Englewood Cliffs, N J: Prentice-Hall

Lorrain P, Corson D R. 1970. Electromagnetic Fields and Waves. 2nd ed. San Francisco, Freeman

Stratton J A. 1941. Electromagnetic Theory. New York: McGraw-Hill

Stratton J A. 1986. translated by He Guoyu. Electromagnetic Theory. Beijing: Beihang University Press

Ulaby F T. 2001. Fundamentals of applied electromagnetics, Media Edition

Xie C F, Rao K J, 1999. Electromagnetic Fields and Waves (3rd edition). Beijing: Higher Education Press

Appendix Common Formula

The Relationship of Three Coordinate Systems

1. Cartesian Coordinates

Unit vectors $\boldsymbol{e}_x$, $\boldsymbol{e}_y$, $\boldsymbol{e}_z$ at Cartesian Coordinates follow the right-hand rule, which is depicted by Figure 1.

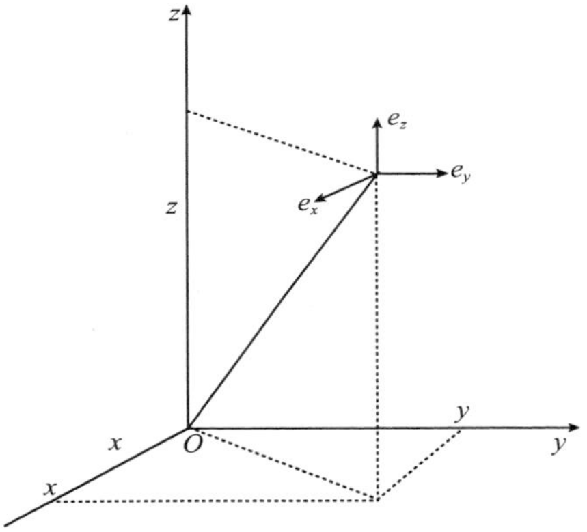

Figure 1 Cartesian coordinates

Cartesian Coordinates

Coordinate range	$-\infty < x < \infty, -\infty < y < \infty, -\infty < z < \infty$
Unit vectors	$\boldsymbol{e}_x$, $\boldsymbol{e}_y$, $\boldsymbol{e}_z$
Relationship among the unit vectors	$\boldsymbol{e}_x \cdot \boldsymbol{e}_x = \boldsymbol{e}_y \cdot \boldsymbol{e}_y = \boldsymbol{e}_z \cdot \boldsymbol{e}_z = 1, \boldsymbol{e}_x \cdot \boldsymbol{e}_y = \boldsymbol{e}_y \cdot \boldsymbol{e}_z = \boldsymbol{e}_z \cdot \boldsymbol{e}_x = 0$ $\boldsymbol{e}_x \times \boldsymbol{e}_y = \boldsymbol{e}_z, \boldsymbol{e}_y \times \boldsymbol{e}_z = \boldsymbol{e}_x, \boldsymbol{e}_z \times \boldsymbol{e}_x = \boldsymbol{e}_y$
Three-dimensional vector expressed by unit vectors	$\boldsymbol{a}(l) = \boldsymbol{a}(x, y, z) = \boldsymbol{e}_x a_x(x, y, z) + \boldsymbol{e}_y a_y(x, y, z) + \boldsymbol{e}_z a_z(x, y, z)$
Vector length unit	$\mathrm{d}\boldsymbol{l} = \boldsymbol{e}_x \mathrm{d}x + \boldsymbol{e}_y \mathrm{d}y + \boldsymbol{e}_z \mathrm{d}z$
Vector volume unit	$\mathrm{d}V = \mathrm{d}x\mathrm{d}y\mathrm{d}z$
Vector area unit	$\mathrm{d}S_x = \mathrm{d}y\mathrm{d}z, \mathrm{d}S_y = \mathrm{d}z\mathrm{d}x, \mathrm{d}S_z = \mathrm{d}x\mathrm{d}y,$ $\mathrm{d}\boldsymbol{S} = \boldsymbol{e}_x \mathrm{d}S_x + \boldsymbol{e}_y \mathrm{d}S_y + \boldsymbol{e}_z \mathrm{d}S_z$
Hamiltonian operator	$\nabla = \boldsymbol{e}_x \frac{\partial}{\partial x} + \boldsymbol{e}_y \frac{\partial}{\partial y} + \boldsymbol{e}_z \frac{\partial}{\partial z}$
Gradient	$\nabla \varphi = \boldsymbol{e}_x \frac{\partial \varphi}{\partial x} + \boldsymbol{e}_y \frac{\partial \varphi}{\partial y} + \boldsymbol{e}_z \frac{\partial \varphi}{\partial z}$

Continued

Divergence	$\nabla \cdot \boldsymbol{a} = \frac{\partial a_x}{\partial x} + \frac{\partial a_y}{\partial y} + \frac{\partial a_z}{\partial z}$
Rotation	$\nabla \times \boldsymbol{a} = \begin{vmatrix} e_x & e_y & e_z \\ \frac{\partial}{\partial x} & \frac{\partial}{\partial y} & \frac{\partial}{\partial z} \\ a_x & a_y & a_z \end{vmatrix}$ $= \boldsymbol{e}_x \left(\frac{\partial a_z}{\partial y} - \frac{\partial a_y}{\partial z} \right) + \boldsymbol{e}_y \left(\frac{\partial a_x}{\partial z} - \frac{\partial a_z}{\partial x} \right) + \boldsymbol{e}_z \left(\frac{\partial a_y}{\partial x} - \frac{\partial a_x}{\partial y} \right)$
Laplace's operation	$\nabla^2 \varphi = \Delta \varphi = \frac{\partial^2 \varphi}{\partial x^2} + \frac{\partial^2 \varphi}{\partial y^2} + \frac{\partial^2 \varphi}{\partial z^2}$

2. Cylindrical Coordinates

Unit vectors $\boldsymbol{e}_r$, $\boldsymbol{e}_\phi$, $\boldsymbol{e}_z$ at cylindrical coordinates follow the right-hand rule, which is depicted by Figure 2.

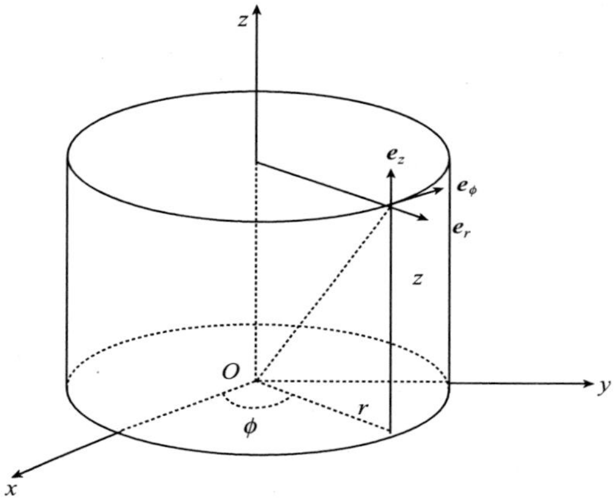

Figure 2 Cylindrical coordinates

Cylindrical Coordinates

Coordinate range	$0 \leqslant r < \infty,\ 0 \leqslant \phi \leqslant 2\pi,\ -\infty < z < \infty$
Coordinate transformation	$r = \sqrt{x^2 + y^2}, \tan \phi = \frac{y}{x}$
Unit vectors	$\boldsymbol{e}_r, \boldsymbol{e}_\phi, \boldsymbol{e}_z$
Relationship among the unit vectors	$\boldsymbol{e}_r \cdot \boldsymbol{e}_r = \boldsymbol{e}_\phi \cdot \boldsymbol{e}_\phi = \boldsymbol{e}_z \cdot \boldsymbol{e}_z = 1, \boldsymbol{e}_r \cdot \boldsymbol{e}_\phi = \boldsymbol{e}_\phi \cdot \boldsymbol{e}_z = \boldsymbol{e}_z \cdot \boldsymbol{e}_r = 0$ $\boldsymbol{e}_r \cdot \boldsymbol{e}_\phi = \boldsymbol{e}_z, \boldsymbol{e}_\phi \times \boldsymbol{e}_z = \boldsymbol{e}_r, \boldsymbol{e}_z \times \boldsymbol{e}_r = \boldsymbol{e}_\phi$
Coordinate transformation of unit vectors	$\boldsymbol{e}_r = \boldsymbol{e}_x \cos \phi + \boldsymbol{e}_y \sin \phi, \boldsymbol{e}_\phi = -\boldsymbol{e}_x \sin \phi + \boldsymbol{e}_y \cos \phi,$ $\boldsymbol{e}_x = \boldsymbol{e}_r \cos \phi - \boldsymbol{e}_\phi \sin \phi, \boldsymbol{e}_y = \boldsymbol{e}_r \sin \phi + \boldsymbol{e}_\phi \cos \phi$
Three-dimensional vector expressed by unit vectors	$\boldsymbol{a}(l) = \boldsymbol{a}(r, \phi, z) = \boldsymbol{e}_r a_r(r, \phi, z) + \boldsymbol{e}_\phi a_\phi(r, \phi, z) + \boldsymbol{e}_z a_z(r, \phi, z)$
Vector length unit	$\mathrm{d}l = \boldsymbol{e}_r \mathrm{d}r + \boldsymbol{e}_\phi r \mathrm{d}\phi + \boldsymbol{e}_z \mathrm{d}z$
Vector volume unit	$\mathrm{d}V = r\mathrm{d}r\mathrm{d}\phi\mathrm{d}z$
Vector area unit	$\mathrm{d}S_r = r\mathrm{d}\phi\mathrm{d}z, \mathrm{d}S_\phi = \mathrm{d}r\mathrm{d}z, \mathrm{d}S_z = r\mathrm{d}r\mathrm{d}\phi,$ $\mathrm{d}\boldsymbol{S} = \boldsymbol{e}_r \mathrm{d}S_r + \boldsymbol{e}_\phi \mathrm{d}S_\phi + \boldsymbol{e}_z \mathrm{d}S_z$
Hamiltonian operator	$\nabla = \boldsymbol{e}_r \frac{\partial}{\partial r} + \boldsymbol{e}_\phi \frac{1}{r} \frac{\partial}{\partial \phi} + \boldsymbol{e}_z \frac{\partial}{\partial z}$

Continued

Gradient	$\nabla\varphi = \boldsymbol{e}_r\frac{\partial\varphi}{\partial r} + \frac{\boldsymbol{e}_\phi}{r}\frac{\partial\varphi}{\partial\phi} + \boldsymbol{e}_z\frac{\partial\varphi}{\partial z}$
Divergence	$\nabla\cdot\boldsymbol{a} = \frac{1}{r}\frac{\partial}{\partial r}(ra_r) + \frac{1}{r}\frac{\partial a_\phi}{\partial\phi} + \frac{\partial a_z}{\partial z}$
Rotation	$\nabla\times\boldsymbol{a} = \begin{vmatrix} \frac{1}{r}\boldsymbol{e}_r & \boldsymbol{e}_\phi & \frac{1}{r}\boldsymbol{e}_z \\ \frac{\partial}{\partial r} & \frac{\partial}{\partial\phi} & \frac{\partial}{\partial z} \\ a_r & ra_\phi & a_z \end{vmatrix}$ $= \boldsymbol{e}_r\left(\frac{1}{r}\frac{\partial a_z}{\partial\phi} - \frac{\partial a_\phi}{\partial z}\right) + \boldsymbol{e}_\phi\left(\frac{\partial a_r}{\partial z} - \frac{\partial a_z}{\partial r}\right)$ $+\boldsymbol{e}_z\left(\frac{1}{r}\frac{\partial}{\partial r}(ra_\phi) - \frac{1}{r}\frac{\partial a_r}{\partial\phi}\right)$
Laplace's operation	$\nabla^2\varphi = \Delta\varphi = \frac{1}{r}\frac{\partial}{\partial r}\left(r\frac{\partial\varphi}{\partial r}\right) + \frac{1}{r^2}\frac{\partial^2\varphi}{\partial\phi^2} + \frac{\partial^2\varphi}{\partial z^2}$

3. Spherical Coordinates

Unit vectors $\boldsymbol{e}_r$, $\boldsymbol{e}_\theta$, $\boldsymbol{e}_\phi$ at spherical coordinates follow the right-hand rule, which is depicted by Figure 3.

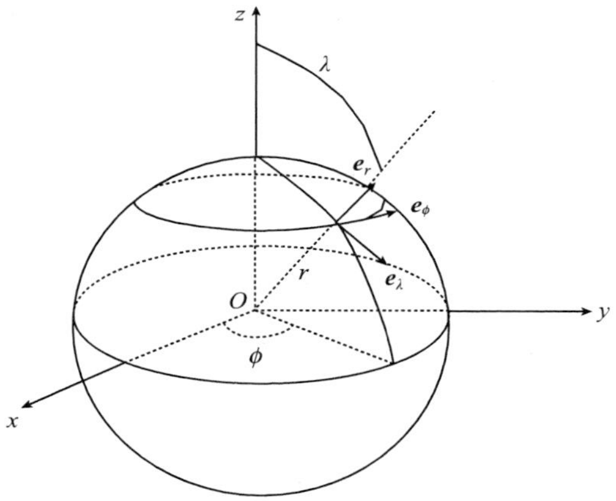

Figure 3 Spherical coordinates

Spherical Coordinates

Coordinate range	$0 \leqslant r < \infty, 0 \leqslant \theta \leqslant \pi, 0 \leqslant \phi < 2\pi$
Coordinate transformation	$r = \sqrt{x^2+y^2+z^2}, \tan\theta = \frac{\sqrt{x^2+y^2}}{z}, \tan\phi = \frac{y}{x}$
Unit vectors	$\boldsymbol{e}_r, \boldsymbol{e}_\theta, \boldsymbol{e}_\phi$
Relationship among the unit vectors	$\boldsymbol{e}_r\cdot\boldsymbol{e}_r = \boldsymbol{e}_\theta\cdot\boldsymbol{e}_\theta = \boldsymbol{e}_\phi\cdot\boldsymbol{e}_\phi = 1, \boldsymbol{e}_r\cdot\boldsymbol{e}_\theta = \boldsymbol{e}_\theta\cdot\boldsymbol{e}_\phi = \boldsymbol{e}_\phi\cdot\boldsymbol{e}_r = 0$ $\boldsymbol{e}_r\times\boldsymbol{e}_\theta = \boldsymbol{e}_\phi, \boldsymbol{e}_\theta\times\boldsymbol{e}_\phi = \boldsymbol{e}_r, \boldsymbol{e}_\phi\times\boldsymbol{e}_r = \boldsymbol{e}_\theta$
Coordinate transformation of unit vectors	$\boldsymbol{e}_r = \boldsymbol{e}_x\sin\theta\cos\phi + \boldsymbol{e}_y\sin\theta\sin\phi + \boldsymbol{e}_z\cos\theta,$ $\boldsymbol{e}_\theta = \boldsymbol{e}_x\cos\theta\cos\phi + \boldsymbol{e}_y\cos\theta\sin\phi - \boldsymbol{e}_z\sin\theta$ $\boldsymbol{e}_\phi = -\boldsymbol{e}_x\sin\phi + \boldsymbol{e}_y\cos\phi,$ $\boldsymbol{e}_x = \boldsymbol{e}_r\sin\theta\cos\phi + \boldsymbol{e}_\theta\cos\theta\cos\phi - \boldsymbol{e}_\phi\sin\phi$ $\boldsymbol{e}_y = \boldsymbol{e}_r\sin\theta\sin\phi + \boldsymbol{e}_\theta\cos\theta\sin\phi - \boldsymbol{e}_\phi\cos\phi$ $\boldsymbol{e}_z = \boldsymbol{e}_r\cos\theta - \boldsymbol{e}_\theta\sin\theta$

Continued

Three-dimensional vector expressed by unit vectors	$\boldsymbol{a}(l)=\boldsymbol{a}(r,\theta,\phi)=\boldsymbol{e}_r a_r(r,\theta,\phi)+\boldsymbol{e}_\theta a_\theta(r,\theta,\phi)+\boldsymbol{e}_\phi a_\phi(r,\theta,\phi)$
Vector length unit	$\mathrm{d}\boldsymbol{l}=\boldsymbol{e}_r\mathrm{d}r+\boldsymbol{e}_\theta r\mathrm{d}\theta+\boldsymbol{e}_\phi r\sin\theta\mathrm{d}\phi$
Vector volume unit	$\mathrm{d}V=r^2\sin\theta\mathrm{d}r\mathrm{d}\theta\mathrm{d}\phi$
Vector area unit	$\mathrm{d}S_r=r^2\sin\theta\mathrm{d}\theta\mathrm{d}\phi,\mathrm{d}S_\theta=r\sin\theta\mathrm{d}r\mathrm{d}\phi,\mathrm{d}S_\phi=r\mathrm{d}r\mathrm{d}\theta,$ $\mathrm{d}\boldsymbol{S}=\boldsymbol{e}_r\mathrm{d}S_r+\boldsymbol{e}_\theta\mathrm{d}S_\theta+\boldsymbol{e}_\phi\mathrm{d}S_\phi$
Hamiltonian operator	$\nabla=\boldsymbol{e}_r\frac{\partial}{\partial r}+\boldsymbol{e}_\theta\frac{1}{r}\frac{\partial}{\partial\theta}+\boldsymbol{e}_\phi\frac{1}{r\sin\theta}\frac{\partial}{\partial\phi}$
Gradient	$\nabla\varphi=\boldsymbol{e}_r\frac{\partial\varphi}{\partial r}+\frac{\boldsymbol{e}_\theta}{r}\frac{\partial\varphi}{\partial\theta}+\frac{\boldsymbol{e}_\phi}{r\sin\theta}\frac{\partial\varphi}{\partial\phi}$
Divergence	$\nabla\cdot\boldsymbol{a}=\frac{1}{r^2}\frac{\partial}{\partial r}(r^2a_r)+\frac{1}{r\sin\theta}\frac{\partial}{\partial\theta}(\sin\theta a_\theta)+\frac{1}{r\sin\theta}\frac{\partial a_\phi}{\partial\phi}$
Rotation	$\nabla\times\boldsymbol{a}=\begin{vmatrix}\frac{1}{r^2\sin\theta}\boldsymbol{e}_r & \frac{1}{r\sin\theta}\boldsymbol{e}_\theta & \frac{1}{r}\boldsymbol{e}_\phi\\ \frac{\partial}{\partial r} & \frac{\partial}{\partial\theta} & \frac{\partial}{\partial\phi}\\ a_r & ra_\theta & r\sin\theta a_\phi\end{vmatrix}$ $=\boldsymbol{e}_r\left(\frac{1}{r\sin\theta}\frac{\partial}{\partial\theta}(\sin\theta a_\phi)-\frac{1}{r\sin\theta}\frac{\partial a_\theta}{\partial\phi}\right)$ $+\boldsymbol{e}_\theta\left(\frac{1}{r\sin\theta}\frac{\partial a_r}{\partial\phi}-\frac{1}{r}\frac{\partial}{\partial r}(ra_\phi)\right)+\boldsymbol{e}_\phi\left(\frac{1}{r}\frac{\partial}{\partial r}(ra_\theta)-\frac{1}{r}\frac{\partial a_r}{\partial\theta}\right)$
Laplace's operation	$\nabla^2\varphi=\Delta\varphi=\frac{1}{r^2}\frac{\partial}{\partial r}\left(r^2\frac{\partial\varphi}{\varphi r}\right)+\frac{1}{r^2\sin\theta}\frac{\partial}{\partial\theta}\left(\sin\theta\frac{\partial\varphi}{\partial\theta}\right)+\frac{1}{r^2\sin^2\theta}\frac{\partial^2\varphi}{\partial\phi^2}$

Answer

Chapter 1 Vector Analysis

1.1 (1) $\boldsymbol{e}_x 0.27 + \boldsymbol{e}_y 0.54 - \boldsymbol{e}_z 0.80$

(2) -11

(3) $-\boldsymbol{e}_x 6 - \boldsymbol{e}_y 15 - \boldsymbol{e}_z 12$

1.4 $\boldsymbol{X} = \dfrac{(P\boldsymbol{A} - \boldsymbol{A} \times \boldsymbol{P})}{|\boldsymbol{A}|^2}$

1.7 $\dfrac{25}{\sqrt{3}}$

1.8 $\nabla \times a = 3\boldsymbol{e}_y - 4\boldsymbol{e}_z$

1.9 $\nabla \varphi = -12\boldsymbol{e}_x - 9\boldsymbol{e}_y + 16\boldsymbol{e}_z$

1.10 $\nabla \left(\dfrac{1}{r}\right) = -\dfrac{r}{r^3}$

1.14 (1) Vector $\boldsymbol{A}$, which is irrotational but has non-zero divergence, can be expressed as the gradient of a scalar function, of which the source distribution is $\nabla \cdot \boldsymbol{A} = 2r \sin \phi$

(2) Vector $\boldsymbol{B}$, which is solenoidal but has non-zero curl, can be expressed as the curl of a vector, of which the source distribution is $\nabla \times \boldsymbol{B} = \boldsymbol{e}_z(2x - 6y)$

1.17 $\displaystyle\int_V \nabla \cdot r \mathrm{d}\boldsymbol{V} = 4\pi a^3$

1.18 $\displaystyle\oint \boldsymbol{A} \cdot \mathrm{d}\boldsymbol{l} = \int_S \nabla \times \boldsymbol{A} \cdot \mathrm{d}\boldsymbol{S} = 8$

1.21 $\nabla \cdot \boldsymbol{A} = 3x^2 + 5y^4x^5 + 5z^4x^2y^3$

1.23 $\left.\dfrac{\partial u}{\partial l}\right|_M = \dfrac{26}{7}$

1.24 (1) $\displaystyle\int \boldsymbol{E} \cdot \mathrm{d}l = 14$

(2) $\displaystyle\int \boldsymbol{E} \cdot \mathrm{d}l = 14$

$\boldsymbol{E}$ is conservative field

Chapter 2 Electrostatic Fields

2.1 ① $\boldsymbol{E} = \boldsymbol{e}_x \dfrac{2q}{4\pi\varepsilon_0 (5^2)} = \boldsymbol{e}_x \dfrac{q}{50\pi\varepsilon_0}$; ② $\varphi = 0$

2.2 $\boldsymbol{E} = -\nabla\varphi = -\nabla \left\{ \dfrac{q}{4\pi\varepsilon_0 [x - (-5)]} + \dfrac{-q}{4\pi\varepsilon_0 [x - 5]} \right\} = \boldsymbol{e}_x \dfrac{q}{50\pi\varepsilon_0}$

2.4 ① $D_n = \boldsymbol{e}_r \rho_s = \boldsymbol{e}_r \dfrac{q}{4\pi a^2}$; ② $C = \dfrac{q}{U} = 4\pi\varepsilon_0 a$

2.5 ① $\rho_l = \dfrac{2\pi\varepsilon_0 U}{\ln\left(\dfrac{b}{a}\right)}$; ② $\rho_{s1} = \dfrac{\varepsilon_0 U}{a \ln\left(\dfrac{b}{a}\right)}$; ③ $\rho_{s2} = \dfrac{-\varepsilon_0 U}{b \ln\left(\dfrac{b}{a}\right)}$

2.6 ① $E_r = 0$; ② $E_r = \dfrac{\rho_0}{3\varepsilon_0}\left(r - \dfrac{a^3}{r^2}\right)$; ③ $E_r = \dfrac{\rho_0}{3\varepsilon_0 r^2}(b^3 - a^3)$

2.7 (1) $E_x = E_y = 0, E_z = \dfrac{a\rho l z}{2\varepsilon_0(a^2 + z^2)^{3/2}}$

(2) $E_z = E_y = 0, E_x = -\dfrac{a^2\rho l}{4\varepsilon_0(a^2 + z^2)^{3/2}}$

(3) $E_z = E_x = 0, E_y = \dfrac{a^2\rho l}{4\varepsilon_0(a^2 + z^2)^{3/2}}$

2.8 $\boldsymbol{E} = \boldsymbol{e}_z\dfrac{\rho_s}{2\varepsilon_0}\left(1 - \dfrac{z}{\sqrt{a^2 + z^2}}\right)$

2.9 $E_x = \dfrac{\rho_s}{\pi\varepsilon}$

2.10 $E_r = 0(r < a), \phi = \text{const.} = \dfrac{\rho_s a}{\varepsilon_0}\ln\left(\dfrac{r_\text{P}}{a}\right)\ (r \leqslant a)$;

$E_r = \dfrac{\rho_s a}{\varepsilon_0 r}(r \geqslant a),\ \phi = \dfrac{\rho_s a}{\varepsilon_0}\ln\left(\dfrac{r_\text{P}}{r}\right)\ (r \geqslant a)$.

2.11 (1) It is the solution of electrostatic field

(2) Potential is $\varphi = x^2 - xyz$

2.12 The minimum of $E_r|_{r=a}$ is $(E_r|_{r=a})_{\min} = \dfrac{U}{b}\cdot\dfrac{1}{\left[\dfrac{a}{b} - \left(\dfrac{a}{b}\right)^2\right]} = \begin{cases} 4U/b \\ 2U/a \end{cases}$

2.13 $\varphi = -\dfrac{\rho_s a}{\pi\varepsilon_0}$

2.14 (1) $\varphi = \dfrac{\rho l}{2\pi\varepsilon_0}\ln\dfrac{\left[\dfrac{L}{2} + \sqrt{\left(\dfrac{L}{2}\right)^2 + r^2}\right]}{r}$

(2) $\boldsymbol{E} = \boldsymbol{e}_r\dfrac{\rho l}{4\pi\varepsilon_0 r}\dfrac{L}{\sqrt{\left(\dfrac{L}{2}\right)^2 + r^2}} = -\nabla\varphi$

2.15 $E_r = \dfrac{abU}{(b - a)r^2}$

2.16 (1) $E_r = \dfrac{q}{4\pi\varepsilon r^2}, a < r < b$

(2) $U = \dfrac{b - a}{4\pi\varepsilon ab}q$

(3) $C = \dfrac{4\pi\varepsilon ab}{b - a}$

2.17 $C = \dfrac{2\pi\varepsilon}{\ln\dfrac{b}{a}}$

2.18 $\begin{cases} A\left(1 - \dfrac{5r^2}{3\varepsilon_0}\right), & r < a \\ B, & r = a \\ 0, & r > a \end{cases}$

2.19 $a = \dfrac{1}{e}b$

2.20 $C = \dfrac{2\pi(\varepsilon_1 + \varepsilon_2)ab}{b - a}$

2.21 $E_y = \dfrac{\sqrt{2}q}{\pi\varepsilon_0 l^2}$

2.23 $-\left(1-\frac{1}{\sqrt{2}}\right)q=-0.293q$

2.24 $\rho(r)=\begin{cases}\varepsilon_0(5r^2+4Ar) & (r\leqslant a)\\ 0 & (r\geqslant a)\end{cases}$, $\rho_s=0$

2.25 (1) $\rho=\frac{6\varepsilon_0 r^3}{a^4}$; (2) $\rho_s=2\varepsilon_0$; (3) $\varphi_a=2a$; (4) $\varphi_0=2.2a$

2.26 (1) $r<a:\boldsymbol{E}=0$; $a<r<b:\boldsymbol{E}=\boldsymbol{e}_r\frac{\rho_{s1}a}{\varepsilon_0 r}$; $r>b:\boldsymbol{E}=\boldsymbol{e}_r\frac{\rho_{s1}a+\rho_{s2}b}{\varepsilon_0 r}$

(2) $\frac{\rho_{s1}}{\rho_{s2}}=-\frac{b}{a}$

2.27 (1) $r<a:\boldsymbol{E}=0:a<r<b:\boldsymbol{E}=\boldsymbol{e}_r\frac{\rho_{s1}a^2}{\varepsilon_0 r^2}$; $r>b:\boldsymbol{E}=\boldsymbol{e}_r\frac{\rho_{s1}a^2+\rho_{s2}b^2}{\varepsilon_0 r^2}$

(2) $U=\frac{\rho_{s1}a^2}{\varepsilon_0}\left(\frac{1}{a}-\frac{1}{b}\right)$

2.28 $C_0=\dfrac{2\pi}{\dfrac{\ln(b/a)}{\varepsilon}+\dfrac{\ln(c/b)}{\varepsilon_0}}$

2.29 $C=4\pi\varepsilon_0\sqrt{a^2+b^2}$

2.30 (1) $E=\frac{abU_0}{(b-a)r^2}\boldsymbol{e}_r$; $D=\frac{\varepsilon_0 bU_0}{(b-a)r}\boldsymbol{e}_r$; $P=\frac{(r-a)\varepsilon_0 bU_0}{(b-a)r^2}\boldsymbol{e}_r$, $\varphi=\frac{a\tau}{2\pi\varepsilon_0}\left(\frac{1}{r}-\frac{1}{b}\right)$

(2) The polarization surface charge density is

$\rho_s=0$ at $r=a$

$\rho_s=\frac{\varepsilon_0 U_0}{b}$ at $r=b$

The polarization volume density of charge is

$$\rho=-\frac{ab\varepsilon_0 U_0}{(b-a)r^3}$$

(3) The capacitance of unit length is

$$C_0=\frac{\tau}{U_0}=\frac{2\pi\varepsilon_0 b}{(b-a)}$$

2.31 (1) The bound charges surface density of each surface is $\rho_s=\frac{L}{2}\boldsymbol{P}_0$

The bound volume charge density is$\rho=-3\boldsymbol{P}_0$

(2) Omitted

2.32 (1) The $\boldsymbol{D}$ and $\boldsymbol{E}$ that are inside and outside the Sphere can be expressed as

Inside the sphere, i.e. $r<a:D_{r(i)}=\rho r/3, E_{r(i)}=\rho r/(3\varepsilon_0)$

Outside the sphere, i.e. $r>a:D_{r(o)}=\rho a^3/(3r^2), E_{r(o)}=\rho a^3/(3\varepsilon_0 r^2)$

(2) The internal and external electric potential of the sphere φ internal electric potential $\phi_{\rm i}$ ($r<a$) : $\phi_{\rm i}=\frac{\rho}{2\varepsilon_0}\left(a^2-\frac{r^2}{3}\right)$ external electric potential $\phi_{\rm o}(r>a):\phi_{\rm o}=\frac{\rho a^3}{3\varepsilon_0 r}$

(3) Electrostatic field energy: $W_e=\frac{4\pi\rho^2a^5}{15\varepsilon_0}$

2.33 (1) $\boldsymbol{E}=\begin{cases}0 & 0<r<a\\ \boldsymbol{e}_r\dfrac{q}{2\pi(\varepsilon_1+\varepsilon_2)r^2} & r>a\end{cases}$

(2) Charge distribution of upper hemisphere

$$\rho_{s_1}=\frac{q}{2\pi^2}\frac{\varepsilon_1}{(\varepsilon_1+\varepsilon 2)}$$

charge distribution of lower hemisphere

$$\rho_{s_2} = \frac{q}{2\pi a^2} \frac{\varepsilon_1}{(\varepsilon_1 + \varepsilon 2)}$$

(3) $W = \dfrac{q^2}{4\pi a(\varepsilon_1 + \varepsilon_{2)})}$

Chapter 3 Constant Magnetic Field

3.2 (1) The As, which are generated by the current of each corresponding points (such as M, N) on the conductive ring at point O, have same magnitude but opposite directions.

(2) The $\boldsymbol{B}$ generated by the current of each point on the conductive ring at point O is in the same direction with each other and can be summed up.

(3) Though $\boldsymbol{A} = 0$ at point O, the $\boldsymbol{A}$ around point O is not. That is to say the rate of change of $\boldsymbol{A}$ around point O is not zero. Therefore it is possible that $\boldsymbol{B} = \nabla \times \boldsymbol{A} \neq 0$

3.3 (1) Area $r \leqslant a$, $\nabla \times \boldsymbol{H} = \boldsymbol{e}_z J_0$

(2) Area $r \geqslant a$, $\nabla \times \boldsymbol{H} = 0$

(3) The curl of constant magnetic field at certain point equals to the current density $\boldsymbol{J}$ of that point. In other words the $\boldsymbol{J}$ is the curl source of that point.

3.4 (1) $\boldsymbol{B}_\phi = \boldsymbol{e}_\phi \dfrac{\mu_0 k r^3}{4}$

(2) omitted

3.5 Inside the cylinder $H_{\phi(\mathrm{i})} = \dfrac{Ir}{2\pi a^2} \quad (r < a)$

Outside the cylinder $H_{\phi(\mathrm{o})} = \dfrac{I}{2\pi r} \quad (r > a)$

3.6 When $0 < r < a$, $\boldsymbol{J} = 0$

When $a < r < b$, $\boldsymbol{J} = \boldsymbol{e}_z \dfrac{I}{\pi(b^2 - a^2)}$

When $r > b$, $\boldsymbol{J} = 0$

3.7 $\boldsymbol{B} = \boldsymbol{e}_x \dfrac{\mu_0 I}{\pi b} \arctan \dfrac{b}{2a}$

3.8 $\boldsymbol{A} = \boldsymbol{e}_z \dfrac{\mu_0 I}{4\pi} \ln \left[\dfrac{\sqrt{(L/2 - z)^2 + r^2} + (L/2 - z)}{\sqrt{(L/2 + z)^2 + r^2} - (L/2 + z)} \right]$

$$\boldsymbol{B} = \boldsymbol{e}_z \frac{\mu_0 I}{4\pi} \ln \left[\frac{L/2 + z}{\sqrt{(L/2 + z)^2 + r^2}} + \frac{L/2 - z}{\sqrt{(L/2 - z)^2 + r^2}} \right]$$

3.9 $\boldsymbol{A} = \boldsymbol{e}_z \dfrac{\mu_0 I}{2\pi} \ln \dfrac{r^-}{r^+} = \boldsymbol{e}_z \dfrac{\mu_0 I}{2\pi} \ln \left(\dfrac{a^2 + r^2 + 2ar\cos\phi}{a^2 + r^2 - 2ar\cos\phi} \right)^{\frac{1}{2}}$

3.10 $\boldsymbol{B} = \begin{cases} \boldsymbol{e}_\phi \mu_0 J_0 \left(\dfrac{r}{2} - \dfrac{2r^2}{3a} \right), & r < a \\ -\boldsymbol{e}_\phi \dfrac{\mu_0 J_0 a^2}{6r}, & r > a \end{cases}$

3.11 (a) $\boldsymbol{B} = \boldsymbol{e}_z \dfrac{\mu_0 I}{2a}$; (b) $\boldsymbol{B} = \boldsymbol{e}_z \dfrac{\mu_0 I}{4a}$; (c) $\boldsymbol{B} = \boldsymbol{e}_z \dfrac{\mu_0 I}{2a} \left(\dfrac{1}{\pi} + \dfrac{1}{2} \right)$

3.12 $r \leqslant a, \boldsymbol{B} = \boldsymbol{e}_{\phi\mu 0} \left(\dfrac{r^3}{4} + \dfrac{4r^2}{3} \right)$; $r \geqslant a, \boldsymbol{B} = \boldsymbol{e}_\phi \dfrac{\mu_0}{r} \left(\dfrac{a^4}{4} + \dfrac{4a^3}{3} \right)$

3.13 $\boldsymbol{H} = \begin{cases} \boldsymbol{e}_\phi \dfrac{\mu_0 I}{\pi r(\mu_0 + \mu)}, & x < 0 \\ \boldsymbol{e}_\phi \dfrac{\mu I}{\pi r(\mu_0 + \mu)}, & x > 0 \end{cases}$

3.14 The distribution of magnetic field between the two plates is uniform: $H_x = J_s$ and it is $\boldsymbol{H} = 0$ outside the plates.

3.15 $r > R_0$, $\boldsymbol{B} = 0$, $\boldsymbol{H} = 0$;

$$R_0 \leqslant r \leqslant R_0 + d, \boldsymbol{H} = \boldsymbol{e}_\phi \frac{I(r^2 - R_0^2)}{2\pi r[(R_0+d)^2 - R_0^2]}, \boldsymbol{B} = \mu_0 \boldsymbol{H}$$

$$R_0 + d \leqslant r \leqslant \infty, \boldsymbol{H} = \boldsymbol{e}_\phi \frac{I}{2\pi r}, \boldsymbol{B} = \mu_0 \boldsymbol{H}$$

3.16 $z > 0, \boldsymbol{B}_1 = \boldsymbol{e}_\phi \dfrac{\mu_0 I}{2\pi r}; z < 0, \boldsymbol{B}_2 = \boldsymbol{e}_\phi \dfrac{\mu I}{2\pi r}$

3.17 when $r < a$

$$\boldsymbol{H}_1 = \frac{Ir}{2\pi a^2}\boldsymbol{e}_\phi, \boldsymbol{B}_1 = \frac{\mu_0 Ir}{2\pi a^2}\boldsymbol{e}_\phi, \boldsymbol{P}_{\mathrm{m}1} = 0$$

when $a < r < b$

$$\boldsymbol{H}_2 = \frac{I}{2\pi r}\boldsymbol{e}_\phi, \boldsymbol{B}_2 = \frac{\mu_1 I}{2\pi r}\boldsymbol{e}_\phi, \boldsymbol{P}_{m2} = \frac{(\mu_{r1} - 1)I}{2\pi r}\boldsymbol{e}_\phi$$

when $b < r < c$

$$\boldsymbol{H}_3 = \frac{I}{2\pi r}\boldsymbol{e}_\phi, \boldsymbol{B}_3 = \frac{\mu_2 I}{2\pi r}\boldsymbol{e}_\phi, \boldsymbol{P}_{m3} = \frac{(\mu_{r2} - 1)I}{2\pi r}\boldsymbol{e}_\phi$$

3.18 $M = \dfrac{\mu_0 a^2}{2d}$

3.19 $\dfrac{\mu_0}{\pi} \ln\left(\dfrac{D}{a}\right) + \dfrac{\mu_0}{4\pi}$

3.20 $L_0 = \dfrac{\mu_0}{8\pi}$

3.21 $M = \dfrac{\mu_0 a}{2\pi} \ln \dfrac{b+c}{b}$

3.22 $\varPhi = \dfrac{\mu_0 I}{\pi}\left[\dfrac{b}{2} - \dfrac{d}{\sqrt{3}} \ln\left(1 + \dfrac{\sqrt{3}b}{2d}\right)\right]$

3.24 $M = \dfrac{\mu_0 \pi a^2 b^2}{2(b^2 + d^2)^{3/2}}$

3.25 $W_m = \dfrac{\mu_0 I^2}{16\pi} + \dfrac{\mu_0 I^2}{4\pi} \ln \dfrac{b}{a}; L_0 = \dfrac{\mu_0}{8\pi} + \dfrac{\mu_0}{2\pi} \ln \dfrac{b}{a}$

3.26 $M = \dfrac{\Psi_{21}}{I_1} = \mu_0 N \pi b^2 \cos\theta$

Chapter 4 Steady Electric Field

4.3 In the Steady electric field, the spatial distribution of charge and the electric charge remain invariance. Therefore, ...

4.4 $R = \dfrac{\ln\left(\dfrac{b}{a}\right)}{2\pi\sigma}$

4.5 (1) 399 A; (2)2 96.1A/m^2; (3) 285A/m^2

4.6 $\boldsymbol{J}_s = \boldsymbol{e}_r \dfrac{I}{2\pi r}$, the current in the $\dfrac{\pi}{3}$ sector is $\dfrac{I}{6}$

4.7 Hints: in the non-uniform medium $\varepsilon, \sigma, \boldsymbol{E}$ are all spatial functions. The solution can be obtained by development based on $\nabla \cdot \boldsymbol{J} = \nabla \cdot \sigma \boldsymbol{E} = 0$ and $\nabla \cdot \boldsymbol{D} = \nabla \cdot \varepsilon \boldsymbol{E} = \rho$.

$$\rho = \sigma \boldsymbol{E} \cdot \left(\frac{\sigma \nabla \varepsilon - \varepsilon \nabla \sigma}{\sigma^2}\right) = \sigma \boldsymbol{E} \cdot \nabla \frac{\varepsilon}{\sigma}$$

4.8 $R = \dfrac{1}{4\pi\sigma_0 K} \ln \dfrac{R_2(R_1 + K)}{R_1(R_2 + K)}$

4.9 $d_2 = \dfrac{\sigma_2 d_1}{\sigma_1}$

4.10 (1) $E_1 = \dfrac{\sigma_2 U_0}{\sigma_2 d_1 + \sigma_1 d_2}, E_2 = \dfrac{\sigma_1 U_0}{\sigma_2 d_1 + \sigma_1 d_2}, J_1 = J_2 = \dfrac{\sigma_1\sigma_2 U_0}{\sigma_2 d_1 + \sigma_1 d_2},$

(2) $U_1 = \dfrac{d_1\sigma_2 U_0}{\sigma_2 d_1 + \sigma_1 d_2}, U_2 = \dfrac{d_2\sigma_1 U_0}{\sigma_2 d_1 + \sigma_1 d_2}$

(3) For the upper polar plate $\rho_{s1} = \dfrac{\varepsilon_1\sigma_2 U_0}{\sigma_2 d_1 + \sigma_1 d_2}$, for the lower polar plate $\rho_{s2} = \dfrac{-\varepsilon_2\sigma_1 U_0}{\sigma_2 d_1 + \sigma_1 d_2}$, for the interface $\rho_s = \dfrac{(\varepsilon_2\sigma_1 - \varepsilon_1\sigma_2)U_0}{\sigma_2 d_1 + \sigma_1 d_2}$

4.11 (1) $R_1 = \dfrac{2d}{\sigma a(r_2^2 - r_1^2)}$

(2) $R_2 = \dfrac{1}{\sigma a d}\ln\dfrac{r_2}{r_1}$

(3) $R_3 = \dfrac{\alpha}{\sigma d \ln(r_2/r_1)}$

4.12 (1) $\boldsymbol{J} = \boldsymbol{e}_r\dfrac{U_0}{r^2 K}; r = a, \rho_s = \dfrac{\varepsilon_1 U_0}{\sigma_1 a^2 K}; r = b, \rho_s = \dfrac{\left(\dfrac{\varepsilon_2}{\sigma_2} - \dfrac{\varepsilon_1}{\sigma_1}\right)U_0}{b^2 K}; r = c, \rho_s = -\dfrac{\varepsilon_2 U_0}{\sigma_2 c^2 K}$

(2) $R = \dfrac{K}{4\pi}$, where, $K = \dfrac{b-a}{\sigma_1 ab} + \dfrac{c-b}{\sigma_2 bc}$

4.13 (1) In the medium 1 $\boldsymbol{E}_1 = \boldsymbol{e}_r\dfrac{U_0}{\left(\ln\dfrac{b}{a} + \dfrac{\sigma_1}{\sigma_2}\ln\dfrac{c}{b}\right)r}$, in the medium 2 $\boldsymbol{E}_2 = \boldsymbol{e}_r\dfrac{U_0}{\left(\dfrac{\sigma_2}{\sigma_1}\ln\dfrac{b}{a} + \ln\dfrac{c}{b}\right)r}$

(2) $\rho_s = \dfrac{U_0}{b}\cdot\dfrac{\sigma_2\varepsilon_1 - \sigma_1\varepsilon_2}{\left(\sigma_2\ln\dfrac{b}{a} + \sigma_1\ln\dfrac{c}{b}\right)}$

(3) $C_0 = \dfrac{2\pi\varepsilon_1\varepsilon_2}{\left(\varepsilon_2\ln\dfrac{b}{a} + \varepsilon_1\ln\dfrac{c}{b}\right)}, G_0 = \dfrac{2\pi\sigma_1\sigma_2}{\left(\sigma_2\ln\dfrac{b}{a} + \sigma_1\ln\dfrac{c}{b}\right)}$

4.14 $R = \dfrac{\ln\cot\dfrac{\theta_0}{2}}{\sigma\pi(b-a)}$

4.15 $R = \dfrac{1}{4\pi\sigma}\left(\dfrac{1}{R_1} + \dfrac{1}{R_2} - \dfrac{1}{d-R_1} - \dfrac{1}{d-R_2}\right)$

Chapter 5 Solutions of Electrostatic Field Boundary Value Problem

5.2
$$\varphi = \frac{U_0}{d}x$$
$$E = -\nabla\varphi = -\boldsymbol{e}_x\frac{U_0}{d}$$

5.3
$$\varphi = -\frac{\rho_0}{2\varepsilon_0}x^2 + \left(\frac{U_0}{d} + \frac{\rho_0 d}{2\varepsilon_0}\right)x$$
$$\boldsymbol{E} = -\nabla\varphi = -\boldsymbol{e}_x\left(\frac{U_0}{d} + \frac{\rho_0}{\varepsilon_0}\left(\frac{d}{2} - x\right)\right)$$

5.4
$$\varphi = \frac{q}{4\pi\varepsilon_0}\left(\frac{1}{x-h} - \frac{1}{x+h}\right)$$
$$\boldsymbol{E} = \boldsymbol{e}_x\left(\frac{q}{4\pi\varepsilon_0}\right)\left(\frac{1}{(x-h)^2} - \frac{1}{(x+h)^2}\right)$$

5.5 The electric potential distribution inside the slot can be expressed as

$$\varphi=\sum_{n=1}^{\infty}\frac{2U_0}{n\pi a\mathrm{sh}\frac{n\pi}{a}b}(1-\cos n\pi)\sin\frac{n\pi}{a}x\mathrm{sh}\frac{n\pi}{a}y$$

5.6 $\varphi=\frac{U_0}{a}(a-x)$

5.7 (1) $\varphi=\sum_{n=1}^{-\infty}(-1)^n\frac{2U_0}{n\pi}\sin\left(\frac{n\pi}{d}x\right)\mathrm{e}^{-\frac{n\pi}{d}y}+\frac{U_0}{d}x$

(2) $\rho_s=-\varepsilon_0\left[\sum_{n=1}^{\infty}(-1)^n\frac{2U_0}{d}\mathrm{e}^{-\frac{n\pi}{d}y}+\frac{U_0}{d}\right],x=0$

$\rho_s=\varepsilon_0\left[\sum_{n=1}^{\infty}\frac{2U_0}{d}\mathrm{e}^{-\frac{n\pi}{d}y}+\frac{U_0}{d}\right],x=d$

$\rho_s=\varepsilon_0\frac{2U_0}{d}\left[\sum_{n=1}^{\infty}(-1)^n\sin\frac{n\pi}{d}x\right],y=0$

5.8 $z\geqslant 0,\varphi=\frac{U_0}{d}y+\sum_{n=1}^{\infty}\frac{2U_0}{(n\pi)^2}\frac{b}{d}\sin\frac{n\pi}{b}d\sin\frac{n\pi}{b}y\mathrm{e}^{-\frac{n\pi}{b}z}$

$z>0,\varphi=\frac{U_0}{d}y+\sum_{n=1}^{\infty}\frac{2U_0}{(n\pi)^2}\frac{b}{d}\sin\frac{n\pi}{b}d\sin\frac{n\pi}{b}y\mathrm{e}^{-\frac{n\pi}{b}z}$

5.9 $\varphi_1=\frac{\rho l}{\varepsilon_0 n\pi}\sum_{n=1}^{\infty}\frac{1}{n}\sin\left(\frac{n\pi}{a}d\right)\sin\left(\frac{n\pi}{a}y\right)\mathrm{e}^{-\frac{n\pi}{a}x}\ (x>0)$

$\varphi_2=\frac{\rho l}{\varepsilon_0 n\pi}\sum_{n=1}^{\infty}\frac{1}{n}\sin\left(\frac{n\pi}{a}d\right)\sin\left(\frac{n\pi}{a}y\right)\mathrm{e}^{\frac{n\pi}{a}x}\ (x<0)$

5.10 The electric potential function between the coaxial lines is $\varphi=\frac{U_0}{\ln\frac{a}{b}}\ln\frac{r}{b}$

The Electric field intensity between the coaxial lines is $\boldsymbol{E}=\boldsymbol{e}_r\frac{-U_0}{r\ln\frac{a}{b}}$

5.11 $\varphi=-\sum_{n=1,3,5,\cdots}\frac{4V_0}{n\pi}\left(\frac{r}{a}\right)^n\sin n\phi$

5.12 The electric potential outside the cylinder is $\varphi(r,\phi)=(a^2r^{-1}-r)E_0\cos\phi+C(r\geqslant a)$, where C is constant.

$\rho_s=2\varepsilon_0E_0\cos\phi$

5.13 $\varphi=\sum_{n=1,3,5,\cdots}^{\infty}\frac{2U_0}{n\pi}\left(\frac{r}{b}\right)^n\sin n\phi+\sum_{n=1,3,5,\cdots}^{\infty}(-1)^{\frac{n-1}{2}}\frac{2U_0}{n\pi}\left(\frac{r}{b}\right)^n\cos n\phi,r\leqslant b$

5.14 $\varphi=\frac{U_0}{\ln\left(\tan\frac{\theta_0}{2}\right)}\cdot\ln\left(\tan\frac{\theta}{2}\right)$

Electric field intensity $\boldsymbol{E}:E_\theta=-\frac{U_0}{r\ln\left(\tan\frac{\theta_0}{2}\right)\cdot\sin\theta}$

5.15 (1) $\varphi=(-E_0r+E_0a^3/r^2)\cos\theta+U_0a/r-U_0$

(2) $\varphi=(-E_0r+E_0a^3/r^2)\cos\theta+\frac{Q}{4\pi\varepsilon_0}\left(\frac{1}{r}-\frac{1}{a}\right)$

5.16 Consider the cylindrical coordinates of which the axis is the axis of the cylinder. Let the bottom surface of the cylinder be on the plane. Thus,

$$\varphi = \sum_{n=1,3,5} \frac{4U_{.0}}{n\pi I_0\left(\frac{n\pi}{l}a\right)} \sin\left(\frac{n\pi}{l}z\right) I_0\left(\frac{n\pi}{l}r\right)$$

where I_0 is the Zeroth Order Bessel Function of the First Kind so that it satisfies the boundary condition of side face.

5.17 (1) The electric energy of each image charge and the coordinates is as follows: $-q$ (0.366, 1.366), q (−1.366, 0.366), $-q$ (−1.366, −0.366), q (0.366, −1.366), -q (1, −1)

(2) $2.62\times10^9 q$ (V)

5.18 The magnitude is $-q$ and the position is at $(-h, 0, 0)$,

$$\varphi = \frac{q}{4\pi\varepsilon_0}\left[\frac{1}{[(x-h)^2+y^2+z^2]^{1/2}} - \frac{1}{[(x+h)^2+y^2+z^2]^{1/2}}\right]$$

$$\rho_s = -\frac{\varepsilon_0 q}{2\pi a^2}(\mathrm{c/m^2})$$

5.19 $C_0 = \dfrac{2\pi\varepsilon_0}{\ln \dfrac{2d_1 d_2}{a\sqrt{d_1^2+d_2^2}}}$

5.20 (1) Omitted

(2) Place another point electric charge q'' at the center of the metal ball and $q'' = 4\pi\varepsilon_0 a U_0$

5.21

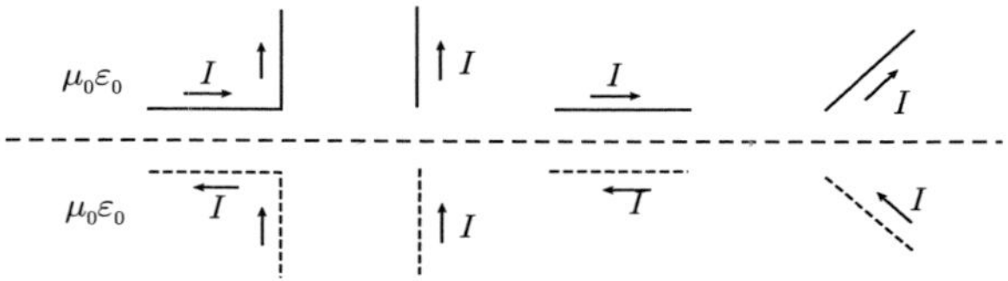

5.22 $\varphi = \dfrac{3q}{4\pi\varepsilon_0}\Big[(9r^2+a^2-6ar\cos\theta)^{-\frac{1}{2}} - (9r^2+a^2+6ar\cos\theta)^{-\frac{1}{2}}$

$-(r^2+9a^2-6ar\cos\theta)^{-\frac{1}{2}} + (r^2+9a^2+6ar\cos\theta)^{-\frac{1}{2}}\Big]$

5.23 $R = \dfrac{1}{2\pi\sigma h}\ln\dfrac{4h}{d}$

Chapter 6 Alternating Electromagnetic Fields

6.1 (1) $\nabla \times E = -\dfrac{\partial \boldsymbol{B}}{\partial t}$

(2) $\nabla \times \boldsymbol{H} = \boldsymbol{J} + \dfrac{\partial \boldsymbol{D}}{\partial t}$

6.2 $\boldsymbol{J}_{\mathrm{d}} = \boldsymbol{e}_x 50\sin(2\pi z)\cos(6\pi\times10^8 t)$

6.6 $E_x = \dfrac{k}{\omega\varepsilon_0}H_0\sin(\omega t - kz)$

6.7 $k = \omega\sqrt{\mu_0\varepsilon_0}$

6.8 0

6.9 $\nabla \times \boldsymbol{B} = \mu\boldsymbol{J} + \mu\varepsilon\dfrac{\partial E}{\partial t} + \dfrac{\nabla\mu\times\boldsymbol{B}}{\mu}$

$\nabla \times \boldsymbol{E} = -\dfrac{\partial \boldsymbol{B}}{\partial t}$

$\nabla \cdot \boldsymbol{B} = 0$

$\nabla\varepsilon \cdot \boldsymbol{E} + \varepsilon\nabla \cdot \boldsymbol{E} = \rho$

6.10 Hints: the formula of ρ in the uniform Conductive medium can be obtained by $\nabla \cdot \boldsymbol{J} = \dfrac{\partial \rho}{\partial t}$ and $\nabla \cdot \boldsymbol{J} = \nabla \cdot \sigma E = \cdots$

(1) The formula of ρ is $\dfrac{\partial \rho}{\partial t} + \dfrac{\sigma}{\varepsilon}\rho = 0$ and the solution is $\rho = \rho_0 e^{-\frac{t}{\tau}}$; (2) $\tau = \dfrac{\varepsilon}{\sigma}$; (3) $\tau = 5.2\times 10^{-20}\mathrm{s}$

6.11 $\boldsymbol{J}_\mathrm{d} = -\boldsymbol{e}_x kA_2 \cos 4x \cos(\omega t - ky) + \boldsymbol{e}_y 4A_2 \sin 4x \sin(\omega t - ky) - \boldsymbol{e}_z kA_1 \sin 4x \sin(\omega t - ky)$

6.12 $\boldsymbol{J}_\mathrm{d} = -\boldsymbol{e}_r \dfrac{\varepsilon U_m \omega}{r \ln \dfrac{b}{a}} \cos \omega t; i_\mathrm{d} = \dfrac{2\pi\varepsilon U_m \omega t}{\ln \dfrac{b}{a}} \cos \omega t$

6.13 $\boldsymbol{H} = -\boldsymbol{e}_x 2.29\times 10^{-4} \sin(10\pi x)\cos(6\pi\times 10^9 t - kz) - \boldsymbol{e}_z 1.33\times 10^{-4} \cos(10\pi x)\sin(6\pi \times 10^9 t - kz)$

6.14 $k = \omega\sqrt{\mu_0\varepsilon_0}$; $\boldsymbol{H} = \boldsymbol{e}_\phi \sqrt{\dfrac{\varepsilon_0}{\mu_0}}\left(\dfrac{E_0}{r}\right)\sin\theta\cos(\omega t - kr)$

6.15 (1) $E = \boldsymbol{e}_r \dfrac{H_\mathrm{m} k}{\omega\varepsilon_0 r} \sin kz \sin \omega t, k = \dfrac{n\pi}{l}(n = 1, 2, 3, \cdots)$

(2) End face: at $z = 0, \boldsymbol{J}_\mathrm{s} = -\boldsymbol{e}_r \dfrac{H_\mathrm{m}}{r} \cos \omega t, \rho_s = 0$; at $z = 1, \boldsymbol{J}_s = \boldsymbol{e}_r \dfrac{H_\mathrm{m}}{r} \cos kl \cos \omega t, \rho_s = 0$

Inner surface of conductor: $\boldsymbol{J}_s = \boldsymbol{e}_y \dfrac{H_\mathrm{m}}{a} \cos kz \cos \omega t, \rho_s = \dfrac{H_\mathrm{m} k}{a\omega} \sin kz \sin \omega t$;

Outer surface of conductor: $\boldsymbol{J}_s = -\boldsymbol{e}_y \dfrac{H_\mathrm{m}}{b} \cos kz \cos \omega t, \rho_s = -\dfrac{H_\mathrm{m} k}{b\omega} \sin kz \sin \omega t$;

6.16 (1) $\boldsymbol{H} = -\boldsymbol{e}_x \dfrac{k}{\omega\mu_0} E_0 \sin(\omega t - kz)$

(2) Omitted

(3) $S_\mathrm{av} = \boldsymbol{e}_z \dfrac{E_0^2}{2} \varepsilon_0 c$

6.17 $\boldsymbol{S} = \boldsymbol{e}_r \dfrac{k}{\omega\varepsilon_0} \dfrac{H_\mathrm{m}^2}{r^2} \sin^2\theta \cos^2(\omega t - kr)$

6.18 $\boldsymbol{P}_\mathrm{av} = \dfrac{4}{3}\pi \dfrac{A_0^2}{\eta_0} = \dfrac{A_0^2}{90}$

6.19 Satisfaction; at the surface of $x = 0$, $\boldsymbol{J}_s = -\boldsymbol{e}_y H_0 \cos(kz - \omega t)$; at the surface of $x = a$, $\boldsymbol{J}_\mathrm{s} = -\boldsymbol{e}_y H_0 \cos(kz - \omega t)$

$$\boldsymbol{S} = \boldsymbol{e}_x \frac{1}{4} H_0^2 \mu\omega \left(\frac{a}{\pi}\right) \sin\left(\frac{2\pi x}{a}\right) \sin 2(kz - \omega t) - \boldsymbol{e}_x \frac{1}{2} H_0^2 \mu\omega k \left(\frac{a}{\pi}\right)^2 \sin^2\left(\frac{\pi x}{a}\right) [1 - \cos 2(kz - \omega t)]$$

$$\boldsymbol{S}_\mathrm{av} = -\boldsymbol{e}_z \frac{1}{2} H_0^2 \mu\omega k \left(\frac{a}{\pi}\right)^2 \sin^2\left(\frac{\pi x}{a}\right)$$

6.20 Omitted

6.21 Use the cylindrical coordinates

(1) $E_z = \dfrac{U}{d}, H_\phi = \dfrac{\sigma U r}{2d}, \boldsymbol{S} = -\boldsymbol{e}_r \dfrac{\sigma U^2 r}{2d^2}$;

(2) The power loss and the power that enters into are both $\dfrac{\sigma\pi a^2 U^2}{d}$

6.22 The mean value of the Poynting vector

$$\boldsymbol{S}_\mathrm{av} = \frac{1}{2}\mathrm{Re}[\dot{\boldsymbol{E}} \times \dot{\boldsymbol{H}}^*] = \frac{k}{2\omega\mu_0}(E_0^2 + E_0'^2)$$

Chapter 7 Propagation of Plane Wave in Infinite Medium

7.1 (1) The wave propagates in the direction of $-z$.

(2) $f = 300\text{MHz},\ \lambda = 1\text{m},\ k = 2\pi\text{rad/m},\ v_{\text{p}} = 3\times 10^8\text{m/s}$
(3) $\boldsymbol{H} = \boldsymbol{e}_x 0.1\cos(6\pi\times 10^8 t + 2\pi z)\text{A/m}$

7.2 (1) $\boldsymbol{H} = -\boldsymbol{e}_x\dfrac{E_0}{\eta_0}\sin(\omega t - kz) + \boldsymbol{e}_y\dfrac{E_0}{\eta_0}\sin(\omega t - kz)$

(2) Instantaneous value: $\boldsymbol{S}(t) = \boldsymbol{e}_z 2\dfrac{E_0^2}{\eta_0}\sin^2(\omega t - kz)$

Mean value: $\boldsymbol{S}_{\text{av}} = \boldsymbol{e}_z\dfrac{E_0^2}{\eta_0}$

7.3 (1) $\boldsymbol{H} = -\boldsymbol{e}_y\dfrac{1}{3\pi}\cos(9\times 10^9 t + 30z)\text{A/m}, \boldsymbol{E} = \boldsymbol{e}_x 40\cos(9\times 10^9 t + 30z)\text{V/m}$

(2) $f = \dfrac{9}{2\pi}\times 10^9\text{Hz}, \lambda = \dfrac{\pi}{15}\text{m}$

7.4 (1) $\boldsymbol{e}_k = \dfrac{2}{3}\left(-\boldsymbol{e}_x + \boldsymbol{e}_y + \dfrac{1}{2}\boldsymbol{e}_z\right)$

(2) $\lambda = \dfrac{4}{3}\text{m}, f = 225\times 10^6\text{Hz}$

(3) $\boldsymbol{E} = 4\pi\times 10^{-5}\left(-\boldsymbol{e}_x - \dfrac{7}{2}\boldsymbol{e}_y + 5\boldsymbol{e}_z\right)\cos\left[\dfrac{9}{2}\pi\times 10^9 t + \pi\left(x - y - \dfrac{1}{2}z\right)\right]\text{V/m}$

(4) $\boldsymbol{S}_{\text{av}} \approx 8\times 10^{-10}\left(\dfrac{-2\boldsymbol{e}_x + 2\boldsymbol{e}_y + \boldsymbol{e}_z}{3}\right)\text{W/m}^2$

7.5 (1) $\boldsymbol{B} = -\boldsymbol{e}_y[\cos(2\pi\times 10^6 t - 2\pi\times 10^{-12}x)]\times 10^{-12}\text{T}$
(2) $\varepsilon_{\text{r}} = 9$

7.6 (1) $f = \dfrac{5}{\pi}\times 10^9\text{Hz}$

(2) $k = \omega\sqrt{\mu\varepsilon} = 50\text{rad/m}$

(3) $\boldsymbol{H}(t) = \boldsymbol{e}_y\dfrac{5}{8\pi}\cos(10^{10}t - 50z)\text{A/m}$

(4) $\boldsymbol{S}_{\text{av}} = \boldsymbol{e}_z\dfrac{125}{8\pi}\text{W/m}^2$

7.7 (1) Levorotatory circular polarization; (2) Line polarization wave; (3) Levorotatory elliptic polarization; (4) Line polarization wave

7.8 (1) Omitted

(2) Let the levorotatory elliptic polarization be expressed as $\boldsymbol{E} = (\boldsymbol{e}_x a + \text{j}\boldsymbol{e}_y b)\text{e}^{\text{j}(\omega t - kz)}$ and thus the radii of the two circular polarization wave are $a_1 = \dfrac{a+b}{2}, a_2 = \dfrac{a-b}{2}$ respectively.

7.9 (1) $\boldsymbol{H} = \dfrac{1}{120\pi}[\boldsymbol{e}_y E_0\sin(\omega t - kz) - \boldsymbol{e}_x E_0'\cos(\omega t - kz)]$

(2) Both electric field and magnetic field are levorotatory elliptic polarization

(3) $\boldsymbol{S}_{\text{av}} = \boldsymbol{e}_z\dfrac{1}{240\pi}(E_0^2 + E_0'^2)$

7.10 $\varepsilon_{\text{r}} = 4.94; v_{\text{r}} = 1.35\times 10^8\text{m/s}$

7.11 $f = 25\times 10^8(\text{Hz}); \mu_{\text{r}} = 1.99; \varepsilon_{\text{r}} = 1.13$

7.12 55.5W

7.13 The instantaneous value of the total power is $28.86\sin^2(\omega t)$mW. The mean value of the total power is 13.43mW

7.14 $\boldsymbol{E} = E_0(\boldsymbol{e}_x - \text{j}\boldsymbol{e}_y)\text{e}^{\text{j}(\omega t - kz)}$; $\boldsymbol{H} = \sqrt{\dfrac{\varepsilon}{\mu}}E_0(\boldsymbol{e}_y + \text{j}\boldsymbol{e}_x)\text{e}^{\text{j}(\omega t - kz)}$; $\boldsymbol{S}_{\text{av}} = \boldsymbol{e}_z\sqrt{\dfrac{\varepsilon}{\mu}}E_0^2$

7.16 (1) $d = 0.76$mm
(2) $d = 1.414$mm

7.17 $h = 8.3\times 10^{-6}$m

7.18 (1) $R_{\mathrm{D}} \approx 2.44 \times 10^{-3}(\Omega)$

(2) $R_{\mathrm{s}} \approx 2.61 \times 10^{-3}(\Omega/\mathrm{m}^2)$

(3) $R_{\mathrm{A}} \approx 0.277(\Omega)$

7.19 $\boldsymbol{E} = \boldsymbol{e}_x 7259\mathrm{e}^{-8.89z}\cos(10^7\pi t - 8.89z + 8.89)\mathrm{V/m}$

$\boldsymbol{H} = \boldsymbol{e}_y 2310\mathrm{e}^{-8.89z}\cos(10^7\pi t - 8.89z + 8.10)\mathrm{V/m}$

$v_{\mathrm{p}} = 3.53 \times 10^6 \mathrm{m/s}$

$\lambda = 0.707\mathrm{m}$

7.20 When $f = 10\mathrm{kHz}$, $\alpha = 0.126\pi$, $\lambda = 15.87\mathrm{m}$, $\eta = 0.0316\pi(1+j)(\Omega)$

When $f = 1\mathrm{MHz}$, $\alpha = 1.26\pi, \lambda = 1.587\mathrm{m}$, $\eta = 0.316\pi(1+j)(\Omega)$

When $f = 10\mathrm{MHz}$, $\alpha = 4\pi, \lambda = 0.5\mathrm{m}$, $\eta = \pi(1+j)(\Omega)$

When $f = 1\mathrm{GHz}$, $\alpha = 24.65\pi, \beta = 2\pi \times 32.4\mathrm{rad/m}$, $\lambda = 0.03\mathrm{m}$, $\eta = 42/\sqrt{1 - j \times 0.89}(\Omega)$

7.21 (1)1.395m

(2) $\eta_{\mathrm{c}} = \dfrac{238.4}{\sqrt{1 - j \times 0.01}}, \lambda = 0.063\mathrm{m}, v_{\mathrm{p}} = 1.897 \times 10^8 \mathrm{m/s}$

(3) $\boldsymbol{H}(x,t) = \boldsymbol{e}_z 0.21\mathrm{e}^{-0.497x}\sin(6\pi \times 10^9 t - 31.6\pi x + \pi/3 - 0.0016\pi)\mathrm{A/m}$

7.22 $\boldsymbol{H}(z,t) = -\boldsymbol{e}_x 2.28 \times 10^3 \sin(3\pi \times 10^6 t - 1.91 \times 10^4 z - \pi/4)\mathrm{e}^{-1.91\times 10^4 z}$

Chapter 8 Reflection and Refraction of Electromagnetic Waves

8.1 (1) $H_y^+ = \dfrac{k}{\omega\mu_0}E_x^+ = \left(\dfrac{1}{\eta_0}\right)E_x^+ = \dfrac{1}{\eta_0}E_0^+\mathrm{e}^{\mathrm{j}(\omega t - kz)}$

(2) $H_y^- = \dfrac{1}{\eta_0}E_0^+\mathrm{e}^{\mathrm{j}(\omega t + kz)}$

(3) $H = 2\left(\dfrac{1}{\eta_0}\right)E_0^+\cos(kz)\,\mathrm{e}^{\mathrm{j}\omega t}$

8.2 (1) k_x represents the phase constant of the TEM wave in the direction of x and it is similar for k_y, k_z.

(2) $k_x^2 + k_y^2 + k_z^2 = k^2$

8.3 (1) $k_x = k\sin\theta = 2, k_z = k\cos\theta = 3$

(2) $\tan\theta = \dfrac{2}{3}$, thus $\theta = 33.69°$

(3) $k = \sqrt{k_x^2 + k_y^2} = \sqrt{13} \approx 3.61$

8.4 (1) From the equation $\dfrac{v_1}{\sin\theta_{\mathrm{i}}} = \dfrac{v_2}{\sin\theta_{\mathrm{r}}}$ we can have $\theta_{\mathrm{i}} = \theta_{\mathrm{r}}$

(2) From the equation $\dfrac{v_1}{\sin\theta_{\mathrm{i}}} = \dfrac{v_2}{\sin\theta_{\mathrm{r}}}$ there are $\dfrac{\sin\theta_{\mathrm{r}}}{\sin\theta_{\mathrm{i}}} = \dfrac{v_2}{v_1} = \dfrac{\sqrt{\varepsilon_{\mathrm{r1}}\varepsilon_0\mu_0}}{\sqrt{\varepsilon_{\mathrm{r2}}\varepsilon_0\mu_0}} = \dfrac{\sqrt{\varepsilon_{\mathrm{r1}}}}{\sqrt{\varepsilon_{\mathrm{r2}}}} = \dfrac{n_1}{n_2}$

8.5 (1) θ_{i} =36.87°

(2) f=477.465 MHz, λ =0.6283m

(3) The complex representation of the reflective electric field is $\boldsymbol{E}^- = -\boldsymbol{e}_y 10 \cdot \mathrm{e}^{\mathrm{j}(\omega t - 6x + 8z)}$

(4) The electric field of composite wave is $\boldsymbol{E} = -\mathrm{j}\boldsymbol{e}_y 20\sin(8z)\cdot\mathrm{e}^{\mathrm{j}(\omega t - 6x)}$

8.6 $0.1161\times10^{-6}\mathrm{W/m}^2$

8.7 ε_{r} =6.85

8.8 (1) $\boldsymbol{e}_k = -\boldsymbol{e}_x\dfrac{2}{\sqrt{5}} + \boldsymbol{e}_z\dfrac{1}{\sqrt{5}}$

(2) $\theta_{\mathrm{i}} = \arctan\dfrac{1}{2} \approx 26.56$

(3) $\boldsymbol{H}^- = \left(\boldsymbol{e}_x\dfrac{1}{\sqrt{5}} - \boldsymbol{e}_z\dfrac{2}{\sqrt{5}}\right)\dfrac{1}{6\pi}\mathrm{e}^{\mathrm{j}(-4x-2z)}$

(4) The mean value of power density in the x direction is 0

8.11 $E_2 \approx 2.356\times10^{-3}\mathrm{V/m}$; $E_3 \approx 7.82 \times 10^{-4}\mathrm{V/m}$; $E_4 \approx 1.564 \times 10^{-3}\mathrm{V/m}$

8.12 $v_{\mathrm{g}} = 4\sqrt{\dfrac{\omega}{2\mu\sigma}}$

8.13 (1) $\boldsymbol{E}^- = -\boldsymbol{e}_x \frac{1}{3} E_0 \sin\omega\left(t + \frac{z}{v_1}\right)$

$\boldsymbol{H}^- = \boldsymbol{e}_y \frac{1}{360\pi} E_0 \sin\omega\left(t + \frac{z}{v_1}\right)$ where $v_1 = \frac{1}{\sqrt{\mu_0\varepsilon_0}} = c$

(2) $\boldsymbol{E}^T = \boldsymbol{e}_x \frac{2}{3} E_0 \sin\omega\left(t - \frac{z}{v_2}\right)$

$\boldsymbol{H}^T = \boldsymbol{e}_y \frac{1}{90\pi} E_0 \sin\omega\left(t - \frac{z}{v_2}\right)$ where $v_2 = \frac{1}{\sqrt{4\mu_0\varepsilon_0}} = \frac{c}{2}$

8.14 (1) $\boldsymbol{E}^- = -\frac{1}{3} E_0(\boldsymbol{e}_x + \mathrm{j}\boldsymbol{e}_y)\mathrm{e}^{\mathrm{j}kz}$

(2) $\boldsymbol{H}^T = \frac{1}{90\pi} E_0(\boldsymbol{e}_y - \mathrm{j}\boldsymbol{e}_x)\mathrm{e}^{-\mathrm{j}2kz}$

(3) The incident wave is levorotatory circular polarization, the reflected wave is dextrorotatory circular polarization and the transmitted wave is levorotatory circular polarization.

8.15 (1) $\varepsilon_r = 9$;

(2) $S_{av}^- = 0.25\text{mW/m}^2$; $S_{av}^T = 0.75\text{mW / m}^2$

8.16 (1) $\varepsilon_r = 4$

(2) $\boldsymbol{E}^T = \boldsymbol{e}_x 10^{-3}\cos(3\times10^8 t - 2z)\text{V/m}$; $\boldsymbol{H}^T = \boldsymbol{e}_y 5.3\times10^{-6}\cos(3\times10^8 t - 2z)\text{A/m}$

8.17 (1) 0.269; (2) 6.935×10^{-4}

8.18 The amplitude of electric field of reflected wave is 33.3V/m; the amplitude of electric field of transmitted wave is 66.7V/m

8.19 (1) $\boldsymbol{E}^- = \boldsymbol{e}_x 2.77\cos(1.8\times10^9 t + 6z + 157°)$

$\boldsymbol{H}^- = -\boldsymbol{e}_y 7.35\times10^{-3}\cos(1.8\times10^9 t + 6z + 157°)$

$\boldsymbol{E}^T = \boldsymbol{e}_x 7.53\mathrm{e}^{-2.29z}\cos(1.8\times10^9 t - 9.76z - 172°)$

$H^T = \boldsymbol{e}_y 0.033\mathrm{e}^{-2.29z}\cos(1.8\times10^9 t - 9.76z - 185°)$

(2) $\boldsymbol{S}_{av1} = \boldsymbol{e}_z 0.122\text{W/m}^2$, $\boldsymbol{S}_{av2} = \boldsymbol{e}_z 0.122\text{W/m}^2$

8.20 (1) $f = \frac{15}{2\pi}\times10^8\text{Hz}$

(2) $\theta_i = \arctan\frac{3}{4}$

(3) $\boldsymbol{E}^- = -\boldsymbol{e}_y E_0 \mathrm{e}^{\mathrm{j}(1.5\times10^9 t - 3x + 4z)}$

(4) $\boldsymbol{E} = -\boldsymbol{e}_y \mathrm{j}2E_0\sin 4z\,\mathrm{e}^{\mathrm{j}(1.5\times10^9 t - 3x)}$

8.21 (1) $\omega = 12\times10^8\text{rad/m}$, $A = -2$

(2) $\boldsymbol{E}^+ = -120\pi\left(\frac{\sqrt{3}}{2}\boldsymbol{e}_x + 2\boldsymbol{e}_y + \frac{1}{2}\boldsymbol{e}_x\right)\sin(12\times10^8 t + 2x - 2\sqrt{3}z)\text{V/m}$

(3) $\theta_i = \frac{\pi}{6}$

8.22 $\varepsilon_r = 3$

8.23 (1) 63.4°; (2) 18%

8.25 $\theta_c = 41.81°$; $\theta_B = 33.69°$

8.26 (1) $\theta_c = 6.38°$; (2) $R_\perp = \mathrm{e}^{\mathrm{j}38°}$; (3) $T_\perp = 1.89\mathrm{e}^{\mathrm{j}19°}$

Chapter 9 Two-Conductor Transmission Lines—Transverse Electromagnetic Wave Guiding System

9.1 (1) $Z_{in} = Z_c$ (2) $Z_{in} = \mathrm{j}Z_c\tan kl$ (3) $Z_{in} = -\mathrm{j}Z_c\cot kl$

9.2 (1) Capacitance; (2) $R_{in} = 0$; $x_{in} = -\frac{Z_c^2}{\omega l}$

9.3 $Z_{in} = (26.32 - \mathrm{j}9.87)\,\Omega$

9.4 (1) $D = 25.5\text{mm}$; (2) $b = 3.91\text{mm}$

9.5 In the case when the terminal is short circuit the length is 0.074λ; In the case when the terminal is open circuit the length is 0.324λ

9.6 $P = \frac{ab}{2}\sqrt{\frac{\varepsilon}{\mu}}|H_0|^2$ (H_0 is the complex amplitude of TEM mode magnetic field), $\alpha = \frac{1}{a\sigma\delta}\sqrt{\frac{\varepsilon}{\mu}}$, $Z_c = \frac{a}{b}\sqrt{\frac{\mu}{\varepsilon}}$

9.8 $Z_c = 74.99\Omega, \gamma = \text{j}6.28\text{rad/m}$

9.9 $Z_c = 120\ln\left(\frac{D}{a}\right)$

9.10 $Z_c = 120\ln\left[\frac{D}{2a} + \sqrt{\left(\frac{D}{2a}\right)^2 - 1}\right]$

9.11 (1) $Z_c = 50\Omega$; (2) The minimal VSWR is 2, and the corresponding voltage reflection coefficient is 1/3

9.12 $Z_c \approx 399.1\Omega$; $\alpha \approx 0.0132\text{Np/m}$; $\beta \approx 20.94\text{rad/m}$; $\lambda = 0.3\text{m}$; $v_p = 3 \times 10^8\text{m/s}$

9.13 (1) Terminal short circuit; (2) Terminal open circuit; (3) Pure capacitance load; (4) Pure inductance load

9.14 For the inductive reactance the shortest length is 0.211λ; for the capacitive reactance, the shortest length is 0.289λ

9.15 (1) $v_p = 1.5 \times 10^8\text{m/s}$, $\lambda = 50\text{m}$; (2) $R_L = 540\Omega$, $P = 9.26\text{W}$; (3) 0.131A

9.16 (1)0.27+j0.168; (2)(466+j206) Ω; (3) The resistance that is $l = 0.3\lambda + \frac{(2n-1)\lambda}{2}, (n = 1, 2, \cdots)$ away from the terminal is 150Ω; The resistance that is $l = 0.3\lambda + \frac{(2n-1)\lambda}{4}, (n = 1, 2, \cdots)$ away from the terminal is 600Ω

9.17 $Z_{02} = 150\Omega$; $f = 500\text{MHz}$

9.19 $L_0 = 2.5 \times 10^{-7}\text{H/m}$, $C_0 = 4.45 \times 10^{-13}\text{F/m}$

9.20 Based on the electric field formula in coaxial line $E_r = \frac{U_0}{r\ln\frac{b}{a}}$, it can be obtained as 2780kW

Chapter 10 TE Mode, TM Mode Transmit System—Waveguide

10.1 (1) $H_y = \frac{\beta}{\omega\mu}E_x = \frac{\beta}{\omega\mu}E_0\sin(\pi y/b)\text{e}^{\text{j}(\omega t - \beta z)}$

(2) The power density in the direction of z

$$\boldsymbol{S}_{av} = \boldsymbol{e}_z\frac{1}{2}\frac{\beta}{\omega\mu}\sin^2(\pi y/b)|E_0|^2$$

(3) $P = \frac{1}{4}\cdot ab\cdot\frac{\beta}{\omega\mu}|E_0|^2$

10.2 (1) 16.36GHz;

(2) 12.86GHz

10.3 (1) $k_x = 50\pi, k_y = 0, k_z = 44\pi$

(2) $$E_x = E_z = 0, E_y = E_0\sin k_x x\text{e}^{\text{j}(\omega t - k_z z)}$$
$$H_x = -\frac{k_z}{\omega\mu_0}E_0\sin k_x x\text{e}^{\text{j}(\omega t - k_z z)}, H_y = 0, H_z = \frac{\text{j}k_x}{\omega\mu_0}E_0\cos k_x\cdot x\text{e}^{\text{j}(\omega t - k_z z)}$$

(3) The wavelength of waveguide is 0.045 m

(4) Phase velocity 3.45×10^8 m/s, group velocity 2.61×10^8m/s

10.4 (1) $\lambda_c = 3\sqrt{2}\text{cm}$, $\lambda_g = 3\sqrt{2}\text{cm}$

(2) $a = 9\text{cm}$, $b = 6\text{cm}$

10.5 (1) When filled with air, λ_c are 14, 7, 6, 5.51, 5.51cm respectively

f_c are 2.14, 4.29, 5.00, 5.44×10^9Hz respectively

When filled with medium, λ_c are 14, 7, 6, 5.51, 5.51cm respectively

f_c are 1.07, 2.14, 2.50, 2.72×10^9Hz respectively

(2) When filled with air, there is only TE_{10} mode; When filled with medium, there are TE_{10}, TE_{20}, TE_{01}, TE_{11}, TM_{11} totally 5 modes.

10.6 When λ=10mm, TE_{01}, TE_{10}, TE_{11}, TM_{11}, TE_{20}, TE_{21}, TM_{21}, TE_{30}, TE_{31}, TM_{31}, TE_{40} can be transmitted; When $\lambda = 30$mm, it can transmit TE_{10}

10.7 $\sqrt{2}\text{cm} < a < 2\text{cm}$

10.8 It can't transmit the signal of wavelength 5cm; Signal of wavelength 3cm can be transmitted by TE_{10} mode; The signal of wavelength 2cm can be transmitted by TE_{10} and TE_{01} mode.

10.9 Hints: Method I: After Δtthere is a phase difference 180° lies between the upper side frequency and the lower side frequency, i.e. $[(\omega+\omega_m)\,\Delta t - (\omega-\omega_m)\,\Delta t] = \pi$. Thus, Δt can be computed ....

Method II: Use $(\beta_+ - \beta_-)\,l = \pi$ to compute the waveguide length l directly, where $\beta_+ = k_+\sqrt{1-\left(\frac{\lambda_+}{\lambda_c}\right)^2} = (\omega+\omega_m)\sqrt{\mu_0\varepsilon_0}\cdot\sqrt{1-\left(\frac{\lambda_+}{\lambda_c}\right)^2}$ and similarly is β_-. 4687.5m

10.11 (1) When $\lambda = 16$cm, it cannot transmit any mode.

When $\lambda = 8$cm, it can transmit TE_{10} mode

When $\lambda = 6.5$cm, it can transmit TE_{10}, TE_{20}, TE_{01} mode

(2) $f = 2.187 \sim 3.959$GHz

10.12 (1) $\beta = 1.58\times10^2$rad/m, $\eta_{TE_{10}} = 499\Omega$

(2) $\beta = 3.49\times10^2$rad/m, $\eta_{TE_{10}} = 201\Omega$

(3) $\alpha = 0.89\times10^2$Np/m, $\eta_{TE_{10}} = \text{j}433.5\Omega, z\approx 0.01\text{m}$

10.13 (1) $\beta = 234$rad/m, $\lambda_g = 0.0268$m, $v_p = 2.68\times10^8$m/s, $\eta_{TE} = 337.4\Omega$

(2) $\alpha = 1.187$dB/m

10.14 (1) The proof is omitted (2) $P_{\max} = 1$MW

10.16 0.096dB/m

10.17 The transmit power of TE_{0n} wave is $P = \frac{\pi a^2}{2Z_{TE}}E_{\phi m}^2 J_0^2(k_c a)$, where $k_c = \frac{u'_{on}}{a}$, $E_{\phi m} = \frac{\omega\mu}{k_c}H_0$

10.19 833km

10.20 (1) Filled with air, 49.97Ω

(2) When $\varepsilon_r = 2.25\eta = 120\pi/\sqrt{2.25}$, characteristic impedance is 33.32Ω

Chapter 11 Electromagnetic Radiation

11.2 (1) $R_r = 0.079\Omega$

(2) $P = 7.11$mW

11.3 0.237 Ω

11.4 1.1×10^{-12}W

11.5 $\theta = 0°$ $|\boldsymbol{E}_\theta| = 0, |\boldsymbol{H}_\phi| = 0$

$\theta = 45°$ $|\boldsymbol{E}_\theta| = 6.66\text{mV/m}, |\boldsymbol{H}_\phi| = 17.7\mu\text{A/m}$

$\theta = 90°$ $|\boldsymbol{E}_\theta| = 9.49\text{mV/m}, |H_\phi| = 25\mu\text{A/m}$

11.6 $H_\phi \approx \text{j}\frac{Idl}{2\lambda r}\sin\theta \text{e}^{\text{j}(\omega t - kr)}$; $E_\theta \approx \text{j}\frac{Idl}{2\lambda r}\left(\frac{k}{\omega\varepsilon}\right)\sin\theta \text{e}^{\text{j}(\omega t-kr)}$

11.7 $\boldsymbol{P}_2 = 40$mW

11.8 (1) $\boldsymbol{E} = \boldsymbol{e}_\theta\frac{\text{j}30kh}{r}I_0\sin\theta^{-\text{j}kr}$, $H = \boldsymbol{e}_\phi\frac{\text{j}30kh}{r\eta^0}I_0\sin\theta\text{e}^{-\text{j}kr}$

(2) $R_r = 20\pi^2\left(\frac{2h}{\lambda}\right)^2$

11.10 (1) Proof is omitted

(2) $\boldsymbol{H} = \boldsymbol{e}_\phi\text{j}\frac{I_m\text{e}^{-\text{j}kr}}{2\pi r_0}\cdot\frac{\cos\left(\frac{\pi}{2}\cos\theta\right)}{\sin\theta}$, $\boldsymbol{E} = \boldsymbol{e}_\theta\text{j}\frac{60I_m\text{e}^{-\text{j}kr}}{r_0}\cdot\frac{\cos\left(\frac{\pi}{2}\cos\theta\right)}{\sin\theta}$

(3) $\boldsymbol{S} = \boldsymbol{e}_r \dfrac{30I_{\mathrm{m}}^2}{\pi^2 r_0^2} \cdot \dfrac{\cos^2\left(\frac{\pi}{2}\cos\theta\right)}{\sin^2\theta}$

11.11 $|E_{\max}| = 60 \times 10^{-3}\mathrm{V/m}$

11.12 $|\boldsymbol{E}| = 1.885 \times 10^{-3}\mathrm{V/m}\,|\boldsymbol{H}| = 5 \times 10^{-6}\mathrm{A/m}$;
$\boldsymbol{P}_{\mathrm{r}} = 78.96 \times 10^{-3}\mathrm{W}, R_{\mathrm{r}} = 78.96 \times 10^{-3}\Omega$

11.13 (1) $R_{\mathrm{r}} = 8.8 \times 10^{-3}\Omega$;
(2) $R_{\mathrm{r}} = 2.44 \times 10^{-8}\Omega$

11.14 45° departure from South

If the receiver stays at South, field intensity will first decrease to zero as the element antenna rotates to the position where the receiving antenna locates at the extended line of its axis and then increase as the element antenna continues to rotate.

If the transmitter and receiver are both element antennas, the maximal and minimal received field intensity appear where the two element antennas are parallel and vertical respectively and the received field intensity lies between the maximal value and minimal value (zero) in other cases.

11.15 Total Radiation Field: $E_\theta = -\dfrac{\mu_0\omega^2 p_0}{4\pi r}\sin\theta\left(\mathrm{e}^{\mathrm{j}k\frac{d}{2}\cos\theta} + j\mathrm{e}^{-\mathrm{j}k\frac{d}{2}\cos\theta}\right)\mathrm{e}^{-\mathrm{j}kr}, \boldsymbol{B}_\phi = \dfrac{1}{c}E_\theta$;

Average energy flux density $\boldsymbol{S} = \boldsymbol{e}_{\mathrm{r}}\dfrac{\mu_0\omega^2 p_0^2}{16\pi^2 c r^2}\sin^2\theta$:

Total radiation power $\dfrac{\mu_0\omega^4 p_0^2}{6\pi c}$

11.16 2.22W

11.17 $\boldsymbol{E} = \dfrac{\eta_0 k^2 I_2 a^2}{4r}\sin\theta\mathrm{e}^{-\mathrm{j}(kr-\frac{\pi}{2})}(\boldsymbol{e}_\theta - \mathrm{j}\boldsymbol{e}_\phi)$. It is dextrorotatory circular polarization.

11.18 $\boldsymbol{E} = \dfrac{\omega^2 p_0}{4\pi\varepsilon_0 r c^3}\mathrm{e}^{-\mathrm{j}(kr+\phi)}(\boldsymbol{e}_\theta\cos\theta - \mathrm{j}\boldsymbol{e}_\phi)\boldsymbol{B} = \dfrac{\omega^2 p_0}{4\pi\varepsilon_0 r c^3}\mathrm{e}^{-\mathrm{j}(kr+\phi)}(\mathrm{j}\boldsymbol{e}_\theta - \boldsymbol{e}_\phi\cos\theta)$

$\boldsymbol{S} = \dfrac{\omega^4\boldsymbol{P}_0^2}{32\pi^2 c^3 r^2}(1+\cos^2\theta)\boldsymbol{e}_{\mathrm{r}}; P = \dfrac{\omega^4 p_0^2}{6\pi c^3}$

11.19 $\boldsymbol{E}_2 = m\boldsymbol{E}_1\mathrm{e}^{\mathrm{j}\Psi}$, $\Psi = \alpha + kd\sin\theta\cos\phi$. Total field is $\boldsymbol{E} = \boldsymbol{E}_1 + \boldsymbol{E}_2 = \boldsymbol{E}_1(1 + m\mathrm{e}^{\mathrm{j}\Psi})$. Proof can be obtained after calculating magnitude.

11.20 $\alpha = \dfrac{\pi}{2}$